U0200263

安长发博士科研论文集

Dr. An, Chang-Fa Scientific Research Papers

安长发　著

學苑出版社

图书在版编目（CIP）数据

安长发博士科研论文集 / 安长发著 .
—北京：学苑出版社，2015. 8

ISBN 978-7-5077-4848-2

Ⅰ . ① 安… Ⅱ . ① 安… Ⅲ . ① 空气动力学—文集 ② 计算流体力学—文集 Ⅳ . ① V211.1-53 ② O35-53

中国版本图书馆 CIP 数据核字（2015）第 208390 号

责任编辑： 郑泽英　李点点
封面设计： 陈四雄
出版发行： 学苑出版社
社　　址： 北京市丰台区南方庄 2 号院 1 号楼
邮政编码： 100079
网　　址： www.book001.com
电子邮箱： xueyuanpress@163.com
联系电话： 010-67601101（销售部）、67603091（总编室）
经　　销： 全国新华书店
印 刷 厂： 北京长阳汇文印刷厂
开本尺寸： 787mm × 1092mm　1 / 16
印　　张： 32
字　　数： 510 千字
版　　次： 2015 年 11 月北京第 1 版
印　　次： 2015 年 11 月北京第 1 次印刷
定　　价： 188. 00 元

简介与领域

Biography and field

姓名：安长发（An Chang-Fa）

出生：1944 年 5 月 16 日

电邮：changfaan@gmail.com

教育背景：

1967 年　哈尔滨军事工程学院 航天工程系 弹道式导弹设计与制造专业 本科毕业

1981 年　中国科学技术大学 近代力学系 流体力学专业 硕士研究生毕业

1992 年　加拿大 温莎大学 (University of Windsor) 应用数学系 博士研究生毕业

工作经历：

1981—1986 年　中国科学技术大学 近代力学系 流体力学专业 讲师

1986—1995 年　加拿大 温莎大学 (University of Windsor) 应用数学系 博士研究生，博士后研究员，兼职教授

1995—1998 年　加拿大 帝国石油资源有限公司 (Imperial Oil Resources Limited) 研发中心 研究专家，加拿大 卡尔加里大学 (University of Calgary) 机械工程系 兼职教授

1998—2008 年　美国 克莱斯勒 (Chrysler) 汽车公司 技术中心 高级工程师

2009—2009 年　中国 长安汽车公司 工程研究院 高级技术专家

2010—2013 年　美国 诺沃 (NOVO）汽车声学系统有限公司 高级工程师

所涉专业领域：

- 航空航天流体空气动力学
- 计算流体动力学

• 应用数学与力学

• 石油工程

• 汽车工程

本人从事科学研究与工程技术工作 30 多年，从航空航天经石油工业到汽车工业，其间多次与国内外大学的教学工作相交叉，既有基础理论研究又有工程实际应用。在完成具体工程项目的同时，如时间允许尽量对所涉及到的基础理论进行深入研究，以提高所承担项目的技术质量与自己的理论水平。

研究成果与工作业绩：

在国内外学术杂志与专业会议上发表科技论文 60 多篇

担任科技论文集审稿与专业会议分会主席 6 次

在国际专业技术会议上获得奖状与奖牌 6 次

接受杰出研究成果媒体专访两次

多年为专业组织与学会的会员：

Ⅰ）美国汽车工程师学会 (SAE)

Ⅱ）美国机械工程师学会 (ASME)

Ⅲ）加拿大计算流体动力学学会 (CFDSC)

Ⅳ) 美国数学学会 (AMS)

序　言

Preface

安长发博士专长空气动力学与计算流体力学，长期以来在这两个领域呕心沥血，做出了卓越的贡献，取得了丰硕的成果。本论文集汇总了安博士多年来在国内外期刊杂志和学术会议上发表的 60 多篇论文，按不同的研究阶段分成了四个部分，即：1. 出国前（1981—1987 年）在中国科技大学攻读硕士研究生及之后的任教期间发表的论文，这期间的学术论文着重于空气动力学方面，特别是机翼的空气动力稳定性的研究；2. 在加拿大温莎大学攻读博士学位及博士后期间（1991—1996 年）的研究成果，着重于计算流体力学的理论问题的研究，特别是对于流函数坐标的网格设计和理论作了较为深入的探讨和研究；3. 在加拿大帝国石油资源有限公司工作期间（1996—2000 年）的研究成果，主要研究了当时世界石油开采和运输公司的一个非常热门的课题——拦油栅拦油失效的流体动力学理论和实践问题。这一阶段是安长发博士科学研究最鼎盛的时期，在这一阶段他所发表的学术论文数量最多，论文既包括了作者在帝国石油公司所做有关拦油栅拦油失效的丰富的实验成果，也包括了应用计算流体力学方法探讨这种失效形态的流体动力学机制和理论，将理论计算的结果与实验数据作了认真细致的比较，两部分数据相一致的结果既考验了作者所掌握的理论方法，又为实验结果的可靠性提供了理论支持，在此基础上进行的一系列不同栏油栅结构的计算流体力学理论分析为设计新型有效的拦油栅结构提供了依据；4. 在美国克莱斯勒汽车公司工作期间（2003—2008 年）的研究成果，其间研究了数种公司正在运行和新开发的汽车车厢结构的空气动力学噪音问题，提出了新车型降噪结构设计，取得了很好的成果，很遗憾由于公司对新车型的保密要求，公开发表的论文未能完全反映这一阶段所取得的丰硕成果。

本论文集的出版实现了安长发博士多年以来的一个心愿，为此我向他表示衷心的祝贺。安长发博士是我近 20 年来志同道合的好朋友。我们相识于 1995 年在新加坡召开的第六届亚洲流体力学会议上，而后我曾经邀请他来北京化工大学讲过学，之后我们又在加拿大温莎大学一起合作共事过近十年时间。

安长发博士对待科学研究工作精益求精，严谨细致，认真负责的科学态度给我留下了极其深刻的印象，经安长发博士整理出来的资料，成果从来都是整齐干净，一丝不苟的。这本论文集的出版将为后人提供一份很好的学习和参考材料。

张政

2015 年 7 月 13 日

前　言
Prologue

多年来我一直有一个心愿，就是把我曾经发表过的科技论文汇编成册印刷出版。一方面可以概括我一生的奋斗作为纪念，另一方面也可以给子孙后代留下一笔精神财富。目前我年事已高且健康欠佳，更加时不我待。今将我在各个时期的论文搜集整理，作为《安长发博士科研论文集》一书出版。在我有生之年完成此项工作，以了却我今生的夙愿。

论文集的编排分为四个部分，以时间先后为序。第 Ⅰ 部分为出国前的研究成果，以硕士论文为起点。第 Ⅱ 部分为出国后在大学里的研究成果，以博士论文为基础。第 Ⅲ 部分为受聘于加拿大帝国石油资源有限公司时期的研究成果。第Ⅳ部分为任职于美国克莱斯勒汽车公司时期的研究成果。全书总共有论文 60 多篇近 500 页。由于本人的主要技术专长为计算流体动力学 CFD，因此文章中图形很多而且有些是彩色的。因为不同颜色有不同意义，所以凡是有彩色图形的页面一律用道林纸彩色印刷。全书出版时用硬皮精装，费用可能高些但是值得。

由于时间跨度 30 多年，各个时期论文的格式都有所不同，可能会对出版工作造成一定的困难。晚期发表的论文格式比较简单，大多为 PDF 或 DOC 文件格式，比较方便。但是早期发表的论文格式比较复杂。由于软件的版本太旧或者早已停版，电子文件无法调出，甚至文件原稿都无法找到。只好从论文抽印本或会议论文集中的每一页用扫描的方法生成 PDF 或 JPG 文件格式。还有一个问题就是，出国前在学术会议上所发表论文的个人存底，因多年的生活动荡而早已遗失。所以只能标明该报告收藏于有关的图书馆或资料室，但没有时间和机会去搜集整理。这些缺欠令本人深感遗憾，也望读者能够体谅。

童秉纲教授是我的硕士论文导师，也是我学术道路上的引路人。他的培养教育之恩，我将铭记在心永志不忘。Ronald M. Barron 教授是我的博士论文导师，也是我在国外科技生涯中亦师亦友的指导者。几十年来不仅在学术上语言上潜移默化滋润无声，而且对多民族多元文化非常友善包容和尊重，这些都令我今生今世难以忘怀。张政教授是我多年在科研工作中的合作者和最可信赖的私人朋友。他能答应为本书作序使我倍感温馨可靠。 Ronald H. Goodman 博士和 Hugh M. Brown 博士都是加拿大帝国石油资源有限公司的资深研究专家，也是我受聘于该公司时期的项目负责人和主要合作者。他们慧眼识人、知人善任，放手让我发挥技术专长，从而把我们的科研工作推向巅峰。Mark E. Gleason 博士是美国克莱斯勒汽车公司的风洞试验室主任。Kanwerdip Singh 博士是该公司 CFD 核心部门的经理。他们思路开阔眼光前瞻，在试验设备的使用与计算机资源的分配方面给予我优先保证和鼎力支持，使我在汽车风躁 (Buffeting) 方面的研究取得了很大成功。另外，我在各个时期的同学、同事和朋友，都曾与我朝夕相处、甘苦与共。本书也是为了与他们共享成果以志纪念。在我奋斗的一生中，我的家人始终与我站在一起，给予理解鼓励和支持，本书也是一件作为回报的珍贵礼物。在本书的出版过程中，学苑出版社的朋友热心协助积极筹划，作了大量艰苦细致的工作，本人表示衷心的感谢。最后，敬请广大读者对于本书可能出现的谬误之处予以批评指正。

安长发

2015 年 4 月 30 日

目　　录
Contents

第 I 部分　出国前硕士学位论文阶段

Ⅰ–01 ………………………………………………………………………………… 2

安长发，非对称转捩对烧蚀钝锥静稳定性的影响，硕士研究生毕业论文，中国科学技术大学，中国，安徽，合肥，1981

Ⅰ–02 ………………………………………………………………………………… 4

安长发，非对称转捩对钝锥静稳定性的影响，空气动力学学报，第 1 期，pp. 100–104, 1983

Ⅰ–03 ………………………………………………………………………………… 9

安长发，庄礼贤，非定常流理论在三角翼动导数计算中的应用，空气动力学杂志，第 4 期，pp. 46–52, 1984

Ⅰ–04 ………………………………………………………………………………… 16

徐守栋，**安长发**，有限元法在跨音速翼型计算中的应用，中国科学技术大学学报，第 16 卷，第 1 期，pp. 74–78, 1986

Ⅰ–05 ………………………………………………………………………………… 22

An, Chang-Fa, Calculation of pressure distribution on 2–D thin airfoil in transonic flows by finite element methods, 34th Canadian Aeronautics and Space Institute Aerodynamics Symposium on Computational Fluid Dynamics, Toronto, Canada, May 25–26, 1987

出国前在专业会议上发表的的技术报告 ……………………………………… 24

第 II 部分　出国后博士学位论文阶段

Ⅱ–01 ………………………………………………………………………………… 26

An, Chang-Fa, Transonic computation using Euler Equations in stream function coordinate system, Ph. D. dissertation, University of Windsor, Windsor, Ontario, Canada, 1992

Ⅱ-02 ·· 28

Barron, R.M. and **An, C.-F.**, Analysis and design of transonic airfoils using streamwise coordinates, Proceedings of 3rd International Conference on Inverse Design Concepts and Optimization in Engineering Sciences (ICIDES-Ⅲ), pp.359-370, Washington, D.C., USA, Oct. 23-25, 1991

Ⅱ-03 ·· 41

An, C.-F. and Barron, R.M., Transonic calculations using Euler equations in stream function coordinates, 10th Canadian Symposium on Fluid Dynamics, Saint John, New Brunswick, Canada, p7, June 4-6, 1992

Ⅱ-04 ·· 43

An, C.-F. and Barron, R.M., Numerical solution of transonic full-potential-equivalent equations in von Mises co-ordinates, International Journal for Numerical Methods in Fluids, Vol.15, pp.925-952, 1992

Ⅱ-05 ·· 71

Barron, R.M., **An, C.-F.** and Zhang, S., Survey of the streamfunction-as-a-coordinate method in CFD, Proceedings of Inaugural Conference of the CFD Society of Canada (CFD93), pp.325-336, Montreal, Canada, June 14-15, 1993

Ⅱ-06 ·· 84

Zhang, S. Barron, R.M. and **An, C.-F.**, A new O-grid system in computational aerodynamics, Proceedings of 2nd Annual Conference of the CFD Society of Canada (CFD94), pp.131-138, Toronto, Canada, June 1-3, 1994

Ⅱ-07 ·· 93

An, C.-F., Numerical simulation in SFC for shock reflection and interaction, 7th International Conference on Boundary and Interior Layers Computational & Asymptotic Methods (BAIL VII), Article 8-2, Beijing, China, Sept. 5-8, 1994

Ⅱ-08 ·· 97

An, C.-F., Barron, R.M. and Zhang, S., Arc length-streamfunction formulation and its application, Appl. Math. Modelling, Vol.18, No.9, pp.478-485, 1994

Ⅱ-09 ·· 105

Barron, R.M., **An, C.-F.** and Zhang, S., Unsteady conservative streamfunction coordinate formulation: 1-D isentropic flow, Appl. Math. Modelling, Vol.18, No.9, pp.486-493, 1994

Ⅱ-10 …… 113

An, C.-F. and Barron, R.M., Transonic Euler computation in streamfunction co-ordinates, International Journal for Numerical Methods in Fluids, Vol.20, pp.75–94, 1995

Ⅱ-11 …… 133

An, C.-F., Barron, R.M. and Zhang, S., Streamfunction coordinate formulation for one-dimensional unsteady flow, Mathematical Models and Methods in Applied Sciences, Vol.5, No.3, pp.401–414, 1995

Ⅱ-12 …… 147

Zhang, S., Barron, R.M. and **An, C.-F.**, Transonic flow computations using streamfunction and potential function, Communications in Numerical Methods in Engineering, Vol.11, pp.585–595, 1995

Ⅱ-13 …… 158

An, C.-F., Streamfunction coordinate formulation vs unsteady inviscid flow, Proceedings of 6th Asian Congress of Fluid Mechanics, Vol.1, pp.481–484, Singapore, May 22–26, 1995

Ⅱ-14 …… 163

An, C.-F., Barron, R.M. and Zhang, S., Stream function coordinate Euler formulation and shocktube application, Applied Mathematical Modelling, Vol.20, pp.421–428, 1996

Ⅱ-15 …… 171

Barron, R.M. and **An, C.-F.**, Numerical simulation of jet impingement turbulent flow with confinement plate, Proceedings of 3rd International Symposium on Experimental & Computational Aerothermodynamics of Internal Flows Ⅲ (ISAIF-Ⅲ), pp.685–690, Beijing, China, Sept. 1–6, 1996

Ⅱ-16 …… 178

Barron, R.M. and **An, C.-F.**, Turbulent jet impingement flow computation, International Journal of Turbo and Jet Engines, Vol.13, No.4, pp.295–306, 1996

Ⅱ-17 …… 191

An, C.-F., Stream function coordinate method and its applications in nonlinear gasdynamics, Proceedings of 3rd International Conference of Differential Equations & Applications, pp.19–20, St. Petersburg, Russia, June 12–17, 2000

Ⅱ-18 …… 194

An, C.-F., Numerical study on 1D unsteady flows in gasdynamics using stream function coordinate method, Mathematical Research, Vol.6, pp.24–33, 2000

第 Ⅲ 部分 受聘于加拿大帝国石油公司时期

Ⅲ -01 ······ 206

An, C.-F., Clavelle, E.J., Brown, H.M. and Barron, R.M., CFD simulation of oil boom failure, Proceedings of 4th Annual Conference of the CFD Society of Canada (CFD96), pp.401–408, Ottawa, Canada, June 2–6, 1996

Ⅲ -02 ······ 215

Goodman, R.H., Brown, H.M., **An, Chang-Fa** and Rowe, Richard, Dynamic modeling of oil boom failure using computational fluid dynamics, Spill Science & Technology Bulletin, Vol.3, No.4, pp.213–216, 1996

Ⅲ -03 ······ 219

An, Chang-Fa, Brown, H.M., Goodman, R.H. and Clavelle, Eric, Animation of boom failure processes, Spill Science & Technology Bulletin, Vol.3, No.4, pp.221–224, 1996

Ⅲ -04 ······ 223

An, C.-F., Brown, H.M., Goodman, R.H., Clavelle, E.J., Rowe, R.D. and Barron, R.M., Hydrodynamic behavior of contained oil slick on flowing water, Proceedings of 5th Annual Conference of the CFD Society of Canada (CFD97), pp.6: 31–36, Victoria, Canada, May 25–27, 1997

Ⅲ -05 ······ 230

An, C.-F., Goodman, R.H., Brown, H.M. and Clavelle, E.J., Computer animation of oil boom drainage failure, Proceedings of 5th Annual Conference of the CFD Society of Canada (CFD97), pp.16: 27–32, Victoria, Canada, May 25–27, 1997

Ⅲ -06 ······ 236

Goodman, R.H., Brown, H.M., **An, C.-F.** and Rowe, R.D. Dynamic modeling of oil boom failure using computational fluid dynamics, Proceedings of 20th Arctic & Marine Oilspill Program (AMOP) Technical Seminar, Vol.1, pp.437–455, Vancouver, Canada, June 11–13, 1997

Ⅲ -07 ······ 256

Brown, H.M., Goodman, R.H., **An, C.-F.** and Bittner, J., Boom failure mechanisms: comparison of channel experiments with computer modeling results, Proceedings of 20th Arctic & Marine Oilspill Program (AMOP) Technical Seminar, Vol.1, pp.457–467, Vancouver, Canada, June 11–13, 1997

Ⅲ -08 ······ 267

An, C.-F., Brown, H.M., Goodman, R.H. and Clavelle, E.J., Animation of boom failure processes, Proceedings of 20th Arctic & Marine Oilspill Program (AMOP) Technical Seminar, Vol.2, pp.1181–1188, Vancouver, Canada, June 11–13, 1997

Ⅲ –09 ······························· 276

程石勇，张政，**安长发**，应用 VOF 方法对水流中拦油栅失效进行数值模拟尝试，全国第七届计算传热会议论文集，pp. 189–197, 中国，北京，1997.10.4–6

Ⅲ –10 ······························· 285

An, C.-F., Brown, H.M., Goodman, R.H., Rowe, R.D. and Zhang, Z., Numerical modeling of the dynamics of an oil slick spilled on flowing water, Proceedings of International Symposium on Multiphase Fluid, Non-Newtonian Fluid & Physico-Chemical Fluid Flows (ISMNP'97), pp.3–45–52, Beijing, China, Oct. 7–10, 1997

Ⅲ –11 ······························· 294

An, C.-F., Goodman, R.H., Brown, H.M., Clavelle, E.J. and Barron, R.M., Oil–water interfacial phenomena behind a boom on flowing water, Proceedings of 10th International Symposium on Transport Phenomena (ISTP–10), Vol.1, pp.13–18, Kyoto, Japan, Nov. 30–Dec. 3, 1997

Ⅲ –12 ······························· 301

An, C.-F., Barron, R.M., Brown, H.M. and Goodman, R.H., Droplet entrainment boom failure and Kelvin–Helmholtz instability, Proceedings of 8th International Offshore & Polar Engineering Conference (ISOPE'98), Vol. II, pp.322–326, Montreal, Canada, May 24–29, 1998

Ⅲ –13 ······························· 307

An, C.-F. and Barron, R.M., Comparative study of models for oil boom simulation, Proceedings of 6th Annual Conference of the CFD Society of Canada (CFD98), pp. IV: 37–42, Quebec City, Canada, June 7–9, 1998

Ⅲ –14 ······························· 315

Brown, H.M., Goodman, R.H. and **An, C.-F.**, Flow around oil containment barriers, Proceedings of 21st Arctic & Marine Oilspill Program (AMOP) Technical Seminar, Vol.1, pp.345–354, Edmonton, Canada, June 10–12, 1998

Ⅲ –15 ······························· 326

Barron, R.M., **An, C.-F.** and Zhang Z., Unstable interface between oil and water with fast relative motion, Proceedings of 3rd International Conference of Fluid Mechanics (ICFM–Ⅲ), pp.247–252, Beijing, China, July 7–10, 1998

Ⅲ -16 ………………………………………………………………………………………… 333

Brown, H.M., Goodman, R.H. and **An, C.-F.**, Development of containment booms for oil spill in fast flowing water, Proceedings of 22nd Arctic & Marine Oilspill Program (AMOP) Technical Seminar, Vol.2, pp.813–823, Calgary, Canada, June 2–4, 1999

Ⅲ -17 ………………………………………………………………………………………… 345

Zhang, Z., **An, C.-F.**, Barron, R.M., Brown, H.M. and Goodman, R.H., Numerical study on (porous) net–boom systems–front net inclined angle effect, Proceedings of 22nd Arctic & Marine Oilspill Program (AMOP) Technical Seminar, Vol.2, pp.903–919, Calgary, Canada, June 2–4, 1999

Ⅲ -18 ………………………………………………………………………………………… 362

Zhang, Z., Barron, R.M. and **An, C.-F.**, Numerical study on new net–boom structures, Proceedings of 8th Asian Congress of Fluid Mechanics, pp.806–809, Shenzhen, China, Dec. 6–10, 1999

Ⅲ -19 ………………………………………………………………………………………… 367

Zhang, Z., Barron, R.M. and **An, C.-F.**, Numerical investigation of 3–D effect of water channel in oil boom experiments, Proceedings of 4th Asian Computational Fluid Dynamics Conference (ACFD4), pp.398–403, Mianyang, China, Sept. 18–22, 2000

Ⅲ -20 ………………………………………………………………………………………… 374

Barron, R.M., **An, Chang-Fa** and Zhang, Z., Effects of float shape and simulation coordinates for the numerical study of oil net–boom structures, Proceedings of International Conference on Applied Computational Fluid Dynamics (ACFD2000), pp.60–67, Beijing, China, Oct. 17–20, 2000

Ⅲ -21 ………………………………………………………………………………………… 383

An, Chang-Fa, CFD analysis helps develop oil containment boom, Pollution Equipment News, Vol.34, No.3, pp.4–5, June 2001

第 Ⅳ 部分　任职于美国克莱斯勒汽车公司时期

Ⅳ -01 ………………………………………………………………………………………… 387

Barron, R.M., Yang, H., El Saheli, A., **An, C.-F.** and Rankin, G.W., Numerical simulation for an automotive engine cooling fan, Proceedings of 8th Annual Conference of the CFD Society of Canada (CFD2K), Vol.2, pp.867–872, Montreal, Canada, June 11–13, 2000

Ⅳ -02 ………………………………………………………………………………………… 394

Zhang, Z, Barron, R. and **An, C.-F.**, CFD study of power spectrum analysis for air flowing over a cavity, Proceedings of 4th Computational Aeroacoustics (CAA) Workshop on Benchmark Problems, pp.75–76, Cleveland, OH, USA, Oct. 20–22, 2003

Ⅳ –03 397

An, Chang-Fa, Alaie, S.M., Sovani, S.D., Scislowicz, M.S. and Singh, K., Side window buffeting characteristics of an SUV, Proceedings of SAE 2004 World Congress: Vehicle Aerodynamics 2004, SP–1874, pp. 43–53, or SAE paper #2004–01–0230, Detroit, MI, USA, March 8–11, 2004

Ⅳ –04 410

An, Chang-Fa, Alaie, S.M. and Scislowicz, M.S., Impact of cavity on sunroof buffeting–a two dimensional CFD study, Proceedings of ASME/JSME Pressure Vessels & Piping Conference: Computational Technologies for Fluid/Thermal/Structural/Chemical Systems with Industrial Applications, Vol. 1, pp. 133–144, San Diego, CA, USA, July 25–29, 2004

Ⅳ –05 423

Zhang, Z., Barron, R. and **An, C.-F.**, Spectral analysis for air flow over a cavity, NASA/CP-2004-212954, pp.197–204, September 2004

Ⅳ –06 432

An, Chang-Fa, Puskarz, M., Singh, K. and Gleason, M.E., Attempts for reduction of rear window buffeting using CFD, Proceedings of SAE 2005 World Congress: Vehicle Aerodynamics 2005, SP–1931, pp. 97–104, or SAE paper #2005–01–0603, Detroit, MI, USA, April 11–14, 2005

Ⅳ –07 442

An, Chang-Fa, Survey of CFD studies on automotive buffeting, Proceedings of 13th Annual Conference of the CFD Society of Canada, pp. 223–230, St. John's, NL, Canada, July 31–Aug. 3, 2005

Ⅳ –08 452

Zhang, Z, Barron, R. and **An, C.-F.**, Post–processing in computational aeroacoustic studies, Proceedings of 13th Annual Conference of the CFD Society of Canada, pp. 274–281, St. John's, NL, Canada, July 31–Aug. 3, 2005

Ⅳ –09 460

Barron, R., Zhang, Z. and **An, C.-F.**, Computational aeroacoustic study of airflow over a cavity, Proceedings of 13th Annual Conference of the CFD Society of Canada, pp. 282–289, St. John's, NL, Canada, July 31–Aug. 3, 2005

Ⅳ –10 468

An, Chang-Fa, Puskarz, M., Singh, K. & Gleason, M.E., Attempts for reduction of rear window buffeting using CFD, SAE 2005 Transactions Journal of Passenger Cars: Mechanical Systems, Vol.

113, Session 6, pp.657–664, March 2006

Ⅳ–11 ··· 471

An, Chang-Fa and Singh, K., Optimization study for sunroof buffeting reduction, Proceedings of SAE 2006 World Congress: Vehicle Aerodynamics 2006, SP–1991, pp. 19–28, or SAE paper #2006–01–0138, Detroit, MI, USA, April 3–6, 2006

Ⅳ–12 ··· 483

An, Chang-Fa and Singh, K., Sunroof buffeting suppression using a dividing bar, Proceedings of SAE 2007 World Congress: Vehicle Aerodynamics 2007, SP–2066, pp. 419–425, or SAE paper#2007–01–1552, Detroit, MI, USA, April 16–19, 2007

Ⅳ–13 ··· 492

An, Chang-Fa, Singh, K., Sunroof buffeting suppression using a dividing bar, SAE 2007 Transactions Journal of Passenger Car: Mechanical Systems, Vol. 116, Session 6, pp. 1501–1507, April 2008

第 I 部分 出国前硕士学位论文阶段

中国科学技术大学

研究生毕业论文

研究生姓名：安長发

指导教师姓名：童秉纲

专　　业：高速空气动力学

研究方向：气动稳定性

论文题目：非对称转捩对烧蚀钝锥静稳定性的影响

一九八一年六月五日

这是硕士研究生毕业论文，中国科学技术大学，中国，安徽，合肥。原文现收藏于校图书馆。

非对称转捩对钝锥静稳定性的影响

安长发

（中国科学技术大学）

一、引　　言

当再入飞行器进入大气层时，在一定的高度范围内会发生边界层转捩。飞行试验的结果表明，在转捩的高度上出现了攻角发散的现象，从而影响了飞行器的性能。因此国外从六十年代起就已注意这个问题了。早期的研究主要是对称转捩问题，但后来发现，所谓“非对称转捩”对稳定性的影响更大，所以七十年代以后逐渐侧重于非对称转捩的研究。代表这些研究成果的主要文献有[1]—[4]等。

本文首先归纳了一些国外的实验结果，对尖锥和钝锥的非对称转捩的空间分布形状进行了数学描述，从而给出经验的逼近公式。然后根据转捩影响稳定性的机理，采用简单实用的方法进行理论分析，从而建立起一组估算非对称转捩对静稳定性影响的工程计算方法，并将计算结果与一些实验数据进行了对比。最后讨论了非对称转捩中一些主要因素的影响，并对某些可能的危险情况作了估计。

本文假定再入体是中小钝度的光滑钝锥；烧蚀的结果只引起边界层内的质量引射而不改变锥的几何外形；整个表面温度相同，有效烧蚀热也相同；来流为小攻角。计算模型的尺寸和基本几何关系见图1。

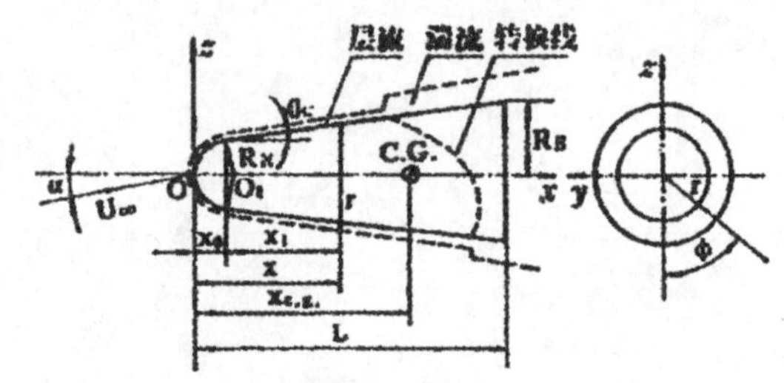

图 1　模型的基本几何关系

二、非对称转捩的空间分布形状

影响边界层转捩的因素很多，但在本文假设下最主要的是来流雷诺数，头部钝度，攻角和表面烧蚀率等。归纳[1]—[4]的实验结果，可以定性地看出各主要因素对转捩位置的影响：

1．当来流雷诺数增大时，转捩位置前移。

2．当钝度比增大时，转捩位置后移，但当钝度比大到一定程度时，后移停止，若再增大钝度比，则转捩位置反而前移。前者称为小钝度区，后者称为大钝度区，两者的分界雷诺数为 $Re_N \approx 8\times10^4$，对于10°左右的钝锥，可粗略地认为分界钝度比为0.1左右。

3．当烧蚀率增大时，转捩位置略向前移。

本文于1981年9月14日收到，1982年7月8日收到修改稿。

4．当来流有攻角时，转捩变成非对称的了，不同钝度的锥，非对称转捩的特征是不同的。对于尖锥，当攻角增大时，背风面转捩位置很快地前移，而迎风面转捩位置则向后移动(如图2a)；对于小钝度锥，随着攻角的增大，背风面转捩位置也是很快地前移，但与尖锥不同的是，迎风面转捩位置也略向前移，从而使得最迟转捩位置不是在迎风面而是在两侧(如图2-b)；对于大钝度锥，迎风面转捩位置甚至比背风面还要提前，所以其非对称转捩的形状与小钝度锥的情况正好相反(如图2-c)。

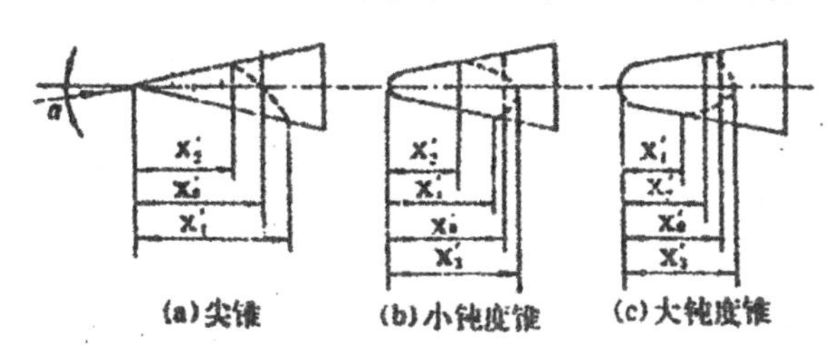

图 2　非对称转捩的基本形式

为了描述转捩的非对称程度，定义转捩畸变系数 $\xi=\frac{x_2'-x_1'}{x_1'}$，$\eta=\frac{x_3'-x_1'}{x_1'}$，其中 x_1', x_2', x_3' 分别是迎风面、背风面和侧面的转捩位置。将图2中的 φ 角展开变成图3，并设小钝度锥的转捩位置由图3-b所示的三次曲线来描述

$$x_{tr}=a_0+a_1\varphi+a_2\varphi^2+a_3\varphi^3 \tag{1}$$

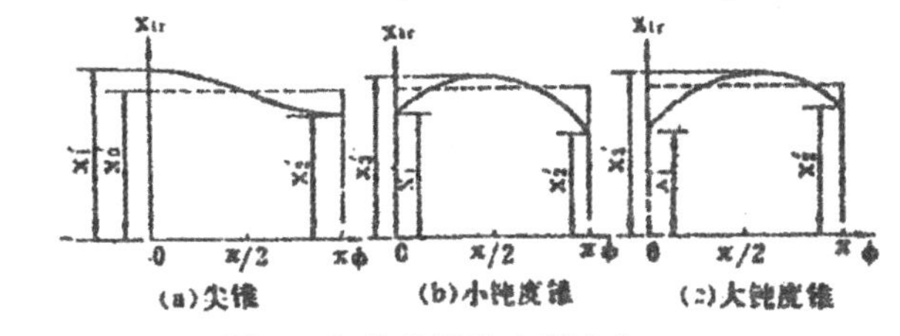

图 3　转捩位置的空间分布

其中四个待定系数由 $\varphi=0,\pi,\pi/2$ 时 $x_{tr}=x_1', x_2', x_3'$ 和 $\varphi=\pi/2, dx_{tr}/d\varphi=0$ 四个条件定出，再把 x_2', x_3' 换成 ξ,η，则

$$x_{tr}=x_1'\left[1+(\xi+4\eta)\frac{\varphi}{\pi}-4(\xi+\eta)\frac{\varphi^2}{\pi^2}+4\xi\frac{\varphi^3}{\pi^3}\right] \tag{2}$$

在小攻角时，可以假定 ξ 与 α 成正比：$\xi=-A_1\alpha$，根据[1]的实验结果，$A_1\approx 0.1—0.2$，$\xi\approx-4\eta$，则小钝度锥的转捩位置可用下式来表示：

$$x_{tr}=x_1'\left\{1+A_1\alpha\left[3\frac{\varphi^2}{\pi^2}-4\frac{\varphi^3}{\pi^3}\right]\right\} \tag{3}$$

对于大钝度锥，根据图2-c和图3-c所示的转捩特征，应以 x_2' 代替 x_1'，$(\pi-\varphi)$ 代替 φ，则

$$x_{tr}=x_2'\left\{1+A_1\alpha\left[3\cdot\frac{(\pi-\varphi)^2}{\pi^2}-4\cdot\frac{(\pi-\varphi)^3}{\pi^3}\right]\right\} \tag{4}$$

对于尖锥，根据图2-a和图3-a，可采用简单的三角函数来逼近

$$x_{tr}=x_1'\left(1-A_1\alpha\frac{1-\cos\varphi}{2}\right) \tag{5}$$

或

$$x_{tr}=x_0'\left(1+A_1\alpha\frac{\cos\varphi}{2}\right) \tag{6}$$

这里 $x_0'\approx\frac{1}{2}(x_1'+x_2')$ 为零攻角时的转捩位置。

三、分析、计算和讨论

当边界层转捩时，其厚度发生了变化，这就相当于等效物形的变化，从而造成表面压力分布的变化，另外，由于两类边界层速度分布规律的不同，所以表面摩阻也会发生变化。将这两个变化量对全物面积分，就得到了力和力矩的变化量，其中俯仰力矩的变化量就涉及到再入体的稳定性。

本文在计算表面压力时，采用了内伏牛顿流公式；在计算边界层位移厚度和摩阻时，参照[5]采用了 Timmer 的工程计算公式；在计算烧蚀率时，首先算出表面热流的分布，然后利用有效烧蚀热换算成烧蚀率；在计算边界层外缘的热力学函数时采用有效比热比的方法。在得到表面

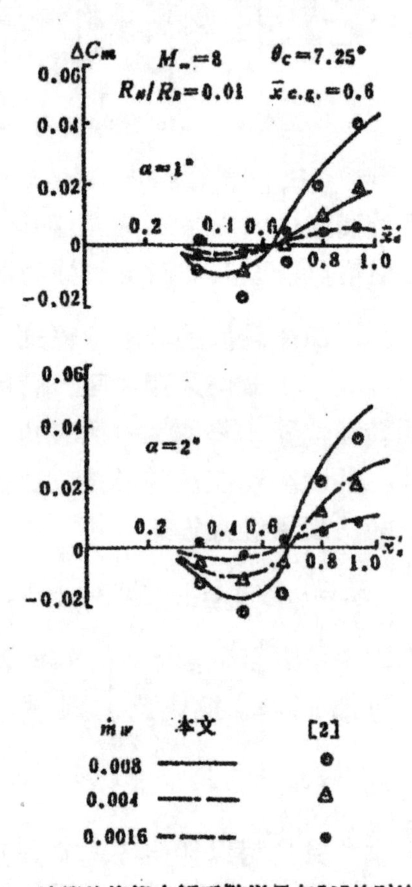

图 5　计算的俯仰力矩系数增量与[2]的对比

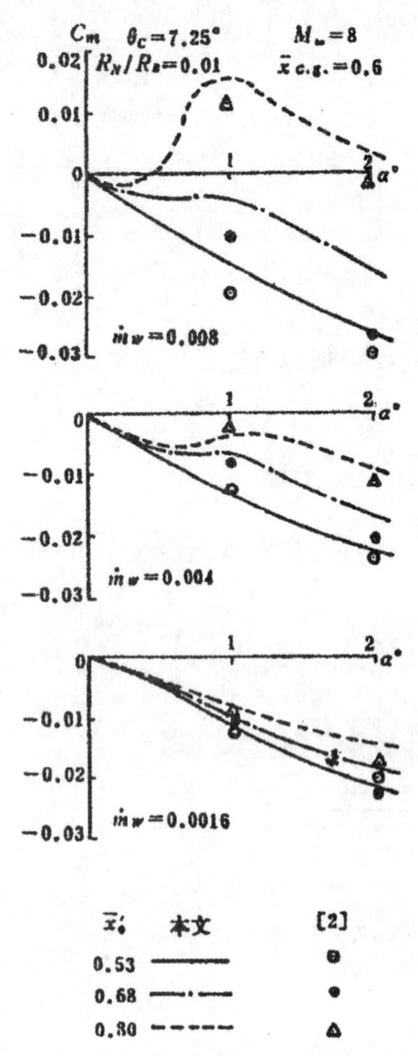

图 4　计算的俯仰力矩系数与[2]的对比

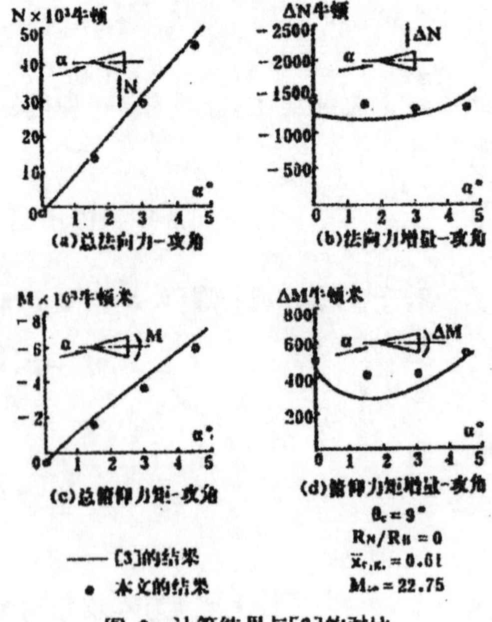

图 6　计算结果与[3]的对比

压力和摩阻的分布之后，采用 Gauss 数值积分即可求出力、力矩和压力中心。分析和计算方法的细节可参见[6]。

图 4、5 是计算结果与[2]中实验数据的对比，图 6 是计算结果与[3]的对比。可以看出，本文计算的结果与[2、3]基本上是一致的。

表 1 是本文计算的结果与[1]中实验数据的对比，可以看出，本文结果与[1]的实验值也是很接近的。

表 1　计算结果与[1]的对比

α	M_∞	R_N/R_B	来　源	C_N	C_m	$\Delta\bar{x}_{cp}$	$\Delta C_m/C_{mi}$
2°	13.2	0	本　文	0.0652	0.103	0.008	−0.014
			[1]	0.0645	0.102	0.013	−0.030
1°	11.3	0.06	本　文	0.0363	0.0562	0.012	−0.011
			[1]	0.0313	0.0466	0.008	−0.010
3°	11.3	0.21	本　文	0.0730	0.0865	−0.011	0.029
			[1]	0.0630	0.0803	−0.009	0.030

为了全面分析非对称转捩的影响，还通过一个小钝度锥模型的典型算例，就非对称转捩中的几个主要因素对静稳定性的影响进行了分析，详细的讨论可参见 [3]。根据这些分析和讨论，可以得出如下简短的结论：

1. 当钝锥后体背风面开始出现非对称转捩时，将产生一个削弱其静稳定性的俯仰力矩增量，且压心前移。这种不稳定的影响随着转捩位置的前移而减弱，随着烧蚀率的增大而加强，随着攻角和钝度比的增大而减弱。

2. 因此，对于大烧蚀率的小钝度锥，在小攻角下飞行时，后体的非对称转捩对静稳定性的影响是不利的，严重时有可能造成攻角发散。

3. 在分析非对称转捩对静稳定性的影响时，如果材料烧蚀率较小，则不仅要考虑压力变化的影响，而且还要考虑摩阻变化的影响；如果材料烧蚀率较大，则可以忽略摩阻变化的影响。

本文作者衷心地感谢童秉纲教授的全面指教；感谢空间研究院计算站李显霖老师在数值计算方面的指导和帮助。

参　考　文　献

[1] Holden, M.S., AD/A-065173, 1978.
[2] Martellucci, A.& Neff, R.S., JSR, vol.8, No.5, 476-482, 1971.
[3] Nickell, J.C.et al., AD/A-040947, 1976.
[4] Uffelman, K.E.& Deffenbaugh, F.D., AIAA paper 79-1626, 1979.
[5] 童秉纲、李显霖，钝锥在质量引射下稳定性导数的分析，中国科技大学内部报告，1980。
[6] 安长发，非对称转捩对烧蚀钝锥静稳定性的影响，中国科技大学研究生毕业论文，1981。

INFLUENCE OF ASYMMETRIC TRANSITION ON STATIC STABILITY OF BLUNTED CONES

An Changfa

(Department of Modern Mechanics, University of Science and Technology of China)

Abstract

In this paper the effect of asymmetric boundary-layer transition on static stability of blunted cones is investigated and an engineering analytical method estimating this effect is proposed. The calculated results are compared well with some available experimental data. By calculating a typical example, the effects of some principal factors in asymmetric transition on static stability of the cone are discussed and estimated.

安长发博士科研论文集

非定常流理论在三角翼动导数计算中的应用

安长发　庄礼贤
(中国科学技术大学)

摘要　在飞行器的预研工作中，需要对各种气动外形的动导数进行快速和准确的估算。本文对三角翼纵向稳定性导数的计算方法及相应的理论进行了综述，并对它们的应用条件、准确程度及优缺点作了简要的述评。为了便于实际应用，本文着重讨论了那些解析或半解析的结果。

一、引　　言

飞行器动导数的研究已开展多年，发表过许多论文，召开过多次国际性专题会议，出版过一些专著、数据图集和手册。但是在现有的理论和计算方法中，解析和半解析的简单结果是不多的，许多方法最后都导致比较繁琐的数值计算。一般说来，数值方法固然精度较好，但将它用于飞行器气动力的初步估算并不很适宜，因为这种方法既耗资费时，又难以反映出流动的内在规律。所以解析的方法和结果仍是十分重要的。

三角翼是一种很典型的机翼和尾翼外形。故本文对其在不同速度范围内的动导数计算方法进行了综述。这些方法有：薄翼在零攻角附近振动问题的非定常线化势流理论，厚翼在大攻角下振动问题的超音速和高超音速统一的非定常流理论，以及近年来发展起来的非定常高超音速流的 Newton-Busemann 理论。同时还将各种理论计算的结果与实验数据进行了比较，并讨论了各种理论的应用范围。为了便于型号研制部门的实际应用，本文主要介绍了一些解析和半解析的结果。

二、非定常小扰动势流的线化理论

假设一薄三角翼在气流中绕零攻角作小振幅的俯仰谐振(见图1)，这时它对气流的扰动是微小的，流场内无强激波因而是近似等熵的。忽略空气的粘性、热传导和质量力并视为完全气体，则流场可以用下列无量纲的非定常小扰动速势方程来描述：

亚：$\beta^2\phi_{xx}+\phi_{yy}+\phi_{zz}-2M^2\phi_{xt}-M^2\phi_{tt}=0$　　(1)

跨：$[1-M^2-(\gamma+1)M^2\phi_x]\phi_{xx}+\phi_{yy}+\phi_{zz}-2M^2\phi_{xt}-M^2\phi_{tt}=0$　　(2)

超：$B^2\phi_{xx}-\phi_{yy}-\phi_{zz}+2M^2\phi_{xt}+M^2\phi_{tt}=0$　　(3)

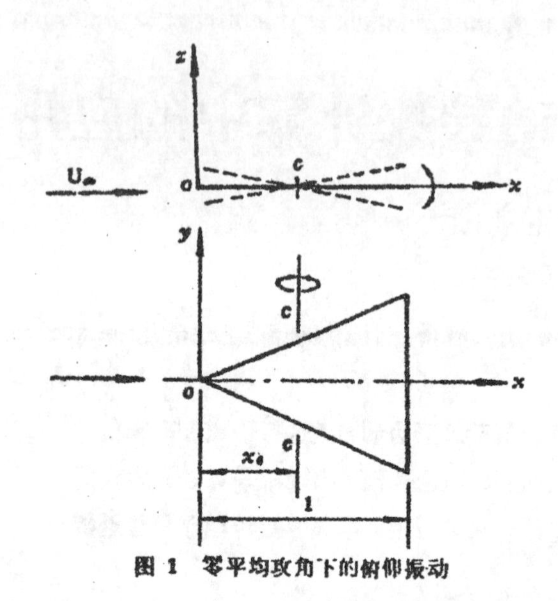

图 1 零平均攻角下的俯仰振动

这里M是来流马赫数，$\beta=\sqrt{1-M^2}$，$B=\sqrt{M^2-1}$。在适当的边界条件下解出上述方程并求出翼面上的压力分布，积分后即可求出动导数。

1．超音速情况

Miles, J. W. 在其专著[1]中系统地阐述了非定常超音速势流的线化理论。对于低频俯仰谐振的平板三角翼，他采用频率展开的方法来解方程(3)，从而得到了稳定性导数的一阶近似计算公式：

$$C_{m_\alpha}=\frac{4}{B}\left(x_0-\frac{2}{3}\right)f_0 \tag{4}$$

$$C_{m_q}=-\frac{4}{B}\left\{\left[\left(x_0-\frac{2}{3}\right)^2+\frac{1}{18}\right]f_0+\left(\frac{1}{2}-\frac{2}{3}x_0\right)\left(f_1-f_0\right)\right\} \tag{5}$$

$$C_{m_{\dot{\alpha}}}=\frac{4}{B}\left[f_0-\frac{3M^2}{B^2}\left(f_0-\frac{2}{3}f_1\right)\right]\left(\frac{1}{3}x_0-\frac{1}{4}\right) \tag{6}$$

其中

$$\left.\begin{aligned}
&f_0=\begin{cases}\dfrac{\pi m}{2E(m')} & m\leqslant 1\\ 1 & m>1\end{cases}\\
&f_1=\begin{cases}\dfrac{3/4\pi m(1-m^2)}{(1-m^2)E(m')+m^2[K(m')-E(m')]} & m\leqslant 1\\ 1 & m>1\end{cases}\\
&m=B\operatorname{tg}\delta,\quad m'=\sqrt{1-m^2},\quad m\leqslant 1
\end{aligned}\right\} \tag{7}$$

式中δ为三角翼的半顶角，x_0为俯仰轴位置到顶点的距离，$K(m')$和$E(m')$分别是模数为m'的第一、二类完全椭圆积分。

当$m<1$时，三角翼具有亚音速前缘，通常称之为“窄翼”，$f_0,f_1<1$；当$m>1$时，具有超音速前缘，称为“宽翼”，$f_0=f_1=1$。图2画出了“窄翼”情况下f_0,f_1与m的关系曲线。图3是按公式(4)－(6)计算的结果与Thompson, J.S.[2]实验结果的对比，图4是计算结果与Ericsson, L.E.[3]实验结果的对比，由这两个图可以看出，无论是“窄翼”还是“宽翼”，线化势流理论都能给出较好的结果。

2．跨音速情况

当来流马赫数接近于1时，由于ϕ_x与$1-M$具有相同的数量级，所以在一般情况下方程(2)是不能线化的。但是当折合频率k满足$|1-M|/k\ll 1$时，Landahl, M.T.[4]证明了速势方程(2)可以线化成：

$$\phi_{yy}+\phi_{zz}-2M^2\phi_{xt}-M^2\phi_{tt}=0 \tag{8}$$

对于小展弦比三角翼，当振动频率较低时，Landahl将Fourier变换后的速势展开成展弦比和频率乘积的级数，并用Adams-Sears迭代法求出了三阶近似解，最后借助广义气动力系数的概念导出了动导数的计算公式：

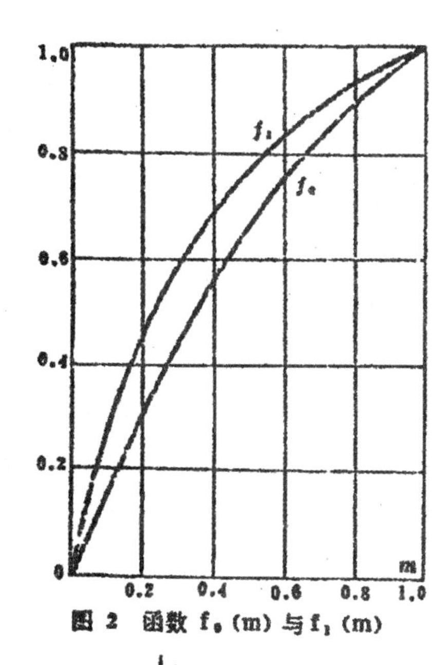

图 2 函数 $f_0(m)$ 与 $f_1(m)$

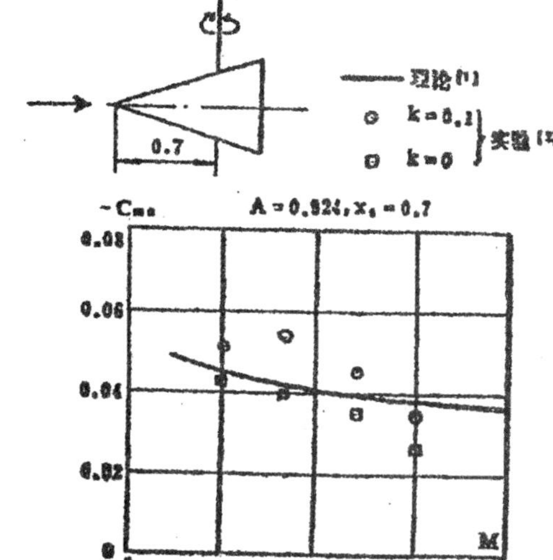

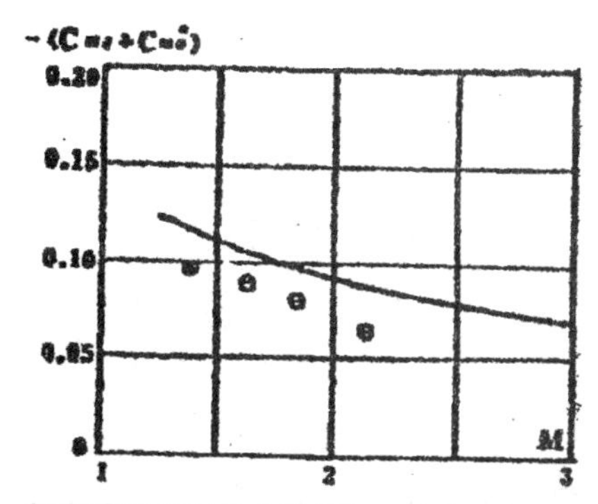

图 3 超音速三角翼稳定性导数理论与实验的比较(一)

$$C_{m_\alpha}=\frac{\pi A}{2}\left[\left(x_0-\frac{2}{3}\right)+\frac{\pi k M^2 A^2}{64}\left(\frac{3}{4}-x_0\right)\right] \quad (9)$$

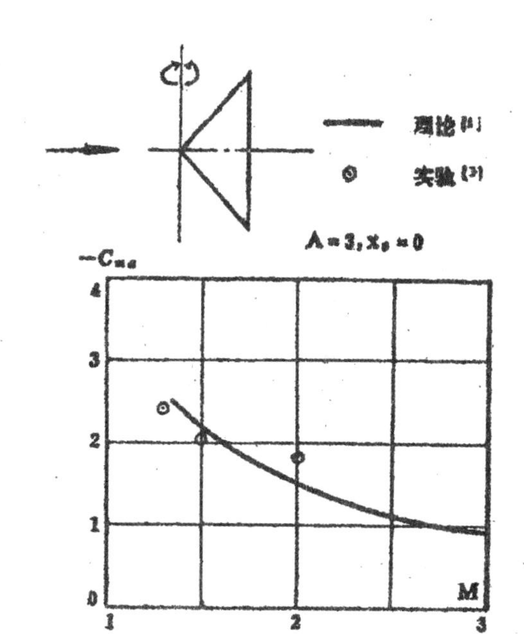

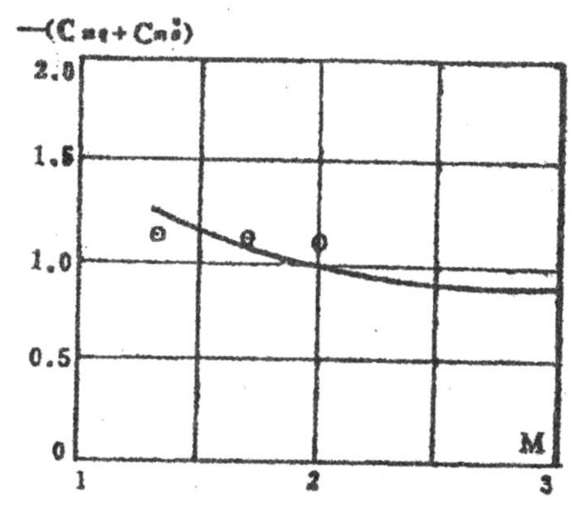

图 4 超音速三角翼稳定性导数理论与实验的比较(二)

$$C_{m_q}=-\frac{\pi A}{2}\left\{\left(\frac{3}{4}-\frac{5}{3}x_0+x_0^2\right)-\frac{M^2A^2}{64}\left[\frac{3}{8}+\pi k\left(\frac{6}{5}-\frac{9}{4}x_0+x_0^2\right)\right]\right\} \quad (10)$$

$$C_{m_{\dot\alpha}}=-\frac{\pi A}{2}\left\{\left(\frac{1}{4}-\frac{1}{3}x_0\right)-\frac{M^2A^2}{64}\left[2\bar{v}\left(\frac{3}{4}-x_0\right)-\frac{1}{2}+\pi k\left(\frac{2}{5}-\frac{1}{2}x_0\right)\right]\right\} \quad (11)$$

其中 $\bar{v}=\ln(128/kA^2)-\gamma$，$\gamma=0.5772\cdots$ 为 Euler 常数，A 为展弦比。

图5给出了按公式(9)—(11)计算的结果与 Orlik-Rückemann[5] 实验结果的对比，结果还是令人满意的。

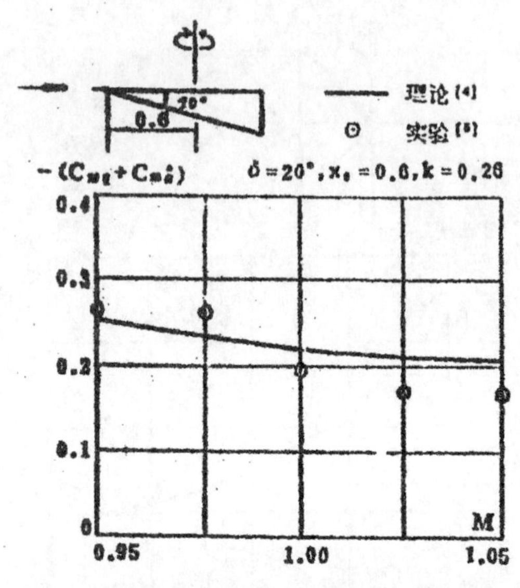

图 5　跨音速三角翼阻尼导数理论与实验的比较

3. 亚音速情况

对于亚音速小扰动方程 (1), Tobak, M. 和 Ressing, H. C. [6] 采用"准定常"的方法得到了计算三角翼动导数的半解析结果。设方程(1)解的形式为

$$\phi=-\frac{M^2}{\beta^2}\dot{\alpha}\phi_1+\dot{\alpha}\left(t+\frac{M^2}{\beta^2}x\right)\phi_2+\phi_3 \tag{12}$$

其中 ϕ_1, ϕ_2 分别是单位俯仰角速度和单位攻角所引起的定常速势，ϕ_3 是为使后缘和尾迹满足 Kutta 条件的附加速势，三者都满足定常小扰动方程，可由定常流理论解出。对于平板三角翼，Tobak 导出了非常简单的动导数计算公式

$$(C_{m_\alpha})_0=-x_a C_{L\alpha} \tag{13}$$

$$(C_{m_q})_0=-\frac{27}{16}x_a^2 C_{L\alpha} \tag{14}$$

$$(C_{m_{\dot{\alpha}}})_0=-\frac{9}{16}x_a^2 C_{L\alpha}+C_{m_c} \tag{15}$$

式中的下标"0"表示俯仰轴位置在三角翼顶点的导数；x_a 是压心位置，$C_{L\alpha}$ 是翼的升力曲线斜率，C_{m_c} 是附加项，它们都是 βA 的函数，具体表达式可参阅[6]。图6给出了 x_a，$C_{L\alpha}$ 与 βA 的关系，图7给出了 C_{m_c} 与 βA 的关系利用这些曲线很容易求出对顶点的动导数，从而不难换算成对任意俯仰轴的动导数。图8是按公式(13)—(15)计算的结果与实验数据[7]的对比，由图可以看出 Tobak 的理论具有一定的精度。

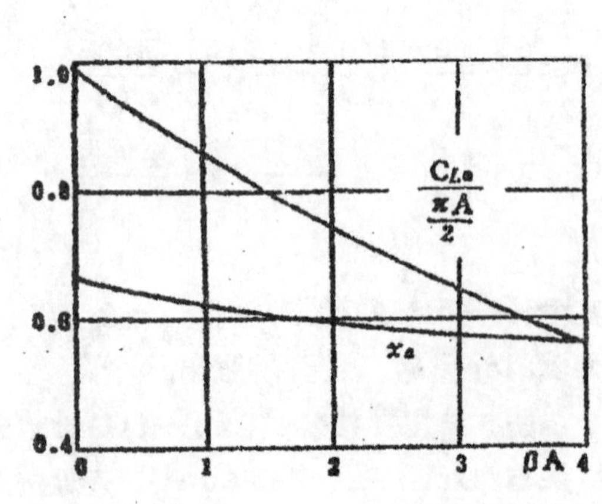

图 6　参数 $C_{L\alpha}$，x_a 随 βA 的变化关系

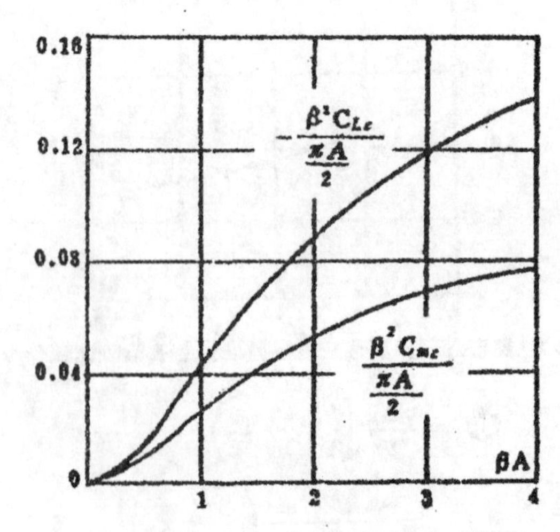

图 7　参数 C_{Lc}, C_{mc} 随 βA 的变化关系

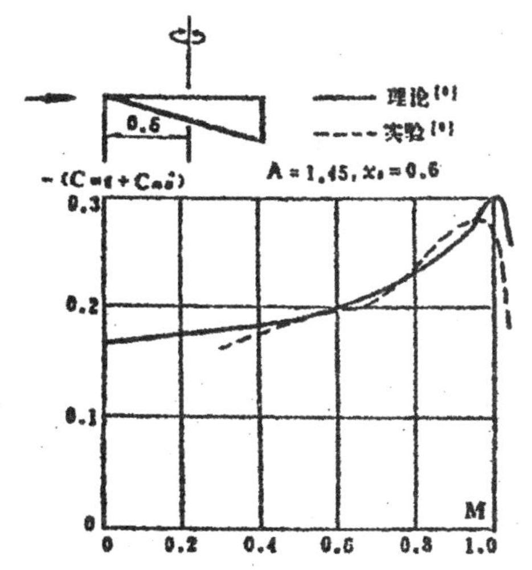

图 8 亚音速三角翼阻尼导数理论与实验的比较

Tobak 的方法具有“半解析半经验”的性质。另外还有一类“半解析半数值”的方法，即非定常流方程基本解迭加的方法。这种方法最后归结为某一下洗积分方程的求解问题。但目前对这种积分方程还只能用数值解法，其中偶极子栅格法是公认的有效方法[7]。对于亚音速大攻角翼的振动问题，最近 Lan，C.E[8] 将“吸力比拟”法推广到非定常情况，但气动力的计算仍归结为积分方程的数值解。限于篇幅，本文就不作详细介绍了。

三、超音速和高超音速的统一理论

势流理论只适合于小攻角情况，当平均攻角较大时，即使振幅很小，小扰动条件也已被破坏，流场中会出现激波，从而使势流理论失效。近年来，许为厚（Hui，W.H.）等人对机翼在大攻角下低频微幅谐振的非定常流场作了一系列研究，提出了超音速和高超音速统一的非定常解析理论。这一理论以平均攻角下的定常流场为基本流动，而非定常振动则被认为是对此基本流动的小扰动。该理论的可用前提是前缘激波必须附体。文献[9]和[10]分别研究了振动楔和大攻角下的振动平板，文献[11]用条带理论把上述结果推广到三维机翼。

对于在大平均攻角 α_m 下慢振动的平板三角翼，[11]给出了计算稳定性导数的解析表达式：

$$C_{m_\alpha}=-A\left(\frac{2}{3}-x_0\right) \tag{16}$$

$$C_{m_{\dot{\alpha}}}+C_{m_q}=-\frac{B+C}{4}+\frac{B+3C}{3}x_0-Cx_0^2 \tag{17}$$

其中

$$\left.\begin{aligned}
&A=\lambda_0\tilde{C}/\tilde{A}+\lambda_1M_1/\sqrt{M_1^2-1}\\
&B=\mu_0(2G-I)+\mu_1M_1(M_1^2-2)/(M_1^2-1)^{3/2}\\
&C=\mu_0I+\mu_1M_1/\sqrt{M_1^2-1}\\
&\lambda_0=2\rho_0u_0^2/M_0,\quad \lambda_1=2\rho_1u_1^2/M_1\\
&\mu_0=2\rho_0u_0/M_0,\quad \mu_1=2\rho_1u_1/M_1
\end{aligned}\right\} \tag{18}$$

式中 M_0，ρ_0，u_0 和 M_1，ρ_1，u_1 分别表示在定常基本流动中气流内转 α_m 角所形成的斜激波和气流外转 α_m 角所形成的扇形膨胀波后面的气流的马赫数、密度和速度，它们都是来流马赫数 M、平均攻角 α_m 和空气比热比 γ 的函数。另外，$\tilde{A}$，$\tilde{C}$，G，I 也都只是 M，α_m，γ 的函数，所以 A、B、C 都只依赖 M、α_m、γ。

特别当 $\alpha_m \to 0$ 时，$A \to 4/\beta$，$B \to 4(\beta^2-1)/\beta^3$，$C \to 4/\beta$，这里的 $\beta=\sqrt{M^2-1}$，则式(16)，(17)退化成线化势流理论的结果，即(4)—(6)($f_0=f_1=1$)，这正是所预期的。

图9是按式(16)、(17)计算的结果与文献[12]按不同方法计算结果的对比，由图可见，不同理论所得的结果大体上是一致的。

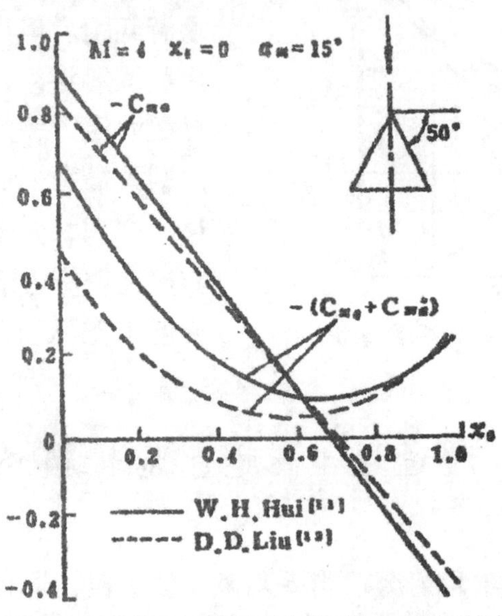

图 9 超音速三角翼在大攻角下的稳定性导数

四、非定常的 Newton-Busemann 理论

实际上，上述的"统一理论"对于高超音速情况并不很适宜，因为这一理论没有考虑激波对膨胀波的反射，而在马赫数很大时这种反射是不能忽略的。在高超音速气流中，通常采用 Newton 碰撞理论计算气动力。为了考虑气流质点轨道的弯曲效应，还应加上 Busemann 离心力修正项。但是在定常情况下，理论和实验都指出[13]，考虑物面曲率效应的修正公式反而不如纯碰撞公式准确，所以处理定常问题往往不用 Busemann 修正公式。对于非定常问题则不同，即使物面是平面，由于振动也会造成气流质点轨道的显著弯曲，从而使离心力修正项不容忽略。

Hui, W.H. 和 Tobak, M.[14] 研究了二维翼型在高超音速气流中作微幅低频谐振的非定常流场，得到了稳定性导数的表达式：

$$C_{m_\alpha}=-4\{\int_0^1 P_1(x)[x+f(x)f'(x)]dx-x_0\int_0^1 P_1(x)dx\} \tag{19}$$

$$\begin{aligned}C_{m_q}+C_{m_{\dot\alpha}}=-4\{&\int_0^1 P_2(x)[x+f(x)f'(x)]dx-\\&-x_0\int_0^1 P_1(x)[x+f(x)f'(x)]dx-\\&-x_0\int_0^1 P_2(x)dx+x_0^2\int_0^1 P_1(x)dx\}\end{aligned} \tag{20}$$

其中

$$\left.\begin{aligned}P_1(x)=&2\mu^2(x)f'(x)+\kappa(x)\int_0^x\mu(\xi)[1-f'^2(\xi)]d\xi\\P_2(x)=&2\mu^2(x)f'(x)[x+f(x)f'(x)]+2f(x)+\\&+\kappa(x)\int_0^x\mu(\xi)\{\xi[1-f'^2(\xi)]+2f(\xi)f'(\xi)\}d\xi-\\&-\kappa(x)\int_0^x d\xi\int_\xi^x ds/\mu(s)\end{aligned}\right\} \tag{21}$$

这里 $f(x)$ 是翼型表面方程，$\mu(x)=1/\sqrt{1+f'^2(x)}$，$\kappa(x)=\mu^3(x)f''(x)$。

文献[15]把上述结果推广到旋成体，[16]进一步推广到任意频率，[17]又推广到任意三维物体。对于三角翼，既可以用条带理论把二维结果加以推广，也可以直接应用[17]的结果。

非定常 Newton-Busemann 理论的正确性有一个简单的佐证，如果用前述的"超音速和高超音速统一理论"来计算二维尖楔并取极限 $M\to\infty$, $\gamma\to 1$，则其结果与非定常 N-B 理论所得的结果完全一致。另外，[15]把尖锥阻尼导数的计算结果和其它理论结果与实验数据[18]进行了对比，见图 10。由图可见，非定常 N-B 理论结果与实验符合得相当好，比单纯Newton碰撞理论的结果有了明显的改进。另外由图10还可以看出，离心力修正项的贡献与碰撞压力项的贡献具有相同的数量级。

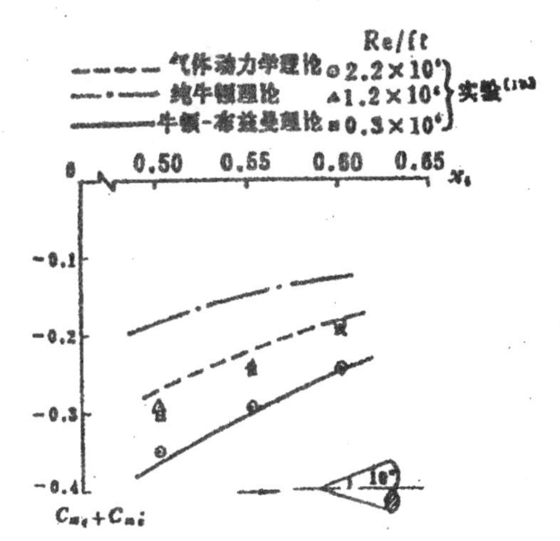

图 10 尖锥在高超音速气流中的俯仰阻尼导数

五、结　　论

1. 三角翼是飞机和导弹上一种常见的翼面形状，所以正确估算其动导数十分必要。在飞行器预研工作中，利用解析理论来估算动导数不仅很快，而且能够看出动导数的变化规律。

2. 对于薄翼在零攻角附近的低频微幅谐振问题，亚、超音速以及满足一定条件的跨音速流动，可以用非定常线化势流理论估算动导数。

3. 当厚翼在超音速气流中在大攻角下作低频微幅谐振时，只要前缘激波保持附体，其动导数的计算就可以采用超音速和高超音速的统一理论。

4. 对于高超音速气流中翼的动导数计算，可以采用非定常的 Newton-Busemann 理论。离心力修正项是非常重要的，不能轻易忽略。

参　考　文　献

[1] Miles, J. W., The Potential Theory of Unsteady Supersonic Flow, Cambridge monographs on mechanics and applied mathematics, Cambridge at the university press (1959).

[2] Thompson, J.S. & Fall, R.A., ARC CP 3355 (1962).

[3] Ericsson, L.E. & Reding, J.P., AIAA 77-667 (1977).

[4] Landahl, M. T., Unsteady Transonic Flow, International series of monographs in aeronautics and astronautics division II; Aerodynamics Vol.2, Oxford, Pergamon press (1961).

[5] Orlik-Rückemann, K. & Olsson, C. O., FFA Rep. 62 (1956).

[6] Tobak, M. & Ressing, H. C., AGARD Rep. 343 (1961).

[7] Albano, E. & Rodden, W. P., AIAA J., 7, 11 (1969).

中国科学技术大学学报

（抽 印 本）

JOURNAL OF CHINA UNIVERSITY OF SCIENCE AND TECHNOLOGY

（OFFPRINT）

第16卷 第1期 1986年3月
中国科学技术大学学报
JOURNAL OF CHINA UNIVERSITY OF SCIENCE AND TECHNOLOGY
Vol.16,No.1 March, 1986

有限元法在跨音速翼型计算中的应用

——定常亚临界初算

徐守栋　　安长发

（近代力学系）

摘　　要

本文用 Galerkin 有限元方法计算了对称双弧翼型在定常亚临界流中的表面压力分布。数值计算的结果与实验数据基本符合。

本文从跨音速小扰动方程出发，用有限元法计算了对称双弧翼型的跨音速流场。参照文［1］，采用矩形网格剖分，Hermite 插值函数，Galerkin 方法控制误差。但在有限元分析中，没有使用文［1］在求单元矩阵元素值时的数值积分方法，而是找到了一个递推公式，把刚度矩阵中全部元素的值都一次积分出来了。通过对 6％对称双弧翼型在亚临界下压力分布的计算表明，这种方法是简易可行的。

一、理 论 分 析

对于二维薄翼的定常跨音速绕流，可用跨音速小扰动方程和相应的边界条件来描述

$$[1-M_\infty^2-M_\infty^2(\gamma+1)\phi_x]\phi_{xx}+\phi_{yy}=0, \tag{1}$$

$$\phi_y-(1+\phi_x)g'=0,\quad \text{（在翼面上）} \tag{2}$$

$$\phi=0,\quad \text{（在无穷远处）} \tag{3}$$

其中 $g(x)$ 为无量纲翼面方程。当 $M_\infty\to1$ 时，文［2］证明了 $M_\infty\sqrt{1+(\gamma+1)\phi_x}=M_l$（当地马赫数），则（1）可写成

$$\phi_{xx}+\phi_{yy}=f,\qquad f=M_l^2\phi_{xx}, \tag{4}$$

作为初步近似，边界条件（3）可认为在有限远处也成立。

若翼型是对称的且来流为零攻角，则可只考虑上半平面。今取出长为5倍、高为2倍弦

1985年6月6日收到。

的长矩形域，剖分成64个单元得85个节点如图1，对于薄翼型，可近似地认为翼面附近的

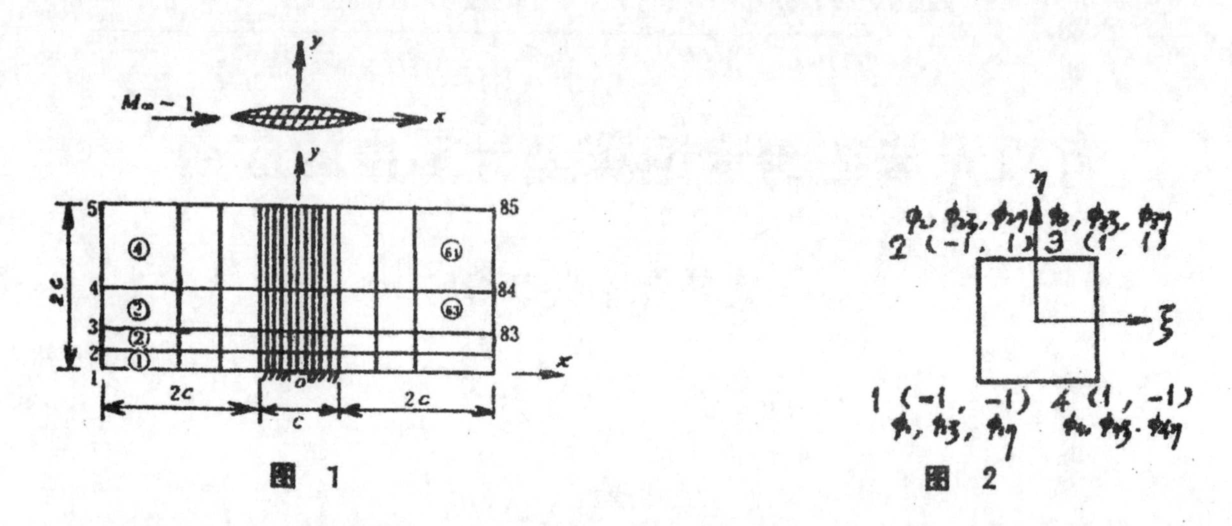

图 1　　　　图 2

单元也是矩形。作变换$\xi=(x-x_c)/a$，$\eta=(y-y_c)/b$，其中a,b，(x_c,y_c)分别为某个单元的长高之半和形心坐标，则每个矩形单元都变成了相同的正方形单元，如图2。

对正方形单元内的函数ϕ作Hermite插值

$$\bar{\Phi}(\xi,\eta)=\phi_i N_i(\xi,\eta),\tag{5}$$

其中$\phi_i(i=1,2,\cdots,12)$代表单元节点上的函数及导数：ϕ_1，$\phi_{1\xi}$，$\phi_{1\eta}$，ϕ_2，$\phi_{2\xi}$，$\phi_{2\eta}$，ϕ_3，$\phi_{3\xi}$，$\phi_{3\eta}$，ϕ_4，$\phi_{4\xi}$，$\phi_{4\eta}$。$N_i(\xi,\eta)(i=1,2,\cdots,12)$是形状函数，具体表达式可参见康纳的书[3]，如

$$\begin{cases}N_1=\frac{1}{8}(\eta-1)(\xi-1)\left[\frac{1}{2}(\eta-1)(\xi-1)-\frac{1}{2}(\eta+1)(\xi+1)\right.\\ \qquad\left.-(\eta-1)(\eta+1)-(\xi-1)(\xi+1)\right],\\ \cdots\cdots\cdots\cdots\\ N_3=-\frac{1}{8}(\eta-1)^2(\eta+1)(\xi-1),\\ \cdots\cdots\cdots\cdots\\ N_6=-\frac{1}{8}(\eta-1)(\eta+1)^2(\xi-1),\\ \cdots\cdots\cdots\cdots\end{cases}\tag{6}$$

根据Galerkin方法，在单元域E和翼面W上有

$$\iint_E(\bar{\Phi}_{xx}+\bar{\Phi}_{yy}-\bar{f})N_i dxdy+\int_W[\bar{\Phi}_y-(1+\bar{\Phi}_x)]g'N_i dl=0\tag{7}$$

应用Green定理，得

$$\iint_E(\bar{\Phi}_x N_{i,x}+\bar{\Phi}_y N_{i,y})dxdy=\oint_{\partial E}\bar{\Phi}_n N_i dl-\iint_E M_l^2\bar{\Phi}_{xx}N_i dxdy$$
$$+\int_W[\bar{\Phi}_y-(1+\bar{\Phi}_x)]g'N_i dl=0\tag{8}$$

式中 $\bar{\Phi}_n$ 为 $\bar{\Phi}$ 在单元边界 ∂E 上的外法向导数。式（8）右端第一项和第三项的前半部分恰好互相抵消，把插值函数（5）代入并变换成(ξ,η)的函数，得出

$$S_{ij}\phi_j=F_i,\quad i=1,2,\cdots,12 \tag{9}$$

其中

$$\begin{cases} S_{ij}=\dfrac{b}{a}A_{ij}+\dfrac{a}{b}B_{ij}, & F_i=-\dfrac{b}{a}\bar{M}^2C_{ij}\phi_j-aX_i, \\ A_{ij}=\displaystyle\int_{-1}^{1}\int_{-1}^{1}N_{i\xi}N_{j\xi}d\xi d\eta, & B_{ij}=\displaystyle\int_{-1}^{1}\int_{-1}^{1}N_{i\eta}N_{j\eta}d\xi d\eta, \\ C_{ij}=\displaystyle\int_{-1}^{1}\int_{-1}^{1}N_iN_{j\xi\xi}d\xi d\eta, & X_i=\displaystyle\int_W N_ig'd\xi,\quad \bar{M}=\dfrac{1}{4}(M_1+M_2+M_3+M_4), \end{cases} \tag{10}$$

把局部有限元方程（9）合成为总体方程

$$S_{kl}\phi_l^{(n)}=F_k^{(n-1)}, \tag{11}$$

其中 $k,l=1,2,\cdots,255$。再用迭代法解方程组（11）即可得到各节点参数（包括导数），从而可以求出节点处的速度 $u=\phi_x$，$v=\phi_y$ 和压力系数 $C_p=-2\phi_x$。

在单元分析中出现的矩阵 A_{ij},B_{ij},C_{ij}，文［1］是在计算机上进行数值积分的，这会带来两个问题：一是增加了机时，二是近似计算引起了新的误差。实际上，这些矩阵中的全部元素都是可以精确积分出来的，只不过矩阵元素太多，而且每个元素又都是二重积分，所以计算起来太繁琐冗长罢了。但是经过分析后发现有如下规律：1）C_{ij} 有许多元素是零；2）B_{ij} 与 A_{ij} 的所有元素都具有某种对应关系；3）A_{ij} 可以找到一个递推公式。这样可以大大简化这些积分计算。

由式（6）可知，插值函数及其导数都是由因子 $(\eta-1)$，$(\eta+1)$，$(\xi-1)$，$(\xi+1)$ 的各次幂的乘积的组合，而 A_{ij} 的各元素都是这些组合在正方形单元域内的二重积分，考虑到这些积分的可分离变量性质，则只需计算如下积分即可：

$$I_{mn}=\int_{-1}^{1}\int_{-1}^{1}(\eta-1)^m(\eta+1)^n d\eta=(-1)^m\cdot 2^{m+n+1}\cdot\frac{m!n!}{(m+n+1)!}, \tag{12}$$

利用公式（12）可以算出 A_{ij} 的全部元素，如

$$A_{3,6}=\int_{-1}^{1}\int_{-1}^{1}N_{3\xi}N_{6\xi}d\xi d\eta=\frac{1}{32}I_{3,3}=-\frac{1}{35},$$

另外，根据 $N_i(\xi,\eta)$ 对 ξ,η 的对称性，发现 B_{ij} 与 A_{ij} 的元素之间有如下对应关系：

A	1	2	3	4	5	6	7	8	9	10	11	12
B	1	3	2	10	12	11	7	9	8	4	6	5

如 $B_{2,11}=A_{3,6}$。这样算出 A_{ij} 同时也得到了 B_{ij}、C_{ij} 的计算类似 A_{ij}，只是更简单些。求出这几个矩阵的全部元素以后可存入计算机备用，从而可以避免进行数值积分了。

二、算例和讨论

为了检验该方法的精度和效率，本文以相对厚度为6％的对称双弧翼型在 $M_\infty=0.806$，

0.861的两种亚临界条件下进行了试算，算得的压力分布与实验数据的对比见图3。图中还画出了文［1］的计算结果。可以看出，虽然本文剖分得比文［1］粗糙些（［1］为120单

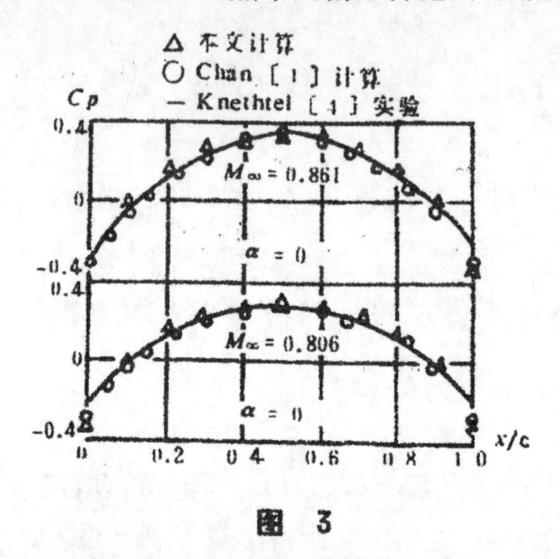

图 3

元150节点），但计算结果的精度差不多，都与实验数据基本符合。

本文计算是在M—140F中型机上完成的，每个状态所用的CPU时间为3分钟。文[1]是在大型机Univac上计算的，每个状态计算需40秒。计及两种机器运算速度上的差别，本文方法在运算次数上约占文［1］的45%，从而证明了本文在基本上保证精度的前提下提高了效率节省了机时。

参考文献

［1］ Chan, S. T. K., Brashears, M. R., Young, V. Y. C., *AIAA* 75—79, 1975.

［2］ 罗时钧等，跨音速定常势流的混合差分法，国防工业出版社，1979。

［3］ 康纳，J. J. 等，流体流动的有限元法，吴望一译，科学出版社，1981。

［4］ Knechtel, E. D., *NASA* TN D—15, 1959.

安长发博士科研论文集

The Application of Finite Element Methods to Prediction of Aerodynamic Forces on Transonic Airfoils

Xu Shoudong　　An Changfa

Department of Modern Mechanics

Abstract

In this paper the Finite Element Methods based on the Galerkin approch from the Transonic Small Perturbation Equations are developed and used to predict surface pressure distributions on the airfoils which are placed in steady subcritical flow. The numerical results are compared well with the experimental data.

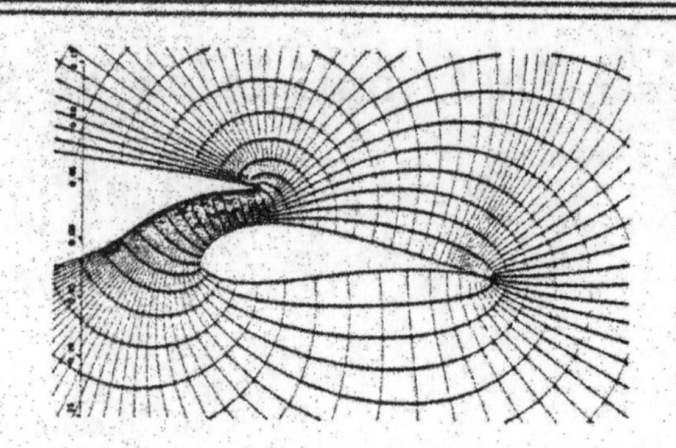

CASI AERODYNAMICS SYMPOSIUM ON COMPUTATIONAL FLUID DYNAMICS

May 25-26, 1987

Loews-Westbury Hotel

Toronto
Ontario

BOOK OF ABSTRACTS

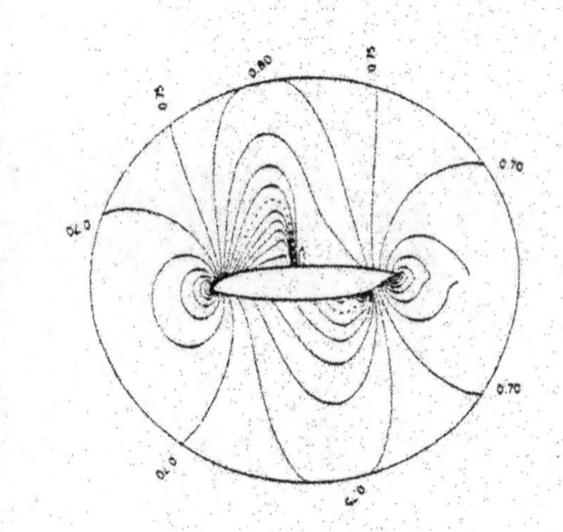

C F AN

University of Windsor

Abstract

Calculation of Pressure Distribution

on 2-D Thin Airfoil in Transonic Flows

by Finite Element Methods

by

Chang-fa An*, Visiting Scholar,

Fluid Dynamics Research Institute,

University of Windsor, Windsor, Ontario, N9B 3P4.

In this paper the Finite Element Method, based on the Galerkin approach is developed and applied to predict the pressure distribution on the surface of airfoils which are placed in steady transonic flows.

Similar to Chan's method (AIAA Paper, 75-79), in this approach rectangular elements are used and Hermite interpolation functions are chosen to determine the values of unknown functions in the elements. The error caused by substituting approximate values of unknown functions into the transonic small-disturbance equation is controlled by a Weighted Residual Method in which the shape functions are chosen as weighted functions.

In Chan's method, a number of numerical integrations have to be performed to evaluate the values of components of the element stiffness matrices. In the present approach, such integrals can be calculated exactly after finding some interesting properties of the element matrices.

This approach is shown to be effective by calculating the surface pressure distribution of a 2-D symmetric 6% biarc airfoil located in subcritical airflow. The advantage of the approach over the "Numerical Integration Method" is mainly that it can significantly decrease the computer time. More than half of the time can be saved using this approach.

One problem is that large computer storage is still needed, although computer time is decreased considerably. In addition, for calculation in the supercritical case, a kind of "up-wind technique" is required to deal with the local supersonic regions and some further efforts are needed.

---

*Permanent Address: Department of Modern Mechanics, Univ. of Science & Technology of China, Hefei, Anhui, People's Republic of China.

出国前在专业会议上发表的的技术报告

Ⅰ-06

安长发, 非对称边界层转淚对烧蚀钝锥静稳定性的影响, 第四届再入宇航飞行器全国会议, 中国, 四川, 绵阳, 1981. 11. 27 – 12. 3

– 本技术报告存档于中国空气动力研究与发展中心资料室, 中国, 四川, 绵阳

Ⅰ-07

安长发, 亚跨超音速三角翼纵向稳定性导数的估算方法, 有翼导弹纵向稳定性导数的全国讨论会, 中国, 北京, 1983. 9. 12 – 15

– 本技术报告存档于北京空气动力研究所资料室

Ⅰ I-08

安长发, 细长翼身组合体在亚 音速气流中的动导数预估, 航天工业部技术评估会, 中国, 安徽, 黄山, 1985. 6. 22 – 25

– 本技术报告存档于航天工业部技术档案室

Ⅰ-09

安长发, 关于机动再入宇航飞行器脉动压力研究的当前发展动态, 机动再入宇航飞行器的全国讨论会, 中国, 广西, 桂林, 1985. 11. 17 – 20

– 本技术报告存档于航天工业部技术档案室

Ⅰ-10

安长发, 李杰忠, 跨超音速风洞的重建和标定, 第六届风洞试验技术全国会议, 中国, 云南, 昆明, 1985. 11. 28 – 12. 2

– 本技术报告存档于航空工业部技术档案室

第II部分
出国后博士学位论文阶段

TRANSONIC COMPUTATION USING EULER EQUATIONS IN STREAM FUNCTION COORDINATE SYSTEM

by

Chang-Fa An

安长发博士科研论文集

A Dissertation
Submitted to the Faculty of Graduate Studies and Research
through the Department of Mathematics and Statistics
in Partial Fulfillment of the Requirements for the
Degree of Doctor of Philosophy
at the University of Windsor

Windsor, Ontario, Canada
1992

这是博士学位论文，加拿大 温莎大学 (University of Windsor, Windsor, Ontario, Canada)。原文现收藏于校图书馆。

Changfa An
Washington, D.C.
Oct. 23, 1991

PROCEEDINGS

THIRD INTERNATIONAL CONFERENCE ON INVERSE DESIGN CONCEPTS AND OPTIMIZATION IN ENGINEERING SCIENCES ICIDES-III

October 23-25, 1991
Washington, D.C.

Sponsored by:

NASA HEADQUARTERS
Aeronautics Directorate
Aerodynamics Division
600 Independence Avenue, SW
Washington, D.C. 20546

NATIONAL SCIENCE FOUNDATION
Division of Electrical and Communications Systems
Communications & Computational Systems Program
1800 G Street, NW
Washington, D.C. 20550

OFFICE OF NAVAL RESEARCH
Department of the Navy
Mechanics Division
800 North Quincy Street
Arlington, VA 222171

PENN STATE UNIVERSITY
College of Engineering
Department of Aerospace Engineering
233 Hammond Building
University Park, PA 16802

Organized and edited by:

George S. Dulikravich

ANALYSIS AND DESIGN OF TRANSONIC AIRFOILS USING STREAMWISE COORDINATES

R.M. BARRON AND C.-F. AN

Department of Mathematics and Statistics
and Fluid Dynamics Research Institute
University of Windsor
Windsor, Ontario, Canada N9B 3P4

Abstract. In this work, a new approach is developed for analysis and design of transonic airfoils. A set of full-potential-equivalent equations in von Mises coordinates is formulated from the Euler equations under the irrotationality and isentropic assumptions. This set is composed of a main equation for the main variable y, and a secondary equation for the secondary variable R. The main equation is solved by type-dependent differencing combined with a shock point operator. The secondary equation is solved by marching from a non-characteristic boundary. Sample computations on NACA 0012 and biconvex airfoils show that, for the analysis problem, the present approach achieves good agreement with experimental C_p distributions. For the design problem, the approach leads to a simple numerical algorithm in which the airfoil contour is calculated as a part of the flow field solution.

1. Introduction

Transonic flow is a widely encountered phenomenon in aeronautics and astronautics but is not easy to calculate because the flow field, and the governing equations as well, are mixed type. Therefore, transonic computation had little progress until 1971 when Murman and Cole developed a type-dependent difference scheme and successfully solved the transonic small disturbance (TSD) equation[1]. Since then, transonic computation has become one of the most upsurging topics for computational fluid dynamicists[2-8]. In 1974, Jameson extended transonic flow computation to the full potential (FP) stage by constructing a rotated difference scheme[4]. Afterwards, papers were published on transonic computation by solving Euler equations[5,6] and their equivalent streamfunction-vorticity formulation[7,8]. Nevertheless, in spite of the recent active efforts on Euler solvers, the full potential calculation is still attractive due to its simplicity, efficiency and sufficient accuracy.

The von Mises transformation is a type of streamline-based transformation which generates a streamwise coordinate system. The von Mises formulation has a number of advantages when applied in CFD. For example, one can resolve the problem of body-fitting coordinates without performing any grid generation. This is because the governing equation (flow physics) and grid generation equation (flow geometry) are combined together in this formulation. Furthermore, the boundary condition on the airfoil for the analysis problem is Dirichlet, and a non-iterative design technique

---

For presentation at the Third International Conference on Inverse Design Concepts and Optimization in Engineering Sciences (ICIDES-III), Washington, D.C., USA, October 23-25, 1991

can be developed for the inverse problem, leading to simplified numerical algorithms and a saving of computer time. Therefore, since Barron[9] connected the von Mises transformation with Martin's approach[10] and solved incompressible 2-D symmetric flow, numerical simulations based on the von Mises transformation have been considerably extended, such as to incompressible lifting[11], axisymmetric[12] and design[13] problems, and to transonic flow[14,15]. In addition, Greywall[16] and Dulikravich[17] obtained a similar formulation for incompressible and compressible flows, respectively.

However, when extending Barron's approach[9] to transonic flow, several problems appear. For compressible flow, apart from the von Mises variable y, another variable, the density ρ, must be updated in each iteration. But in the transonic range, the classical difficulty of double value density-massflux relation still exists. Besides, shock waves are not easy to handle in von Mises coordinates either by the artificial density technique or by type-dependent differencing. Recently, the authors[18] developed a new approach to overcome these difficulties by solving so-called full-potential-equivalent equations in von Mises coordinates. The principal advances over the previous transonic work[14,15] are as follows: 1) To update density, instead of solving the non-linear algebraic Bernoulli equation, a first order partial differential equation is solved, thereby avoiding the double density problem; 2) To handle shock waves properly, a shock point operator in von Mises coordinates is proposed and combined with the type-dependent difference scheme so that shock waves can be captured correctly; 3) Introducing a concept of generalized density linearizes the density equation.

In the next section, an outline of the mathematical formulation is given. The numerical algorithms for analysis and design problems are constructed in sections 3 and 4. In section 5, sample computations are performed to test the approach, and conclusions are given in section 6.

2. Flow Equations in Streamwise Coordinates

Two dimensional, steady, inviscid fluid flows are governed by the Euler equations

$$\begin{pmatrix} \rho u \\ \rho u^2 + p \\ \rho u v \\ \rho u H \end{pmatrix}_x + \begin{pmatrix} \rho v \\ \rho u v \\ \rho v^2 + p \\ \rho v H \end{pmatrix}_y = 0 \tag{2-1}$$

where ρ is density, u and v are velocity components in Cartesian coordinates, p is pressure, $H = \frac{\gamma}{\gamma-1} p/\rho + (u^2 + v^2)/2$ is total enthalpy and γ is the ratio of specific heats. ρ, u, v and p are normalized by free stream density ρ_∞, speed q_∞ and dynamic pressure head $\rho_\infty q_\infty^2$ while x and y are scaled by the airfoil chord length.

Introducing streamfunction ψ, such that $\psi_y = \rho u, \psi_x = -\rho v$ and substituting into equation $(2-1)$, one gets

$$\begin{pmatrix} \psi_y^2/\rho + p \\ -\psi_x\psi_y/\rho \\ \psi_y H \end{pmatrix}_x + \begin{pmatrix} -\psi_x\psi_y/\rho \\ \psi_x^2/\rho + p \\ -\psi_x H \end{pmatrix}_y = 0. \tag{2-2}$$

Streamfunction $\psi = \psi(x,y)$ can be rewritten in an implicit form, $F(x,y,\psi) = 0$, or in an explicit form, $y = y(x,\psi)$. This process is equivalent to introducing von Mises transformation: $x \equiv \phi, y = y(\phi,\psi)$. If the Jacobian $J = \partial(x,y)/\partial(\phi,\psi) \neq 0, \infty$, then the transformation is single-valued and $(2-2)$ becomes

$$\begin{pmatrix} 1/(\rho y_\psi) + p y_\psi \\ y_\phi/(\rho y_\psi) \\ H \end{pmatrix}_\phi + \begin{pmatrix} -p y_\phi \\ p \\ 0 \end{pmatrix}_\psi = 0 \tag{2-3}$$

where the total enthalpy $H = \frac{\gamma}{\gamma-1}p/\rho + (1+y_\phi^2)/(2\rho^2 y_\psi^2)$. The streamline ordinate y, called von Mises variable, is viewed as a function of ϕ and ψ. The velocity components can be easily calculated from $u = 1/(\rho y_\psi), v = y_\phi/(\rho y_\psi)$, after y and ρ are solved.

It is known that the entropy increase across a shock wave is of third order of the shock strength. So, if the shock is not strong, transonic flow can be assumed isentropic and irrotational. Replacing the energy equation in $(2-3)$ by the isentropic relation and keeping in mind that $\phi \equiv x$, we reduce $(2-3)$ to

$$(\frac{1}{\rho y_\psi} + p y_\psi)_x - (p y_x)_\psi = 0, \tag{2-4a}$$

$$(\frac{y_x}{\rho y_\psi})_x + p_\psi = 0, \tag{2-4b}$$

$$p = \frac{\rho^\gamma}{\gamma M_\infty^2}, \tag{2-4c}$$

$$(\frac{y_x}{\rho})_x - (\frac{1+y_x^2}{\rho y_\psi})_\psi = 0 \tag{2-4d}$$

where M_∞ is free stream Mach number and the last equation is the irrotationality condition, $\omega = 0$, expressed in von Mises coordinates. Substituting $(2-4c)$ into $(2-4a)$ and $(2-4b)$ and expanding $(2-4d)$, we get

$$-y_{x\psi} + y_\psi(y_\psi^2 \frac{\rho^{\gamma+1}}{M_\infty^2} - 1)\frac{\rho_x}{\rho} - y_x y_\psi^2 \frac{\rho^{\gamma+1}}{M_\infty^2}\frac{\rho_\psi}{\rho} = 0, \tag{2-5a}$$

$$y_\psi y_{xx} - y_x y_{x\psi} - y_x y_\psi \frac{\rho_x}{\rho} + y_\psi^2 \frac{\rho^{\gamma+1}}{M_\infty^2}\frac{\rho_\psi}{\rho} = 0, \tag{2-5b}$$

$$y_\psi^2 y_{xx} - 2y_x y_\psi y_{x\psi} + (1+y_x^2)y_{\psi\psi} - y_x y_\psi^2 \frac{\rho_x}{\rho} + y_\psi(1+y_x^2)\frac{\rho_\psi}{\rho} = 0. \qquad (2\text{-}5c)$$

Properly manipulating the above three equations can produce several sets of equations. Each set has two independent equations for two variables. To make the formulation more compact, define generalized density $R = \rho^{\gamma+1}$ as an alternative to density ρ. Solving for ρ_x/ρ and ρ_ψ/ρ from $(2-5a)$and $(2-5b)$, and substituting into $(2-5c)$, one gets

$$(y_\psi^2 - \frac{M_\infty^2}{R})y_{xx} - 2y_x y_\psi y_{x\psi} + (1+y_x^2)y_{\psi\psi} = 0. \qquad (2\text{-}6a)$$

Eliminating $y_{x\psi}$ from $(2-5a)$ and $(2-5b)$ gives

$$y_x y_\psi^2 R_x - y_\psi(1+y_x^2)R_\psi = (\gamma+1)M_\infty^2 y_{xx}. \qquad (2\text{-}6b)$$

Equation $(2-5a)$ can be rewritten as

$$y_\psi(y_\psi^2 - \frac{M_\infty^2}{R})R_x - y_x y_\psi^2 R_\psi = (\gamma+1)M_\infty^2 y_{x\psi}. \qquad (2\text{-}6c)$$

Substituting the above $y_{xx}, y_{x\psi}$ into $(2-5c)$ and replacing ρ by R, one obtains

$$y_x(y_\psi^2 - \frac{M_\infty^2}{R})R_x + y_\psi(1 - y_x^2 - \frac{M_\infty^2}{R}\frac{1+y_x^2}{y_\psi^2})R_\psi = (\gamma+1)M_\infty^2 \frac{1+y_x^2}{y_\psi^2} y_{\psi\psi}. \qquad (2\text{-}6d)$$

It is important to note that $(2-6b)$ is linear after introducing the new variable R. The term M_∞^2/R is usually called compressibility factor.

In principle, any two of the above four equations could be combined as a set of equations to solve for y and R. But, in practice, equation $(2-6a)$ is always selected to solve for y and one of the remaining three equations is selected to solve for R. Equation $(2-6a)$ is a second order, non-linear, partial differential equation of mixed type depending on the local flow property. If the flow is subsonic/supersonic, then $(2-6a)$ is elliptic/hyperbolic. In other words, the mathematical classification of the equation is consistent with the physical nature of the local flow. Therefore, $(2-6a)$ is named the main equation for the corresponding main variable y. Equations $(2-6b)-(2-6d)$ are called secondary equations for the secondary variable R. Among the three secondary equations, $(2-6b)$ appears simpler because it is linear and hence priority is given to it to accompany the main equation. The main equation $(2-6a)$ and one of the secondary equations $(2-6b)-(2-6d)$ constitute a set of so-called full-potential-equivalent equations. They are coupled with each other and solved in an alternating and iterative manner. More details and other forms of full-potential-equivalent equations can be found in [18].

3. Analysis Problem

For a symmetric airfoil placed in a transonic stream at zero angle of attack, the governing equations $(2-6a)$ and $(2-6b)$ can be rewritten as

$$A_1 y_{xx} + A_2 y_{x\psi} + A_3 y_{\psi\psi} = 0, \tag{3-1}$$

$$B_1 R_x + B_2 R_\psi = B_3 \tag{3-2}$$

where $A_1 = y_\psi^2 - M_\infty^2/R, A_2 = -2y_x y_\psi, A_3 = 1 + y_x^2, B_1 = y_x y_\psi^2, B_2 = -y_\psi(1 + y_x^2), B_3 = (\gamma+1)M_\infty^2 y_{xx}$. The boundary conditions on y are Dirichlet: $y = f(x)$ on the airfoil, $y = \psi$ at infinity, $y = 0$ on the symmetry line and $R = 1$ at infinity, where $f(x)$ is the airfoil shape function. The computational domain and boundary conditions are shown in Fig.1.

Since the mathematical character of $(3-1)$ depends on the local flow property, it is necessary to apply Murman and Cole's type-dependent scheme[1] to solve for y. Applying the type-dependent difference scheme to $(3-1)$ gives

$$Ay_{i,j-1} + By_{i,j} + Cy_{i,j+1} = RHS \tag{3-3}$$

where $A = \beta^2 A_3 - \frac{1-\nu}{2}\beta A_2, B = -2\beta^2 A_3 + (1-3\nu)A_1, C = \beta^2 A_3 + \frac{1-\nu}{2}\beta A_2,$

$$\begin{aligned} RHS = & -\nu A_1(y_{i+1,j} + y_{i-1,j}) + (1-\nu)A_1(2y_{i-1,j} - y_{i-2,j}) \\ & - \nu\beta A_2(y_{i+1,j+1} - y_{i+1,j-1} - y_{i-1,j+1} + y_{i-1,j-1})/4 \\ & + (1-\nu)\beta A_2(y_{i-1,j+1} - y_{i-1,j-1})/2, \end{aligned}$$

and $\beta = \Delta x/\Delta\psi$, for $i = 2,3,\ldots,I_{max}-1, j = 2,3,\ldots,J_{max}-1$. The switch parameter $\nu = 1$ for a subsonic point, $\nu = 0$ for a supersonic point. The resulting system of difference equations $(3-3)$ has a tridiagonal coefficient matrix so that SLOR can be applied by relaxing along vertical lines, sweeping from left to right and iterating up to convergence. (see Fig.1)

After $y(x,\psi)$ is solved from the main equation $(3-1)$ and y_x, y_ψ, y_{xx} are properly differenced, the secondary equation $(3-2)$ can be solved for $R(x,\psi)$ by marching from an initial line other than its characteristic curve. The slope of its characteristic curve is $d\psi/dx = -(1+y_x^2)/(y_x y_\psi)$. At infinity, $d\psi/dx = \infty$. Thus, left and right far field boundaries are characteristic curves and hence cannot serve as initial lines. Fortunately, the horizontal boundary is not a characteristic and we can march equation $(3-2)$ from the top boundary to the airfoil using the condition $R=1$ at infinity.

The Crank-Nicolson scheme for $(3-2)$ gives

$$\widetilde{A}R_{i-1,j} + \widetilde{B}R_{i,j} + \widetilde{C}R_{i+1,j} = \widetilde{RHS} \tag{3-4}$$

where $\widetilde{RHS} = \widetilde{C}R_{i-1,j+1} + \widetilde{B}R_{i,j+1} + \widetilde{A}R_{i+1,j+1} + 4\Delta x B_3, \widetilde{A} = -B_1, \widetilde{B} = -4\beta B_2,$ $\widetilde{C} = B_1, \beta = \Delta x/\Delta\psi$ for $j = J_{max} - 1, ..., 3, 2, 1, i = 2, 3, ..., I_{max} - 1$. The system of difference equations $(3-4)$ can be solved row by row from the horizontal far field to the airfoil using SLOR, but no iteration is needed because $(3-2)$ is linear. After R is solved, the pressure coefficient is calculated from

$$C_p = \frac{2}{\gamma M_\infty^2}(R^{\frac{\gamma}{\gamma+1}} - 1). \tag{3-5}$$

However, it has been found after numerical tests that this procedure is efficient only for flow in which the shock is weak. For flow with a stronger shock, the iterations fail to converge. To overcome this difficulty, a special treatment of the shock wave is proposed following the ideas of Murman's shock structure analysis[2,3]. For usual transonic flows, the shock wave is approximately normal and the shock jump conditions are given by

$$[\frac{y_x}{\rho}] = 0, \qquad [y_\psi] = 0. \tag{3-6}$$

where [...] represents a jump across the shock. Based on this analysis, the difference quotient approximations to $y_{xx}, y_{x\psi}$ at a shock point, i.e. grid point just behind the shock, are constructed as below:

$$(y_{xx})_{i,j} = \frac{1}{\Delta x^2}(y_{i+1,j} - y_{i,j} - \alpha_j y_{i-1,j} + \alpha_j y_{i-2,j}) \tag{3-7a}$$

$$(y_{x\psi})_{i,j} = \frac{1}{4\Delta x\Delta\psi}(y_{i+1,j+1} - y_{i+1,j-1} + y_{i,j+1} - y_{i,j-1} - 3y_{i-1,j+1} + 3y_{i-1,j-1} + y_{i-2,j+1} - y_{i-2,j-1}). \tag{3-7b}$$

where α_j is the density jump factor on j^{th} streamline and given by the Rankine-Hugoniot relation of a normal shock. $(3-7a)$ and $(3-7b)$ are called shock point operator (SPO) in von Mises coordinates. The difference equations $(3-3)$ for y and $(3-4)$ for R are modified using SPO. Numerical experimentation has shown that SPO must be applied in the $y_{xx}, y_{x\psi}$ terms of the main equation $(3-1)$ and in the B_3 term of the secondary equation $(3-2)$. SPO is a crucial tool to capture shock waves in supercritical transonic flows.

4. Design Problem

Similar to the analysis problem, the main equation $(2-6a)$ and secondary equations $(2-6b)$ or $(2-6c)$ can be solved for y and R alternatively:

$$A_1 y_{xx} + A_2 y_{x\psi} + A_3 y_{\psi\psi} = 0, \tag{4-1a}$$

$$B_1 R_x + B_2 R_\psi = B_3, \tag{4-1b}$$

$$D_1 R_x + D_2 R_\psi = D_3 \tag{4-1c}$$

where $A_1 = y_\psi^2 - M_\infty^2/R, A_2 = -2y_x y_\psi, A_3 = 1 + y_x^2, B_1 = y_x y_\psi^2, B_2 = -y_\psi(1 + y_x^2), B_3 = (\gamma+1)M_\infty^2 y_{xx}, D_1 = y_\psi(y_\psi^2 - M_\infty^2/R), D_2 = -y_x y_\psi^2, D_3 = (\gamma+1)M_\infty^2 y_{x\psi}$. The boundary conditions are the same as in the analysis problem, except on the airfoil, which is unknown. There, the pressure coefficient C_{ps} is specified, hence, the generalized density is also specified:

$$R_s = (1 + \gamma M_\infty^2 C_{ps}/2)^{(\gamma+1)/\gamma} \tag{4-2}$$

On the airfoil surface, the Bernoulli equation in von Mises coordinates leads to

$$F(x)y_\psi^2 - y_x^2 = 1 \tag{4-3}$$

where

$$F(x) = \frac{2}{(\gamma-1)M_\infty^2}[(1 + \frac{\gamma-1}{2}M_\infty^2)R_s^{\frac{-2}{\gamma+1}} - R_s].$$

This is a Neumann boundary condition on the airfoil when solving $(4-1a)$ for y. $(4-2)$ is a Dirichlet boundary condition on the airfiol when solving $(4-1c)$ for R. In addition, on a symmetry line off the airfoil, $R_\psi = 0$.

If streamlines do not intesect each other on the airfoil, then $y_\psi > 0$, and if, furthermore, $F(x) \neq 0$ on the airfoil, then equation $(4-3)$ gives $y_\psi = \sqrt{(1 + y_x^2)/F(x)}$. For most practical transonic flows the required conditions are easily satisfied as long as C_{ps} is reasonably specified. Differencing y_ψ, we get

$$y_{i,1} = [4y_{i,2} - y_{i,3} - 2G(x_i)]/3 \tag{4-4}$$

where $G(x_i) = \Delta\psi\sqrt{[1 + (y_x^2)_{i,1}]/F(x_i)}$. Considering this new boundary condition, we modify system $(3-3)$ as follows:

For $j = 2$, equation $(3-3)$ reads $Ay_{i,1} + By_{i,2} + Cy_{i,3} = RHS$. Substituting $(4-4)$ into it, we have

$$(B + 4A/3)y_{i,2} + (C - A/3)y_{i,3} = RHS + 2AG(x_i)/3 \tag{4-5}$$

Replacing the first equation in system $(3-3)$ by $(4-5)$, solving the resulting system and applying $(4-4)$, we can obtain the desired airfoil contour $f(x_i) = y_{i,1}$ without further iteration of the airfoil shape. The computational domain and boundary conditions are shown in Fig. 2.

To solve for the secondary variable R, two secondary equations $(4-1b)$ and $(4-1c)$ are available. For equation $(4-1b)$, the marching procedure is the same as in the analysis problem, while for equation $(4-1c)$, the marching procedure is different. The slope of its characteristic curve is $d\psi/dx = -(y_x y_\psi)/(y_\psi^2 - M_\infty^2/R)$. At infinity, $d\psi/dx = 0$. So, the horizontal far field boundary is a characteristic curve, but the vertical boundaries are not. Therefore, the marching process can be carried out from left to right.

Crank-Nicolson scheme for $(4-1c)$ gives

$$\bar{A}R_{i,j-1} + \bar{B}R_{i,j} + \bar{C}R_{i,j+1} = R\bar{H}S \tag{4-6}$$

where $R\bar{H}S = \bar{C}R_{i-1,j-1} + \bar{B}R_{i-1,j} + \bar{A}R_{i-1,j+1} + 4\Delta x D_3, \bar{A} = -\beta D_2, \bar{B} = 4D_1, \bar{C} = \beta D_2, \beta = \Delta x/\Delta\psi$, for $i = 2,3,...,I_{max}-1, j = 2,3,...,J_{max}-1$.

For the first equation in system $(4-6)$, the boundary conditions $R_{i,1} = R_s(x_i)$ on the airfoil and $R_{i,1} = R_{i,2}$ on symmetry line should be imposed. It is noted that $y_{x\psi}$ in D_3 should be type-dependent differenced with SPO to keep consistency with the main equation.

Both $(4-1b)$ and $(4-1c)$ have been coupled with $(4-1a)$. Numerical experiments have shown that $(4-1c)$ gives better accuracy than $(4-1b)$. This is reasonable because the boundary condition on the airfoil is considered not only in the main equation $(4-1a)$, but also in the secondary equation $(4-1c)$, while it is not suitably considered in the secondary equation $(4-1b)$. However, the price to pay is more iterations because $(4-1c)$ is non-linear.

5. Sample Computations

The approach developed here is applied to calculated transonic flows for both analysis and design problems. Only symmetric airfoils at zero angle of attack are considered, but both subcritical and supercritical Mach numbers are included. In the computational domain, a 65x33 uniform mesh covers $-2 \le x \le 3, 0 \le \psi \le 2.5$ and the airfoil is placed between 0 and 1. For higher Mach numbers, a 80x31 mesh has been used. The computational domain and boundary conditions are shown in Figures 1 and 2.

Figures 3 and 4 are comparisons of calculated C_p distributions of NACA 0012 with experimental data at NAE[19] for $M_\infty = 0.490$ and at ONERA[19] for $M_\infty = 0.803$. Figure 5 indicates the calculated C_p distribution of a 6 percent biconvex airfoil at $M_\infty = 0.909$ compared with experimental data at NASA[20]. From these plots we can see that the present approach is able to accurately predict C_p distributions on airfoils in transonic flows. The agreement between computed pressure and available experimental data is quite satisfactory. For supercritical transonic flows, the shock wave can be captured by the presently proposed type-dependent scheme with SPO.

Figure 6 shows the designed contour of a 6 percent biconvex airfoil compared with the exact shape[21]. The specified C_p distribution on the airfoil comes from experiments at NASA[20] for $M_\infty = 0.909$. Figures 7 and 8 give designed NACA 0012 contours compared with the exact shape[21]. The specified C_p is from NAE[19] for $M_\infty = 0.490$ and ONERA[19] for $M_\infty = 0.803$. Here, we can see that the present approach is capable of designing airfoil contours with satisfactory accuracy.

6. Conclusions

1) The newly developed approach based on the full-potential-equivalent equations in von Mises coordinates is able to solve transonic flows for both analysis and design problems.

2) The full-potential-equivalent equations are composed of a main equation for the corresponding main variable, streamline ordinate y, and a secondary equation for the related secondary variable, generalized density R.

3) The type-dependent difference scheme with shock point operator is effective to solve the main equation for y and the shock point operator is crucial to capture shock waves in supercritical transonic flows.

4) The secondary equation can be solved for R by marching from a certain non-characteristic, density-specified boundary. Crank-Nicolson scheme proves to be useful to march such a equation.

5) For analysis problems, the boundary condition on the airfoil is Dirichlet, which is easy to implement.

6) For design problems, the airfoil contour can be obtained in a non-iterative manner because it is a part of the solution of the main equation.

References

1. E. M. Murman and J. D. Cole, Calculation of Plane Steady Transonic Flows, AIAA J., Vol.9, 114-121 (1971)

2. E. M. Murman and J. A. Krupp, Solution of the Transonic Potential Equation Using a Mixed Finite Difference System, Lecture Notes in Physics, Vol.8, pp199-206, Springer-Verlag, Berlin, 1974

3. E. M. Murman, Analysis of Embedded Shock Waves Calculated by Relaxation Methods, AIAA J., Vol.12, 626-633 (1974)

4. A. Jameson, Iterative Solution of Transonic Flows over Airfoils and Wings, Including Flows at Mach 1, Communications on Pure and Applied Mathematics, Vol.27, 283-309 (1974)

5. R. M. Beam and R. F. Warming, An Implicit Finite Differnce Algorithm for Hyperbolic System in Conservation Law Form, J.of Comp. Phys., Vol.22, 87-110 (1976)

6. J. C. Steger, Implicit Finite Difference Simulation of Flow around Arbitrary Two Dimensional Geometries, AIAA J., Vol.16, 676-686 (1978)

7. M. Hafez and D. Lovell, Numerical Solution of Transonic Stream Function Equation, AIAA J., Vol.21, 327-335 (1983)

8. H. L. Atkins and H. A. Hassan, A New Stream Function Formulation for the Steady Euler Equations, AIAA J., Vol.23, 701-706 (1985)

9. R. M. Barron, Computation of Incompressible Potential Flow Using von Mises Coordinates, J. of Math. and Comp. in Simulation, Vol.31, 177-188 (1989)

10. M. H. Martin, The Flow of a Viscous Fluid I, Archives for Rational Mechanics and Analysis, Vol.41, 266-286 (1971)

11. R. K. Naeem and R. M. Barron, Lifting Airfoil Calculations Using von Mises Variables, Communications in Applied Numerical Methods, Vol. 5, 203-210(1989)

12. R. M. Barron, S. Zhang, A. Chandna and N. Rudraiah, Axisymmetric Potential Flow Calculations, Part 1: Analysis Mode, Communications in Applied Numerical Methods, Vol. 6, 437-445(1990)

13. R. M. Barron, A Non-Iterative Technique for Design of Aerofoils in Incompressible Potential Flow, Communications in Applied Numerical Methods, Vol. 6, 557-564(1990)

14. R. M. Barron and R. K. Naeem, Numerical Solution of Transonic Flows on a Streamfunction Coordinate System, Intl. J. for Num. Methods in Fluids, Vol.9, 1183-1193 (1989)

15. R. K. Naeem and R. M. Barron, Transonic Computations on a Natural Grid, AIAA J., Vol.28, 1836-1838(1990)

16. M. S. Greywall, Streamwise Computation of 2-D Incompressible Potential Flows, J. of Comp. Phys., Vol.59, 224-231 (1985)

17. G. S. Dulikravich, A Stream-Function-Coordinate (SFC) Concept in Aerodynamic Shape Design, AGARD VKI Lecture Series, May 14-18, 1990

18. C.-F. An and R. M. Barron, Numerical Solution of Transonic Full-Potential-Equivalent Equations in von Mises Coordinates, to appear

19. J. J. Thibert, M. Grandjacques and L. H. Ohman, Experimental Data Base for Computer Program Assessment, AGARD AR-138, pp.A1-1 — A1-36, 1979

20. E. D. Knetchtel, Experimental Investigation at Transonic Speeds of Pressure Distributions over Wedge and Circular-Arc-Airfoil Sections and Evaluation of Perforated Wall Interferance, NASA TN D-15, 1959

21. I. Abbott and A. E. von Doenhoff, Theory of Airfoil Section, Dover Publ. Inc., N.Y., 1959

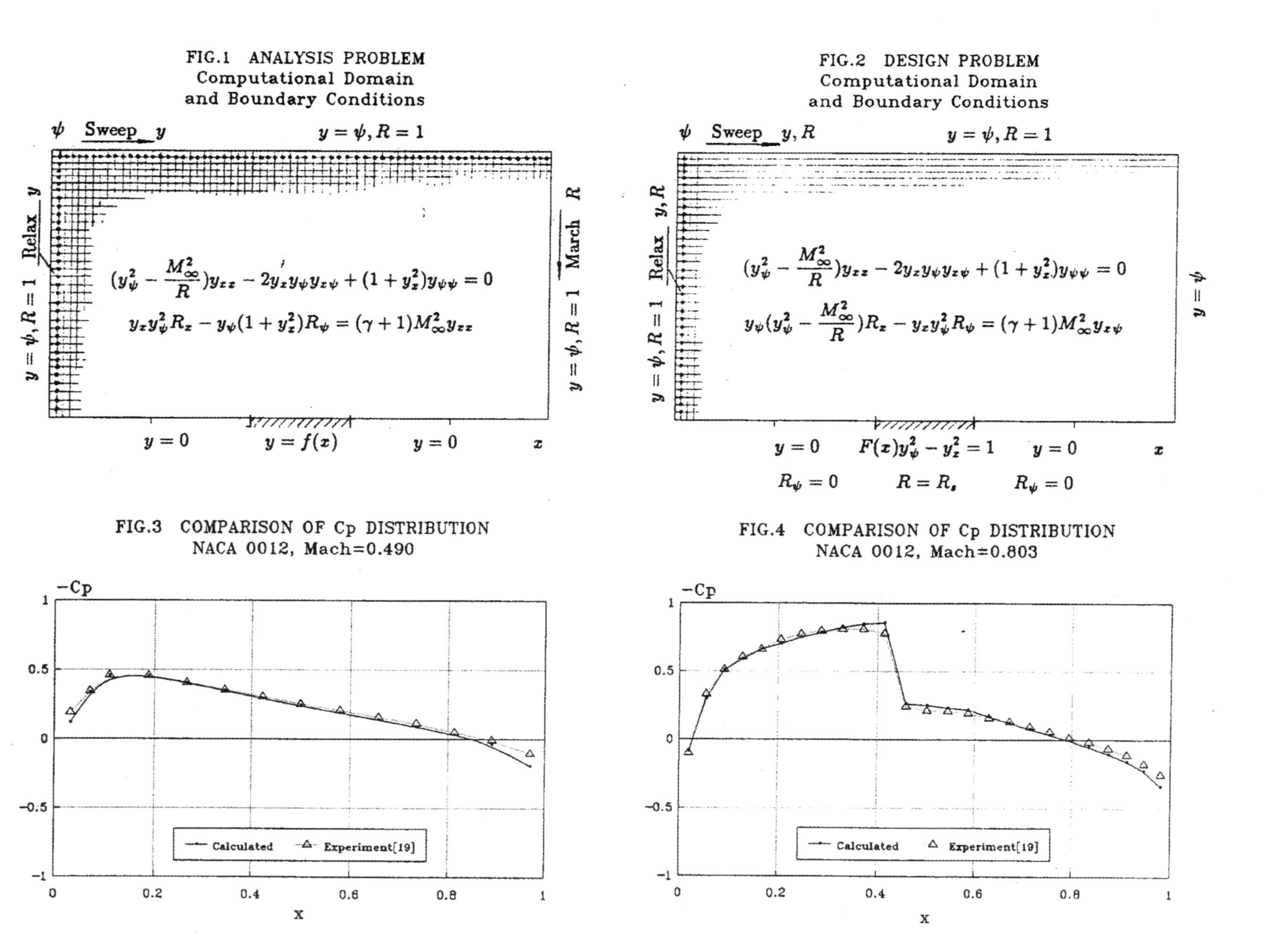

FIG.1 ANALYSIS PROBLEM
Computational Domain
and Boundary Conditions

FIG.2 DESIGN PROBLEM
Computational Domain
and Boundary Conditions

FIG.3 COMPARISON OF Cp DISTRIBUTION
NACA 0012, Mach=0.490

FIG.4 COMPARISON OF Cp DISTRIBUTION
NACA 0012, Mach=0.803

FIG.5 COMPARISON OF Cp DISTRIBUTION
Biconvex(6%) Airfoil, Mach=0.909

FIG.6 DESIGNED BICONVEX(6%) AIRFOIL
Mach=0.909

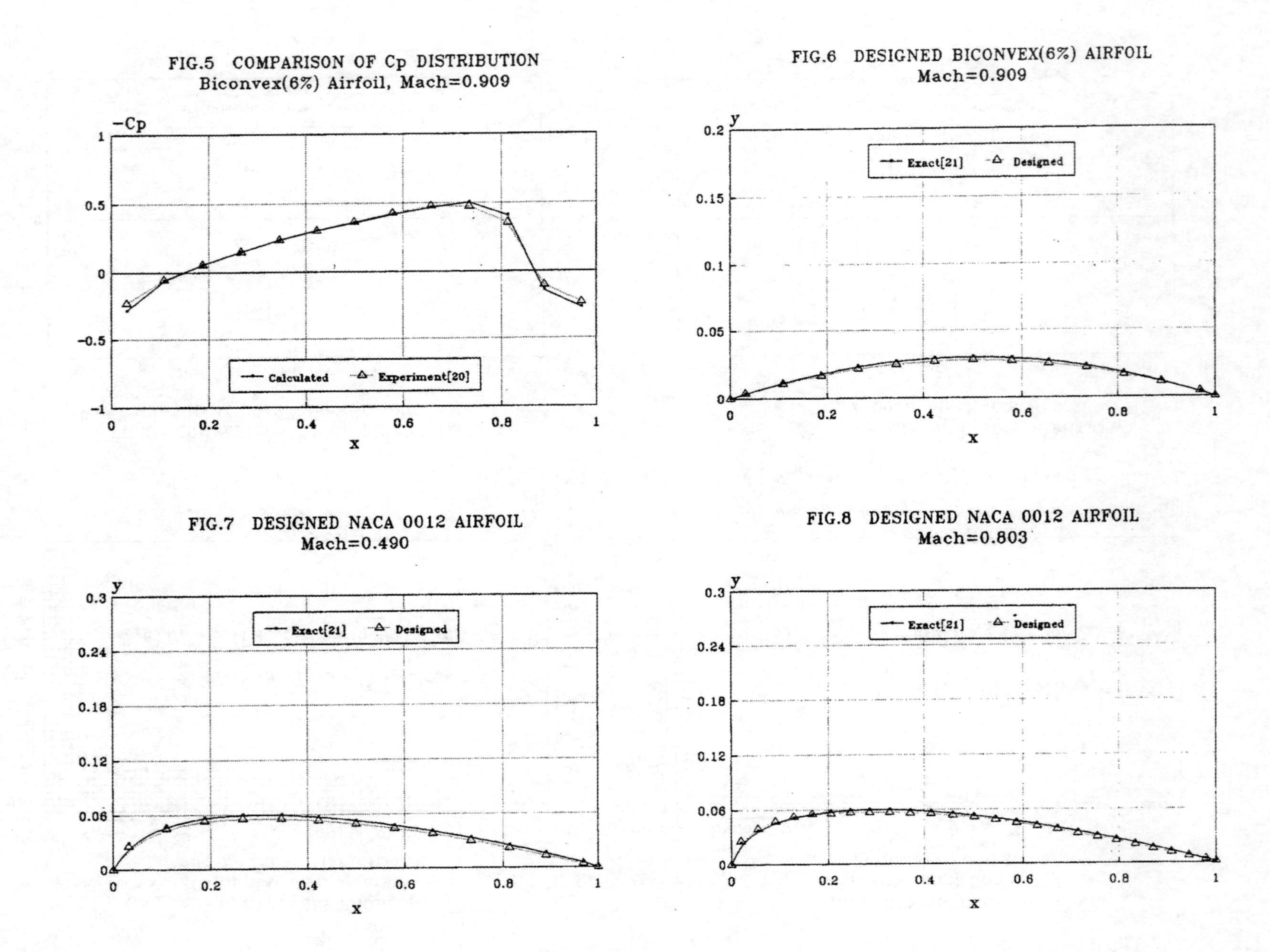

FIG.7 DESIGNED NACA 0012 AIRFOIL
Mach=0.490

FIG.8 DESIGNED NACA 0012 AIRFOIL
Mach=0.803

LIVRE DES RESUMES
BOOK OF ABSTRACTS

Tenth Canadian Symposium on Fluid Dynamics

Dixième Symposium Canadien sur la Dynamique des Fluides

Theoretical, Numerical, Experimental and Industrial Fluid Mechanics

Mécanique des Fluides Théorique, Numérique, Expérimentale et Industrielle

4 - 6 June / juin 1992
The University of New Brunswick
Saint John, New Brunswick

TRANSONIC CALCULATIONS USING EULER EQUATIONS IN STREAM FUNCTION COORDINATES

C.-F. An and R.M. Barron

Department of Mathematics and Statistics
and Fluid Dynamics Research Institute
University of Windsor
Windsor, Ontario, Canada N9B 3P4

Abstract

In this paper, a new approach has been developed to calculate two dimensional steady transonic flow past airfoils using the Euler equations in stream function coordinates.

Due to the importance of transonic flow phenomenon in aeronautics and aerospace engineering, transonic calculations have been a major topic of research for the past two decades. Most existing transonic codes require the use of a grid generator to determine a suitable distribution of grid points. Although simple in concept, the grid generation phase of the calculation takes a significant proportion of the CPU time and storage requirements. However, this time-consuming step can be completely avoided by introducing the von Mises transformation and the corresponding stream function coordinates. The von Mises transformation combines the flow physics and flow geometry together and produces a formulation in streamwise and body-fitting coordinates without performing any conventional grid generation.

In the present work, a set of Euler equivalent equations in stream function coordinates is formulated through the introduction of the stream function and the von Mises transformation. This set consists of three equations with three unknowns, one geometric and two physical quantities. To solve these equations, several numerical techniques are applied, including type-dependent differencing, shock point operator, marching from a non-characteristic boundary, successive line over relaxation, etc. Particular attention has to be paid to the supersonic case where a careful treatment of the shock is essential. It is shown that the shock point operator is crucial to accurately capture shock wave in this formulation. The computed results show good agreement with existing experimental data. The limitations of the approach and further investigations will also be discussed.

INTERNATIONAL JOURNAL FOR NUMERICAL METHODS IN FLUIDS, VOL. 15, 925–952 (1992)

NUMERICAL SOLUTION OF TRANSONIC FULL-POTENTIAL-EQUIVALENT EQUATIONS IN VON MISES CO-ORDINATES

C.-F. AN AND R. M. BARRON

Department of Mathematics and Statistics and Fluid Dynamics Research Institute, University of Windsor, Windsor, Ontario, Canada N9B 3P4

SUMMARY

In this paper a new approach to calculate transonic flows is developed. A set of full-potential-equivalent equations in the von Mises co-ordinate system is formulated under the irrotationality and isentropic assumptions. The emphasis is placed on supercritical flow, in which the treatment of embedded shock waves is crucial to get convergent solutions. Shock jump conditions are employed and shock point operators (SPOs) are constructed in the body-fitting streamline co-ordinate system. SPOs and a type-dependent difference scheme are applied to solve the 'main' equation for the 'main' variable, the streamline ordinate y. A number of 'secondary' equations are deduced for the corresponding 'secondary' variables. An optimal combination for the 'secondary' variable, its equation and related difference scheme is selected to be the generalized density R, its linear equation and the Crank–Nicolson scheme. Numerical results show that the present approach gives good agreement with experimental data and other computational work for NACA0012 and biconvex aerofoils in both subcritical and supercritical ranges.

KEY WORDS CFD Potential transonic flow Streamwise co-ordinates Shock point operator

1. INTRODUCTION

Transonic flows are very often encountered in compressible flows, such as the flows around aerofoils, wings, through nozzle throats, cascade blades or past blunt bodies, etc. Thus, in the past two decades, transonic flow computation has been an upsurging topic for CFD workers in the aeronautical and applied mathematical communities. The history of its development may be broken down into three stages.

TSD stage. In 1971 Murman and Cole[1] first developed a type-dependent finite difference relaxation method and successfully solved the mixed-type transonic small-disturbance (TSD) equation. Subsequently Murman and Krupp[2] and Murman[3] improved the scheme, analysed the shock jump conditions and proposed the concept of shock point operator. After that there were many advances on the TSD equation (see e.g. References 4 and 5).

FP stage. In 1974 Jameson[6] extended Murman's[3] scheme and constructed his rotated difference scheme to solve the full-potential (FP) equation in the transonic regime. Since then transonic full-potential calculations have been developed considerably.[7–10]

Euler stage. In the early 1970s Magnus and Yoshihara[11] solved the Euler equations in the transonic range. However, the computer time was too much for the method to be useful. After

0271–2091/92/200925–28$19.00

Received February 1991
Revised May 1992

1976 many other researchers[12–16] attempted to solve the steady Euler equations by seeking the time-asymptotic solution of the unsteady Euler equations. The approach is now referred to as a time-dependent technique. On the other hand, quite a few researchers attacked the Euler equations from another side, namely the streamfunction–vorticity formulation, and solved it as an alternative to the steady Euler equations.[17–20]

In general, Euler solvers are more accurate, but more complicated and consume more computer time than the full-potential solvers. Therefore, in spite of the recent active efforts on Euler solvers, the full-potential calculations are still more attractive and widely used in practical transonic computation owing to their simplicity, efficiency and accuracy. The purpose of this paper is to report some transonic calculations at the FP stage.

The von Mises transformation is a classical transformation in fluid mechanics. After the transformation the co-ordinates become body-fitting ones and it is easy to obtain streamlines, so this transformation is especially convenient for aerodynamic design problems. Moreover, some boundary conditions after the transformation are Dirichlet-type, which is believed to provide a faster convergence rate for an iterative solution process.

Barron[21] connected Martin's approach[22] with the von Mises transformation and successfully solved the resulting elliptic equation to simulate the incompressible potential flow past an aerofoil. Later on[23,24] Barron's method[21] was extended to compressible flow with some promising results, especially for subcritical transonic flows.

In this paper a new approach to calculate transonic flows is developed. Emphasis is placed on the supercritical case, in which the treatment of an embedded supersonic pocket, partly bounded by a shock wave, is crucial to get acceptable convergent solutions.

Starting from the 2D steady Euler equations, through the introduction of the streamfunction and the von Mises transformation, a set of Euler-equivalent equations in the von Mises co-ordinate system is deduced. With the assumptions of irrotationality and the isentropic condition, a number of sets of full-potential-equivalent equations are formulated.

In each set of equations the 'main' equation is a second-order, non-linear partial differential equation for a 'main' geometric variable, namely the streamline ordinate y, as a function of abscissa x and streamfunction ψ. This 'main' equation has Dirichlet boundary conditions (for the analysis problem) and is coupled with one of the following 'secondary' variables: density ρ, generalized density R, squared Mach number M^2, x-velocity component u or reciprocal of density σ. All these 'secondary' variables as functions of x and ψ are solved using a corresponding 'secondary' equation of the proper set.

To get a numerical solution of the 'main' equation which is mathematically and physically consistent and well-classified, Murman and Cole's type-dependent scheme[1] is applied. To handle the embedded shock wave properly, a shock jump condition and a shock point operator (SPO) in von Mises co-ordinates are deduced using Murman's[3] ideas as a guide.

The 'secondary' equations are either first-order partial differential equations or first-order ordinary differential equations for the corresponding 'secondary' variables. Various difference schemes have been tested, including explicit, implicit, Crank–Nicolson, type-dependent and type-dependent with SPO, to solve the 'secondary' equation. The optimal 'secondary' variable, equation and scheme are selected after some computer experimentation. Generalized density R, its linear equation and the Crank–Nicolson scheme are recommended as the optimal combination.

The computed results from the recommended combination are compared with experimental data and other computations. Some discussion, suggestion and concluding remarks are given at the end of this paper.

2. MATHEMATICAL FORMULATION

2.1. Euler-equivalent equations

The governing equations of 2D, steady, inviscid flows are the Euler equations

$$\begin{pmatrix} \rho u \\ \rho u^2+p \\ \rho uv \\ \rho uH \end{pmatrix}_x + \begin{pmatrix} \rho v \\ \rho uv \\ \rho v^2+p \\ \rho vH \end{pmatrix}_y = 0, \tag{1a}$$

$$H=\frac{\gamma}{\gamma-1}\frac{p}{\rho}+\frac{u^2+v^2}{2}, \tag{1b}$$

where ρ is the density, u and v are the velocity components in the x- and y-directions respectively, p is the pressure, H is the total enthalpy per unit mass and γ is the ratio of specific heats. The dependent variables are normalized by the freestream density ρ_∞, speed q_∞ and dynamic pressure head $\rho_\infty q_\infty^2$. The independent variables x and y are scaled by a characteristic length, e.g. the chord length of the aerofoil.

Introducing the streamfunction ψ such that

$$\psi_y=\rho u, \qquad \psi_x=-\rho v \tag{2}$$

and substituting (2) into (1), one gets

$$\begin{pmatrix} \psi_y^2/\rho+p \\ -\psi_x\psi_y/\rho \\ \psi_y H \end{pmatrix}_x + \begin{pmatrix} -\psi_x\psi_y/\rho \\ \psi_x^2/\rho+p \\ -\psi_x H \end{pmatrix}_y = 0, \tag{3a}$$

$$H=\frac{\gamma}{\gamma-1}\frac{p}{\rho}+\frac{\psi_x^2+\psi_y^2}{2\rho^2}. \tag{3b}$$

The first component equation in (1a), the continuity equation, is satisfied automatically by introduction of the streamfunction. Thus only three equations are left, with three unknowns ψ, ρ and p.

If we interchange the roles of dependent variable ψ and independent variable y and keep the role of independent variable x invariant, then we have in fact introduced the von Mises transformation:

$$\phi=x, \quad \psi=\psi(x, y) \qquad \text{or} \qquad x=\phi, \quad y=y(\phi, \psi). \tag{4}$$

The Jacobian of the transformation is

$$J=\left.\frac{\partial(x, y)}{\partial(\phi, \psi)}\right|_{\phi=x}=y_\psi \tag{5}$$

and the differential operators are

$$\frac{\partial}{\partial x}=\frac{\partial}{\partial \phi}+\psi_x\frac{\partial}{\partial \psi} \quad \text{and} \quad \frac{\partial}{\partial y}=\psi_y\frac{\partial}{\partial \psi} \quad \text{when } \phi=x. \tag{6}$$

Acting (6) on y gives

$$\psi_x = -\frac{y_\phi}{y_\psi}, \qquad \psi_y = \frac{1}{y_\psi}. \tag{7}$$

Thus the differential operators become

$$\frac{\partial}{\partial x} = \frac{\partial}{\partial \phi} - \frac{y_\phi}{y_\psi}\frac{\partial}{\partial \psi}, \qquad \frac{\partial}{\partial y} = \frac{1}{y_\psi}\frac{\partial}{\partial \psi}. \tag{8}$$

Using (7) and (8) in (3) and simplifying, one gets

$$\begin{pmatrix} 1/\rho y_\psi + p y_\psi \\ y_\phi/\rho y_\psi \\ H \end{pmatrix}_\phi + \begin{pmatrix} -p y_\phi \\ p \\ 0 \end{pmatrix}_\psi = 0, \tag{9a}$$

$$H = \frac{\gamma}{\gamma-1}\frac{p}{\rho} + \frac{1+y_\phi^2}{2\rho^2 y_\psi^2}, \tag{9b}$$

where $\phi = x$.

The third equation in (9a), $H_\phi = 0$, means that the total enthalpy H is invariant along a streamline. For a flow problem with uniform freestream the total enthalpy H is also invariant along a line other than a streamline in the freestream. Therefore H = constant throughout the flow field. This is the *isoenergetic* or *homoenergetic* assumption made by most Euler solver researchers. The constant can be evaluated at the freestream condition as

$$H = H_\infty = \frac{1}{\gamma-1}\frac{1}{M_\infty^2} + \frac{1}{2}. \tag{10}$$

Finally, for 2D, steady, inviscid flows the governing *Euler-equivalent equations* in the von Mises co-ordinate system are

$$\left(\frac{1}{\rho y_\psi} + y_\psi p\right)_x - (y_x p)_\psi = 0, \tag{11a}$$

$$\left(\frac{y_x}{\rho y_\psi}\right)_x + p_\psi = 0, \tag{11b}$$

$$\frac{\gamma}{\gamma-1}\frac{p}{\rho} + \frac{1+y_x^2}{2\rho^2 y_\psi^2} = H_\infty. \tag{11c}$$

Equations (11a) and (11b) are x-momentum and y-momentum equations respectively and the energy equation is replaced by an algebraic equation (11c) for p and ρ. These three equations contain three unknowns, i.e. density ρ, pressure p and streamline ordinate y, as functions of abscissa x and streamfunction ψ.

2.2. *Full-potential-equivalent equations*

The Euler-equivalent equations (11) are not easy to solve even though they are kept in conservative form and the energy equation has been simplified to an algebraic equation via the homoenergetic assumption. However, for many practical transonic problems the flow can be assumed irrotational and isentropic. In this case the irrotationality condition

$$\omega = v_x - u_y = 0 \tag{12}$$

and the isentropic relation

$$p=\rho^{\gamma}/\gamma M_{\infty}^{2} \tag{13}$$

can be introduced.

Using the von Mises transformation, (12) becomes

$$(y_{\psi}v)_{x}-(y_{x}v+u)_{\psi}=0.$$

From (2) and (7),

$$\frac{v}{u}=-\frac{\psi_{x}}{\psi_{y}}=y_{x}\quad (\text{note that } \phi=x),$$

so the irrotationality condition becomes

$$(y_{x}y_{\psi}u)_{x}-[(1+y_{x}^{2})u]_{\psi}=0 \tag{14}$$

or

$$\left(\frac{y_{x}}{\rho}\right)_{x}-\left(\frac{1+y_{x}^{2}}{\rho y_{\psi}}\right)_{\psi}=0, \tag{15}$$

where $\rho u y_{\psi}=1$ has been employed.

Therefore a set of so-called full-potential-equivalent equations in the von Mises co-ordinate system is composed of the x-momentum equation (11a), the y-momentum equation (11b), the isentropic relation (13) and the irrotationality condition (15).

To simplify the formulation further, let us differentiate the isentropic relation (13) w.r.t. x and ψ:

$$p_{x}=\frac{\rho^{\gamma-1}}{M_{\infty}^{2}}\rho_{x},\qquad p_{\psi}=\frac{\rho^{\gamma-1}}{M_{\infty}^{2}}\rho_{\psi}. \tag{16}$$

Substituting (16) into (11a), the x-momentum equation becomes

$$-y_{x\psi}+y_{\psi}\left(y_{\psi}^{2}\frac{\rho^{\gamma+1}}{M_{\infty}^{2}}-1\right)\frac{\rho_{x}}{\rho}-y_{x}y_{\psi}^{2}\frac{\rho^{\gamma+1}}{M_{\infty}^{2}}\frac{\rho_{\psi}}{\rho}=0. \tag{17}$$

Similarly, the y-momentum equation (11b) becomes

$$y_{\psi}y_{xx}-y_{x}y_{x\psi}-y_{x}y_{\psi}\frac{\rho_{x}}{\rho}+y_{\psi}^{2}\frac{\rho^{\gamma+1}}{M_{\infty}^{2}}\frac{\rho_{\psi}}{\rho}=0. \tag{18}$$

The irrotationality condition (15) itself gives

$$y_{\psi}^{2}y_{xx}-2y_{x}y_{\psi}y_{x\psi}+(1+y_{x}^{2})y_{\psi\psi}-y_{x}y_{\psi}^{2}\frac{\rho_{x}}{\rho}+y_{\psi}(1+y_{x}^{2})\frac{\rho_{\psi}}{\rho}=0 \tag{19}$$

after expansion.

Proper combinations of these three equations (17)–(19) produce a number of sets of equations which can be used to solve for y and ρ. For example, solving for ρ_{x}/ρ and ρ_{ψ}/ρ from (17) and (18), plugging them into (19) and simplifying the resulting equation, one gets

$$(y_{\psi}^{2}-K)y_{xx}-2y_{x}y_{\psi}y_{x\psi}+(1+y_{x}^{2})y_{\psi\psi}=0, \tag{20a}$$

where

$$K=M_{\infty}^{2}/\rho^{\gamma+1} \tag{20b}$$

is referred to as the 'compressibility parameter'. Eliminating the $y_{x\psi}$-term from (17) and (18), one

第II部分 出国后博士学位论文阶段

gets

$$y_x y_\psi^2 (\rho^{\gamma+1})_x - y_\psi (1+y_x^2)(\rho^{\gamma+1})_\psi = (\gamma+1) M_\infty^2 y_{xx} \tag{21}$$

and equation (17) itself can be rewritten as

$$y_\psi \left(y_\psi^2 - \frac{M_\infty^2}{\rho^{\gamma+1}} \right) (\rho^{\gamma+1})_x - y_x y_\psi^2 (\rho^{\gamma+1})_\psi = (\gamma+1) M_\infty^2 y_{x\psi}. \tag{22}$$

Substituting y_{xx} in (21) and $y_{x\psi}$ in (22) into (19) gives

$$y_x \left(y_\psi^2 - \frac{M_\infty^2}{\rho^{\gamma+1}} \right) (\rho^{\gamma+1})_x + y_\psi \left((1-y_x^2) - \frac{M_\infty^2}{\rho^{\gamma+1}} \frac{1+y_x^2}{y_\psi^2} \right) (\rho^{\gamma+1})_\psi = (\gamma+1) M_\infty^2 \frac{1+y_x^2}{y_\psi^2} y_{\psi\psi}. \tag{23}$$

The 'main' equation (20) is a second-order non-linear partial differential equation for the streamline ordinate y, and it is coupled with another unknown, the density ρ. The type of this 'main' equation is mixed depending on the local flow behaviour. In fact, the discriminant of equation (20) is

$$\Delta = 4 y_\psi^2 (M^2 - 1), \tag{24}$$

where the squared local Mach number is

$$M^2 = \frac{M_\infty^2}{\rho^{\gamma+1}} \frac{1+y_x^2}{y_\psi^2}. \tag{25}$$

Hence, if the local flow is supersonic, then $M^2 > 1$ and $\Delta > 0$, so that equation (20) must be hyperbolic. In contrast, if the local flow is subsonic, equation (20) is elliptic. In other words, the mathematical classification of equation (20) is consistent with the physical meaning of the flow. This significant property of the equation for y creates the possibility of applying the type-dependent difference scheme originally developed by Murman and Cole[1] for TSD problems.

It should be mentioned that equation (20) was first obtained and classified by Naeem and Barron.[24] Dulikravich[25] also got a similar equation to (20) with a different-scaled compressibility parameter. If $K \to 0$, then equation (20) reduces to the incompressible flow equation

$$y_\psi^2 y_{xx} - 2 y_x y_\psi y_{x\psi} + (1+y_x^2) y_{\psi\psi} = 0. \tag{26}$$

This equation was obtained by Barron.[21] Greywall[26] deduced the same equation by a different method.

In order to find a proper 'secondary' equation to solve for the 'secondary' variable ρ, any of the above three equations (21)–(23) can be chosen to construct a complete set of equations. However, to simplify the formulation, define the generalized density

$$R = \rho^{\gamma+1}. \tag{27}$$

Then equations (20)–(23) and (25) become

$$(y_\psi^2 - K) y_{xx} - 2 y_x y_\psi y_{x\psi} + (1+y_x^2) y_{\psi\psi} = 0, \tag{28a}$$

where

$$K = M_\infty^2 / R, \tag{28b}$$

$$y_x y_\psi^2 R_x - y_\psi (1+y_x^2) R_\psi = (\gamma+1) M_\infty^2 y_{xx}, \tag{29}$$

$$y_\psi \left(y_\psi^2 - \frac{M_\infty^2}{R} \right) R_x - y_x y_\psi^2 R_\psi = (\gamma+1) M_\infty^2 y_{x\psi}, \tag{30}$$

$$y_x\left(y_\psi^2-\frac{M_\infty^2}{R}\right)R_x+y_\psi^2\left(1-y_x^2-\frac{M_\infty^2}{R}\frac{1+y_x^2}{y_\psi^2}\right)R_\psi=(\gamma+1)M_\infty^2\frac{1+y_x^2}{y_\psi^2}y_{\psi\psi},\tag{31}$$

$$M^2=\frac{M_\infty^2}{R}\frac{1+y_x^2}{y_\psi^2}.\tag{32}$$

Equation (29) can also be rewritten in the 'conservative' form

$$(y_xR)_x-\left(\frac{1+y_x^2}{y_\psi}R\right)_\psi=(\gamma+2)M_\infty^2\frac{y_{xx}}{y_\psi^2}.\tag{33}$$

All the above 'secondary' equations (29)–(31) and (33) are first-order partial differential equations and can be solved for the 'secondary' variable R. However, only equation (29) seems to be the simplest one because it is linear. Thus why do we not choose it first to solve for R? Equation (29) can be solved (marched) from an initial data line other than its characteristic curve as long as y and its derivatives are known. In fact, the slope of the characteristics of equation (29) is

$$\frac{d\psi}{dx}=-\frac{1+y_x^2}{y_xy_\psi}.\tag{34}$$

At infinity, $y_x\to0$ and $y_\psi\to1$, so $d\psi/dx\to\infty$. Thus both the left and right boundaries at infinity are characteristic curves and therefore cannot serve as initial data lines. Fortunately, the horizontal boundary at infinity meets the requirement of such an initial line. Hence we can march (29) to the aerofoil from the top boundary for the upper half-plane and from the bottom boundary for the lower half-plane.

The boundary conditions for the 'main' equation (28) are very simple Dirichlet ones. For a symmetric aerofoil at zero incidence,

$$y=f(x)\quad\text{on the aerofoil,}\tag{35a}$$

$$y=\psi\quad\text{at infinity,}\tag{35b}$$

$$y=0\quad\text{on the symmetry line,}\tag{35c}$$

where $f(x)$ is the shape function of the aerofoil. For the 'secondary' equation (29), both initial and boundary conditions are

$$R=1\quad\text{at infinity.}\tag{36}$$

2.3. Alternative 'secondary' equations

Apart from equations (29)–(31) and (33), a number of other 'secondary' equations can be found to solve for the corresponding 'secondary' variables. For example, if we eliminate the R_ψ-term from (29) and (30), we can get a first-order ordinary differential equation for R:

$$R_x=\frac{\gamma+1}{2}M_\infty^2\left(\frac{M_\infty^2}{R}\frac{1+y_x^2}{y_\psi^2}-1\right)^{-1}\left(\frac{1+y_x^2}{y_\psi^2}\right)_x.\tag{37}$$

Similarly, eliminating the R_x-term from (30) and (31), we have

$$R_\psi=\frac{\gamma+1}{2}M_\infty^2\left(\frac{M_\infty^2}{R}\frac{1+y_x^2}{y_\psi^2}-1\right)^{-1}\left(\frac{1+y_x^2}{y_\psi^2}\right)_\psi.\tag{38}$$

It is interesting that (37) and (38) are astonishingly similar and symmetric.

At a shock wave it is better to replace (37) by the Rankine–Hugoniot relation

$$R^{+}=\left[\frac{1}{R}\left(\frac{(\gamma+1)\,(1+y_x^2)}{2(\gamma-1)H_\infty y_\psi^2}\right)^{\gamma+1}\right]^{-} \tag{39}$$

where the superscripts '+' and '−' represent the downstream and upstream sides of the shock wave respectively.

Equation (37) accompanied by (39) and equation (38) can be marched along the x- and ψ-directions step-by-step.

Furthermore, if we differentiate (29) w.r.t. ψ and (30) w.r.t. x and subtract them to eliminate third-order y-derivatives $y_{x\psi x}$ and $y_{x\psi x}$, we can get a second-order, but non-homogeneous, partial differential equation for R with the same differential operator as in the y-equation (28):

$$\left(y_\psi^2-\frac{M_\infty^2}{R}\right)R_{xx}-2y_x y_\psi R_{x\psi}+(1+y_x^2)R_{\psi\psi}=G, \tag{40a}$$

where

$$G=\frac{\gamma+1}{2}M_\infty^2\,\frac{\left\{(y_x y_\psi^2)_\psi-\left[y_\psi\left(y_\psi^2-\frac{M_\infty^2}{R}\right)\right]_x\right\}\left(\frac{1+y_x^2}{y_\psi^2}\right)_x+\{(y_x y_\psi^2)_x-[y_\psi(1+y_x^2)]_\psi\}\left(\frac{1+y_x^2}{y_\psi^2}\right)_\psi}{y_\psi\left(\frac{M_\infty^2}{R}\frac{1+y_x^2}{y_\psi^2}-1\right)}. \tag{40b}$$

The boundary condition for R on the aerofoil is

$$R=\left[1+\frac{\gamma-1}{2}M^2\left(1-\frac{1+f'^2(x)}{R^{2/(\gamma+1)}y_\psi^2}\right)\right]^{(\gamma+1)/(\gamma-1)}. \tag{41}$$

We can see that both differential equation (40) and boundary condition (41) are non-linear.

All the above R-equations (37), (38) and (40) can be used as 'secondary' equations to solve for R to update the compressibility parameter $K=M_\infty^2/R$ in the 'main' equation (28).

On the other hand, using the R–M^2 relation (32) in equation (37), we get a first-order ordinary differential equation for M^2:

$$(M^2)_x=\frac{M^2\{1+[(\gamma-1)/2]M^2\}}{1-M^2}\left[\ln\left(\frac{1+y_x^2}{y_\psi^2}\right)\right]_x. \tag{42}$$

However, at a shock wave equation (42) should be replaced by the Rankine–Hugoniot relation

$$(M^2)^{+}=\frac{1+[(\gamma-1)/2]\,(M^2)^{-}}{\gamma(M^2)^{-}-(\gamma-1)/2}. \tag{43}$$

Accordingly, the compressibility parameter in equations (28) should be expressed as

$$K=M^2\frac{y_\psi^2}{1+y_x^2}. \tag{44}$$

Superficially, this set of equations seems to be a good formulation, but it is too early to conclude its success before performing some computations.

Next we will just list some of the other alternative sets of full-potential-equivalent equations for reference without giving any derivations. In each set the 'main' equation for y takes the form of

(28) and the compressibility K takes different forms in different sets:

y–ρ set
$$K=M_\infty^2/\rho^{\gamma+1}, \tag{45a}$$
$$y_\psi^2\left(y_\psi^2\frac{\rho^{\gamma+1}}{M_\infty^2}-1+(\gamma-1)y_x^2\right)\rho_x-y_xy_\psi\left(y_\psi^2\frac{\rho^{\gamma+1}}{M_\infty^2}+(\gamma-1)(1+y_x^2)\right)\rho_\psi$$
$$-\frac{\gamma-1}{2}\left[y_\psi^2(y_x^2)_x+\left(\frac{1}{\gamma-1}-2y_x^2\right)(y_\psi^2)_x+\frac{y_x}{y_\psi}(1+y_x^2)(y_\psi^2)_\psi\right]\rho=0; \tag{45b}$$

y–σ set
$$K=M_\infty^2\sigma^{\gamma+1}, \tag{46a}$$
$$(y_x\sigma)_x-\left(\frac{1+y_x^2}{y_\psi}\sigma\right)_\psi=0, \tag{46b}$$
$$\sigma=1/\rho; \tag{46c}$$

y–u set
$$K=M_\infty^2y_\psi^{\gamma+1}u^{\gamma+1}, \tag{47a}$$
$$(y_xy_\psi u)_x-[u(1+y_x^2)]_\psi=0; \tag{47b}$$

y–u–ρ set
$$K=M_\infty^2/\rho^{\gamma+1}, \tag{48a}$$
$$(y_xy_\psi u)_x-[(1+y_x^2)u]_\psi=0, \tag{48b}$$
$$\rho=\left(1+\frac{\gamma-1}{2}M_\infty^2[1-(1+y_x^2)u^2]\right)^{1/(\gamma-1)}. \tag{48c}$$

It can be pointed out that although these formulations are theoretically rational, the computer calculations may fail to give acceptable results. However, it is difficult to undergo the tedious and time-consuming work of running all possible combinations of these sets of equations with various schemes and relaxation parameters on a computer. Thus we have to select an optimal combination of a 'best' secondary variable, a 'best' secondary equation and a 'best' difference scheme. Regardless, it is natural and logical to choose the simplest set of equations (28) and (29) to first attack the problem.

3. NUMERICAL ALGORITHM

3.1. Type-dependent scheme

Because equation (28) is conveniently classified to hyperbolic or elliptic type depending on the local supersonic or subsonic flow properties, it is permissible to apply Murman and Cole's type-dependent scheme to solve (28) for y as long as R is known.

Equation (28) can be rewritten as

$$A_1 y_{xx}+A_2 y_{x\psi}+A_3 y_{\psi\psi}=0, \tag{49a}$$

where

$$A_1=y_\psi^2-K, \qquad K=M_\infty^2/R, \tag{49b}$$

$$A_2=-2y_x y_\psi, \tag{49c}$$

$$A_3=1+y_x^2. \tag{49d}$$

The type-dependent scheme reads

$$\{A_1[\nu\Delta_x+(1-\nu)\nabla_x]\nabla_x+A_2[\nu\delta_x+(1-\nu)\nabla_x]\delta_\psi+A_3\delta_{\psi\psi}\}y_{i,j}=0, \tag{50a}$$

where Δ, ∇ and δ are forward, backward and central difference operators and

$$\nu=\begin{cases}1 & \text{if local flow is subsonic,}\\ 0 & \text{if local flow is supersonic,}\end{cases} \tag{50b}$$

is a switch parameter.

Expanding (50a) and rearranging the terms, one gets

$$Ay_{i,j-1}+By_{i,j}+Cy_{i,j+1}=\text{RHS}, \quad i=2,3,\ldots,I_{\max-1}, \quad j=2,3,\ldots,J_{\max-1} \tag{51a}$$

where

$$A=\beta^2 A_3-\frac{1-\nu}{2}\beta A_2, \qquad \beta=\frac{\Delta x}{\Delta\psi}, \tag{51b}$$

$$B=-2\beta^2 A_3+(1-3\nu)A_1, \tag{51c}$$

$$C=\beta^2 A_3+\frac{1-\nu}{2}\beta A_2, \tag{51d}$$

$$\begin{aligned}\text{RHS}=&-\nu A_1(y_{i+1,j}+y_{i-1,j})+(1-\nu)A_1(2y_{i-1,j}-y_{i-2,j})\\ &-\frac{\nu}{4}\beta A_2(y_{i+1,j+1}-y_{i+1,j-1}-y_{i-1,j+1}+y_{i-1,j-1})\\ &+\frac{1-\nu}{2}\beta A_2(y_{i-1,j+1}-y_{i-1,j-1}).\end{aligned} \tag{51e}$$

From (35) the boundary conditions are

$$y_{i,1}=\begin{cases}f(x_i) & \text{if } i_{LE}\leqslant i\leqslant i_{TE},\\ 0 & \text{if } i<i_{LE} \text{ or } i>i_{TE},\end{cases} \tag{52a}$$

$$y_{i,j}=\psi_j \quad \text{if } i=1 \text{ or } I_{\max} \text{ or } j=J_{\max}. \tag{52b}$$

The system of difference equations (51) with boundary conditions (52) has a tridiagonal coefficient matrix, so the successive line overrelaxation (SLOR) procedure can be used. Along a vertical i fixed line, system (51) is relaxed and then a sweep from left to right is made. A simple tridiagonal solver is available to use, but an iterative procedure has to be applied owing to the non-linearity of the equations. The stencils of the type-dependent schemes and boundary conditions are shown in Figure 1.

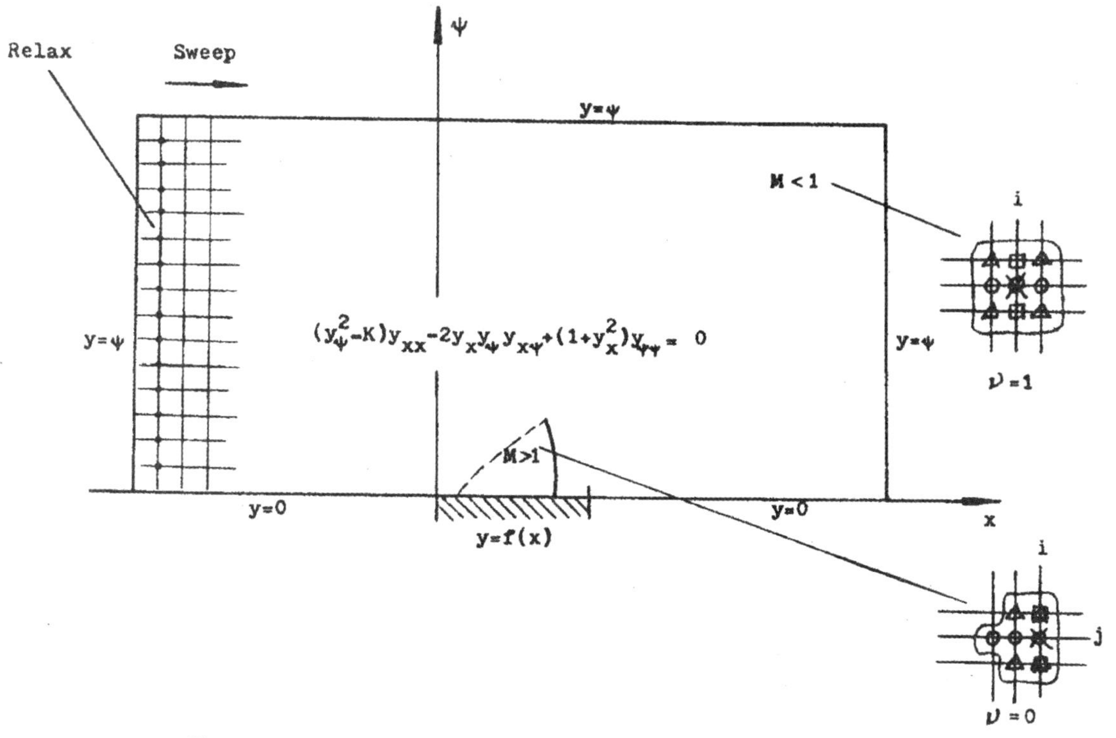

Figure 1. Type-dependent scheme and boundary conditions for y-equation

3.2. Crank–Nicolson scheme

As mentioned earlier, after $y(x, \psi)$ has been solved from the 'main' equation and y_x, y_ψ and y_{xx} have been properly differenced from y, the 'secondary' equation (29) can be solved for $R(x, \psi)$ by marching from the horizontal far-field boundary to the aerofoil.

Equation (29) can be rewritten as

$$B_1 R_x + B_2 R_\psi = B_3, \tag{53a}$$

where

$$B_1 = y_x y_\psi^2, \tag{53b}$$

$$B_2 = -y_\psi(1+y_x^2), \tag{53c}$$

$$B_3 = (\gamma+1) M_\infty^2 y_{xx}. \tag{53d}$$

The Crank–Nicolson difference scheme is an implicit, second-order-accurate and unconditionally stable difference scheme. Applying it to equation (53a) at point $(i, j+\frac{1}{2})$ gives

$$B_1 \frac{R_{i+1,j+1/2} - R_{i-1,j+1/2}}{2\Delta x} + B_2 \frac{R_{i,j+1} - R_{i,j}}{\Delta\psi} = B_3.$$

Evaluating each variable at level $j+\frac{1}{2}$ by the average on levels j and $j+1$, one gets

$$\tilde{A} R_{i-1,j} + \tilde{B} R_{i,j} + \tilde{C} R_{i+1,j} = \widetilde{\mathrm{RHS}}, \quad j = J_{\max-1}, \ldots, 3, 2, 1, \quad i = 2, 3, \ldots, I_{\max-1} \tag{54a}$$

where

$$\tilde{A} = -B_1, \tag{54b}$$

$$\tilde{B} = -4\beta B_2, \qquad \beta = \frac{\Delta x}{\Delta \psi}, \tag{54c}$$

$$\tilde{C} = B_1, \tag{54d}$$

$$\widetilde{\mathrm{RHS}} = \tilde{C}\,R_{i-1,j+1} + \tilde{B}\,R_{i,j+1} + \tilde{A}\,R_{i+1,j+1} + 4\Delta x B_3, \tag{54e}$$

with

$$B_1 = \tfrac{1}{2}[(y_x y_\psi^2)_{i,j} + (y_x y_\psi^2)_{i,j+1}], \tag{54f}$$

$$B_2 = -\tfrac{1}{2}\{[y_\psi(1+y_x^2)]_{i,j} + [y_\psi(1+y_x^2)]_{i,j+1}\}, \tag{54g}$$

$$B_3 = \frac{\gamma+1}{2} M_\infty^2 [(y_{xx})_{i,j} + (y_{xx})_{i,j+1}]. \tag{54h}$$

The system of difference equations (54) can be solved line-by-line horizontally from the far-field boundary to the aerofoil using SLOR with a tridiagonal solver, but no iterative procedure is needed at this stage owing to its linearity. The difference scheme stencil and boundary conditions are shown in Figure 2. The following points should be emphasized.

(i) The coefficient matrix of the system (54) is diagonally dominant at most grid lines. This fact guarantees the numerical solution of the system. In fact, for most grid points off the aerofoil surface, $y_x \approx 0$, $y_\psi \approx 1$ and β can be chosen equal to or greater than unity, then

$$|\tilde{B}| - (|\tilde{A}| + |\tilde{C}|) = 4\beta|B_2| - 2|B_1| \geqslant 2|y_\psi|[2(1+y_x^2) - |y_x y_\psi|] \geqslant 0. \tag{55}$$

This inequality is easily satisfied for most points. Only for points close to the aerofoil or shock wave may the inequality be violated. We should pay particular attention to those situations.

(ii) The marching process can only be carried out from the horizontal far-field boundary instead of the vertical far-field boundary or aerofoil surface, because R is easy to specify at infinity and, more importantly, the horizontal boundary is not a characteristic curve, while the left and right far-field boundaries are such curves.

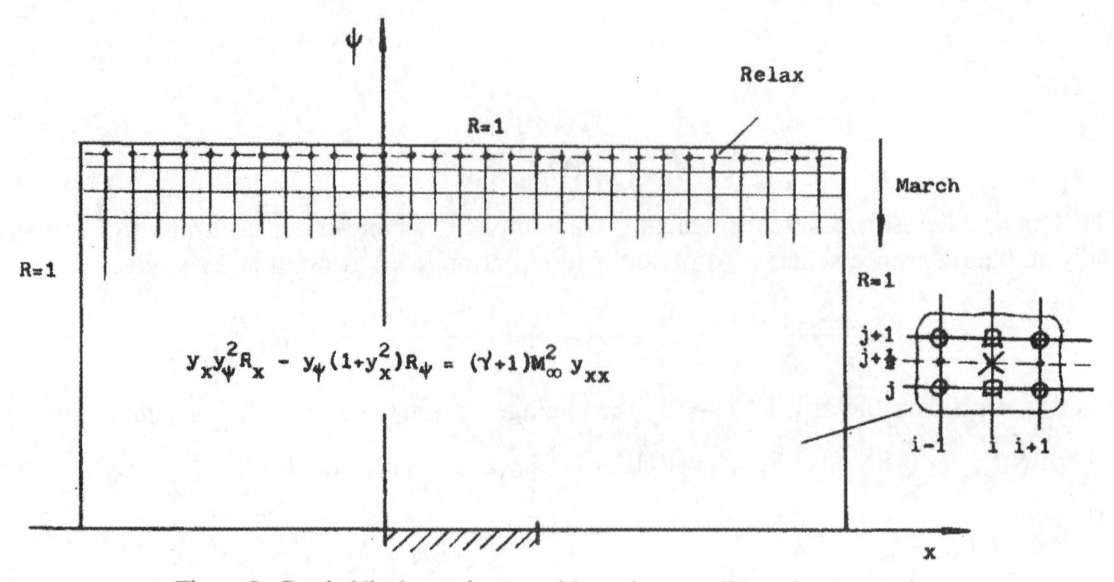

Figure 2. Crank–Nicolson scheme and boundary conditions for R-equation

(iii) The right-hand-side term in equation (53a) includes y_{xx}, which must be type-dependent differenced with SPOs (explained in the next subsection) to keep consistent with the case in the y-equation. However, for R_x and y_x, whether type-dependent differencing with SPOs is needed or not can be determined only after the numerical tests. A number of alternative schemes have been tested, including explicit, implicit, first-order and second-order type-dependent with and without SPOs, etc. It was found that the Crank–Nicolson scheme is the optimal one.

After R has been solved, the squared Mach number is calculated from (32) and the pressure coefficient is calculated from

$$C_p=\frac{2}{\gamma M_\infty^2}(R^{\gamma/(\gamma+1)}-1). \tag{56}$$

3.3. Shock wave treatment

After some numerical tests it was found that the type-dependent scheme is effective to achieve good solutions for the subcritical case of transonic flows and for the supercritical case with very weak shock waves. However, for a supercritical transonic flow with moderate or strong shock wave the iterative procedure either fails to converge or is forced to be stopped owing to inaccurate intermediate values of the unknowns y and R.

In the early work of Murman and Cole[1] the shock jump conditions are automatically contained in their TSD formulation in the integral form. The sonic line and shock waves evolve naturally during the course of iteration. The shock is captured as part of the continuous solution and is smeared out over several grid points. Thus no special shock wave treatment is needed for their computation. Murman, in his later paper,[3] proposed the concept of shock point operator (SPO) for the TSD equations, but his SPO cannot be applied directly here. However, his analysis of the shock structure provides a useful hint for modification of the FP-equivalent equations. Next let us analyse our shock jump conditions, propose an SPO and apply it to solve the problem of supercritical FP transonic flow.

Shock jump conditions. Suppose an oblique shock with velocity V makes an angle β with the x-axis, u and v are Cartesian components of V, V_n and V_t are normal and tangential components of V and α is the angle of V with the x-axis. Superscripts '$-$' and '$+$' represent upstream and downstream of the shock. Then (Figure 3) the tangential shock wave relation $V_t^-=V_t^+$ gives

$$(v^+-v^-)\sin\beta+(u^+-u^-)\cos\beta=0$$

or

$$\frac{v^+-v^-}{u^+-u^-}=-\cot\beta=-\left(\frac{dx}{dy}\right)_s. \tag{57}$$

The normal shock wave relation $\rho^-V_n^-=\rho^+V_n^+$ gives

$$(\rho^+u^+-\rho^-u^-)\sin\beta-(\rho^+v^+-\rho^-v^-)\cos\beta=0$$

or

$$\frac{\rho^+u^+-\rho^-u^-}{\rho^+v^+-\rho^-v^-}=\cot\beta=\left(\frac{dx}{dy}\right)_s, \tag{58}$$

where $(dx/dy)_s=\cot\beta$ is the slope of the shock wave. Equations (57) and (58) can be expressed in

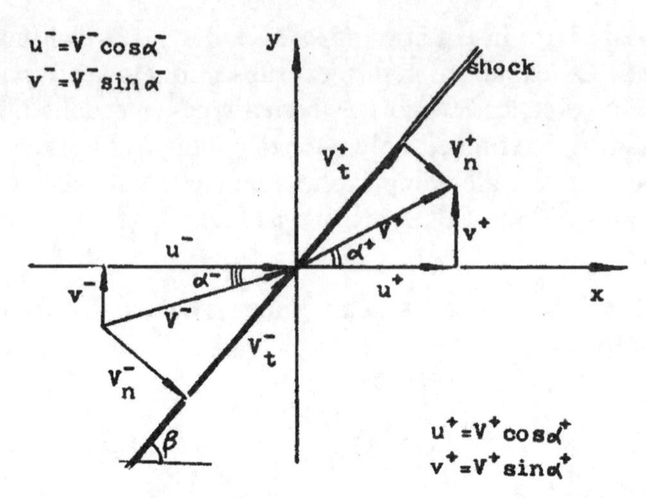

Figure 3. Shock jump conditions

the compact form

$$[v]+\left(\frac{\mathrm{d}x}{\mathrm{d}y}\right)_s[u]=0, \qquad [\rho u]-\left(\frac{\mathrm{d}x}{\mathrm{d}y}\right)_s[\rho v]=0,$$

where $[\cdot]$ represents the jump across the shock. Considering $\rho u y_\psi=1$ and $v=y_x u$, we get the oblique shock jump conditions

$$\left[\frac{y_x}{\rho y_\psi}\right]+\left(\frac{\mathrm{d}x}{\mathrm{d}y}\right)_s\left[\frac{1}{\rho y_\psi}\right]=0, \tag{59a}$$

$$\left[\frac{1}{y_\psi}\right]-\left(\frac{\mathrm{d}x}{\mathrm{d}y}\right)_s\left[\frac{y_x}{y_\psi}\right]=0. \tag{59b}$$

For a normal shock, i.e. a shock whose plane is perpendicular to the x-axis, $(\mathrm{d}x/\mathrm{d}y)_s=0$. The shock jump conditions are then reduced to

$$\left[\frac{y_x}{\rho y_\psi}\right]=0, \qquad \left[\frac{1}{y_\psi}\right]=0.$$

Finally, we get the normal shock jump conditions of the simplest form

$$\left[\frac{y_x}{\rho}\right]=0, \qquad [y_\psi]=0. \tag{60}$$

Shock point operator (SPO). The shock wave appearing in many transonic flow regions is approximately normal and we assume that it is an infinitely thin discontinuous surface located at point $(i-\frac{1}{2}, j)$ and perpendicular to the x-axis. From the previous paragraph, only y_x/ρ and y_ψ are continuous across this discontinuity, but y_x, y_{xx}, $y_{x\psi}$, ρ, R, M^2, etc. are not (see Figure 4). The normal shock jump conditions (60) are

$$y_\psi^+=y_\psi^-, \qquad y_x^+=y_x^-\frac{\rho^+}{\rho^-}. \tag{61}$$

The Rankine–Hugoniot relation for a normal shock is

$$\frac{\rho^+}{\rho^-}=\frac{[(\gamma+1)/2](M^2)^-}{1+[(\gamma-1)/2](M^2)^-}. \tag{62}$$

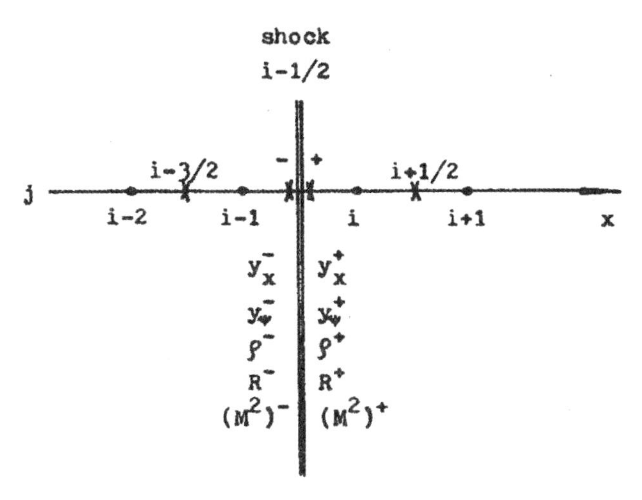

Figure 4. Shock point operator

The squared Mach number at $(i-\frac{1}{2}, j)^-$ can be evaluated by extrapolation from the two upstream points as

$$M_{i-1/2,j}^2 = \tfrac{3}{2} M_{i-1,j}^2 - \tfrac{1}{2} M_{i-2,j}^2. \tag{63}$$

Thus the shock jump conditions (61) can be rewritten as

$$y_\psi^+ = y_\psi^-, \qquad y_x^+ = \alpha_j y_x^-, \tag{64a}$$

where

$$\alpha_j = \left(\frac{\rho^+}{\rho^-}\right)_j = \frac{[(\gamma+1)/4](3M_{i-1,j}^2 - M_{i-2,j}^2)}{1+[(\gamma-1)/4](3M_{i-1,j}^2 - M_{i-2,j}^2)} \tag{64b}$$

is the density jump factor on the jth streamline. Furthermore, we can construct a difference scheme for y_x at a shock point $i=i_s$, i.e. the grid point just behind the shock:

$$\begin{aligned}(y_x)_{i,j} &= \tfrac{1}{2}[(y_x)_{i+1/2,j} + (y_x)_{i-1/2,j}^+] \\ &= \tfrac{1}{2}[(y_x)_{i+1/2,j} + \alpha_j (y_x)_{i-1/2,j}^-] \\ &= \frac{1}{2\Delta x}(y_{i+1,j} - y_{i,j} + \alpha_j y_{i-1,j} - \alpha_j y_{i-2,j}),\end{aligned} \tag{65}$$

where second-order difference formulae

$$(y_x)_{i+1/2,j} = \frac{1}{\Delta x}(y_{i+1,j} - y_{i,j}) \quad \text{and} \quad (y_x)_{i-1/2,j}^- = \frac{1}{\Delta x}(y_{i-1,j} - y_{i-2,j})$$

have been used. Similarly, for $i=i_s$,

$$(y_{xx})_{i,j} = \frac{1}{\Delta x^2}(y_{i+1,j} - y_{i,j} - \alpha_j y_{i-1,j} + \alpha_j y_{i-2,j}), \tag{66}$$

$$\begin{aligned}(y_{x\psi})_{i,j} = \frac{1}{4\Delta x \Delta\psi}(&y_{i+1,j+1} - y_{i+1,j-1} + y_{i,j+1} - y_{i,j-1} - 3y_{i-1,j+1} + 3y_{i-1,j-1} \\ &+ y_{i-2,j+1} - y_{i-2,j-1}).\end{aligned} \tag{67}$$

Equations (65)–(67) define so-called shock point operators in von Mises co-ordinates. Owing to the special treatment of the grid point at a shock wave, we have to revise the type-dependent scheme (51) for the y-equation and (54) for the R-equation.

Type-dependent scheme with SPOs. Based on the SPOs derived above, the system of difference equations corresponding to the 'main' equation (49a) for y is still of the same form as (51a), i.e.

$$Ay_{i,j-1}+By_{i,j}+Cy_{i,j+1}=\mathrm{RHS}, \tag{68a}$$

but the coefficients and right-hand-side term are revised as

$$A=\beta^2 A_3-\begin{Bmatrix}(1-\nu)/2\\ \frac{1}{4}\end{Bmatrix}\beta A_2 \quad \begin{cases}\text{if } i\neq i_s,\\ \text{if } i=i_s,\end{cases} \tag{68b}$$

$$B=-2\beta^2 A_3+\begin{Bmatrix}1-3\nu\\ -1\end{Bmatrix}A_1 \quad \begin{cases}\text{if } i\neq i_s,\\ \text{if } i=i_s,\end{cases} \tag{68c}$$

$$C=\beta^2 A_3+\begin{Bmatrix}(1-\nu)/2\\ \frac{1}{4}\end{Bmatrix}\beta A_2 \quad \begin{cases}\text{if } i\neq i_s,\\ \text{if } i=i_s,\end{cases} \tag{68d}$$

$$\mathrm{RHS}=\begin{cases}-\nu A_1(y_{i+1,j}+y_{i-1,j})+(1-\nu)A_1(2y_{i-1,j}-y_{i-2,j})\\ \quad -\dfrac{\nu}{4}\beta A_2(y_{i+1,j+1}-y_{i+1,j-1}-y_{i-1,j+1}+y_{i-1,j-1})\\ \quad +\dfrac{1-\nu}{2}\beta A_2(y_{i-1,j+1}-y_{i-1,j-1}) \quad \text{if } i\neq i_s,\\ -A_1(y_{i+1,j}-\alpha_j y_{i-1,j}+\alpha_j y_{i-2,j})\\ \quad -\frac{1}{4}\beta A_2(y_{i+1,j+1}-y_{i+1,j-1}-3y_{i-1,j+1}\\ \quad +3y_{i-1,j-1}+y_{i-2,j+1}-y_{i-2,j-1}) \quad \text{if } i=i_s.\end{cases} \tag{68e}$$

For the 'secondary' equation (53a) the corresponding difference equation, its coefficients and its RHS term take the same form as in equations (54a)–(54h), but the second derivative y_{xx} in equation (54h) should be approximated by a type-dependent difference scheme with shock point operator

$$(y_{xx})_{ij}=\begin{cases}\dfrac{\nu}{\Delta x^2}(y_{i+1,j}-2y_{i,j}+y_{i-1,j})+\dfrac{1-\nu}{\Delta x^2}(y_{i,j}-2y_{i-1,j}+y_{i-2,j}) & \text{if } i\neq i_s,\\ \dfrac{1}{\Delta x^2}(y_{i+1,j}-y_{i,j}-\alpha_j y_{i-1,j}+\alpha_j y_{i-2,j}) & \text{if } i=i_s,\end{cases} \tag{69}$$

where the switch parameter ν is defined in equation (50b) and the density jump factor α_j is given by equation (64b).

Criterion of classifying points. In order to determine the local flow type at grid point (i,j), we have to check the flow property at two adjacent points, namely the current point (i,j) and the upstream point $(i-1,j)$. The criteria are shown in Table I. In computational practice the sonic point does not need to be distinguished because there is no jump across it, but the shock point must be identified carefully, and this is a key step to get a convergent solution.

Table I

$M^2_{i-1,j}$	$M^2_{i,j}$	Local flow type at (i, j)
<1	<1	Subsonic point
<1	>1	Sonic point
>1	>1	Supersonic point
>1	<1	Shock point

4. CLUSTERING TRANSFORMATION

In order to improve the accuracy by increasing the number of grid points on the surface of the aerofoil without consuming too much computer time, some kind of clustering (stretching) transformation can be applied to pack mesh points for larger-gradient regions and spread out mesh points for smaller-gradient regions.

Jones' algebraic stretching transformation[5] is a simple and effective one:

$$x = a\,\mathrm{e}^{-b\xi^2}\tan\xi, \tag{70a}$$

$$\psi = d\tan\eta. \tag{70b}$$

A new variable Y can be introduced by

$$y = Y + \psi. \tag{71}$$

Substituting (70) and (71) into (28), we get

$$A_1 Y_{\xi\xi} + A_2 Y_{\xi\eta} + A_3 Y_{\eta\eta} + A_4 Y_\xi + A_5 Y_\eta = 0, \tag{72a}$$

where

$$A_1 = (Y_\eta + \psi_\eta)^2 - \tilde{K}, \qquad \tilde{K} = K\psi_\eta^2 = \frac{M_\infty^2}{R}\psi_\eta^2, \tag{72b}$$

$$A_2 = -2Y_\xi(Y_\eta + \psi_\eta), \tag{72c}$$

$$A_3 = Y_\xi^2 + x_\xi^2, \tag{72d}$$

$$A_4 = -\frac{x_{\xi\xi}}{x_\xi}A_1, \tag{72e}$$

$$A_5 = -\frac{\psi_{\eta\eta}}{\psi_\eta}A_3 \tag{72f}$$

and the squared Mach number is

$$M^2 = \frac{M_\infty^2}{R}\frac{\psi_\eta^2}{x_\xi^2}\frac{Y_\xi^2 + x_\xi^2}{(Y_\eta + \psi_\eta)^2}. \tag{73}$$

The normal shock jump conditions (60) become

$$\left[\frac{Y_\xi}{\rho x_\xi}\right] = 0, \qquad \left[\frac{Y_\eta}{\psi_\eta} + 1\right] = 0.$$

Because x_ξ and ψ_η are continuous everywhere, the shock jump conditions have the same form as (60), i.e.

$$\left[\frac{Y_\xi}{\rho}\right] = 0, \qquad [Y_\eta] = 0. \tag{74}$$

第二部分　出国后博士学位论文阶段

Similarly, we get SPOs for clustered co-ordinates:

$$Y_\eta^+ = Y_\eta^-, \qquad Y_\xi^+ = \alpha_j Y_\xi^-, \tag{75a}$$

where

$$\alpha_j = \left(\frac{\rho^+}{\rho^-}\right)_j = \frac{[(\gamma+1)/4]\,(3M_{i-1,j}^2 - M_{i-2,j}^2)}{1+[(\gamma-1)/4]\,(3M_{i-1,j}^2 - M_{i-2,j}^2)} \tag{75b}$$

is the density jump factor on the jth streamline and

$$(Y_\xi)_{i,j} = \frac{1}{2\Delta\xi}(Y_{i+1,j} - Y_{i,j} + \alpha_j Y_{i-1,j} - \alpha_j Y_{i-2,j}), \tag{76a}$$

$$(Y_{\xi\xi})_{i,j} = \frac{1}{\Delta\xi^2}(Y_{i+1,j} - Y_{i,j} - \alpha_j Y_{i-1,j} + \alpha_j Y_{i-2,j}), \tag{76b}$$

$$(Y_{\xi\eta})_{i,j} = \frac{1}{4\Delta\xi\Delta\eta}(Y_{i+1,j+1} - Y_{i+1,j-1} + Y_{i,j+1} - Y_{i,j-1} - 3Y_{i-1,j+1} + 3Y_{i-1,j-1} + Y_{i-2,j+1} - Y_{i-2,j-1}). \tag{76c}$$

Using the type-dependent scheme with SPOs on equation (72a), we get the following system of difference equations with tridiagonal coefficient matrix:

$$AY_{i,j-1} + BY_{i,j} + CY_{i,j+1} = \text{RHS}, \quad i=2,3,\ldots,I_{\max-1}, \quad J=2,3,\ldots,J_{\max-1}, \tag{77a}$$

where

$$A = \beta^2 A_3 - \tfrac{1}{2}\beta\Delta\xi A_5 - \begin{Bmatrix}(1-v)/2\\ \tfrac{1}{4}\end{Bmatrix}\beta A_2 \quad \begin{cases}\text{if } i\neq i_s,\\ \text{if } i=i_s,\end{cases} \tag{77b}$$

$$B = -2\beta^2 A_3 + \begin{Bmatrix}1-3v\\ -1\end{Bmatrix}A_1 + \begin{Bmatrix}1-v\\ -\tfrac{1}{2}\end{Bmatrix}\Delta\xi A_4 \quad \begin{cases}\text{if } i\neq i_s,\\ \text{if } i=i_s,\end{cases} \tag{77c}$$

$$C = \beta^2 A_3 + \tfrac{1}{2}\beta\Delta\xi A_5 + \begin{Bmatrix}(1-v)/2\\ \tfrac{1}{4}\end{Bmatrix}\beta A_2 \quad \begin{cases}\text{if } i\neq i_s,\\ \text{if } i=i_s,\end{cases} \tag{77d}$$

$$\text{RHS} = \begin{cases} -vA_1(Y_{i+1,j}+Y_{i-1,j}) + (1-v)A_1(2Y_{i-1,j} - Y_{i-2,j}) \\ \quad -\dfrac{v}{4}\beta A_2(Y_{i+1,j+1} - Y_{i+1,j-1} - Y_{i-1,j+1} + Y_{i-1,j-1}) \\ \quad +\dfrac{1-v}{2}\beta A_2(Y_{i-1,j+1} - Y_{i-1,j-1}) \\ \quad -\dfrac{v}{2}\Delta\xi A_4(Y_{i+1,j} - Y_{i-1,j}) + (1-v)\Delta\xi A_4\, Y_{i-1,j} & \text{if } i\neq i_s, \\ -A_1(Y_{i+1,j} - \alpha_j Y_{i-1,j} + \alpha_j Y_{i-2,j}) \\ \quad -\tfrac{1}{4}\beta A_2(Y_{i+1,j+1} - Y_{i+1,j-1} - 3Y_{i-1,j+1} + 3Y_{i-1,j-1} \\ \quad + Y_{i-2,j+1} - Y_{i-2,j-1}) \\ \quad -\tfrac{1}{2}\Delta\xi A_4(Y_{i+1,j} + \alpha_j Y_{i-1,j} - \alpha_j Y_{i-2,j}) & \text{if } i=i_s, \end{cases} \tag{77e}$$

$$\beta = \Delta\xi/\Delta\eta.$$

The boundary conditions are Dirichlet-type:

$$Y_{i,1}=\begin{cases} f(\xi_i) & \text{if } i_{LE}\leqslant i\leqslant i_{TE}, \\ 0 & \text{if } i<i_{LE} \text{ or } i>i_{TE}, \end{cases} \tag{77f}$$

$$Y_{i,j}=0 \quad \text{if } j=J_{max} \text{ or } i=1 \text{ or } i=I_{max}.$$

Applying Jones' transformation to equation (29), we get

$$B_1 R_\xi + B_2 R_\eta = B_3, \tag{78a}$$

where

$$B_1 = Y_\xi (Y_\eta + \psi_\eta)^2, \tag{78b}$$

$$B_2 = -(Y_\eta + \psi_\eta)(Y_\xi^2 + x_\xi^2), \tag{78c}$$

$$B_3 = (\gamma+1) M_\infty^2 \psi_\eta^2 \left(Y_{\xi\xi} - \frac{x_{\xi\xi}}{x_\xi} Y_\xi \right). \tag{78d}$$

The Crank–Nicolson scheme for equation (78a) produces the same form of difference equation, its coefficients and its RHS term as in equations (54a)–(54e), but the B's take a different form, i.e.

$$B_1 = \tfrac{1}{2}\{[Y_\xi(Y_\eta+\psi_\eta)^2]_{i,j} + [Y_\xi(Y_\eta+\psi_\eta)^2]_{i,j+1}\}, \tag{79a}$$

$$B_2 = -\tfrac{1}{2}\{[(Y_\eta+\psi_\eta)(Y_\xi^2+x_\xi^2)]_{i,j} + [(Y_\eta+\psi_\eta)(Y_\xi^2+x_\xi^2)]_{i,j+1}\}, \tag{79b}$$

$$B_3 = \frac{\gamma+1}{2} M_\infty^2 \left\{ \left[\psi_\eta^2 \left(Y_{\xi\xi} - \frac{x_{\xi\xi}}{x_\xi} Y_\xi \right) \right]_{i,j} + \left[\psi_\eta^2 \left(Y_{\xi\xi} - \frac{x_{\xi\xi}}{x_\xi} Y_\xi \right) \right]_{i,j+1} \right\}, \tag{79c}$$

and the ξ-derivatives Y_ξ and $Y_{\xi\xi}$ in equation (79c) should be approximated by a type-dependent difference scheme with SPOs

$$(Y_\xi)_{i,j} = \begin{cases} \dfrac{\nu}{2\Delta\xi}(Y_{i+1,j} - Y_{i-1,j}) + \dfrac{1-\nu}{\Delta\xi}(Y_{i,j} - Y_{i-1,j}) & \text{if } i \neq i_s, \\ \dfrac{1}{2\Delta\xi}(Y_{i+1,j} - Y_{i,j} + \alpha_j Y_{i-1,j} - \alpha_j Y_{i-2,j}) & \text{if } i = i_s, \end{cases} \tag{80a}$$

$$(Y_{\xi\xi})_{i,j} = \begin{cases} \dfrac{\nu}{\Delta\xi^2}(Y_{i+1,j} - 2Y_{i,j} + Y_{i-1,j}) + \dfrac{1-\nu}{\Delta\xi^2}(Y_{i,j} - 2Y_{i-1,j} + Y_{i-2,j}) & \text{if } i \neq i_s, \\ \dfrac{1}{\Delta\xi^2}(Y_{i+1,j} - Y_{i,j} - \alpha_j Y_{i-1,j} + \alpha_j Y_{i-2,j}) & \text{if } i = i_s. \end{cases} \tag{80b}$$

5. RESULTS AND DISCUSSION

The method developed here is used to calculate the transonic flows past symmetric aerofoils at zero angle of attack. Both subcritical and supercritical Mach numbers are considered. For unstretched co-ordinates (x, ψ) a 65×33 uniform mesh covers the domain $-2\leqslant x\leqslant 3$, $0\leqslant\psi\leqslant 2{\cdot}5$ and the aerofoil is located between 0 and 1 with 13 grid points on the surface. For stretched co-ordinates (ξ, η) a 65×33 uniform grid covers the computational domain $-1{\cdot}54\leqslant\xi\leqslant 1{\cdot}54$, $0\leqslant\eta\leqslant 1{\cdot}54$, corresponding to $-7{\cdot}04\leqslant x\leqslant 7{\cdot}04$, $0\leqslant\psi\leqslant 19{\cdot}47$. The aerofoil is located between $-0{\cdot}5$ and $0{\cdot}5$ and has 25 grid points on it (see Figure 5).

Most of the computations have been carried out on the y–R set of equations (28) and (29) for (x, ψ) co-ordinates and on the Y–R set of equations (72) and (78) for (ξ, η) co-ordinates. Other sets

第二部分 出国后博士学位论文阶段

of equations were only used for comparison with the y–R set. The relaxation parameters were taken as $\omega=1{\cdot}7$–$1{\cdot}8$ for the subcritical case and $\omega=0{\cdot}8$–$0{\cdot}9$ for the supercritical case.

Figure 6 shows the comparison of the calculated C_p-distribution of a 6% biconvex aerofoil at $M_\infty=0{\cdot}909$ with experimental data.[27] The type-dependent difference scheme with SPO is used for the y-equation. We can see that the shock wave is very weak and accurately captured by the present scheme.

Figures 7–9 are comparisons of calculated NACA0012 C_p-distributions in (x, ψ) co-ordinates with experimental data at NAE[28] for $M_\infty=0{\cdot}490$ and at ONERA[28] for $M_\infty=0{\cdot}756$ and $0{\cdot}803$. In

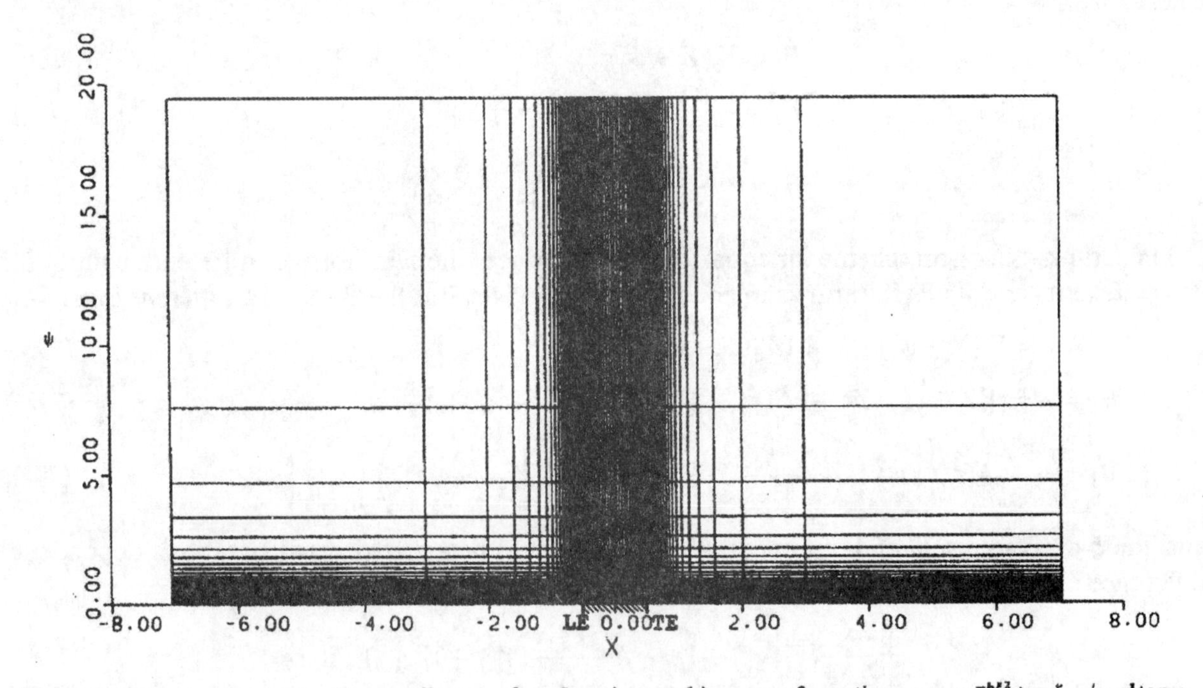

Figure 5. Grid system in (x, ψ) co-ordinates after Jones' stretching transformation; $x=ae^{-b\xi^2}\tan\xi$, $\psi=d\tan\eta$, $-1{\cdot}54\leqslant\xi\leqslant1{\cdot}54$, $0\leqslant\eta\leqslant1{\cdot}54$, $a=0{\cdot}9$, $b=0{\cdot}6$, $d=0{\cdot}6$, $-7{\cdot}04\leqslant x\leqslant7{\cdot}04$, $0\leqslant\psi\leqslant19{\cdot}47$, $x_{LE}=-0{\cdot}5$, $x_{TE}=0{\cdot}5$

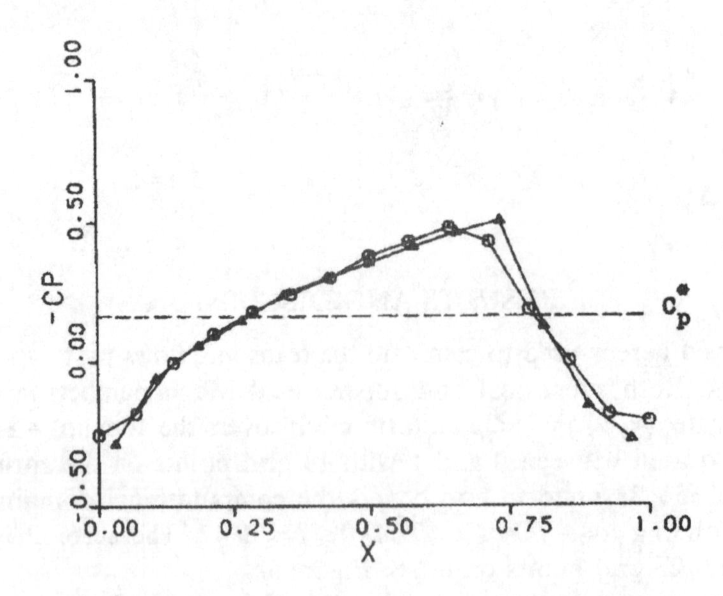

Figure 6. Comparison of pressure coefficient distribution with experimental data, 6% biconvex, $M_\infty=0{\cdot}909$, y–R set: —△—, present; —○—, experimental data at NASA[27]

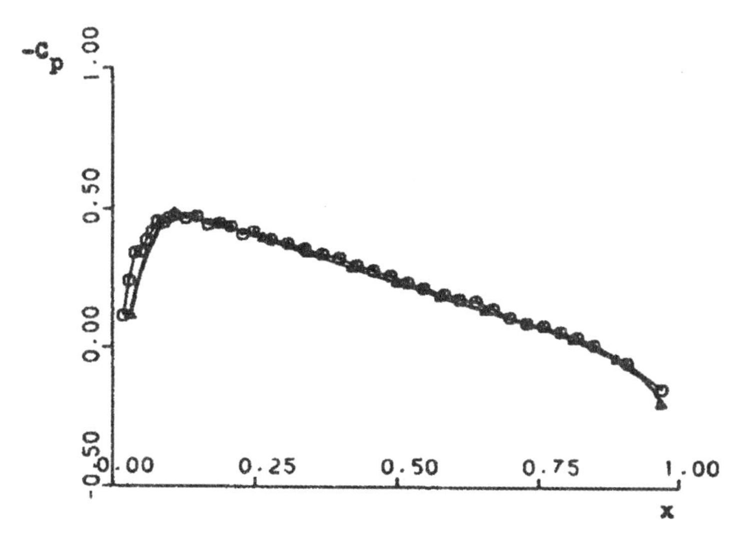

Figure 7. Comparison of pressure coefficient distribution with experimental data, NACA0012, $M_\infty = 0{\cdot}490$, *y–R* set: —△—, present; —○—, experimental data at NAE[28]

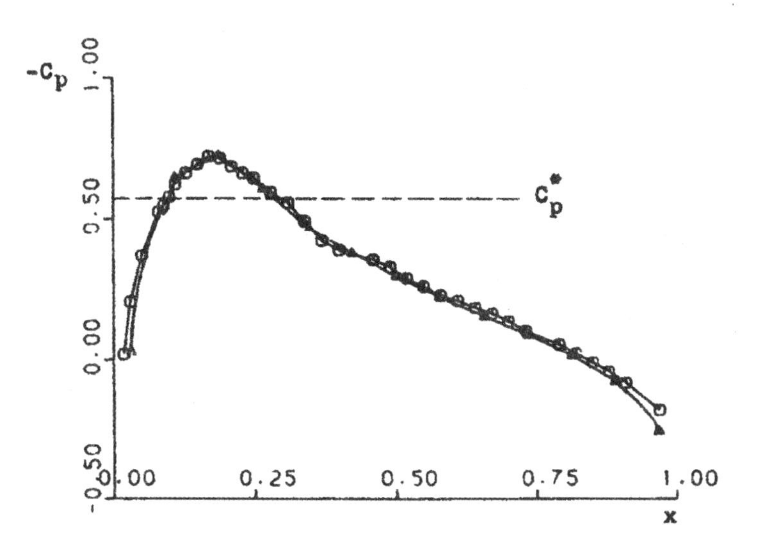

Figure 8. Comparison of pressure coefficient distribution with experimental data, NACA0012, $M_\infty = 0{\cdot}756$, *y–R* set: —△—, present; —○—, experimental data at ONERA[28]

order to show the improvement in accuracy using stretched co-ordinates, Figures 10 and 11 are comparisons of calculated C_p-distributions of NACA0012 with ONERA[28] for $M_\infty = 0{\cdot}803$ and NAE[28] for $M_\infty = 0{\cdot}817$. From these plots we can see that for both subcritical and supercritical cases the present approach gives good agreement with available experiments. For supercritical flow the shock wave location and strength are well predicted, but the shock is smeared out over two or three grid points sometimes and slight oscillation occurs after the shock wave.

Figure 12 shows the effect of the shock point operator in the *y*-equation on shock wave capturing for NACA0012 at $M_\infty = 0{\cdot}803$. Obviously, the type-dependent (TD) scheme with no SPO cannot correctly capture the shock wave and the C_p-distribution is incorrect, while the TD scheme with SPO is able to do so. Hence the SPO is a crucial tool to capture shock waves automatically.

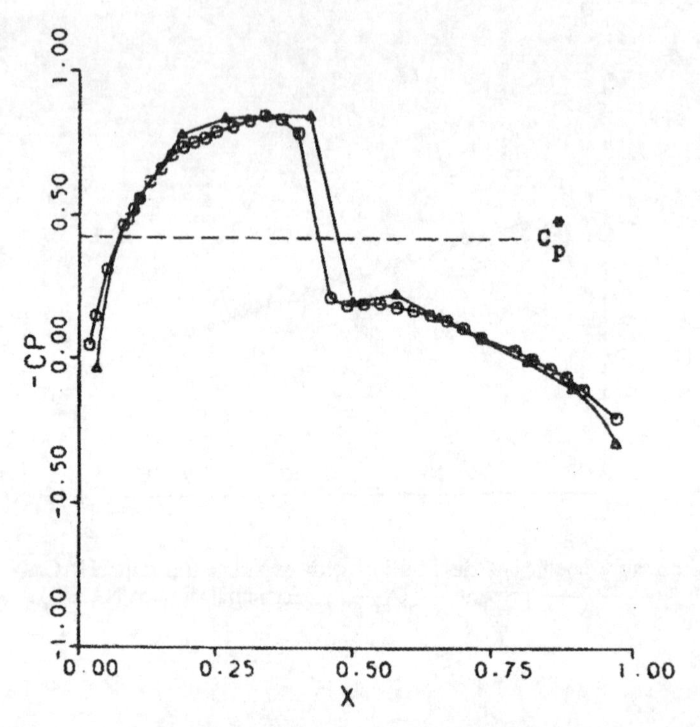

Figure 9. Comparison of pressure coefficient distribution with experimental data, NACA0012, $M_\infty = 0{\cdot}803$, y–R set: —Δ—, present, uniform grid; —○—, experimental data at ONERA[28]

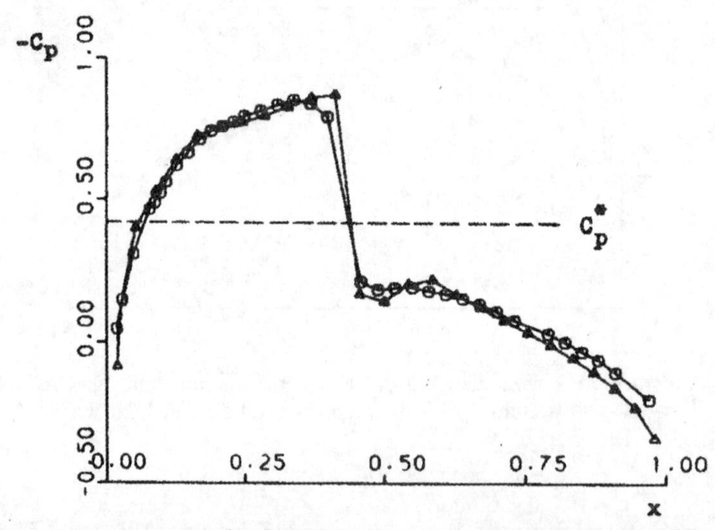

Figure 10. Comparison of pressure coefficient distribution with experimental data, NACA0012, $M_\infty = 0{\cdot}803$, Y–R set: —Δ—, present, clustered grid; —○—, experimental data at ONERA[28]

Figures 13–16 show the C_p-distributions for NACA0012 using various sets of equations: y–R with R_x given by (37), y–R with $L(y)=0$ and $L(R)=G$, y–M^2 and y–u–ρ. It was found that all the sets are capable of producing good results for subcritical flows. However, some difficulties are encountered in the computation for supercritical flows. Therefore the y–R set of equations (28) and (29) are recommended as an optimal set.

Figure 17 illustrates the convergence history of the R-equation. The horizontal axis gives the total number of iterations for y and the vertical axis is the logarithm of the error of R between two

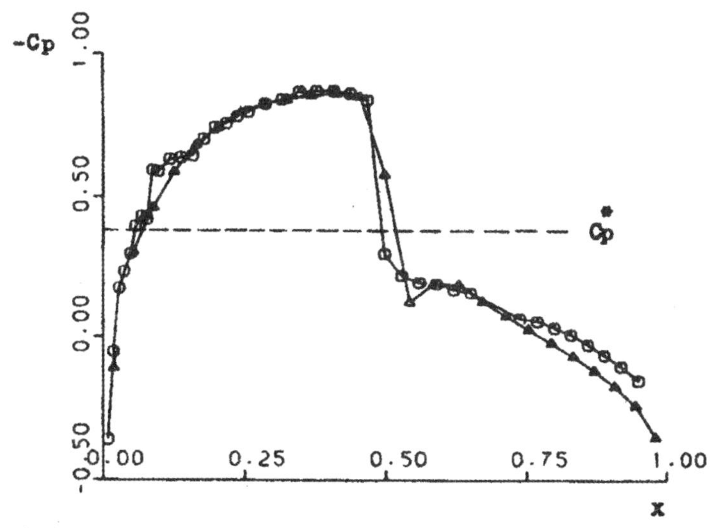

Figure 11. Comparison of pressure coefficient distribution with experimental data, NACA0012, $M_{\infty}=0{\cdot}817$, Y–R set: ——△——, present, clustered grid; ——○——, experimental data at NAE[28]

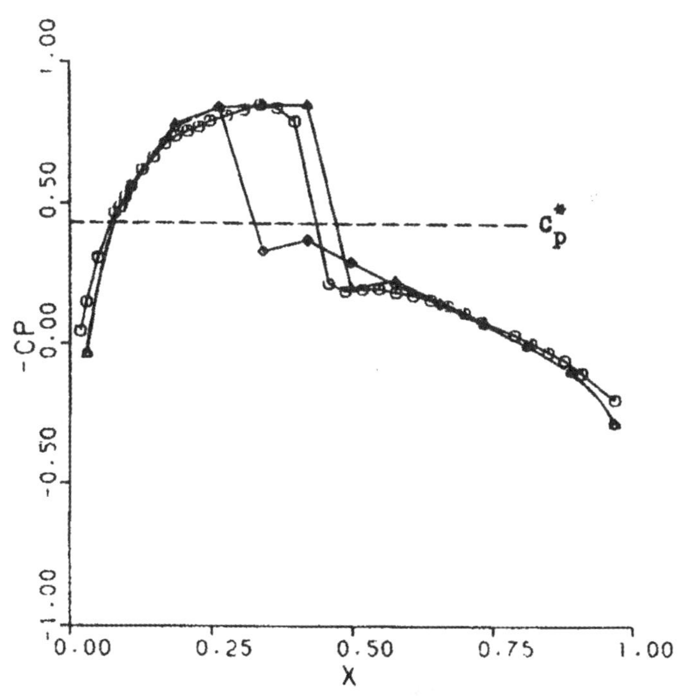

Figure 12. Effect of shock point operator (SPO) for y_{xx}, $y_{x\psi}$ on shock strength and location, NACA0012, $M_{\infty}=0{\cdot}803$, y–R set: ——△——, present; ——○——, experimental data at ONERA;[28] ——◇——, present without SPO

successive global iterations. For subcritical Mach number the convergence rate is fast and the maximum error decreases steeply, but for supercritical flow the convergence rate is much slower and the maximum error decreases slowly with violent oscillation.

Figure 18 indicates the effect of artificial density methods on the capturing of shock waves. Central differencing is used for both the y- and R-equations and a conventional artificial density $\tilde{R}=R-\mu R_x \Delta x$ is added. The result shows that the C_p-distribution is totally incorrect and the

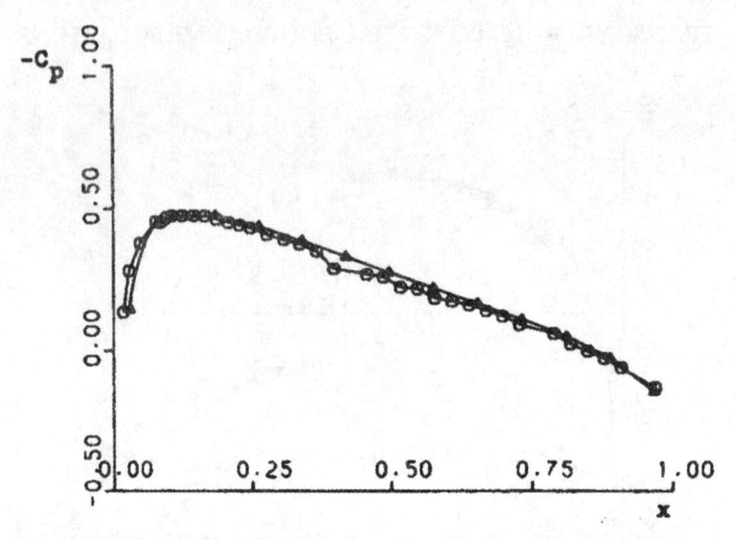

Figure 13. Comparison of pressure coefficient distribution with experimental data, NACA0012, $M_\infty = 0{\cdot}502$, y–R set, R_x given by (37): ——Δ——, present; ——O——, experimental data at ONERA[28]

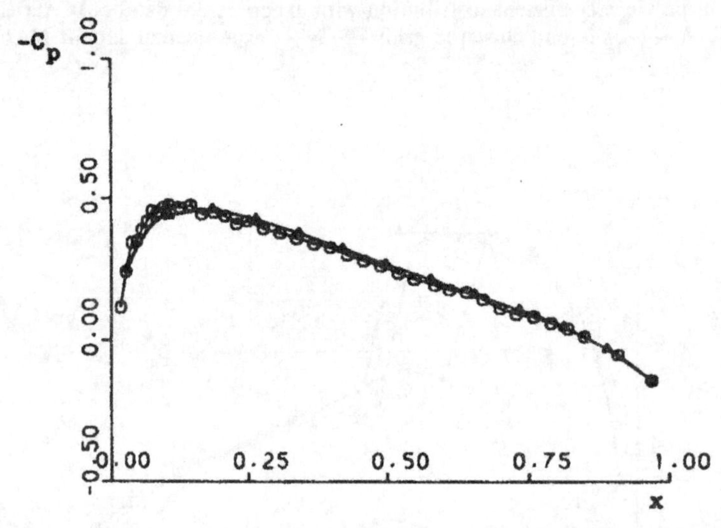

Figure 14. Comparison of pressure coefficient distribution with experimental data, NACA0012, $M_\infty = 0{\cdot}490$, y–R set, $L(y) = 0$, $L(R) = G$: ——Δ——, present; ——O——, experimental data at NAE[28]

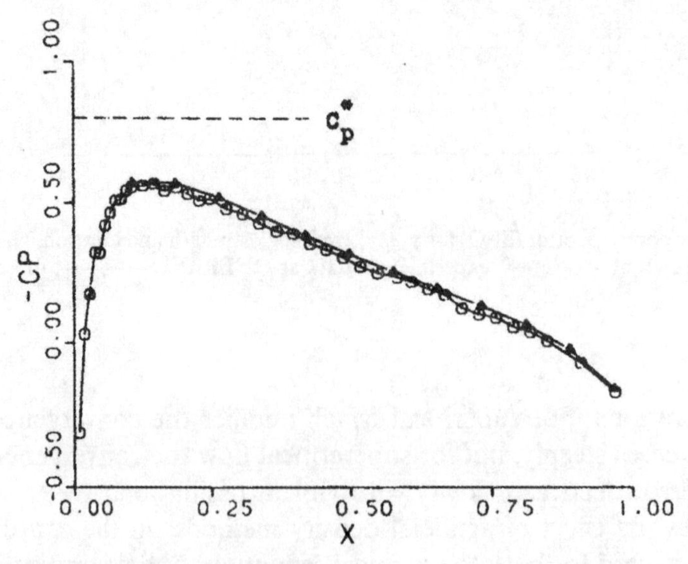

Figure 15. Comparison of pressure coefficient distribution with experimental data, NACA0012, $M_\infty = 0{\cdot}693$, y–M^2 set, $(M^2)_x$ given by (42): ——Δ——, present; ——O——, experimental data at NAE[28]

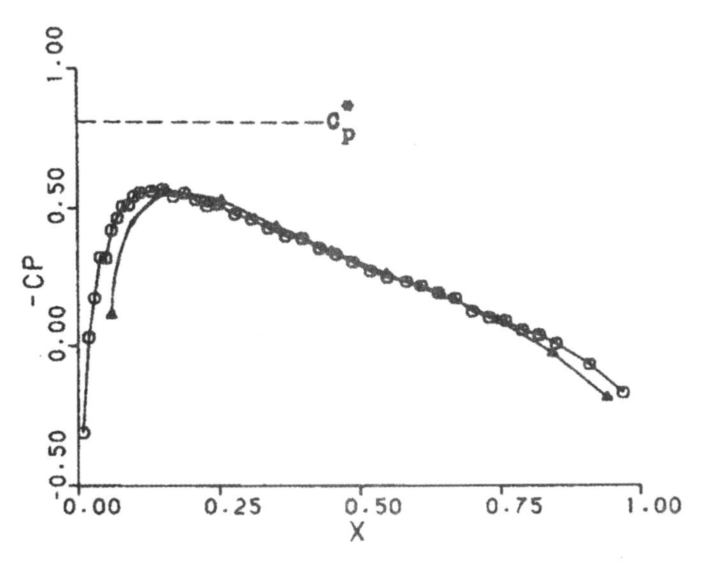

Figure 16. Comparison of pressure coefficient distribution with experimental data, NACA0012, $M_\infty=0{\cdot}693$, y–u–ρ set: ——△——, present; ——○——, experimental data at NAE[28]

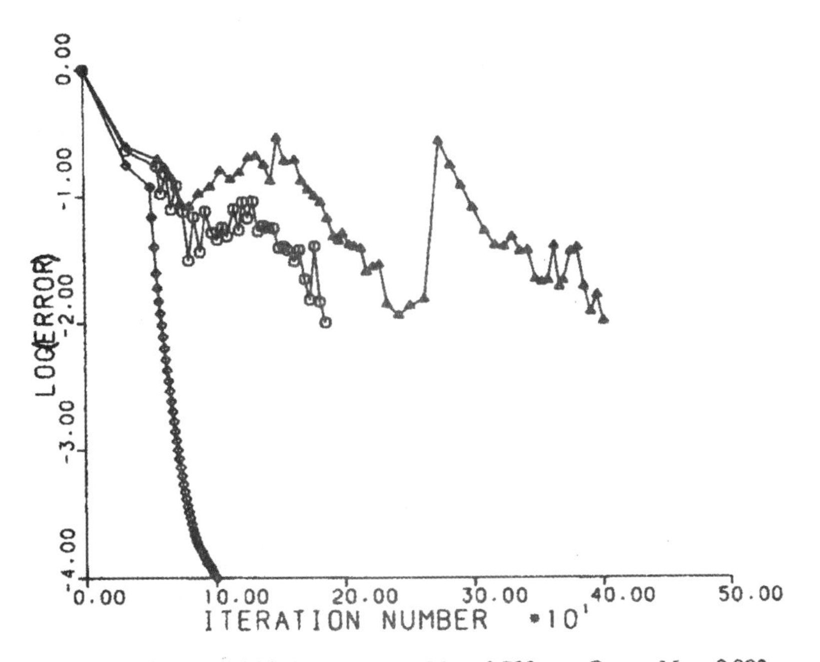

Figure 17. Convergence history for NACA0012: ——◇——, $M_\infty=0{\cdot}703$; ——○——, $M_\infty=0{\cdot}803$; ——△——, $M_\infty=0{\cdot}835$

shock wave is not captured. Hence it seems that the widely used artificial density technique in velocity potential computation is not easily extendable for our formulation.

Figure 19 is the comparison of the C_p-distribution for NACA0012 at $M_\infty=0{\cdot}803$ for two difference schemes using the R-equation (29). One of them is the Crank–Nicolson scheme with TD and SPO for the y_{xx}-term in the R-equation. The other is TD with SPO for all y_{xx}-, y_x- and R_x-terms in the R-equation. It is clear that in the latter case the shock wave position and strength are highly oscillatory, while the former scheme gives accurate shock position and strength and less oscillation. Hence the first scheme is recommended.

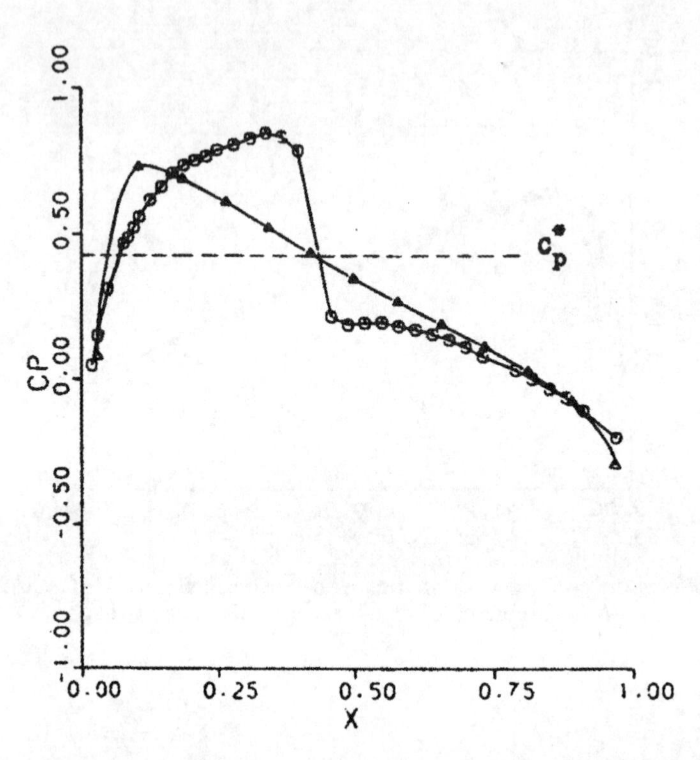

Figure 18. Inability of artificial density method to capture shock waves, NACA0012, $M_\infty=0{\cdot}803$, y–R–$\tilde{R}$ set: ——△——, artificial density method with central differencing; ——○——, experimental data at ONERA[28]

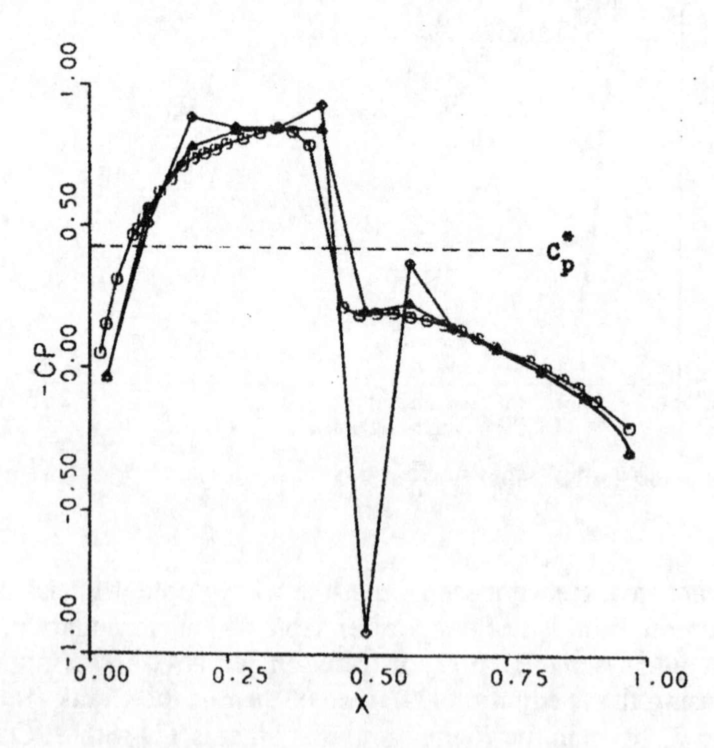

Figure 19. Effect of type-dependent (TD) differencing with shock point operator (SPO) for y_x, R_x, NACA0012, $M_\infty=0{\cdot}803$, y–R set: ——△——, present; ——○——, experimental data at ONERA;[28] ——◇——, present with y_x, R_x, TD+SPO

6. CONCLUDING REMARKS

1. The newly developed approach based on the full-potential-equivalent equations in von Mises co-ordinates, type-dependent scheme with shock point operator and successive line overrelation procedure is able to obtain accurate numerical solutions for both subcritical and supercritical transonic flows.
2. Embedded shock waves in supercritical flows can be captured automatically in the process of iteration and are smeared out over two or three grid points.
3. The full-potential-equivalent equations consist of a 'main' equation for the corresponding 'main' variable, which is the streamline ordinate y, and a 'secondary' equation for one of the following 'secondary' variables: density ρ, generalized density R, squared Mach number M^2, x-velocity component u, etc.
4. The type-dependent difference scheme with shock point operator (SPO) is effective to solve the 'main' equation for y and the SPO is crucial to capture shock waves, especially for moderate or strong ones.
5. Among various 'secondary' variables, their equations and a number of difference schemes, the generalized density R, its linear equation (29) and the Crank–Nicolson scheme are selected as an optimal combination to accompany the 'main' variable y and its equation.
6. The method developed here accurately captures shock waves in supercritical transonic flow without employing artificial dissipation schemes. This is advantageous, since the use of artificial density or viscosity requires a certain amount of tuning of the code, i.e. a number of parameters must be adjusted to achieve converged solutions. The present method, in which the only adjustable quantities are the relaxation parameters, is very robust and relatively easier to use.

ACKNOWLEDGEMENT

The authors thank the Natural Science and Engineering Research Council of Canada for financially supporting this project.

REFERENCES

1. E. M. Murman and J. D. Cole, 'Calculation of plane steady transonic flows', *AIAA J.*, **9**, 114–121 (1971).
2. E. M. Murman and J. A. Krupp, 'Solution of the transonic potential equation using a mixed finite difference system', *Lecture Notes in Physics*, Vol. 8, Springer, Berlin, 1971, pp. 199–206.
3. E. M. Murman, 'Analysis of embedded shock waves calculated by relaxation methods', *AIAA J.*, **12**, 626–633 (1974).
4. M. Hafez and H. K. Cheng, 'Shock fitting applied to relaxation solutions of transonic small disturbance equations', *AIAA J.*, **5**, 786–793 (1977).
5. D. J. Jones and R. G. Dickinson, 'A description of the NAE two-dimensional transonic small disturbance computer method', *NAE Laboratory Technical Report LTR-HA-39*, January 1980.
6. A. Jameson, 'Iterative solution of transonic flows over airfoils and wings, including flow at Mach 1', *Commun. Pure Appl. Math.*, **27**, 283–309 (1974).
7. A. Jameson, 'Numerical solution of transonic flows with shock waves', *Symp. Transsonicum II*, Springer, Berlin, 1976, pp. 384–414.
8. M. Hafez, J. South and E. M. Murman, 'Artificial compressibility methods for numerical solution of transonic full potential equations', *AIAA J.*, **17**, 838–844 (1979).
9. W. G. Habashi and M. Hafez, 'Finite element solutions of transonic flow problems', *AIAA J.*, **20**, 1368–1378 (1982).
10. M. Hafez and D. Lovell, 'Improved relaxation schemes for transonic potential calculations', *Int. j. numer. methods fluids*, **8**, 1–6 (1988).
11. R. Magnus and H. Yoshihara, 'Inviscid transonic flow over airfoils', *AIAA J.*, **8**, 2157–2162 (1970).
12. R. M. Beam and R. F. Warming, 'An implicit finite difference algorithm for hyperbolic system in conservation law form', *J. Comput. Phys.*, **22**, 87–110 (1976).
13. J. C. Steger, 'Implicit finite difference simulation of flow around arbitrary two-dimensional geometries', *AIAA J.*, **16**, 676–686 (1978).

14. T. H. Pulliam and D. S. Chaussee, 'A diagonal form of an implicit approximate factorization algorithm', *J. Comput. Phys.*, **39**, 343–383 (1981).
15. A. Jameson, W. Schmidt and E. Turkel, 'Numerical solutions of the Euler equations by finite-volume methods using Runge–Kutta time-stepping schemes', *AIAA Paper 81-1259*, 1981.
16. R. H. Ni, 'A multi-grid scheme for solving the Euler equations', *AIAA J.*, **20**, 1565–1571 (1982).
17. W. G. Habashi and M. Hafez, 'Finite element stream function solutions for transonic turbomachinery flows', *AIAA Paper 82-1268*, 1982.
18. M. Hafez and D. Lovell, 'Numerical solution of transonic stream function equation', *AIAA J.*, **21**, 327–335 (1983).
19. W. G. Habashi, P. L. Kotiuga and L. A. McLean, 'Finite element simulation of transonic flows by modified potential and stream function methods', *Eng. Anal.*, **2**, 150–154 (1985).
20. H. L. Atkins and H. A. Hassan, 'A new stream function formulation for the steady Euler equations', *AIAA J.*, **23**, 701–706 (1985).
21. R. M. Barron, 'Computation of incompressible potential flow using von Mises coordinates', *J. Math. Comput. Simul.*, **31**, 177–188 (1989).
22. M. H. Martin, 'The flow of a viscous fluid I', *Arch. Rat. Mech. Anal.*, **41**, 266–286 (1971).
23. R. M. Barron and R. K. Naeem, 'Numerical solution of transonic flows on a streamfunction coordinate system', *Int. j. numer. methods fluids*, **9**, 1183–1193 (1989).
24. R. K. Naeem and R. M. Barron, 'Transonic computation on a natural grid', *AIAA J.*, **28**, 1836–1838 (1990).
25. G. S. Dulikravich, 'A stream-function-coordinate (SFC) concept in aerodynamic shape design', *AGARD VKI Lecture Series* (1990).
26. M. S. Greywall, 'Streamwise computation of 2-D incompressible potential flows', *J. Comput. Phys.*, **59**, 224–231 (1985).
27. E. D. Knetchtel, 'Experimental investigation at transonic speeds of pressure distributions over wedge and circular-arc-airfoil sections and evaluation of perforated wall interference', *NASA TN D-15*, 1959.
28. J. J. Thibert, M. Grandjacques and L. H. Ohman, 'Experimental data base for computer program assessment', *AGARD AR-138*, 1979, pp. A1-1–A1-36.

PROCEEDINGS

CONFERENCE OF THE CFD

SOCIETY OF CANADA

MONTREAL

JUNE 14 - 15, 1993

COMPTES RENDUS

CONFÉRENCE DE LA SOCIÉTÉ

CANADIENNE DE CFD

MONTRÉAL

14 - 15 JUIN 1993

Changfa An
Montreal, PQ, Canada
July 20, 1993

SURVEY OF THE STREAMFUNCTION-AS-A-COORDINATE METHOD IN CFD*

R.M. BARRON, C.-F. AN AND S. ZHANG

Fluid Dynamics Research Institute
University of Windsor
Windsor, Ontario, Canada N9B 3P4

Abstract. The idea of streamfunction as an independent variable in fluid dynamics problems was initiated by von Mises about 65 years ago. However, the approach found limited direct application until the work of Stanitz in the early 1950's. A more general approach was developed by Martin. This formulation is now known as the streamfunction-as-a-coordinate (SFC) method. During the last decade, the SFC method has been applied in CFD. Flow regimes up to high transonic Mach numbers have been considered with application to both aerodynamic analysis and design. In this paper the fundamental concepts and the general formulation of the SFC method will be described and earlier work will be reviewed. Recent results obtained by the present authors, from the viewpoint of a unified methodology for full-potential and Euler models, will be presented. Other approaches within the general SFC formulation will be discussed.

1. INTRODUCTION

Use of the streamfunction as an independent variable in the study of fluid dynamics problems can be traced back to 65 years ago when von Mises[1] proposed a coordinate transformation in which the roles of streamfunction and one of the geometric coordinates, usually the transverse coordinate to the main flow direction, are interchanged so that the streamfunction ψ serves as a coordinate and the transverse coordinate y serves as an dependent variable in the transformed system of governing equations. Therefore, this approach may be called a 'streamfunction-as-a-coordinate' method. However, the approach found limited direct application for several decades, used primarily in boundary layer studies.

Practical application of the streamfunction-as-a-coordinate (SFC) method was pioneered by Stanitz in the 1950's[2-6]. Stanitz used the streamlines and potential curves as a set of curvilinear coordinates to design 2D stationary channels (ducts). He solved for velocity q and its direction θ as unknown functions of velocity potential ϕ and streamfunction ψ and, furthermore, obtained x and y coordinates of the

*For presentation at Inaugural Conference of the CFD Society of Canada, June 14-15, 1993, Montreal, Canada

channel walls. In 1971, a general formulation of the flow equations was developed by Martin[7], who formally applied results from differential geometry to transform the equations for 2D viscous incompressible flows from Cartesian coordinates to curvilinear coordinates. Like some earlier work, Martin used the streamfunction to define one of the coordinate curves. But unlike others, he left the second family of coordinate curves arbitrary so that it could be specified by the user depending on the particular problem of interest. If the second family of curves remains the original Cartesian coordinate in the main flow direction, the approach becomes the one that von Mises developed. If the second family of curves is chosen as velocity potential, the approach is the same as Stanitz's method.

Paralleling the rapid development of modern electronic computers, CFD, as a vital branch of fluid mechanics, has advanced dramatically. At the same time, the use of the SFC method in CFD has received more attention since the late 1960's and 1970's because of a number of attractive advantages of the approach. For instance, the conventional grid generation procedure, a necessary first step of most CFD applications, can be omitted due to the fact that the equations to be solved in the transformed plane play a double role of governing equations (physical laws of the flow) and grid generation equations (geometric characteristics of the flow). Needless to say, a considerable portion of the computational time and storage requirement can be reduced. Some typical works are reviewed below.

For incompressible potential flow, Jeppson[8] solved an axisymmetric flow from a large reservoir through a circular orifice using a streamline coordinate system in which the velocity potential and Stokes streamfunction are independent variables and the radial and axial dimensions are unknowns. Chan[9] solved the vertical coordinate y of large-amplitude solitary waves in the compex potential plane in which velocity potential ϕ and streamfunction ψ serve as coordinates. Yang and Nelson[10] employed a similar method to solve a 2D incompressible potential flow through the Griffith diffuser. Breeze-Stringfellow and Burggraf[11] treated the interference flow of a propeller and a nacelle by solving the axisymmetric potential flow in 'stream tube coordinates' where the streamfunction and axial distance are coordinates. Using a similar streamline coordinates, Greywall[12-14] solved 2D and axisymmetric incompressible potential flows.

For viscous incompressible fluid flows, Duda and Vrentas[15] developed a so-called 'Protean coordinate system', whose fundamental feature is the use of stream function as an independent variable in the radial direction of an axisymmetric flow. Clermont and Lande[16] and Andre et al.[17] developed a 'stream tube method' in which the flow field is solved in a mapped domain where the transformed streamlines are rectilinear or circular. Their method is used to investigate axisymmetric viscous flows including the flow through a convergent duct and a jet flow at the exit of a cylindrical tube. To handle hydraulic power machinery problems, Takahashi[18] derived the governing equations for 2D and axisymmetric laminar flows in an orthogonal curvilinear coordinate system where one coordinate is a streamline and the other is perpendicular to it and the metrics are to be solved as unknowns. Using this method, Takahashi solved the 2D liquid jet flow from a channel with parallel walls into the atmosphere, Takahashi and Tsukiji[19] solved a 2D laminar jet issuing from a skewed-symmetrical orifice in the spool valve mechanism generally used in hydraulic power systems, Tsukiji and Takahashi[20] solved an axisymmetric laminar jet leaving a Poiseuille tube, Zarbi and Takahashi[21] studied the laminar jet flow through a convergent nozzle into the atmosphere by solving velocity and its direction.

For compressible fluid flows, in the mid-1960's Honda[22-23] studied the inverse problem of a shock layer around a blunt-nosed body in a hypersonic stream. He solved 2D and axisymmetric inviscid rotational flow in an orthogonal streamfunction coordinate system. Stanitz[24-26] extended his earlier work to the 3D case. Independently, Pearson[27] proposed a 3D streamline coordinate method in which the streamline geometry is expressed in terms of two parameters (for 2D flow only one parameter is needed) and is corrected iteratively by the transformed governing equations in the streamline coordinates. In fact, this approach is the same as the SFC method if the two parameters are taken as two streamfunctions. Using this method, he calculated a 2D jet flow from an orifice into the air. Owen and Pearson[28] extended the method to solve an axisymmetric actuator disc flow in a turbomachine and 3D flows through ducts with variable cross sections and the flow past a 3D corner. In the case of unsteady flow, Srinivasan and Spalding[29] solved the shock tube flow using 1D unsteady gas dynamics equations for primitive variables in streamfunction coordinates. Huang and Dulikravich[30] applied the SFC concept to obtain an explicit formulation for 3D inviscid steady compressible flow and solved the incompressible flow around a cylinder and the subsonic flow past an airfoil. Dulikravich[31] emphasized some advantageous features of the SFC concept for aerodynamic shape design. Bonatak, Chaviaropoulos and Papailiou[32] solved velocity modulus and flow angle in potential-streamfunction coordinates to complete a cascade design problem.

In turbomachinery analysis and design, since Wu[33] proposed the general theory of the stream surfaces to solve cascade problems, the applications of this theory have been extensive. A typical work of this kind is the image-plane method used in the inverse problem of cascade flows by Liu and Tao[34] and Chen et al.[35]. Their method leads to a problem of solving an integro-differential equation for the meridian angle of cylindrical coordinates in the coordinates composed of the axial distance and the streamfunction, which construct an image-plane. Liu and Zhang[36] improved their approach by introducing another unknown, the moment function, which represents a generalization of the Kutta-Joukowski lift theorem for 2D flow. The above approaches and applications are summarized by Liu[37].

Barron[38] first connected Martin's[7] approach with the SFC method in CFD. Treating the von Mises transformation as a specialized SFC method, he successfully solved the resulting elliptic equation to simulate incompressible potential flow past an airfoil. Since then, numerical simulations based on the SFC method have been considerably extended, such as to incompressible lifting flows[39], axisymmetric flows[40] and design problems[41-42], as well as to transonic analysis and design problems[43-50].

2. FLOW EQUATIONS IN SFC SYSTEM

The governing flow equations can be transformed from physical space, with Cartesian coordinates (x, y), to a computational space, with general curvilinear coordinates (ξ, η), by the transformation equations

$$\tau = t, \quad \xi = \xi(x, y, t), \quad \eta = \eta(x, y, t). \tag{1}$$

The transformed equations for unsteady 2D flow, in strong conservation law form, are[51]

$$\frac{\partial Q}{\partial \tau} + \frac{\partial}{\partial \xi}(E - E_\nu) + \frac{\partial}{\partial \eta}(F - F_\nu) = 0 \tag{2}$$

where

$$Q = J\begin{bmatrix} \rho \\ \rho u \\ \rho v \\ e \end{bmatrix}, \quad E = J\begin{bmatrix} \rho U \\ \rho u U + \xi_x p \\ \rho v U + \xi_y p \\ (e+p)U - \xi_t p \end{bmatrix}, \quad F = J\begin{bmatrix} \rho V \\ \rho u V + \eta_x p \\ \rho v V + \eta_y p \\ (e+p)V - \eta_t p \end{bmatrix},$$

$$E_\nu = \frac{J}{Re}\begin{bmatrix} 0 \\ \xi_x \tau_{11} + \xi_y \tau_{12} \\ \xi_x \tau_{21} + \xi_y \tau_{22} \\ \xi_x \beta_1 + \xi_y \beta_2 \end{bmatrix}, \quad F_\nu = \frac{J}{Re}\begin{bmatrix} 0 \\ \eta_x \tau_{11} + \eta_y \tau_{12} \\ \eta_x \tau_{21} + \eta_y \tau_{22} \\ \eta_x \beta_1 + \eta_y \beta_2 \end{bmatrix}. \tag{3}$$

In these equations, $J = \partial(x, y)/\partial(\xi, \eta)$ is the transformation Jacobian, U and V are contravariant velocity components given by

$$U = \xi_t + \xi_x u + \xi_y v, \quad V = \eta_t + \eta_x u + \eta_y v \tag{4}$$

and τ_{ij} and β_i are the components of the shear-stress tensor and heat flux vector, respectively.

Defining the streamfunction $\psi(x,y,t)$ such that

$$\rho u = \psi_y, \quad \rho v = -\psi_x \tag{5}$$

one can obtain the flow equations in general SFC by identifying ψ with the function η. A special case of the general SFC is the von Mises system which is obtained by choosing the coordinate ξ to be x. In this case, it is easy to show that

$$\xi_x = 1, \quad \xi_y = 0, \quad \xi_t = 0, \quad J = y_\psi,$$

$$\eta_x = -\frac{y_x}{y_\psi} = -\rho v, \quad \eta_y = \frac{1}{y_\psi} = \rho u, \quad \eta_t = -\frac{y_\tau}{y_\psi}. \tag{6}$$

The equation (2) can then be expressed as

$$\frac{\partial Q}{\partial \tau} + \frac{\partial}{\partial x}(E - E_\nu) + \frac{\partial}{\partial \psi}(F - F_\nu) = 0 \tag{7}$$

where

$$Q = \begin{bmatrix} \rho y_\psi \\ 1 \\ y_x \\ e y_\psi \end{bmatrix}, \quad E = \begin{bmatrix} 1 \\ 1/(\rho y_\psi) + p y_\psi \\ y_x/(\rho y_\psi) \\ (e+p)/\rho \end{bmatrix}, \quad F = -\begin{bmatrix} \rho y_\tau \\ p y_x + y_\tau / y_\psi \\ -p + y_\tau y_x / y_\psi \\ e y_\tau \end{bmatrix},$$

$$E_\nu = \frac{1}{Re}\begin{bmatrix} 0 \\ y_\psi \tau_{11} \\ y_\psi \tau_{21} \\ y_\psi \beta_1 \end{bmatrix}, \quad F_\nu = \frac{1}{Re}\begin{bmatrix} 0 \\ -y_x \tau_{11} + \tau_{12} \\ -y_x \tau_{21} + \tau_{22} \\ -y_x \beta_1 + \beta_2 \end{bmatrix} \tag{8}$$

and

$$\tau_{11} = \frac{2}{3}[2(u_x - \frac{y_x}{y_\psi}u_\psi) - \frac{1}{y_\psi}v_\psi]$$

$$\tau_{12} = \tau_{21} = v_x - \frac{y_x}{y_\psi}v_\psi + \frac{1}{y_\psi}u_\psi$$

$$\tau_{22} = \frac{2}{3}[\frac{2}{y_\psi}v_\psi - u_x + \frac{y_x}{y_\psi}u_\psi]$$

$$\beta_1 = \frac{\gamma}{P_r}[\frac{\partial}{\partial x} - \frac{y_x}{y_\psi}\frac{\partial}{\partial \psi}]e_I + u\tau_{11} + v\tau_{12}$$

$$\beta_2 = \frac{\gamma}{P_r}\frac{1}{y_\psi}\frac{\partial}{\partial\psi}e_I + u\tau_{12} + v\tau_{22}$$

$$e_I = \frac{e}{\rho} - \frac{u^2+v^2}{2} = \frac{e}{\rho} - \frac{1+y_x^2}{2\rho^2 y_\psi^2}.$$

If, furthermore, the ideal gas law is assumed, then the total energy per unit volume is given by

$$e = \frac{1}{\gamma-1}p + \frac{1+y_x^2}{2\rho y_\psi^2} \tag{9}$$

and, therefore,

$$e_I = \frac{1}{\gamma-1}\frac{p}{\rho}. \tag{10}$$

It is often useful to work with the vorticity function which can be expressed as

$$y_\psi \omega = [\frac{y_x}{\rho}]_x - [\frac{1+y_x^2}{\rho y_\psi}]_\psi. \tag{11}$$

In the remainder of this paper, we restrict our discussions to steady flows for which $\partial/\partial\tau = 0$, and the first equation in (7) is identically satisfied. Equation (9) can be used to eliminate e, and hence equations (7) are a system of three equations for the three unknowns y, ρ, p as functions of x and ψ. Having solved for y, ρ, p, the velocity components are given by

$$u = \frac{1}{\rho y_\psi}, \quad v = y_x u. \tag{12}$$

3. BOUNDARY CONDITIONS

In this section we discuss typical boundary conditions for the function $y(x,\psi)$. The boundary conditions for other unknowns do not change from physical to computational space, or can be derived following standard procedures.

3.1 Far field condition:

An incoming flow at angle of incidence α and free stream Mach number M_∞ is considered as a combination of a uniform stream and a vortex. In this case, y is given implicitly by

$$y(x,\psi) = xtan\alpha + \psi sec\alpha - \frac{\Gamma}{2\pi}sec\alpha Ln[x^2 + (1-M_\infty^2)y^2(x,\psi)] \tag{13}$$

where Γ is the circulation.

3.2 Solid boundary condition:

Let the solid boundary be described in the physical domain by the curve $y = f(x)$.

Inviscid case: The condition of flow tangency in the physical domain can be expressed as $v(x,y) = f'(x)u(x,y)$ on $y = f(x)$ where $f(x)$ is a known function for the analysis problem and an unknown function for the design problem. In the computational domain, for the analysis problem,

$$y(x,\psi) = f(x) \quad on \quad \psi = \psi_b \tag{14}$$

where ψ_b is known. For the design problem, a derivative boundary condition is imposed, e.g., for potential flow

$$G(x)y_\psi^2 - y_x^2 = 1 \quad on \quad \psi = \psi_b \tag{15}$$

where $G(x)$ is a known function of the prescribed surface speed or pressure.

Viscous case: The no-slip condition for the analysis problem is $v = u = 0$ on $y = f(x)$. In this computational domain, this condition is replaced by $y(x,\psi) = f(x)$ on $\psi = \psi_b$. No work seems to have been done using the SFC formulation for design problems in viscous flow.

3.3 Inlet condition:

Various inlet conditions can be transformed to the computational domain. For example, for a Poiseuille flow, y may be given implicitly as a function of ψ by $\psi = -2y^3 + 3y^2$, or a condition of parallel flow ($v = 0$) is expressed by $y_x = 0$.

4. INVISCID MODEL

For inviscid flow the fluxes E_ν and F_ν vanish and the Euler equations are obtained. Eliminating e and p and using (8), equation (11) reduces to

$$(y_\psi^2 - \frac{1}{\rho^2 a^2})y_{xx} - 2y_x y_\psi y_{x\psi} + (1 + y_x^2)y_{\psi\psi} = \rho\omega y_\psi^3[1 + (\gamma - 1)M^2] \tag{16}$$

where

$$a = \sqrt{\frac{1}{M_\infty^2} + \frac{\gamma - 1}{2}(1 - \frac{1 + y_x^2}{\rho^2 y_\psi^2})} \quad \text{and} \quad M = \frac{\sqrt{1 + y_x^2}}{a\rho y_\psi}$$

are speed of sound and the local Mach number, respectively. It can be shown that this equation is elliptic, hyperbolic or parabolic as the local flow is subsonic, supersonic or sonic. For isentropic full-potential flows, the right hand side of (16) vanishes ($\omega = 0$) and it is convenient to express the equation in terms of the free Mach number using

$$\frac{1}{\rho^2 a^2} = \frac{M_\infty^2}{\rho^{\gamma+1}}. \tag{17}$$

Incompressible potential flows can be recovered by setting $M_\infty = 0$.

Typical results of calculations based on the Euler model (16), and for compressible full-potential and incompressible potential flows are shown in figures 1—4.

5. CONCLUDING REMARKS

One of the major weakness of von Mises coordinates is that the Jacobian becomes infinite at points where the flow reverses direction. This leads to non-uniqueness of $y(x, \psi)$ and is particularly troublesome in the leading edge region of an airfoil calculation if the angle of attack is too high. Similarly, separated viscous flows cannot be directly treated with this method. Several possibilities are currently being explored to address these issues.

For inviscid flows, the more general SFC method may be applicable, in which ξ is not x, but is chosen in some other fashion. For example, ξ might be chosen to be arc length, in which case the functions $x(\xi, \psi)$ and $y(\xi, \psi)$ must satisfy the constraint $x_\xi^2 + y_\xi^2 = 1$, or chosen such that the $\xi(x, y) = constant$ curves are orthogonal to the streamlines. An interesting alternative is to calculate the $\xi = constant$ curves from a specified distribution of the Jacobian function, in a manner similar to that done in hyperbolic grid generation. This has the advantage of allowing for explicit or implicit cell area control in the computation.

For viscous flows, it is necessary to devise a procedure to handle separation. Currently under investigation is a zonal procedure in which the main flow is first solved using von Mises coordinates. The location of the separation and re-attachment points and the shape of the separation streamline are obtained as part of the main flow solution. The recirculation region can then be solved using conventional methods.

To conclude, the SFC formulation has been successful for aerodynamic analysis and design for inviscid flows over a wide range of flow conditions. As work progresses, it is expected that the key limitations of the SFC method can be overcome and the formulation will be useful for practical analysis and design problems.

NOMENCLATURE

t, τ	time
x, y	Cartesian coordinates
ξ, η	curvilinear coordinates
J	Jacobian
Q	vector of primitive variables
E, F, E_ν, F_ν	flux vectors
Re, Pr	Reynolds and Prandtl numbers
ρ, p, e	density, pressure, total energy per unit volume
u, v	velocity components
U, V	contravariant velocity components
τ_{ij}, β_i	shear-stress tensor, heat-flux vector
ψ, ψ_b	streamfunction, value of ψ on body
γ	adiabatic exponent
ω	vorticity
α	angle of attack
Γ	circulation
M, M_∞	local and free stream Mach numbers
$f(x)$	airfoil boundary shape function
a	sound speed

安长发博士科研论文集

REFERENCES

1. VON MISES, R., *Bemerkungen zur Hydrodynamik*, ZAMM **7** (1927), 425.
2. STANITZ, J.D., *Design of two-dimensional channels with prescribed velocity distributions along the channel walls, Part I — Relaxation solutions*, NACA TN 2595 (1952).
3. STANITZ, J.D., *Design of two-dimensional channels with prescribed velocity distributions along the channel walls, Part II — Solution by Green's function*, NACA TN 2595 (1952).
4. STANITZ, J.D. and SHELDRAKE, L.J., *Application of a channel design method to high solidity cascades and tests of an impulse cascade with 90° turning*, NACA TN 2652 (1952).
5. STANITZ, J.D., OSBORN, W.M. and MIZISIN, J., *An experimental investigation of secondary flow in an accelerating, rectangular elbow with 90° of turning*, NACA TN 3015 (1953).
6. STANITZ, J.D., *Aerodynamic design of efficient two-dimensional channels*, Trans. ASME **75** (1953), 7.
7. MARTIN, M.H., *The flow of a viscous fluid. I*, Archives for Rational Mechanics and Analysis **41** (1971), 266-286.
8. JEPPSON, R.W., *Inverse formulation and finite difference solution for flow from a circular orifice*, J. Fluid Mech. **40 Part I** (1970), 215-223.
9. CHAN, R.K.-C., *A discretized solution for the solitary wave*, J. Comp. Phys. **16** (1974), 32-48.
10. YANG, T. and NELSON, C.D., *Griffith diffusers*, J. Fluids Eng. **101** (1979), 473-477.

11. BREEZE-STRINGFELLOW, A. and BURGGRAF, O.R., *Computation of propeller nacelle interference flows using streamtube co-ordinates*, "Proc. Appl. Mech. Bioeng. and Fluids Eng. Conf.," ed. by K. N. Ghia and U. Ghia, 1983, pp. 107-116.
12. GREYWALL, M.S., Comp. Methods Appl. Mech. Eng. **21** (1980), 231.
13. GREYWALL, M.S., Comp. Methods Appl. Mech. Eng. **36** (1983), 71.
14. GREYWALL, M.S., *Streamwise computation of 2-D incompressible potential flows*, J. Comp. Phys. **59** (1985), 224-231.
15. DUDA, J.L. and VRENTAS, J.S., *Fluid mechanics of laminar liquid jets*, Chem. Eng. Sci. **22** (1967), 855-869.
16. CLERMONT, L.R. and LANDE, M.E., *A method for the simulation of plane or axisymmetric flows of incompressible fluids based on the concept of the stream function*, Eng. Computations **3** (1986), 339-347.
17. ANDRE, P., CLERMONT, J.R. and LANDE, M.E., *Analysis of incompressible flows by the stream tube method: computation of axisymmetric converging flows and free surface flows*, "Numerical Method for Laminar and Turbulent Flows," Proc. 6th Intl. Conf. held in Swansea, UK, ed. by C. Taylor et al., Pineridge Press, 1989, pp. 1439-1448.
18. TAKAHASHI, K., *A numerical analysis of flow using streamline coordinates (The case of two-dimensional steady incompressible flow)*, Bulletin of the JSME **25** (1982), 1696-1702.
19. TAKAHASHI, K. and TSUKIJI, T., *Numerical analysis of a laminar jet using a streamline coordinate system*, Transactions of the CSME **9** (1985), 165-170.
20. TSUKIJI, T. and TAKAHASHI, K., *Numerical analysis of an axisymmetric jet using a streamline coordinate system*, JSME Intl. J. **30** (1987), 1406-1413.
21. ZARBI, G. and TAKAHASHI, K., *Prediction of the laminar two-dimensional jet flow through a convergent channel*, JSME Intl. J., Series II **34** (1991), 115-121.
22. HONDA, M., *Stream-function co-ordinates in rotational flow and an analysis of the flow in a shock layer, Part I: Two-dimensional flow,*, J. Inst. Maths. Applics. **1** (1965), 127-148.
23. HONDA, M., *Stream-function co-ordinates in rotational flow and an analysis of the flow in a shock layer, Part II: Axisymmetric flow*, J. Inst. Maths. Applics. **2** (1966), 55-75.
24. STANITZ, J.D., *General design method for three-dimensional, potential flow fields, Part I — Theory*, NASA CR 3288 (1980).
25. STANITZ, J.D., *General design method for three-dimensional, potential flow fields, Part II — Computer program DIN3D1 for simple, unbranched ducts*, NASA CR 3926 (1985).
26. STANITZ, J.D., *A review of certain inverse methods for the design of ducts with 2- or 3-dimensional potential flow*, Appl. Mech. Rev. **41** (1988), 217-238.
27. PEARSON, C.E., *Use of streamline coordinates in the numerical solution of compressible flow problems*, J. Comp. Phys. **42** (1981), 257-265.
28. OWEN, D.R. and PEARSON, C.E., *Numerical solution of a class of steady-state Euler equations by a modified streamline method*, AIAA 88-0625 (1988).
29. SRINIVASAN, K. and SPALDING, D.B., *The stream function coordinate system for solution of one-dimensional unsteady compressible flow problems*, Appl. Math. Modelling **10** (1986), 278-283.
30. HUANG, C.-Y. and DULIKRAVICH, G.S., *Stream function and stream function coordinate (SFC) formulation for inviscid flow field calculations*, Computer Methods Appl. Mech. Eng. **59** (1986), 155-177.
31. DULIKRAVICH, G.S., *A stream-function-coordinate (SFC) concept in aerodynamic shape design*, "AGARD-R-780," Brussels, Belgium, 1990, pp. 6.1-6.6.
32. BONATAKI, E., CHAVIAROPOULOS, P. and PAPAILIOU, K.D., *An inverse inviscid method for the design of quasi-three dimensional rotating turbomachinery cascades*, "Proc. 3rd Intl. Conf. on Inverse Design Concepts and Optimization in Eng. Sci.," held in Washington, D.C., USA, ed. by G. S. Dulikravich, 1991, pp. 189-200.

33. WU, C.-H., *A general theory of 3-D flow in subsonic and supersonic turbomachines of axial-, radial- and mixed-flow types*, NACA TN 2604 (1952).
34. LIU, G.-L. and TAO, C., *A universal image-plane method for inverse and hybrid problems of compressible cascade flow on arbitrary stream sheet of revolution: part I — theory*, "Numerical Method for Laminar and Turbulent Flows," Proc. 6th Intl. Conf. held in Swansea, UK, ed. by C. Taylor et al., Pineridge Press, 1989, pp. 1343-1354.
35. CHEN, K.-M., ZHANG, D.-F. and ZHU, Z.-G., *A universal image-plane method for direct-, inverse- and hybrid problems of compressible cascade flow on arbitrary stream sheet of revolution: part II — numerical solution*, "Numerical Method for Laminar and Turbulent Flows," Proc. 6th Intl. Conf. held in Swansea, UK, ed. by C. Taylor et al., Pineridge Press, 1989, pp. 1553-1365.
36. LIU, G.-L. and ZHANG, D.-F., *The moment function formulation of inverse and hybrid problems for blade-to-blade compressible viscous flow along axisymmetric stream sheet*, "Numerical Method for Laminar and Turbulent Flows," Proc. 6th Intl. Conf. held in Swansea, UK, ed. by C. Taylor et al., Pineridge Press, 1989, pp. 1289-1300.
37. LIU, G.-L., *Research on inverse, hybrid and optimization problems in engineering sciences with emphasis on turbomachine aerodynamics: review of Chinese advances*, "Proc. 3rd Intl. Conf. on Inverse Design Concepts and Optimization in Eng. Sci.," held in Washington, D.C., USA, ed. by G. S. Dulikravich, 1991, pp. 145-163.
38. BARRON, R.M., *Computation of incompressible potential flow using von Mises coordinates*, Mathematics and Computers in Simulation 31 (1989), 177-188.
39. NAEEM, R.K. and BARRON, R.M., *Lifting airfoil calculations using von Mises variables*, Comm. Appl. Num. Methods 5 (1989), 203-210.
40. BARRON, R.M., ZHANG, S., CHANDNA, A. and RUDRAIAH, N., *Axisymmetric potential flow calculations. part 1: analysis mode*, Comm. Appl. Num. Methods 6 (1990), 437-445.
41. BARRON, R.M., *A non-iterative technique for design of aerofoils in incompressible potential flow*, Comm. Appl. Num. Methods 6 (1990), 557-564.
42. ZHANG, S., BARRON, R.M. and RUDRAIAH, N., *Axisymmetric potential flow calculations. part 2: design mode*, Comm. Appl. Num. Methods 7 (1991), 563-567.
43. BARRON, R.M. and NAEEM, R.K., *Numerical solution of transonic flows on a streamfunction co-ordinate system*, Intl. J. Num. Methods in Fluids 9 (1989), 1183-1193.
44. NAEEM, R.K. and BARRON, R.M., *Transonic computations on a natural grid*, AIAA J. 28 (1990), 1836-1838.
45. BARRON, R.M. and NAEEM, R.K., *2-D transonic calculations on a flow-based grid system*, Mathematics and Computers in Simulation 33 (1991), 65-67.
46. BARRON, R.M. and AN, C.-F., *Analysis and design of transonic airfoils using streamwise coordinates*, "Proc. 3rd Intl. Conf. on Inverse Design Concepts and Optimization in Eng. Sci.," held in Washington, D.C., USA, ed. by G. S. Dulikravich, 1991, pp. 359-370.
47. AN, C.-F. and BARRON, R.M., *Numerical solution of transonic full-potential-equivalent equations in von Mises coordinates*, Intl. J. Num. Methods in Fluids 15 (1992), 925-952.
48. AN, C.-F. and BARRON, R.M., *Transonic Euler computation in stream function coordinates*, to appear.
49. AN, C.-F., *Transonic computation using Euler equations in stream function coordinate system*, Ph.D. Dissertation, University of Windsor, 1992.
50. ZHANG, S., *Streamwise transonic computations*, Ph.D. Dissertation, University of Windsor, 1993.
51. PEYRET, R. and VIVIAND, H., *Computation of viscous compressible flows based on the Navier-stokes equations*, AGARD-AG-212.

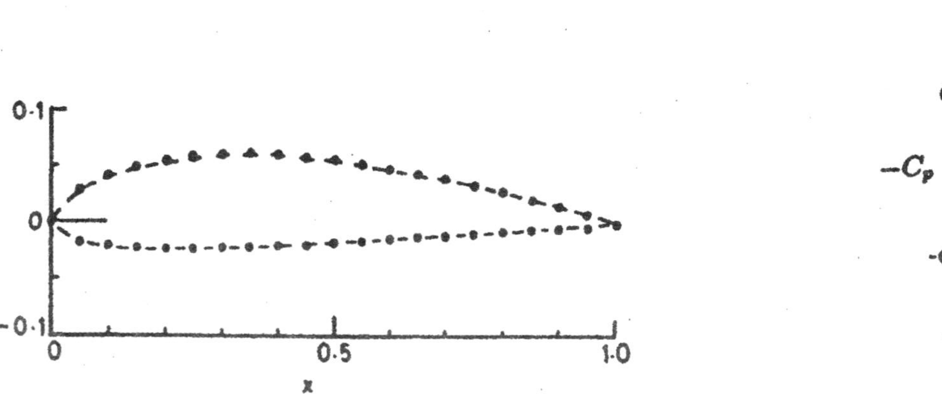

Fig. 1 NACA 2408 airfoil design, $M_\infty = o, \alpha = 5°$[41]

- - - exact; . . . computed

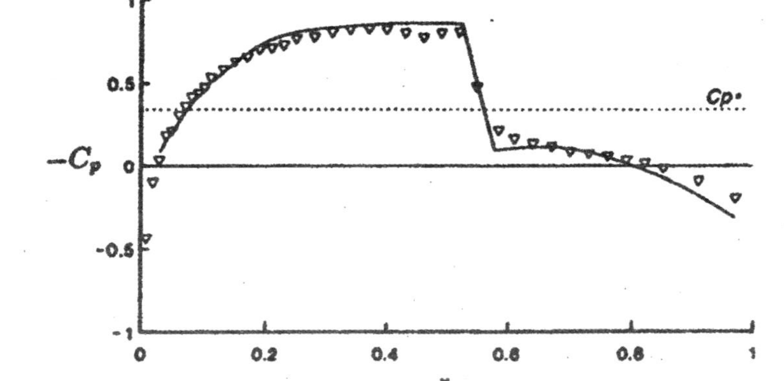

Fig. 2 Pressure coefficient for NACA 0012, $M_\infty = 0.835, \alpha = 0°$[49]

▽▽▽ experimental; — full-potential

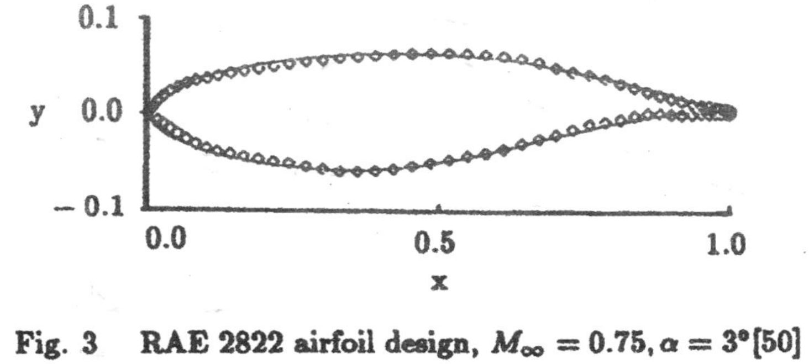

Fig. 3 RAE 2822 airfoil design, $M_\infty = 0.75, \alpha = 3°$[50]

— exact; ◇◇◇ full-potential

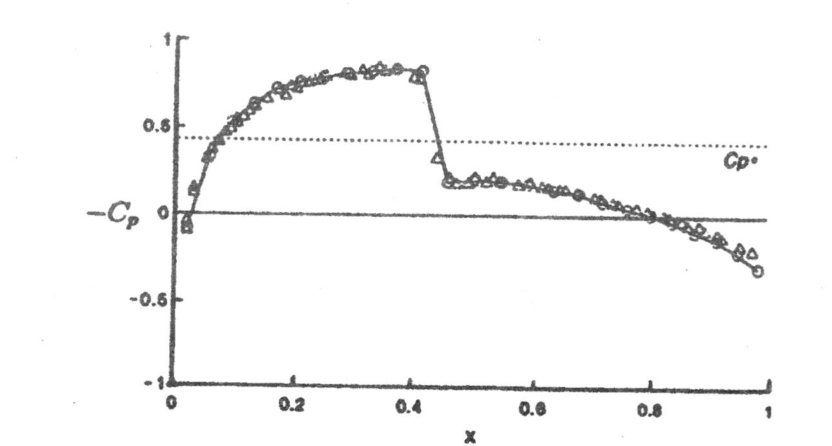

Fig. 4 Pressure coefficient for NACA 0012, $M_\infty = 0.803, \alpha = 0°$[49]

△△△ experimental; — Euler

Proceedings
Comptes rendus

CFD 94

Second Annual Conference of the CFD Society of Canada

Société canadienne de

Society of Canada

Deuxième congrès annuel de la société canadienne de CFD

Toronto, Ontario

June 1-3, 1994 1er au 3 juin 1994

Editors/Editeurs: James J. Gottlieb and/et C. Ross Ethier

A NEW O-GRID SYSTEM IN COMPUTATIONAL AERODYNAMICS

S. ZHANG, R. M. BARRON AND C.-F. AN

Department of Mathematics and Statistics
and Fluid Dynamics Research Institute
University of Windsor
Windsor, Ontario, Canada N9B 3P4

第二部分 出国后博士学位论文阶段

Abstract

Purpose: To create an efficient O-grid system for 2D and 3D flow computations.
Methods: Transonic flows past airfoils and wings are calculated by combining streamfunction coordinate formulations and new O-grid systems.
Results: Calculations are performed for several airfoils and an ONERA M6 wing. The results compare favorably with other available data.
Conclusions: The proposed O-grid system is efficient, allows for very accurate specification of boundary conditions, and eliminates some of the primary deficiencies of other formulations.

1. Introduction

There are a number of difficulties which must be resolved, such as nonlinearity and mixed type of partial differential equations, boundary conditions and coordinate systems, in order to compute transonic flows.

In computational fluid dynamics, it is essential to use grid generation to generate an effective computational mesh for a specific problem. A very successful coordinate system, first introduced by Sells [1] for subcritical flow calculations, is obtained by using complex conformal mapping for each airfoil calculation, which maps the interior of the unit disk to the exterior of the airfoil. These coordinates, which automatically distribute the grid points near the leading and trailing edges, have been extensively used in [2] and [3]. But, there are defects in this coordinate system. Near the center of the circle, the modulus of the complex conformal mapping approaches $1/r^2$, where r is the radial component of the coordinate system, possibly leading to large errors, infinity cannot be chosen as the outer boundary, and the complex conformal mapping is needed for each airfoil computation.

In some earlier works (eg. [4-6]) a streamfunction-coordinate method has been formulated. All of these streamfunction-coordinate formulations suffer from two difficulties for external flow computations. One difficulty is that the grid usage is not efficient since too many grid points are off the airfoil or wing surfaces. The second problem stems from the singularities at the leading and trailing edges and multivaluedness of the streamline ordinate if the flow has a moderate or higher angle of attack.

In this paper, a simple technique is used to introduce a grid system which maintains the advantages of the streamfunction-coordinate formulation and eliminates the defects of the complex conformal mapping method and streamfunction formulation mentioned above.

2. Streamfunction-Coordinate Equations

It can be shown [7] that the equations for perturbed potential function $\Phi(x,\psi)$ and streamline ordinate function $Y(x,\psi)$ are

$$A_1 Y_{xx} + A^2 Y_{xx} - 2ABY_{x\psi} + B^2 Y_{\psi\psi} = 0 \quad (1)$$

$$A_1 \Phi_{xx} + A^2 \Phi_{xx} - 2AB\Phi_{x\psi} + B^2 \Phi_{\psi\psi} = 0 \quad (2)$$

and the associated boundary conditions are

$$Y(x,\psi) = 0 \quad at \quad x = \pm\infty \quad or \quad \psi = \pm\infty$$

$$Y(x,0^{\pm}) = f^{\pm}(x) - x tan\alpha, \quad for \quad 0 \le x \le 1 \quad (3)$$

$$\Phi(x,\psi) = \frac{\Gamma}{2\pi} tan^{-1}[\sqrt{(1-M_{\infty}^2)} tan(\beta-\alpha)] \quad at \ x = \pm\infty \quad or \quad \psi = \pm\infty$$

$$\frac{\partial\Phi(x,\psi)}{\partial\psi}|_{\psi=0^{\pm}} = \frac{(f^{\pm})'[Y_{\psi} + \frac{1}{cos\alpha}]}{g_{11}^{\pm}}(\frac{\partial\Phi(x,\psi)}{\partial x}|_{\psi=0^{\pm}} + \frac{1}{cos\alpha}) - tan\alpha \quad (4)$$

where

$$Y(x,\psi) = y(x,\psi) - (\frac{\psi}{cos\alpha} + x tan\alpha), \qquad \Phi(x,\psi) = \phi(x,\psi) - (\frac{x}{cos\alpha} + \psi tan\alpha)$$

$$A_1 = (\frac{1}{cos\alpha} + Y_{\psi})^2(1-M^2) \qquad A = (\frac{1}{cos\alpha} + Y_{\psi})(tan\alpha + Y_x)$$

$$B = 1 + (tan\alpha + Y_x)^2 \quad (5)$$

with ϕ and $y(x,\psi)$ as potential and streamline ordinate functions respectively, M is the local Mach number, and the circulation Γ can be determined by the jump in potential across the trailing edge, i.e.,

$$\Gamma = \phi_{TE}^{+} - \phi_{TE}^{-} \quad (6)$$

in each iteration, where ϕ_{TE}^{+} and ϕ_{TE}^{-} are the values of ϕ at the upper and lower trailing edge of the airfoil. The Kutta condition is satisfied by adjusting Γ until q^2 is continuous at the trailing edge. The airfoil upper and lower surfaces are defined by $f^{+}(x)$ and $f^{-}(x)$ respectively, $tan\beta = y/x$ and $g_{11}^{\pm} = 1 + ((f^{\pm})')^2$. The chord length of the airfoil has been taken as characteristic length. Equation (4) express the condition of flow tangency at the airfoil surface, $v = uf'(x)$ (u and v are velocity components).

3. New O-grid System

An O-grid system is introduced by the transformation

$$x = \frac{1}{4}(\frac{1}{r} + r)cos\theta + \frac{1}{2}, \qquad \psi = -\frac{1}{4}(\frac{1}{r} - r)sin\theta \tag{6}$$

which transforms the rectangle $(r, \theta) \in [0,1] \times [0, 2\pi]$ to the whole (x, ψ) plane, with $r = 0$ corresponding to the infinity of the (x, ψ) plane. The line segments $l_1 : (r = 1, 0 \leq \theta \leq \pi)$ and $l_2 : (r = 1, \pi \leq \theta \leq 2\pi)$ are transformed to the line segments L_1: $(0 \leq x \leq 1, \psi = 0^-)$ and $L_2 : (0 \leq x \leq 1, \psi = 0^+)$ respectively (see Figures 1-2). Thus the images of l_1 and l_2, in the physical plane, are lower and upper airfoil surfaces respectively.

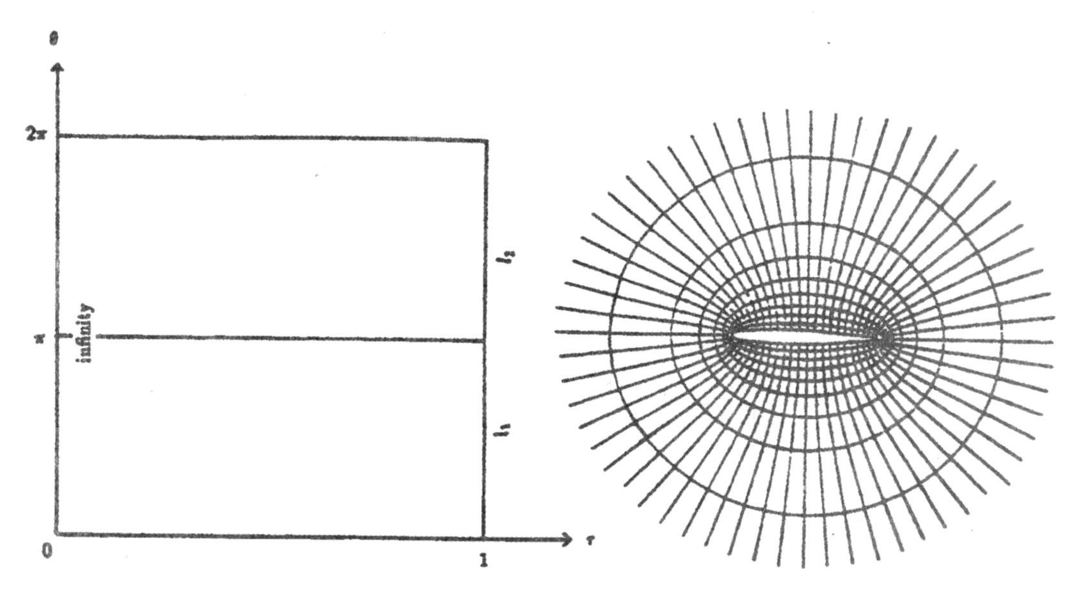

Figure 1. The Computational Domain $(r, \theta) \in [0,1] \times [0, 2\pi]$ Figure 2. A Typicial Grid in The Physical Plane $(x(r,\theta), y(r,\theta)) \in R^3$

In (r, θ) plane, equations (1) and (2) can be expressed as

$$R_1 Y_{\theta\theta} + R_2 Y_{\theta r} + R_3 Y_{rr} + Q_1 Y_{\theta\theta} + Q_2 Y_{\theta r} + Q_3 Y_{rr} = P_1(r) Y_r + P_1(\theta) Y_\theta \tag{7}$$

$$R_1 \Phi_{\theta\theta} + R_2 \Phi_{\theta r} + R_3 \Phi_{rr} + Q_1 \Phi_{\theta\theta} + Q_2 \Phi_{\theta r} + Q_3 \Phi_{rr} = P_1(r) \Phi_r + P_1(\theta) \Phi_\theta \tag{8}$$

where

$$P_1(r) = -(A_1 r_{xx} + A^2 r_{xx} - 2AB r_{x\psi} + B^2 r_{\psi\psi})$$

$$P_1(\theta) = -(A_1 \theta_{xx} + A^2 \theta_{xx} - 2AB\theta_{x\psi} + B^2 \theta_{\psi\psi}) \tag{9}$$

$$R_1 = A_1 \theta_x^2 \qquad R_2 = 2A_1 r_x \theta_x \qquad R_3 = A_1 r_x^2 \qquad Q_1 = [A\theta_x - B\theta_\psi]^2$$

$$Q_2 = 2[Ar_x - Br_\psi][A\theta_x - B\theta_\psi] \qquad Q_3 = [Ar_x - Br_\psi]^2 \tag{10}$$

and A_1, A and B are evaluated by using (5).

The boundary conditions for $Y(r,\theta)$ are

$$Y(0,\theta)=0 \quad for \quad 0\le\theta\le 2\pi$$

$$Y(1,\theta)=G^{+}(\theta)-\frac{1}{2}(1+cos\theta)tan\alpha, \qquad for \quad \pi\le\theta\le 2\pi$$

$$Y(1,\theta)=G^{-}(\theta)-\frac{1}{2}(1+cos\theta)tan\alpha, \qquad for \quad 0\le\theta\le \pi \tag{11}$$

where $G^{\pm}(\theta)=f^{\pm}(\frac{1}{2}+\frac{1}{2}cos\theta)$. The boundary conditions for $\Phi(r,\theta)$ are

$$\Phi(0,\theta)=\frac{\Gamma}{2\pi}tan^{-1}[\sqrt{(1-M_\infty^2)}tan(\beta-\alpha)]$$

$$\frac{\partial\Phi}{\partial r}|_{r=1}=\frac{g_{12}^{+}}{g_{11}^{+}}|_{r=1}(\frac{sin\theta}{2cos\alpha}-\frac{\partial\Phi}{\partial\theta}|_{r=1})-\frac{sin\theta}{2}tan\alpha \qquad for \quad \pi\le\theta\le 2\pi$$

$$\frac{\partial\Phi}{\partial r}|_{r=1}=\frac{g_{12}^{-}}{g_{11}^{-}}|_{r=1}(\frac{sin\theta}{2cos\alpha}-\frac{\partial\Phi}{\partial\theta}|_{r=1})-\frac{sin\theta}{2}tan\alpha \qquad for \quad 0\le\theta\le \pi \tag{12}$$

where

$$\frac{g_{12}^{\pm}}{g_{11}^{\pm}}|_{r=1}=-\frac{2G_\theta^{\pm}(\frac{sin\theta}{cos\alpha}+2Y_r)}{sin^2\theta+4(G_\theta^{\pm})^2} \tag{13}$$

and

$$tan\beta=\frac{y}{x}|_{\infty}=lim_{r\to 0}\frac{y(r,\theta)}{x(r,\theta)}=-\frac{1}{cos\alpha}tan\theta+tan\alpha \tag{14}$$

4. Extension to 3D

In 3D, the potential and streamfunctions are introduced as

$$\vec{u}=\vec{\nabla}\phi=(\phi_x,\phi_y,\phi_z), \qquad \rho\vec{u}=\vec{\nabla}\psi\times\vec{\nabla}\chi. \tag{15}$$

It can be shown [7,8] that the equations for ϕ, ψ and χ are

$$\vec{\nabla}\bullet(\rho\vec{u})=\vec{\nabla}\bullet(\rho\vec{\nabla}\phi)=0 \tag{16}$$

$$\vec{\omega}\bullet\vec{\nabla}\psi=0, \qquad \vec{\omega}\bullet\vec{\nabla}\chi=0 \tag{17}$$

where $\vec{\omega}=\vec{\nabla}\times\vec{u}$ is the vorticity vector.

The streamfunction-coordinate (SFC) formulation has not been well-developed or implemented for 3D external transonic flows. Huang and Dulikravich [5] have derived an explicit vector operator form of the 3D streamfunction equations and obtained the transformed equations based on the streamfunction coordinates. Greywall [6] and Zhang [9] have formulated the 3D SFC equations using two equations for two unknowns $y(x,\psi,\chi)$ and $z(x,\psi,\chi)$. Sherif and Hafiz [10] have used the two streamfunctions as dependent variables for the numerical calculation of 3D transonic flows, but the two streamfunctions they used do not represent physical streamsurfaces. The variational principle of the streamfunction formulations for 2D and 3D flows has been extensively studied by Guderley [8].

Two streamfunctions in 3D flows correspond to two families of streamsurfaces. The use of a streamfunction coordinate will transform the body surface from the physical domain into a rectangle (a line segment in 2D). As in 2D, the boundary conditions in the SFC formulation will be simplified. However, in external transonic flow calculations, the body surface ordinate $z(x,\psi,0)$ (suppose $\chi=0$ corresponds to the wing surface) is a function of $y(x,\psi,0)$. In fact $y(x,\psi,0)$ is an unknown if both ordinates $y(x,\psi,\chi)$ and $z(x,\psi,\chi)$ are treated as dependent variables in the SFC formulations. This causes a major difficulty for numerical computations, since the actual location at which the surface boundary condition must be applied is not known [9]. In the present work, only one streamfunction χ is treated as an independent variable, and the second streamfunction ψ and streamline ordinate z are solved as unknowns. In this way, the boundary conditions are much easier to handle. All flow equations are obtained and solved in the (x,y,χ) coordinates.

An extended O-grid system in 3D can be defined using

$$x=c(y)\left[\frac{1}{4}(\frac{1}{r}+r)cos\theta+\frac{1}{2}\right]+\bar{c}(y),\qquad y=y,\qquad \chi=c(y)\left[\frac{1}{4}(r-\frac{1}{r})sin\theta\right]\tag{18}$$

which transforms the rectangular block $(r,y,\theta)\in[0,1]\times[-\infty,\infty]\times[0,2\pi]$ to the whole (x,y,χ) space, where c(y) and $\bar{c}$(y) are local chord length and local leading edge position at wing spanwise station y. If the wing is positioned so that the y-coordinate is parallel to the wing spanwise direction, then $r=0$ corresponds to the infinity of the (x,y,χ) space at each spanwise station $y=const$. The surfaces s_1: $(r=1,y,0\le\theta\le\pi)$ and $s_2:(r=1,$ y, $\pi\le\theta\le 2\pi)$ are transformed to the surfaces $(\bar{c}(y)\le x\le\bar{c}(y)+c(y)$, y, $\chi=0^-)$ and $(\bar{c}(y)\le x\le\bar{c}(y)+c(y)$, y, $\chi=0^+)$ respectively. The images of s_1 and s_2, in the physical coordinates are lower and upper wing surfaces.

The flow equations (16) and (17) are solved for the unknowns ϕ, ψ and stream-surface ordinate $z(x,y,\chi)$ (for details, cf.[7]).

5. Numerical Scheme, Results and Discussions

Equations (7) and (8) in 2D or (17) and (18) in 3D are solved by the finite difference method with a successive line over-relaxation method applied in the usual way ([3,7]).

The test calculations are performed for various airfoils and an ONERA M6 wing. The results compare well with other available data and are presented in Figs 3-6.

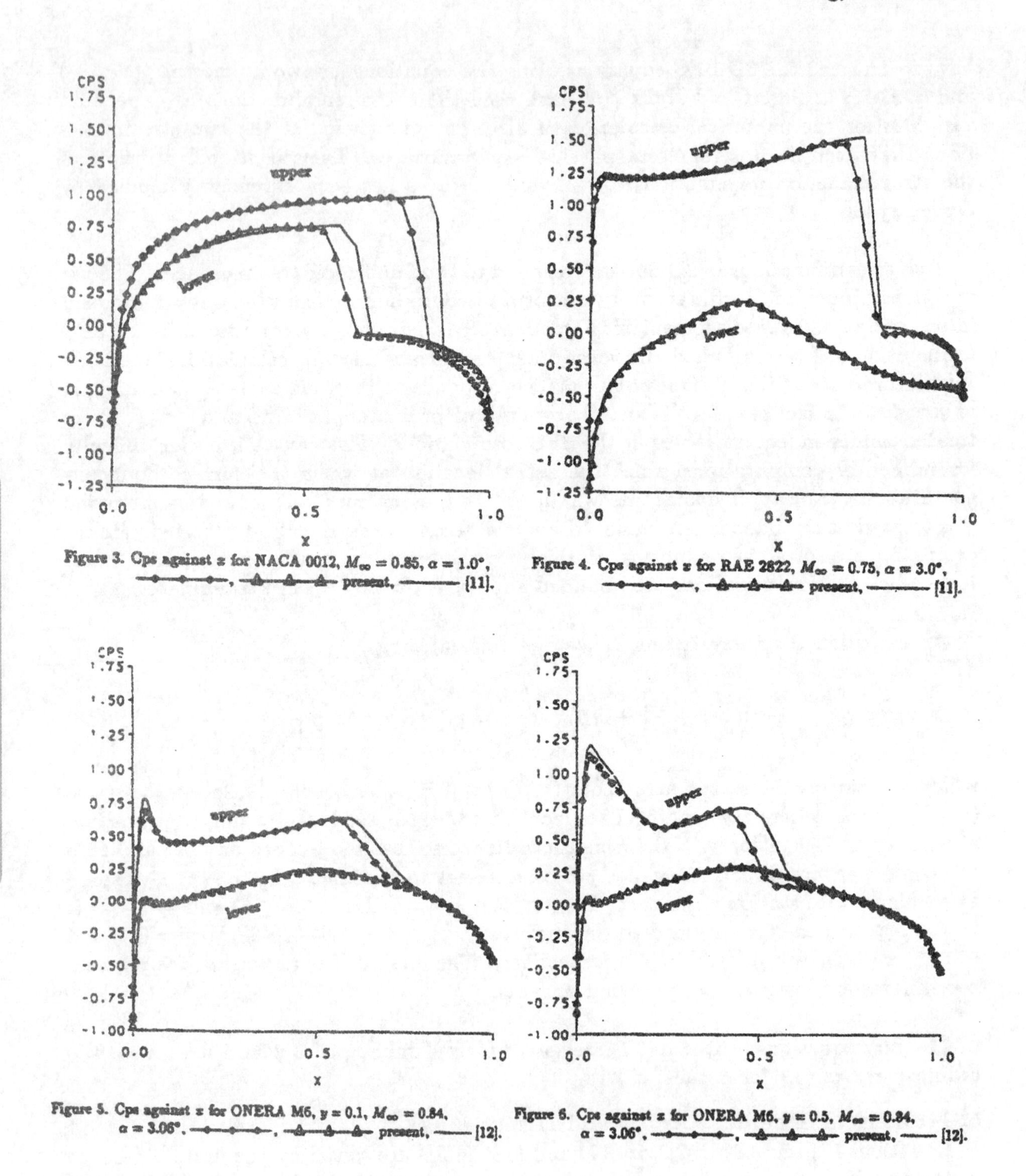

Figure 3. Cps against x for NACA 0012, $M_\infty = 0.85$, $\alpha = 1.0°$, —◆—◆—◆—, —△—△—△— present, ——— [11].

Figure 4. Cps against x for RAE 2822, $M_\infty = 0.75$, $\alpha = 3.0°$, —◆—◆—◆—, —△—△—△— present, ——— [11].

Figure 5. Cps against x for ONERA M6, $y = 0.1$, $M_\infty = 0.84$, $\alpha = 3.06°$, —◆—◆—◆—, —△—△—△— present, ——— [12].

Figure 6. Cps against x for ONERA M6, $y = 0.5$, $M_\infty = 0.84$, $\alpha = 3.06°$, —◆—◆—◆—, —△—△—△— present, ——— [12].

There are several important features of the method described in this work which should be emphasized:

(a) The grids naturally conform to, and cluster, near the airfoil or wing (Figs 1-2). This is achieved without any complicated conformal mappings.

(b) In the (r,θ) coordinate system, $\Gamma\theta/2\pi$ is subtracted from Φ to get a function $\overline{\Phi}$

$$\phi(x,\psi) = \overline{\Phi}(r,\theta) + \frac{\Gamma\theta}{2\pi} + (\frac{x}{cos\alpha} + \psi tan\alpha) \tag{19}$$

which is single-valued (similarly in 3D).

(c) Transformation Jacobian $J = det\frac{\partial(x,\psi)}{\partial(r,\theta)}$ becomes zero at $r = 1, \theta = 0$ and at $r = 1, \theta = \pi$, which are the leading and trailing edges. The Kutta condition requires that the tangential speed

$$\left[(u,v)^t \bullet \begin{pmatrix} 1 \\ f_x \end{pmatrix}\right]_{r=1} = (u + y_x v)_{r=1} = sec\alpha - \frac{2}{sin\theta}(\overline{\Phi}_\theta + \frac{\Gamma}{2\pi})$$

at the trailing edge must be finite at $\theta = 0$, requiring that

$$\Phi_\theta = (\overline{\Phi}_\theta + \frac{\Gamma}{2\pi})|_{r=1,\theta=0} = 0. \tag{20}$$

Equation (20) gives an equation for calculating the circulation Γ

$$\Gamma = -2\pi\overline{\Phi}_\theta|_{r=1,\theta=0} \tag{21}$$

(d) One of the advantages of using the O-type grid is that in equations (7) and (8), the terms involving R_3Y_{rr} and $R_3\Phi_{rr}$ become zero on the airfoil surface. Hence, it is not neccessary to introduce a dummy grid line as in Jameson's formulation [3], in order to use an upwind difference scheme. This advantage also applies to 3D equations [7].

(e) At the leading edge, the boundary condition for Φ in (12) can be written as

$$\frac{\partial\Phi}{\partial r}|_{r=1} = \left(\frac{Y_r}{Y_\theta}\frac{\partial\Phi}{\partial\theta}\right)|_{r=1,\theta=\pi} \tag{22}$$

In general Y_θ is not zero at the leading edge if the airfoil does not have a cusped leading edge. Formula (22) is a well-behaved boundary condition in the computations and it is superior to other formulas generally used in SFC formulations.

(f) As has been mentioned before, if the uniform flow is at moderate or higher angle of attack, $y(x,\psi)$ ($z(x,y,\chi)$ in 3D) is multivalued with respect to the variable x. In the (r,θ) coordinate system, the function $y(r,\theta)$ ($z(r,y,\theta)$ in 3D) is unique with respect to (r,θ). This may be a solution to the problem of multivaluedness for the streamfunction-coordinate formulation [7].

6. Conclusions

A new O-type coordinate system (r,θ) ((r,y,θ) in 3D) is introduced for the SFC formulations. In this new coordinate system, the airfoil analysis (and design) problems can be easily formulated when combined with SFC formulations. There are three major developments in this work. First, true infinity boundary conditions are achieved without creating any singularities in the partial differential equations. Second, in the (r,θ) system, $y(r,\theta)$ is unique with respect to (r,θ) and this may be a solution for the difficulty of multivaluedness in the SFC formulations. The last is that, in the (r,θ) system, the boundary condition at the leading edge is well-behaved.

The computational results show that the new O-grid system is a simple, effective and innovative coordinate system for the SFC formulations and computational aerodynamics.

7. References

1. Sells, C. C. "Plane Subcritical Flow Past a Lifting Airfoil". *Proceeding of The Royal Society of London.* **308A**, 377-401 (1968).
2. Garabedian, P. and Korn, D. "Analysis of Transonic Airfoils". *Comm. on Pure and Appl Math.* **24**, 841-851 (1972).
3. Jameson, A. "Iterative Solution of Transonic Flows over Airfoils and Wings, Including Flows at Mach 1". *Comm. on Pure and App Math.* **27**, 283-309 (1974).
4. Barron, R. "Computation of Incompressible Potential Flow Using von Mises Coordinates". *Math and Computers in Simul.* **31** No **3**, 177-188 (1989).
5. Huang, C-Y. and Dulikravich, G. S. "Stream Function and Stream-Function-Coordinate (SFC) Formulation for Inviscid Flow Field Calculations". *Computer Meth. in Appl Mech and Eng.* **59**, 155-177 (1986).
6. Greywall, G. S. "Streamwise Computations of Three Dimensional Incompressible Potential Flows". *J. of Comp Phys.* **78**, 178-193 (1988).
7. Zhang, S. Streamwise Transonic Computations. Ph.D Dissertation, University of Windsor, Windsor, Canada (October, 1993).
8. Guderley, K. G. "An Extremum Principle for Three Dimensional Compressible Inviscid Flows". *SIAM J. on Appl Math.* **23** No **2**, 259-275 (1972).
9. Zhang, S. Application of Stream Function Coordinates in Three Dimensional Fluid Flows. MSc Major Paper, University of Windsor, Windsor, Canada (September, 1987).
10. Sherif, A. and Hafez, M. "Computation of Three-Dimensional Transonic Flows Using Two Stream Functions". *Int. J. for Numer Meth in Fluids.* **8**, 17-29 (1988).
11. Viviand, H. "Numerical Solutions of Two-Dimensional Reference Test Cases" in *AGARD Advisory Report No 211* (1985).
12. Sacher, P. "Numerical Solutions for Three-Dimensional Cases-Swept Wings", in *AGARD Advisory Report No 211* (1985).

The Seventh International Conference
on
Boundary and Interior Layers
Computational and Asymptotic Methods

BOOK OF ABSTRACTS

September 5~8 1994
Beijing, CHINA

C. R. [illegible]
University of Windsor
Canada

CHINESE AERODYNAMICS RESEARCH SOCIETY

第二部分 出国后博士学位论文阶段

NUMERICAL SIMULATION IN SFC FOR SHOCK REFLECTION AND INTERACTION*

C.-F. An
(Dept. of Mathematics & Statistics
and Fluid Dynamics Research Institute
Univ. of Windsor, Windsor, Ont., Canada N9B 3P4)

One dimensional unsteady compressible inviscid flow is a basic and important phenomenon in fluid mechanics. The Euler equations is the most general model for this type of flow. Therefore, numerical solutions of the Euler equations for one dimensional unsteady flow have been paid an intensive attention for several decades. However, most Euler computations are carried out in Cartesian coordinates or in transformed coordinates through grid generation. Alternatively, some researchers deal with the problem in streamfunction coordinates. The streamfunction coordinate(SFC) formulation has a number of advantages. Particularly, for one dimensional unsteady flow, the Jacobian of the transformation is always nonzero and finite so that the transformation is always permissible. Another prominent advantage of the SFC formulation is the ease of drawing particle path lines which makes the post-processing more expressive and less expensive.

For one dimensional unsteady compressible flow of an inviscid fluid, the governing Euler equations for primitive variables in Cartesian coordinates are

$$\vec{Q}_t^* + \vec{F}_x^* = 0 \tag{1}$$

where $\vec{Q}^* = [\rho, \rho u, e]^t$, $\vec{F}^* = [\rho u, \rho u^2 + p, (e+p)u]^t$, $p = (\gamma - 1)(e - \rho u^2/2)$ is pressure, e is total energy per unit volume, ρ is density, u is velocity and γ is the ratio of specific heats. Defining streamfunction ψ such that $\psi_x = \rho$, $\psi_t = -\rho u$, making von Mises transformation $x = x(\psi, \tau)$, $t \equiv \tau$ and introducing the compatibility condition $x_{\psi\tau} = x_{\tau\psi}$, the Euler equations (1) can be re-formulated to the Euler equtions for general variables in streamfunction coordinates(SFC),

$$\vec{Q}_\tau + \vec{F}_\psi = 0 \tag{2}$$

where $\vec{Q} = [q_1, q_2, q_3]^t$, $\vec{F} = [-q_2, (\gamma-1)(q_3 - q_2^2/2)/q_1, (\gamma-1)(q_3 - q_2^2/2)q_2/q_1]^t$ and the general variables are $q_1 = x_\psi$, $q_2 = x_\tau$, $q_3 = x_\psi e$.

Classification of equations (2) is important to decide what type of numerical scheme to choose. The flux Jacobian matrix, $\partial\vec{F}/\partial\vec{Q}$, has three real eigenvalues, $\sigma_{1,2} = \pm\sqrt{\gamma p \rho}$ and $\sigma_3 = 0$. Hence, (2) is always hyperbolic for which the Godunov's predictor-corrector scheme is one of the optimal solvers,

$$\begin{cases} \overline{\vec{Q}}_{j+1/2} = \frac{1}{2}(\vec{Q}_{j+1}^n + \vec{Q}_j^n) - \frac{\Delta\tau}{\Delta\psi}(\vec{F}_{j+1}^n - \vec{F}_j^n) \\ \vec{Q}_j^{n+1} = \vec{Q}_j^n - \frac{\Delta\tau}{\Delta\psi}(\overline{\vec{F}}_{j+1/2} - \overline{\vec{F}}_{j-1/2}). \end{cases} \tag{3}$$

*For presentation at the Seventh International Conference on Boundary and Interior Layers Computational and Asymptotic Methods, September 5-8, 1994, Beijing, China

The scheme is stable when $CFL \leq 1$. Having solved for $\vec{Q}$, the Cartesian coordinate $x = x(\psi, \tau)$ can be calculated by marching $x_j^n \doteq x_{j-1}^n + \Delta\psi[(q_1)_{j-1}^n + (q_1)_j^n]/2$ and the physical variables can be evaluated by $\rho = 1/x_\psi = 1/q_1$, $u = x_\tau = q_2$, $e = q_3/q_1$.

Suppose a shocktube (with pressure ratio 2:1) is initially separated by a diaphragm at $x_d = 0.8$. Both chambers are filled with stationary air ($\gamma = 1.4$) of the same temperature. At moment $t = 0$, the diaphragm is suddenly broken. Then, the high pressure air rushes to the low pressure chamber. As a result, three waves are produced: a rarefaction wave travelling to the left, a shock wave travelling to the right and a contact discontinuity also travelling to the right behind the shock. Because the diaphragm is located so far from the left end wall that the left-travelling rarefaction wave will not reach the wall even after the right-travelling shock wave hits the right end wall, reflects from there and meets the contact discontinuity, so the left boundary conditons are very simple but the right boundary conditions are more complicated to consider the shock reflection.

The test computation is performed in the computational domain $0 \leq \psi \leq 0.9$, $0 \leq \tau \leq 0.4$. The ψ coordinate is divided into 100 uniform intervals. The τ intervals are not uniform due to constant $CFL = 0.5$. Three moments, $t_1 = 0.11$, $t_2 = 0.1826$ and $t_3 = 0.3385$ are selected to demonstrate typical flow situations. Figure 1 shows the particle path lines in the physical plane. Each line describes the path line of a certain air particle in the tube. Figures 2 shows the comparison of density distribution between the present computation and exact solution. The simulation accuracy is very good, especially for the contact discontinuity. At $t_1 = 0.11$, the three waves are completely formed. The transition of the contact discontinuity needs only one interval. The transition of the shock occupies about 3–4 intervals but there is no oscillation. At $t_2 = 0.1826$, the shock has been reflected from the wall but has not yet met the contact discontinuity. At $t_3 = 0.3385$, the reflected shock has already penetrated the contact discontinuity. The simulation of rarefaction wave is less accurate and the reason is not clear at present. Figures 3 shows the historic evolutions of density. The processes of three wave propagations, shock reflection and its interaction with the contact discontinuity are three dimensionally exhibited.

Fig. 1
Particle
Path Lines

t

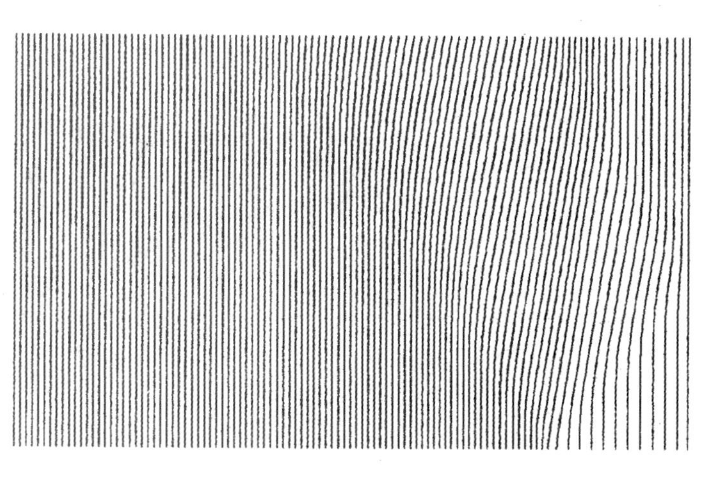

x

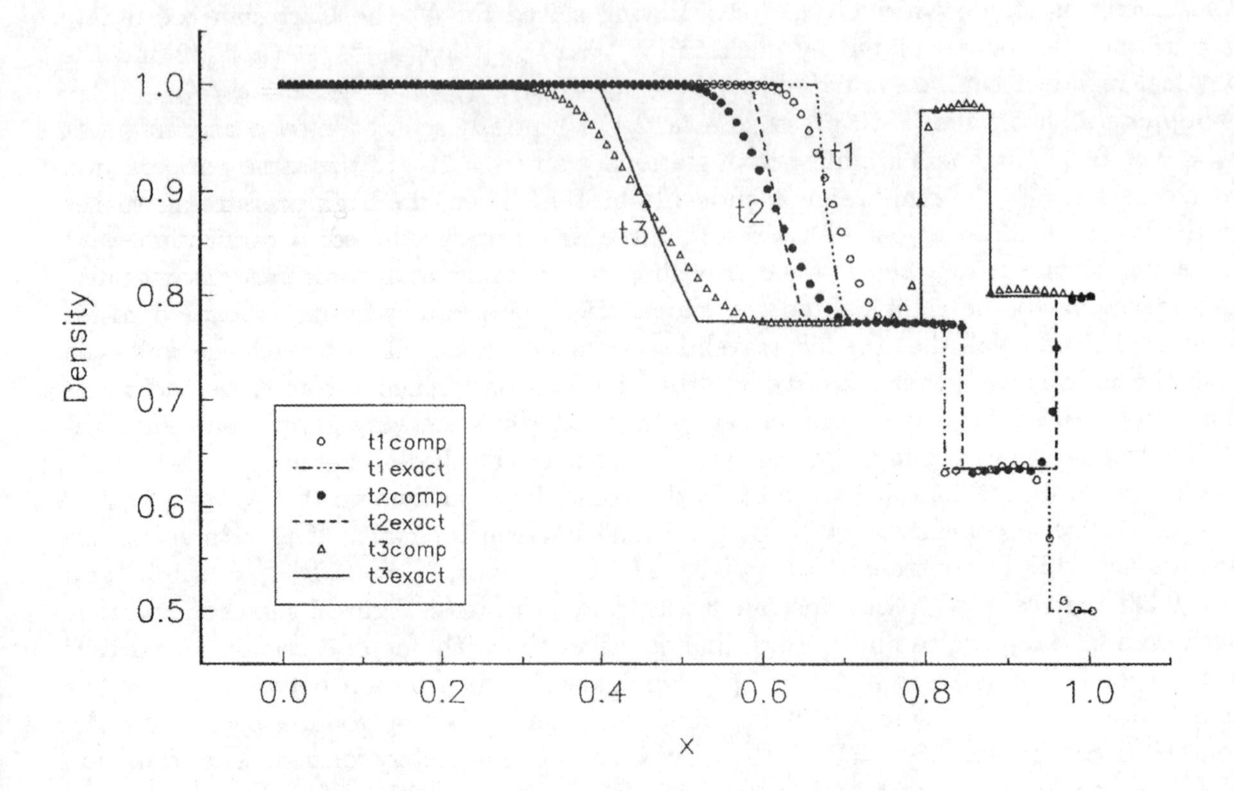

Fig. 2　Density Distribution

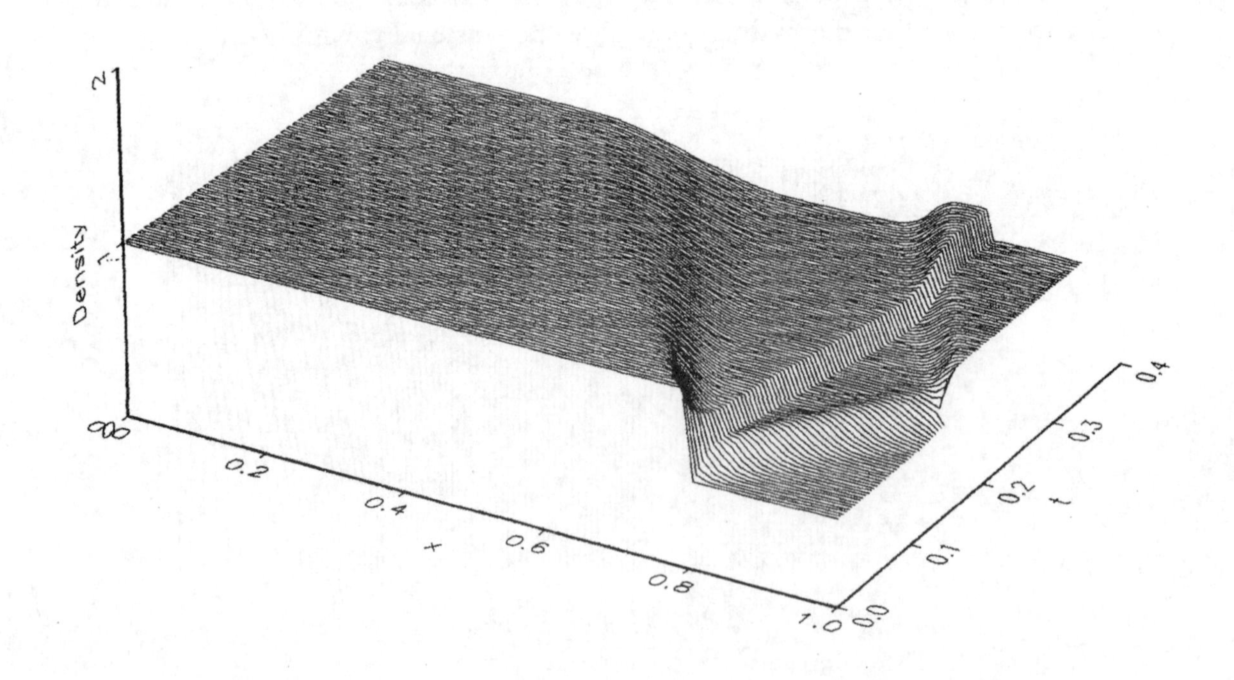

Fig. 3　Historic Evolution of Density

Arc length–streamfunction formulation and its application

C-F. An, R. M. Barron, and S. Zhang

Department of Mathematics and Statistics and Fluid Dynamics Research Institute, University of Windsor, Windsor, Ontario, Canada

In this article, a new approach has been developed to calculate two-dimensional steady incompressible potential internal and external flows using an arc length–streamfunction formulation. The computational domain is obtained by a coordinate transformation that specifies the curvilinear coordinates s, ψ as the arc length along a streamline and streamfunction. For incompressible potential flow, instead of solving the Laplace equation for streamfunction in cartesian coordinates, we solve two elliptic differential equations for the cartesian coordinate x and y, as unknown functions of s and ψ. In this approach, a single set of equations plays a double role. Physically, it serves as the governing equations of fluid flow while, geometrically, it also acts as grid generation equations. Thus, when the equations are solved, the grid and the flow field are obtained simultaneously, reducing computer time and storage requirements. Most boundary conditions are of the Dirichlet type, which simplifies the numerical procedure. Several examples have been computed, and the results demonstrate that this formulation has significant promise. Possible extensions and restrictions of the proposed approach have been addressed.

Keywords: CFD, arc length–stream function, potential flow

1. Introduction

Since Martin[1] derived a general formulation of the Navier-Stokes equations in curvilinear coordinates, one of which is streamfunction, based on the theory of differential geometry, the streamfunction-as-a-coordinate method and its application in CFD have gained considerable attention.[2–8] A survey of the streamfunction-as-a-coordinate method has been given by the present authors.[9] However, Martin specified only one family of coordinate curves and left the second arbitrary so that the user has an option to specify it depending on the particular problem of interest. For potential flows, a typical choice for the second coordinate is the velocity potential,[7] so that the two families of coordinate curves are orthogonal. But it is difficult to extend this choice to more complex flows, rotational flow for example. Although one can assume orthogonality between the two coordinates, the second family of curves loses its physical meaning. Another good choice is that the second curvilinear coordinate remains as the cartesian coordinate x itself, i.e., the von Mises transformation.[4,6,8] This von Mises formulation has been successfully applied to many cases.[8–13] However, when a reverse flow region appears in the flow field, a multivaluedness problem arises.

In this paper, the arc length along a streamline is chosen as the second family of curvilinear coordinates. The coordinate net is not orthogonal, but the multivaluedness difficulty can be avoided. The introduction of a new variable requires an additional condition, i.e., the arc length constraint condition. In Sections 2 and 3, the arc length–streamfunction formulation and the governing equations of the two-dimensional (2-D) steady incompressible potential flows are derived from the basic principles of differential geometry. Section 4 describes the numerical techniques involved in the solution procedure. Several examples are given and the computed results are discussed in Section 5. In the last section, some concluding remarks are given, including strengths and weaknesses of the method, its possible extensions to more complicated flow situations, and practical restrictions.

Address reprint requests to Dr. An at the Dept. of Mathematics and Statistics and Fluid Dynamics Research Institute, University of Windsor, Windsor, Ontario, Canada N9B 3P4.

Received 22 July 1993; revised 14 February 1994; accepted 8 March 1994

2. Curvilinear coordinate transformation

Suppose two families of curves, $s = const.$ and $\psi = const.$, overlay the (x, y) plane and form a regular coordinate net in which each s-curve intersects with a ψ-curve at only one point and neither an s-curve nor a ψ-curve intersects with itself (see *Figure 1*). For the time being, s and ψ are arbitrary curvilinear coordinates and do not yet have any specific physical meaning. Consider a coordinate transformation

$$x = x(s, \psi), \quad y = y(s, \psi) \tag{1}$$

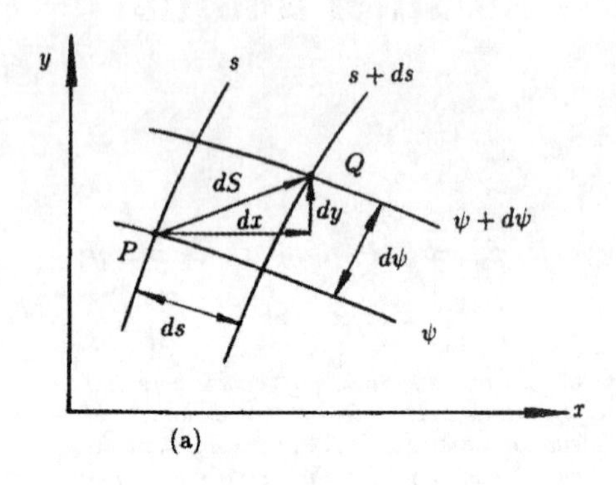

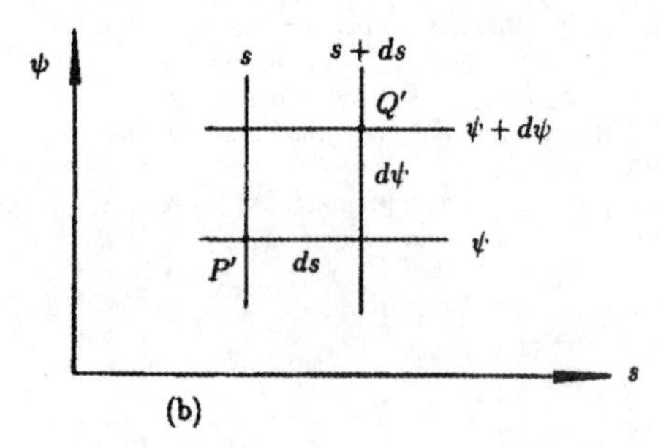

Figure 1. Curvilinear coordinate transformation. (a) Physical plane, (b) computational plane.

The differentials of (1) are

$$dx = x_s ds + x_\psi d\psi, \quad dy = y_s ds + y_\psi d\psi \tag{2}$$

From the theory of differential geometry, the squared length of an element arc in (x, y) plane can be expressed by the first fundamental form

$$dS^2 = dx^2 + dy^2 = E ds^2 + 2F ds d\psi + G d\psi^2 \tag{3}$$

where

$$E = x_s^2 + y_s^2, \quad F = x_s x_\psi + y_s y_\psi, \quad G = x_\psi^2 + y_\psi^2 \tag{4}$$

are the metrics of the transformation. By the chain rule,

$$\frac{\partial}{\partial x} = s_x \frac{\partial}{\partial s} + \psi_x \frac{\partial}{\partial \psi}, \quad \frac{\partial}{\partial y} = s_y \frac{\partial}{\partial s} + \psi_y \frac{\partial}{\partial \psi} \tag{5}$$

Applying these operators to x and y, respectively, we have

$$x_s s_x + x_\psi \psi_x = 1, \quad x_s s_y + x_\psi \psi_y = 0,$$
$$y_s s_x + y_\psi \psi_x = 0, \quad y_s s_y + y_\psi \psi_y = 1 \tag{6}$$

Solving for s_x, ψ_x, s_y, ψ_y gives

$$s_x = \frac{y_\psi}{J}, \quad \psi_x = -\frac{y_s}{J}, \quad s_y = -\frac{x_\psi}{J}, \quad \psi_y = \frac{x_s}{J} \tag{7}$$

where J is the Jacobian of the transformation

$$J = \frac{\partial(x, y)}{\partial(s, \psi)} = \det \begin{bmatrix} x_s & x_\psi \\ y_s & y_\psi \end{bmatrix} = x_s y_\psi - x_\psi y_s \tag{8}$$

It is easy to verify that

$$J^2 = EG - F^2 \tag{9}$$

If the Jacobian is nonzero and finite, then the transformation (1) is one to one, and there are no singularities throughout the domain of interest. In this case, the differential operators (5) are transformed as

$$\frac{\partial}{\partial x} = \frac{y_\psi}{J}\frac{\partial}{\partial s} - \frac{y_s}{J}\frac{\partial}{\partial \psi}, \quad \frac{\partial}{\partial y} = -\frac{x_\psi}{J}\frac{\partial}{\partial s} + \frac{x_s}{J}\frac{\partial}{\partial \psi} \tag{10}$$

and second-order derivatives become

$$\begin{aligned} \frac{\partial^2}{\partial x^2} = \frac{1}{J^2}&\left\{ y_\psi^2 \frac{\partial^2}{\partial s^2} - 2 y_s y_\psi \frac{\partial^2}{\partial s \partial \psi} + y_s^2 \frac{\partial^2}{\partial \psi^2} \right\} \\ &+ \left\{ \frac{y_\psi}{J}\left(\frac{y_\psi}{J}\right)_s - \frac{y_s}{J}\left(\frac{y_\psi}{J}\right)_\psi \right\} \frac{\partial}{\partial s} \\ &+ \left\{ -\frac{y_\psi}{J}\left(\frac{y_s}{J}\right)_s + \frac{y_s}{J}\left(\frac{y_s}{J}\right)_\psi \right\} \frac{\partial}{\partial \psi}, \end{aligned} \tag{11}$$

$$\begin{aligned} \frac{\partial^2}{\partial y^2} = \frac{1}{J^2}&\left\{ x_\psi^2 \frac{\partial^2}{\partial s^2} - 2 x_s x_\psi \frac{\partial^2}{\partial s \partial \psi} + x_s^2 \frac{\partial^2}{\partial \psi^2} \right\} \\ &+ \left\{ \frac{x_\psi}{J}\left(\frac{x_\psi}{J}\right)_s - \frac{x_s}{J}\left(\frac{x_\psi}{J}\right)_\psi \right\} \frac{\partial}{\partial s} \\ &+ \left\{ -\frac{x_\psi}{J}\left(\frac{x_s}{J}\right)_s + \frac{x_s}{J}\left(\frac{x_s}{J}\right)_\psi \right\} \frac{\partial}{\partial \psi} \end{aligned} \tag{12}$$

3. Arc length–streamfunction formulation

In the coordinate transformation (1), if s is taken to be arc length along a streamline and ψ to be a stream function, then equation (3) and the second-order differential operators can be significantly reduced.

Applying (11) and (12) to streamfunction ψ yields

$$\psi_{xx} = -\frac{y_\psi}{J}\left(\frac{y_s}{J}\right)_s + \frac{y_s}{J}\left(\frac{y_s}{J}\right)_\psi, \tag{13}$$

$$\psi_{yy} = -\frac{x_\psi}{J}\left(\frac{x_s}{J}\right)_s + \frac{x_s}{J}\left(\frac{x_s}{J}\right)_\psi \tag{14}$$

The first two terms on the right-hand sides of equations (11) and (12) vanish because ψ is independent of s and the second derivative of ψ with respect to ψ itself is also zero. Only the last terms in equations (11) and (12) involving the first derivative of ψ with respect to ψ, which is unity, remain.

The 2-D steady incompressible potential flow is governed by the Laplace equation for streamfunction ψ

$$\nabla^2 \psi = \psi_{xx} + \psi_{yy} = 0 \tag{15}$$

Substituting (13) and (14) into (15) gives

$$y_\psi\left(\frac{y_s}{J}\right)_s - y_s\left(\frac{y_s}{J}\right)_\psi + x_\psi\left(\frac{x_s}{J}\right)_s - x_s\left(\frac{x_s}{J}\right)_\psi = 0 \tag{16}$$

Expanding it and using (8) for J, one gets

$$x_s(G y_{ss} - 2F y_{s\psi} + E y_{\psi\psi}) = y_s(G x_{ss} - 2F x_{s\psi} + E x_{\psi\psi}) \tag{17}$$

where the metrics E, F, and G are defined in (4).

Equation (17) is a second-order nonlinear partial differential equation for y and x as functions of s and ψ. In this equation, the unknown variables $y(s, \psi)$ and

$x(s, \psi)$ are coupled with each other and have a symmetric or interchangeable format. So, this equation can be used to solve for y if x has been obtained, and vice versa. This equation is elliptic for y as well as for x because the discriminants for both y and x are less than zero, i.e.,

$$\Delta_y = 4x_s^2(F^2 - EG) = -4x_s^2 J^2 < 0, \quad (18)$$

$$\Delta_x = 4y_s^2(F^2 - EG) = -4y_s^2 J^2 < 0 \quad (19)$$

Nevertheless, one equation (17) is not enough to solve for two unknowns y and x. Another independent equation is needed to complete the formulation. It is known that along a streamline, $\psi = const.$ or $d\psi = 0$; hence equation (3) reduces to

$$dS^2 = Eds^2 \quad (20)$$

On the other hand, along a stream line, $S = s$, or

$$dS = ds \quad (21)$$

Comparing equations (20) and (21) and recalling (4), we get

$$E = x_s^2 + y_s^2 = 1 \quad (22)$$

This first-order partial differential equation is also symmetric or interchangeable for x and y.

Equations (17) and (22) constitute the arc length–streamfunction formulation for incompressible potential flows and can be solved simultaneously for y and x under appropriate boundary conditions. In principle, the choice of which equation should be solved for y and which for x is completely arbitrary. However, for conventional flow problems, for which the x-axis roughly represents the main flow direction and the y-axis represents the transverse direction, it is recommended that (17) be used as the y-equation and (22) as the x-equation. The second-order elliptic equation (17) can be solved by successive line overrelaxation (SLOR) numerically in the whole region, and the first-order equation (22) can be solved by marching along a streamline and proceeding streamline by streamline.

From the authors' experiences, however, the system of equations (17) and (22) does not always work well, especially for the flows with rapid change in direction. In such a case, equation (22) cannot be solved because the approximate values of y, obtained during the process of iteration, may lead to $y_s^2 > 1$. Then, from (22),

$$x_s = \pm\sqrt{1 - y_s^2} \quad (23)$$

is imaginary. Also, it is not obvious which sign should be chosen before the square root, even when y_s^2 remains less than one.

To overcome this difficulty, these two equations can be manipulated to get a more favorable form. Differentiating equation (22) with respect to s and ψ, respectively, we have

$$x_s x_{ss} + y_s y_{ss} = 0, \qquad x_s x_{s\psi} + y_s y_{s\psi} = 0 \quad (24)$$

Substituting x_{ss}, $x_{s\psi}$ given in (24) into (17) and rearranging terms, we get an elliptic equation for y

$$Gy_{ss} - 2Fy_{s\psi} + y_{\psi\psi} = x_s y_s x_{\psi\psi} + y_s^2 y_{\psi\psi} \quad (25)$$

Similarly, eliminating y_{ss}, $y_{s\psi}$ from equations (24) and (17) gives

$$Gx_{ss} - 2Fx_{s\psi} + x_{\psi\psi} = y_s x_s y_{\psi\psi} + x_s^2 x_{\psi\psi} \quad (26)$$

The discriminants of equations (25) and (26) are negative, and therefore both equations are elliptic if the nonhomogeneous right-hand-side terms are not considered. The system of equations (25) and (26) does not exhibit the above-mentioned calculation difficulty.

For the arc length–streamfunction formulation, most boundary conditions are of the Dirichlet type on body surfaces, on walls, on symmetry lines, and at the far field. For example, the boundary condition on a body surface or wall $\psi = const.$ is

$$y = f[x(s)] \quad (27)$$

where $f(x)$ is the prescribed shape function of the body or wall. The boundary conditions at an inlet and outlet may be of the Neumann type depending on the specific problem. Such cases will be discussed later.

4. Numerical techniques

The governing equations for 2-D incompressible potential flow in the arc length–streamfunction coordinates, i.e., (25) and (26), can be expressed as

$$A_1\begin{bmatrix} y \\ x \end{bmatrix}_{ss} + A_2\begin{bmatrix} y \\ x \end{bmatrix}_{s\psi} + A_3\begin{bmatrix} y \\ x \end{bmatrix}_{\psi\psi} = A_4 \quad (28)$$

where

$$A_1 = G, \quad A_2 = -2F, \quad A_3 = 1,$$

$$A_4 = \begin{bmatrix} x_s y_s x_{\psi\psi} \\ y_s x_s y_{\psi\psi} \end{bmatrix} + \begin{bmatrix} y_s^2 y_{\psi\psi} \\ x_s^2 x_{\psi\psi} \end{bmatrix} \quad (29)$$

Central differencing all second derivatives in equation (28) with respect to s and ψ and rearranging the resulting difference equation in tridiagonal form, one gets

$$A\begin{bmatrix} y \\ x \end{bmatrix}_{i,j-1} + B\begin{bmatrix} y \\ x \end{bmatrix}_{i,j} + C\begin{bmatrix} y \\ x \end{bmatrix}_{i,j+1} = RHS \quad (30)$$

where

$$A = \beta^2 A_3$$

$$B = -2(A_1 + \beta^2 A_3), \quad \beta = \Delta s/\Delta\psi$$

$$C = \beta^2 A_3$$

$$RHS = \Delta s^2 A_4 - A_1\left\{\begin{bmatrix} y \\ x \end{bmatrix}_{i+1,j} + \begin{bmatrix} y \\ x \end{bmatrix}_{i-1,j}\right\} - \frac{\beta}{4}A_2\left\{\begin{bmatrix} y \\ x \end{bmatrix}_{i+1,j+1} - \begin{bmatrix} y \\ x \end{bmatrix}_{i+1,j-1} - \begin{bmatrix} y \\ x \end{bmatrix}_{i-1,j+1} + \begin{bmatrix} y \\ x \end{bmatrix}_{i-1,j-1}\right\} \quad (31)$$

where $i = 2, 3, \ldots, I_{max} - 1$ and $j = 2, 3, \ldots, J_{max} - 1$. The system of difference equations (31) can be solved by a tridiagonal solver and the SLOR algorithm can be applied.

5. Boundary conditions

The boundary conditions on a body surface or wall are the Dirichlet type (27). But for the arc length–streamfunction formulation, it is necessary to express the condition as

$$x_i = x(s_i), \quad y_i = f(x_i) \tag{32}$$

where ψ takes its value on the boundary and i takes all possible values. Usually, the second condition in (32) is given explicitly. But the first condition in (32) is usually an implicit function. As a matter of fact, the inverse function of it is nothing but the arc length formula of a given curve $y = f(x)$ between x_0 and x_i (see *Figure 2*)

$$s_i = \int_{x_0}^{x_i} \sqrt{1 + [f'(\xi)]^2}\, d\xi \tag{33}$$

Solving equation (33) for x_i when s_i is given is equivalent to finding a zero of the following function

$$T(x_i) = \int_{x_0}^{x_i} \sqrt{1 + [f'(\xi)]^2}\, d\xi - s_i \tag{34}$$

Differentiating (34) with respect to x_i gives

$$R(x_i) = \frac{dT(x_i)}{dx_i} = \sqrt{1 + [f'(x_i)]^2} \tag{35}$$

Using the Newton–Raphson iteration requires that

$$x_i^{(n)} = x_i^{(n-1)} - \frac{T(x_i)}{R(x_i)} \qquad n = 1, 2, \ldots \tag{36}$$

where $x_i^{(0)}$ should be guessed initially. At convergence, the value of x_i for the given s_i is known. When i runs through all possible values, the discretized boundary condition $x_i = x(s_i)$ can be obtained point by point.

Special care must be taken to impose boundary conditions at an inlet or outlet because the boundary conditions may be the Neumann type. Suppose the flow at the inlet is uniform and parallel to the x-axis, i.e., $u = 1$, $v = 0$ (see *Figure 3(a)*). By the conventional

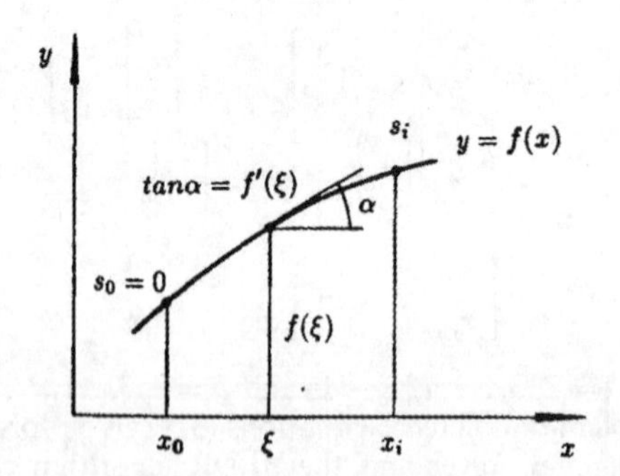

Figure 2. Arc length evaluation and inverse calculation.

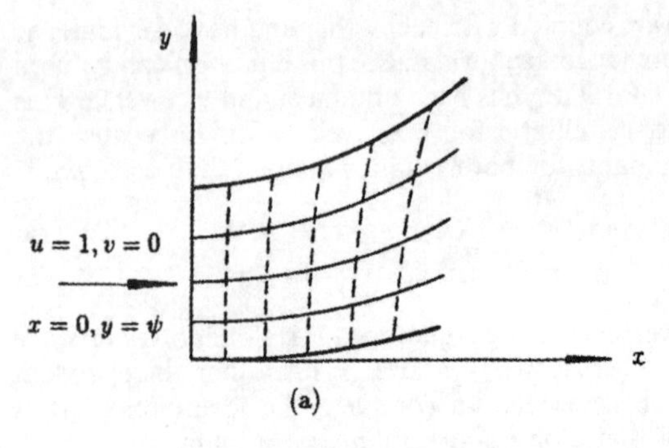

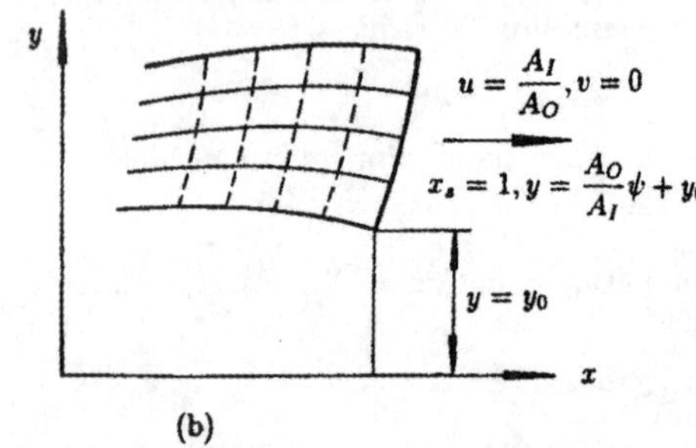

Figure 3. Inlet and outlet boundary conditions. (a) Inlet, (b) outlet.

definition of streamfunction and equation (7),

$$u = \psi_y = \frac{x_s}{J}, \quad v = -\psi_x = \frac{y_s}{J} \tag{37}$$

Hence, $x_s = J$, $y_s = 0$. Recalling $x_s^2 + y_s^2 = 1$, we get $x_s = 1$, $y_s = 0$ and, therefore, $J = y_\psi = 1$. Integrating implies $x = s + c_1$ and $y = \psi + c_2$. Setting $s = 0$ when $x = 0$ and $\psi = 0$ when $y = 0$ leads to $c_1 = 0$ and $c_2 = 0$. Finally, the boundary conditions at the inlet ($s = 0$) are

$$x = 0, \qquad y = \psi \tag{38}$$

This is a Dirichlet boundary condition. However, the outlet boundary conditions are slightly different (see *Figure 3(b)*). Suppose the flow at the outlet is also uniform and parallel to the x-axis so that $v = 0$. But, u cannot be specified as 1 because the mass flux conservation law has to be observed. So, $u = A_I/A_O$ where A_I and A_O are the widths of the inlet and outlet sections, respectively. Based on an analysis similar to the inlet, we get $x_s = 1$, $y_s = 0$, and $J = y_\psi = A_O/A_I$; therefore,

$$x_s = 1, \quad y = \frac{A_O}{A_I}\psi + y_0 \tag{39}$$

where y_0 is the height of the zero streamline at the outlet. It should be noted that only the y equation can be integrated to get a Dirichlet specification. The x equation should not be integrated because there is no information to determine constant c_1. Hence, the boundary condition for the x equation is of the Neumann type at the outlet. Using $x_s = 1$ or $x_{i+1,j} = x_{i,j} + \Delta s$ in the difference expression of x_{ss} and a one-sided difference for $x_{s\psi}$, the coefficient B and the right-hand-side term RHS of the

安长发博士科研论文集

difference equation (31) should be replaced by

$$B = -(A_1 + 2\beta^2 A_3)$$
$$RHS = \Delta s^2 A_4 - A_1(x_{I-1,j} + \Delta s) - \frac{\beta}{2} A_2(x_{I,j+1} - x_{I,j-1} - x_{I-1,j+1} + x_{I-1,j-1}) \quad (40)$$

where $I = I_{max}$ and $j = 2, 3, \ldots, J_{max} - 1$. If, furthermore, the starting point ($s = 0$) is not set at the inlet but somewhere else, then the boundary condition for the x equation at the inlet is also the Neumann type, $x_s = 1$. In this case, the coefficient B and right-hand-side term RHS should be replaced by

$$B = -(A_1 + 2\beta^2 A_3)$$
$$RHS = \Delta s^2 A_4 - A_1(x_{2,j} - \Delta s) - \frac{\beta}{2} A_2(x_{2,j+1} - x_{2,j-1} - x_{1,j+1} + x_{1,j-1}) \quad (41)$$

where $j = 2, 3, \ldots, J_{max} - 1$.

6. Examples

1. *Flow through a tank.* The first example is an incompressible potential flow through a tank (see *Figure 4*). The numbers of grid points in s and ψ directions are $I_{max} = 57$ and $J_{max} = 17$, respectively. For this tank, s takes values from 0 to $s_{max} = 1.75$ while ψ varies from 0 to $\psi_{max} = 0.25$. As described in the previous section, the boundary conditions on all walls are prescribed as the Dirichlet type. The inlet and outlet boundary conditions are

$$\begin{cases} x = 0, \quad y = \psi & \text{at inlet} \\ x_s = 1, \quad y = \psi + 0.75 & \text{at outlet} \end{cases} \quad (42)$$

It should be noted that the outlet boundary condition for the y equation is of the Dirichlet type, but the outlet boundary condition for the x equation is of the Neumann type. Hence, equation (40) should be used. The calculated streamlines are shown in *Figure 4*, and the velocity vector plot is shown in *Figure 5*.

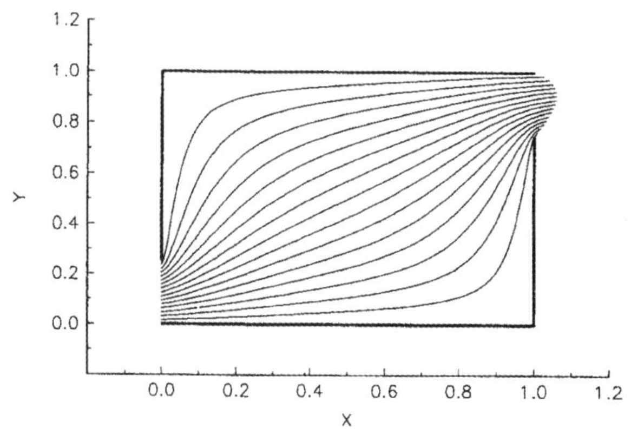

Figure 4. Streamlines of flow through tank I.

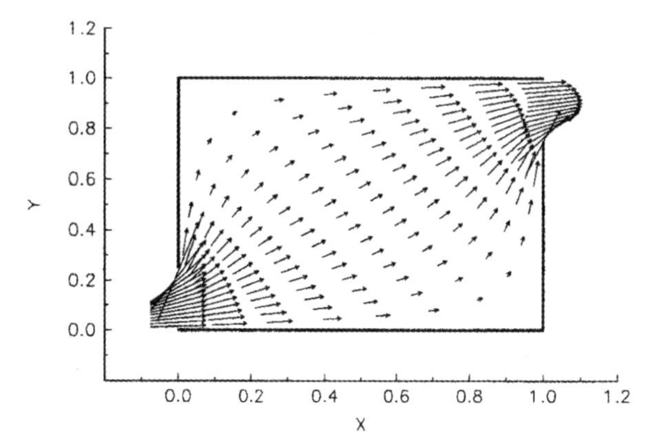

Figure 5. Velocity vector plot of flow through tank I.

2. *Flow through a tank with sharp turn.* The second example differs from the first tank in that the streamlines approach the vertical asymptotic line in the outlet region (see *Figure 6*). This kind of flow cannot be handled by the von Mises formulation because the Jacobian of the transformation approaches infinity in the exit region. The inlet and outlet boundary conditions are

$$\begin{cases} x = 0, \quad y = \psi & \text{at inlet} \\ x = 2.0 - \psi, \quad y_s = 1 & \text{at outlet} \end{cases} \quad (43)$$

The mesh is 49 × 17. The velocity vector plot is shown in *Figure 6*.

3. *Nozzle flow.* The third example is flow through a nozzle (see *Figure 7*). Due to symmetry of the nozzle, only the upper half is considered. The wall shape function is given by

$$f(x) = \begin{cases} 1.25 & x \le 0 \\ 1.25 - 6x^2 & 0 < x \le 0.25 \\ 2 - 6x + 6x^2 & 0.25 < x \le 0.5 \\ 1 - 2x + 2x^2 & 0.5 < x \le 0.75 \\ -1.25 + 4x - 2x^2 & 0.75 < x \le 1 \\ 0.75 & x > 1 \end{cases} \quad (44)$$

The total length of the curved part of the wall ($0 \le x \le 1$) is divided into 40 intervals, and each of

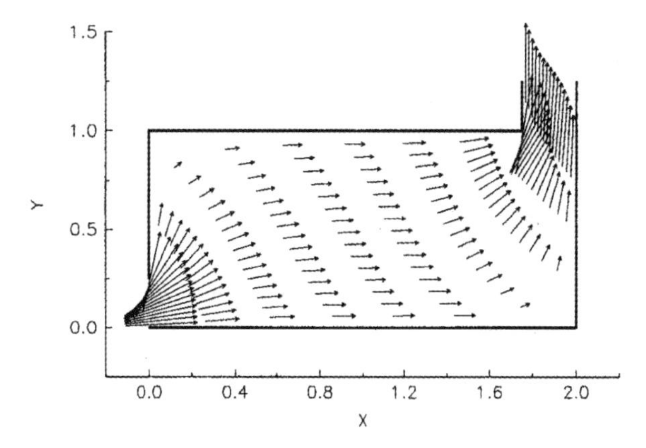

Figure 6. Velocity vector plot of flow through tank II.

第II部分 出国后博士学位论文阶段

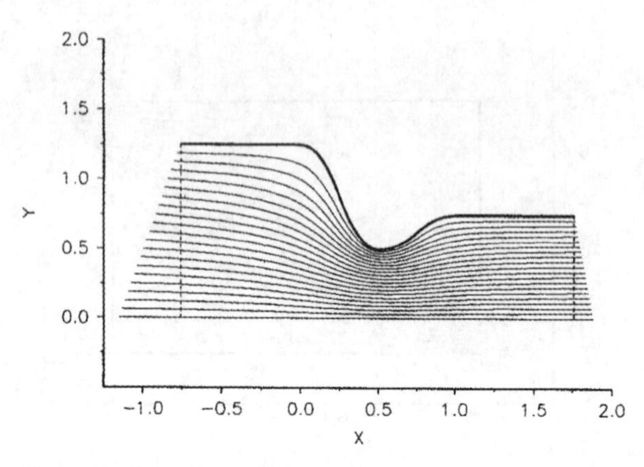

Figure 7. Streamlines of a nozzle flow.

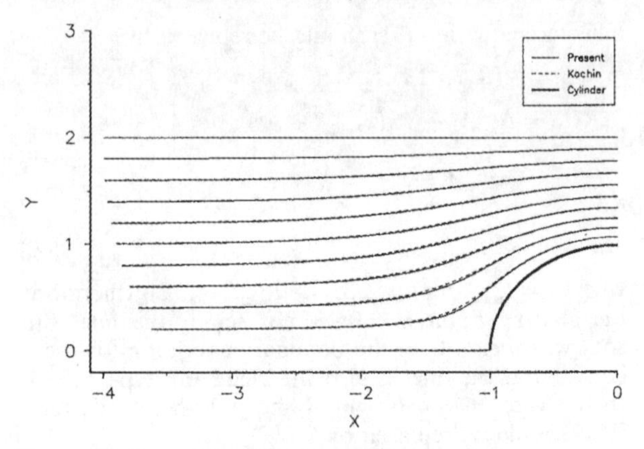

Figure 8. Streamlines of flow around a cylinder.

the left ($x < 0$) or right ($x > 1$) straight lines contains 20 segments. The maximum arc length is 3.032 and $\Delta s = 0.03679$. The mesh is 81×21. The starting point ($s = 0$) is chosen to be the middle point of the curved part of the wall corresponding to $x_0 = 0.3636$, and $s_{max} = 1.516$, $s_{min} = -1.516$, $\psi_{max} = 1.25$, and $\psi_{min} = 0$. The inlet and outlet boundary conditions differ from those in examples 1 and 2 because the position of $s = 0$ does not correspond to the inlet but to the line $x = x_0$. The inlet and outlet boundary conditions are

$$\begin{cases} x_s = 1, \quad y = \psi & \text{at inlet} \\ x_s = 1, \quad y = \dfrac{0.75}{1.25}\psi & \text{at outlet} \end{cases} \tag{45}$$

In this example, both inlet and outlet boundary conditions are the Dirichlet type for the y equation and the Neumann type for the x equation. Hence, both (40) and (41) should be used to solve the x equation. The calculated streamlines are drawn in *Figure 7*.

4. *Flow around a cylinder.* The fourth example is an incompressible potential flow around a cylinder placed in the middle of a channel composed of two parallel walls. The half distance between the walls is 2 and the radius of the cylinder is 1 (see *Figure 8*). Due to symmetry of the flow, only a quarter of the field with length of 4 is considered. The starting point $s = 0$ is chosen at the outlet. The boundary conditions at the inlet and outlet are

$$\begin{cases} x_s = 1 \quad y = \psi & \text{at inlet} \\ x = 0, \quad y_s = 0 & \text{at outlet} \end{cases} \tag{46}$$

The boundary conditions on the cylinder surface, on the wall, and on the symmetry line are the Dirichlet type for both y and x equations:

$$\begin{cases} x = s, \quad y = 2 \quad \text{on wall} \\ x = r\sin\dfrac{s}{r}, \quad y = r\cos\dfrac{s}{r} \quad \text{on cylinder} \\ x = s + r(\pi/2 - 1), \quad y = 0 \quad \text{on symmetry line} \end{cases} \tag{47}$$

where $r = 1$ is the radius of the cylinder. The mesh is 101×51, $s_{min} = -4$, $s_{max} = 0$, $\psi_{min} = 0$, $\psi_{max} = 2$, and $\Delta s = \Delta\psi = 0.04$. There are 40 points on the cylinder surface. *Figure 8* gives the calculated streamlines of the flow compared with the streamlines of the analytical solution:[14]

$$\psi = y - \frac{a}{\pi} sh^2\left[\frac{\pi r}{a}\right] \frac{\sin(2\pi y/a)}{ch(2\pi x/a) - \cos(2\pi y/a)} \tag{48}$$

where $a = 4$ is the width of the channel and $r = 1$ is the cylinder radius. *Figure 9* is the horizontal velocity component distribution along the zero streamline compared with the result of the same analytical solution.[14] These comparisons confirm that the present formulation provides accurate results.

5. *Flow past an airfoil.* The last example is flow past a NACA 0012 airfoil with angle of incidence $\alpha = 0°$ and $4°$ (see *Figure 10*). The mesh is 81×81. There are 21 points on each side of the airfoil, whose total arc length is 1.0182. We have taken $s_{max} = 2.0363$, $s_{min} = -2.0363$, $\psi_{max} = 2$, $\psi_{min} = -2$, $\Delta s = 0.051$, and $\Delta\psi = 0.05$. The boundary condition on the airfoil, as in (32), is $x_i = x(s_i)$, $y_i = f_{\pm}(x_i)$ where $f_{\pm}(x)$ represents upper and lower shape functions of the airfoil and $x(s)$ is the inverse function of the arc length formula (33). To specify $x_i = x(s_i)$, Newton–Raphson iteration is

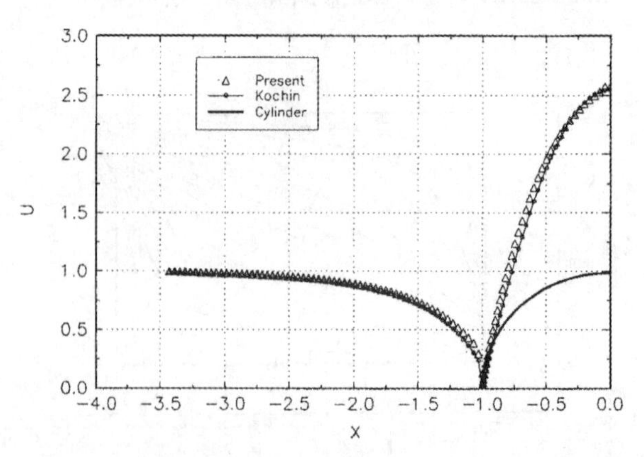

Figure 9. Velocity distribution on cylinder and symmetry line $\psi = 0$.

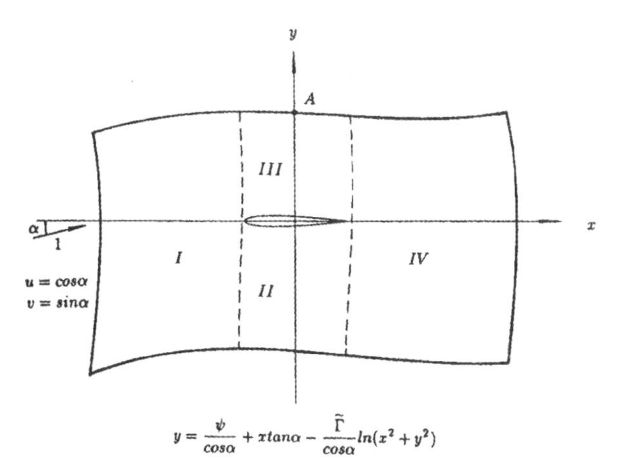

Figure 10. Flow past an airfoil.

needed for each airfoil point. The boundary condition in the far field can be specified from the following expression:

$$y = \frac{\psi}{\cos\alpha} + x\tan\alpha - \frac{\tilde{\Gamma}}{\cos\alpha}\ln(x^2 + y^2) \tag{49}$$

where $\tilde{\Gamma}$ is the modified circulation, $\tilde{\Gamma} = \pi\alpha = 0$, and 0.198 for $\alpha = 0°$ and 4° from thin airfoil theory. The boundary condition specification is more complicated than previous examples. To specify the top boundary condition we first determine the y value at the top middle point A where $s = 0$, $x = 0$, $\psi = 2$, using equation (49) with Newton–Raphson iteration. The coordinates x and y of the next point along the top streamline can be calculated by the arc length constraint condition (22), or its difference form

$$(x_i - x_{i-1})^2 + (y_i - y_{i-1})^2 = \Delta s^2 \tag{50}$$

Starting from point A and solving (49) and (50) for x and y point by point, we can get the coordinates of all points along the top boundary by sweeping in both directions. The same procedure can be applied for the bottom boundary. After determining top and bottom boundary conditions, simple interpolation can be used to give an initial guess of the flow field. The left and right boundaries are like an inlet and outlet. Assuming they are far enough from the airfoil, both inlet and outlet boundary conditions are $u = \cos\alpha$, $v = \sin\alpha$. From equations (37) and (22), $u^2 + v^2 = 1/J^2 = 1$. Therefore $J = 1$ and the inlet and outlet boundary conditions are

$$x_s = \cos\alpha, \qquad y_s = \sin\alpha \tag{51}$$

It can be seen that the boundary conditions at both inlet and outlet for both y and x equations are the Neumann type. Hence, equations (40) and (41) with minor modification to consider the effect of the angle of attack should be applied. The whole domain is divided into four subregions. In each subregion, the SLOR algorithm is invoked to sweep from left to right, and all four subregions are swept in the order I, II, III, IV for both y and x equations until the solution converges. *Figure 11* shows the predicted streamlines of flow past a NACA 0012 airfoil with angle of attack $\alpha = 4°$. *Figure 12* shows the Cp distribution on the airfoil for the case of $\alpha = 0°$ compared with an analytical result.[15] *Figure 13* shows the Cp distribution on the upper and lower surfaces of the airfoil at $\alpha = 4°$ compared with a calculation using the boundary element method.[16] These results confirm the applicability of the present approach.

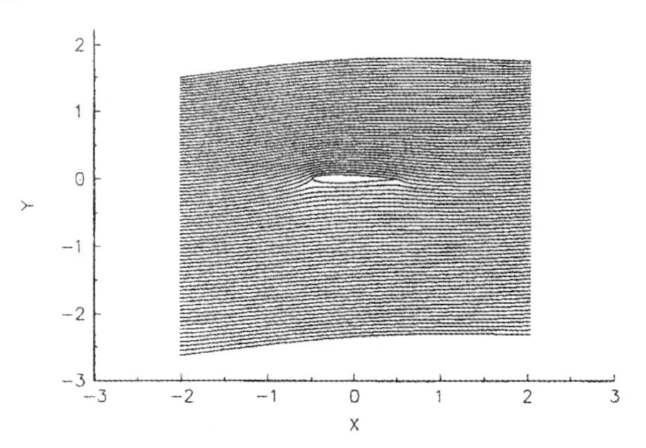

Figure 11. Streamlines of flow past NACA 0012 at $\alpha = 4°$.

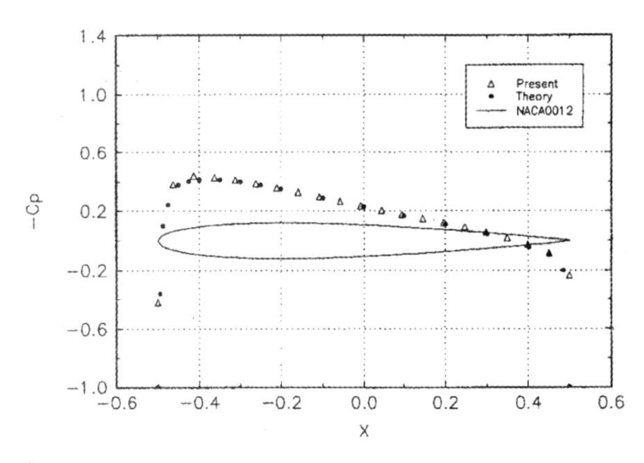

Figure 12. Cp distribution on NACA 0012 at $\alpha = 0°$.

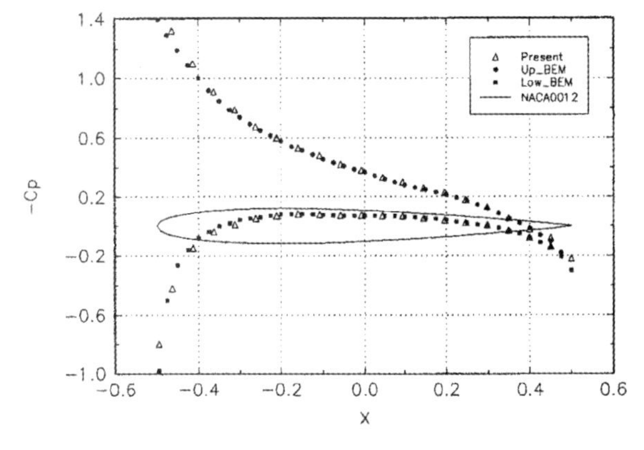

Figure 13. Cp distribution on NACA 0012 at $\alpha = 4°$.

第二部分 出国后博士学位论文阶段

7. Concluding remarks

One of the major weakness of von Mises coordinates is that the Jacobian becomes infinite at points where the flow reverses direction. This leads to nonuniqueness of $y(x, \psi)$ and is particularly troublesome in the region where the flow changes its direction sharply. The proposed arc length–streamfunction formulation is one of the methods to overcome this problem.

For a 2-D inviscid incompressible potential flow, the two curvilinear coordinates can be chosen as the arc length along a streamline and streamfunction s, ψ. After this coordinate transformation, the governing equations of the flow are a set of elliptic equations for cartesian coordinates x and y, accompanied by an arc length constraint condition $x_s^2 + y_s^2 = 1$. Several computed examples demonstrate that the present approach is effective and robust enough to solve incompressible potential flows.

The proposed approach can also be extended to more complex forms. For example, it can be extended to compressible flow using either the full potential equation or Euler equations. It may be possible to treat viscous flow without separation by including additional terms in the governing equations and developing an equation for the vorticity. Extension to steady 3-D and unsteady 2-D flows may also be possible, but will undoubtedly be challenging. As with the von Mises approach, the present formulation should also lead to an efficient procedure for pure inverse design of airfoils and channels.

However, this approach has some restrictions. In particular, for a flow in which the flow direction changes rapidly, the total arc length is different for each streamline within the physical domain. As a result, some streamlines may extend beyond the original physical domain. Furthermore, the skewness of the cells may be large, which may lead to slow convergence or even divergence of the computation. This deficiency may be overcome by using a stretched arc length to replace the real arc length. In addition, the present method is unable to handle separated flows and other flows with closed streamlines. This technique is still in the exploratory stage, and much work is required before it is useful in industrial or commercial applications.

Nomenclature

A_1, A_2, A_3, A_4	defined in (29)
A, B, C	defined in (31), (40), or (41)
a	width of channel
Cp	pressure coefficient
E, F, G	metrics of coordinate transformation
$f(x)$	airfoil boundary shape function
J	Jacobian
S	general arc length
s	arc length along a streamline
$R(x)$	defined in (35)
r	radius of cylinder
$T(x)$	defined in (34)
u, v	velocity components
x, y	cartesian coordinates
α	angle of attack
β	ratio of Δs to $\Delta\psi$
$\tilde{\Gamma}$	modified circulation
ψ	streamfunction
$\Delta s, \Delta\psi$	increments of s and ψ, respectively
Δ_x, Δ_y	discriminants of x and y equations, respectively
∇^2	Laplace operator

References

1 Martin, M. H. The flow of a viscous fluid. I. *Arch. Rational Mech. Anal.* 1971, **41**, 266–286

2 Pearson, C. E. Use of streamline coordinates in the numerical solution of compressible flow problems. *J. Comp. Phys.* 1981, **42**, 257–265

3 Takahashi, K. A numerical analysis of flow using streamline coordinates (The case of two-dimensional steady incompressible flow). *Bull. JSME* 1982, **25**, 1696–1702

4 Greywall, M. S. Streamwise computation of 2-D incompressible potential flows. *J. Comp. Phys.* 1985, **59**, 224–231

5 Srinivasan, K. and Spalding, D. B. The stream function coordinate system for solution of one-dimensional unsteady compressible flow problems. *Appl. Math. Modelling* 1986, **10**, 278–283

6 Huang, C-Y. and Dulikravich, G. S. Stream function and stream function coordinate (SFC) formulation for inviscid flow field calculations. *Comput. Meth. Appl. Mech. Eng.* 1986, **59**, 155–177

7 Stanitz, J. D. A review of certain inverse methods for the design of ducts with 2- or 3-dimensional potential flow. *Appl. Mech. Rev.* 1988, **41**, 217–238

8 Barron, R. M. Computation of incompressible potential flow using von Mises coordinates. *Math. Comput. Simulation* 1989, **31**, 177–188

9 Barron, R. M., An, C-F., and Zhang, S. Survey of the streamfunction-as-a-coordinate method in CFD. *Proceedings of the Inaugural Conference of the CFD Society of Canada*, ed. R. Camarero. Montreal, Canada, 1993

10 Barron, R. M. and Naeem, R. K. Numerical solution of transonic flows on a streamfunction co-ordinate system. *Intl. J. Numer. Methods Fluids* 1989, **9**, 1183–1193

11 Barron, R. M., Zhang, S., Chandna, A., and Rudraiah, N. Axisymmetric potential flow calculations. Part 1: Analysis mode. *Comm. Appl. Numer. Methods* 1990, **6**, 437–445

12 Barron, R. M. and An, C-F. Analysis and design of transonic airfoils using streamwise coordinates. *Proceedings of the 3rd International Conference on Inverse Design Concepts and Optimization in Engineering Science*, ed. G. S. Dulikravich. Washington, DC, 1991, pp. 359–370

13 An, C-F. and Barron, R. M. Numerical solution of transonic full-potential-equivalent equations in von Mises coordinates. *Int. J. Numer. Methods Fluids* 1992, **15**, 925–952

14 Kochin, N. E., Kibel, I. A., and Rose, N. V. *Theoretical Hydrodynamics*, Translated from Russian by D. Boyanovitch. Interscience Publishers, New York, 1964, pp. 289–290

15 Abbott, I. H. and von Doenhoff, A. E. *Theory of Wing Sections.* Dover Publications, New York, 1959, p. 321

16 Carey, G. F. and Kim, S. W. Lifting airfoil calculation using the boundary element method. *Int. J. Numer. Methods Fluids* 1983, **3**, 481–492

Unsteady conservative streamfunction coordinate formulation: 1-D isentropic flow

R. M. Barron, C-F. An, and S. Zhang

Department of Mathematics and Statistics and Fluid Dynamics Research Institute, University of Windsor, Windsor, Ontario, Canada

In this paper the conservative formulation for the numerical simulation of one-dimensional (1-D) unsteady flow in streamfunction coordinates has been established. Instead of solving a single nonlinear nonconservative second-order partial differential equation for the cartesian coordinate x in the computational plane (ψ, τ), a conservative system of two equations for x_ψ and x_τ has been formulated in the computational plane and solved numerically applying Godunov's explicit, first-order, two-step predictor–corrector scheme. This scheme is dissipative and nondispersive, which is favorable for dealing with discontinuities and capable of suppressing possible oscillations. Some sample computations have been carried out using both conservative and nonconservative formulations. The advantages of the conservative form become obvious, especially when the flow is compressed so strongly that a discontinuity, or a shock wave, is about to appear.

Keywords: CFD, unsteady flow, conservative streamfunction coordinates

1. Introduction

Only a few papers,[1–4] dealing with the numerical solution of unsteady flow using stream function coordinates (SFC), have appeared in the literature. However, the SFC method, showing significant advantages when applied to 1-D unsteady flow, has an attractive potential in CFD. In earlier exploratory work,[4] the von Mises transformation was applied to the equations governing 1-D unsteady isentropic flows. The Jacobian of the transformation is always a finite, positive, and nonzero quantity and, hence, the transformation never fails. As a result, the governing equations are simplified to a nonlinear second-order hyperbolic partial differential equation, a wave equation with a variable propagation speed, for the cartesian coordinate x as a function of streamfunction ψ and time τ. Using a three-step explicit leapfrog scheme, this equation can be successfully solved to simulate 1-D unsteady isentropic flows. However, difficulties arise when the approach is applied to a strong compression wave problem, especially when the flow is compressed so strongly that a discontinuity, or a shock wave, is about to appear. In these cases, oscillations occur, and the accuracy decreases. It is believed that these difficulties are partly attributable to the nonconservative nature of the governing equation.

Address reprint requests to Dr. Barron at the Dept. of Mathematics and Statistics and Fluid Dynamics Research Institute, University of Windsor, Windsor, Ontario, Canada N9B 3P4.

Received 15 November 1993; revised 2 March 1994; accepted 8 March 1994

To overcome the above-mentioned difficulties, in this work the governing equations for 1-D unsteady isentropic flow are re-formulated as a conservative system of two first-order equations for two variables, x_ψ and x_τ, in the computational plane (ψ, τ). The system of equations is of the hyperbolic type. Therefore, a numerical scheme that explicitly marches the solution from initial time can be applied to solve the system. Comparisons of the present conservative formulation with the nonconservative one show that a significant improvement is realized.

In Section 2 of this paper, the conservative system of equations in streamfunction coordinates for 1-D unsteady isentropic flow is derived from 1-D Euler equations with the isentropic assumption. Section 3 analyzes the eigenvalues of the flux vector Jacobian of the system and describes the numerical algorithm. The initial and boundary conditions for some selected examples are described in Section 4. In Section 5, the calculated results using both conservative and nonconservative formulations are compared. Section 6 offers some concluding remarks.

2. Mathematical formulation

For one-dimensional unsteady isentropic flow, the governing equations are

$$\frac{\partial \tilde{\rho}}{\partial \tilde{t}} + \frac{\partial(\tilde{\rho}\tilde{u})}{\partial \tilde{x}} = 0, \tag{1}$$

$$\frac{\partial \tilde{u}}{\partial \tilde{t}} + \tilde{u}\frac{\partial \tilde{u}}{\partial \tilde{x}} + \frac{1}{\tilde{\rho}}\frac{\partial \tilde{p}}{\partial \tilde{x}} = 0, \tag{2}$$

$$\tilde{p} = \text{const.}(\tilde{\rho})^\gamma \tag{3}$$

where the dimensional density $\tilde{\rho}$, velocity $\tilde{u}$, and pressure $\tilde{p}$ are unknown functions of time $\tilde{t}$ and coordinate $\tilde{x}$, and γ is the ratio of specific heats. Choosing stagnation density $\tilde{\rho}_0$ and pressure $\tilde{p}_0$ as reference properties and some characteristic length $\tilde{L}$ as reference length, i.e.,

$$\rho = \frac{\tilde{\rho}}{\tilde{\rho}_0}, \; p = \frac{\tilde{p}}{\tilde{p}_0}, \; u = \frac{\tilde{u}}{\sqrt{\tilde{p}_0/\tilde{\rho}_0}}, \; x = \frac{\tilde{x}}{\tilde{L}}, \; t = \frac{\tilde{t}\sqrt{\tilde{p}_0/\tilde{\rho}_0}}{\tilde{L}} \tag{4}$$

the governing equations take the following nondimensional form,

$$\rho_t + (\rho u)_x = 0, \tag{5}$$

$$\rho u_t + \rho u u_x + p_x = 0, \tag{6}$$

$$p = \rho^\gamma \tag{7}$$

Substituting p from (7) into (6), the momentum equation becomes

$$\rho u_t + \rho u u_x + \gamma \rho^{\gamma-1} \rho_x = 0 \tag{8}$$

Introducing a streamfunction ψ such that

$$\psi_x = \rho, \; \psi_t = -\rho u \tag{9}$$

we have

$$\rho = \psi_x, \; u = -\frac{\psi_t}{\psi_x} \tag{10}$$

Furthermore, introducing the von Mises-like transformation,

$$t \equiv \tau, \; x = x(\psi, \tau) \tag{11}$$

the roles of x as an independent variable and ψ as a dependent variable can be switched. From the chain rule,

$$\frac{\partial}{\partial t} = \frac{\partial}{\partial \tau} + \psi_t \frac{\partial}{\partial \psi}, \; \frac{\partial}{\partial x} = \psi_x \frac{\partial}{\partial \psi} \tag{12}$$

Applying these differential operators to x leads to

$$\psi_t = -\frac{x_\tau}{x_\psi}, \; \psi_x = \frac{1}{x_\psi} \tag{13}$$

Comparing (13) with (10), one gets

$$\rho = \frac{1}{x_\psi}, \; u = x_\tau \tag{14}$$

From (13), differential operators (12) become

$$\frac{\partial}{\partial t} = \frac{\partial}{\partial \tau} - \frac{x_\tau}{x_\psi}\frac{\partial}{\partial \psi}, \; \frac{\partial}{\partial x} = \frac{1}{x_\psi}\frac{\partial}{\partial \psi} \tag{15}$$

The Jacobian of the transformation (11)

$$J = \frac{\partial(x, t)}{\partial(\psi, \tau)} = x_\psi \tag{16}$$

is always positive due to the fact that it is the reciprocal of density, which is always nonzero, positive, and finite. This simple fact guarantees that the transformation (11) is always one-to-one and invertible. Substituting the expressions for ρ and u in (14) and differential operators in (15) into the momentum equation (8), we obtain

$$x_{\tau\tau} = \gamma \frac{x_{\psi\psi}}{x_\psi^{\gamma+1}} \tag{17}$$

This hyperbolic nonlinear second-order partial differential equation was solved using the three-step midpoint leapfrog scheme in a previous study.[4] A piston-driven expansion wave problem and a density disturbance propagation problem were successfully simulated. For the expansion part of these problems, the method gives excellent results but for the compression part it is not as accurate. Some oscillations appear as the flow is compressed strongly. For a piston-driven compression wave problem, the computed density distribution has unacceptable errors, especially when the compression becomes increasingly stronger and a shock wave is about to occur. In this case, it is beneficial and perhaps necessary to write the governing equations in conservative form and to use a dissipative numerical scheme. The momentum equation (17) can be written as

$$x_{\tau\tau} = -\left[\frac{1}{x_\psi^\gamma}\right]_\psi \tag{18}$$

If x_τ and x_ψ are considered to be independent of each other, this single equation is not enough to solve for these two variables. An additional equation is needed to close the system. Substituting ρ and u from (14) and differential operators from (15) into the continuity equation (5) and simplifying, a complementary equation is obtained,

$$x_{\psi\tau} = x_{\tau\psi} \tag{19}$$

Combining (19) and (18) together, a conservative system for the unknowns x_ψ, x_τ can be formed,

$$\begin{bmatrix} x_\psi \\ x_\tau \end{bmatrix}_\tau + \begin{bmatrix} -x_\tau \\ x_\psi^{-\gamma} \end{bmatrix}_\psi = 0 \tag{20}$$

or in vector form,

$$\vec{Q}_\tau + \vec{F}_\psi = 0 \tag{21}$$

where

$$\vec{Q} = \begin{bmatrix} q_1 \\ q_2 \end{bmatrix} = \begin{bmatrix} x_\psi \\ x_\tau \end{bmatrix}, \vec{F} = \begin{bmatrix} f_1 \\ f_2 \end{bmatrix} = \begin{bmatrix} -x_\tau \\ x_\psi^{-\gamma} \end{bmatrix} = \begin{bmatrix} -q_2 \\ q_1^{-\gamma} \end{bmatrix} \tag{22}$$

3. Numerical algorithm

In order to decide what type of numerical scheme is appropriate to solve system of equations (20), classification of the system is important. The Jacobian of the flux vector is defined as

$$A = \frac{\partial \vec{F}}{\partial \vec{Q}} = \frac{\partial(f_1, f_2)}{\partial(q_1, q_2)} \tag{23}$$

Substituting q_1, q_2, f_1, and f_2 from (22) into (23) gives

$$A = \begin{bmatrix} 0 & -1 \\ -\gamma q_1^{-(\gamma+1)} & 0 \end{bmatrix} \tag{24}$$

The eigenvalues σ of matrix A satisfy

$$det[\sigma I - A] = det\begin{bmatrix} \sigma & 1 \\ \gamma q_1^{-(\gamma+1)} & \sigma \end{bmatrix} = 0 \quad (25)$$

Solving this quadratic algebraic equation for σ, the roots are

$$\sigma_{1,2} = \pm\sqrt{\gamma q_1^{-(\gamma+1)}} = \pm\sqrt{\gamma \rho^{\gamma+1}} \quad (26)$$

Because density ρ is always positive, both eigenvalues are real and the system of equations (21) is hyperbolic.[5] The system can be solved by marching from some initial conditions.[6]

The rectangular domain in (ψ, τ) plane can be discretized into $J_{max} \times N_{max}$ grid points (*Figure 1*). The difference equation, discretized from (21), can be marched level by level from initial time $t = 0$ ($n = 1$). According to Sod's survey[7] of finite-difference methods for systems of nonlinear hyperbolic equations, Godunov's explicit first-order, two-step predictor–corrector scheme is one of the most successful methods available. It can be written as

$$\begin{cases} \bar{\vec{Q}}_{j+1/2} = \frac{1}{2}(\vec{Q}_{j+1}^n + \vec{Q}_j^n) - \dfrac{\Delta\tau}{\Delta\psi}(\vec{F}_{j+1}^n - \vec{F}_j^n) \\ \vec{Q}_j^{n+1} = \vec{Q}_j^n - \dfrac{\Delta\tau}{\Delta\psi}(\bar{\vec{F}}_{j+1/2} - \bar{\vec{F}}_{j-1/2}) \end{cases} \quad (27)$$

where the variables with overbars have intermediate values between time levels n and $n+1$. From equation (22), vector $\vec{F}$ is a function of vector $\vec{Q}$, i.e.,

$$\vec{F} = \vec{F}(\vec{Q}) \quad \text{or} \quad f_1 = -q_2,\ f_2 = q_1^{-\gamma} \quad (28)$$

Therefore, $\vec{F}_j^n$ can be evaluated from the solved (if $n > 1$) or initial (if $n = 1$) value of $\vec{Q}_j^n$ at any time level n. $\bar{\vec{F}}_{j+1/2}$ can be evaluated from the predicted value $\bar{\vec{Q}}_{j+1/2}$ at the "overbarred" intermediate time level. The marching process is sketched in *Figure 1*. Godunov's scheme is stable when

$$CFL = \sigma_{max}\frac{\Delta\tau}{\Delta\psi} \leq 1 \quad (29)$$

where CFL is the Courant–Friedrichs–Lewy number and $\sigma_{max} = max(\sigma_j)$ for all j's. At each time level n, σ_j is different for each point j and σ_{max} is maximum over all j's. However, σ_{max} varies from level to level, and hence

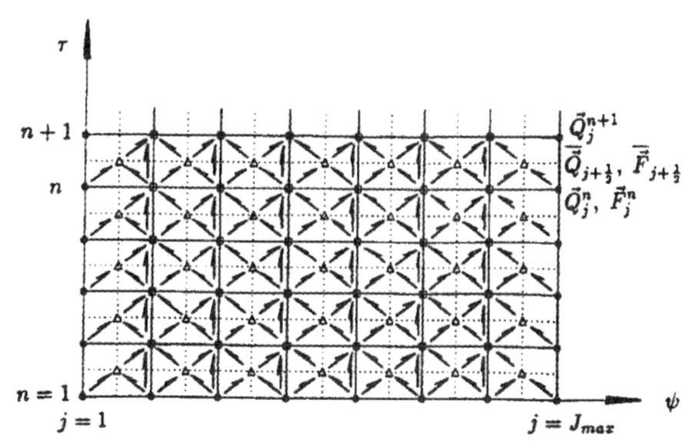

Figure 1. Godunov's scheme.

$\Delta\tau$ also varies from level to level if the CFL number is specified as a constant. The optimal CFL number has to be chosen through computational experimentation. The two steps at each time level are staggered to eliminate accumulated numerical errors and to increase accuracy. The first predictor step, containing the average of values at two adjacent grid points, like the Lax scheme, makes the scheme dissipative, enabling it to deal with possible discontinuities.

Having solved for $\vec{Q} = \begin{bmatrix} q_1 \\ q_2 \end{bmatrix}$, the cartesian coordinate $x = x(\psi, \tau)$ of particles can be calculated as follows. For a given time level $\tau = \tau^n$,

$$x(\psi_j, \tau^n) = \int_{\psi_{j-1}}^{\psi_j} q_1\, d\psi = \frac{\Delta\psi}{2}[(q_1)_{j-1} + (q_1)_j] \quad (30)$$

for $j = 2, 3, \ldots, J_{max}$ and $n = 1, 2, \ldots$ Then, taking ψ_j as constant, particle path lines of the flow are given by $x(\psi_j, \tau^n)$, $n = 1, 2, \ldots$

4. Initial and boundary conditions

The initial and boundary conditions are discussed in this section for selected examples, which will be computed later.

The first example is the same as example 2 in Ref. 4, i.e., the density disturbance propagation problem. The initial and boundary conditions are cited here for convenience. The initial conditions in the physical plane are

$$\rho(x, 0) = g(x) = \begin{cases} 1, & 0 \leq x \leq 0.45 \\ 1 + 0.2\sin\left[\dfrac{x - 0.45}{0.1}\pi\right], & 0.45 < x < 0.55 \\ 1, & 0.55 \leq x \leq 1 \end{cases} \quad (31)$$

$$u(x, 0) = 0 \quad (32)$$

In the computational plane,

$$q_1(\psi, 0) = f(\psi) = \begin{cases} 1, & 0 \leq x \leq 0.45 \\ 1/\left[1 + 0.2\sin\left(\dfrac{x - 0.45}{0.1}\pi\right)\right], & 0.45 < \psi < 0.563 \\ 1, & 0.563 \leq \psi \leq 1.013 \end{cases} \quad (33)$$

$$q_2(\psi, 0) = 0 \quad (34)$$

In (33), x is calculated iteratively by

$$x = \psi - \frac{0.02}{\pi}\left\{1 - \cos\left[\frac{x - 0.45}{0.1}\pi\right]\right\} \tag{35}$$

The left and right boundary conditions in the physical plane are

$$\rho(0, t) = 1, \qquad u(0, t) = 0 \tag{36}$$

and

$$\rho(1, t) = 1, \qquad u(1, t) = 0 \tag{37}$$

In the computational plane, these become

$$q_1(0, \tau) = 1, \qquad q_2(0, \tau) = 0 \tag{38}$$

and

$$q_1(\psi_d, \tau) = 1, \qquad q_2(\psi_d, \tau) = 0 \tag{39}$$

where $\psi_d = 1.013$.

The second example in this work is a piston-driven compression wave, similar to example 1 in Ref. 4, but with the piston moving in the opposite direction. The piston position is given by

$$X(t) = 2t^2 \tag{40}$$

The initial conditions in the physical and computational planes are

$$\rho(x, 0) = 1, \qquad u(x, 0) = 0 \tag{41}$$

and

$$q_1(\psi, 0) = 1, \qquad q_2(\psi, 0) = 0 \tag{42}$$

respectively. The left boundary conditions can be specified as follows. For a given piston location $X(t)$, the piston speed is

$$u[X(t), t] = X'(t) \tag{43}$$

which also represents the velocity of the fluid particles adjacent to the piston. From the isentropic relation, the density can be evaluated as

$$\rho[X(t), t] = \left[1 + \frac{\gamma - 1}{2}X'(t)\right]^{2/(\gamma-1)} \tag{44}$$

In the computational plane, the left boundary conditions are

$$q_1(0, \tau) = \left[1 + \frac{\gamma - 1}{2}X'(\tau)\right]^{-2/(\gamma-1)} \tag{45}$$

$$q_2(0, \tau) = X'(\tau) \tag{46}$$

The right boundary conditions in the physical and computational planes are

$$\rho(1, t) = 1, \qquad u(1, t) = 0 \tag{47}$$

and

$$q_1(1, \tau) = 1, \qquad q_2(1, \tau) = 0 \tag{48}$$

respectively.

The third example is also a piston-driven compression problem, but the piston movement is different from the previous one. Suppose a cylindrical tube is filled with air and the piston is at rest in the left end of the tube. At

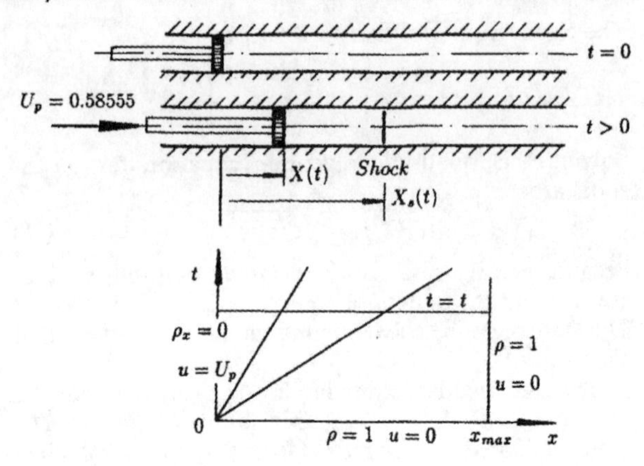

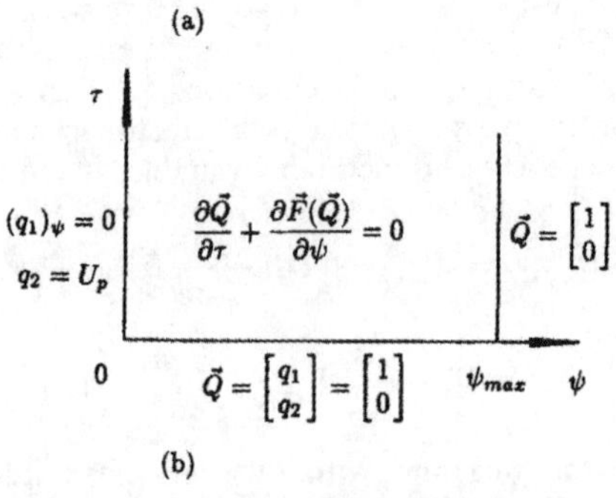

Figure 2. Piston-driven shock wave. (a) Physical plane, (b) computational plane.

$t = 0$, the piston is instantaneously accelerated to a constant speed $U_p = 0.58555$ (*Figure 2*). Our task is to investigate the subsequent unsteady flow in the tube. The initial conditions in the physical and computational planes are the same as in (41) and (42), respectively. The left boundary conditions in the physical plane are

$$\rho_x[X(t), t] = 0, \tag{49}$$

$$u[X(t), t] = U_p \tag{50}$$

where $X(t) = U_p t$ is the piston location at time t. In the computational plane, the left boundary conditions are

$$q_{1\psi}(0, \tau) = 0, \tag{51}$$

$$q_2(0, \tau) = U_p \tag{52}$$

To obtain (51), it is noted from the second of the differential operators (15), the first equation of (14), and the definition of $\vec{Q}$ in (22) that

$$\rho_x = \frac{1}{x_\psi}\rho_\psi = \frac{1}{x_\psi}\left[\frac{1}{x_\psi}\right]_\psi = \frac{1}{q_1}\left[\frac{1}{q_1}\right]_\psi = -\frac{q_{1\psi}}{q_1^3} \tag{53}$$

The right boundary conditions in the physical and computational planes are the same as (47) and (48), respectively.

5. Computed results and discussion

Density disturbance propagation

For the computational domain, the segment $0 \le \psi \le \psi_d$ ($\psi_d = 1.013$) is divided into 100 intervals and the total

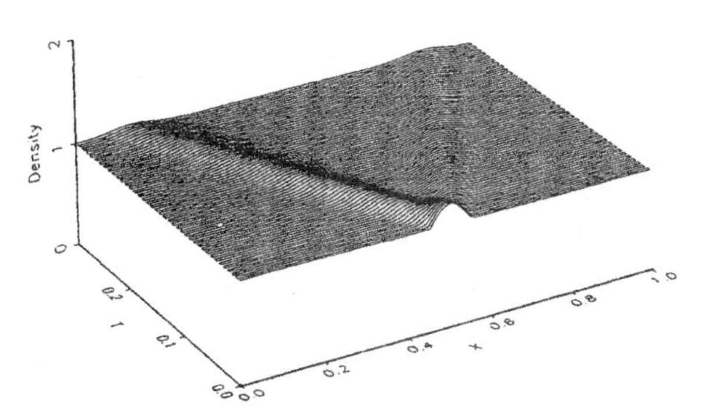

Figure 3. Density surface of example 1.

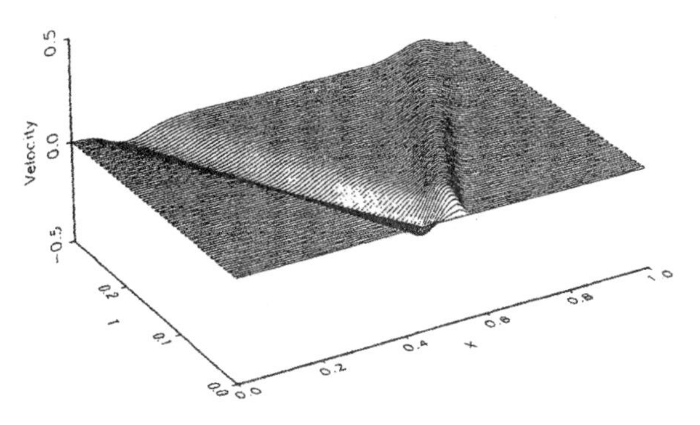

Figure 4. Velocity surface of example 1.

number of time steps is 50, i.e., $J_{max} = 101$ and $N_{max} = 51$. After marching 50 steps, $\tau = \tau_{max} = 0.3$. This corresponds to the physical domain $0 \le x \le 1$, $0 \le t \le 0.3$. The *CFL* number is chosen as 0.5. *Figures 3* and *4* give the surfaces of density and velocity distributions above the physical domain. *Figures 5* and *6* show the density and velocity distributions for both conservative and nonconservative formulations at $t = 0.3$, compared with a standard primitive variable Euler computation under the isentropic assumption. From these graphs we observe that the present conservative computation is much more accurate than the nonconservative one. The oscillations that occur in the nonconservative solution are efficiently

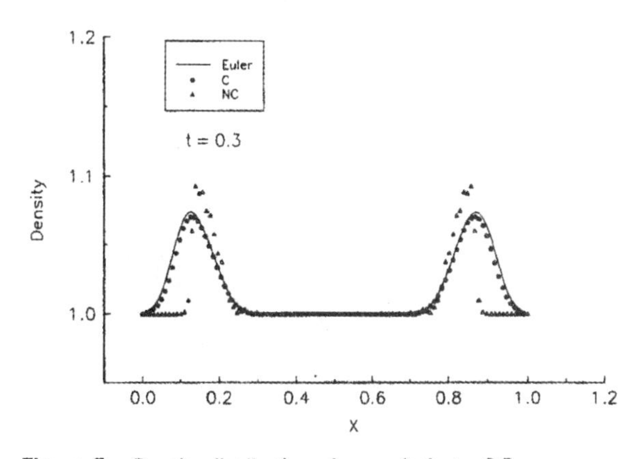

Figure 5. Density distribution of example 1: $t = 0.3$.

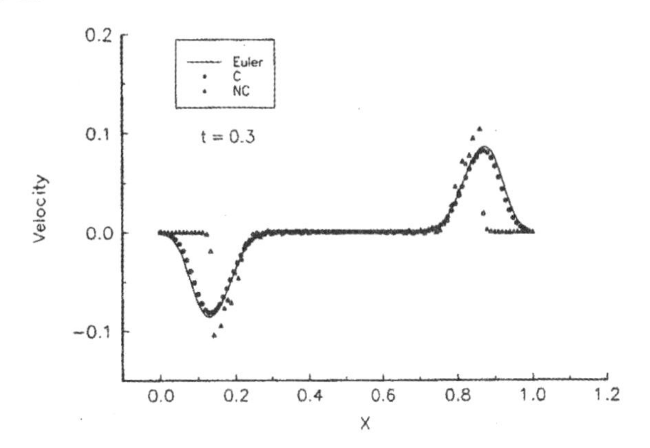

Figure 6. Velocity distribution of example 1: $t = 0.3$.

removed, and a smoother solution is obtained. This is not surprising because Godunov's scheme is dissipative, so the solution is smeared and the oscillations are suppressed.

Piston-driven compression wave

The computational domain and its discretization are the same as in the first example. The physical domain is not rectangular for this example because the left boundary is not a straight line perpendicular to the x-axis, but a curve described by a function of time, i.e., $X(t)$. *Figures 7* and *8* give the surfaces of density and velocity distributions, respectively, above the physical domain. *Figure 9* depicts the particle path lines. Here,

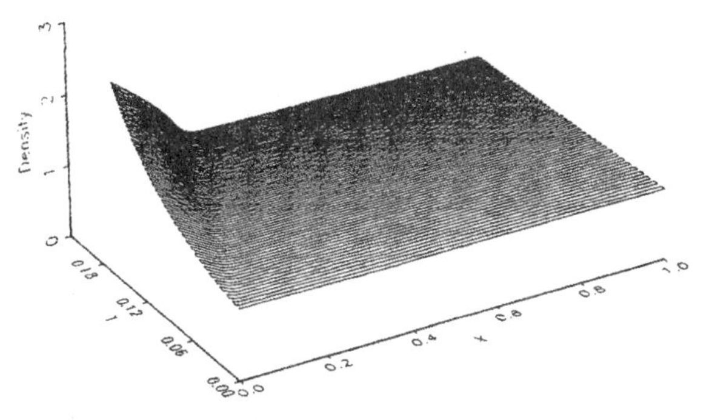

Figure 7. Density surface of example 2.

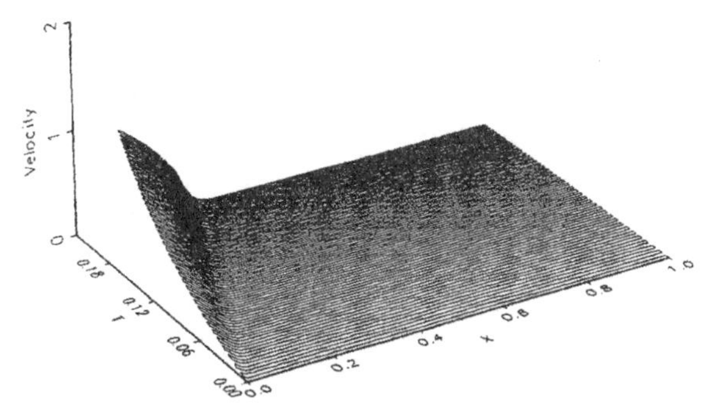

Figure 8. Velocity surface of example 2.

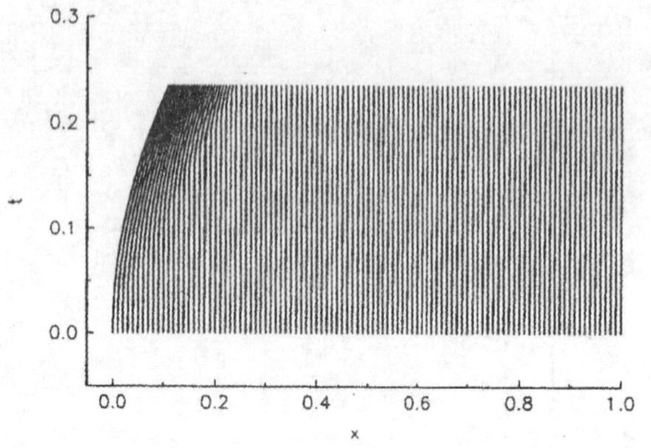

Figure 9. Particle path lines of example 2.

once again, we take advantage of the streamfunction formulation to easily draw the path lines. *Figures 10* and *11* show the calculated density and velocity distributions at $t = 0.226$ using both conservative and nonconservative formulations and compared with an exact solution. The exact solution in the compression region is obtained from the characteristics theory:[8]

$$u_k = X'(t_k),\ \rho_k = \left[1 + \frac{\gamma - 1}{2} X'(t_k)\right]^{2/(\gamma - 1)} \tag{54}$$

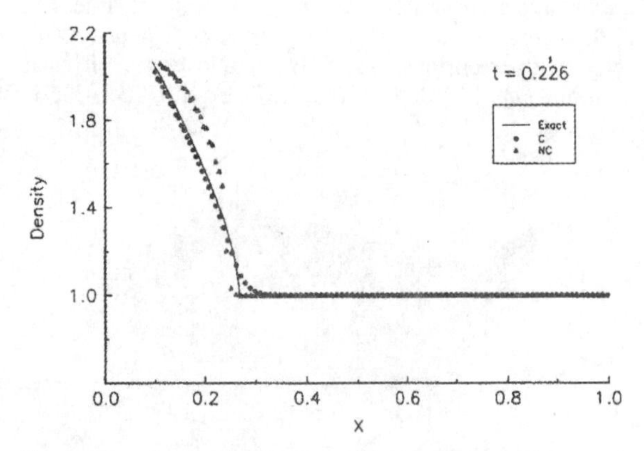

Figure 10. Density distribution of example 2: $t = 0.226$.

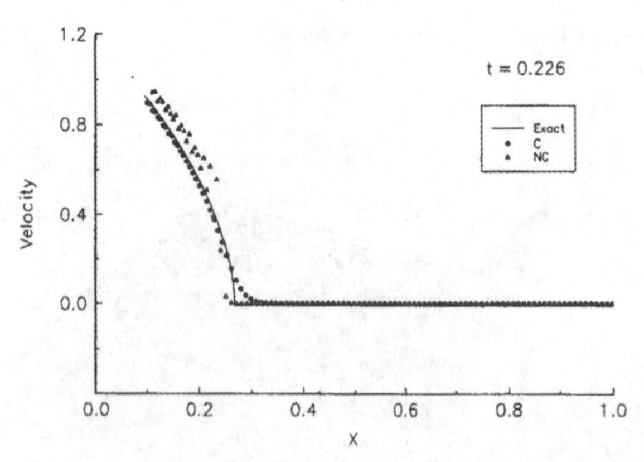

Figure 11. Velocity distribution of example 2: $t = 0.226$.

and

$$x_k = X(t_k) + \left[1 + \frac{\gamma + 1}{2} X'(t_k)\right](t_{max} - t_k) \tag{55}$$

From these graphs we can see that for the compression wave problem, the nonconservative formulation is not satisfactory whereas the conservative formulation gives excellent results compared with the exact solution.

Piston-driven shock wave

Figures 12 and *13* show the surfaces of density and velocity distributions for the piston-driven shock wave problem. *Figures 14* and *15* indicate density and velocity distributions at $t = 0.6$ compared with an exact solution. *Figure 16* gives the particle path lines. The exact solution is extracted from Landau and Lifshitz's book:[9]

$$u = \begin{cases} \dfrac{\gamma + 1}{4} U_p + \sqrt{\left[\dfrac{\gamma + 1}{4} U_p\right]^2 + c_0^2}, & x \le X_s(t) \\ 0, & x > X_s(t) \end{cases} \tag{56}$$

$$\rho = \left[1 + \frac{\gamma - 1}{2} u\right]^{2/(\gamma - 1)} \tag{57}$$

The shock wave and piston locations are

$$X_s(t) = \left\{\frac{\gamma + 1}{4} U_p + \sqrt{\left[\frac{\gamma + 1}{4} U_p\right]^2 + c_0^2}\right\} t \tag{58}$$

$$X(t) = U_p t \tag{59}$$

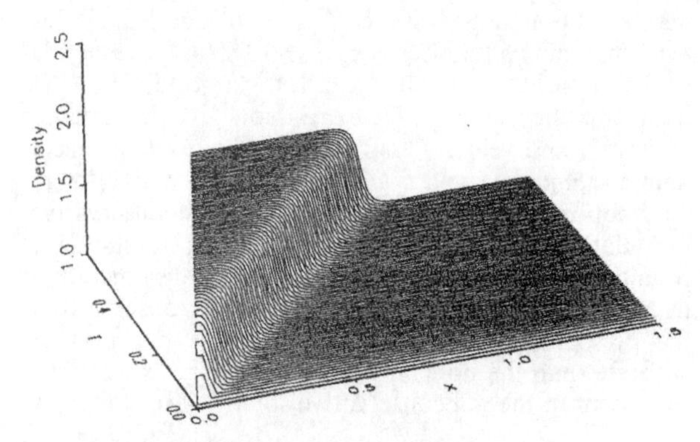

Figure 12. Density surface of example 3.

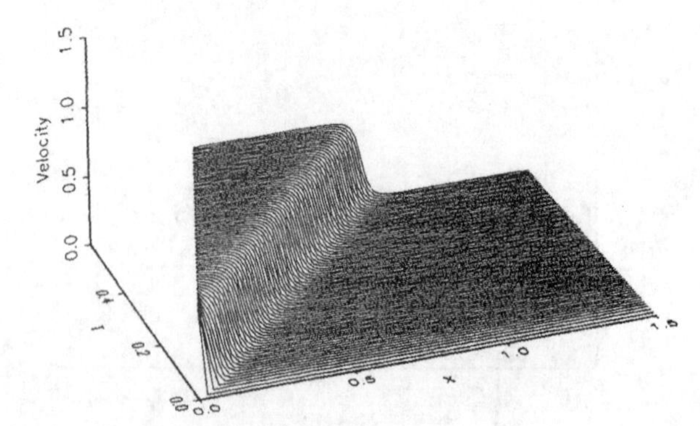

Figure 13. Velocity surface of example 3.

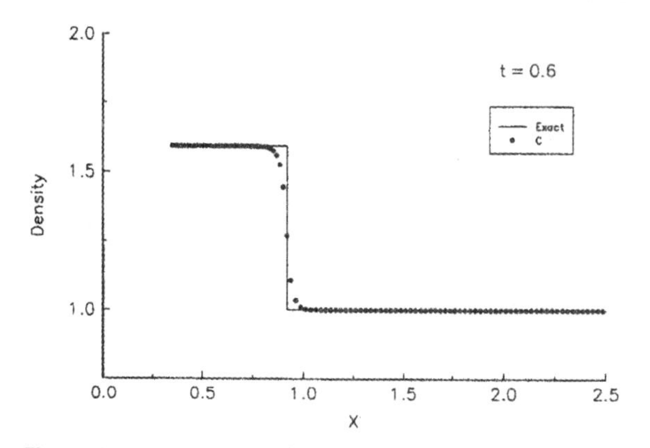

Figure 14. Density distribution of example 3: $t = 0.6$.

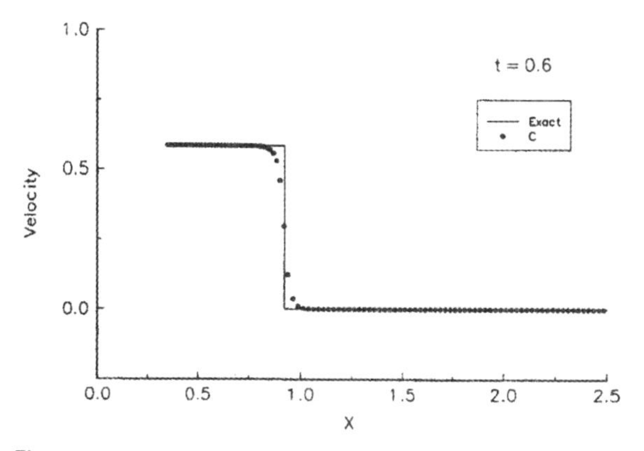

Figure 15. Velocity distribution of example 3: $t = 0.6$.

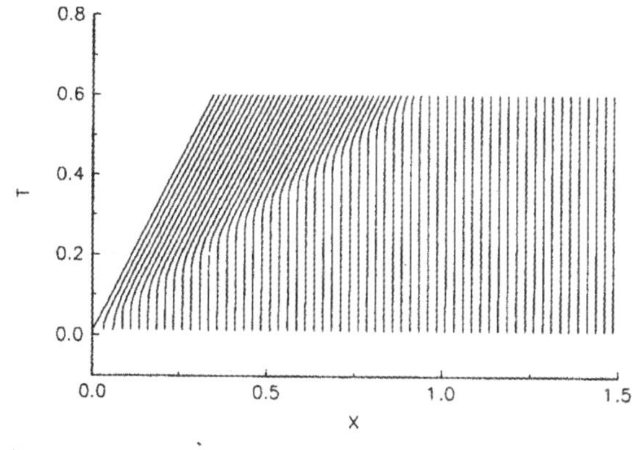

Figure 16. Particle path lines of example 3.

where $c_0 = \sqrt{\gamma}$ is stagnation speed of sound. The nonconservative formulation cannot give an acceptable solution for this piston-driven shock wave problem. Therefore, it is necessary to use the conservative formulation to deal with a compression wave problem, especially when the compression is so strong that a discontinuity, or a shock wave, is about to appear.

6. Concluding remarks

For a 1-D unsteady isentropic flow, the governing equations in streamfunction coordinates can be formulated as a conservative system of two equations for two unknowns, x_ψ and x_τ, as functions of ψ and τ.

The derived system can be solved numerically using Godunov's explicit first-order, two-step, predictor–corrector scheme under relevant initial and boundary conditions. Density disturbance propagation, piston-driven compression wave, and piston-driven shock wave problems have been successfully simulated using the conservative formulation, and the calculated results are excellent compared with the exact solutions.

The proposed conservative formulation is more accurate and powerful than the nonconservative one in dealing with compression flow problems, expecially when the compression is so strong that a discontinuity, or a shock wave, is about to appear.

However, when the present method is applied to solve a shock tube problem, the contact discontinuity cannot be simulated. This is because the isentropic assumption is not valid and entropy increase across a strong shock cannot be ignored. A further investigation is under way to extend the present method to treat shock tube problems using the conservative Euler equations in streamfunction coordinates.

Nomenclature

A	Jacobian of flux vector
CFL	Courant–Friedrichs–Lewy number
$\vec{F}$	flux vector
f	initial condition in computational plane (33)
f_1, f_2	components of $\vec{F}$
g	initial condition in physical plane (31)
J	Jacobian of von Mises transformation
p	pressure
Q	vector of unknown variables
q_1, q_2	components of $\vec{Q}$
t	time
U_p	piston speed in example 3
u	velocity
X	piston location
X_s	shock wave location
x	cartesian coordinate
γ	ratio of specific heats
ρ	density
σ	eigenvalue of A
τ	time in computational plane
ψ	streamfunction
$\Delta\tau, \Delta\psi$	increments of τ and ψ, respectively
Superscript $(\tilde{\ })$	dimensional quantities
Subscript $(.)_0$	stagnation quantities

References

1. Srinivasan, K., and Spalding, D. B. The stream function coordinate system for solution of one-dimensional unsteady compressible flow problems. *Appl. Math. Modelling* 1986, **10**, 278–283

第二部分 出国后博士学位论文阶段

2. An, C-F. Transonic computation using Euler equations in stream function coordinate system. Ph.D. Thesis, University of Windsor, 1992
3. Barron, R. M., An, C-F., and Zhang, S. Survey of the streamfunction-as-a-coordinate method in CFD. *Proceedings of the Inaugural Conference of the CFD Society of Canada*, Montreal, Canada, 1993
4. An, C-F., Barron, R. M., and Zhang, S. Streamfunction coordinate formulation for 1-dimensional unsteady flow. *Math. Models and Methods in Appl. Sci.* 1994, in press
5. Courant, R., and Hilbert, D. *Methods of Mathematical Physics*, Vol. II. Interscience Publishers, New York, 1962, Chapter 2
6. Hoffmann, K. A. *Computational Fluid Dynamics for Engineers*. Eng. Education System, Wichita, KS, USA, 1989, Chapter 6
7. Sod, G. A. A survey of several finite difference methods for system of nonlinear hyperbolic conservation laws. *J. Comp. Phys.* 1978, **27**, 1–31
8. Shapiro, A. H. *The Dynamics and Thermodynamics of Compressible Fluid Flow*, Vol. II. The Ronald Press Company, New York, 1953, Chapter 24
9. Landau, L. D. and Lifshitz, E. M. *Fluid Mechanics*, translated from Russian by J. B. Sykes and W. H. Reid. Pergamon Press, London, 1959, Chapter 10

INTERNATIONAL JOURNAL FOR NUMERICAL METHODS IN FLUIDS, VOL. 20, 75–94 (1995)

TRANSONIC EULER COMPUTATION IN STREAMFUNCTION CO-ORDINATES

C.-F. AN AND R. M. BARRON*

Department of Mathematics and Statistics and Fluid Dynamics Research Institute, University of Windsor, Windsor, Ontario Canada, N9B 3P4

SUMMARY

A new approach has been developed to calculate two-dimensional steady transonic flows past aerofoils using the Euler equations in streamfunction co-ordinates. Most existing transonic computation codes require the use of a grid generator to determine a suitable distribution of grid points. Although simple in concept, the grid generation may take a considerable proportion of the CPU time and storage requirements. However, this grid generation step can be avoided by introducing the von Mises transformation, which produces a formulation in streamwise and natural body-fitting co-ordinates. In this work a set of Euler equivalent equations in streamfunction co-ordinates is formulated, consisting of three equations with three unknowns; one is a geometric variable, the streamline ordinate y, and the other two are physical quantities, the density ρ and the vorticity ω. To solve these equations, type-dependent differencing, development of a shock point operator, marching from a non-characteristic boundary and successive line overrelaxation are applied. Particular attention has been paid to the supercritical case where a careful treatment of the shock is essential. It is shown that the shock point operator is crucial to accurately capture shock waves. The computed results show excellent agreement with existing experimental data and other computations.

KEY WORDS Euler equations Streamfunction co-ordinates

INTRODUCTION

Transonic flow is a widely encountered phenomenon in aeronautics and astronautics, occurring in flows past aerofoils, wings, through nozzle throats, cascade blades or around blunt bodies, etc. Transonic flow is more difficult to solve compared with pure subsonic or supersonic flows, because the flow fields have mixed zones and shock waves. Owing to these difficulties, there was little progress in transonic computations until the early 1970s. Since then the numerical simulation of transonic flow has been an active research topic for computational fluid dynamicists working in applied mathematics and aeronautical and aerospace engineering. The earliest efforts on transonic computation used the transonic small-disturbance (TSD) equation[1–4] and subsequently methods were developed for the full potential equations.[5–9] During the last decade attention has turned to the most accurate model for inviscid transonic computation, the Euler equations. Typical approaches include the implicit finite difference scheme,[10] the implicit approximate factorization method,[11] the finite volume scheme with explicit Runge–Kutta time stepping,[12] the flux-vector-splitting method,[13] the multigrid scheme,[14] the total-variation-diminishing scheme[15] and the finite element method.[16]

As an alternative to the above approaches, several researchers have replaced the primitive variable formulation of the Euler equations with a streamfunction–vorticity formulation and have successfully computed transonic flows.[17–22]

* Author to whom correspondence should be addressed.

CCC 0271–2091/95/010075–20

Received 28 August 1992
Revised 31 August 1993

第二部分 出国后博士学位论文阶段

In most CFD applications grid generation is a necessary first step in order to provide a body-fitting mesh system and this process may exhaust a considerable portion of CPU time. Hence the degree to which a numerical method can reduce this portion of time is an important index of its efficiency and applicability. Conventional numerical grid generation can be completely avoided by introducing the von Mises transformation and the corresponding streamfunction co-ordinates (SFCs). The von Mises transformation is a streamline-based co-ordinate transformation which analytically produces a body-fitting co-ordinate system. The transformation allows a single set of equations to play a double role, i.e. simultaneously serving as governing equations (flow physics) and grid generation equations (flow geometry). Therefore in recent years streamfunction (or streamline) co-ordinates have been exploited for the computation of 2D and axisymmetric incompressible potential flows,[23–28] incompressible viscous flows[29–31] and compressible potential flows.[32–40] Similar ideas have been used in turbomachinery analysis and design.[41–43]

In this study a technique is developed to calculate two-dimensional steady transonic flows past aerofoils using the Euler equations in streamfunction co-ordinates. Introducing the streamfunction and the von Mises transformation, a set of Euler equivalent equations in streamfunction co-ordinates is formulated. It consists of three coupled equations with three unknowns; one is a geometrical variable, the streamline ordinate y, and the other two are physical quantities, the density ρ and the vorticity ω. To solve the "main equation' for y, which is a second-order partial differential equation, a type-dependent difference scheme is applied. To treat the embedded shock wave, the shock jump conditions are analysed and a shock point operator is constructed in streamfunction co-ordinates. In order to solve for the density ρ, researchers traditionally use the Bernoulli equation. In the transonic range, however, the classical double-density problem exists in the new streamfunction co-ordinate formulation. Even if the artificial density technique is applied, in conjunction with the use of upwind differencing in supersonic regions, the supersonic pocket and shock waves are still difficult to handle. This is perhaps because there is no obvious mechanism by which the artificial density can provide dissipation to the y-equation, in the sense explained by Jameson[6] or Hafez *et al.*[7] for the potential equation and by Habashi and Hafez[17] or Hafez and Lovell[18] for the streamfunction equation. To overcome this difficulty, instead of solving the algebraic Bernoulli equation, a first-order partial differential equation called the 'secondary equation' is solved to avoid the double-density problem. Once y and ρ are obtained, ω can be easily calculated.

In Section 2 the Euler equivalent equations are formulated in streamfunction co-ordinates. In Section 3 the related numerical methodologies are discussed. Particular attention has been paid to the shock wave treatment, including the analysis of the shock jump conditions and the construction of the shock point operator (SPO). In Section 4 sample computations are conducted and the calculated results are compared with available experimental data and other computations. In the last section brief conclusions are given and the advantages and limitations of the present approach are discussed.

MATHEMATICAL FORMULATION

For a two-dimensional, steady, inviscid flow around an aerofoil the most accurate mathematical model is the Euler equations

$$\begin{pmatrix} \rho u \\ \rho u^2 + p \\ \rho uv \\ \rho uH \end{pmatrix}_x + \begin{pmatrix} \rho v \\ \rho uv \\ \rho v^2 + p \\ \rho vH \end{pmatrix}_y = 0, \qquad (1)$$

where $H = [\gamma/(\gamma - 1)]p/\rho + (u^2 + v^2)/2$ is the total enthalpy per unit mass, ρ is the density, u and v are the velocity components in Cartesian co-ordinates (x, y), p is the pressure and γ is the ratio of specific heats. The dependent variables ρ, u, v and p have been normalized by the quantities at freestream condition: density ρ_∞, speed V_∞ and dynamic pressure head $\rho_\infty V_\infty^2$. The independent variables x and y have been scaled by the aerofoil chord length.

Introducing the streamfunction ψ such that

$$\psi_y = \rho u, \qquad \psi_x = -\rho v, \tag{2}$$

the continuity equation in (1) is automatically satisfied. The explicit form of the streamfunction $\psi = \psi(x, y)$ can be considered in an implicit form $F(x, y; \psi) = 0$ or in an alternative explicit form as $y = y(x, \psi)$. This process is equivalent to the introduction of the von Mises transformation[44]

$$x \equiv x, \qquad y = y(x, \psi). \tag{3}$$

If the Jacobian $J = \partial(x, y)/\partial(x, \psi) = y_\psi \neq 0, \infty$, then the transformation (3) is one-to-one and the differential operators are transformed to

$$\left.\frac{\partial}{\partial x}\right|_y = \left.\frac{\partial}{\partial x}\right|_\psi - \frac{y_x}{y_\psi}\frac{\partial}{\partial \psi}, \qquad \frac{\partial}{\partial y} = \frac{1}{y_\psi}\frac{\partial}{\partial \psi}. \tag{4}$$

Therefore the Euler equations in streamfunction co- ordinates become

$$\begin{pmatrix} 1/\rho y_\psi + p y_\psi \\ y_x/\rho y_\psi \\ H \end{pmatrix}_x + \begin{pmatrix} -p y_x \\ p \\ 0 \end{pmatrix}_\psi = 0, \tag{5}$$

where

$$H = \frac{\gamma}{\gamma - 1}\frac{p}{\rho} + \frac{1 + y_x^2}{2\rho^2 y_\psi^2}.$$

The co-ordinates (x, ψ) are referred to as the streamfunction co-ordinates (SFCs)[34] and the Euler equations in streamfunction co-ordinates, (5), are completely equivalent to the Euler equations in Cartesian co-ordinates, (1), as long as the von Mises transformation (3) is valid in the problem under consideration. In this new formulation there are three dependent variables: streamline ordinate y, density ρ and pressure p; as unknown functions of two independent variables: abscissa x and streamfunction ψ.

The last equation in (5), $H_x = 0$, means that the total enthalpy H is invariant along a streamline. However, for a flow with uniform freestream H is invariant along any line and hence H is a constant throughout the flow field. This is the so-called homoenergetic condition which is satisfied in most practical problems. The constant can be evaluated at freestream condition as

$$H = H_\infty = \tfrac{1}{2} + \frac{1}{(\gamma - 1)M_\infty^2}. \tag{6}$$

Thus equations (5) can be rewritten as

$$\left[\frac{1}{\rho y_\psi} + p y_\psi\right]_x - [p y_x]_\psi = 0, \tag{7}$$

$$\left[\frac{y_x}{\rho y_\psi}\right]_x + p_\psi = 0, \tag{8}$$

$$\frac{\gamma}{\gamma - 1}\frac{p}{\rho} = H_\infty - \frac{1 + y_x^2}{2\rho^2 y_\psi^2}. \tag{9}$$

Here the energy equation has been reduced to an algebraic equation for p, ρ and derivatives of y owing to the homoenergetic condition. The velocity components can be calculated from

$$u = \frac{1}{\rho y_\psi}, \qquad v = \frac{y_x}{\rho y_\psi} \tag{10}$$

in streamfunction co-ordinates.

Differentiating (9) with respect to x and ψ and substituting p_x and p_ψ into (7) and (8) yields

$$\begin{aligned} 2y_x y_\psi^2 y_{xx} &+ 2y_\psi\left(\frac{1}{\gamma-1} - 2y_x^2\right)y_{x\psi} + 2y_x(1+y_x^2)y_{\psi\psi} \\ &= y_\psi^2\left(2H_\infty \rho^2 y_\psi^2 - \frac{\gamma+1}{\gamma-1} + y_x^2\right)\frac{\rho_x}{\rho} - y_x y_\psi(2H_\infty\rho^2 y_\psi^2 + 1 + y_x^2)\frac{\rho_\psi}{\rho}, \end{aligned} \tag{11}$$

$$\frac{2\gamma}{\gamma-1}y_\psi^2 y_{xx} - \frac{2(2\gamma-1)}{\gamma-1}y_x y_\psi y_{x\psi} + 2(1+y_x^2)y_{\psi\psi} = \frac{2\gamma}{\gamma-1}y_x y_\psi^2\frac{\rho_x}{\rho} - y_\psi(2H_\infty\rho^2 y_\psi^2 + 1 + y_x^2)\frac{\rho_\psi}{\rho}. \tag{12}$$

Then, using the definition of vorticity ($\omega = v_x - u_y$) in streamfunction co-ordinates,[36]

$$\omega = \frac{1}{y_\psi}\left(\left[\frac{y_x}{\rho}\right]_x - \left[\frac{1+y_x^2}{\rho y_\psi}\right]_\psi\right), \tag{13}$$

and eliminating ρ_x/ρ and ρ_ψ/ρ using (11) and (12), we get

$$(y_\psi^2 - Z_1)y_{xx} - 2y_x y_\psi y_{x\psi} + (1+y_x^2)y_{\psi\psi} = Z_2, \tag{14}$$

where

$$Z_1 = \frac{[2/(\gamma-1)]y_\psi^2}{2H_\infty\rho^2 y_\psi^2 - (1+y_x^2)}, \qquad Z_2 = \rho\omega y_\psi^3\frac{2H_\infty\rho^2 y_\psi^2 + 1 + y_x^2}{2H_\infty\rho^2 y_\psi^2 - (1+y_x^2)}$$

are the terms representing compressibility and rotational effects respectively. Equation (14), which can be solved for y if ρ and ω are known, is a second-order non-linear non-homogeneous partial differential equation. To classify this equation, one can show that its discriminant is

$$\Delta = 4y_\psi^2(M^2 - 1), \tag{15}$$

where M is the local Mach number. Thus we observe that if the local flow is supersonic (or subsonic), then the governing equation must be hyperbolic (or elliptic) and vice versa. Therefore the mathematical classification of the governing equation in streamfunction co-ordinates is consistent with the physical nature of the local flow. This feature provides the possibility of applying the type-dependent difference scheme originally proposed by Murman and Cole[1] to numerically solve equation (14) for y.

It is obvious that equation (14) for y is coupled with ρ and ω through Z_1 and Z_2. Therefore, to solve equation (14) for y iteratively, ρ and ω must be updated from iteration to iteration. The density can be updated from a first-order non-linear partial differential equation obtained by eliminating the term $y_\psi^2 y_{xx} - 2y_x y_\psi y_{x\psi} + (1+y_x^2)y_{\psi\psi}$ from (13) and (14),

$$y_x y_\psi^2 \rho_x - y_\psi(1+y_x^2)\rho_\psi = (Z_1 y_{xx} + Z_3)\rho, \tag{16}$$

where

$$Z_3 = \frac{2\rho\omega y_\psi^3(1+y_x^2)}{2H_\infty\rho^2 y_\psi^2 - (1+y_x^2)}$$

is the rotational term. The slope of the characteristic curve for this equation is

$$\frac{d\psi}{dx} = -\frac{1+y_x^2}{y_x\, y_\psi}. \tag{17}$$

At infinity $y_x \to 0$, $y_\psi \to 1$ and hence $d\psi/dx \to \infty$. Therefore the characteristic curve of (16) at infinity is a vertical line in the (x, ψ)-plane.

After y is solved from (14) and ρ is updated from (16), the vorticity ω can be updated from its definition (13). These equations constitute a complete set of 'Euler equivalent equations' in streamfunction co-ordinates. For convenience equation (14) is referred to as the 'main equation' for the corresponding 'main variable' y and the other equations are referred to as the "secondary equations' for the related "secondary variables' ρ and ω. Having obtained y, ρ and ω, the local Mach number can be obtained from

$$M^2 = \frac{[2/(\gamma-1)](1+y_x^2)}{2H_\infty\rho^2 y_\psi^2 - (1+y_x^2)} \tag{18}$$

and the pressure coefficient from

$$C_p = 2\left(p - \frac{1}{\gamma M_\infty^2}\right), \tag{19}$$

where

$$p = \frac{\gamma-1}{\gamma}\left(H_\infty\rho - \frac{1+y_x^2}{2\rho y_\psi^2}\right).$$

NUMERICAL METHODOLOGIES

Suppose an aerofoil is placed in a two-dimensional air flow with freestream Mach number M_∞ at an angle of attack α. From the previous section the governing Euler equivalent equations in streamfunction co-ordinates are composed of (14), (16) and (13)

$$(y_\psi^2 - Z_1)y_{xx} - 2y_x y_\psi y_{x\psi} + (1+y_x^2)y_{\psi\psi} = Z_2, \tag{20}$$

$$y_x y_\psi^2 \rho_x - y_\psi(1+y_x^2)\rho_\psi = (Z_1\, y_{xx} + Z_3)\rho, \tag{21}$$

$$\omega = \frac{1}{y_\psi}\left(\left[\frac{y_x}{\rho}\right]_x - \left[\frac{1+y_x^2}{\rho y_\psi}\right]_\psi\right). \tag{22}$$

On the aerofoil the boundary condition for y is Dirichlet,

$$y = f_\pm(x), \tag{23}$$

where $f_+(x)$ and $f_-(x)$ represent the shape functions of the upper and lower surfaces of the aerofoil respectively. In the far field the streamfunction can be expressed by the sum of a uniform flow, a doublet and a vortex.[20] In most cases the doublet term is sufficiently small and can be ignored.

第二部分 出国后博士学位论文阶段

Therefore the boundary condition at infinity is given in an explicit form for ψ,

$$\psi(x\ y) = y\cos\alpha - x\sin\alpha + \frac{\Gamma}{2\pi}\ln[x^2 + (1 - M_\infty^2)y^2],$$

in the physical domain or in an implicit form for y,

$$y(x,\ \psi) = \frac{\psi}{\cos\alpha} + x\tan\alpha - \frac{\Gamma}{2\pi\cos\alpha}\ln\{x^2 + (1 - M_\infty^2)[y(x,\ \psi)]^2\}, \tag{24}$$

in the computational domain. This algebraic equation for $y = y(x, \psi)$ is non-linear and an iteration algorithm (e.g. Newton's iteration) must be applied. In addition, the Kutta condition must be satisfied, i.e. the pressures

$$p_{\mathrm{TE}} = \frac{\gamma - 1}{\gamma}\left(H_\infty\rho - \frac{1 + y_x^2}{2\rho y_\psi^2}\right)_{\mathrm{TE}} \tag{25}$$

calculated from the upper and lower surfaces at the trailing edge must be equal to each other. Sketches of the physical and computational domains and the boundary conditions are shown in Figures 1(a) and 1(b) respectively.

Type-dependent scheme for the main equation

Since the main equation (20) is well classified as hyperbolic- or elliptic-type depending on whether the local flow is supersonic or subsonic, it is possible to apply the type-dependent scheme to solve for y. Equation (20) can be rewritten as

$$A_1 y_{xx} + A_2 y_{x\psi} + A_3 y_{\psi\psi} = A_4, \tag{26}$$

where

$$A_1 = y_\psi^2 - Z_1, \qquad A_2 = -2y_x y_\psi, \qquad A_3 = 1 + y_x^2, \qquad A_4 = Z_2.$$

The type-dependent scheme reads

$$\{A_1[\nu\Delta_x + (1-\nu)\nabla_x]\nabla_x + A_2[\nu\delta_x + (1-\nu)\nabla_x]\delta_\psi + A_3\delta_{\psi\psi}\}y_{i,j} = A_4, \tag{27}$$

where Δ, ∇ and δ are forward, backward and central difference quotient operators respectively, and the switch parameter

$$\nu = \begin{cases} 1 & \text{for subsonic points,} \\ 0 & \text{for supersonic points.} \end{cases} \tag{28}$$

In this formulation, upwinding in the x-direction at supersonic points has the effect of a rotated difference scheme, since the backward differencing is actually in the direction of the streamline.

Expanding the type-dependent scheme (27) and rearranging the terms to express it in a tridiagonal coefficient matrix form, one gets

$$Ay_{i,j-1} + By_{i,j} + Cy_{i,j+1} = \mathrm{RHS}, \tag{29}$$

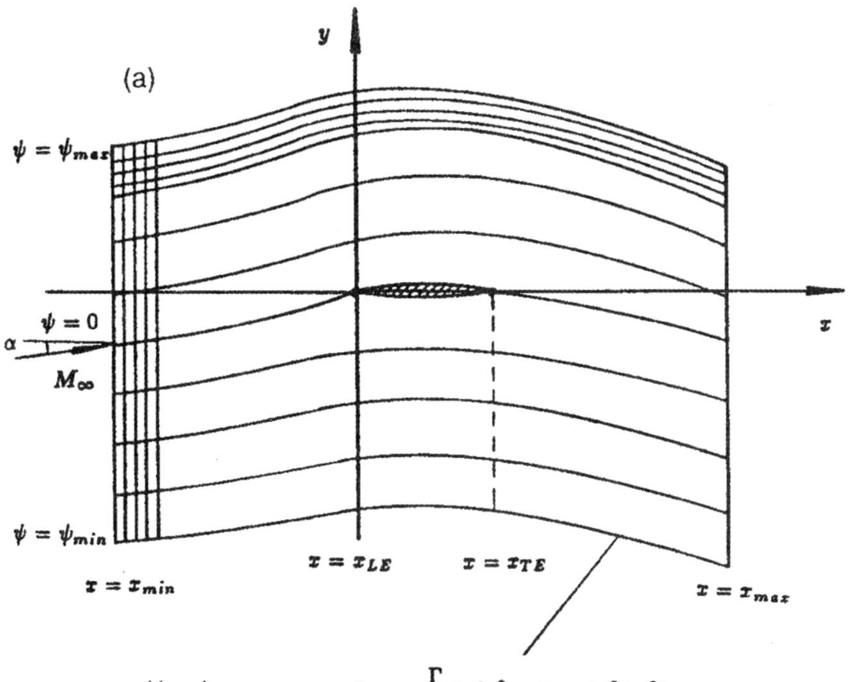

$$\psi(x,y) = y\cos\alpha - x\sin\alpha + \frac{\Gamma}{2\pi}\ln\{x^2 + (1 - M_\infty^2)y^2\}$$

Figure 1(a). Physical plane

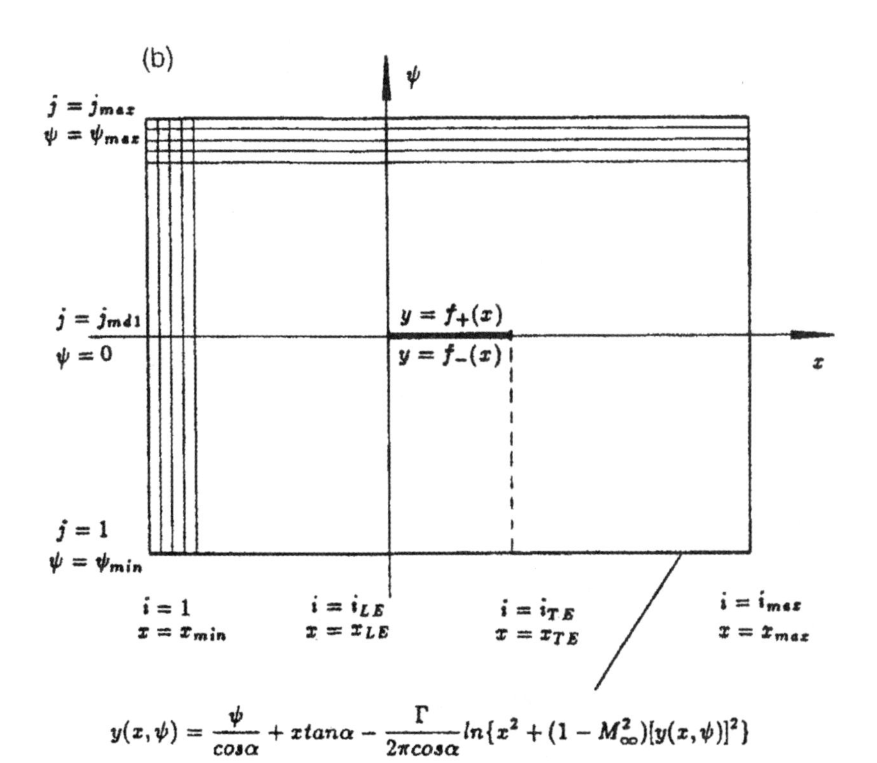

$$y(x,\psi) = \frac{\psi}{\cos\alpha} + x\tan\alpha - \frac{\Gamma}{2\pi\cos\alpha}\ln\{x^2 + (1 - M_\infty^2)[y(x,\psi)]^2\}$$

Figure 1(b). Computational plane

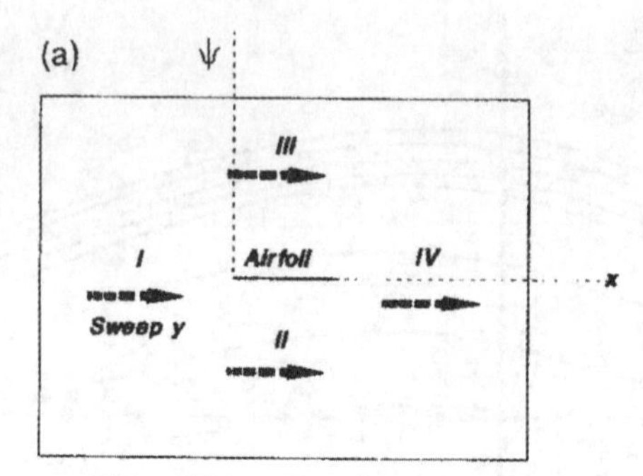

Figure 2(a). y-Equation sweeping

where

$$A = \beta^2 A_3 - (1-\nu)\beta A_2/2, \qquad B = -2\beta^2 A_3 + (1-3\nu)A_1, \qquad C = \beta^2 A_3 + (1-\nu)\beta A_2/2,$$

$$\begin{aligned} \text{RHS} = & -\nu A_1(y_{i+1,j} + y_{i-1,j}) + (1-\nu)A_1(2y_{i-1,j} - y_{i-2,j}) \\ & - \nu\beta A_2(y_{i+1,j+1} - y_{i+1,j-1} - y_{i-1,j+1} + y_{i-1,j-1})/4 \\ & + (1-\nu)\beta A_2(y_{i-1,j+1} - y_{i-1,j-1})/2 + \Delta x^2 A_4 \end{aligned}$$

for $i = 2, 3, \ldots, I_{\max-1}, j = 2, 3, \ldots, J_{\max-1}$ and $\beta = \Delta x/\Delta\psi$.

The computational domain is divided into four subdomains as shown in Figure 2(a) and each subdomain is swept sequentially by SLOR from left to right. The whole process should be iterated up to convergence owing to the non-linearity of the equation.

Marching the secondary equation

Since the secondary equation (21) has vertical characteristics in the far field, it can be solved by marching line-by-line from the horizontal far-field boundary, which is a non-characteristic curve at which ρ=1. In our case equation (21) can be marched to the aerofoil from lower and upper boundaries for lower and upper half-planes respectively (Figure 2(b)). Equation (21) can be rewritten as

$$B_1\rho_x + B_2\rho_\psi + B_3\rho = 0, \tag{30}$$

where

$$B_1 = y_x y_\psi^2, \qquad B_2 = -y_\psi(1+y_x^2), \qquad B_3 = -(Z_1 y_{xx} + Z_3).$$

Applying the Crank–Nicholson scheme to (30) at point $(i, j-\frac{1}{2})$ for the lower half-plane, evaluating the density ρ at level $j-\frac{1}{2}$ by the average at levels j and $j-1$ and rearranging the equations in a

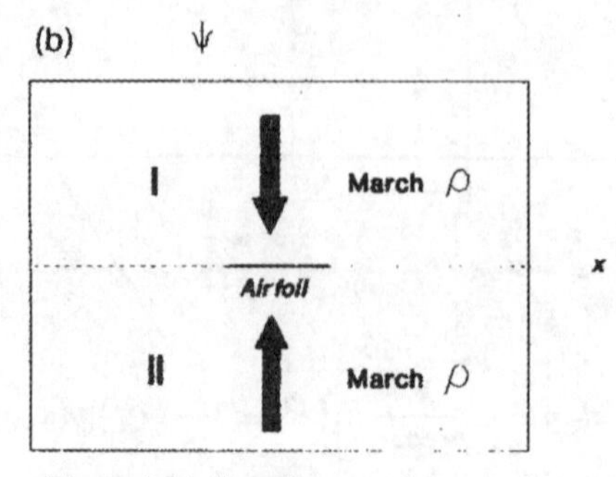

Figure 2(b). Density equation marching

tridiagonal form, one gets

$$\tilde{A}\rho_{i-1,j} + \tilde{B}\rho_{i,j} + \tilde{C}\rho_{i+1,j} = \widetilde{\mathrm{RHS}}, \tag{31}$$

where

$$\begin{aligned} \tilde{A} &= -B_1, \qquad \tilde{B} = 4\beta B_2 + 2\Delta x B_3, \qquad \tilde{C} = B_1, \\ \widetilde{\mathrm{RHS}} &= B_1\rho_{i-1,j-1} + (4\beta B_2 - 2\Delta x B_3)\rho_{i,j-1} - B_1\rho_{i+1,j-1} \end{aligned} \tag{32}$$

for $i = 2, 3, \ldots, I_{\max-1}$ and $j = 2, 3, \ldots, J_{\mathrm{md1}}$. Here J_{md1} represents the streamline coinciding with the aerofoil surface. It should be noted that B_1, B_2 and B_3 are taken as the averages of the corresponding quantities at j and $j-1$. A similar expression holds for the upper half-plane. Along the zero streamline upstream and downstream of the aerofoil the density ρ should be evaluated by averaging the values from upper and lower half-planes after each iteration. At convergence these values will be the same.

The system of difference equations (31) has a tridiagonal coefficient matrix and can be solved line-by-line horizontally using SLOR from the far- field boundaries to the aerofoil. An iterative procedure is used, because (30) is non-linear. It should also be pointed out that the second derivative y_{xx} in B_3 should be type-dependently differenced to keep consistency with the y-equation. After ρ is solved, the vorticity can be updated from (22) and the local Mach number can be calculated from (18). Finally, the Kutta condition requires that the pressures at the trailing edge calculated from the upper surface, p_{TE}^{+} and from the lower surface, p_{TE}^{-} must be equal to each other. That is,

$$(\Delta p)_{\mathrm{TE}} = p_{\mathrm{TE}}^{+} - p_{\mathrm{TE}}^{-} = 0, \tag{33}$$

where the pressures $p_{\mathrm{TE}}^{\pm}$ are given by (25). If (23) is not satisfied, the circulation Γ around the aerofoil can be corrected from the expression

$$\Gamma^{(n+1)} = \Gamma^{(n)} + \beta_0(\Delta p)_{\mathrm{TE}}, \tag{34}$$

where the superscripts (n) and $(n+1)$ indicate the iteration levels and the relaxation parameter β_0 can be determined from numerical tests. The numerical solution process can be described as below.

First, the tridiagonal system of algebraic equations (29) is solved for y along a vertical line. Each vertical line is then successively relaxed from left to right in each subdomain and the subdomains are swept in the order I, II, III and IV. Each time after y is relaxed, the error between the current and previous iterations is checked for all grid points. If the error is less than the prescribed tolerance, the iterations are considered to be converged, otherwise the iterations are repeated until convergence. After y is converged, ρ can be solved from equation (31). The tridiagonal system of algebraic equations (31) is solved along a horizontal line and each horizontal line is marched upwards or downwards to the aerofoil. After ρ is converged, the Mach number M and vorticity ω are calculated and the Kutta condition (33) is checked. If it is not satisfied, Γ is updated using (34) and the procedure is repeated again up to convergence. The Mach number is used to distinguish the grid point type: subsonic, supersonic or shock wave. The computational flowchart is shown in Figure 3.

Shock jump condition

As might be expected, numerical tests indicated that the type-dependent scheme is effective only for subcritical flows and for supercritical flows with weak shock waves. For a supercritical flow with moderate or strong shock waves the computation either fails to converge or is forced to stop owing to inaccurate intermediate values of the unknowns y and ρ during the process of iteration. Occasionally the computation converges but gives inaccurate pressure distributions and incorrect shock wave

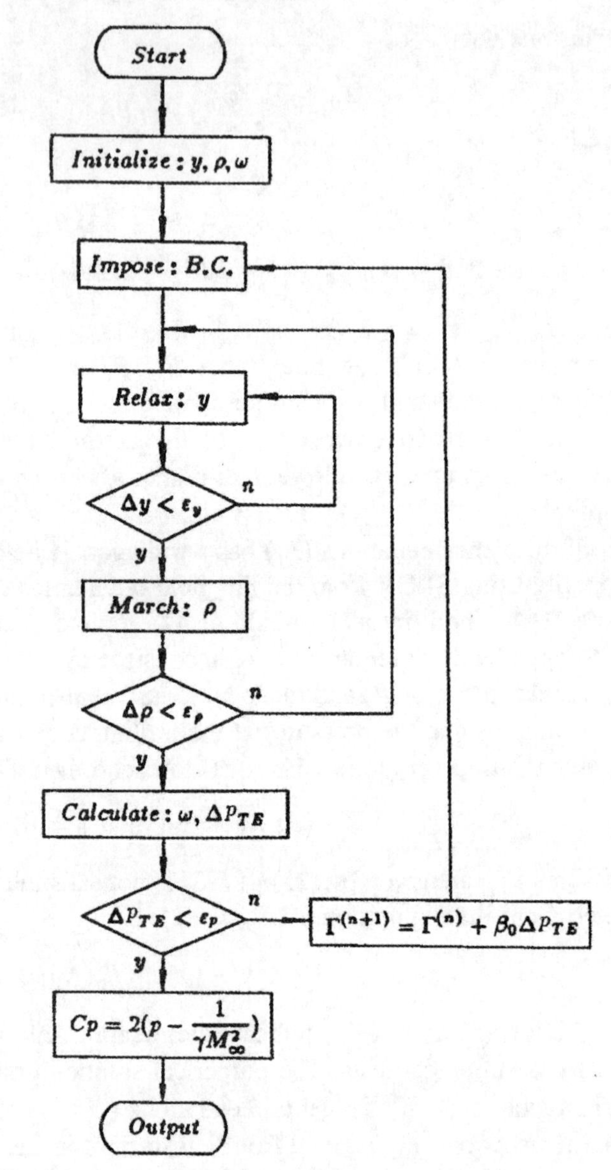

Figure 3. Computational flow chart

positions. This means that the shock waves are not being properly handled. Therefore a special treatment of shock waves is necessary. In the early work of Murman and Cole[1] the shock jump conditions are automatically incorporated in their scheme for the TSD equation. The sonic line and weak shock waves develop naturally during the course of iteration. No special shock wave treatment has been made in their computation. To improve this approach, Murman[2] proposed the concept of shock point operator (SPO) for the TSD equation and was able to achieve an improved solution. Although Murman's SPO cannot be applied here directly, the basic idea and analysis of the shock wave structure provide a useful hint for extension to more accurate models such as full potential or Euler equations.

Suppose an oblique shock wave makes an angle β with the x-axis. Let V and α be the velocity and its angle with the x-axis, u and v be the x- and y-velocity components and V_n and V_t be the normal and tangential velocity components to the shock. Superscripts '$\mp$' represent upstream and downstream of the shock (Figure 4(a)). The tangential shock relation $V_t^+ = V_t^-$ gives

$$V^+ \cos(\beta - \alpha^+) = V^- \cos(\beta - \alpha^-).$$

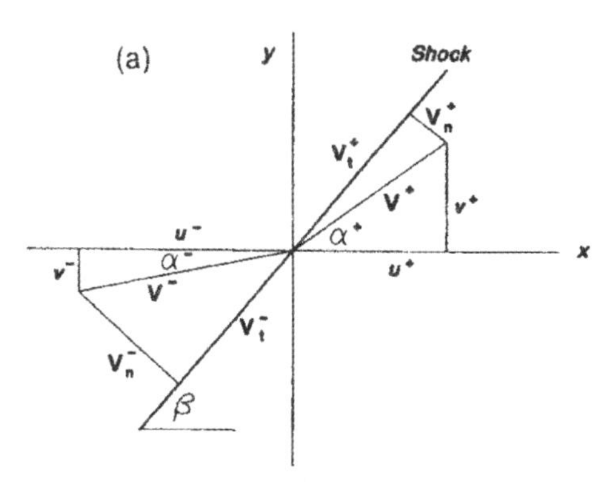

Figure 4(a). Shock jump condition

Expanding this equation and using the relations $u^{\pm} = V^{\pm} \cos \alpha^{\pm}$ and $v^{\pm} = V^{\pm} \sin \alpha^{\pm}$ yields

$$\frac{v^+ - v^-}{u^+ - u^-} = -K, \tag{35}$$

where $K = 1/\tan \beta$ is the reciprocal of the shock wave slope. Similarly the normal shock relation $\rho^+ V_n^+ = \rho^- V_n^-$ gives

$$\frac{\rho^+ u^+ - \rho^- u^-}{\rho^+ v^+ - \rho^- v^-} = K. \tag{36}$$

Equations (35) and (36) can be expressed in a compact form

$$[v] + K[u] = 0, \qquad [\rho u] - K[\rho v] = 0, \tag{37}$$

where [...] represents the jump of the corresponding quantities across the shock wave. Recalling that $\rho u y_\psi = 1$ and $v = y_x u$, the oblique shock jump conditions in streamfunction co-ordinates are

$$\left[\frac{y_x}{\rho y_\psi}\right] + K\left[\frac{1}{\rho y_\psi}\right] = 0, \qquad \left[\frac{1}{y_\psi}\right] - K\left[\frac{y_x}{y_\psi}\right] = 0. \tag{38}$$

For a shock wave which is perpendicular to the x-axis, $\beta = \pi/2$, $K = 0$ and the shock jump conditions reduce to

$$\left[\frac{y_x}{\rho}\right] = 0, \qquad [y_\psi] = 0. \tag{39}$$

These are the normal shock jump conditions in streamfunction co-ordinates.

Shock point operator

For moderate transonic Mach number the shock wave is approximately normal and can therefore be assumed to be an infinitely thin discontinuity surface located at point $(i-\frac{1}{2}, j)$ and perpendicular to the x-axis (Figure 4(b)). The shock jump conditions (39) can be written as

$$y_x^+ = \mu y_x^-, \qquad y_\psi^+ = y_\psi^-, \tag{40}$$

where

$$\mu = \frac{\rho^+}{\rho^-} = \frac{[(\gamma+1)/2](M^2)^-}{1 + [(\gamma-1)/2](M^2)^-}$$

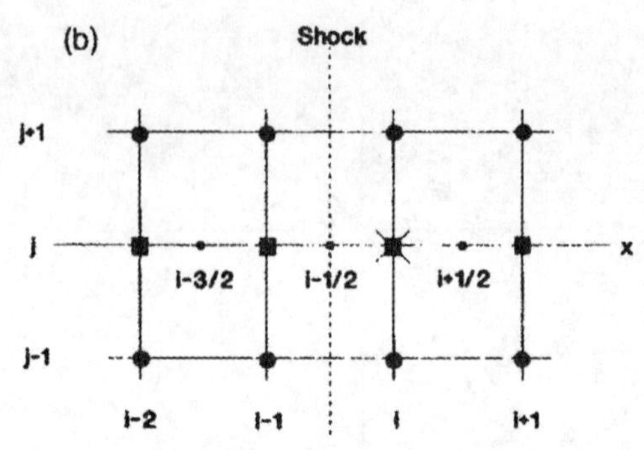

Figure 4(b). Shock point operator

is the density jump factor and can be evaluated from the Rankine–Hugoniot relation. The squared Mach number at point $(i-\frac{1}{2},\ j)^-$ can be evaluated by extrapolation from the two upstream grid points $(i-1,\ j)$ and $(i-2,\ j)$ as

$$(M^2)^-_{i-1/2,j} = \tfrac{3}{2}M^2_{i-1,j} - \tfrac{1}{2}M^2_{i-2,j}\,.$$

Thus the shock jump conditions (40) can be rewritten as

$$(y_x)^+_{i-1/2,j} = \mu_j (y_x)^-_{i-1/2,j}, \qquad (y_\psi)^+_{i-1/2,j} = (y_\psi)^-_{i-1/2,j}, \tag{41}$$

where

$$\mu_j = \frac{[(\gamma+1)/4](3M^2_{i-1,j} - M^2_{i-2,j})}{1 + [(\gamma-1)/4](3M^2_{i-1,j} - M^2_{i-2,j})}$$

is the density jump factor on the jth streamline. Finally we construct the difference approximation to y_{xx} at a shock point $i = i_s$, i.e. the grid point just behind the shock, so that for i_s we have

$$\begin{aligned}(y_{xx})_{i,j} &= \frac{1}{\Delta x}[(y_x)_{i+1/2,j} - (y_x)^+_{i-1/2,j}] \\ &= \frac{1}{\Delta x}[(y_x)_{i+1/2,j} - \mu_j (y_x)^-_{i-1/2,j}] \\ &= \frac{1}{\Delta x}\left(\frac{1}{\Delta x}(y_{i+1,j} - y_{i,j}) - \frac{\mu_j}{\Delta x}(y_{i-1,j} - y_{i-2,j})\right) \\ &= \frac{1}{\Delta x^2}(y_{i+1,j} - y_{i,j} - \mu_j y_{i-1,j} + \mu_j y_{i-2,j}).\end{aligned} \tag{42}$$

Similarly the cross-derivative can be approximated by

$$\begin{aligned}(y_{x\psi})_{i,j} &= \frac{1}{\Delta x}[(y_\psi)_{i+1/2,j} - (y_\psi)^+_{i-1/2,j}] \\ &= \frac{1}{\Delta x}[(y_\psi)_{i+1/2,j} - (y_\psi)^-_{i-1/2,j}] \\ &= \frac{1}{4\Delta x \Delta\psi}(y_{i+1,j+1} - y_{i+1,j-1} + y_{i,j+1} - y_{i,j-1} - 3y_{i-1,j+1} \\ &\qquad + 3y_{i-1,j-1} + y_{i-2,j+1} - y_{i-2,j-1}).\end{aligned} \tag{43}$$

Equations (42) and (43) define the shock point operator in streamfunction co-ordinates. Numerical tests for both full potential[40] and the present Euler calculations show that the first derivatives y_x, ρ_x,

Table I

$M^2_{i-1,j}$	$M^2_{i,j}$	Local flow type at (i,j)
<1	<1	Subsonic point
<1	>1	Sonic point
>1	>1	Supersonic point
>1	<1	Shock point

etc. do not need special treatment across shock waves, although they do have jumps across the shock waves.

Owing to the special treatment of the grid point at a shock wave, the type-dependent difference scheme (29) and the Crank–Nicolson scheme (31) must be revised. The system of difference equations for y is of the same form as (29), but the coefficients A, B, C and the RHS term have to be modified to

$$A=\beta^2A_3-\begin{cases}(1-\nu)\beta A_2/2 & \text{if } i\neq i_s,\\ \beta A_2/4 & \text{if } i=i_s.\end{cases}$$

$$B=-2\beta^2A_3+\begin{cases}(1-3\nu)A_1 & \text{if } i\neq i_s,\\ -A_1 & \text{if } i=i_s.\end{cases}$$

$$C=\beta^2A_3+\begin{cases}(1-\nu)\beta A_2/2 & \text{if } i\neq i_s,\\ \beta A_2/4 & \text{if } i=i_s.\end{cases}$$

$$\text{RHS}=\begin{cases}-\nu A_1(y_{i+1,j}+y_{i-1,j})+(1-\nu)A_1(2y_{i-1,j}-y_{i-2,j}) & \\ -\nu\beta A_2(y_{i+1,j+1}-y_{i+1,j-1}-y_{i-1,j+1}+y_{i-1,j-1})/4 & \\ \quad+(1-\nu)\beta A_2(y_{i-1,j+1}-y_{i-1,j-1})/2+\Delta x^2A_4 & \text{if } i\neq i_s,\\ -A_1(y_{i+1,j}-\mu_j y_{i-1,j}+\mu_j y_{i-2,j}) & \\ -\beta A_2(y_{i+1,j+1}-y_{i+1,j-1}-3y_{i-1,j+1} & \\ \quad+3y_{i-1,j-1}+y_{i-2,j+1}-y_{i-2,j-1})/4+\Delta x^2A_4 & \text{if } i=i_s,\end{cases}\tag{44}$$

where $\beta=\Delta x/\Delta\psi$, $i=2, 3, \ldots, I_{\max-1}$ and $j=2, 3, \ldots, J_{\max-1}$.

Similar changes occur in the equation for ρ, i.e. (31). In particular, the second derivative y_{xx} in the expression for B_3 should be approximated by the type-dependent difference with shock point operator.

In order to determine the type of local flow at grid point (i,j), the Mach number at two adjacent grid points is checked, i.e. the current point (i,j) and the immediately upstream point $(i-1,j)$. The criteria are given in Table I.

In computational practice the sonic points need not be distinguished, because there is no jump across the sonic line. However, the shock points must be identified carefully and this is a key step in supercritical transonic flow computation.

SAMPLE COMPUTATIONS

The Euler equivalent equations in streamfunction co-ordinates have been used to calculate transonic flows past aerofoils at subcritical and supercritical Mach numbers. The computations are executed for the set of equations (29) for y and (31) for ρ. The vorticity ω is calculated from (22) using central differences for all x- and ψ-derivatives. The iteration involves three levels of loops. The internal loop is for the y-iteration, the intermediate loop is for ρ and the external iteration loop is for ω and Γ. For

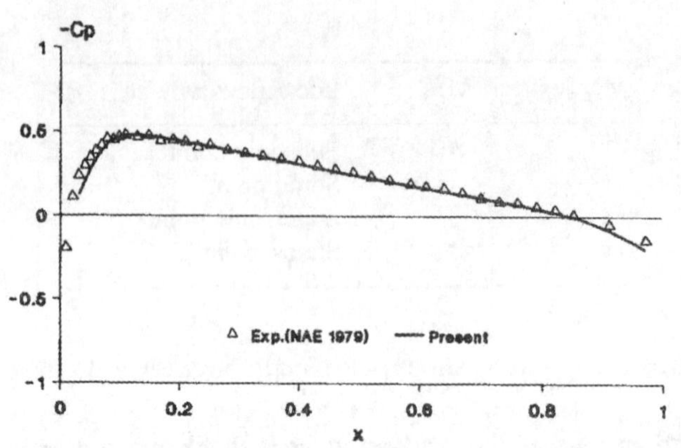

Figure 5. C_p comparison: NACA 0012, $\alpha = 0°$, $M_\infty = 0{\cdot}490$

subcritical and shock-free or weak shock supercritical flows, accurate solutions are obtained on a 65×65 uniform mesh in the (x, ψ)-domain, which is truncated as $-2 \leq x \leq 3$, $-2{\cdot}5 \leq \psi \leq 2{\cdot}5$ with the aerofoil located between 0 and 1. For flows with moderate or strong shock waves, clustering transformations[45] are required to place a sufficient number of grid points on the aerofoil to accurately predict the locations and strengths of the shocks.

Figures 5–10 show comparisons between the calculated results and experimental data or other computations. The experimental data are extracted from work at ONERA (France),[46] NAE (Canada)[47] and NASA (U.S.A).[48] Figure 5 is the comparison of the C_p-distribution for NACA 0012 for purely subsonic flow at Mach number $M_\infty = 0{\cdot}490$ at zero angle of attack. Figure 6 is for NACA 0012 at $M_\infty = 0{\cdot}503$ and $\alpha = 6{\cdot}05°$. Figure 7 gives the C_p-distribution for NACA 0012 at slightly supercritical Mach number $M_\infty = 0{\cdot}756$ and $\alpha = 0°$. These shock-free calculations, performed on a uniform grid, show excellent agreement between computed and experimental results. Figure 8 gives the results of a supercritical calculation on a clustered grid. This example shows that the computational method described here can be used to accurately capture shock waves. Figure 9 illustrates the C_p-distribution on a 6% biconvex aerofoil at $M_\infty = 0{\cdot}909$ and $\alpha = 0°$ and shows excellent agreement with experiments. Excellent agreement is also seen in Figure 10, where the present calculations carried out on a clustered grid are compared with earlier test case computations[49] for $M_\infty = 0{\cdot}8$ and $\alpha = 1{\cdot}25°$. Figures 11 and 12 show the Mach number and entropy contours respectively for the same test case of NACA 0012 at $M_\infty = 0{\cdot}8$ and $\alpha = 1{\cdot}25°$.

Figure 13 demonstrates the evolution of iterations and convergence process for a typical supercritical calculation on a clustered grid: NACA 0012, $M_\infty = 0{\cdot}803$ and $\alpha = 0°$. The C_p-distributions

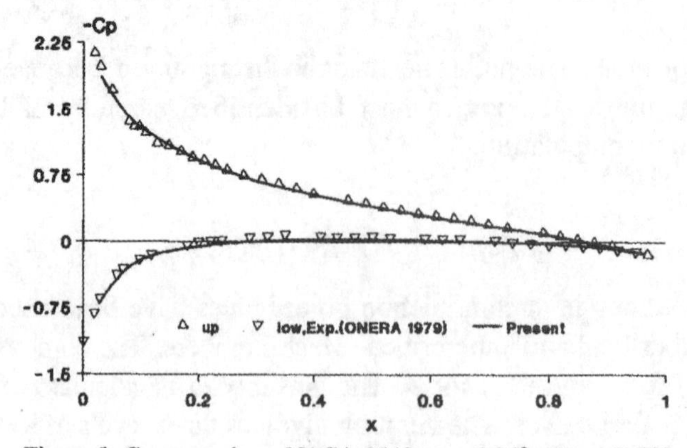

Figure 6. C_p comparison: NACA 0012, $\alpha = 6{\cdot}05°$, $M_\infty = 0{\cdot}503$

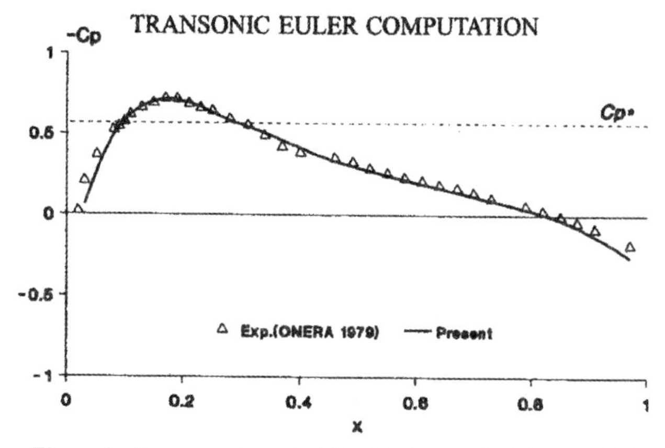

Figure 7. C_p comparison: NACA 0012, $\alpha = 0°$, $M_\infty = 0{\cdot}756$

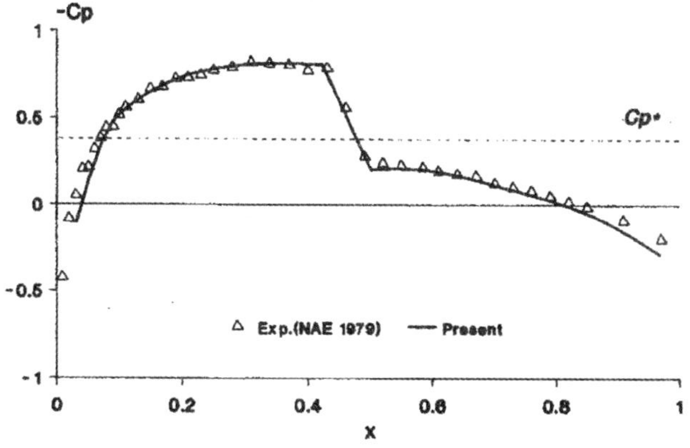

Figure 8. C_p comparison: NACA 0012, $\alpha = 0°$, $M_\infty = 0{\cdot}814$

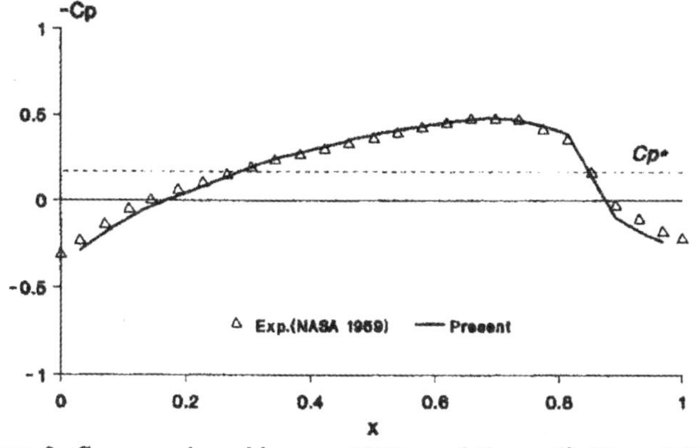

Figure 9. C_p comparison: biconvex (6%) aerofoil, $\alpha = 0°$, $M_\infty = 0{\cdot}909$

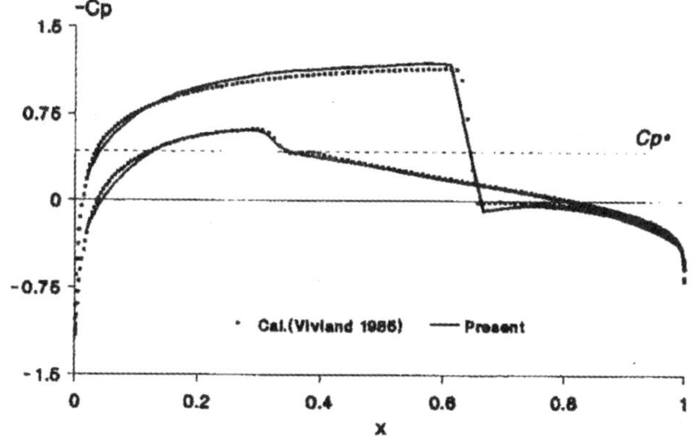

Figure 10. C_p comparison: NACA 0012, $\alpha = 1{\cdot}25°$, $M_\infty = 0{\cdot}800$

第II部分 出国后博士学位论文阶段

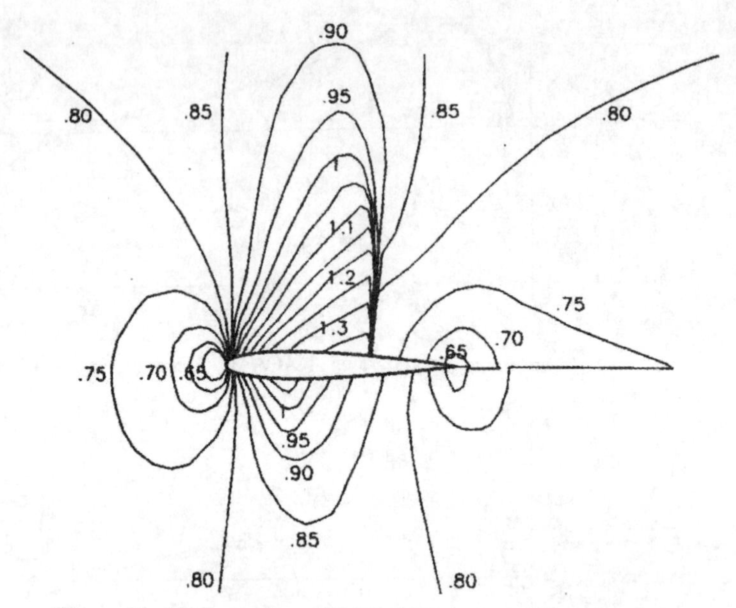

Figure 11. Mach contour: NACA 0012, $\alpha = 1{\cdot}25°$, $M_\infty = 0{\cdot}8$

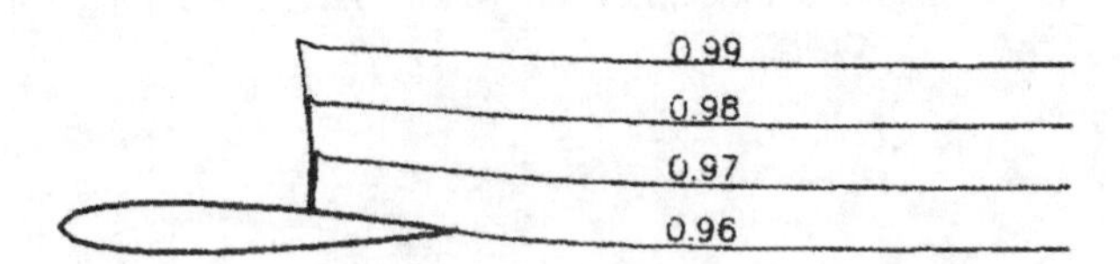

Figure 12. Total pressure contour: NACA 0012, $\alpha = 1{\cdot}25°$, $M_\infty = 0{\cdot}8$

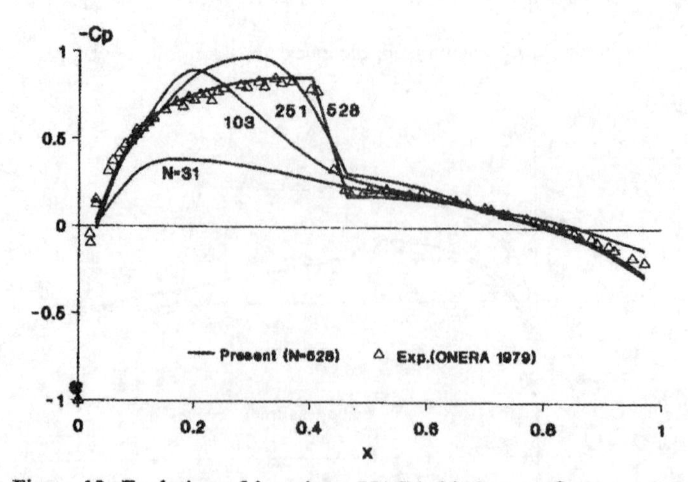

Figure 13. Evolution of iterations: NACA 0012, $\alpha = 0°$, $M_\infty = 0{\cdot}803$

安长发博士科研论文集

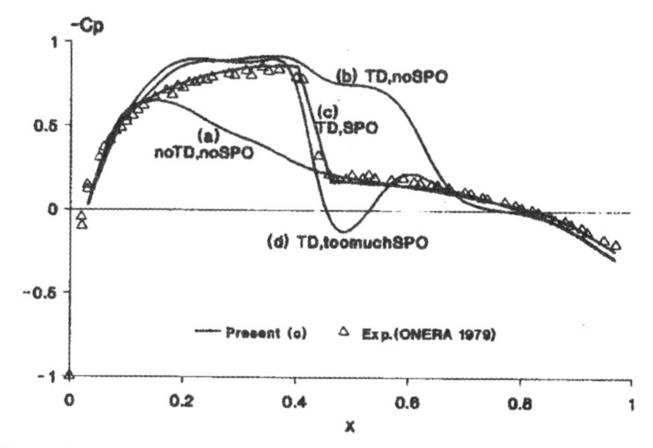

Figure 14. Effect of TD and SPO: NACA 0012, $\alpha = 0°$, $M_\infty = 0{\cdot}803$

on the aerofoil are plotted after the following numbers of iterations: 31, 103, 251 and 528. These plots show that before 251 iterations the calculated flow field has no shock wave. After 251 iterations a shock wave is formed and pushed backwards, with the accuracy improving as the iteration proceeds. The solution converges after 528 iterations and gives excellent results.

Figure 14 shows the effect of type-dependent (TD) differencing and shock point operator (SPO) in a typical supercritical calculation: NACA 0012, $M_\infty = 0{\cdot}803$ and $\alpha = 0°$. Curve (a) gives the result of central differencing with neither TD nor SPO. The C_p-distribution is totally unacceptable, completely missing the supercritical region and shock wave. Curve (b) shows the result of TD differencing only, without SPO. The C_p-distribution is inaccurate and the shock wave location is too far downstream. Curve (c) shows the result of TD differencing plus appropriate SPO, i.e. SPO is used in terms y_{xx} and $y_{x\psi}$ only. The calculation gives an accurate C_p-distribution and shock wave location. Curve (d) shows the result of TD differencing plus too much SPO, i.e. SPO is used not only in y_{xx} and $y_{x\psi}$ but also in terms y_x and ρ_x. The C_p-distribution is unacceptable again and a severe oscillation occurs in the shock wave region. Comparing these curves, one can conclude that the TD differencing plus the appropriate SPO is an effective scheme to calculate transonic flows and the SPO is a crucial tool to accurately capture the embedded shock waves.

CONCLUSIONS

The Euler equivalent equations in streamfunction co-ordinates consist of a main equation for the corresponding main (geometric) variable, the streamline ordinate y, and a secondary equation for the secondary (physical) variable, the density ρ, and an equation for the vorticity ω. These three equations are coupled together and must be solved simultaneously or iteratively.

The main equation for y is a second-order non-linear partial differential equation with Dirichlet boundary conditions. It is solved using type-dependent differencing plus a shock point operator. The shock point operator is a crucial numerical tool to accurately capture the embedded shock waves.

The secondary equation for ρ is a first-order partial differential equation and is solved by marching vertically from far-field boundaries to the aerofoil. The Crank–Nicolson scheme is effective in solving this equation.

The approach based on the Euler equivalent equations in streamfunction co- ordinates is able to simulate transonic flows past two-dimensional aerofoils. The embedded shock waves in supercritical flows can be accurately captured in the process of iteration. The calculated results compare favourably with existing experimental data and other computations.

APPENDIX: NOMENCLATURE

C_p	pressure coefficient
H	total enthalpy
J	Jacobian of a transformation
K	$1/\tan\beta$
M	local Mach number
p	pressure
u, v	velocity components in directions x and y respectively
V	speed
(x, y)	Cartesian co-ordinates
(x, ψ)	streamfunction co-ordinates

Greek letters

α	angle of attack or velocity angle with x- axis
β	$\Delta x/\Delta\psi$ or shock wave angle with x-axis
β_0	relaxation parameter for Γ
γ	ratio of specific heats
Γ	circulation
μ	density jump factor
ν	switch parameter
ρ	density
ψ	streamfunction
ω	vorticity

Subscripts

i, j	grid points
LE, TE	leading and trailing edges respectively
x, y, ψ	partial derivatives
∞	freestream value

安长发博士科研论文集

REFERENCES

1. E. M. Murman and J. D. Cole, 'Calculation of plane steady transonic flows', *AIAA J.*, **9**, 114–121 (1971).
2. E. M. Murman, 'Analysis of embedded shock waves calculated by relaxation methods', *AIAA J.*, **12**, 626–633 (1974).
3. S. T. K. Chan, M. R. Brashears and V. Y. C. Young, 'Finite element analysis of transonic flow by the method of weighted residuals', *AIAA Paper 75-79*, 1975.
4. M. Hafez and H. K. Cheng, 'Shock-fitting applied to relaxation solutions', *AIAA J.*, **15**, 786–793 (1977).
5. A. Jameson, 'Iterative solution of transonic flows over airfoils and wings, including flows at Mach 1', *Commun. Pure Appl. Math.*, **27**, 283–309 (1974).
6. A. Jameson, "Numerical computation of transonic flows with shock waves', in K. Oswatitsch and D. Rues (eds), *Symp. Transsonicum II*, Springer, New York, 1976, pp. 384–414.
7. M. Hafez, J. South and E. M. Murman, 'Artificial compressibility methods for numerical solutions of transonic full potential equation', *AIAA J.*, **17**, 838–844 (1979).
8. T. L. Holst and W. F. Ballhaus, 'Fast conservative schemes for the full potential equation applied to transonic flows', *AIAA J.*, **17**, 145–152 (1979).
9. W. G. Habashi and M. Hafez, "Finite element solutions of transonic flow problems', *AIAA J.*, **20**, 1368–1376 (1982).
10. J. L. Steger, "Implicit finite-difference simulation of flow about arbitrary two-dimensional geometries', *AIAA J.*, **16**, 679–686 (1978).
11. T. H. Pulliam and D. S. Chaussee, 'A digital form of an implicit approximate factorization algorithm', *J. Comput. Phys.*, **39**, 347–363 (1981).

12. A. Jameson, W. Schmidt and E. Turkel, 'Numerical solution of the Euler equations by finite volume methods using Runge–Kutta time stepping schemes', *AIAA Paper 81-1259*, 1981.
13. J. L. Steger and R. F. Warming, 'Flux vector splitting of the inviscid gasdynamic equations with application of finite-difference methods', *J. Comput. Phys.*, **40**, 263–293 (1981).
14. R.-H. Ni, "A multi-grid system for solving the Euler equations', *AIAA J.*, **20**, 1565–1571 (1982).
15. A. Harten, "A high resolution scheme for the computation of weak solutions of hyperbolic conservation laws', *J. Comput. Phys.*, **49**, 357–393 (1983).
16. A. Ecer and H. U. Akay, 'A finite element formulation for steady transonic Euler equations', *AIAA J.*, **21**, 343–350 (1983).
17. W. G. Habashi and M. Hafez, Finite element stream function solutions for transonic turbomachinery flows', *AIAA Paper 82-1268*, 1982.
18. M. Hafez and D. Lovell, 'Numerical solution of transonic stream function equation', *AIAA J.*, **21**, 327–335 (1983).
19. W. G. Habashi, P. L. Kotiuga and L. A. McLean, 'Finite element simulation of transonic flows by modified potential and stream function methods', *Eng. Anal.* **2**, 150–154 (1985).
20. H. L. Atkins and H. A. Hassan, 'A new stream function formulation for the steady Euler equations', *AIAA J.*, **23**, 701–706 (1985).
21. M. Hafez and J. Ahmad, 'Numerical simulation of rotational flows', in J. Zierep and H. Oertel (eds), *Symp. Transsonicum III*, Springer, New York, 1988, pp. 339–354.
22. M. Hafez, C. Yam, K. Tang and H. Dwyer, 'Calculations of rotational flows using stream function', *AIAA Paper 89-0474*, 1989.
23. R. W. Jeppson, 'Inverse formulation and finite difference solution for flow from a circular orifice', *J. Fluid Mech.*, **40**, 215–223 (1970).
24. T. Yang and C. D. Nelson, 'Griffith diffusers', *J. Fluids Eng.*, **101**, 473–477 (1979).
25. M. S. Greywall, 'Streamwise computation of 2-D incompressible potential flows', *J. Comput. Phys.*, **59**, 224–231 (1985).
26. R. M. Barron, 'Computation of incompressible potential flow using von Mises coordinates', *Math. Comput. Simul.*, **31**, 177–188 (1989).
27. R. M. Barron, S. Zhang, A. Chandna and N. Rudraiah, 'Axisymmetric potential flow calculations. Part 1: Analysis mode', *Commun. Appl. Numer. Methods*, **6**, 437–445 (1990).
28. R. M. Barron, 'A non-iterative technique for design of aerofoils in incompressible potential flow', *Commun. Appl. Numer. Methods*, **6**, 557–564 (1990).
29. J. L. Duda and J. S. Vrentas, 'Fluid mechanics of laminar liquid jets', *Chem. Eng. Sci.*, **22**, 855–869 (1967).
30. L. R. Clermont and M. E. Lande, 'A method for the simulation of plane or axisymmetric flows of incompressible fluid based on the concept of the stream function', *Eng. Comput.*, **3**, 339–347 (1986).
31. K. Takahashi, 'A numerical analysis of flow using streamline coordinates (the case of two-dimensional steady incompressible flow)', *Bull. JSME*, **25**, 1696–1702 (1982).
32. C. E. Pearson, 'Use of streamline coordinates in the numerical solution of compressible flow problems', *J. Comput. Phys.*, **42**, 257–265 (1981).
33. D. R. Owen and C. E. Pearson, 'Numerical solution of a class of steady state Euler equations by a modified streamline method', *AIAA Paper 88-0625*, 1988.
34. C.-Y. Huang and G. S. Dulikravich, 'Stream function and stream function coordinate (SFC) formulations for inviscid flow field calculations', *Comput. Methods Appl. Mech. Eng.*, **59**, 155–177 (1986).
35. G. S. Dulikravich, 'A stream-function-coordinate (SFC) cncept in aerodynamic shape design', *AGARD VKI Lecture Series*, 1990.
36. R. M. Barron and R. K. Naeem, 'Numerical solution of transonic flows on a streamfunction co-ordinate system', *Int. j. numer. methods fluids*, **9**, 1183–1193 (1989).
37. R. K. Naeem and R. M. Barron, 'Transonic computations on a natural grid', *AIAA J.*, **28**, 1836–1838 (1990).
38. R. M. Barron and R. K. Naeem, '2-D transonic calculations on a flow- based grid system', *Math. Comput. Simul.*, **33**, 65–67 (1991).
39. R. M. Barron and C.-F. An, 'Analysis and design of transonic airfoils using streamwise coordinates', in G. S. Dulikravich (ed), *Proc. 3rd Int. Conf. on Inverse Design Concepts and Optimization in Engineering Science*, Washington, DC, 1991, pp. 359–370.
40. C.-F. An and R. M. Barron, 'Numerical solution of transonic full- potential-equivalent equations in von Mises coordinates', *Int. j. numer. methods fluids*, **15**, 925–952 (1992).
41. G.-L. Liu and C. Tao, 'A universal image-plane method for inverse and hybrid problems of compressible cascade flow on arbitrary stream sheet of revolution: part I—theory', in C. Taylor *et al.* (eds), *Numerical Methods in Laminar and Turbulent Flows*, Vol. 6, Pt. II, Pineridge, Swansea, 1989, pp. 1343–1354.
42. K.-M. Chen, D.-F. Zhang, and Z.-G. Zhu, 'A universal image-plane method for direct-, inverse- and hybrid problems of compressible cascade flow on arbitrary stream sheet of revolution: part II—numerical solution', in C. Taylor *et al.* (eds), *Numerical Methods in Laminar and Turbulent Flows*, Vol. 6, Pt. II, Pineridge, Swansea, 1989, pp. 1355–1365.
43. G.-L. Liu and D.-F. Zhang, 'The moment function formulation of inverse and hybrid problems for blade-to-blade compressible viscous flow along axisymmetric stream sheet', in C. Taylor *et al.* (eds), *Numerical Methods in Laminar and Turbulent Flows*, Vol. 6, Pt. II, Pineridge, Swansea, 1989, pp. 1289–1300.
44. R. von Mises, 'Bemerkungen zur Hydrodynamik', *ZAMM*, **7**, 425 (1927).
45. D. J. Jones and R. G. Dickinson, 'A description of the NAE two-dimensional transonic small disturbance computer method', *NAE Tech. Rep. LTR-HA-39*, 1980.

46. J. J. Thibert and M. Grandjacques, 'Experimental data base for computer program assessment', *AGARD AR-138,* 1979, pp. A1-1–A1-19.
47. L. H. Ohman, 'Experimental data base for computer program assessment', *AGARD AR-138*, 1979, pp. A1-20–A1-36.
48. E. D. Knetchtel, 'Experimental investigation at transonic speeds of pressure distributions over wedge and circular-arc-airfoil sections and evaluation of perforated wall interference', *NASA TN D-15,* 1959.
49. H. Viviand, 'Numerical solutions of two-dimensional reference test cases', *AGARD AR-211,* 1985, pp. 6-1–6-68.

Mathematical Models and Methods in Applied Sciences
Vol. 5, No. 3 (1995) 401–414

STREAMFUNCTION COORDINATE FORMULATION FOR ONE-DIMENSIONAL UNSTEADY FLOW

C.-F. AN, R. M. BARRON and S. ZHANG

Department of Mathematics and Statistics, and Fluid Dynamics Research Institute, University of Windsor, Windsor, Ontario, Canada N9B 3P4

Communicated by L. M. de Socio
Received 25 March 1994

In this paper, the streamfunction-as-a-coordinate concept has been extended to 1-D unsteady flow. The governing equation in streamfunction coordinates in this case is a second order nonlinear hyperbolic partial differential equation or a wave equation with variable propagation speed for the Cartesian coordinate $x = x(\psi, \tau)$ where ψ is streamfunction and τ is time. The equation has been solved numerically by the leapfrog scheme. A piston–driven expansion wave and a density disturbance propagation have been successfully simulated and the agreement between the present method and exact solution is excellent. The streamfunction coordinate formulation for 1-D unsteady flow is particularly advantageous due to the fact that the Jacobian is always a positive finite quantity so that there is no singularity in the transformation. An additional benefit of the streamfunction formulation for 1-D unsteady flow is the ability to easily draw particle path lines.

第二部分 出国后博士学位论文阶段

1. Introduction

The concept of streamfunction-as-a-coordinate (SFC) and its applications in computational fluid dynamics (CFD) has received considerable attention in the last decade.[1–6,8,10,11,13,14] A survey of the SFC method in CFD has recently been given by the present authors.[7] Most applications of this method are to 2-D steady flow. Among these SFC applications, the von Mises formulation plays an important role due to a number of advantages. For example, it combines the grid generator and equation solver together without performing conventional grid generation, thereby reducing computer time. The boundary conditions are mostly Dirichlet, which simplifies the algorithm. Another impressive feature of the von Mises formulation is that it is easy to draw streamlines and hence is an excellent tool for flow visualization. But this formulation, when applied to 2-D steady flow, has a troublesome restriction, i.e. inability to treat reverse flows. In that case, the Jacobian becomes infinite and the transformation fails.

However, this restriction does not exist if the von Mises transformation is applied to 1-D unsteady flow because the Jacobian of the transformation in this case has a physical meaning, viz. density, which is always positive and finite. This simple fact guarantees the validity of the transformation and provides an opportunity to use SFC in a variety of 1-D unsteady flows, such as disturbance propagation, expansion and compression wave, wave reflection and interaction, shock tube flow, 1-D section variant nozzle flow, etc.

In this paper, an exploratory effort has been made to initiate the abovementioned series of studies. In Sec. 2, the governing equation for 1-D unsteady flow is derived in streamfunction coordinates. The boundary and initial conditions for two selected examples are given in Sec. 3. Section 4 describes the numerical techniques used in this work. Section 5 discusses and summarizes the calculated results and Sec. 6 gives some concluding remarks.

2. Governing Equations

For 1-D unsteady isentropic flow of an inviscid gas, the governing equations are

$$\rho_t + (\rho u)_x = 0\,, \tag{2.1}$$

$$\rho u_t + \rho u u_x + p_x/\gamma = 0\,, \tag{2.2}$$

$$p = \rho^\gamma\,, \tag{2.3}$$

where density ρ, velocity u and pressure p have been non-dimensionalized by some reference density ρ_0, speed of sound c_0 and $\rho_0 c_0^2/\gamma$, where γ is the ratio of specific heats. The Cartesian coordinate x and time t have been scaled by some characteristic length L and L/c_0. Eliminating p from (2.3) and (2.2), introducing 'streamfunction' ψ such that

$$\psi_x = \rho, \quad \psi_t = -\rho u \tag{2.4}$$

and substituting it in (2.1) and (2.2), the continuity equation (2.1) is automatically satisfied and the momentum equation (2.2) becomes

$$(\psi_t^2 - \psi_x^{\gamma+1})\psi_{xx} - 2\psi_t\psi_x\psi_{xt} + \psi_x^2\psi_{tt} = 0. \tag{2.5}$$

Introducing a von Mises' type transformation

$$x = x(\psi, \tau), \quad t \equiv \tau, \tag{2.6}$$

we have

$$\begin{aligned} \psi_x &= \frac{1}{x_\psi}, \quad \psi_t = -\frac{x_\tau}{x_\psi}\,, \\ \psi_{xx} &= -\frac{x_{\psi\psi}}{x_\psi^3}, \quad \psi_{xt} = -\frac{x_\psi x_{\tau\psi} - x_\tau x_{\psi\psi}}{x_\psi^3}\,, \\ \psi_{tt} &= -\frac{x_\psi^2 x_{\tau\tau} - 2x_\tau x_\psi x_{\tau\psi} + x_\tau^2 x_{\psi\psi}}{x_\psi^3}\,. \end{aligned} \tag{2.7}$$

Therefore, the governing equation for 1-D unsteady flow in Cartesian coordinates (2.5) can be transformed to the streamfunction coordinates, yielding

$$x_{\psi\psi} = x_{\psi}^{\gamma+1} x_{\tau\tau}. \tag{2.8}$$

The Jacobian of the transformation (2.6)

$$J = \frac{\partial(x,t)}{\partial(\psi,\tau)} = x_{\psi} = \frac{1}{\rho} \tag{2.9}$$

is obviously neither zero nor infinite due to the physical meaning of density. This simple fact guarantees that the transformation (2.6) is always one-to-one and invertible, with no singularities. Equation (2.8) is a nonlinear second order partial differential equation and is always hyperbolic because the discriminant

$$\Delta = 4x_{\psi}^{\gamma+1} = \frac{4}{\rho^{\gamma+1}} \tag{2.10}$$

is always positive. The dependent variable $x = x(\psi, \tau)$, as unknown function of streamfunction ψ and time τ, is to be solved from (2.8) under appropriate boundary and initial conditions. After x is obtained, the density, velocity and pressure can be easily calculated from

$$\rho = \frac{1}{x_{\psi}}, \quad u = x_{\tau}, \quad p = \frac{1}{x_{\psi}^{\gamma}}. \tag{2.11}$$

3. Boundary and Initial Conditions

In order to provide an explanation of boundary and initial conditions in the computational plane (ψ, τ), let us consider some examples for which the computation will be carried out.

The first example is an expansion wave propagation (Fig. 1). Suppose the air on the right of the piston placed in a cylinder is at rest. When $t > 0$, the piston is drawn out from the cylinder and accelerated to the left with a given function for its position

$$x = X(t). \tag{3.1}$$

The flow picture in the physical plane is illustrated in Fig. 1a. As the piston moves to the left, a sequence of expansion waves are induced and travel to the right. Since the waves are traveling to the right, the left boundary condition must be specified. The velocity of the air close to the piston is given based on the assumption that the air is traveling with the piston. Hence, the left boundary condition for velocity is

$$u(0,t) = X'(t). \tag{3.2}$$

The left boundary condition for density ρ can be specified through the speed of sound c using the isentropic relation and, in turn, the speed of sound c is related

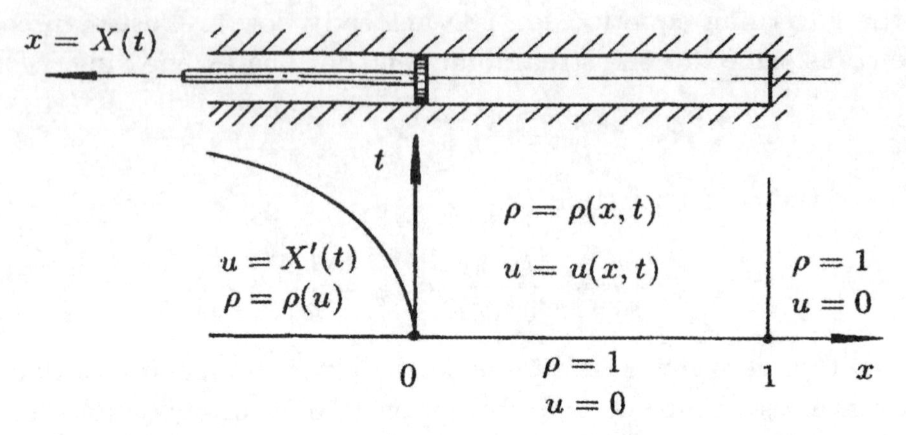

(a) Physical plane

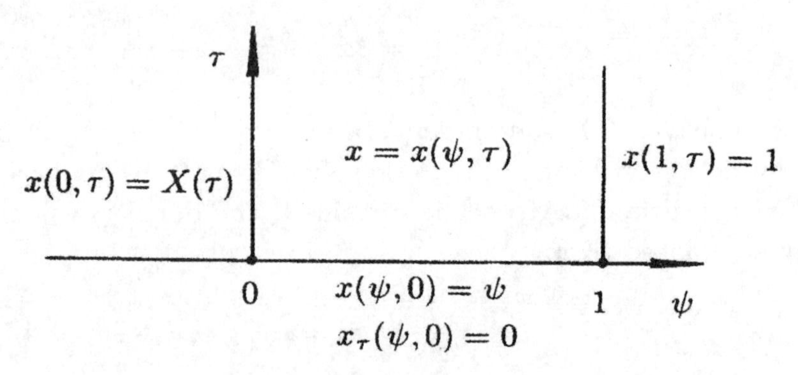

(b) Computational plane

Fig. 1. Piston–driven expansion wave.

to the velocity u in a simple wave region. The left boundary condition for density can be written as

$$\rho(0,t) = [1 + \frac{\gamma - 1}{2} X'(t)]^{2/(\gamma-1)}. \tag{3.3}$$

Considering only the process before the wave front reaches the right end of the cylinder, the condition at the right boundary is not needed. However, from a numerical point of view, the right boundary condition must be specified.[15] It can be expressed as

$$\rho(1,t) = 1, \qquad u(1,t) = 0. \tag{3.4}$$

The initial conditions in the physical plane are

$$\rho(x,0) = 1, \qquad u(x,0) = 0. \tag{3.5}$$

In the computational plane (ψ, τ), the left and right boundary conditions, equivalent to (3.2), (3.3) and (3.4), are

$$x(0,\tau) = X(\tau) \quad \text{and} \quad x(1,\tau) = 1. \tag{3.6}$$

To write these initial conditions in the computational plane, consider Eq. (2.11) and the initial conditions in the physical plane (3.5). Then, $x_\psi = 1/\rho = 1, x_\tau = u = 0$. Therefore, the initial conditions in the computational plane are

$$x(\psi, 0) = \psi, \qquad x_\tau(\psi, 0) = 0. \tag{3.7}$$

Figure 1b shows the boundary and initial conditions in the computational plane.

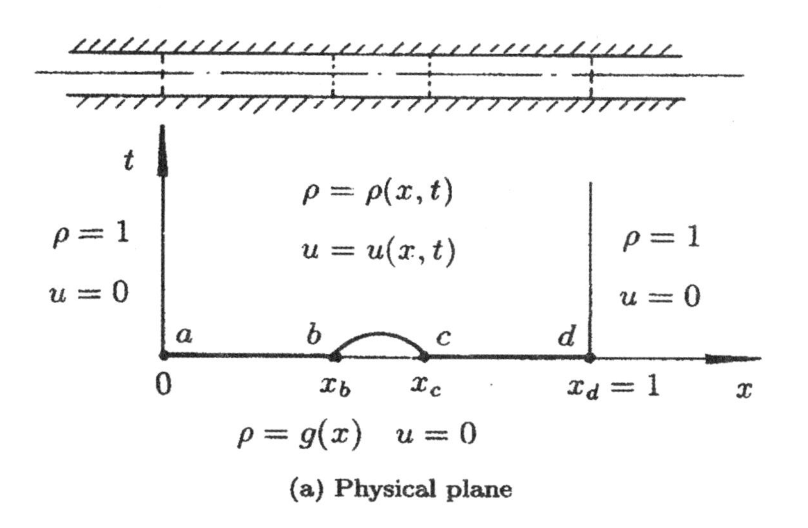

(a) Physical plane

(b) Computational plane

Fig. 2. Density disturbance propagation.

The second example is a 1-D disturbance propagation problem. Suppose an infinitely long tube is filled with stationary air (Fig. 2). When $t = 0$, a density disturbance is imposed in segment *bc*. Our task is to investigate the propagation of this density disturbance. We choose a segment *ad* from $x = 0$ to $x = 1$ as the object of study. Let $x_b = 0.35$ and $x_c = 0.45$. The density disturbance distribution is imposed as

$$\rho(x, 0) = \begin{cases} 1, & 0 \le x \le x_b \\ 1 + 0.2\sin[\frac{x - x_b}{x_c - x_b}\pi], & x_b < x < x_c \\ 1, & x_c \le x \le 1. \end{cases} \tag{3.8}$$

There is no initial velocity disturbance,

$$u(x,0)=0. \tag{3.9}$$

The initial conditions in the computational plane can be specified as follows:
On segment ab: $\psi=\int_0^x \rho(x,0)dx=x$, therefore,

$$x=\psi, \quad 0\le\psi\le\psi_b=0.35\,.$$

On segment bc: $\psi=\psi_b+\int_{x_b}^x \rho(x,0)dx$, therefore,

$$x=\psi-\frac{0.2(x_c-x_b)}{\pi}\{1-\cos[\frac{x-x_b}{x_c-x_b}\pi]\}, \quad \psi_b\le\psi\le\psi_c=0.463\,.$$

On segment cd: $\psi=\psi_c+\int_{x_c}^x \rho(x,0)dx$, therefore,

$$x=\psi-\psi_c+x_c, \qquad \psi_c\le\psi\le\psi_d=1.013\,.$$

To summarize, the initial conditions in the computational plane are

$$x(\psi,0)=\begin{cases}\psi, & 0\le\psi\le\psi_b\\ \psi-\frac{0.2(x_c-x_b)}{\pi}\{1-\cos[\frac{x-x_b}{x_c-x_b}\pi]\}, & \psi_b<\psi<\psi_c\\ \psi-\psi_c+x_c, & \psi_c\le\psi\le\psi_d\end{cases} \tag{3.10}$$

and

$$x_\tau(\psi,0)=0. \tag{3.11}$$

For this example, neither left nor right boundary conditions are required for the partial differential equation (2.8). But, for the purpose of numerical computation, both left and right numerical boundary conditions are needed. A natural consideration is to take

$$\rho=1, \qquad u=0. \tag{3.12}$$

In the computational plane,

$$x(0,\tau)=0, \qquad x(\psi_d,\tau)=1. \tag{3.13}$$

Although this example may not correspond to some direct physical problem, it is useful to validate the numerical scheme since it can be solved exactly using the theory of characteristics. The implementation of numerical boundary conditions is also tested through this example. Figures 2a and 2b show the boundary and initial conditions in the physical and computational planes, respectively.

4. Numerical Techniques

In the last two sections we have deduced governing Eq. (2.8) and, for example 1, initial conditions (3.7) and boundary conditions (3.6), which are collected here for convenience.

$$x_{\tau\tau} = x_\psi^{-(\gamma+1)} x_{\psi\psi}\,, \tag{4.1}$$

$$x(\psi, 0) = \psi\,, \tag{4.2}$$

$$x_\tau(\psi, 0) = 0\,, \tag{4.3}$$

$$x(0, \tau) = X(\tau)\,, \tag{4.4}$$

$$x(1, \tau) = 1\,. \tag{4.5}$$

The solution domain is $0 \le \psi \le 1, \tau \ge 0$. The propagation speed of the nonlinear wave equation (4.1) is

$$a = a(\psi, \tau) = x_\psi^{-(\gamma+1)/2}. \tag{4.6}$$

To solve this problem, the three-step explicit midpoint leap frog scheme can be applied as suggested by Hoffmann,[9] resulting in the following finite difference equation,

$$x_i^{n+1} = 2x_i^n - x_i^{n-1} + C^2(x_{i+1}^n - 2x_i^n + x_{i-1}^n) \tag{4.7}$$

for $i = 2, 3, ..., I_{\max} - 1$ and $n = 2, 3, ...$ where

$$C = C(\psi, \tau) = \frac{a(\psi, \tau)\Delta\tau}{\Delta\psi} \tag{4.8}$$

is the Courant–Friedrichs–Lewy (CFL) number. Here, it is obvious that a numerical boundary condition on the right boundary must be provided to solve this finite difference equation, while it is not needed for the original partial differential equation. Furthermore, because Eq. (4.7) has dependent variable x at three time levels, $n-1, n$ and $n+1$, a starter equation is required. To do this, consider the second initial condition (4.3) and use central differencing

$$\frac{x_i^{n+1} - x_i^{n-1}}{2\Delta\tau} = 0 \quad \text{or} \quad x_i^{n-1} = x_i^{n+1}. \tag{4.9}$$

Substituting (4.9) in (4.7), one gets an equation for the first time step:

$$x_i^{n+1} = x_i^n + \frac{C^2}{2}(x_{i+1}^n - 2x_i^n + x_{i-1}^n) \tag{4.10}$$

for $i = 2, 3, ..., I_{\max} - 1$ and $n = 1$. From stability analysis,[9] the scheme is stable if

$$C_{\max} = \frac{a_{\max}\Delta\tau}{\Delta\psi} \le 1\,, \tag{4.11}$$

where $a_{\max} = \max(a_i)$ for all i's. At each time level n, a_i is different at each point i and $a_{\max}$ is the maximum among all i's. However, $a_{\max}$ is usually different from level to level, and hence $\Delta\tau$ is also different from level to level. The solution accuracy analysis[9] indicates that the best solution is always obtained for C equal or very close to 1. Hence, the calculation procedure is as follows. Set $C_{\max} = 0.99$, i.e.

$$\Delta\tau = 0.99\frac{\Delta\psi}{a_{\max}}. \tag{4.12}$$

After solving x at a particular time level, $a_{\max}$ is evaluated immediately and $\Delta\tau$ is calculated for the next time level from (4.12). In this way, the CFL condition is always observed and the CFL number remains close to 1 for all time levels.

For the second example, the governing equation (2.8), initial and boundary conditions (3.10), (3.11) and (3.13) are gathered as follows:

$$x_{\tau\tau} = x_{\psi}^{-(\gamma+1)} x_{\psi\psi}\,, \tag{4.13}$$

$$x(\psi, 0) = x(\psi, 0)\,, \tag{4.14}$$

$$x_{\tau}(\psi, 0) = 0\,, \tag{4.15}$$

$$x(0, \tau) = 0\,, \tag{4.16}$$

$$x(\psi_d, \tau) = 1\,, \tag{4.17}$$

with $x(\psi, 0)$ given by (3.10).

These equations are similar to example 1, except for the second and the fourth equations. The source of disturbance is the left boundary condition for the first example while it is the initial condition for the second example. The rest of the procedure, such as starter equation, CFL condition and $\Delta\tau$ calculation, etc., are the same as the first example. The only difference is that the disturbance in the second example propagates in both directions while the expansion wave in the first example only travels to the right.

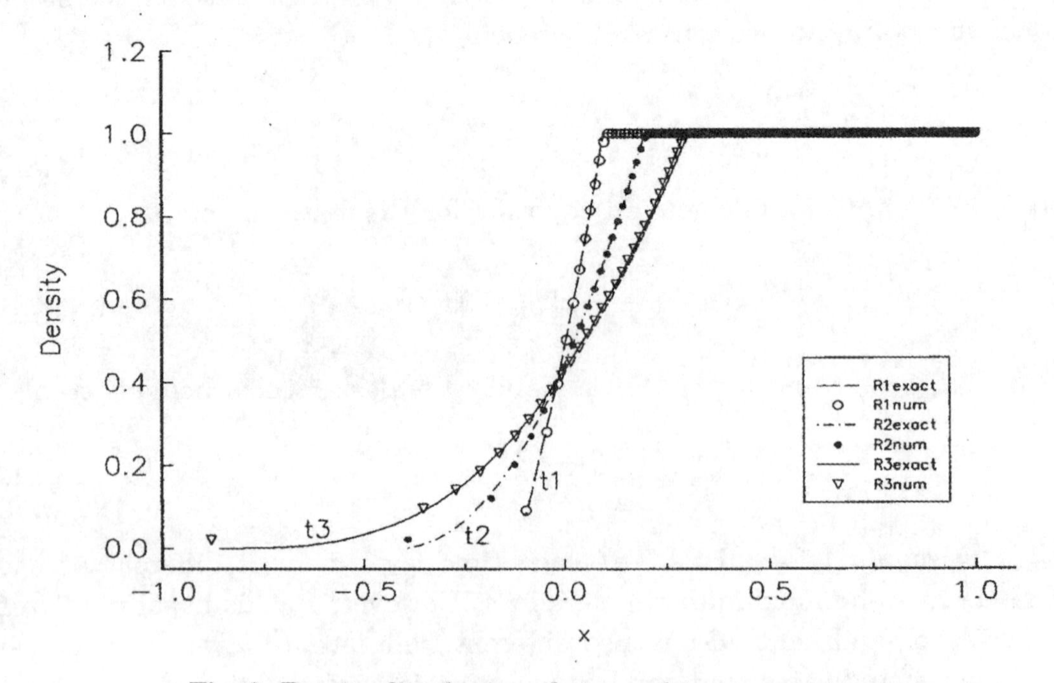

Fig. 3. Density distribution of piston–driven expansion wave.

5. Results and Discussion

5.1. *Piston–driven expansion wave*

The interval $0 \leq \psi \leq 1$ is divided into 100 segments and the total number of time steps is taken as 50, i.e. $I_{\max} = 101$ and $N_{\max} = 51$. The piston motion is described by (3.1) as $X(t) = -At^2$ and the acceleration constant A is chosen to be 10. Figures 3 and 4 compare density and velocity, respectively, between the present computation and the exact solution obtained using characteristics theory. The time levels are chosen to be $t_1 = 0.099, t_2 = 0.198$, and $t_3 = 0.297$, and R_1 refers to the density value at t_1, etc. It is observed that the agreement is excellent. Figure 5 shows the particle path lines in the physical plane. Each line, drawn for $x = x(\psi_i, \tau), i = 1, 2, ..., I_{\max}$, describes the path of a certain air particle in the tube. Physically, the streamfunction $\psi(x,t) = \int_0^x \rho(x,t)dx$ depicts the total mass in the segment from $x = 0$ to $x = x$ at a certain time t. Hence, $x(\psi_i, \tau)$ for a constant ψ_i represents the position of the right end of an imaginary air cylinder originally located in $x(\psi_i, 0)$ with total mass ψ_i. Therefore, $x(\psi_i, \tau)$ represents the location history of the right end of the air cylinder, i.e. the path of the air particle. This expressive picture illustrates a major advantage of the streamfunction coordinates, that is, it is very easy to draw path lines of the air particles in physical plane by fixing $\psi = \psi_i$ and evaluating $x = x(\psi_i, \tau)$. Figures 6 and 7 illustrate the density distribution $\rho(x,t)$ and velocity distribution $u(x,t)$, respectively, in the physical plane. One can easily see that the right traveling expansion wave propagation is successfully simulated. In both figures, the first curve is obviously different from the others. This indicates that the piston moves leftward too quickly for the air to follow. The air particle is almost able to keep its "fleeing speed", i.e. when the density is equal to zero, to follow the piston far behind it. Between the piston and the neighboring air particle there is a vacuum region. This is consistent with the classical analysis of one-dimensional unsteady compressible flow.[12]

5.2. *Density disturbance propagation*

As in the previous example, $I_{\max} = 101$ and $N_{\max} = 51$ is chosen. Figure 8 shows the wave shapes at several time levels, $t_1 = 0, t_2 = 0.0339, t_3 = 0.1694$. The initial wave shape for t_1 is a sinusoidal function. At t_2, the wave becomes lower and fatter but still remains one wave shape. At t_3, two separate waves are formed, distorted so that the fronts are steeper and the tails are flatter. The agreement between the present computation and the exact solution from characteristics theory is excellent. Figure 9 gives the iso-density contours and also shows that the wave fronts become steeper and steeper as time increases. Figure 10 shows the propagation of the density disturbance, represented by the surface $\rho = \rho(x,t)$ above the physical plane. The density disturbance propagates in both directions. Due to the nonlinear nature, the wave shape changes with time. The wave fronts become increasingly steep while the wave

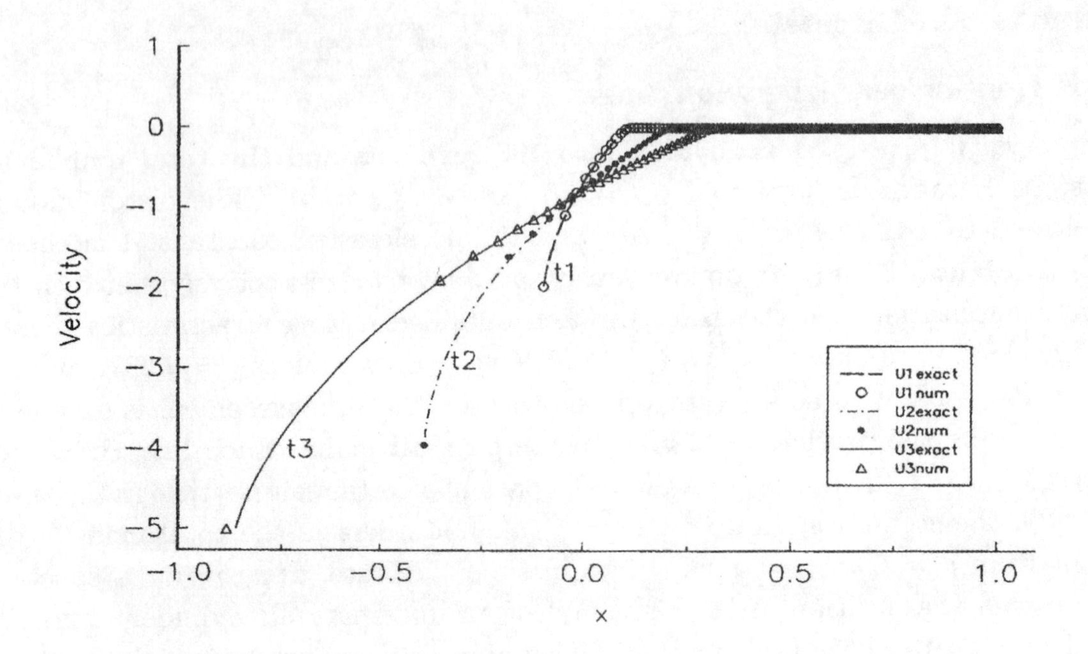

Fig. 4. Velocity distribution of piston–driven expansion wave.

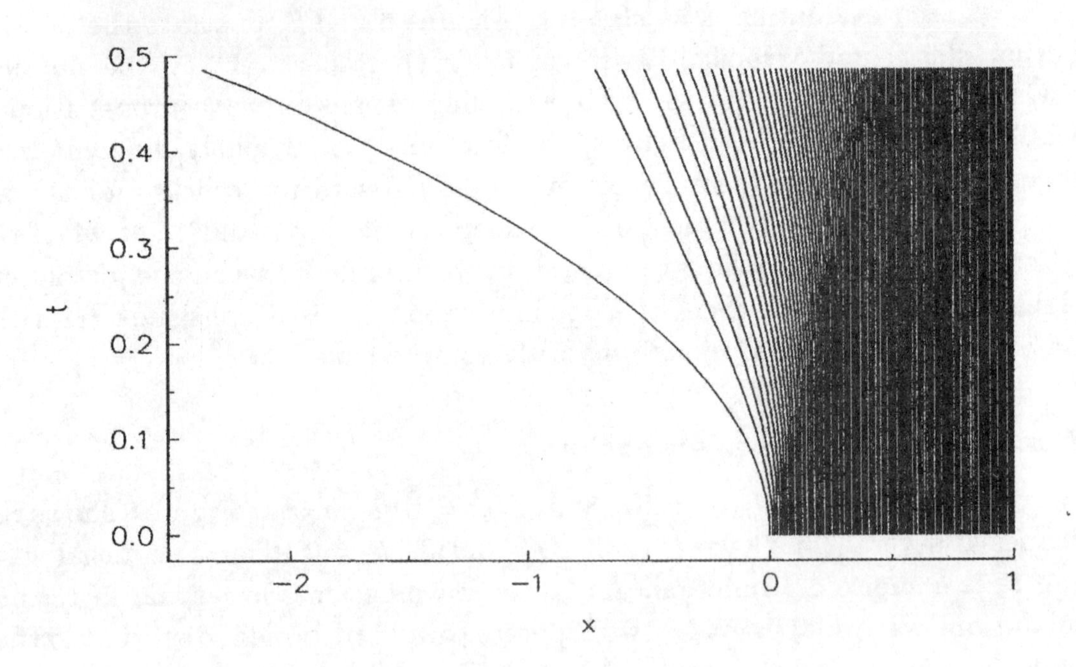

Fig. 5. Particle path lines of piston–driven expansion wave.

tails become increasingly flat. As the disturbance propagates, the wave shape becomes broader and its amplitude decreases, eventually, it splits into two equal parts. Each part propagates in its own direction independently.

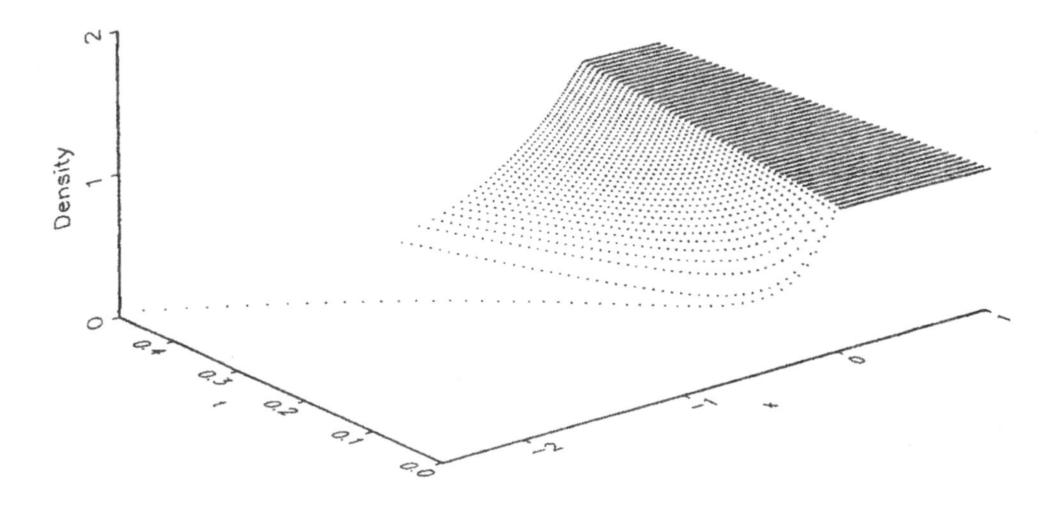

Fig. 6. Density surface of piston–driven expansion wave.

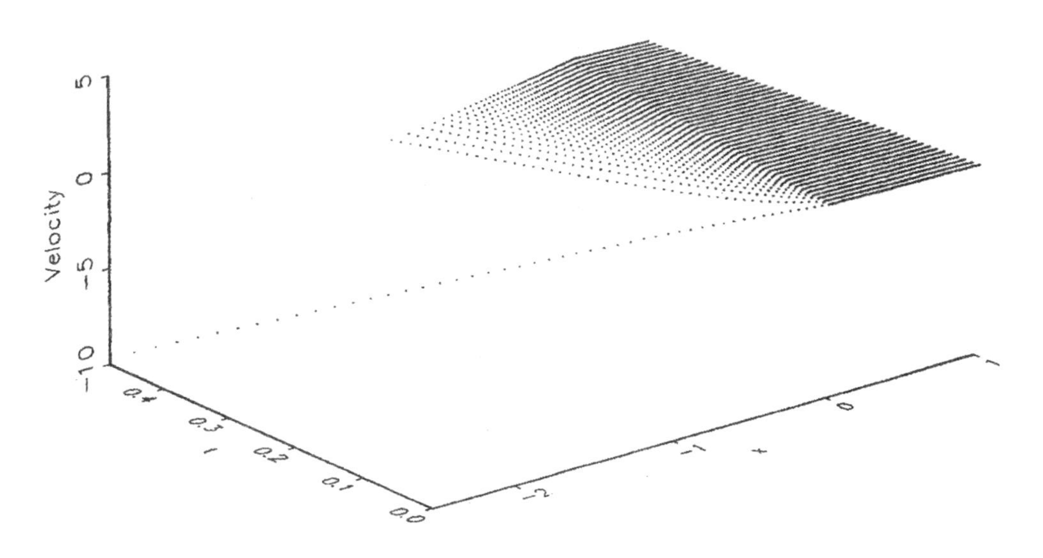

Fig. 7. Velocity surface of piston–driven expansion wave.

6. Concluding Remarks

For a 1-D unsteady flow of an inviscid isentropic gas, the governing equation in streamfunction coordinates (von Mises' type) is a second order nonlinear hyperbolic partial differential equation, i.e. a wave equation with a variable propagation speed, for the Cartesian coordinate $x = x(\psi, \tau)$ as unknown function of streamfunction ψ and time τ.

The derived wave equation can be solved numerically using the leap frog scheme under the appropriate boundary and initial conditions. A piston–driven expansion

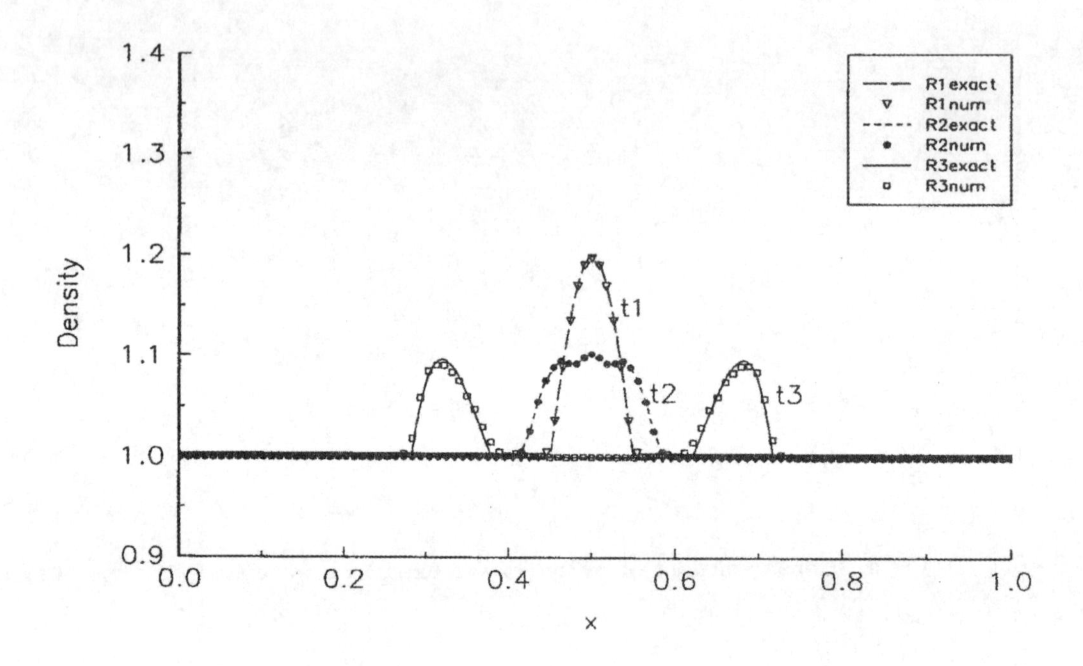

Fig. 8. Density distribution of disturbance propagation.

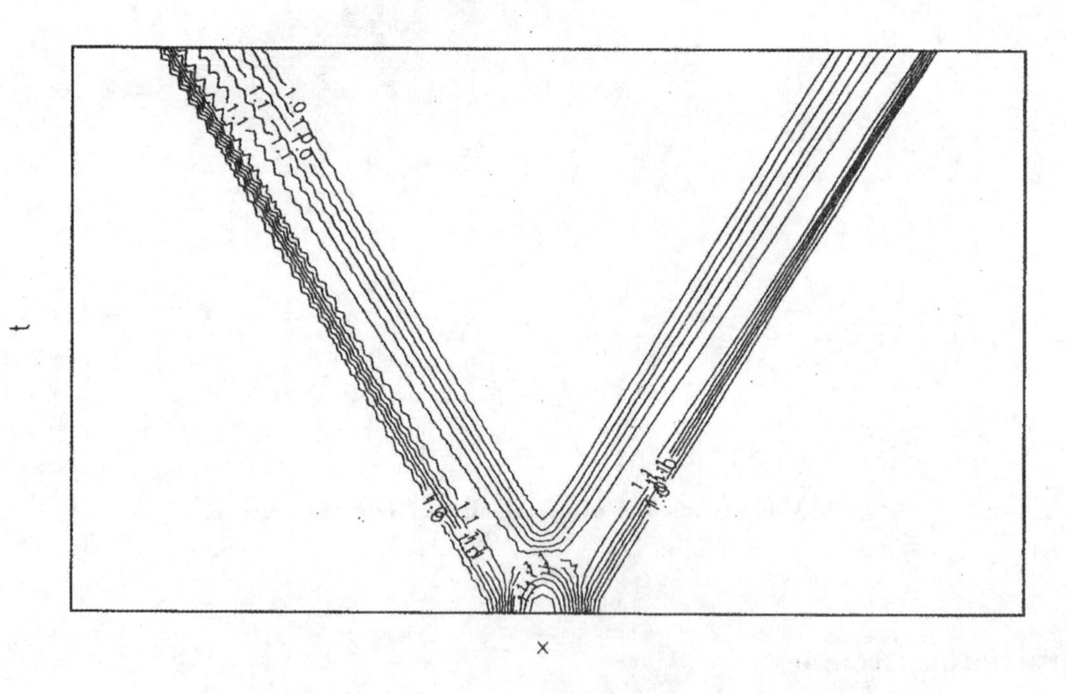

Fig. 9. Iso-density contours of disturbance propagation.

wave and a density disturbance propagation have been successfully simulated and numerical results compare well with the exact solution of the characteristics theory.

An attractive advantage of the streamfunction formulation shown in this study is that the gas particle path lines are easily drawn. This benefit is also known in the 2-D steady flow case in which streamlines are required to be drawn.

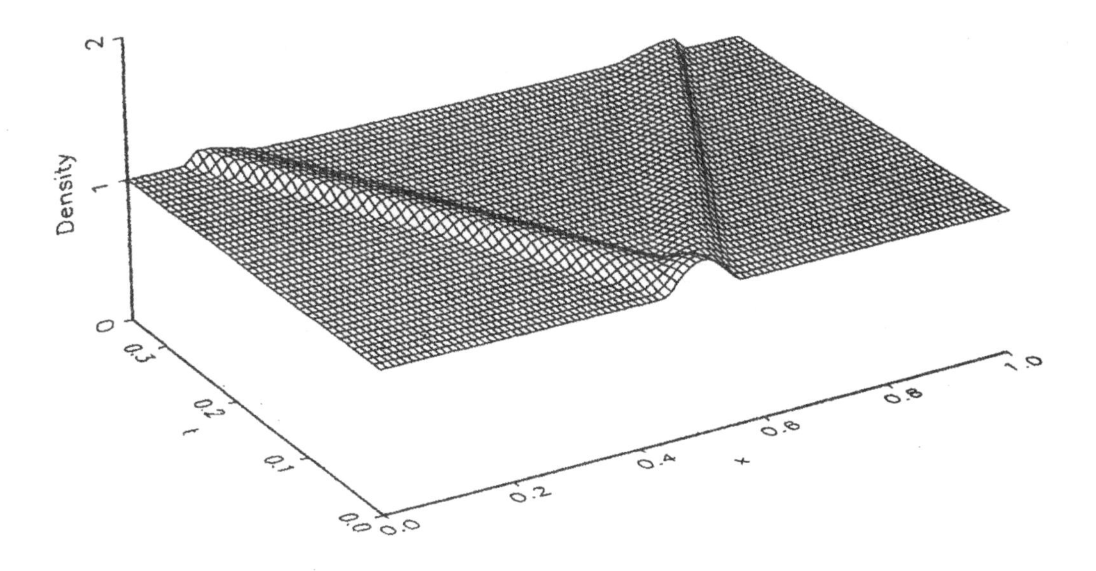

Fig. 10. Density disturbance propagation.

The major drawback of von Mises' coordinates for 2-D steady flow is that the Jacobian becomes infinite at points where the flow reverses the direction. However, this difficulty does not exist for 1-D unsteady flow since the Jacobian has an explicit physical meaning, viz. density, which is always a nonzero and finite quantity. This simple fact guarantees the validity of the von Mises' type transformation for 1-D unsteady flow.

A limitation of the present approach is the inability to simulate a flow with discontinuity or shock. To overcome this limitation, further investigation will be directed to the conservative form of the governing equations in streamfunction coordinates.

References

1. C.-F. An and R. M. Barron *Numerical solution of transonic full-potential-equivalent equations in von Mises' coordinates, Internat. J. Numer. Methods Fluids* **15** (1992) 925–952.
2. C.-F. An, R. M. Barron and S. Zhang, *Arc length-streamfunction formulation and its application, Appl. Math. Modelling* **18** (1994) 478–485.
3. R. M. Barron, *Computation of incompressible potential flow using von Mises' coordinates, Math. Comput. Simulation* **31** (1989) 177–188.
4. R. M. Barron and R. K. Naeem, *Numerical solution of transonic flows on a streamfunction co-ordinate system, Internat. J. Numer. Methods Fluids* **9** (1989) 1183–1193.
5. R. M. Barron, S. Zhang, A. Chandna and N. Rudraiah, *Axisymmetric potential flow calculations. Part 1: analysis mode, Comm. Appl. Numer. Methods* **6** (1990) 437–445.
6. R. M. Barron and C.-F. An, *Analysis and design of transonic airfoils using streamwise coordinates*, in **Proc. of the 3rd Int. Conf. on Inverse Design Concepts and Optimization in Engineering Sciences**, ed. G. S. Dulikravich (1991), pp. 359–370.

7. R. M. Barron, C.-F. An and S. Zhang, *Survey of the streamfunction-as-a-coordinate method in CFD*, in **Proc. of the Inaugural Conference of the CFD Society of Canada**, ed. R. Camarero (1993), pp. 325–333.
8. M. S. Greywall, *Streamwise computation of 2-D incompressible potential flows*, *J. Comput. Phys.* **59** (1985) 224–231.
9. K. A. Hoffmann, **Computational Fluid Dynamics for Engineers** (Engineering Education System, 1989), Chaps. 4–6.
10. C.-Y. Huang and G. S. Dulikravich, *Stream function and streamfunction coordinate (SFC) formulation for inviscid flow field calculations*, *Comput. Methods Appl. Mech. Engrg.* **59** (1986) 155–177.
11. C. E. Pearson, *Use of streamline coordinates in the numerical solution of compressible flow problems*, *J. Comput. Phys.* **42** (1981) 257–265.
12. A. H. Shapiro, **The Dynamics and Thermodynamics of Compressible Fluid Flow** (The Ronald Press Co., 1953), Vol. II, Chap. 24.
13. K. Srinivasan and D. B. Spalding, *The streamfunction coordinate system for solution of one-dimensional unsteady compressible flow problems*, *Appl. Math. Modelling* **10** (1986) 278–283.
14. K. Takahashi, *A numerical analysis of flow using streamline coordinates (The case of two-dimensional steady incompressible flow)*, *Bull. JSME* **25** (1982) 1696–1702.
15. H. C. Yee, R. M. Beam and R. F. Warming, *Boundary approximations for implicit schemes for one-dimensional inviscid equations of gasdynamics*, *AIAA J.* **20** (1982) 1203–1211.

COMMUNICATIONS IN NUMERICAL METHODS IN ENGINEERING, Vol. 11, 585–595 (1995)

TRANSONIC FLOW COMPUTATIONS USING STREAMFUNCTION AND POTENTIAL FUNCTION

S. ZHANG, R. M. BARRON* AND C.-F. AN

Department of Mathematics and Statistics and Fluid Dynamics Research Institute, University of Windsor, Windsor, Ontario, Canada N9B 3P4

SUMMARY

A new formulation is proposed for two dimensional transonic potential flow calculations. The present approach takes advantage of both streamfunction and potential function formulations. A modified form of the partial differential equations and simplified boundary conditions are derived. Numerical examples show that a reliable method has been developed, capable of calculating highly transonic potential flows.

KEY WORDS CFD; transonic flows; von Mises coordinates

第二部分 出国后博士学位论文阶段

1. INTRODUCTION

Significant analytical and computational work has been done in compressible flows using the potential function formulation. Considering the flow to be irrotational, inviscid and isentropic, the momentum equations reduced to the Bernoulli relation between the density and the speed, and the continuity equation provides a second-order partial differential equation for the velocity potential. Many successful transonic codes are based on the potential function formulation.[1-4]

On the other hand, the streamfunction formulation is an attractive approach since it is also applicable to the steady Euler equations governing rotational inviscid flows. The Euler equations can be replaced by second-order partial differential equations in terms of the streamfunction,[5-8] and the boundary conditions on the streamfunction are Dirichlet. Unfortunately, as is well known, the Bernoulli equation yields a non-unique relation between the density and flux, and additional information is needed to make the transition between the subsonic and supersonic solutions.

Considering the steady full-potential or Euler equations, the major difficulties come from the fact that the equations have a mixed elliptic–hyperbolic nature allowing for the presence of discontinuities. The initial breakthrough was made in 1970 by Murman and Cole,[1] who developed the first numerical solution for the steady transonic small-disturbance equation by using a type-dependent difference scheme. Later, this idea was extended by Garabedian and Korn,[2] and Jameson[3] to the transonic full-potential equation. Since then, most research on numerical methods for the steady-state full-potential equation has focused on accelerating iterative methods.

The original foundation of the present work is due to Martin,[9] who transformed the flow equations into curvilinear co-ordinates (ϕ, ψ) where ψ is the streamfunction and ϕ defines an

* Author to whom correspondence should be addressed.

CCC 1069–8299/95/070585–11

Received 2 March 1994
Revised 1 November 1994

arbitrary family of curves. Greywall,[10] Huang and Dulikravich,[11] and Barron[12] proposed a streamfunction-co-ordinate (SFC) formulation for solving two- and three-dimensional fluid flow problems. One very important advantage is that the SFC formulation simultaneously solves a flow problem and determines the flow pattern, thereby bypassing conventional grid generation.

All streamfunction formulations for potential flow suffer from the fact that the density is not uniquely determined in terms of mass flux, making it difficult to obtain numerical solutions. The physically feasible solution may not be obtained in the calculation of transonic external flows if only the conventional iterative procedure is employed.[13,14] There is no difficulty if only incompressible and compressible subsonic flows are considered.[10,11] So far there has been very little work on supercritical transonic external flow using streamfunction-co-ordinate formulations.[5,15–17]

In this paper, a stable and reliable method is presented which overcomes the difficulty of obtaining the density in streamfunction formulations. The basic equations are derived and rewritten in a new form that is similar to Jameson's work.[3] This is followed by some discussion on the boundary conditions of the streamfunction-co-ordinate formulation and the details of the present method. Our discussions are limited to inviscid irrotational two-dimensional flows, and all variables are non-dimensionalised using the flow quantities at infinity.

BASIC EQUATIONS

Consider the 2D steady flow of a perfect gas past an aerofoil. Let (u, v) be the velocity vector and ρ be the density. The equation of conservation of mass is

$$(\rho u)_x + (\rho v)_y = 0 \tag{1}$$

and, for inviscid irrotational flow, the condition of irrotationality is

$$v_x - u_y = 0 \tag{2}$$

Integration of the momentum equations gives the Bernoulli equation

$$\rho = \left[1 - \frac{\gamma - 1}{2} M_\infty^2 (q^2 - 1)\right]^{1/(\gamma-1)} \tag{3}$$

Equations (1)–(3) are to be solved in the region exterior to the aerofoil, with boundary conditions of zero mass flow through the aerofoil and uniform flow at infinity. From equation (1), there exists a streamfunction $\psi(x, y)$ such that $\rho u = \psi_y$, $\rho v = -\psi_x$, and consequently equation (2) can be written as[18]

$$(\psi_x/\rho)_x + (\psi_y/\rho)_y = 0 \tag{4}$$

Also, from equation (2), there exists a potential function $\phi(x, y)$ such that $u = \phi_x$, $v = \phi_y$, and equation (1) can be written as

$$(\rho\phi_x)_x + (\rho\phi_y)_y = 0 \tag{5}$$

The von Mises co-ordinates are now introduced, where the roles of y and ψ are interchanged, i.e. consider $x = x$, $y = y(x, \psi)$. One can show that the irrotationality condition (2) transforms to the conservative form[17]

$$\left[\frac{1 + y_x^2}{\rho y_\psi}\right]_\psi - \left[\frac{y_x}{\rho}\right]_x = 0 \tag{6}$$

Expanding this equation, and using the Bernoulli equation to eliminate derivatives of ρ, equation (6) can be expressed in the quasilinear non-conservative form[5,13,16]

$$\left(y_\psi^2 - \frac{M_\infty^2}{\rho^{\gamma+1}}\right)y_{xx} - 2y_x y_\psi y_{x\psi} + (1 + y_x^2)y_{\psi\psi} = 0 \tag{7}$$

Now consider the potential function $\phi(x, y)$ in the (x, ψ)-plane. Using the definition of ϕ, the velocity components can be expressed as

$$u = \phi_x - \frac{y_x}{y_\psi}\phi_\psi \qquad v = \frac{1}{y_\psi}\phi_\psi \tag{8}$$

Substituting (8) into the conservation of mass equation (1) and using the Bernoulli equation (3) gives

$$\left(y_\psi^2 - \frac{M_\infty^2}{\rho^{\gamma+1}}\right)\phi_{xx} - 2y_x y_\psi \phi_{x\psi} + (1 + y_x^2)\phi_{\psi\psi} = 0 \tag{9}$$

The local Mach number is given by

$$M^2 = \frac{M_\infty^2 q^2}{\rho^{\gamma-1}} = \frac{M_\infty^2(1 + y_x^2)}{y_\psi^2 \rho^{\gamma+1}} \tag{10}$$

and the speed is given by

$$q^2 = \phi_x^2 - \frac{2y_x}{y_\psi}\phi_x\phi_\psi + \frac{1 + y_x^2}{y_\psi^2}\phi_\psi^2 \tag{11}$$

Multiplying equation (7) by $1 + y_x^2$ and using (10), the equation for $y(x, \psi)$ can be recast in the form

$$y_\psi^2(1 - M^2)y_{xx} + y_\psi^2 y_x^2 y_{xx} - 2y_x y_\psi(1 + y_x^2)y_{x\psi} + (1 + y_x^2)^2 y_{\psi\psi} = 0 \tag{12}$$

Similarly, the equation for $\phi(x, \psi)$ becomes

$$y_\psi^2(1 - M^2)\phi_{xx} + y_\psi^2 y_x^2 \phi_{xx} - 2y_x y_\psi(1 + y_x^2)\phi_{x\psi} + (1 + y_x^2)^2 \phi_{\psi\psi} = 0 \tag{13}$$

Equations (12) and (13) are similar to those obtained from the formulation in the (s, n) co-ordinates, where s is the arc length along a co-ordinate curve and n is in the direction normal to the tangent.[3]

Equations (12) and (13) are solved numerically in the (x, ψ)-plane, subject to appropriate boundary conditions, and the density is evaluated from equation (3) after the speed is obtained from equation (11). The significant advantages associated with using equations (12) and (13) will become apparent in the discussion of boundary conditions and numerical procedures.

BOUNDARY CONDITIONS

In the streamline ordinate form, since the aerofoil surface is considered to be a streamline, the Kutta–Joukowsky condition is satisfied automatically, and hence there is no need to explicitly impose the Kutta–Joukowsky condition at the trailing edge.

Secondly, if the far field boundary is far enough away, a conventional (i.e. without the circulation term) Dirichlet boundary condition on the streamfunction can be used, producing better results than a Neumann boundary condition.[19]

Considering the above comments on the streamfunction formulation, one observes that complete Dirichlet boundary conditions can be imposed on $y(x, \psi)$ in the streamfunction formulation of the analysis problem, leading to easy application of second-order difference schemes for numerical approximation. In the far field, the Dirichlet boundary condition is

$$\psi = y\cos\alpha - x\sin\alpha \tag{14}$$

where α is the angle of attack. This equation places a constraint on the influx and outflux of the mass, but not on the lift.

Boundary condition (14) and the absence of an explicit Kutta–Joukowsky condition at the trailing edge are of remarkable importance. Firstly, as a linear function, $y(x, \psi)$ can be obtained explicitly at the far field. Secondly, condition (14) places a constraint on mass flux and serves as the correct boundary condition for the Euler equations governing rotational flows as well as the potential flows considered here.

Now consider the aerofoil lying along $\psi = 0$, at an angle of attack α, where $0 \leqslant \alpha \leqslant \pi/2$, in an unbounded domain. The boundary conditions for $y(x, \psi)$ are:

$$y(x, \psi) = \frac{\psi}{\cos\alpha} + x\tan\alpha \quad \text{at} \quad x^2 + \psi^2 = \infty$$

$$y(x, 0^{\pm}) = f^{\pm}(x) \quad \text{for} \quad 0 \leqslant x \leqslant 1 \tag{15}$$

and for $\phi(x, \psi)$ are:[20]

$$\phi(x, \psi) = \frac{x}{\cos\alpha} + \psi\tan\alpha + \frac{\Gamma}{2\pi}\tan^{-1}[\sqrt{(1 - M_\infty^2)}\tan(\beta - \alpha)] \quad \text{at} \quad x^2 + \psi^2 = \infty$$

$$\frac{\partial\phi(x, \psi)}{\partial\psi}\bigg|_{\psi=0^{\pm}} = \frac{g_{12}^{\pm}}{g_{11}^{\pm}}\frac{\partial\phi(x, \psi)}{\partial x}\bigg|_{\psi=0^{\pm}} \quad \text{for} \quad 0 \leqslant x \leqslant 1 \tag{16}$$

where $g_{11}^{\pm} = 1 + ((f^{\pm})')^2$, $g_{12}^{\pm} = (f^{\pm})'y_{\psi}$, $f^{+}(x)$ and $f^{-}(x)$ define the given aerofoil upper and lower surfaces respectively, $\tan\beta = y/x$ and Γ is the circulation. The chord length of the aerofoil has been taken as the characteristic length. The last equation expresses the condition of flow tangency at the aerofoil surface, $v = uf'(x)$.

For computational purposes it is convenient to introduce the transformations:

$$Y(x, \psi) = y(x, \psi) - \left(\frac{\psi}{\cos\alpha} + x\tan\alpha\right) \tag{17}$$

$$\Phi(x, \psi) = \phi(x, \psi) - \left(\frac{x}{\cos\alpha} + \psi\tan\alpha\right) \tag{18}$$

$$x = a_1\exp(-a_2\xi^2)\tan\xi \tag{19}$$

$$\psi = \psi(\eta) = \ln\left(\frac{\eta}{1 - \eta}\right) \tag{20}$$

so that the interval $\psi \in [-\infty, \infty]$ is mapped to $\eta \in [0, 1]$, with the aerofoil along $\psi = 0$ now corresponding to $\eta = \frac{1}{2}$ and $x \in [-\infty, \infty]$ has been mapped to $\xi \in [-\frac{1}{2}\pi, \frac{1}{2}\pi]$ with a dense clustering of mesh points for properly chosen a_1 and a_2.[21] Then, the equations for Y and Φ

become:

$$A_1\xi_x^2 Y_{\xi\xi} + A_2\xi_x^2 Y_{\xi\xi} + B\xi_x Y_{\xi\eta} + CY_{\eta\eta} + (A_1 + A_2)\xi_{xx} Y_\xi + DY_\eta = 0 \quad (21)$$

$$A_1\xi_x^2 \Phi_{\xi\xi} + A_2\xi_x^2 \Phi_{\xi\xi} + B\xi_x \Phi_{\xi\eta} + C\Phi_{\eta\eta} + (A_1 + A_2)\xi_{xx} \Phi_\xi + D\Phi_\eta = 0 \quad (22)$$

and the boundary conditions are:

$$Y(\xi, 0) = 0 \qquad Y(\xi, 1) = 0 \qquad Y(\pm\pi/2, \eta) = 0$$

$$Y(\xi, 1/2^\pm) = f^\pm(x) - x\tan\alpha \qquad \text{for} \qquad 0 \leqslant \xi \leqslant \xi_{TE}$$

$$\Phi(\xi, 0) = \frac{\Gamma}{4} \qquad \Phi(\xi, 1) = \frac{3\Gamma}{4} \qquad \Phi\left(\pm\frac{\pi}{2}, \eta\right) = \frac{\Gamma}{2\pi}\tan^{-1}[\sqrt{(1 - M_\infty^2)}\tan(\beta - \alpha)]$$

$$\frac{\partial\Phi(\xi, \eta)}{\partial\eta}\Big|_{\eta = 1/2^\pm} = \frac{(f^\pm)'\left[Y_\eta + \dfrac{4}{\cos\alpha}\right]}{(1 + ((f^\pm)')^2)}\left(\xi_x \frac{\partial\Phi(\xi, \eta)}{\partial\xi}\Big|_{\eta = 1/2^\pm} + \frac{1}{\cos\alpha}\right) - 4\tan\alpha$$

$$\text{for} \qquad 0 \leqslant \xi \leqslant \xi_{TE}$$

where

$$A_1 = \left[\frac{1}{\cos\alpha} + Y_\eta\eta(1 - \eta)\right]^2(1 - M^2) \qquad A_2 = \left[\frac{1}{\cos\alpha} + Y_\eta\eta(1 - \eta)\right]^2(\tan\alpha + \xi_x Y_\xi)^2$$

$$B = -2\left[\frac{1}{\cos\alpha} + Y_\eta\eta(1 - \eta)\right]\eta(1 - \eta)(\tan\alpha + \xi_x Y_\xi)[1 + (\tan\alpha + \xi_x Y_\xi)^2]$$

$$C = \eta^2(1 - \eta)^2[1 + (\tan\alpha + \xi_x Y_\xi)^2]^2 \qquad D = [1 + (\tan\alpha + \xi_x Y_\xi)^2]^2\eta(1 - \eta)(1 - 2\eta)$$

The circulation Γ can be determined by the jump in potential across the trailing edge, i.e.

$$\Gamma = \Phi_{TE}^+ - \Phi_{TE}^- \quad (23)$$

where Φ_{TE}^+ and Φ_{TE}^- are the values of Φ at the upper and lower sides of the trailing edge of the aerofoil. The Kutta–Joukowsky condition for the potential function is satisfied by adjusting Γ until q^2 is continuous at the trailing edge. This means that the right amount of circulation must be determined in order that the fluid particles leave the trailing edge smoothly along the streamline.

In the present work, a standard function

$$\Theta(x, \psi) = \frac{\Gamma}{2\pi}\tan^{-1}[\sqrt{(1 - M_\infty^2)}\tan(\beta - \alpha)]$$

is constructed and is subtracted from the potential function Φ to obtain a function which is single-valued.

Finally, the pressure coefficient is evaluated from

$$C_p = 2(p - p_\infty) = \frac{2(\rho^\gamma - 1)}{\gamma M_\infty^2} \quad (24)$$

NUMERICAL SCHEME

Equations (21) and (22) are solved by the finite difference method. A successive line over-relaxation method has been applied in the usual way, using a modification of the numerical scheme which was introduced by Jameson.[3]

At the subsonic points, all the second-order derivatives are approximated by the central finite difference ($F = Y$ or Φ):

$$F_{\xi\xi} = \left(F_{i-1j}^{n+1} - \frac{2}{\omega}F_{ij}^{n+1} - \left(2 - \frac{2}{\omega}\right)F_{ij}^{n} + F_{i+1j}^{n}\right)/\Delta\xi^2$$

$$F_{\xi\eta} = (F_{i+1j+1}^{n} - F_{i+1j-1}^{n} - F_{i-1j+1}^{n+1} + F_{i-1j-1}^{n+1})/4\Delta\xi\Delta\eta \tag{25}$$

$$F_{\eta\eta} = \left(F_{ij-1}^{n+1} - 2F_{ij}^{n+1} + F_{ij+1}^{n+1} - \frac{2-\omega}{\omega}\left(\frac{\Delta\eta}{\Delta\xi}\right)^2(F_{ij}^{n+1} - F_{ij}^{n})\right)/\Delta\eta^2$$

At the supersonic points, the first term in each of equations (21) and (22) is approximated by

$$F_{\xi\xi} = (2F_{ij}^{n+1} - F_{ij}^{n} - 2F_{i-1j}^{n+1} + F_{i-2j}^{n+1})/\Delta\xi^2 \tag{26}$$

and all other terms are approximated by the central difference

$$F_{\xi\xi} = (F_{i-1j}^{n+1} - 2F_{ij}^{n+1} + F_{i+1j}^{n})/\Delta\xi^2 \qquad F_{\eta\eta} = (F_{ij+1}^{n+1} - 2F_{ij}^{n+1} + F_{ij-1}^{n+1})/\Delta\eta^2$$
$$F_{\xi\eta} = (F_{i+1j+1}^{n} - F_{i-1j+1}^{n+1} - F_{i+1j-1}^{n} + F_{i-1j-1}^{n+1}))/4\Delta\xi\Delta\eta \tag{27}$$

Here $1 \leq \omega \leq 2$ is the over-relaxation parameter, and F_{ij}^{n+1} stands for the updated values of $F(\xi, \eta)$ at mesh point (i, j). The numerical difference approximations introduced above are based on analysis of artificial times.[3,22]

Generally, transonic flow computations require a large number of iterations in order to obtain a converged solution. To accelerate the convergence a simple version of the multigrid method is applied in the streamwise direction. Solutions are obtained by performing 10–20 cycles on a 141 × 42 grid, followed by 30–50 cycles on a 71 × 42 grid, and 30–50 cycles on a 36 × 42 grid. After interpolating on the midpoints, a final solution is obtained by 40 cycles on the 71 × 42 grid, followed by 60 cycles on the 141 × 42 grid. In our numerical tests, converged solutions are achieved within 200–400 cycles regardless of M_∞. The build-up of the supersonic zone is employed as the measure of convergence.

RESULTS AND DISCUSSIONS

Test calculations were performed for the NACA 0012, Chiocchia–Nocilla and RAE 2822 aerofoils. All results are obtained on the final 141 × 42 grid.

Figure 1 gives the pressure distribution on a NACA 0012 aerofoil at $M_\infty = 0{\cdot}50$, $\alpha = -0{\cdot}02°$, a pure subsonic flow problem, and comparison is made with available experimental results.[23] Only the upper surface is plotted since the angle of attack is very small and the computational results are almost symmetrical. For purely subsonic flows, the solution converges within 80 iterations, and our results show excellent agreement with the experiments.

Figure 2 shows the pressure distribution on a NACA 0012 aerofoil at $M_\infty = 0{\cdot}803$, $\alpha = 0{\cdot}05°$, a slightly supercritical flow problem. Our present results, obtained within 150 iterations, agree well with the experimental data.

Figure 3 shows the surface pressure distribution for the transonic flow over a NACA 0012 aerofoil at $M_\infty = 0{\cdot}85$, $\alpha = 1.00°$, in comparison with the computational results of Viviand.[24] Good agreement is obtained, although the shock position is slightly different because of the use of a non-conservative form of equations. Within 200 iterations, the present approach provides accurate solutions.

Figure 4 illustrates the pressure distribution on a NACA 0012 aerofoil at $M_\infty = 0{\cdot}95$, $\alpha = 0.00°$, a highly supercritical transonic flow problem. In this strong shock case, the shock-wave position is almost right on the trailing edge, and our results are in good agreement with the

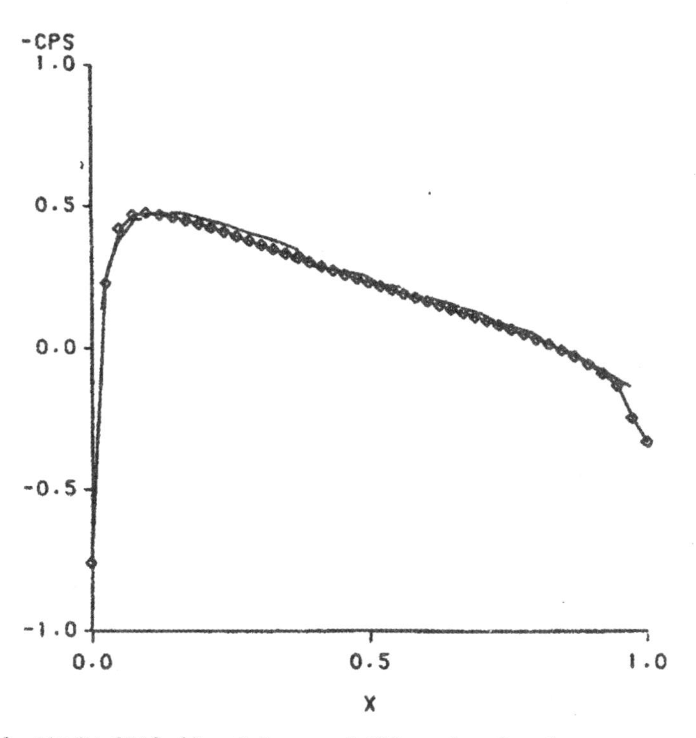

Figure 1. CPS against x for NACA 0012, $M_{\infty}=0{\cdot}5$, $\alpha=-0{\cdot}02°$; —◇—◇—◇— present, ———— Reference 23

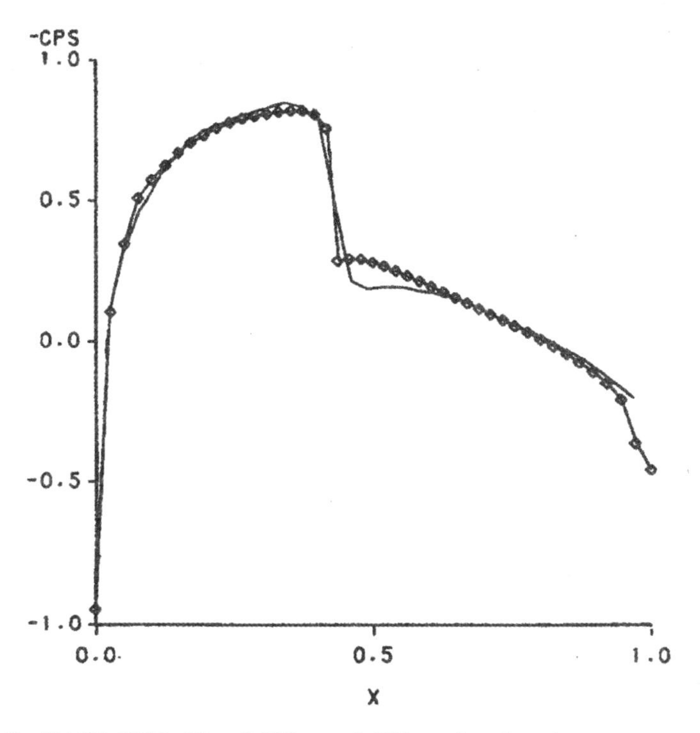

Figure 2. CPS against x for NACA 0012, $M_{\infty}=0{\cdot}803$, $\alpha=0{\cdot}05°$; —◇—◇—◇— present, ———— Reference 23

previous computations.[24] It is interesting to note that nine different codes have been previously tested for this case, and the shock positions vary from the trailing edge at $x=1{\cdot}0$ up to $x=1{\cdot}4$ (for details, see the AGARD Report of Reference 24).

Figure 5 shows the pressure distribution on a Chiocchia–Nocilla aerofoil[24] at $M_{\infty}=0{\cdot}769$, $\alpha=0.00°$, a difficult transonic flow problem. Comparison is given with the computational results

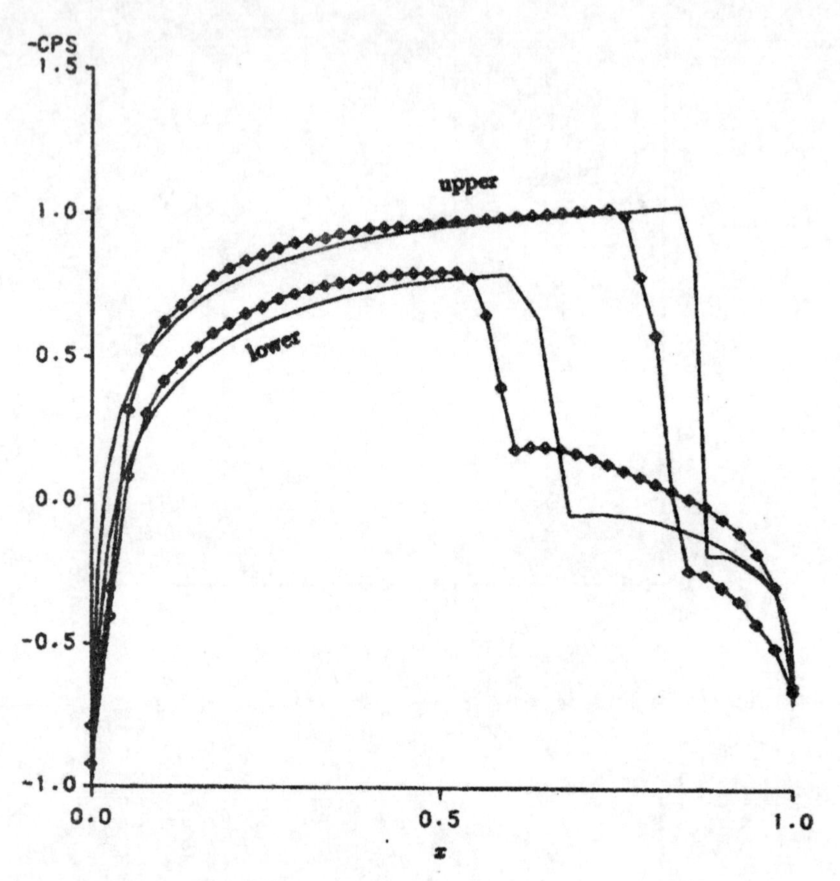

Figure 3. CPS against x for NACA 0012, $M_\infty = 0{\cdot}85$, $\alpha = 1{\cdot}0°$; —◇—◇—◇— present, ———— Reference 24

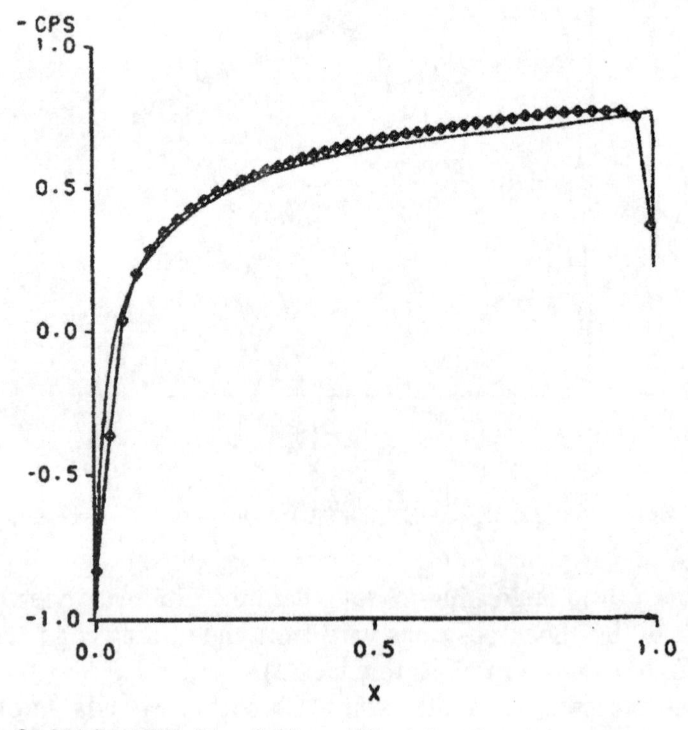

Figure 4. CPS against x for NACA 0012, $M_\infty = 0{\cdot}95$, $\alpha = 0°$; —◇—◇—◇— present, ———— Reference 24

安长发博士科研论文集

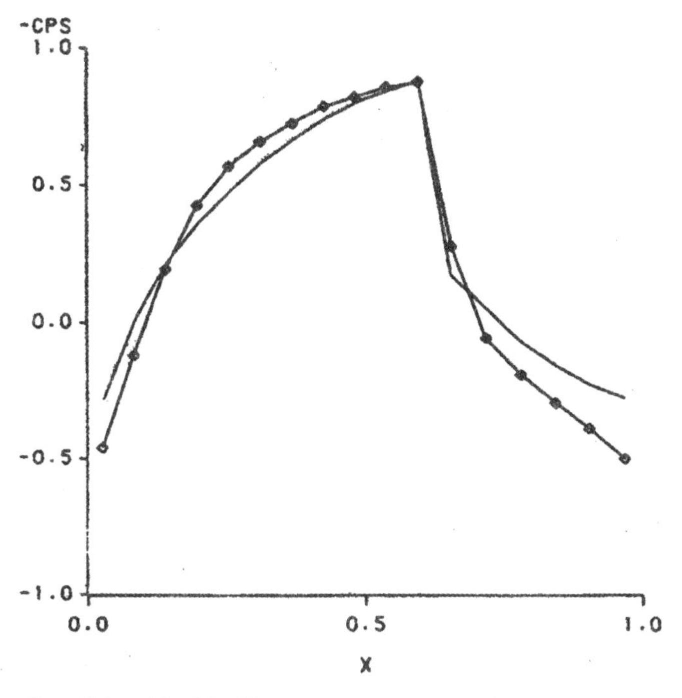

Figure 5. CPS against x for Chiocchia–Nocilla, $M_{\infty}=0\cdot769$, $\alpha=0°$; —◇—◇—◇— present, ———— Reference 24

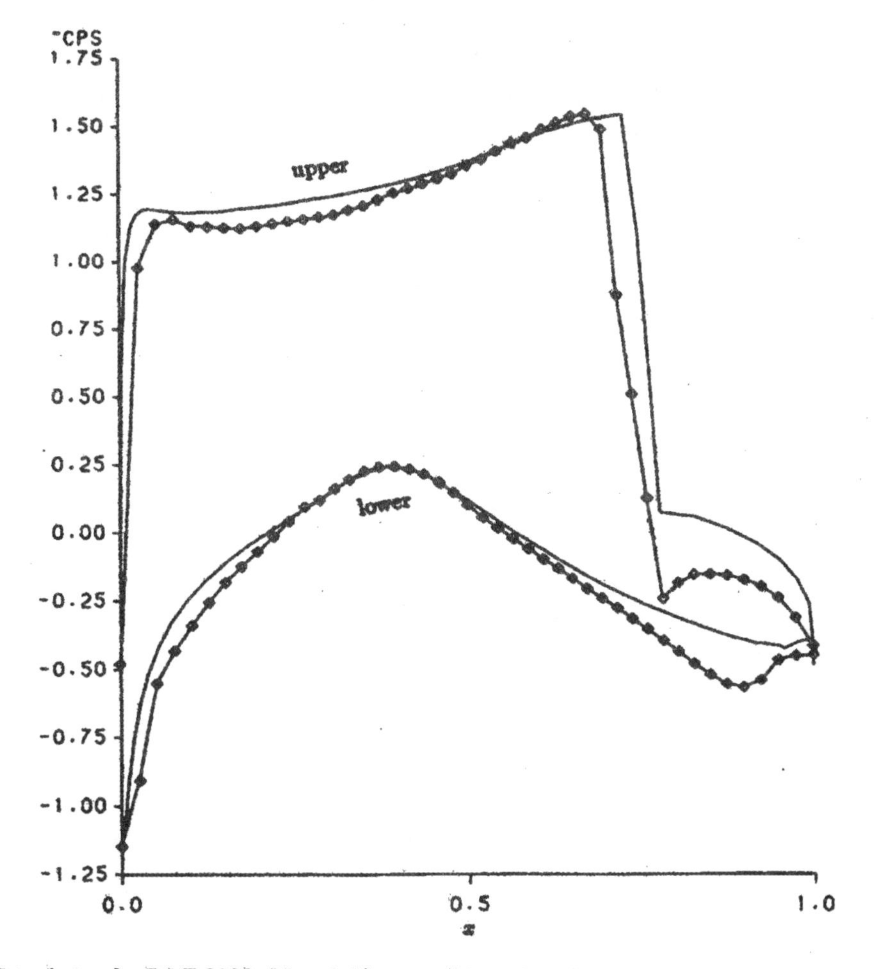

Figure 6. CPS against x for RAE 2822, $M_{\infty}=0\cdot75$, $\alpha=3\cdot0°$; —◇—◇—◇— present, ———— Reference 24

from the AGARD Report of Reference 24 where nine different codes have been tested. Many of the codes were not able to correctly calculate this flow, particularly in the shock-wave region. Our results are in reasonable agreement with one of the best results.[24] In particular, the shock-wave position, which is difficult to capture, is correct.

Figure 6 shows the pressure distribution on an RAE 2822 aerofoil at $M_\infty = 0{\cdot}75$, $\alpha = 3.00°$, and comparison is given with the Euler computational results.[24] There are slight inaccuracies at the leading and trailing edges, but the overall results of the present calculations show good agreement with earlier results.[24]

CONCLUSIONS

As mentioned by Huang and Dulikravich,[11], singular points in the SFC formulation occur where the transformation Jacobian becomes infinity ($u = 0$) or vanishes. Furthermore, for certain other cases, such as for flow with moderate or higher angle of attack, $y(x, \psi)$ becomes double-valued or triple-valued. These problems are associated with the transformation, and these fluid flows, unfortunately, cannot be solved using (x, ψ) co-ordinates.

Other researchers have attempted to extend the streamfunction-co-ordinate methodology to three-dimensional flows, with limited success. However, such flows can in fact be handled using the ideas presented in this paper, provided the right combination of streamfunction and rectangular co-ordinates is chosen.[13]

The numerical method used in this work falls in the category of a shock-capturing scheme. The differencing formulae originally introduced by Jameson[3] allow one to capture accurately the shock locations and strengths, even though the equations are solved in non-conservative form.

There are several key features of the present external transonic potential flow calculation method:

1. Since both streamfunction and potential function are used, the continuity equation and the irrotationality condition have each been satisfied twice.
2. The time-dependent differencing scheme is effective as long as outgoing disturbances do not reflect back into the computational domain. This is achieved by applying boundary conditions at true infinity and adjusting the over-relaxation parameter ω. Normally, ω close to 1 is used at the grid points which are in the subsonic zone and far away from the aerofoil.
3. The equation for $y(x, \psi)$ has served as a grid generation equation, but it is quite different from conventional grid generation. The present method is a natural type of adaptive grid generation (grids are adapted by solving flow equations), since the flow pattern is obtained as part of the solution and used for the grid.
4. Other methods used to compute these types of flows, such as artificial viscosity and density methods, contain some parameters which must often be tuned in order to achieve convergence and also may require adjustment as the Mach number increases. The present method avoids these difficulties.

REFERENCES

1. E. M. Murman and J. D. Cole, 'Calculation of plane steady transonic flows', *AIAA J.*, **9**, 114–121 (1971).
2. P. Garabedian and D. Korn, 'Analysis of transonic airfoils', *Comm. Pure Appl. Math.*, **24**, 841–851 (1972).
3. A. Jameson, 'Iterative solution of transonic flows over airfoils and wings, including flows at Mach 1', *Comm. Pure Appl. Math.*, **27**, 283–309 (1974).

4. M. Hafez, J. South and E. Murman, 'Artificial compressibility method for numerical solutions of transonic full potential equation', AIAA paper 78–1148, 1978, pp. 838–844.
5. R. Barron and C.-F. An, 'Analysis and design of transonic airfoils using streamwise coordinates', in G. S. Dulikravich (Ed.), *Proc. 3rd Int. Conf. on Inverse Design Concepts and Optimization in Eng. Sci.*, Washington, D.C., USA, 1991, pp. 359–370.
6. W. G. Habashi, P. Kotiuga and L. McLean, 'Finite element simulation of transonic flows by modified potential and stream function methods', *Eng. Anal.*, **2**, 150–154 (1985).
7. M. Hafez and D. Lovell, 'Numerical solution of transonic stream function equation', *AIAA J.*, **21**, 327–335 (1983).
8. C. C. L. Sells, 'Plane subcritical flow past a lifting airfoil', *Proc. R. Soc. London A*, **308**, 377–401 (1968).
9. M. H. Martin, 'The flow of a viscous fluid I', *Arch. Ration. Mech. Analy.*, **41**, 266–286 (1971).
10. G. S. Greywall, 'Streamwise computations of two dimensional incompressible potential flows', *J. Comput. Phys.*, **59**, 224–231 (1985).
11. C.-Y. Huang and G. S. Dulikravich, 'Stream function and stream-function-coordinate (SFC) formulation for inviscid flow field calculations', *Comput. Methods Appl. Mech. Eng.*, **59**, 155–177 (1986).
12. R. Barron, 'Computation of incompressible potential flow using von Mises coordinates', *Math. Comput. Simul.*, **31**, 177–188 (1989).
13. S. Zhang, 'Streamwise transonic computations', Ph.D. Dissertation, University of Windsor, Canada, 1993.
14. S. Zhang, R. M. Barron and C.-F. An, 'On streamfunction formulations for transonic computations', to be published.
15. R. Barron and R. K. Naeem, 'Numerical solution of transonic flows on a streamfunction coordinate system', *Int. J. Numer. Methods Fluids*, **9**, 1183–1193 (1989).
16. R. K. Naeem and R. Barron, 'Transonic computations on a natural grid', *AIAA J.*, **28**, 1836–1838 (1990).
17. C.-F. An and R. Barron, 'Numerical solution of transonic full-potential-equivalent equations in von Mises coordinates', *Int. J. Numer. Methods Fluids*, **15**, 925–952 (1992).
18. R. Courant and K. O. Friedrichs, *Supersonic Flow and Shock Waves*, Interscience Publisher, Inc., New York, 1948.
19. H. L. Atkins and H. A. Hassan, 'A new stream function formulation for the steady Euler equations', *AIAA J.*, **23**, 701–706 (1985).
20. G. S. S. Ludford, 'The behavior at infinity of the potential function of a two dimensional subsonic compressible flow', *J. Math. Phys.*, **30**, 131–159 (1951).
21. D. Jones and R. Dickinson, 'A description of the NAE two dimensional transonic small disturbance computer method', NAE Laboratory Technical Report LTR-HA-39, 1980.
22. P. Garabedian, 'Estimation of the relaxation factor for small mesh size', *Math. Tables Aids Comput.*, **10**, 183–185 (1956).
23. J. J. Thibert, M. Grandjacques and L. H. Ohman, 'NACA 0012 airfoil', in AGARD Advisory Report No. 138, 1979.
24. H. Viviand, 'Numerical solutions of two-dimensional reference test cases', in AGARD Advisory Report No. 211, 1985.

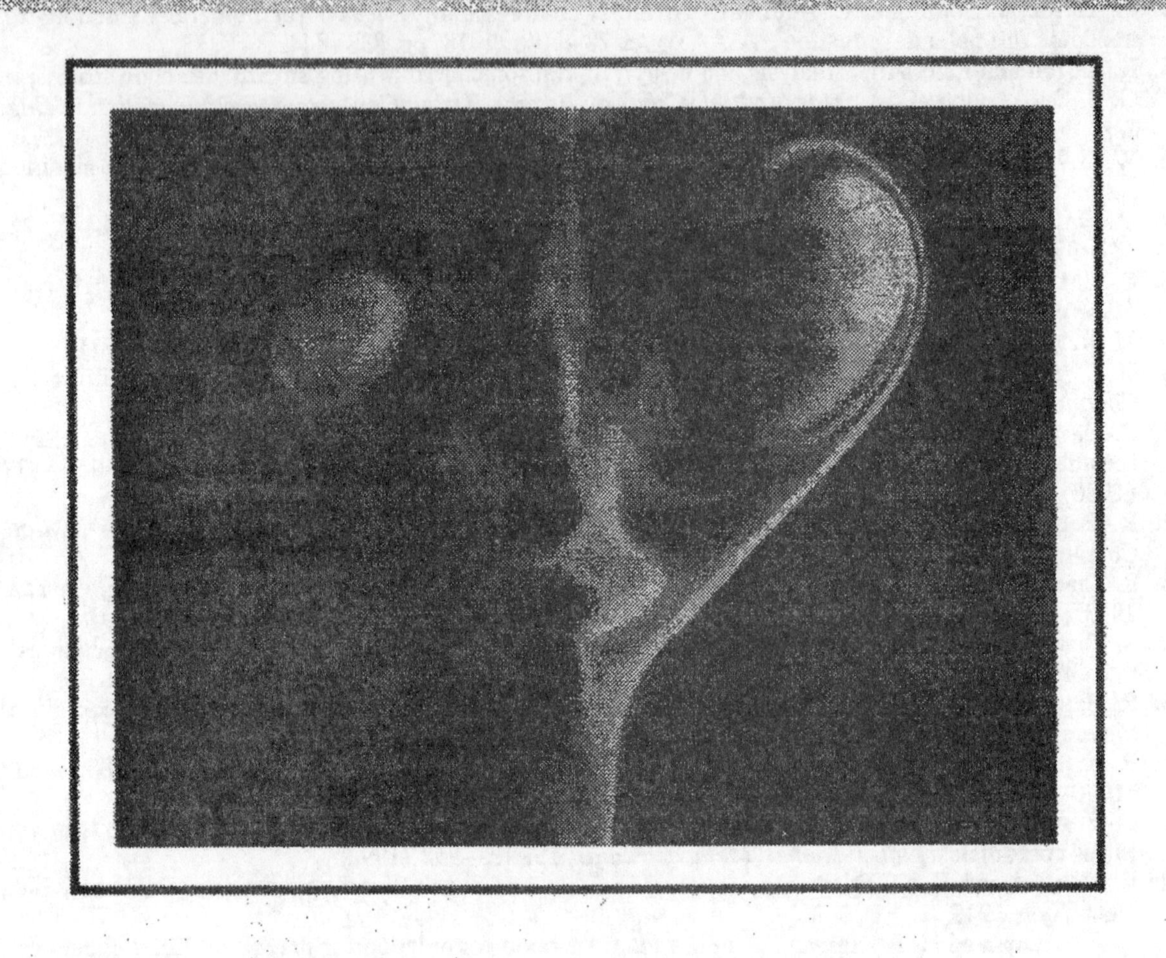

安长发博士科研论文集

THE SIXTH ASIAN CONGRESS OF FLUID MECHANICS

MAY 22 – 26, 1995, SINGAPORE

PROCEEDINGS

VOLUME I

THE SIXTH ASIAN CONGRESS OF FLUID MECHANICS
May 22–26, 1995, Singapore

STREAMFUNCTION COORDINATE FORMULATION VS UNSTEADY INVISCID FLOW*

C.-F. An

Dept. of Mathematics & Statistics
and Fluid Dynamics Research Institute
Univ. of Windsor, Windsor, Ont., Canada

Extended Abstract

The concept of streamfunction coordinates(SFC) and its applications in CFD have received considerable attention in recent years. Among the various SFC formulations, von Mises transformation plays an important role due to a number of advantages. Firstly, this transformation combines flow solver and mesh generator together to form a single set of equations without performing conventional grid generation. Secondly, it is easy to draw streamlines or particle path lines which makes post-processing more expressive and less expensive. Thirdly, and most importantly, when the transformation is applied to one and quasi-one dimensional unsteady flow, the Jacobian is always nonzero and finite. This fact guarantees the validity of the transformation and provides an opportunity to apply the SFC formulation in a variety of one or quasi-one dimensional unsteady flows. In this paper, some recent studies of the SFC formulation and applications by the present author and collaborators have been summarized from the point of view of a unified methodology. The mathematical formulations, numerical algorithms and computed results for several test examples are reported. The advantages and disadvantages of the proposed formulation are also discussed.

(a) If the flow is continuous and isentropic, the governing equations can be reduced to a nonlinear second order hyperbolic partial differential equation, i.e. a nonlinear wave equation:

$$x_{\psi}^{\gamma+1} x_{\tau\tau} = x_{\psi\psi} \tag{1}$$

after introducing streamfunction ψ and von Mises transformation: $x = x(\psi, \tau)$, $t \equiv \tau$. Here x is Cartesian coordinate, τ is time and γ is the ratio of specific heats. There are many numerical schemes available to solve this hyperbolic equation, a three-step leapfrog scheme for example. Having solved for $x(\psi, \tau)$, density and velocity can be evaluated by $\rho = 1/x_{\psi}$, $u = x_{\tau}$.

(b) If the flow has weak discontinuity but it is isentropic, a conservative system of equations in streamfunction coordinates can be derived as below:

$$\begin{bmatrix} x_{\psi} \\ x_{\tau} \end{bmatrix}_{\tau} + \begin{bmatrix} -x_{\tau} \\ x_{\psi}^{-\gamma} \end{bmatrix}_{\psi} = 0. \tag{2}$$

The eigenvalue analysis confirms that this system remains hyperbolic type. Therefore, any marching scheme can be selected to solve it. One of the optimal algorithms is Godunov's predictor-corrector scheme.

(c) If the flow is generally inviscid, possibly containing strong discontinuities (shocks, contact discontinuities, etc. as in the case of shocktube flow), the governing equations can be re-formulated to the following Euler equations in streamfunction coordinates:

$$\begin{bmatrix} x_\psi \\ x_\tau \\ x_\psi e \end{bmatrix}_\tau + \begin{bmatrix} -x_\tau \\ (\gamma-1)(e-x_\tau^2/2x_\psi) \\ (\gamma-1)(e-x_\tau^2/2x_\psi)x_\tau \end{bmatrix}_\psi = 0 \tag{3}$$

where e is total energy per unit volume. It can be verified that the system still remains hyperbolic type and a similar scheme can be utilized. Having solved for the general variables x_ψ, x_τ and e, pressure can be evaluated by $p=(\gamma-1)(e-x_\tau^2/2x_\psi)$.

(d) If the flow is inviscid quasi-one dimensional, a section-variant nozzle flow for example, the above Euler equations in streamfunction coordinates can be extended to the following non-homogeneous equations:

$$\begin{bmatrix} x_\psi \\ x_\tau \\ x_\psi e \end{bmatrix}_\tau + \begin{bmatrix} -x_\tau \\ (\gamma-1)(e-x_\tau^2/2x_\psi) \\ (\gamma-1)(e-x_\tau^2/2x_\psi)x_\tau \end{bmatrix}_\psi = \begin{bmatrix} 0 \\ S^\# \\ 0 \end{bmatrix} \tag{4}$$

where $S^\# = (\gamma-1)(ex_\psi - x_\tau^2/2)S'(x)/S(x)$. $S(x)$ and $S'(x)$ are prescribed section area and its changing rate of the nozzle.

Figure 1 shows the calculated density/velocity distributions compared with the exact solution of characteristics theory for an expansion wave induced by a piston gradually extracted from a tube. The simulation accuracy is excellent. The top part of the figure is particle path lines in the physical plane (x,t). Each line illustrates the locus of a certain particle in the tube. The ease of drawing particle path lines is one of the major advantages of the SFC formulation. One can clearly see from these particle path lines that the air particles move to the left while the rarefaction wave moves to the right.

Figure 2 gives the calculated density/velocity distributions compared with the exact solution and particle path lines for the flow with a shock caused by a piston suddenly pushed in a tube and accelerated to a constant speed. The shock wave is accurately captured and its propagation is well simulated shown in the particle path lines.

Figure 3 shows the calculated density/velocity distributions at two moments compared with the exact solution and particle path lines of a shocktube flow. The shock wave and the contact discontinuity are accurately simulated, especially for the latter. Only one interval is needed to capture this discontinuity. It is a little strange that the simulation accuracy of the rarefaction wave is not as good as the other two waves. The reason is not clear at present.

Figure 4 is a three dimensional demonstration of the density/pressure propagation histories of the shocktube flow. They can give us an intuitive impression of flow evolution with time.

Figure 5 shows the calculated density distributions at two typical moments and particle path lines of a shocktube flow with rarefaction wave reflection on the left end wall. The simulation for the reflected wave is more accurate than for the incidence wave.

Figure 6 shows the density distributions at three typical moments and particle path lines of a shocktube flow including shock reflection on the right end wall and interaction between the reflected shock and the contact discontinuity.

From these figures, one can see that the proposed streamfunction coordinate formulations is powerful and robust to simulate one dimensional unsteady inviscid flows including the flows with shock wave, contact discontinuity, wave reflection and interaction. A dissatisfaction of low simulation accuracy of the rarefaction wave in the shocktube flow is unpleasant. More work is needed to improve this problem. Having obtained these encouraging results, we are confident that the SFC formulation can be extended to more applications, for example, two and three dimensional unsteady flows, the unsteady flow with mass/heat transfer, and so on.

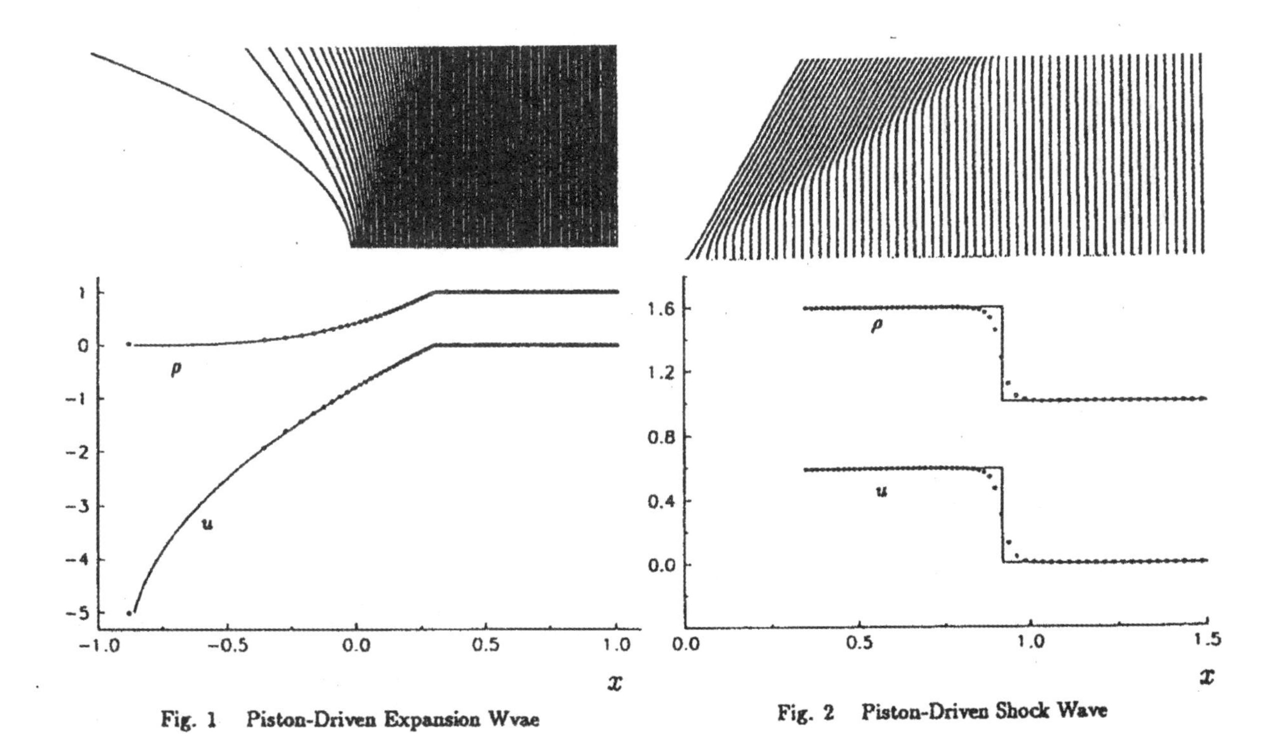

Fig. 1 Piston-Driven Expansion Wvae

Fig. 2 Piston-Driven Shock Wave

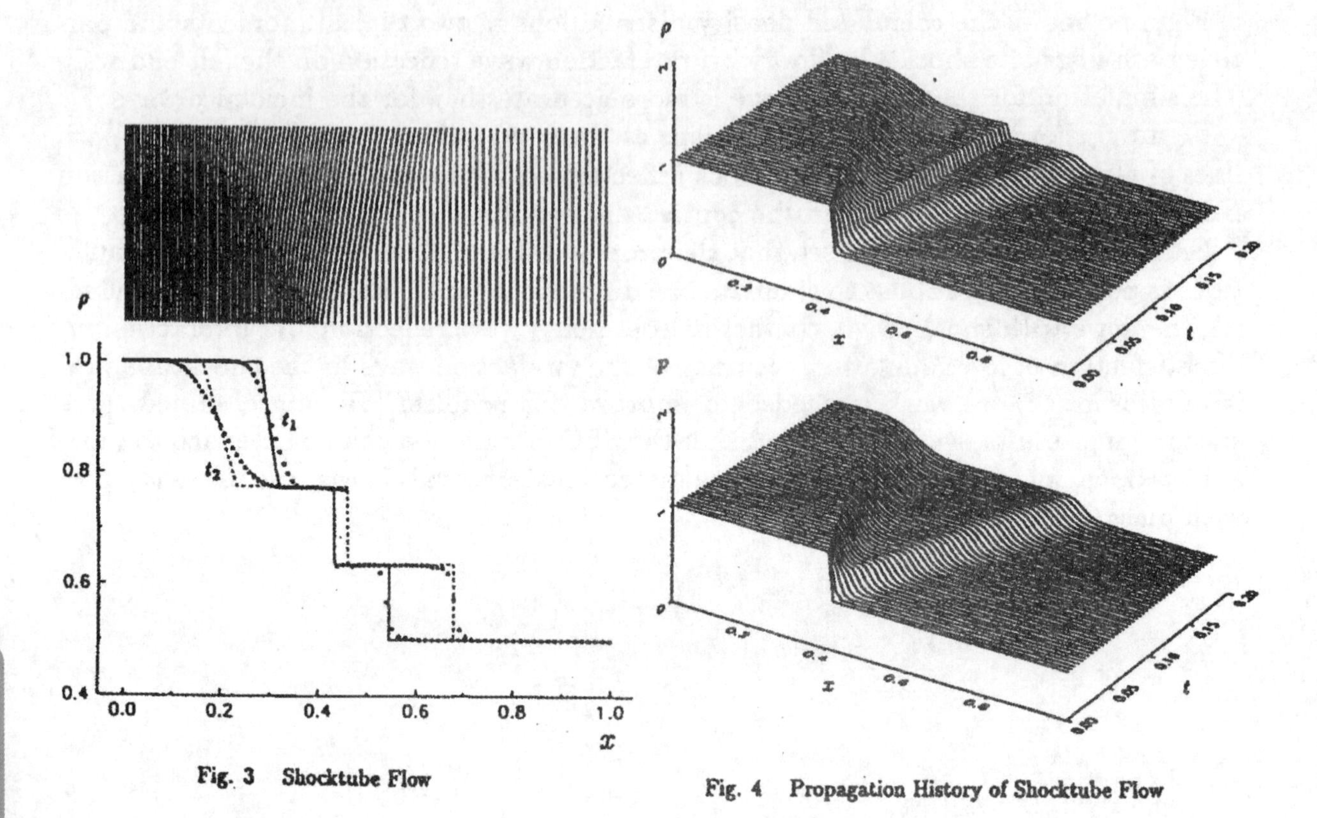

Fig. 3 Shocktube Flow

Fig. 4 Propagation History of Shocktube Flow

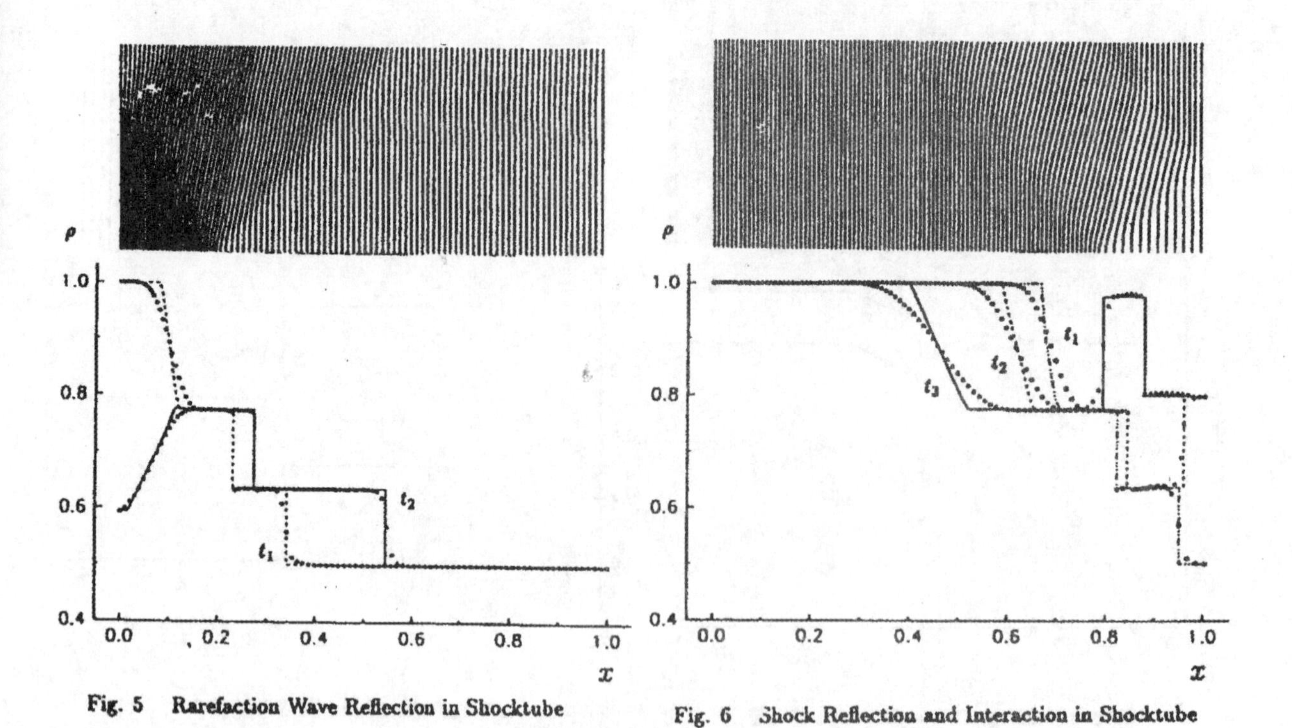

Fig. 5 Rarefaction Wave Reflection in Shocktube

Fig. 6 Shock Reflection and Interaction in Shocktube

安长发博士科研论文集

Stream function coordinate Euler formulation and shocktube application

C.-F. An, R. M. Barron and S. Zhang

Fluid Dynamics Research Institute and Department of Mathematics & Statistics, University of Windsor, Windsor, Ontario, Canada

In this paper, the stream function coordinate SFC Euler formulation for one-dimensional unsteady compressible flow with strong discontinuities has been developed. A conservative system of three equations has been formulated and solved numerically using Godunov's scheme in the computational plane. To illustrate the applicability of the method, a finite length shocktube flow problem has been successfully simulated. The calculated results show good agreement with the exact solution. The present SFC Euler formulation is more powerful than the SFC isentropic formulation and is able to locate exactly the positions and accurately predict the strengths of the shock wave and the contact discontinuity. In fact, only a single grid interval is needed to capture the contact discontinuity. Reflection of the shock wave and rarefaction wave from the tube ends and interaction between the reflected shock wave and the contact discontinuity are also accurately simulated.

Keywords: computational fluid dynamics, Euler, stream function coordinates, shocktube

1. Introduction

In the past few decades, numerous papers have been published on the numerical solutions of nonlinear hyperbolic conservation laws, e.g., the unsteady Euler equations. The states-of-the-art in 1960s, 1970s and 1980s have been summarized by three survey papers.[1-3] However, only a few papers[4-7] have appeared in the literature that deal with unsteady flows using a stream function coordinate (SFC) formulation. Instead some researchers have tackled the Euler equations using the Lagrangian formulation.[8-14] Colella and Woodward presented a piecewise parabolic method in Lagrangian coordinates to solve the Euler equations.[8] Lappas et al.[9] developed an adaptive Lagrangian method for computing one-dimensional flow with or without chemical reaction. Liu et al.[10] proposed an implicit Lagrangian algorithm to solve unsteady gas dynamic equations. Hui and coworkers[11-14] developed the Lagrangian formulation with applications to steady supersonic flow. In principle, the Lagrangian formulation and the SFC formulation possess the same essential features but stem from different considerations. The Lagrangian formulation is based on a reference frame that is fixed in the fluid particle while the SFC formulation focuses on the conservation of physical properties of the flow via the introduction of stream function. Therefore one can say that the Lagrangian formulation and the SFC formulation are equivalent counterparts and the former is more "geometric" while the latter is more "physical."

The SFC formulation has a number of attractive advantages. One of them is that the von Mises–like transformation, generating the SFCs, is always one-to-one when applied to one-dimensional unsteady flow because the Jacobian of the transformation is always positive and finite. Another major advantage is that it is easy to draw the particle pathlines that make the postprocessing more expressive and less expensive. If the flow field is continuous, the governing equations can be simplified to a nonlinear second-order hyperbolic partial differential equation, and many marching schemes are available to compute the flow field.[6] If discontinuities exist in the flow field but they are not very strong, an isentropic conservative SFC formulation can be utilized.[7] However if the discontinuities are so strong that a shock wave and a contact discontinuity appear, as in the case of a shocktube flow, the conservative isentropic SFC formulation fails to capture the contact discontinuity.

In this paper, to overcome this obstacle, the SFC Euler formulation is extended from the SFC isentropic conservative formulation.[7] The governing Euler equations for one-dimensional unsteady flow are reformulated as a conservative system of three equations for three unknown general variables in SFCs. The deduced system remains hyperbolic after the von Mises–like transformation and, therefore, can be solved by marching from initial time. In order to treat

Address reprint requests to Dr. Barron at the Fluid Dynamics Research Institute, University of Windsor, 401 Sunset, Windsor, Ontario, N9B 3P4 Canada.

Received 12 September 1994; revised 21 August 1995; accepted 12 October 1995.

Appl. Math. Modelling 1996 Vol. 20, June

655 Avenue of the Americas, New York, NY 10010

0307-904X/96/$15.00
SSDI 0307-904X(95)00161-1

第二部分 出国后博士学位论文阶段

boundary conditions properly, the rarefaction wave reflection and the shock reflection on the end walls are analyzed. Comparisons of the calculated results with the exact solution confirm the validity and applicability of this approach.

In Section 2, the conservative system of Euler equations in SFCs for one-dimensional unsteady flow is derived from the Euler equations for the primitive variables in Cartesian coordinates via the von Mises–like transformation. Also in this section the eigenvalues of the flux Jacobian matrix of the system are analyzed and the corresponding numerical algorithm is described. In Section 3, the initial and boundary conditions in the computational plane for a selected shocktube flow problem are prescribed considering the wave reflections on the end walls. Sections 4 and 5 outline the theoretical background of the rarefaction wave reflection on the left wall, the shock reflection on the right wall, and the interaction between the reflected shock and the contact discontinuity. This background is the basis of the boundary condition specifications in Section 3. In Section 6, the calculated results are compared with the exact solution and discussed. Some concluding remarks are addressed in Section 7.

安长发博士科研论文集

2. Governing equations and numerical algorithm

One-dimensional unsteady compressible flow for an inviscid fluid is governed by the Euler equations

$$\begin{bmatrix} \rho \\ \rho u \\ e \end{bmatrix}_t + \begin{bmatrix} \rho u \\ \rho u^2 + p \\ u(e+p) \end{bmatrix}_x = 0 \tag{1}$$

where $p=(\gamma-1)(e-\rho u^2/2)$ is pressure, e is total energy per unit volume, ρ is density, u is velocity, and γ is the ratio of specific heats. The variables ρ, e, and u have been nondimensionalized by a reference density ρ_0, reference pressure p_0, and reference velocity $\sqrt{p_0/\rho_0}$, respectively. The independent variables x and t have been nondimensionalized by some characteristic length L and characteristic time $L/\sqrt{p_0/\rho_0}$. For a shocktube flow, it is convenient to choose the stagnation state in the high pressure chamber as the reference state.

Referring to previous work,[6,7] through the introduction of a "stream function" ψ such that

$$\psi_x = \rho \qquad \psi_t = -\rho u \tag{2}$$

and a von Mises–like transformation

$$x = x(\psi, \tau) \qquad t \equiv \tau \tag{3}$$

the Euler equations in Cartesian coordinates (1) can be transformed to those in SFCs, taking the conservative form

$$\begin{bmatrix} x_\psi \\ x_\tau \\ x_\psi e \end{bmatrix}_\tau + \begin{bmatrix} -x_\tau \\ p \\ x_\tau p \end{bmatrix}_\psi = 0 \tag{4}$$

where $p=(\gamma-1)(e-x_\tau^2/2x_\psi)$. Letting

$$q_1 = x_\psi \qquad q_2 = x_\tau \qquad q_3 = x_\psi e \tag{5}$$

be general variables, system (4) can be written in a compact form

$$\mathbf{Q}_\tau + \mathbf{F}_\psi = 0 \tag{6}$$

where

$$\mathbf{Q} = \begin{bmatrix} q_1 \\ q_2 \\ q_3 \end{bmatrix}$$

$$\mathbf{F} = \begin{bmatrix} f_1 \\ f_2 \\ f_3 \end{bmatrix} = \begin{bmatrix} -q_2 \\ (\gamma-1)(q_3 - q_2^2/2)/q_1 \\ (\gamma-1)(q_3 - q_2^2/2)q_2/q_1 \end{bmatrix}$$

The relations between the physical unknowns and the general variables are

$$\rho = 1/x_\psi = 1/q_1 \qquad u = x_\tau = q_2 \qquad e = q_3/q_1$$
$$p = (\gamma-1)(q_3 - q_2^2/2)/q_1 \tag{7}$$

The Jacobian of transformation (3), $J = x_\psi = 1/\rho$, is always positive and finite. The flux Jacobian matrix of the Euler equations (6), defined as $\partial\mathbf{F}/\partial\mathbf{Q}$ possesses three real eigenvalues

$$\sigma_1 = 0 \qquad \sigma_{2,3} = \pm\sqrt{\gamma p \rho} \tag{8}$$

Therefore, system (6) is always hyperbolic.[15] Both Sod's survey[2] and the authors' experience[7] suggest that Godunov's predictor-corrector marching scheme[16] is one of the optimal solvers for such a system. That is

$$\begin{cases} \overline{\mathbf{Q}}_{j+1/2} = \dfrac{1}{2}(\mathbf{Q}_{j+1}^n + \mathbf{Q}_j^n) - \dfrac{\Delta\tau}{\Delta\psi}(\mathbf{F}_{j+1}^n - \mathbf{F}_j^n) \\ \mathbf{Q}_j^{n+1} = \mathbf{Q}_j^n - \dfrac{\Delta\tau}{\Delta\psi}(\overline{\mathbf{F}}_{j+1/2} - \overline{\mathbf{F}}_{j-1/2}) \end{cases} \tag{9}$$

where the vectors with overbars have intermediate values between time levels n and $n+1$. The scheme is stable when the Courant–Friedrichs–Lewy number $(CFL) = \sigma_{max}(\Delta\tau/\Delta\psi) \le 1$. Having solved for $\mathbf{Q} = [q_1\ q_2\ q_3]^t$, the Cartesian coordinate $x = x(\psi, \tau)$ can be calculated by

$$x_j^n \approx x_{j-1}^n + \frac{\Delta\psi}{2}[(q_1)_{j-1}^n + (q_1)_j^n] \tag{10}$$

where $x_j^n = x(\psi_j, \tau^n)$ for $j = 2, 3, \ldots, J_{max}$ and $n = 1, 2, \ldots$

3. Initial and boundary conditions

In order to describe initial and boundary conditions and to test the proposed formulation, let us consider the example of a shocktube flow shown in *Figures 1a* and *1b*. Suppose a tube $(0 \le x \le 1)$ is separated by a diaphragm located at x_d. Both chambers are filled by stationary air $(\gamma = 1.4)$ at the same temperature so that setting $p_1 = 1$ and $p_5 = 0.5$ leads to $\rho_1 = 1$ and $\rho_5 = 0.5$. At moment $t = 0$, the diaphragm is broken and the air in the high pressure cham-

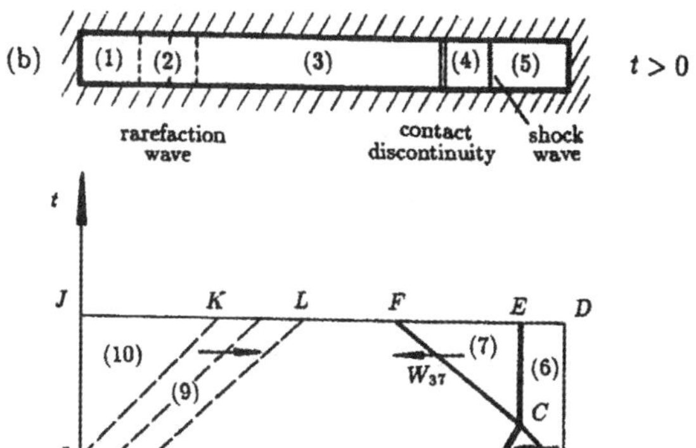

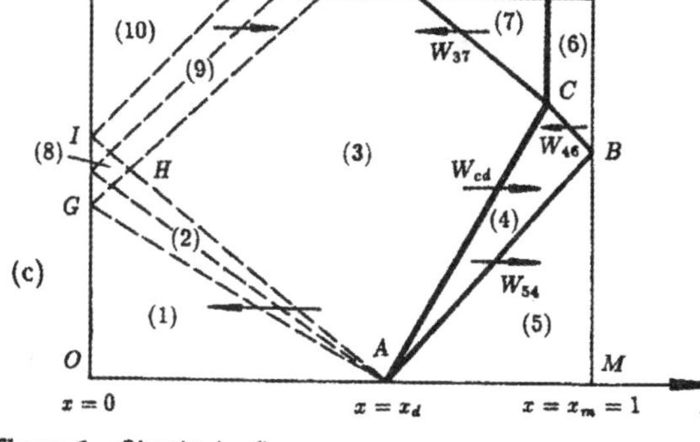

Figure 1. Shocktube flow.

ber rushes to the low pressure chamber. As a result, three waves are produced when $t > 0$: a rarefaction wave travelling to the left (AGI), a shock wave travelling to the right (AB), and a contact discontinuity behind the shock also travelling to the right (AC). Furthermore if the time is long enough, the left-travelling rarefaction wave will hit the left wall (at GI) and reflect back ($GIKL$), and the right-travelling shock will hit the right wall (at B) and reflect back (BC). In this case, both left and right boundary conditions must be treated deliberately to consider these reflections. Furthermore the left-travelling shock reflected from the right wall will encounter the right-travelling contact discontinuity (at C), penetrate through it, and continue to move leftward (CF). The contact discontinuity, however, will stay there or move to the right (CE). *Figure 1c* demonstrates the wave propagations, reflections, and interactions in the physical plane. The treatment of wave reflections and interactions will be discussed in the next two sections.

3.1. Initial conditions

In the physical plane (x, t), the initial conditions for the primitive variables are

$$\rho = \begin{cases} 1 & 0 \le x \le x_d \\ 0.5 & x_d < x \le 1 \end{cases}$$

$$u = 0$$

$$e = \begin{cases} 2.5 & 0 \le x \le x_d \\ 1.25 & x_d < x \le 1 \end{cases} \tag{11}$$

In the computational plane (ψ, τ), the initial conditions for the general variables are

$$q_1 = \begin{cases} 1 & 0 \le \psi \le \psi_d \\ 2 & \psi_d < \psi \le \psi_m \end{cases}$$

$$q_2 = 0$$

$$q_3 = 2.5 \tag{12}$$

where the diaphragm location, ψ_d, and the maximum coordinate, ψ_m, can be determined from

$$\psi = \int_0^x \rho dx = \begin{cases} x & 0 \le x \le x_d \\ (x + x_d)/2 & x_d < x \le x_m = 1 \end{cases} \tag{13}$$

3.2. Boundary conditions

Because the rarefaction wave will eventually hit the left wall, the left boundary conditions should be specified in a piecewise manner (*Figure 1c*). Before the first rarefaction wave reaches the wall (at $t = t_G$), all variables on the boundary are constant and equal to those in region 1. After the last rarefaction wave reaches the wall (at $t = t_I$), all variables on the boundary are also constant and equal to those in region 10. Between times t_G and t_I, all variables on the boundary vary with time except for the velocity which remains zero. Therefore the general variables on the left boundary can be specified as below.

Left ($\psi = 0$);

$$\begin{array}{llll} q_1 = 1 & q_2 = 0 & q_3 = 2.5 & \text{if } t \le t_G \\ q_1 = f(t) & q_2 = 0 & q_3 = g(t) & \text{if } t_G < t < t_I \\ q_1 = \dfrac{1}{\rho_{10}} & q_2 = 0 & q_3 = \dfrac{e_{10}}{\rho_{10}} & \text{if } t \ge t_I \end{array} \tag{14}$$

where functions $f(t)$ and $g(t)$, density ρ_{10} and energy e_{10}, and the times t_G and t_I are evaluated in Section 4.

For the right boundary, piecewise specification of conditions is also required (*Figure 1c*). Before the shock reaches the wall, all variables on the boundary keep their initial values. After the shock leaves from the wall, all variables on the boundary are changed except for the velocity which remains zero. The variables on the boundary are the same as those in region 6. Considering the shock reflection, the right boundary conditions for the general variables in the computational plane can be prescribed as below.

Right ($\psi = \psi_m$):

$$\begin{array}{llll} q_1 = 2 & q_2 = 0 & q_3 = 2.5 & \text{if } t \le t_B \\ q_1 = \dfrac{1}{\rho_6} & q_2 = 0 & q_3 = \dfrac{e_6}{\rho_6} & \text{if } t > t_B \end{array} \tag{15}$$

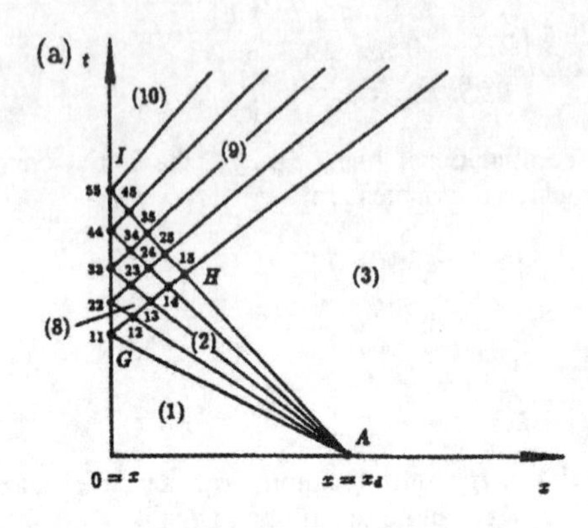

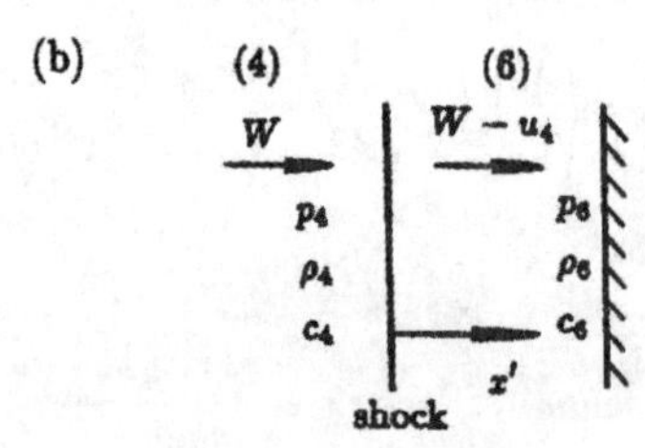

Figure 2. Wave reflection on end walls. (a) Rarefaction wave reflection. (b) Shock wave reflection.

安长发博士科研论文集

where e_6 and ρ_6 are determined from Section 5. The time of shock reflection t_B is given by

$$t_B = \frac{x_m - x_d}{W_{54}} \tag{16}$$

where W_{54} is the speed of the right-travelling shock which can be evaluated using shocktube theory.[17]

4. Rarefaction wave reflection

As stated in the previous section, the left-travelling rarefaction wave will hit the left wall and reflect back, so analysis of the flow field in the reflection region is required. Suppose the region AGH in *Figure 2a* is divided into $(K-1)$ triangular subregions and the velocity u is linearly distributed on the nodes along GH:

$$u_{1,i} = u_{1,1} + \frac{i-1}{K-1}(u_{1,K} - u_{1,1}) \tag{17}$$

for $i = 1, 2, \ldots, K$ where $u_{1,1} = u_1$ and $u_{1,K} = u_3$ are known velocities in regions 1 and 3, respectively. The characteristics theory[17] states that along the first family of characteristics GH, $[dc/du]_I = -(\gamma-1)/2$, or

$$c_{1,i} = c_{1,i-1} - \frac{\gamma-1}{2}(u_{1,i} - u_{1,i-1}) \tag{18}$$

for $i = 2, 3, \ldots, K$. However along each second family of characteristics emitting from point A, all flow properties are invariant and the slope of characteristics is $[dx/dt]_{II} = u - c$ while along the first family of characteristics GH the slope of characteristics is $[dx/dt]_I = u + c$. Using these relations at each node on GH and solving for the coordinates of node $(1, i)$, $i = 2, 3, \ldots, K$, lead to

$$t_{1,i} = \frac{x_d - x_{1,i-1} + (u_{1,i} + c_{1,i})t_{1,i-1}}{2c_{1,i}}$$

$$x_{1,i} = x_d + (u_{1,i} - c_{1,i})t_{1,i} \tag{19}$$

where $u_{1,i}$ and $c_{1,i}$ are obtained from (17) and (18), respectively. Equations (17–19) determine the properties and coordinates of each node on GH. Similarly, the properties and coordinates on GI are

$$u_{j,j} = 0 \qquad c_{j,j} = c_{j-1,j} - \frac{\gamma-1}{2}u_{j-1,j}$$

$$x_{j,j} = 0 \qquad t_{j,j} = \frac{x_d}{c_{j,j}} \tag{20}$$

for $j = 2, 3, \ldots, K$. For the nodes in reflection region GHI, but not on GH nor GI, the properties and coordinates are

$$u_{j,i} = \frac{1}{2}(u_{j,i-1} + u_{j-1,i}) + \frac{1}{\gamma-1}(c_{j,i-1} - c_{j-1,i})$$

$$c_{j,i} = \frac{1}{2}(c_{j,i-1} + c_{j-1,i}) + \frac{\gamma-1}{4}(u_{j,i-1} - u_{j-1,i})$$

$$t_{j,i} = \left[x_{j-1,i} - x_{j,i-1} + (u_{j,i} + c_{j,i})t_{j,i-1} - (u_{j,i} - c_{j,i})t_{j-1,i}\right]/(2c_{j,i})$$

$$x_{j,i} = \left[t_{j-1,i} - t_{j,i-1} + \frac{x_{j,i-1}}{u_{j,i} + c_{j,i}} - \frac{x_{j-1,i}}{u_{j,i} - c_{j,i}}\right] \Bigg/ \left[\frac{1}{u_{j,i} + c_{j,i}} - \frac{1}{u_{j,i} - c_{j,i}}\right] \tag{21}$$

for $j = 2, 3, \ldots, K$ and $i = j+1, \ldots, K$. It is easy to specify other properties, e.g., density, pressure, energy, etc., through the use of isentropic relation. All properties in region 10 are the same as those at node I. An interpolation technique may be used to provide properties at a point other than a node on segment GI. In this way, the functions $f(t)$ and $g(t)$ in the left boundary conditions (14) can be easily evaluated.

5. Shock reflection and interaction with contact discontinuity

5.1. Shock reflection on a wall

Suppose a shock moves to and reflects from a solid wall. W_{54} and W_{46} in *Figure 1* represent the travelling speeds of the incidence and the reflected shocks, respectively, relative to the wall. Let W represent the speed of the reflected shock relative to the air upstream of it, then the "absolute" velocity of the reflected shock relative to the wall is $u_4 - W$. If the reference frame is fixed on the shock, then the velocities of the air upstream and downstream of the shock are W and $W-u_4$, respectively, as shown in *Figure 2b*.

The air flow across the reflected shock is governed by the mass, momentum, and energy conservation laws,

$$\rho_4 W = \rho_6 (W - u_4) \tag{22}$$

$$\rho_4 W^2 + p_4 = \rho_6 (W - u_4)^2 + p_6 \tag{23}$$

$$c_4^2 + \frac{\gamma-1}{2} W^2 = c_6^2 + \frac{\gamma-1}{2}(W-u_4)^2 \tag{24}$$

where $c=\sqrt{\gamma p/\rho}$ is the speed of sound. The Mach number of the reflected shock relative to the upstream air is $M = W/c_4$. Substituting c and M into equations (22)–(24), eliminating p and ρ, and solving for M gives

$$M = \frac{\gamma+1}{4}\frac{u_4}{c_4} + \sqrt{\left[\frac{\gamma+1}{4}\frac{u_4}{c_4}\right]^2 + 1} \tag{25}$$

The reflected shock strength can be evaluated from M

$$\tilde{p}_{46} = 1 + \frac{2\gamma}{\gamma+1}[M^2 - 1] \tag{26}$$

Therefore the properties downstream of the reflected shock are

$$p_6 = \tilde{p}_{46} p_4 \qquad \rho_6 = \frac{(\gamma+1)\tilde{p}_{46} + (\gamma-1)}{(\gamma-1)\tilde{p}_{46} + (\gamma+1)} \rho_4 \tag{27}$$

$$c_6 = \sqrt{\gamma \frac{p_6}{\rho_6}} \qquad e_6 = \frac{p_6}{\gamma - 1} \tag{28}$$

It is noted that the velocity of the air downstream of the reflected shock is $u_6 = 0$. From equations (25)–(28), all physical properties downstream of the reflected shock, i.e., region 6, can be calculated from those upstream of the reflected shock, i.e., region 4. Having calculated the properties in region 6, the boundary conditions on the right wall are easily specified and have been cited in Section 3.

5.2. Shock interaction with contact discontinuity

If the time is long enough, the left-travelling reflected shock will encounter the right-travelling contact discontinuity and interact with it. In fact, this interaction has nothing to do with the numerical solution, e.g., initial and boundary conditions, marching scheme, *CFL* constraint, etc. However a clear theoretical analysis can give a better understanding of the interacted flow field and provide an exact solution, which can be used to validate the numerical simulation. Consider *Figure 1c* and focus on the shock reflection and interaction regions. The contact discontinuity, as a wave, is travelling to the right with speed $W_{cd} = u_4$ while the reflected shock is travelling to the left with speed W_{46}. At moment $t = t_C$, the two waves meet at some point $x = x_C$. After this meeting, the two waves penetrate each other and move apart in their own directions with possibly changed speeds. However we know that the air particles in region 6 remain stationary. Hence the contact discontinuity must be standing at x_C after the moment t_C. Moreover a contact discontinuity can only provide jumps for density and energy, rather than for pressure and velocity. It follows that the pressures and velocities on both sides of the contact discontinuity remain the same, i.e.,

$$u_7 = u_6 = 0 \qquad p_7 = p_6 \tag{29}$$

The rest of the physical quantities in region 7 can be obtained using similar formulas as (25)–(28) with subscripts 4 and 6 replaced by 3 and 7.

6. Test computations and discussion

In order to examine the applicability and accuracy of the SFC Euler formulation, a shocktube flow problem has been computed. In the computational domain $0 \le \psi \le \psi_m$, the ψ coordinate is divided into 100 intervals. This corresponds to a physical domain $0 \le x \le 1$. To investigate the propagation characteristics of the shocktube flow with and without reflection and interaction, three cases have been considered. These cases have the same pressure ratio but different diaphragm locations: $\psi_d = x_d = 0.4$, 0.2, and 0.8. The corresponding computational domains are $0 \le \psi \le \psi_m = 0.7$, 0.6, and 0.9, respectively. The *CFL* number is chosen to be 0.5. Several times have been selected to demonstrate typical flow spectrums for each case.

Figure 3 shows the comparisons of density distribution between the present calculation and the exact solution for the case $x_d = 0.4$ at two typical time levels, $t_1 = 0.101$ and $t_2 = 0.203$. The exact solution is obtained using the shocktube theory.[17] It is observed from these plots that the shock wave and the contact discontinuity have been correctly captured, especially the latter. The position of the contact discontinuity has been exactly located, the constant states on both sides of it have been fully recognized, and its strength has been accurately predicted. The transition of the contact discontinuity needs only a single interval, i.e., it is really a discontinuity jump. However the smearing feature of capturing shock waves in most numerical algorithms also exists for the present calculation. The transition

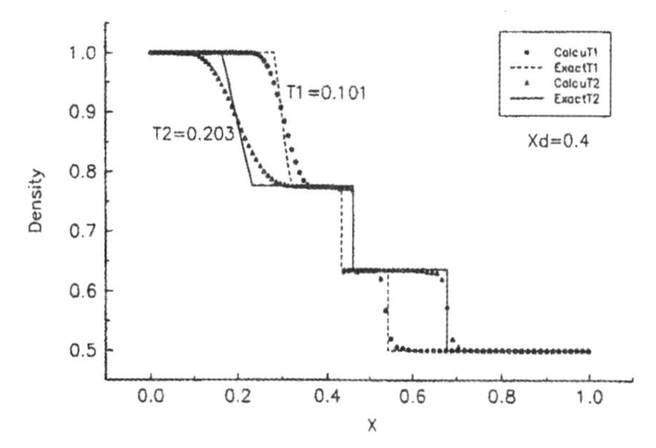

Figure 3. Density distribution ($x_d = 0.4$).

第二部分 出国后博士学位论文阶段

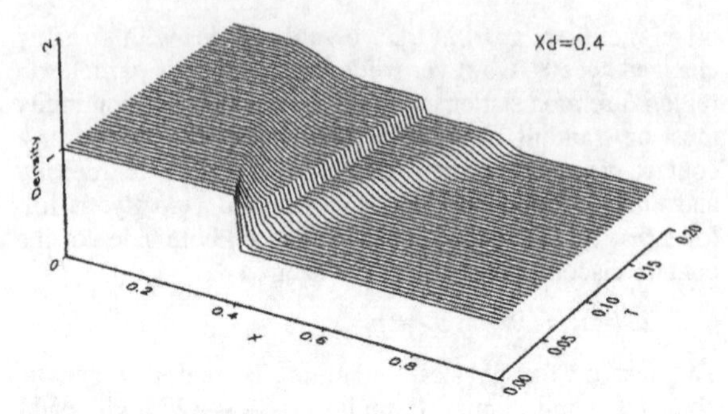

Figure 4. Density evolution ($x_d = 0.4$).

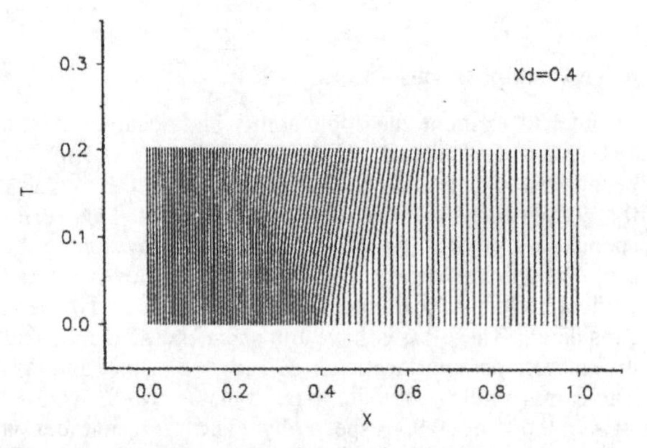

Figure 5. Particle pathlines ($x_d = 0.4$).

of the shock wave occupies four to five intervals, and the corners are rounded. However there is no oscillation either upstream or downstream of the shock. As far as accuracy of the rarefaction wave simulation is concerned, it is not very good, and the reason is unclear at the present time. *Figure 4* is density evolution history in the physical plane for the case $x_d = 0.4$. The whole propagation process of the three waves in the shocktube flow can be three-dimensionally expressed. Once again, the sharp jump of the contact discontinuity can be observed. *Figure 5* shows the particle pathlines in the physical plane for the case $x_d = 0.4$. Each line, drawn for $x = x(\psi_j, t)$, $j = 1, 2, \ldots, J_{max}$, describes the pathline of a certain air particle in the tube. This expressive picture shows a major advantage of the SFC formulation to easily draw the particle pathlines in the physical plane by fixing $\psi = \psi_j$ and evaluating $x = x(\psi_j, t)$. The rarefaction wave, the shock wave, and the contact discontinuity can be identified from the texture of this picture.

Figure 6 shows the comparisons of density distribution between the present calculation and the exact solution for the case $x_d = 0.2$ at two typical time levels, $t_1 = 0.1$ and $t_2 = 0.25$. At time t_1, the rarefaction wave has not yet reached the left wall (see solid dots and dash line). At time t_2, the rarefaction wave has already been reflected from the wall and travels to the right (see triangles and solid line). It can be seen that simulation of the reflected rarefaction wave is considerably improved. Except for the rounded corners, the accuracy is very satisfactory. *Figure 7* is density evolution history in the physical plane for the case

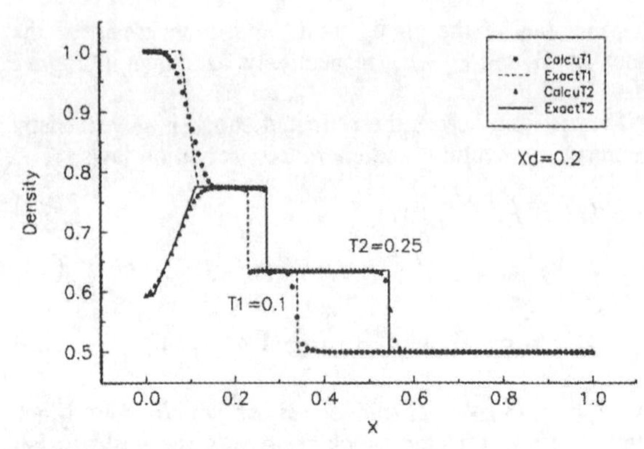

Figure 6. Density distribution ($x_d = 0.2$).

$x_d = 0.2$. It is observed, by paying attention to the top-left corner, that the rarefaction wave reflection has been simulated in a three-dimensional manner.

Figure 8 shows the comparisons of density distribution between the present calculation and the exact solution for the case $x_d = 0.8$ at three typical time levels, $t_1 = 0.11$, $t_2 = 0.1826$, and $t_3 = 0.3385$. At time t_1, three waves have been completely formed and the shock has not yet reached the right wall (see empty circles and double dot dash line). At time t_2, the shock has already been reflected from the wall and moves to the left, but it has not yet met the oncoming contact discontinuity (see solid dots and dash

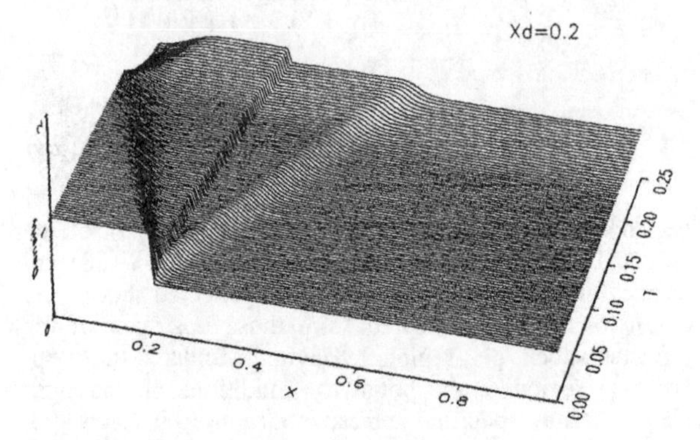

Figure 7. Density evolution ($x_d = 0.2$).

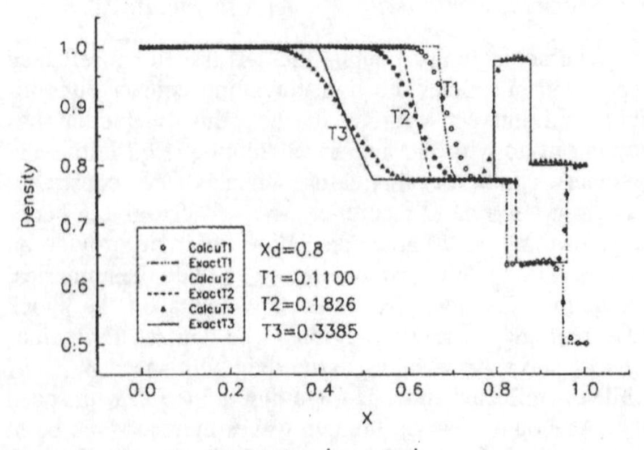

Figure 8. Density distribution ($x_d = 0.8$).

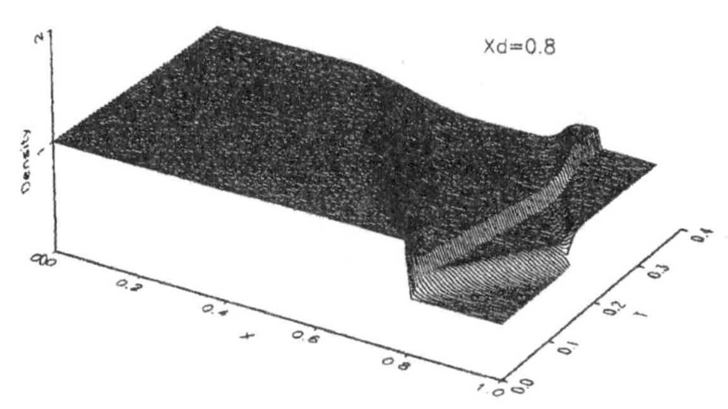

Figure 9. Density evolution ($x_d = 0.8$).

line). At time t_3, the left-travelling reflected shock has already penetrated the right-travelling contact discontinuity and keeps moving leftward (see triangles and solid line). These plots confirm that the present method is capable of simulating the shock reflection on a wall and its interaction with the contact discontinuity. The simulation of the shock reflection remains the same order of accuracy as in the case of an incidence shock wave. The simulation of the shock interaction with the contact discontinuity is satisfactory although the accuracy is not as good as that before the interaction. The maximum relative error of the calculated results is about 3% after the interaction. *Figure 9* is density evolution history in the physical plane for the case $x_d = 0.8$. The propagation of the three waves, the shock wave reflection, and its interaction with the contact discontinuity have been three-dimensionally exhibited.

7. Concluding remarks

For a one-dimensional unsteady compressible flow of an inviscid fluid, the governing Euler equations have been reformulated in SFCs (ψ, τ) as a conservative system of three equations for three unknown general variables. The derived system of equations has been solved numerically using Godunov's predictor-corrector scheme for a shock-tube flow problem including the rarefaction wave reflection, the shock wave reflection, and its interaction with the contact discontinuity.

The calculated results using the present SFC Euler formulation show good agreement with the exact solution, especially for the contact discontinuity. The whole evolution of three wave propagations, the rarefaction wave reflection on the left wall, the shock wave reflection on the right wall, and the interaction between the reflected shock wave and the contact discontinuity have been successfully simulated and graphically demonstrated in a three-dimensional manner.

The advantages of the SFC formulation for one-dimensional unsteady isentropic flow found in previous work, such as positive and finite Jacobian and the ease of drawing particle pathlines, are still retained and play an important role in the present SFC Euler formulation. A new advantage, found in this formulation, is that the contact discontinuity can be exactly predicted. Only a single grid spacing is needed to capture the contact discontinuity. The main disadvantage of the SFC Euler formulation is the low accuracy of the rarefaction wave simulation, although this may be improved by refining the mesh near the left end of the tube.

Nomenclature

c	speed of sound
CFL	Courant–Friedrichs–Lewy number
e	total energy per unit volume
$\mathbf{F}$	flux vector
f_1, f_2, f_2	components of $\mathbf{F}$
J	Jacobian of von Mises–like transformation
M	Mach number
p	pressure
$\tilde{p}$	pressure ratio of downstream to upstream of a shock
$\mathbf{Q}$	vector of unknowns in general form
q_1, q_2, q_3	components of $\mathbf{Q}$
t	time
u	velocity
W	shock wave speed
x	Cartesian coordinate
γ	ratio of specific heats
ρ	density
σ	eigenvalue of matrix $\partial\mathbf{F}/\partial\mathbf{Q}$
τ	time in computational plane
ψ	stream function

References

1. Richtmyer, R. A survey of difference methods for non-steady fluid dynamics. NCAR Tech. Note 63-2, National Center for Atmospheric Research, Boulder, CO, USA, 1962
2. Sod, G. A. A survey of several finite difference methods for system of nonlinear hyperbolic conservation laws, *J. Comp. Phys.*, 1978, 27, 1–31
3. Dervieux, A. and Vijayasundaram, G. On numerical schemes for solving the Euler equations of gas dynamics. *Numerical Methods for the Euler Equations of Fluid Dynamics*, ed. F. Angrand et al., 1985, pp. 121–144
4. Srinivasan, K. and Spalding, D. B. The stream function coordinate system for solution of one-dimensional unsteady compressible flow problems. *Appl. Math. Modelling* 1986, 10, 278–283
5. Barron, R. M., An, C.-F. and Zhang, S. Survey of the streamfunction-as-a-coordinate method in CFD. *Proceedings of the Inaugural Conference of the CFD Society of Canada*. Montreal, Canada, 1993, pp. 325–336
6. An, C.-F., Barron, R. M. and Zhang, S. Streamfunction coordinate formulation for 1-dimensional unsteady flow. *Math. Models Meth. Appl. Sci.* 1995, 5, 401–414
7. Barron, R. M., An, C.-F. and Zhang, S. Unsteady conservative streamfunction coordinate formulation: 1-D isentropic flow. *Appl. Math. Modelling* 1994, 18, 486–493
8. Colella, P. and Woodward, P. R. The piecewise parabolic method (PPM) for gas-dynamical simulations. *J. Comp. Phys.* 1984, 54, 174–201
9. Lappas, T., Leonard, A. and Dimotakis, P. E. An adaptive Lagrangian method for computing 1D reacting and non-reacting flows. *J. Comp. Phys.* 1993, 104, 361–376
10. Liu, F., McIntosh, A. C. and Brindley, J. An implicit Lagrangian method for solving one- and two-dimensional gasdynamic equations. *J. Comp. Phys.* 1994, 110, 112–133

11. Hui, W. H. and Van Roessel, H. J. Unsteady three-dimensional flow theory via material functions. *AGARD-CP-386*, 1985
12. Loh, C. Y. and Hui, W. H. A new Lagrangian method for steady supersonic flow computation. I. Godonov scheme. *J. Comp. Phys.* 1990, 89, 207–240
13. Hui, W. H. and Loh, C. Y. A new Lagrangian method for steady supersonic flow computation. II. Slip-line resolution. *J. Comp. Phys.* 1992, 103, 450–464
14. Hui, W. H. and Loh, C. Y. A new Lagrangian method for steady supersonic flow computation. III. Strong shock. *J. Comp. Phys.* 1992, 103, 465–471
15. Hoffmann, K. A. *Computational Fluid Dynamics for Engineers*, Chapter 6. Engineering Education System, Wichita, KS, USA, 1989
16. Godunov, S. K. *Mathematical Spornik* 1959, 47, 271–306 (in Russian)
17. Shapiro, A. H. *The Dynamics and Thermodynamics of Compressible Fluid Flow*, volume II, Chapter 24. The Ronald Press Company, New York, 1953

AEROTHERMODYNAMICS OF INTERNAL FLOWS III

PROCEEDINGS OF THE THIRD INTERNATIONAL SYMPOSIUM ON EXPERIMENTAL AND COMPUTATIONAL AEROTHERMODYNAMICS OF INTERNAL FLOWS

September 1 - 6, 1996

China Hall of Science and Technology, Beijing, China

第二部分 出国后博士学位论文阶段

Shen YU, Naixing CHEN, Yinming BAI (Eds.)

Changfa An
Beijing, China
Sept. 1, 1996

World Publishing Corporation
Beijing · Shanghai · Xian · Guangzhou

Numerical Simulation of Jet Impingement Turbulent Flow with Confinement Plate

R.M. Barron

Professor, Department of Mathematics and Statistics, and Department of Mechanical and Material Engineering, and Director, Fluid Dynamics Research Institute, University of Windsor, Windsor, Ontario, N9B 3P4, Canada

C.-F. An

Imperial Oil Resources Limited, Calgary, Alberta, T2L 2K8, Canada

A numerical study of confined turbulent jet impingement flow has been performed using a commercial CFD package, FIDAP. Simulation of fully developed turbulent pipe flow at Re_D=70,000 is used to validate the code. Predicted values of skin friction and Nusselt number are within 2.09% and 1.35% of experimental values, respectively. The standard $k-\varepsilon$ turbulent model is implemented. The flow issuing from this pipe impinges on a flat plate located 6 diameters away. Results of the jet impingement simulation are analyzed by examining the flow variables and turbulent quantities at several cross-section along the plate.

Keywords: jet impingement, turbulent flow, numerical simulation, FIDAP

INTRODUCTION

Jet impingement flows have been found to be useful in industrial practice because of their excellent heat and mass transfer characteristics. Industrial situations where jets are used include drying of paper and textiles, heat treatment of non-ferrous metals, cooling of sensitive electronic components, turbine blade cooling, tempering of glass, industrial washers, etc. Physical measurements of these flows are difficult, and the use of numerical simulation can complement experimental testing. Some typical work on this topic can be found in the literature [1-4].

In the present study a numerical simulation of a confined turbulent jet impingement flow has been performed using a commercial CFD package FIDAP [5]. The problem is two-fold: (a) solve fully developed turbulent flow in a round pipe, calculated values of skin friction coefficient and Nusselt number to validate the code and produce profiles of primitive variables at the exit of the pipe; (b) predict the turbulent jet flow issuing from the above pipe exit and impinging onto a plate. Around the exit of the pipe, there is a confinement plate bounding the jet flow region.

Calculations were performed on an SGI IRIS workstation. FIDAP is a complete CFD package comprised of three main parts: pre-processor (mesh generator and initial or boundary conditions), processor (flow solver) and post-processor (graphical visualizer). The code can simulate 2D or 3D, steady of transient, incompressible or compressible, inviscid or viscous (laminar or turbulent) flows, multi-phase, porous media flows, Newtonian or non-Newtonian flows, flows with heat and mass transfer and chemical reactions, etc. The flow solver is based on the finite element method. The package is well documented including theory, algorithms, boundary condition and initial condition specification, mesh generation, tutorials and examples.

FULLY DEVELOPED TURBULENT PIPE FLOW

In order to specify the inlet boundary conditions for the jet impingement flow considered in the next section, and to validate the computational code, a fully developed turbulent flow of air in a round pipe with constant heat flux at the pipe wall is solved.

Relevant parameters are:
pipe diameter = D = 1m
pipe length = L = 100m
density = ρ = 1.205 kg/m^3
viscosity = μ = 1.81 x 10^{-5} kg/(m.s)
Reynolds number = Re_D = 70,000
Prandtl number = Pr = 0.71
thermal conductivity
k = 0.0255346 kg.m/(s^3.K)
specific heat at constant pressure
C_p = 1004.64 m^2/(s^2.K)
wall heat flux = q" = 1000.

The computed results for skin friction coefficient C_f and Nusselt number Nu_H can be compared with those obtained from the empirical formulae

$$1/\sqrt{C_f/2} = 2.46\ln(\mathrm{Re}_D\sqrt{C_f/2}) + 0.30,$$

$$Nu_H = 0.022\,\mathrm{Re}_D^{0.8}\,\mathrm{Pr}^{0.6}$$

which give C_f=0.004837 and Nu_H=134.66.

The standard $k-\varepsilon$ turbulence model given in FIDAP implements the following wall function

$$u^+ = \kappa^{-1}\ln(1+\kappa y^+) + 7.8\left[1 - e^{-y+/11} - y^+ e^{-0.33y+}/11\right]$$

where

$u^+ = u/u^*$, $y^+ = \rho u^* y/\mu$, $u^* = (\tau_w/\rho)^{1/2}$ is friction velocity and κ= 0.41 is the von Karman constant. A uniform 100 x 40 mesh with 9-noded quadrilateral elements has been used to discretize the flow region. The value of y^+ in the first cell near the wall for this mesh is 43.03 which falls in the range 30 – 200 recommended for the standard $k-\varepsilon$ model with the above wall function.

To reduce storage requirement and computational time, a separate algorithm is also available in which an isothermal computation is first performed and the energy equation is then solved for temperature.

With the convergence tolerance of 1 x 10^{-3}, FIDAP required 22 iterations, using the successive substitution algorithm, and a total CPU time of 2800.48 sec. The computed results for the wall friction coefficient and Nusselt number are C_f = 4.938 x 10^{-3} and Nu_H=136.48, representing relative errors of 2.09% and 1.35%, respectively. The results also indicate that the flow is nearly fully developed after about 50% of the total pipe length. The end effect of the pipe causes some ripples in the numerical solution near the exit. Hence, the profiles at the section which is 90m from the inlet are used as the exit profiles and imported as entrance conditions for the jet impingement flow.

JET IMPINGMENT FLOW

The jet impingement flow is illustrated in Fig. 1. The inlet boundary condition is taken from the outlet date (at 90m) of the pipe. On the symmetry line u_r=0, and on the impingement and confinement plates $u_r = u_z = 0$. A clustered mesh system covers the physical domain $0 \le r \le 7$, $0 \le z \le 6$, with 5841 cells. The standard $k-\varepsilon$ turbulence model is used.

FIDAP provides several options for the solution algorithm for the discretized equations. We have used successive substitution. For a tolerance of 3×10^{-3}, the number of iterations for convergence is 22, requiring 10375.56 sec. of CPU time. The convergence history is shown in Fig. 2. Figure 3 - 10 show the plots of u_z, u_r, k and ε as a function of z at r = 0.5 and r = 2.0. Other results have been documented in a technical report [6].

At r = 0.5 (end of pipe wall), u_z rises rapidly in the region close to the inlet, then maintains a nearly constant value until near the plate, where it falls sharply to zero (Fig. 3). Figure 4 indicates that u_r remains almost zero across two-thirds of the region, increasing dramatically near the plate due to the sudden turning of the flow along the plate. The numerical oscillation seen in this figure can be reduced by refining the mesh in this region. The turbulent kinetic energy k and dissipation rate ε, shown in Figs 5 and 6, have peak near the inlet and large increase close to the plate.

At r = 2.0, u_z changes sign from positive to negative at approximately z = 5.0 (Fig. 7). Figure 8 shows that u_r has a maximum at this point. This indicates the existence of a recirculation region in the flow, as can also be seen from streamline plots. k and ε maintain constant values across two-thirds of the region, with sharp increases near the impingement plate, as seen in Figs. 9 & 10.

CONCLUSIONS

FIDAP has the capability to simulate confined turbulent flow impinging on a flat plate. The free jet boundary can be clearly observed and there is a large recirculation between the confinement plate and the impingement plate. The highest speed occurs in the jet region and extends across most of the layer between the two plates. Maximum vorticity appears near the inlet. Pressure contours show that it is maximum at the stagnation point (r = 0, z = 6).

REFERENCES

[1] B.E. Launder and W. Rodi, "The turbulent wall jet - measurements and modeling," Ann. Rev. Fluid Mech. **15**, pp. 429-459 (1983).
[2] S.-H. Chuang, "Numerical simulation of an impinging jet on a flat plate," Int. J. Num. Methods in Fluids, **9**. pp. 1413-1426 (1989).
[3] D. Cooper, D.C. Jackson, B.E. Launder and G.X. Liao, "Impinging jet studies for turbulence model assessment – I, Flow field experiments," Int. J. Heat Mass Transfer, **36**, pp. 2675 – 2684 (1993).
[4] T.J. Craft, L.J.W. Graham and B.E. Launder, "Impinging jet studies for turbulence model assessment – II, An examination of the performance of four turbulence models," Int. J. Heat Mass Transfer, **36**, pp. 2685 - 2697 (1993).
[5] FIDAP Manual, Version 7.0, Fluid Dynamics International, Inc., Chicago, IL, USA (1993).
[6] C.-F. An and R.M. Barron, "Jet impingement flow simulation using FIDAP," Fluid Dynamics Research Institute Report FDRI- TR-95-02, University of Windsor, Canada (1995).

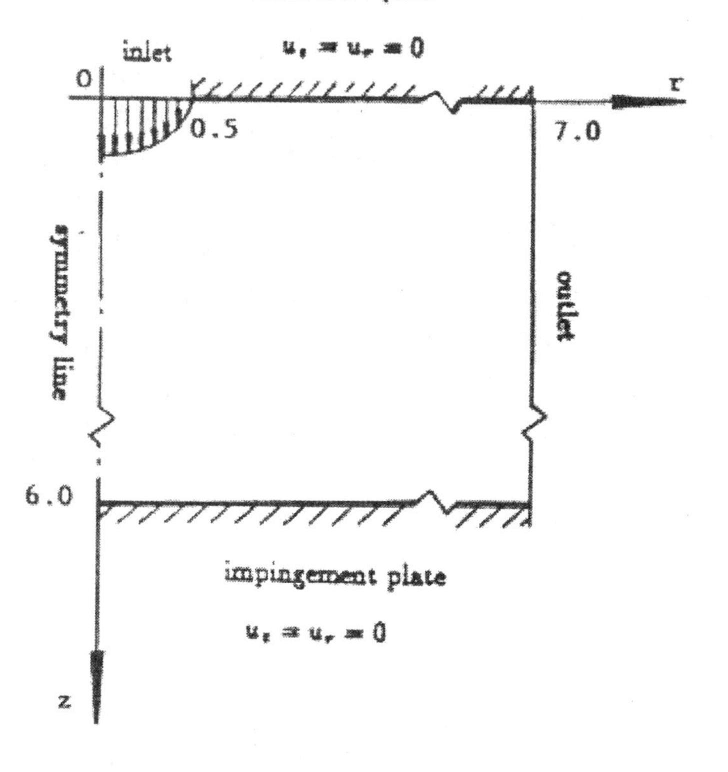

Fig. 1 Turbulent Jet Impingement Flow

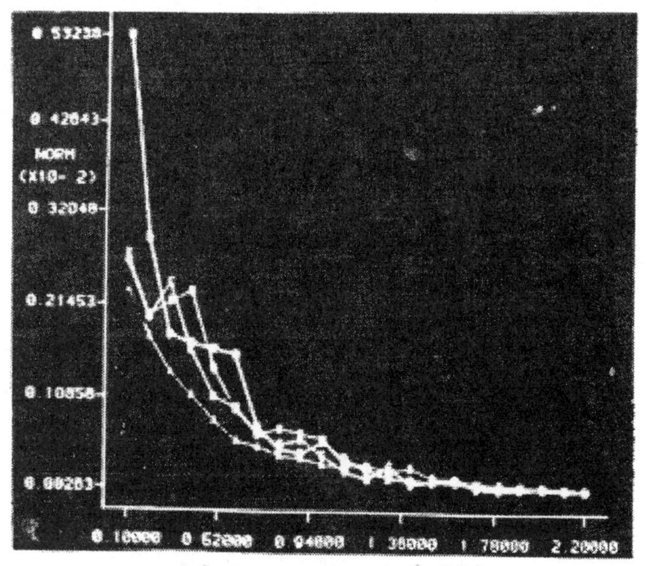

no. of iterations (x10)

Fig. 2 Convergence History for Jet Flow

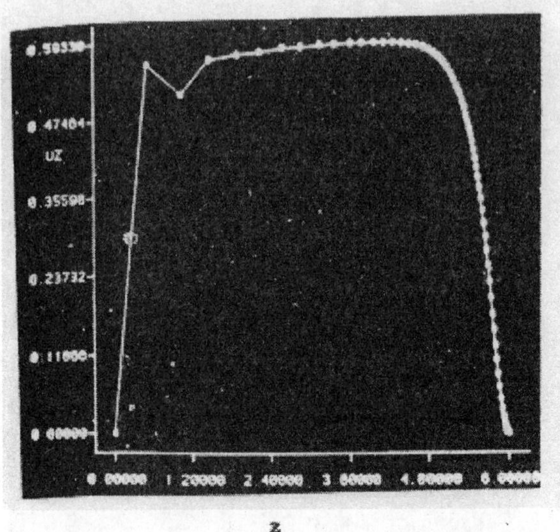

Fig. 3 Velocity Component u_z(r=0.5)

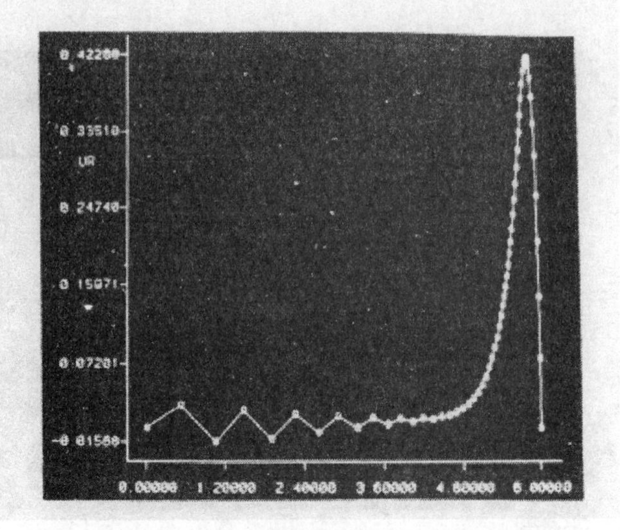

Fig. 4 Velocity Component u_r(r=0.5)

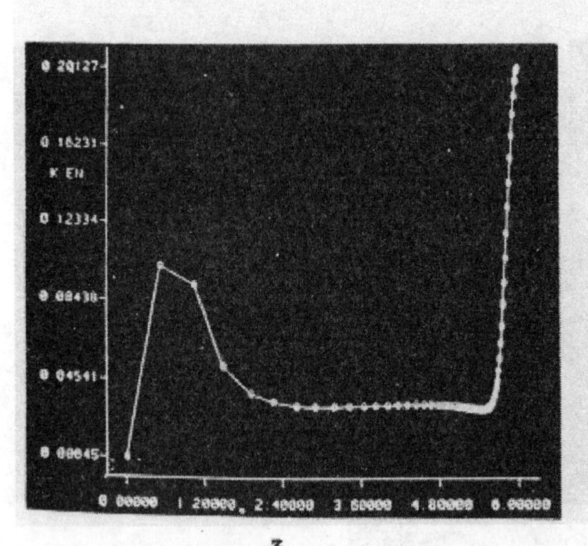

Fig. 5 Turbulent Kinetic Energy (r=0.5)

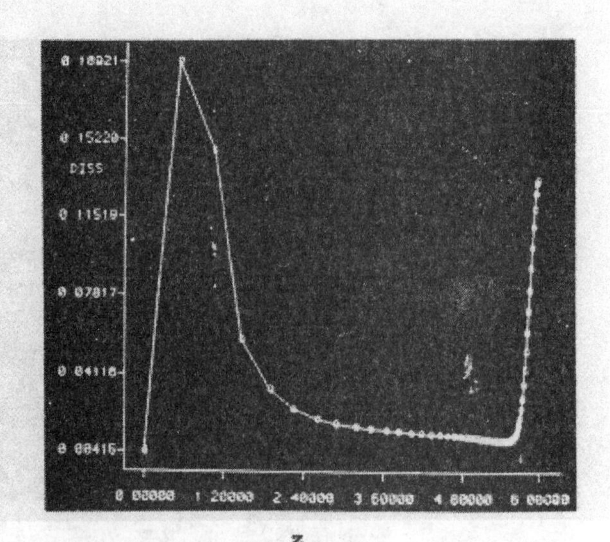

Fig. 6 Dissipation Rate (r=0.5)

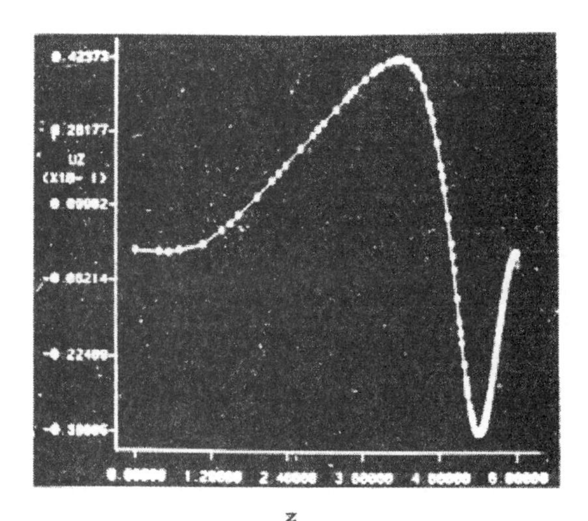

Fig. 7 Velocity Component u_z(r=2.0)

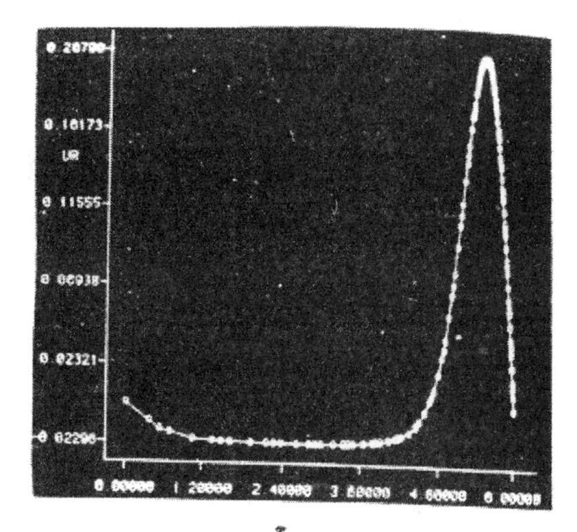

Fig. 8 Velocity Component u_r(r=2.0

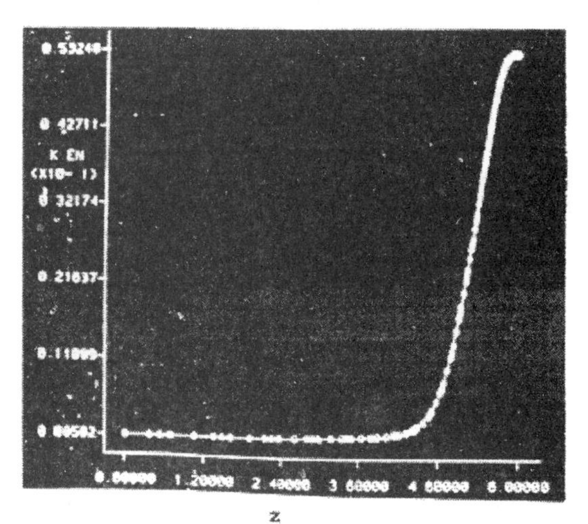

Fig. 9 Turbulent Kinetic Energy (r=2.0)

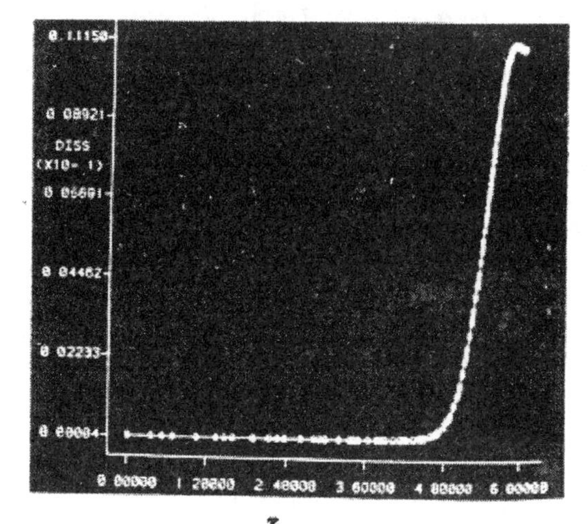

Fig. 10 Dissipation Rate (r=2.0)

ISSN 0334-0082

INTERNATIONAL JOURNAL OF TURBO & JET-ENGINES

CONTENTS VOL. 13, NO. 4, 1996

Page

Measurement of the Flow Inside the Tip Clearance of a Rotating Gas Turbine Rotor
D. Adler, M. Kleiner and S. Balaban 227

A Review on Heat Recovery Exchangers for Very Small Engines
C.F. McDonald 239

A New Pseudo-Potential Model for Rotational Turbo-Flow: (I) Variational Formulation and Finite Element Solution for Transonic Blade-to-Blade Flow
G-L. Liu and H-G. Wang 263

Vortical Structures in Channel Flows with a Backward-Facing Step
T.W.H. Sheu, T.P. Chiang and S.F. Tsai 277

Turbulent Jet Impingement Flow Computation
R.M. Barron and C.F. An 295

© Freund Publishing House, Ltd., London International Journal of Turbo and Jet Engines, 13, 295–306 (1996)

Turbulent Jet Impingement Flow Computation

R.M. Barron*
Fluid Dynamics Research Institute, University of Windsor, Windsor, Ontario, Canada N9B 3P4

and

C.-F. An
Imperial Oil Resources Limited, 3535 Research Road, N.W., Calgary, Alberta, Canada T2L 2K8.

Abstract

Computational studies of a confined turbulent jet impingement flow have been performed using a commercial CFD software – FIDAP. The computation is divided into two parts: (i) a fully developed pipe flow at $Re_D = 70,000$ and (ii) a consequent jet impingement flow. For both the pipe flow and jet impingement flow, the standard $K-\varepsilon$ turbulence model has been implemented. The fully developed pipe flow computation is used to validate the CFD code and to provide the inlet condition of the jet impingement flow. Predicted values of skin friction coefficient C_f and Nusselt number Nu_H are within 2.09% and 1.35%, respectively, of the theoretical results. The flow issuing from the pipe impinges normal to a flat plate located at 6 pipe diameters away. Results of the jet impingement flow computation are analyzed by examining velocity vector, contours and profiles of flow variables and turbulent quantities.

Key words: Jet impingement, turbulent flow, CFD

1. Introduction

Jet impingement flows have been found to be useful in practice because of their excellent heat/mass transfer and reaction characteristics. Industrial situations where jets can be used are: drying of paper and textiles, heat treatment of non-ferrous metals, cooling of sensitive electronic components, turbine blade cooling, tempering of glass, etc. It is well known that jet propulsion plays an essential role in aeronautics and astronautics. During lift-off of aerospace vehicles, the resulting jet impingement flow involves complex physics and attracts multi-disciplinary interests. Because physical measurements of jet flows are sometimes difficult to obtain, the use of numerical simulation can complement experimental testing. Realizing the importance of jet impingement flow, many researchers /1-4/ have dealt with this topic and the 3rd Annual Conference of the CFD Society of Canada (CFD95) set it as a challenging benchmark problem /5/ to provide an opportunity for CFD groups and individuals to validate their formulations, algorithms and codes, to exchange information about this topic and compare their results and experiences.

The computational code used in this project is a commercial software, FIDAP, version 7.0 /6/. It is a complete CFD package comprised of three main parts: pre-processor (mesh generator), processor (flow solver) and post-processor (graphical visualizer). The code can solve 2D and 3D, steady and transient, incompressible and compressible, inviscid and viscous (laminar and turbulent) flows, multi-phase and porous media flows, Newtonian and non-Newtonian flows, flows with heat transfer, mass transfer and chemical reaction and so on. The applications of the software include aerospace, automotive, machinery, electronic, chemical, plastic, metal, glass, paper, food and other industries.

* to whom correspondence should be addressed.

The software can be implemented by batch commands and/or interactive menu-drive commands. The mesh generator can produce both structured (mapped) and unstructured (paved) meshes. It can import geometry from and export results to other software, such as I-DEAS, PATRAN, ANSYS and ICEM-CFD. The flow solver is based on the finite element method (FEM). The post-processor can give a variety of visualizations of flow fields: vector plots, contours, profiles, convergence history and so on. In addition, the package is well documented. The users' manuals contain all aspects of the software: theory, algorithm, boundary condition and initial condition specification, mesh generation, tutorials, examples and others. It also gives descriptions on some important topics in fluid mechanics: turbulence modeling, free-surface flow modeling, compressible flow modeling, nondimensionalization and so on.

Section 2 is dedicated to the first part of the study, computation of fully developed turbulent pipe flow which provides the inlet conditions of the jet impingement flow. The main part of the study, jet impingement flow computation, is reported in section 3 and some remarks and comments are given in section 4.

2. Fully Developed Turbulent Pipe Flow

In order to validate the computational code and provide inlet conditions for the jet impingement flow, a fully developed turbulent flow (Re_D = 70,000) of air (ρ = 1.205 kg/m^3, μ = 1.81 x 10^{-5} kg/m.s, Pr = 0.71) in a round pipe (diameter D = 1 m, length L = 100 m) with constant heat flux at the wall ($\dot{q}''$ = 1,000) is solved. The computed results for skin friction coefficient C_f and Nusselt number Nu_H can be compared with values obtained from the theoretical formulas /7,8/:

$$1/\sqrt{C_f/2} = 2.46ln(Re_D\sqrt{C_f/2}) + 0.30 \quad (1)$$

$$Nu_H = 0.022Re_D^{0.8}Pr^{0.6} \quad (2)$$

which give C_f = 0.004837 and Nu_H = 134.66.

To solve turbulent flow, a standard $K-\varepsilon$ model available in FIDAP is implemented. For an axi-symmetric flow, velocity components, u_z, u_r, turbulent kinetic energy K, its dissipation rate ε, pressure p and temperature T are solved numerically. To reduce storage requirement and computational time, a separate algorithm is used in which an isothermal computation is first performed to solve for u_z, u_r, K and ε (p is calculated afterwards from u_z, u_r). The energy equation is then solved for temperature T.

The boundary conditions are shown in Fig. 1. Based on Re_D = 70,000, the inlet boundary conditions are $u_z = V_{in}$ = 1.0514523 m/s, u_r = 0, K_{in} = 0.004146, ε_{in} = 0.005339, T_{in} = 273K. Here, turbulence intensity I = 0.05 is assumed and the following formulas are used:

$$u_z = V_{in} = \frac{\mu}{\rho D}Re_D, \quad K_{in} = 1.5(IV_{in})^2,$$
$$\epsilon_{in} = \frac{K_{in}^{3/2}}{0.1(D/2)}. \quad (3)$$

For the wall boundary, $u_z = u_r$ = 0, $\dot{q}''$ = 1,000. For the symmetry line boundary, u_r = 0, $\dot{q}''$ = 0. By the convention of FIDAP, K and ε are not prescribed explicitly on the wall and symmetry line and nothing need be specified at the outlet boundary. It is strongly suggested by FIDAP that a good initial condition (rather than the default value of zero) be imposed to accelerate convergence. In this problem, the same values as the inlet boundary condition are chosen as initial condition.

Summary of the pipe flow data:

- Length: L = 100m
- Diameter: D = 1m
- Reynolds Number: Re_D = 70,000
- Inlet Velocity: V_{in} = 1.0514523 m/s, based on ρ, μ, D and Re_D
- Inlet Temperature: T_{in} = 273K
- Inlet Kinetic Energy: K_{in} = 0.004146, based on Intensity I = 0.05
- Inlet Dissipation Rate: ε_{in} = 0.005339
- Wall Heat Flux: $\dot{q}''$ = 1,000

Physical properties of air:

Density: $\rho = 1.205$ kg/m^3

Viscosity: $\mu = 1.81 \times 10^{-5}$ kg/m.s

Thermal Conductivity: $k = 0.0255346$ kg.m/s^3.K

Specific Heat: $C_p = 1{,}004.64$ m^2/s^2.K

Prandtl Number: $Pr = 0.71$

The standard K - ε turbulence model given in FIDAP provides the following wall function:

$$u^+ = \frac{1}{\kappa} ln(1 + \kappa y^+) + 7.8[1 - e^{-y^+/11} - \frac{y^+}{11} e^{-0.33y^+}] \tag{4}$$

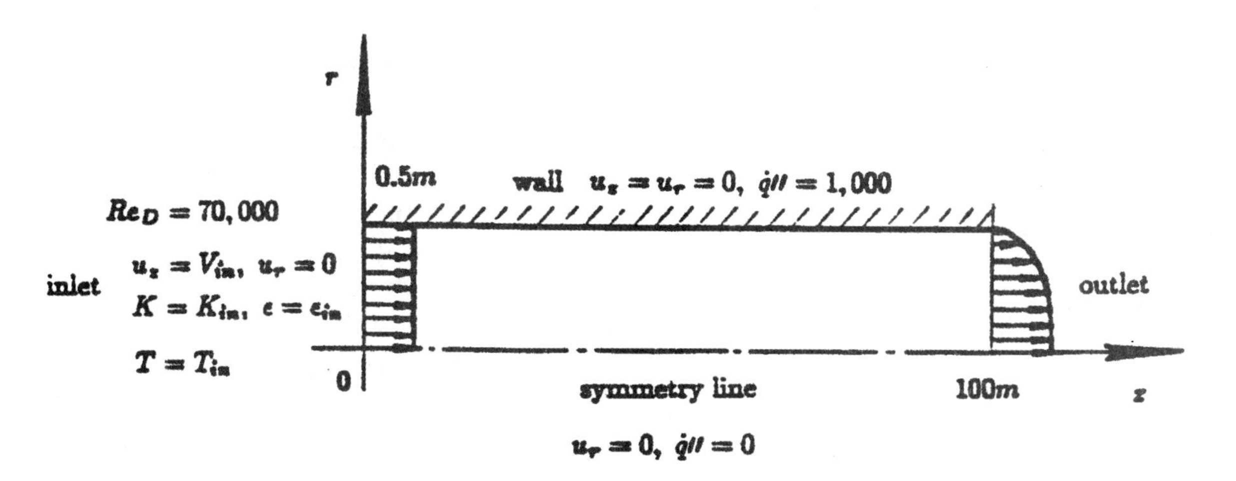

Fig. 1: Fully developed turbulent pipe flow

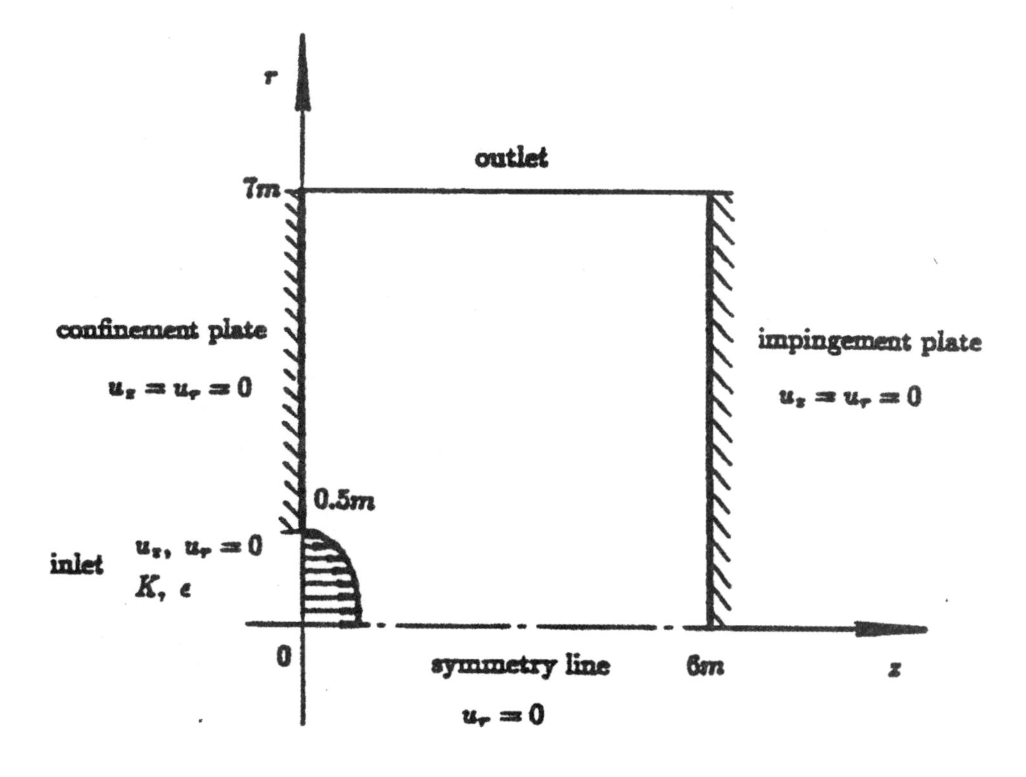

Fig. 2: Axi-symmetric jet impingement flow

where $u^+ = u/u^*$, $y^+ = \rho u^* D/\mu$, $u^* = \sqrt{\tau_w/\rho}$ is friction velocity and $\kappa = 0.41$ is the von Karman constant. The values of y^+ in the first cell near the wall for the two meshes used in this study are 86.06 and 43.03, respectively. Both fall in the range of 30-200 which is recommended for the standard $K-\varepsilon$ turbulence model with the above wall function.

After the flow field is solved, the wall skin friction coefficient and the Nusselt number can be calculated from

$$C_f = \frac{D}{2\rho V_j^2}\frac{\Delta p}{\Delta l}, \qquad Nu_H = \frac{\dot{q}''D}{k(T_w - T_j)} \tag{5}$$

where $\Delta p/\Delta l$ is the calculated pressure gradient (constant), V_j is bulk velocity ($= V_{in}$), T_w is the computed wall temperature and T_j is bulk temperature evaluated from

$$T_j = \frac{2}{V_j(D/2)^2}\int_0^{D/2} ruTdr. \tag{6}$$

Two uniform meshes with 9-noded quadrilateral elements have been used to discretize the flow region: a coarse mesh 100 x 20 and a fine mesh 100 x 40. For the coarse mesh, total solution time is 451.27 sec, total memory requirement is 1.60 MB, number of iterations is 29 for the tolerance of convergence 3×10^{-3} (Fig. 11). For the fine mesh, total solution time is 2,800.48 sec, total memory requirement is 16.85 MB, number of iterations is 22 for the tolerance of convergence 10^{-3} (Fig. 12).

The calculated results are:

For the 100 x 20 coarse mesh,

$C_f = 4.246 \times 10^{-3}$, error 12.2%

$Nu_H = 153.99$, error 14.4%

For the 100 x 40 fine mesh,

$C_f = 4.938 \times 10^{-3}$, error 2.09%

$Nu_H = 136.48$, error 1.35%

The computed velocity, pressure and temperature distributions along the center-line of the pipe and temperature distribution along the wall are shown in Figures 3-6. From these plots we observe that after about 40% of the total length of the pipe, the flow is almost fully developed (Fig. 3); pressure distribution is linear (Fig. 4); temperature distributions on both centerline (Fig. 5) and wall (Fig. 6) of the pipe are nearly linear after about 20% of the total length of the pipe, for the case of constant heat flux on the wall.

The computed profiles of velocity, temperature, K and ε at section 90 m from the pipe inlet are given in Figures 7-10. Because of the pipe end effect on the numerical calculation, the outlet data has some ripples (Figures 3 and 6). Hence, the data at the 90 m section is chosen as the outlet data of the pipe flow. The shapes of these profiles confirm that the turbulent flow has really become fully developed. These data will be used as inlet boundary conditions of the jet impingement flow computation.

The convergence histories for the coarse mesh and fine mesh are shown in Figures 11 and 12, respectively. Appendix A of reference /9/ gives the FIDAP input file for the fine mesh pipe flow computation.

3. Turbulent Jet Impingement Flow

For the axi-symmetric jet impingement flow (Fig. 2), only an isothermal computation has been carried out, i.e. continuity, momentum, K and ε equations have been solved numerically. The inlet boundary condition is specified from the outlet data of the fully developed pipe flow. At the inlet of the jet impingement flow, profiles of velocity component u_z, kinetic energy K and dissipation rate ε are shown in Figures 7, 9 and 10, respectively, and velocity component $u_r = 0$. On the symmetry line, $u_r = 0$. On the impingement plate and confinement plate, $u_z = u_r = 0$. The initial condition is taken as the same data as the inlet boundary condition.

The mesh system is shown in Fig. 13. In z direction (horizontal), 60 packed intervals are chosen. For $0 \le z \le 5.4$ m, the ratio of last to first is 0.05 and for $5.4 \text{ m} \le z \le 6$ m (near the impingement plate), uniform intervals are taken. In r direction (vertical), 20 uniform intervals are chosen for $0 \le r \le 0.5$ m which cover the inlet of the jet impingement flow and are consistent with the outlet of the pipe flow. In this way, it is easy to transfer the outlet data of the pipe flow to specify the boundary condition of the inlet of the jet impingement flow. For 0.5 m $\leq r \leq$ 7 m, 80 clustered intervals are chosen. The ratio of last to first is 5. For the purpose of finite element analysis, all 1,500 quadrilateral elements have 9 nodes. FIDAP provides several options for the solution strategy; in this case we use successive substitution. The computation data are summarized as below: total solution time is 10,375.546 sec, total storage requirement is 27.15 MB, number of iterations is 22 for the tolerance of convergence 3 x 10^{-3}. Fig. 14 shows the convergence history of the jet impingement flow computation. Appendix B of reference /9/ gives the FIDAP input file for this jet flow computation.

Figures 15-22 show the computed jet impingement flow field: velocity vector field (Fig. 15), close up of velocity vector field near the stagnation region (Fig. 16), streamlines (Fig. 17), speed contours (Fig. 18), turbulent kinetic energy contours (Fig. 19), dissipation rate contours (Fig. 20), vorticity contours (Fig. 21) and pressure contours (Fig. 22). From these pictures we can see that the free jet boundary can be clearly distinguished (Figures 15, 18, 19, 20 and 21); there is a large recirculation in the chamber (Fig. 17); the highest speed appears in the jet region (Figures 15 and 18); the largest K and ε appear at the plate and near the inlet (Figures 19 and 20); the maximum vorticity appears near the inlet (Fig. 21) and the highest pressure appears at the stagnation point (Fig. 22).

Figures 23-38 show the profiles of u_z, u_r, K and ε at r = 0, 0.5m, 2.0m, and 7.0m, respectively.

At r = 0, u_z, falls from its value at the inlet to zero at the stagnation point on the plate (Fig. 23) while u_r, remains zero all the way (Fig. 24). K and ε have little change until near the plate where considerable increases are found (Figures 25 and 26).

At r = 0.5m, u_z rises rapidly in the region near the inlet, then maintains a nearly constant value until near the plate, where it falls rapidly to zero (Fig. 27). u_r has some oscillations from inlet to about half the distance towards the plate, probably due to the coarse mesh arrangement in this region (Fig. 13). Near the plate, it increases dramatically due to the turnaround of the flow (Fig. 28). K and ε have similar trends as the case of r = 0 near the plate, but there is a peak near the inlet region. This is consistent with the K and ε contours (Figures 19 and 20).

At r = 2.0 m, u_z has different signs before and after z = 5 m (Fig. 31) and u_r has a maximum at this point (Fig. 32). This indicates the existence of a recirculation region in the flow (cf. Fig. 17). K and ε have a similar trend as the case of r = 0, but the increases are not as sharp (Figures 33 and 34).

At r = 7.0 m (outlet of the jet impingement flow region), u_z is negative (Fig. 35) and u_r changes direction (Fig. 36) due to recirculation (cf. Fig. 17). K and ε increase slowly, reaching a peak value around z = 4.2 m and then decreasing up to the impingement plate (Figures 37 and 38).

4. Remarks and Comments

1. The commercial CFD code FIDAP is able to generate a mesh system in the domain of interest, solve the governing equations of fluid flow and visualize the results in a variety of ways.
2. Jet impingement turbulent flow is an important phenomenon of industrial interest and extensive studies have been performed in academic institutions as well as industries. In principle, turbulent jet impingement flow can be simulated using FIDAP.
3. For the fully developed turbulent pipe flow, which gives the inlet condition of the jet impingement flow, the simulation accuracy is quite good compared to the theoretical results of skin friction coefficient C_f and Nusselt number Nu_H.
4. For the jet impingement turbulent flow with confinement plate, post-processing shows that the calculated results are reasonable.

References

1. B.E. Launder and W. Rodi, "The turbulent wall

jet – measurements and modeling", *Ann. Rev. Fluid Mech.*, **15**, 429-459 (1983).

2. S.-H. Chuang, "Numerical simulation of an impinging jet on a flat plate", *Int. J. Num. Methods in Fluids*, **9**, 1413-1426 (1989).
3. D. Cooper, D.C. Jackson, B.E. Launder and G.X. Liao, "Impinging jet studies for turbulence model assessment – I. Flow-field experiments", *Int. J. Heat Mass Transfer*, **36**, 2675-2684 (1993).
4. T.J. Craft, L.J.W. Graham and B.E. Launder, "Impinging jet studies for turbulence model assessment – II. An examination of the performance of four turbulence models, *Int. J. Heat Mass Transfer*, **36**, 2685-2697 (1993).
5. A. Pollard, S. McIlwain, R.K. Avva, R.M. Barron, A.Y. Boglaev, A.M. Latypov and M.R. Malin, "Validation exercise CFD95: Impinging turbulent round jet with heat transfer", CFDSC/V/95-3, May, 1996.
6. FIDAP Manual, version 7.0, Fluid Dynamics International, Inc., Chicago, IL, USA, 1993.
7. W.M. Kays and M.E. Crawford, *Convective Heat and Mass Transfer*, McGraw-Hill, New York, 1980.
8. W.M. Rohsenow and J.P. Hartnett, eds., *Handbook of Heat Transfer*, McGraw-Hill, New York, 1973.
9. C.-F. An and R.M. Barron, "Jet impingement flow simulation using FIDAP", Fluid Dynamics Research Institute Technical Report, FDRI-TR-95-02, University of Windsor, Windsor, Ontario, Canada, June, 1995.

Fully Developed Pipe Flow

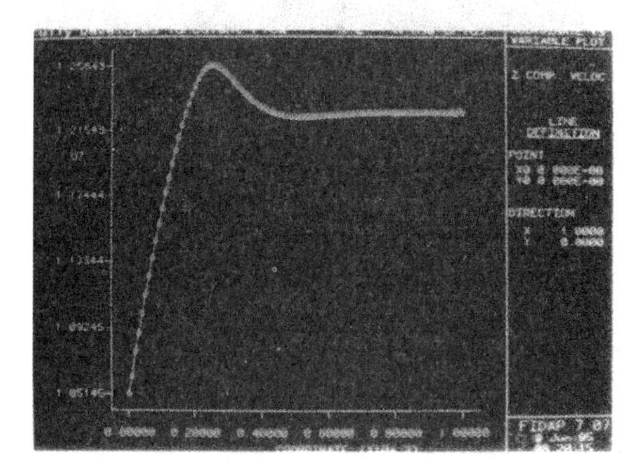

Fig. 3: Velocity distribution on pipe centerline

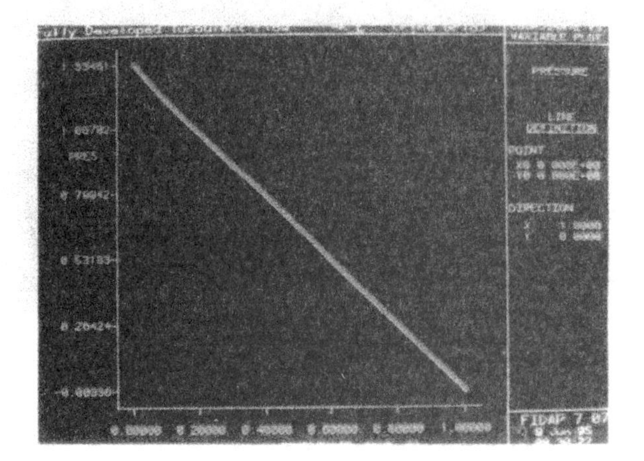

Fig. 4: Pressure distribution on pipe centerline

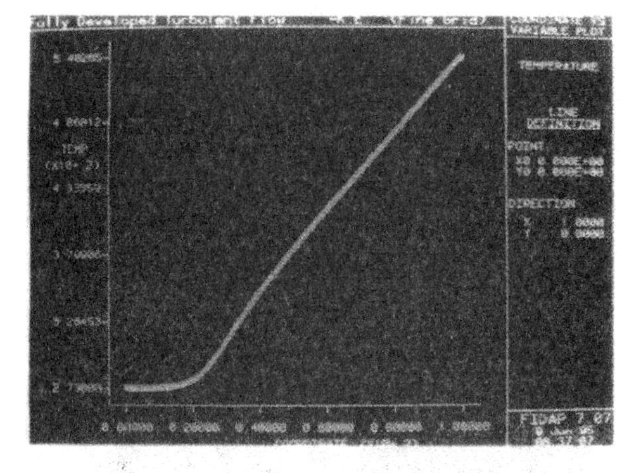

Fig. 5: Temperature distribution on pipe centerline

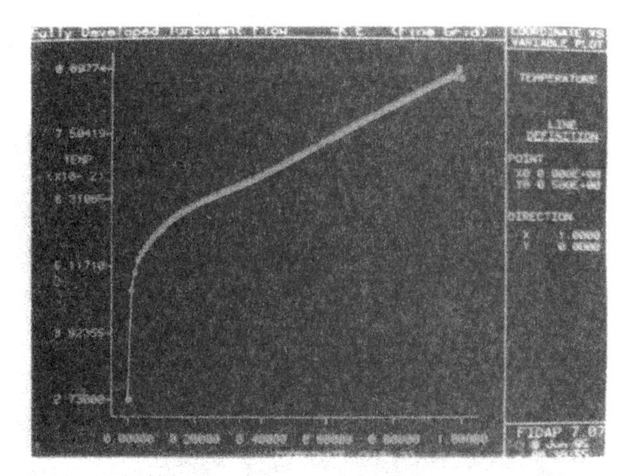

Fig. 6: Temperature distribution on pipe wall

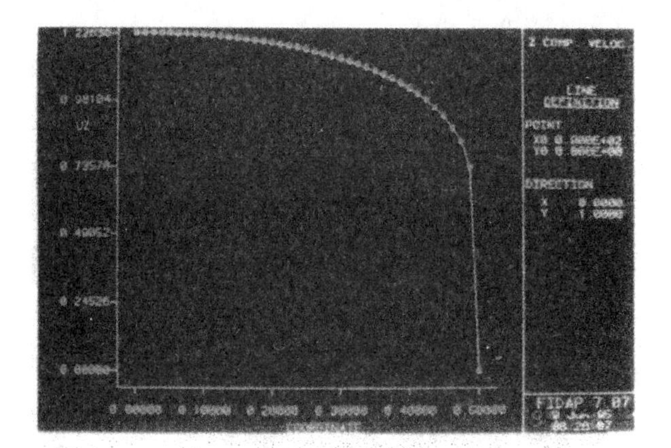

Fig. 7: Velocity profile at section 90m

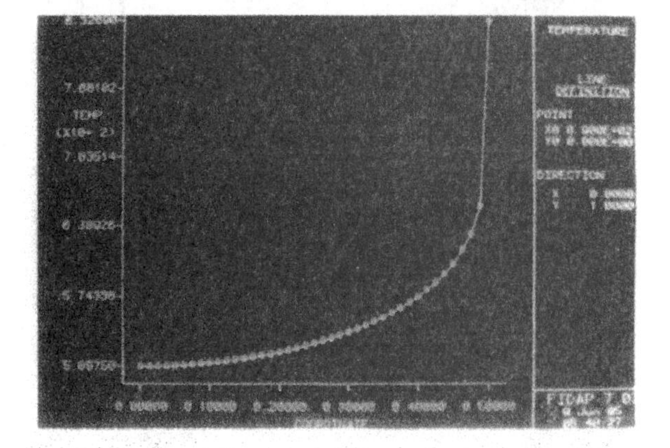

Fig. 8: Temperature profile at section 90m

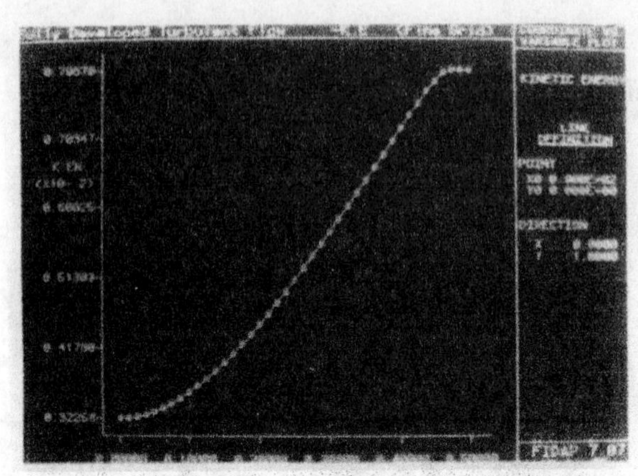

Fig. 9: Kinetic energy profile at section 90m

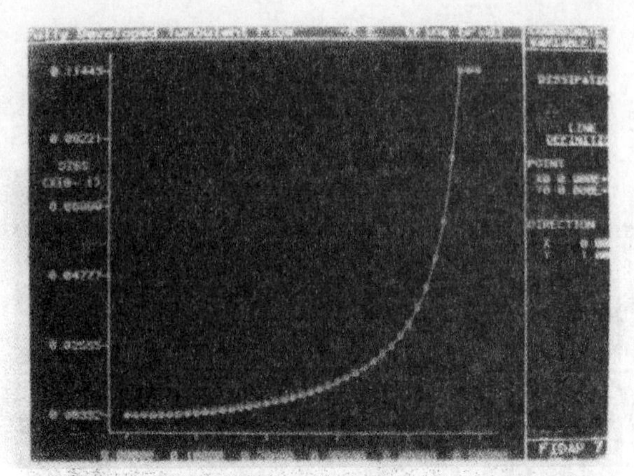

Fig. 10: Dissipation rate profile at section 90m

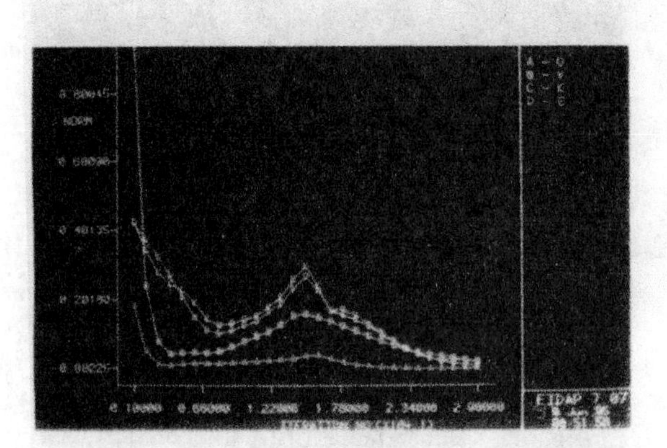

Fig. 11: Convergence history of coarse mesh pipe

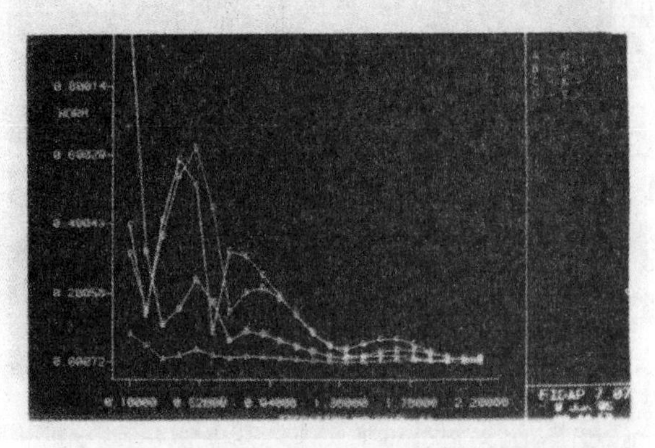

Fig. 12: Convergence history of fine mesh pipe

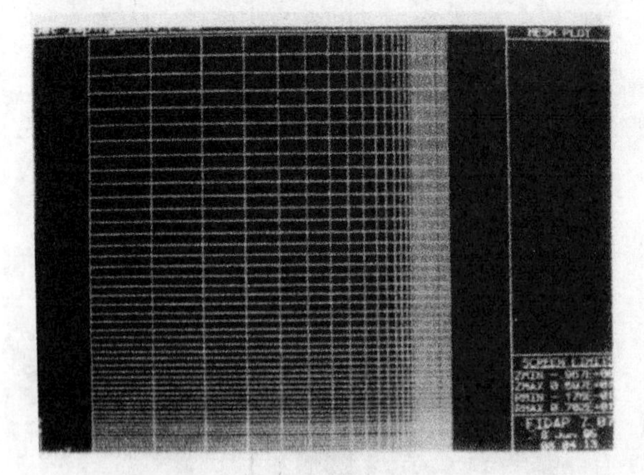

Fig. 13: Mesh system of jet flow

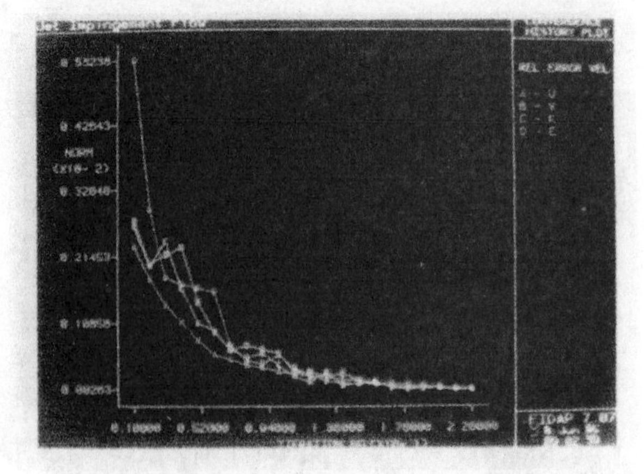

Fig. 14: Convergence history of jet flow

Fig. 15: Velocity vector field of jet flow

Fig. 16: Close up of velocity vector field

Fig. 17: Streamlines of jet flow

Fig. 18: Speed contours of jet flow

Fig. 19: Kinetic energy contours of jet flow

Fig. 20: Dissipation rate contours of jet flow

Fig. 21: Vorticity contours of jet flow

Fig. 22: Pressure contours of jet flow

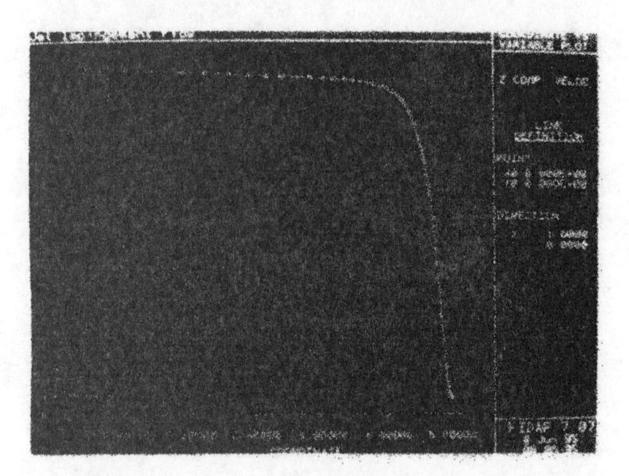

Fig. 23: u_z profile at $r = 0$

Fig. 24: u_r profile at $r = 0$

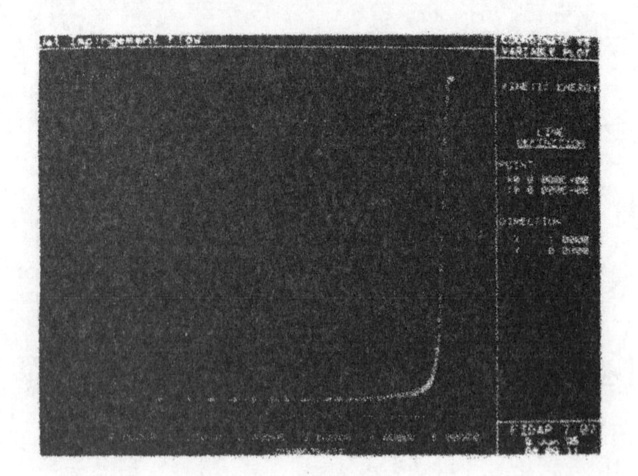

Fig. 25: K profile at $r = 0$

Fig. 26: ε profile at $r = 0$

Fig. 27: u_z profile at $r = 0.5$

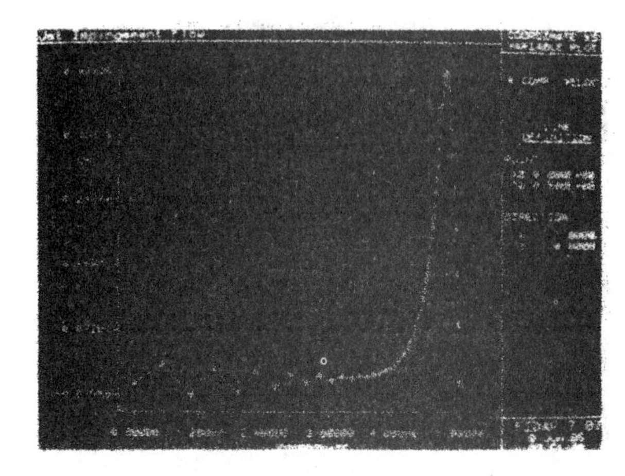

Fig. 28: u_r profile at $r = 0.5$

Fig. 29: K profile at $r = 0.5$

Fig. 30: ε profile at $r = 0.5$

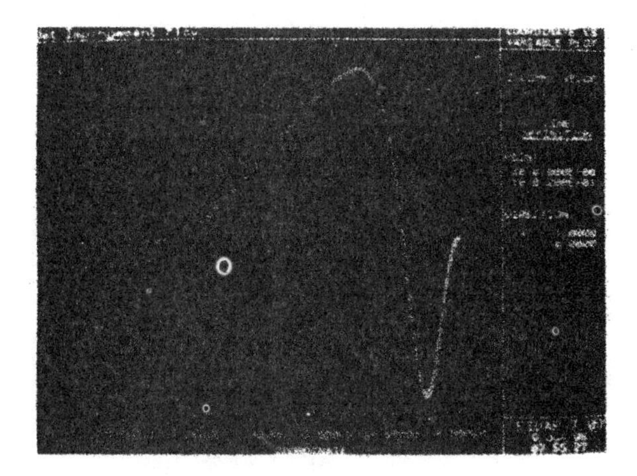

Fig. 31: u_z profile at $r = 2.0$

Fig. 32: u_r profile at $r = 2.0$

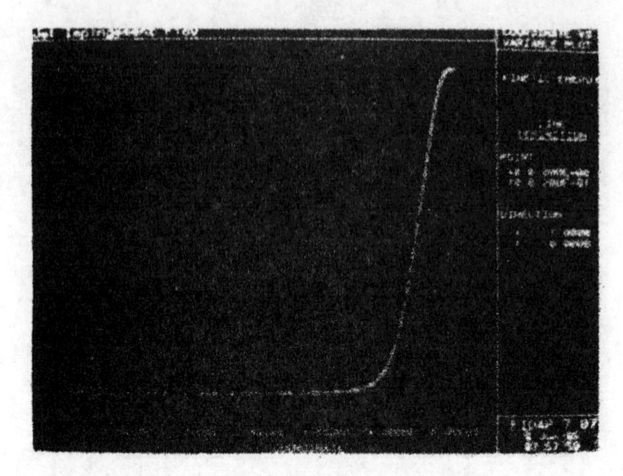

Fig. 33: K profile at $r = 2.0$

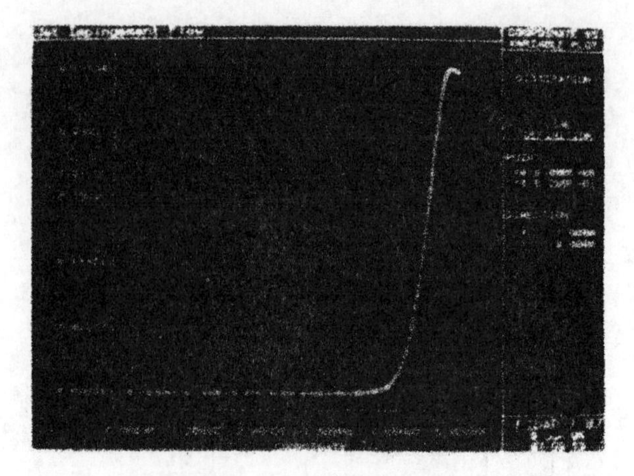

Fig. 34: ε profile at $r = 2.0$

Fig. 35: u_z profile at $r = 7.0$

Fig. 36: u_r profile at $r = 7.0$

Fig. 37: K profile at $r = 7.0$

Fig. 38: ε profile at $r = 7.0$

The Third International Conference

DIFFERENTIAL EQUATIONS

AND

APPLICATIONS

June 12-17 2000

BOOK OF ABSTRACTS

STREAM FUNCTION COORDINATE METHOD AND ITS APPLICATIONS IN NONLINEAR GASDYNAMICS*

C.-F. An
DaimlerChrysler Corporation
Aero-Thermal Development Department
800 Chrysler Drive, CIMS: 481-33-01
Auburn Hills, MI, USA 48326-2757
e-mail: ca56@daimlerchrysler.com

Objectives The stream function coordinate (SFC) method was given considerable attention in the last two decades [1-4]. From numerical point of view, the method has several advantages. Firstly, the method combines flow solver and mesh generator together and formulates a single set of partial differential equations without performing conventional grid generation. Another advantage is that streamlines or particle path lines can be easily traced so that post-processing becomes more expressive and less expensive. Furthermore, when the method is applied to solve one dimensional unsteady flow in gasdynamics, the Jacobian of the transformation is always positive and finite. Therefore, the validity of the transformation is always guaranteed. This provides an opportunity to solve a variety of problems in gasdynamics. In this paper, studies of SFC method on nonlinear flows in gasdynamics by the present author and his colleagues in recent years are summarized. Several examples of one dimensional unsteady problems are presented. Advantages and disadvantages of the method are also addressed. Computer animations based on the simulated examples will be demonstrated at the conference.

Methods/Techniques For one dimensional unsteady compressible flow of an inviscid fluid, the governing Euler equation in Cartesian coordinates is $\mathbf{Q}^*_t + \mathbf{F}^*_x = \mathbf{0}$...(1) where $\mathbf{Q}^*=(\rho, \rho u, e)^t$, $\mathbf{F}^*=[\rho u, \rho u^2+p, (e+p)u]^t$, $p=(\gamma-1)(e-\rho u^2/2)$ is pressure, e is total energy per unit volume, ρ is density, u is velocity and γ is the ratio of specific heat. In these equations, superscript "t" stands for transpose. Defining stream function ψ such that $\psi_x=\rho$, $\psi_t=-\rho u$ and making von Mises-like transformation $t\equiv\tau$, $x=x(\psi,\tau)$, equation (1) can be re-written as $\mathbf{Q}_\tau + \mathbf{F}_\Psi = \mathbf{0}$...(2) where $\mathbf{Q}=(q_1, q_2, q_3)^t$, $\mathbf{F}=[-q_2, (\gamma-1)(q_3-q_2^2/2)/q_1, (\gamma-1)(q_3-q_2^2/2)q_2/q_1]^t$ and $q_1=x_\psi$, $q_2=x_\tau$, $q_3=x_\psi e$. The flux Jacobian matrix of equation (2), $\partial\mathbf{Q}/\partial\mathbf{F}$, has real eigenvalues $\sigma_{1,2} = \pm\sqrt{(\gamma p \rho)}$ and $\sigma_3 = 0$. Therefore, the SFC equation (2) is of hyperbolic type. This means that the Godunov's predictor-corrector scheme is an effective solver. Having solved for **Q**, the geometric/physical variables x, u, p and e can be calculated from q_1, q_2 and q_3.

Results Figure 1(a) shows the comparison of density and velocity between calculation and theoretical solution for a piston-driven expansion wave. Here, R and U represent density and velocity, respectively, and subscripts c and t represent calculation and theoretical solution, respectively. Figure 1(b) shows particle path lines in (x, t) plane and demonstrates the propagation of the wave. It is clear that the accuracy of the calculation is excellent compared to the theory. Figure 2(a) shows the comparison of density between calculation and theory for a shocktube flow at the moment t=0.101 after the diaphragm at Xd=0.4 is broken. It can be seen that the shock wave, the contact discontinuity and the rarefaction wave are well captured in the simulation. The shock jump is smeared out to several grid points and the rarefaction wave spreads over a wider range compared to the theoretical solution. The contact discontinuity, however, is predicted exactly between two grid points. Figure 2(b) shows the propagation of shock wave, contact discontinuity and rarefaction wave in the shocktube. Figure 3(a) shows the comparison of density for the shocktube flow at the moment when the rarefaction wave reaches the left wall and reflects from there. Figure 3(b) shows the propagation and reflection of the rarefaction wave from the left wall. Figure 4(a) shows the comparison of density for the shocktube flow at a later moment when the shock wave has reached the right wall, reflected from there, traveled backward and, finally, meets the incoming contact discontinuity. Figure 4(b) shows the propagation of these three waves, shock wave reflection on the right wall and its interaction with the contact discontinuity.

Conclusions From the above results, the proposed SFC method is able to solve many nonlinear problems in gasdynamics including isentropic/non-isentropic flows, continuous/discontinuous flows, wave reflection on a wall, wave interaction with each other and so on. The contact discontinuity can be captured exactly, but the shock jump is smeared out to several grid points. Further study is needed to improve the simulation accuracy of shock waves.

References:

1. Greywall, M.S., Streamwise computation of 2-D incompressible potential flows, J. Comp. Phys., Vol. 59, pp. 224-231, 1985
2. Takahashi, K. and Tsukiji, T., Numerical analysis of a laminar jet using a streamline coordinate system, Transactions of the CSME, Vol. 9, pp. 165-170, 1985
3. Dulikravich, G.S., A stream-function-coordinate concept in aerodynamic shape design, AGARD-R-780, pp. 6.1-6.6, 1990
4. Barron, R.M., An, C.-F. and Zhang, S., Survey of streamfunction-as-a-coordinate method in CFD, Proc. of Inaugural Conference of the CFD Society of Canada, pp.325-336, Montreal, Canada, June 13-14, 1993

* For presentation at the Third International Conference of Differential Equations and Applications, St. Petersburg, Russia, June 12-17, 2000

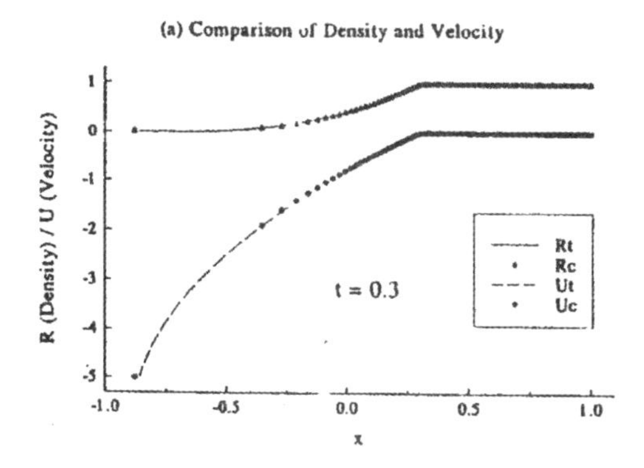

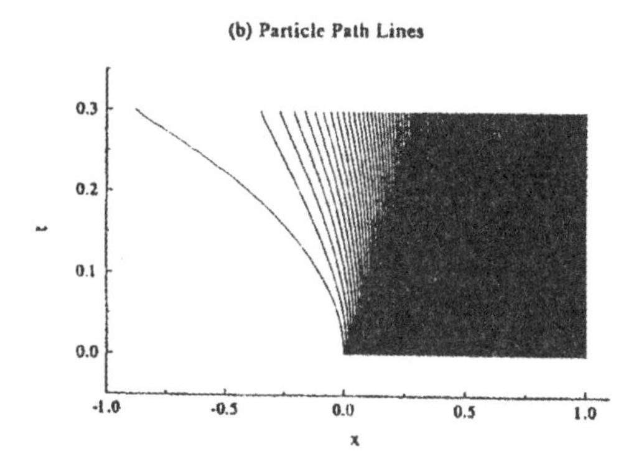

Figure 1. Piston-Driven Expansion Wave

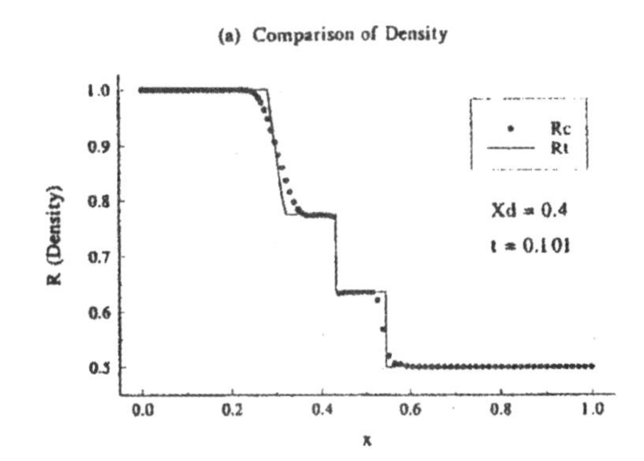

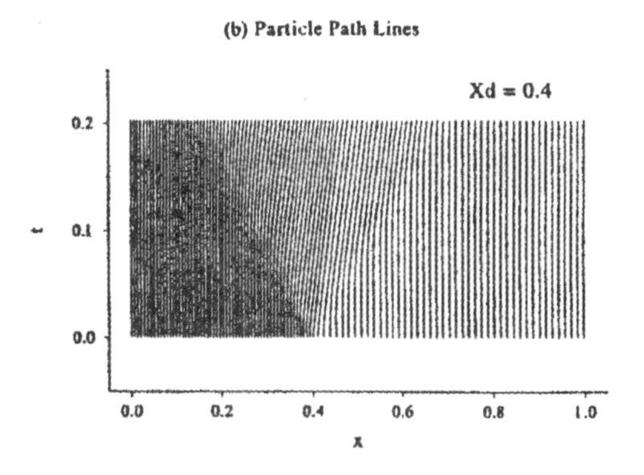

Figure 2. Propagation of Three Waves in Shocktube Flow

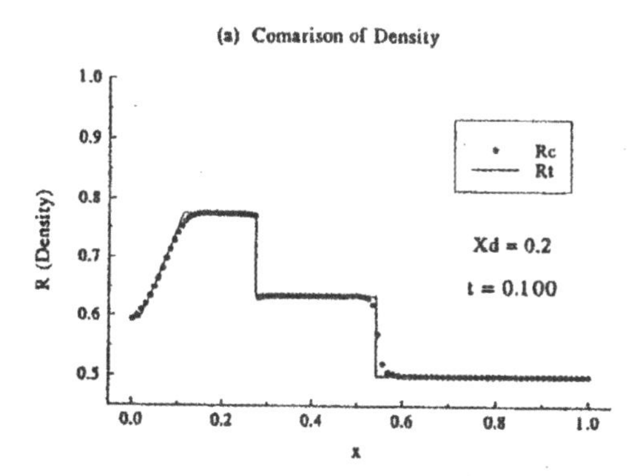

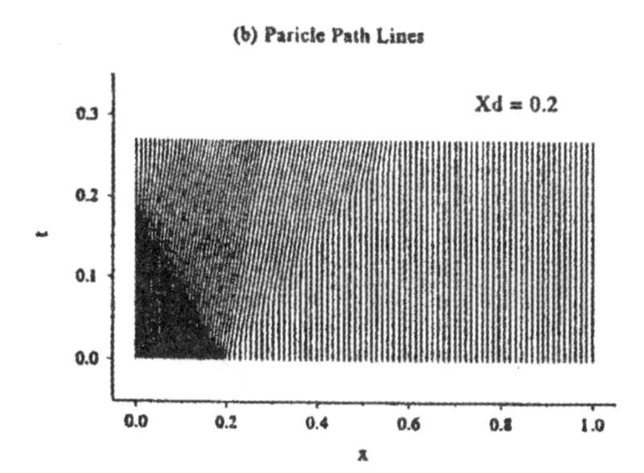

Figure 3. Reflection of Rarefaction Wave In Shocktube Flow

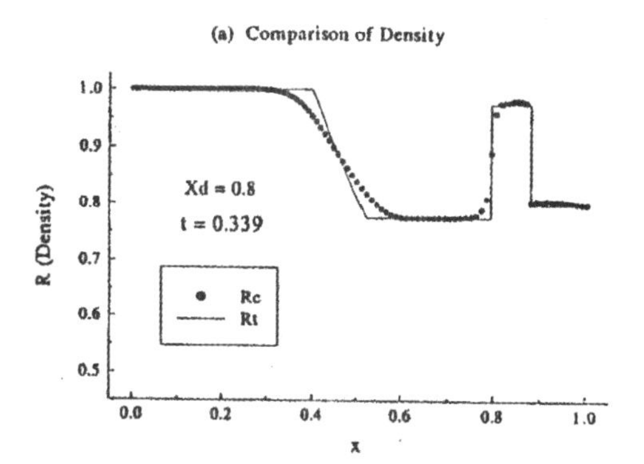

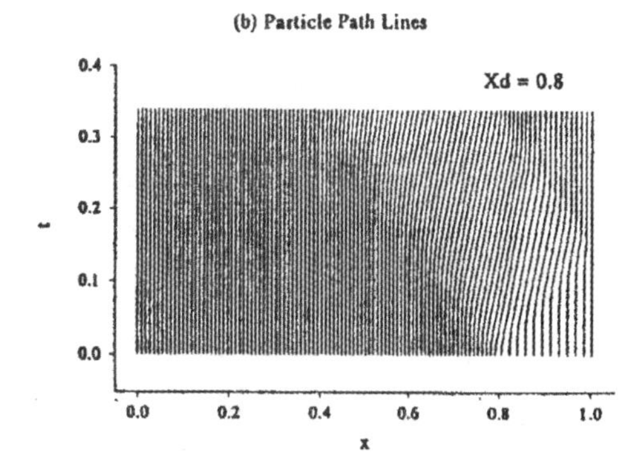

Figure 4. Reflection and Interaction of Shock Wave with Contact Discontinuity

安长发博士科研论文集

MATHEMATICAL RESEARCH

VOLUME 6

THEORY AND PRACTICE OF DIFFERENTIAL EQUATIONS

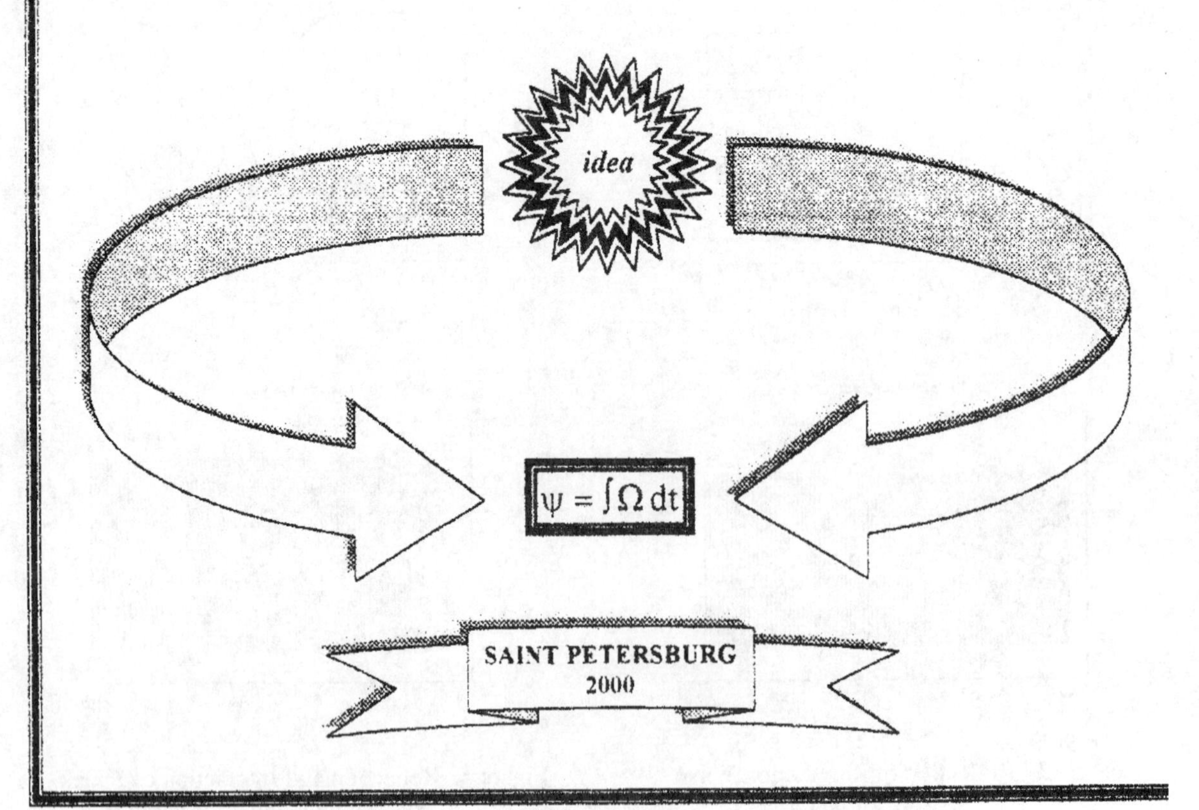

NUMERICAL STUDY ON 1D UNSTEADY FLOWS IN GASDYNAMICS USING STREAM FUNCTION COORDINATE METHOD

C.-F. An
DaimlerChrysler Corporation
Aero-Thermal Development Department
800 Chrysler Drive, CIMS: 481-33-01
Auburn Hills, MI, USA 48326-2757
e-mail: ca56@daimlerchrysler.com

Abstract

In this paper, the study on 1D unsteady flows in gasdynamics is summarized. Mathematical formulation and numerical algorithms for various flow situations are derived and several computational examples are presented. Advantages and disadvantages of the SFC method are also addressed. Computer animations of the examples are available for interested readers.

1. Introduction

The method of stream function coordinate (SFC) was given considerable attention in the last two decades [1-3]. From numerical point of view, the method has several advantages. Firstly, it combines flow solver (flow physics) and grid generator (flow geometry) together and formulates a single set of partial differential equation(s) without performing conventional grid generation. Therefore, the computer resources can be saved significantly. Another advantage is that due to the Lagrangian nature of the method, streamlines or particle path lines can be easily traced so that post-processing becomes more expressive and less expensive. Furthermore, when the method is applied to solve 1D unsteady flows in gasdynamics, the Jacobian of the transformation is always non-zero and finite. Therefore, the validity of the transformation is guaranteed. This fact provides an opportunity to solve a variety of 1D unsteady gasdynamics problems.

The concept of stream function coordinate (SFC) was developed in solving 2D steady flows [4-6]. [4] applied differential geometry theory to the 2D steady incompressible potential flow, derived a second-order equation using SFC method and successfully solved some problems. [5] extended the SFC method to the full potential transonic flows. [6] extended the SFC method to the general inviscid transonic flows. The basic idea in [4] is cited here to describe the concept of the SFC method. For a 2D steady incompressible potential flow, the governing equation in Cartesian coordinate system *(x, y)* is the Laplace equation for stream function $\psi(x,y)$: $\psi_{xx}+\psi_{yy}=0$. Making von Mises transformation $x\equiv\phi, y=y(\phi,\psi)$ and assuming the Jacobian is non-zero and finite, the transformation is permissible and the Laplace equation for $\psi(x, y)$ can be transformed to a second-order elliptic nonlinear partial differential equation for $y(x, \psi)$: $y_\psi^2 y_{xx}-2y_x y_\psi y_{x\psi}+(1+y_x^2)y_{\psi\psi}=0$. Here ϕ is replaced by x in the equation because they are equivalent during the transformation. Although the equation becomes a little bit more complicated, the computational domain of the flow is much more simplified, cf. Figure 1(a) and (b). For a 1D unsteady flow, the governing equation(s) in SFC system can be derived similarly as shown in section 2.

In this paper, the study on 1D unsteady flows in gasdynamics using SFC method by the present author and his colleagues in recent years is summarized. In section 2, the governing equations for various 1D unsteady gasdynamics flows are formulated, eigenvalue analysis is performed and numerical algorithms are constructed. In section 3, several computational examples are presented. In section 4, concluding remarks with advantages and disadvantages of the method are addressed.

2. Governing Equations and Numerical Algorithms

(1) Continuous Isentropic Flow

For a 1D unsteady continuous isentropic flow of an inviscid compressible gas, the governing equations are the continuity equation, momentum conservation and isentropic relation:

$$\tilde{\rho}_{\tilde{t}} + (\tilde{\rho}\tilde{u})_{\tilde{x}} = 0, \tag{2-1-1}$$
$$\tilde{\rho}\tilde{u}_{\tilde{t}} + \tilde{\rho}\tilde{u}\tilde{u}_{\tilde{x}} + \tilde{p}_{\tilde{x}} = 0, \tag{2-1-2}$$
$$\tilde{p} = const \cdot \tilde{\rho}^{\gamma} \tag{2-1-3}$$

where dimensional density $\tilde{\rho}$, velocity $\tilde{u}$ and pressure $\tilde{p}$ are unknown functions of time $\tilde{t}$ and Cartesian coordinate $\tilde{x}$, and γ is the ratio of specific heats of the gas. Choosing stagnation density $\tilde{\rho}_0$ and stagnation speed of sound $\tilde{c}_0$ as primary reference variables and some $\tilde{l}$ as reference length, equations (2-1-1) – (2-1-3) can be non-dimensionalized to

$$\rho_t + (\rho u)_x = 0, \tag{2-1-4}$$
$$\rho u_t + \rho u u_x + p_x / \gamma = 0, \tag{2-1-5}$$
$$p = \rho^{\gamma} \tag{2-1-6}$$

where the non-dimensional variables are: $\rho = \tilde{\rho}/\tilde{\rho}_0$, $u = \tilde{u}/\tilde{c}_0$, $p = \tilde{p}/(\tilde{\rho}_0\tilde{c}_0^2/\gamma)$, $x = \tilde{x}/\tilde{l}$ and $t = \tilde{t}/(\tilde{l}/\tilde{c}_0)$. It is noted that the relation $\tilde{c}_0^{\,2} = \gamma \cdot \tilde{p}_0/\tilde{\rho}_0$ is used to get (2-1-6) from (2-1-3) where $\tilde{p}_0$ is dimensional stagnation pressure. Eliminating p from (2-1-5) and (2-1-6), defining stream function ψ such that

$$\psi_x = \rho, \qquad \psi_t = -\rho u \tag{2-1-7}$$

and substituting (2-1-7) into (2-1-4) and (2-1-5), the continuity equation (2-1-4) is satisfied automatically and the momentum equation (2-1-5) leads to

$$\psi_x^2\psi_{tt} - 2\psi_t\psi_x\psi_{tx} + (\psi_t^2 - \psi_x^{\gamma+1})\psi_{xx} = 0. \tag{2-1-8}$$

Introducing von Mises-like transformation

$$t \equiv \tau, \qquad x = x(\tau, \psi), \tag{2-1-9}$$

the stream function equation (2-1-8) can be simplified to

$$x_{\tau\tau} - x_{\psi}^{-(\gamma+1)} x_{\psi\psi} = 0. \tag{2-1-10}$$

This is a second-order hyperbolic nonlinear partial differential equation, or a wave equation with variable propagation speed, in stream function coordinate system (τ, ψ). Once the unknown variable $x(\tau, \psi)$ is solved, the physical variables density ρ, velocity u and pressure p can be calculated from

$$\rho = x_{\psi}^{-1}, \qquad u = x_{\tau}, \qquad p = x_{\psi}^{-\gamma}. \tag{2-1-11}$$

The Jacobian of the transformation (2-1-9), $J = \partial(t,x)/\partial(\tau,\psi,) = x_{\psi} = 1/\rho$, is obviously non-zero and finite due to the physical meaning of density. Therefore, the transformation (2-1-9) is always valid without any singularities. Moreover, it is obvious that the determinant of equation (2-1-10), $\Delta = 4x_{\psi}^{\gamma+1} = 4/\rho^{\gamma+1}$, is always a real quantity so that the equation must be of hyperbolic type. To solve this hyperbolic equation, the three-step explicit midpoint leapfrog scheme can be used [7]:

$$x_i^{n+1} = 2x_i^n - x_i^{n-1} + C^2(x_{i+1}^n - 2x_i^n + x_{i-1}^n) \tag{2-1-12}$$

for $i = 2,3,\ldots,I_{max}-1$, $n = 2,3,\ldots$ Here, $C = x_{\psi}^{-(\gamma+1)/2}\Delta\tau/\Delta\psi$ is the Courant-Friedrichs-Lewy (CFL) number. The subscripts $i+1$, i, $i-1$ represent the ψ locations and the superscripts $n+1$, n, $n-1$ represent the levels of time τ. Difference equation (2-1-12) can be marched with time if proper initial and boundary conditions are specified for a specific flow. The scheme is stable when $C<1$. To get a good solution, C is usually taken 0.5-0.6. The details of the governing equation and numerical scheme for this continuous isentropic flow can be found in [8].

(2) Discontinuous Isentropic Flow

As indicated in the previous subsection, the 1D unsteady continuous isentropic flow is governed by equation (2-1-10) in SFC system. However, when the flow field has a discontinuity for density or velocity, the non-conservative governing equation (2-1-10) is unable to capture the discontinuity. In order to solve the flow with a discontinuity, a conservative system of equations has to be formulated. If stagnation density $\tilde{\rho}_0$ and stagnation pressure $\tilde{p}_0$ are chosen as primary reference variables, continuity equations (2-1-1) and isentropic relation (2-1-3) can be non-dimensionalized to the same equations as (2-1-4) and (2-1-6), while momentum equation (2-1-2) can be non-dimensionalized to

$$\rho u_t + \rho u u_x + p_x = 0 \qquad (2-2-1)$$

where $\rho = \tilde{\rho}/\tilde{\rho}_0$, $p = \tilde{p}/\tilde{p}_0$, $u = \tilde{u}/\sqrt{\tilde{p}_0/\tilde{\rho}_0}$, $x = \tilde{x}/\tilde{l}$ and $t = t/(\tilde{l}/\sqrt{\tilde{p}_0/\tilde{\rho}_0})$. It is noted that the slight difference between equations (2-2-1) and (2-1-5) is due to different primary reference variables. Similar to the previous procedure, i.e. defining stream function (2-1-7) and introducing von Mises-like transformation (2-1-9), the momentum equation (2-2-1) can be reduced to

$$(x_\tau)_\tau + (x_\psi^{-\gamma})_\psi = 0. \qquad (2-2-2)$$

If the first derivatives x_ψ and x_τ are chosen as two individual unknown variables, one more equation is necessary to relate them to each other. That is, $x_{\psi\tau} = x_{\tau\psi}$, or

$$(x_\psi)_\tau + (-x_\tau)_\psi = 0. \qquad (2-2-3)$$

Combining equations (2-2-2) and (2-2-3) together, one gets

$$\begin{bmatrix} x_\psi \\ x_\tau \end{bmatrix}_\tau + \begin{bmatrix} -x_\tau \\ x_\psi^{-\gamma} \end{bmatrix}_\psi = 0, \qquad (2-2-4)$$

or in a vector form,

$$\vec{Q}_\tau + \vec{F}_\psi = 0 \quad where \quad \vec{Q} = \begin{bmatrix} x_\psi \\ x_\tau \end{bmatrix} = \begin{bmatrix} q_1 \\ q_2 \end{bmatrix} \ and \ \vec{F} = \begin{bmatrix} -x_\tau \\ x_\psi^{-\gamma} \end{bmatrix} = \begin{bmatrix} -q_2 \\ q_1^{-\gamma} \end{bmatrix} = \begin{bmatrix} f_1 \\ f_2 \end{bmatrix} \qquad (2-2-5)$$

are the unknown vector and the flux vector, respectively. Equation (2-2-5) is in a conservative form and, therefore, is more powerful to capture discontinuity than equation (2-1-10). Once equations (2-2-5) are solved, the physical variables can be calculated from

$$\rho = q_1^{-1}, \qquad u = q_2, \qquad p = q_1^{-\gamma}. \qquad (2-2-6)$$

The Jacobian matrix of the flux vector in equation (2-2-5) is

$$J = \frac{\partial \vec{F}}{\partial \vec{Q}} = \frac{\partial(f_1, f_2)}{\partial(q_1, q_2)} = \begin{bmatrix} 0 & -1 \\ -\gamma q_1^{-(\gamma+1)} & 0 \end{bmatrix} = \begin{bmatrix} 0 & -1 \\ -\gamma \rho^{\gamma+1} & 0 \end{bmatrix}. \qquad (2-2-7)$$

Because the eigenvalues of matrix (2-2-7), $\pm\sqrt{\gamma\rho^{\gamma+1}}$, are real quantities, equations (2-2-5) must be of hyperbolic type. For these hyperbolic equations, the Godunov's first-order two-step predictor-corrector scheme can be used [9]:

$$\bar{\vec{Q}}_{j+1/2} = \frac{1}{2}(\vec{Q}^n_{j+1} + \vec{Q}^n_j) - \frac{\Delta\tau}{\Delta\psi}(\vec{F}^n_{j+1} - \vec{F}^n_j), \qquad \vec{Q}^{n+1}_j = \vec{Q}^n_j - \frac{\Delta\tau}{\Delta\psi}(\bar{\vec{F}}_{j+1/2} - \bar{\vec{F}}_{j-1/2}) \qquad (2-2-8)$$

for $j = 2,3,\ldots,J_{max}-1,\ n = 2,3,\ldots$ Here, the vector variables with overbar, $\bar{\vec{Q}}_{j+1/2}$ and $\bar{\vec{F}}_{j\pm1/2}$, have intermediate values between time levels n and $n+1$. Figure 2 draws a sketch of Godunov's predictor-corrector scheme. The details of the mathematical formulation for this conservative system of equations in SFC, eigenvalue analysis and the Godunov's scheme can be found in [10].

(3) General Inviscid Flow

For a general 1D unsteady flow of an inviscid compressible gas, the isentropic assumption does not apply, especially for the flow in which strong discontinuities exist. Under these circumstances, accurate solutions cannot be obtained using either equation (2-1-10) or equation (2-2-5). Even worse, the computation may not be converged at all. In order to overcome this difficulty the governing equations in SFC have to be derived directly from the conservative Euler equations in Cartesian coordinates (t, x):

$$\begin{bmatrix} \rho \\ \rho u \\ e \end{bmatrix}_t + \begin{bmatrix} \rho u \\ \rho u^2 + p \\ u(e+p) \end{bmatrix}_x = 0 \qquad (2-3-1)$$

where stagnation density $\tilde{\rho}_0$ and stagnation pressure $\tilde{p}_0$ are chosen as primary reference variables. Density ρ and velocity u have been non-dimensionalized by $\tilde{\rho}_0$ and $\sqrt{\tilde{p}_0/\tilde{\rho}_0}$, respectively, while both total energy per unit volume e and pressure $p = (\gamma-1)(e - \rho u^2/2)$ have been non-dimensionalized by $\tilde{p}_0$. The independent variables x and t have been non-dimensionalized by $\tilde{l}$ and $\tilde{l}/\sqrt{\tilde{p}_0/\tilde{\rho}_0}$, respectively. Referring to the previous two subsections, through the definition of stream function (2-1-7) and von Mises-like transformation (2-1-9), the Euler equations (2-3-1) can be transformed to the SFC system (τ, ψ):

$$\begin{bmatrix} x_\psi \\ x_\tau \\ x_\psi e \end{bmatrix}_\tau + \begin{bmatrix} -x_\tau \\ (\gamma-1)(e - x_\tau^2 x_\psi^{-1}/2) \\ (\gamma-1)(e - x_\tau^2 x_\psi^{-1}/2)x_\tau \end{bmatrix}_\psi = 0. \qquad (2-3-2)$$

Equations (2-3-2) can be re-written symbolically in a general form,

$$\vec{Q}_\tau + \vec{F}_\psi = 0 \quad where \quad \vec{Q} = \begin{bmatrix} x_\psi \\ x_\tau \\ x_\psi e \end{bmatrix} = \begin{bmatrix} q_1 \\ q_2 \\ q_3 \end{bmatrix} \quad and \quad \vec{F} = \begin{bmatrix} f_1 \\ f_2 \\ f_3 \end{bmatrix} = \begin{bmatrix} -q_2 \\ (\gamma-1)(q_3 - q_2^2/2)q_1^{-1} \\ (\gamma-1)(q_3 - q_2^2/2)q_2 q_1^{-1} \end{bmatrix} \qquad (2-3-3)$$

are the general unknown vector and the flux vector, respectively. The relation between the physical unknowns and the general unknowns is

$$\rho = q_1^{-1},\ u = q_2,\ e = q_3 q_1^{-1},\ p = (\gamma-1)(q_3 - q_2^2/2)q_1^{-1}. \qquad (2-3-4)$$

The Jacobian matrix of the flux vector in equation (2-3-3), $J = \partial\vec{F}/\partial\vec{Q} = \partial(f_1, f_2, f_3)/\partial(q_1, q_2, q_3)$, has three real eigenvalues, $\pm\sqrt{\gamma p\rho}$ and 0, so it must be of hyperbolic type. Similarly, the equations can be numerically solved using the Godunov's predictor-corrector scheme (2-2-8). The details of the mathematical formulation for the general Euler equations in SFC can be found in [11].

3. Computational Examples

i. Piston-driven expansion wave: In this example, the air in a cylindrical tube was at rest at $t = 0$. When $t>0$, the piston on the left side of the air is drawn leftward with the position function $x = -10t^2$. As the piston moves leftward, a sequence of expansion waves is induced and travels rightward. Figure 3(a) shows the comparison of density and velocity between the calculated results using equation (2-1-10) and a theoretical solution [12] at $t = 0.3$ after the beginning of the piston movement. In this figure, R and U represent density and velocity, respectively, and subscripts t and c represent the theoretical solution and the calculation, respectively. Figure 3(b) shows the particle path lines in the physical domain (x, t) and demonstrates the left-moving piston and the right-travelling expansion wave. It is clear that the accuracy of the present calculation is excellent compared to the theoretical solution.

ii. Piston-driven compression wave: Similar to the previous example, the air was at rest at $t = 0$. However, instead of drawing the piston leftward, the piston is pushed rightward with the position function $x = 2t^2$. Figure 4(a) shows the comparison of density and velocity between the calculation using equations (2-2-5) and a theoretical solution [12] at $t = 0.226$. Figure 4(b) shows the particle path lines in the physical domain (x, t) and demonstrates the right-moving piston and the right-travelling compression wave. It is seen that the agreement between the calculation and the theoretical solution is excellent.

iii. Piston-driven shock wave: In the third example, the piston is also pushed rightward, but it is accelerated instantaneously to a constant speed. The position function is $x = 0.58555t$. Figure 5(a) shows the comparison of density and velocity between the calculation using equations (2-2-5) and a theoretical solution [13] at $t = 0.6$. Figure 5(b) shows the particle path lines in the physical domain (x, t) and demonstrates the right-moving piston and the right-travelling shock wave. It is observed that the shock wave can be captured within 3-4 grid points without any oscillation.

iv. Propagation of three waves in a shock tube: This is an example of a shock tube flow. At $t = 0$, the stationary air in a cylindrical tube was separated by a diaphragm located at $x_d = 0.4$. The left chamber has high-pressure air with density $\rho = 1$, while the right chamber has low-pressure air with density $\rho = 0.5$. When $t > 0$, the diaphragm is suddenly broken and the air rushes from the left to the right. As a result, three waves are generated: a rarefaction wave travelling to the left, a shock wave travelling to the right and a contact discontinuity also travelling to the right behind the shock wave. Figure 6(a) shows the comparison of density between the calculation using equations (2-3-3) and a theoretical solution [13] at the moment $t = 0.101$ after the diaphragm breakage. It is obvious that the three waves are well captured in the calculation. The shock wave is smeared out to 3-4 grid points without any oscillation and the rarefaction wave spreads over a wider range compared to the theoretical solution. The contact discontinuity, however, is captured exactly between two grid points. The strengths of all three waves are accurately predicted. Figure 6(b) shows the particle path lines of the shock tube flow in the physical domain (x, t). All three waves in the shock tube flow can be distinguished from different denseness of the particle path lines in the four regions.

v. Reflection of rarefaction wave in a shock tube: This is also a shock tube flow, but the diaphragm is put at $x_d = 0.2$ in order to simulate the reflection of the rarefaction wave on the left wall. Figure 7(a) shows the comparison of density between the calculation using equations (2-3-3) and the theoretical solution [13] at $t = 0.25$. At this moment, the originally left-travelling rarefaction wave has already been reflected from the left wall and is travelling back to the right. The accuracy of the calculation seems to be better for the reflected rarefaction wave than that for the incidence rarefaction wave. The strength of the reflected rarefaction wave is accurately predicted. Figure 7(b) shows the particle path lines in the physical domain (x, t) in which the rarefaction wave reflection can be identified if careful attention is paid to the upper left corner of the graph.

vi. Shock wave reflection and interaction with contact discontinuity in a shock tube: The last example is also a shock tube flow. In order to study the behavior of the shock wave near the right wall, the diaphragm is put at $x_d = 0.8$. Figure 8(a) shows the comparison of density between the calculation using equations (2-3-3) and a theoretical solution [13] at $t = 0.339$. At this moment, the originally right-travelling shock wave has already been reflected from the right wall, has moved back to the left, has penetrated through the oncoming contact discontinuity and keeps moving leftward. It can be seen that the reflected shock wave is also smeared out to 3-4 grid points. The contact discontinuity, which was penetrated by the reflected shock wave, is also exactly captured and is standing there where it met the shock wave. The left-travelling rarefaction wave still spreads over a wider range. Figure 8(b) shows the particle path lines in the physical domain (x, t). All the processes of shock wave behavior, reflecting from the right wall, moving back to the left, penetrating through the contact discontinuity and keeping movement leftward, can be recognized from this shadow graph.

In addition, computer animations are produced to demonstrate the example flows and available from the author to any interested readers.

4. Concluding Remarks

From the present study, the following brief conclusions may be drawn. (i) The proposed SFC method is able to solve many 1D unsteady flows, including continuous isentropic flow, discontinuous isentropic flow and general inviscid flow. (ii) Discontinuities in the flow, such as shock wave, rarefaction wave and contact discontinuity, can be captured using the conservative equations. The location and strengths of all these discontinuities can be accurately predicted. Particularly, the contact discontinuity can be captured exactly with two grid points. (iii) The shock wave is usually smeared out to 3-4 grid points and the rarefaction wave spreads over a wider range compared to the theoretical solutions. These drawbacks may be improved by adding more grid points or selecting less-dissipative schemes in the future. (iv) The major weakness of the method is that it is difficult to extend to the multi-dimensional unsteady flows due to the complication of stream function definition.

Acknowledgement

The author acknowledges his colleagues, Drs. R.M. Barron and S. Zhang from University of Windsor, for their collaborative efforts in the development of the SFC method in the 2D steady flows in aerodynamics and the extension to the 1D unsteady flows in gasdynamics.

References

1. Greywall, M.S., Streamwise computation of 2-D incompressible potential flows, J. Comp. Phys., **59**, 224-231, 1985
2. Dulikravich, G.S., A stream-function-coordinate concept in aerodynamic shape design, AGARD-R-780, 6.1-6.6, 1990
3. Barron, R.M., An, C.-F. and Zhang, S., Survey of streamfunction-as-a-coordinate method in CFD, Proc. of Inaugural Conference of the CFD Society of Canada, 325-336, Montreal, Canada, June 13-14, 1993
4. Barron, R.M., Computation of incompressible potential flow using von Mises coordinates, Math. & Comp. in Simulation, **31**, 177-188, 1989
5. Barron, R.M. and Naeem, R.K., Numerical solution of transonic flows on a streamfunction co-ordinate system, Int'l J. Num. Methods in Fluids, **9**, 1183-1193, 1989
6. An, C.-F., Barron, R.M., Transonic Euler computation in stream function coordinates, Int'l J. Num. Methods in Fluids, **20**, 75-94, 1995
8. Hoffmann, K.A., Computational Fluid Dynamics for Engineers, Chapt. 4-6, Engineering Education System, 1989
9. An, C.-F., Barron, R.M. and Zhang, S., Streamfunction coordinate formulation for one-dimensional unsteady flow, Math. Models & Methods in Appl. Sci., **5**, 401-414, 1995
10. Sod. G.A., Survey of several finite difference methods for system of nonlinear hyperbolic conservation laws, J. of Comp. Phys., **27**, 1-31, 1978
10. Barron, R.M., An, C.-F. and Zhang, S., Unsteady conservative streamfunction coordinate formulation: 1-D isentropic flow, Appl. Math. Modelling, **18**, 486-493, 1994
11. An, C.-F., Barron, R.M. and Zhang, S., Stream function coordinate Euler formulation and shocktube application, Appl. Math. Modelling, **20**, 421-428, 1996
12. Sharpiro, A.H., The Dynamics and Thermodynamics of Compressible Fluid Flow, Vol. II, Chapt. 24, Ronald Press, 1953
13. Landau. L.D. and Lifshits, E.M., Fluid Mechanics, Chapt. 10, translated from Russian by J.B. Sykes and W.H. Reid, Pergamon Press. London, 1959

von Mises
transformation
$\begin{pmatrix} x \\ y \end{pmatrix} \Longrightarrow \begin{pmatrix} x \\ \psi \end{pmatrix}$
$\psi(x,y) \Longrightarrow y(x,\psi)$
$\psi_{xx} + \psi_{yy} = 0$
$y_\psi^2 y_{xx} - 2y_x y_\psi y_{x\psi} + (1 + y_x^2) y_{\psi\psi} = 0$
y
x
ψ
x

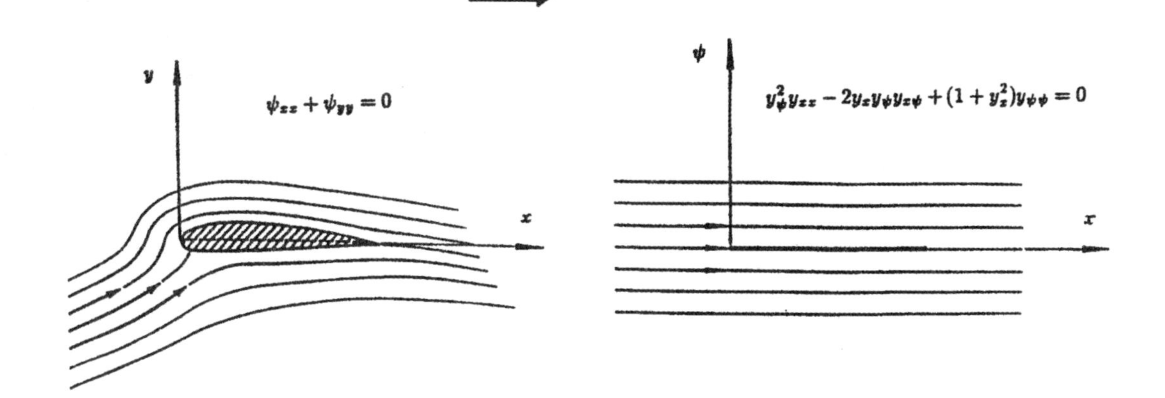

(a) Physical Domain

(b) Computational Domain

Figure 1. Concept of Stream Function Coordinate (SFC)

$$
\begin{cases}
\overline{\vec{Q}}_{j+1/2} = \frac{1}{2}(\vec{Q}_{j+1}^n + \vec{Q}_j^n) - \frac{\Delta\tau}{\Delta\psi}(\vec{F}_{j+1}^n - \vec{F}_j^n) \\
\vec{Q}_j^{n+1} = \vec{Q}_j^n - \frac{\Delta\tau}{\Delta\psi}(\overline{\vec{F}}_{j+1/2} - \overline{\vec{F}}_{j-1/2})
\end{cases}
$$

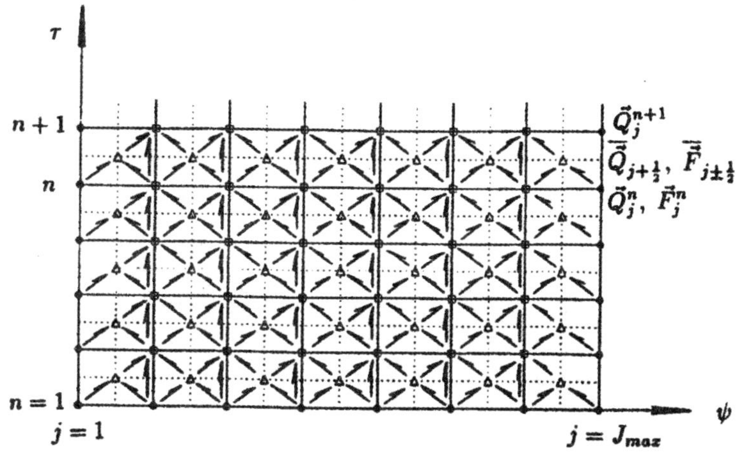

τ
$n+1$
n
$n=1$
$j=1$
$j=J_{max}$
ψ
$\vec{Q}_j^{n+1}$
$\overline{\vec{Q}}_{j+\frac{1}{2}}$, $\overline{\vec{F}}_{j\pm\frac{1}{2}}$
$\vec{Q}_j^n$, $\vec{F}_j^n$

Figure 2. Godunov's Predictor-Corrector Scheme

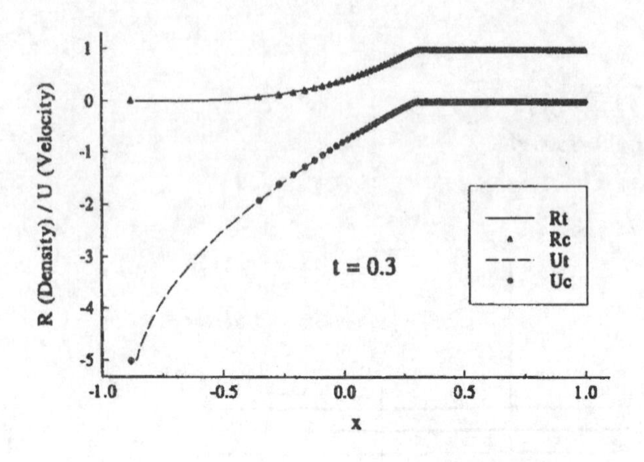

(a) Comparison of Density and Velocity

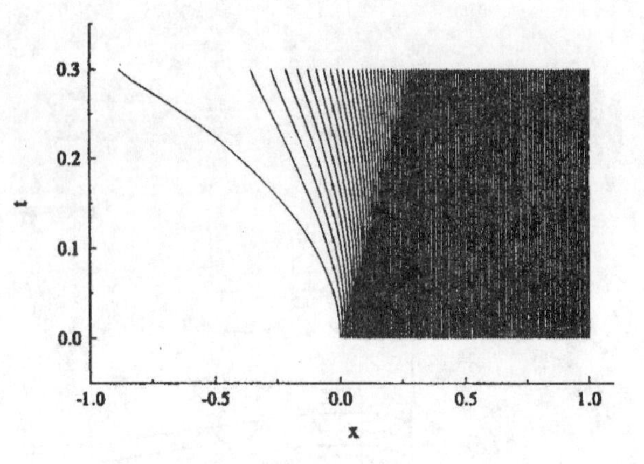

(b) Particle Path Lines

Figure 3. Piston-Driven Expansion Wave

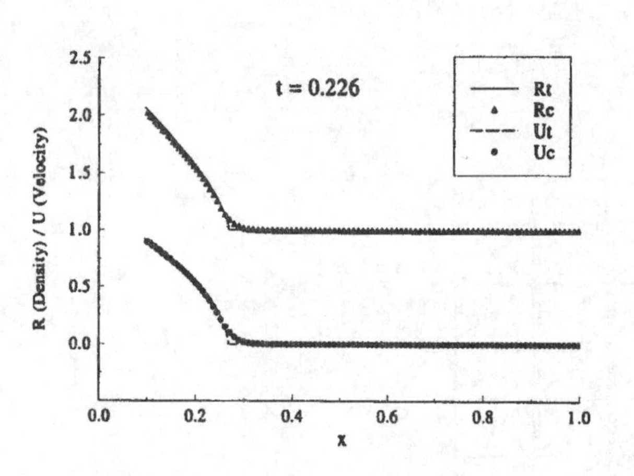

(a) Comparison of Density and Velocity

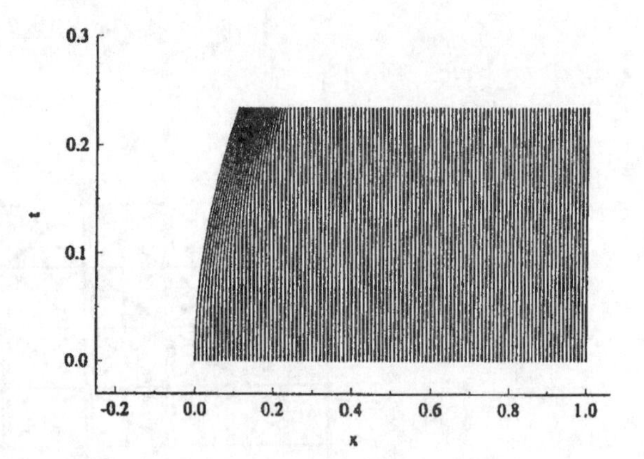

(b) Particle Path Lines

Figure 4. Piston-Driven Compression Wave

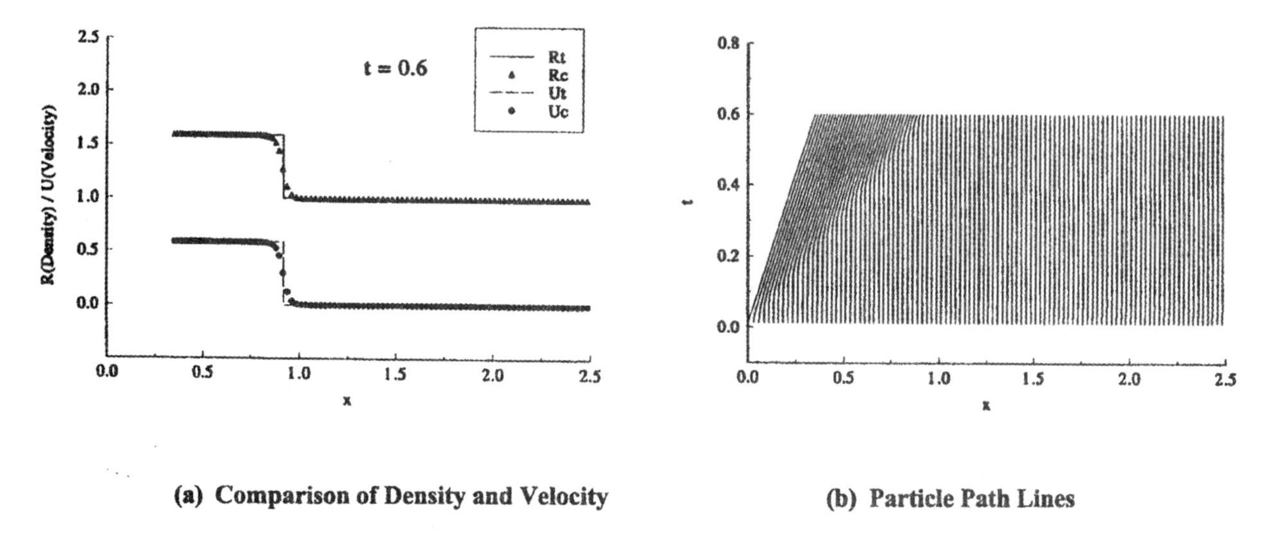

(a) Comparison of Density and Velocity

(b) Particle Path Lines

Figure 5. Piston-Driven Shock Wave

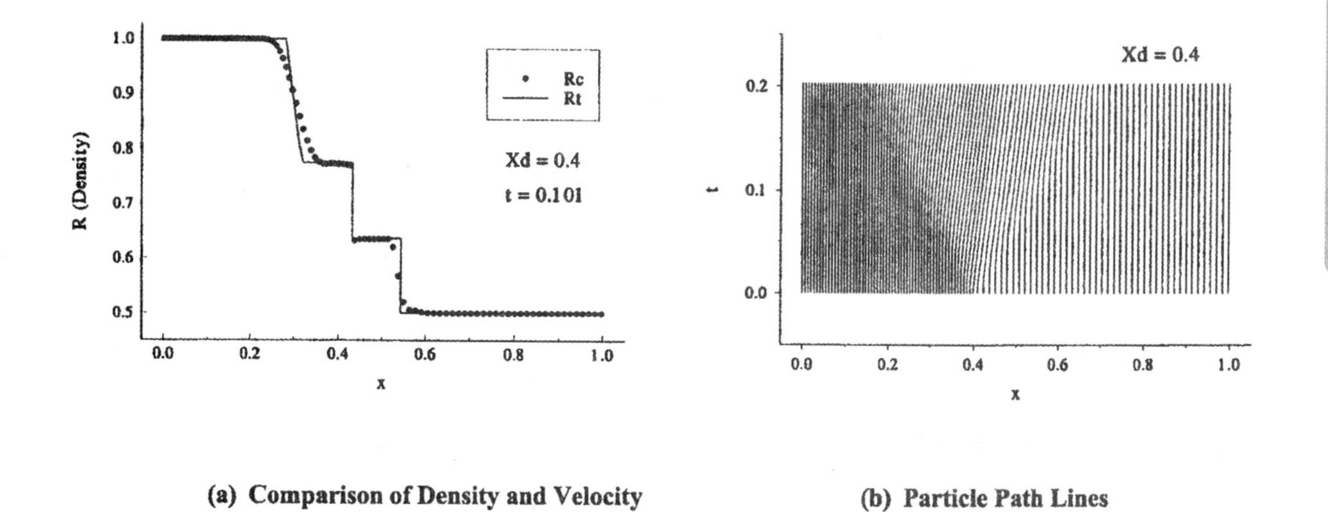

(a) Comparison of Density and Velocity

(b) Particle Path Lines

Figure 6. Propagation of Three Waves in a Shock Tube

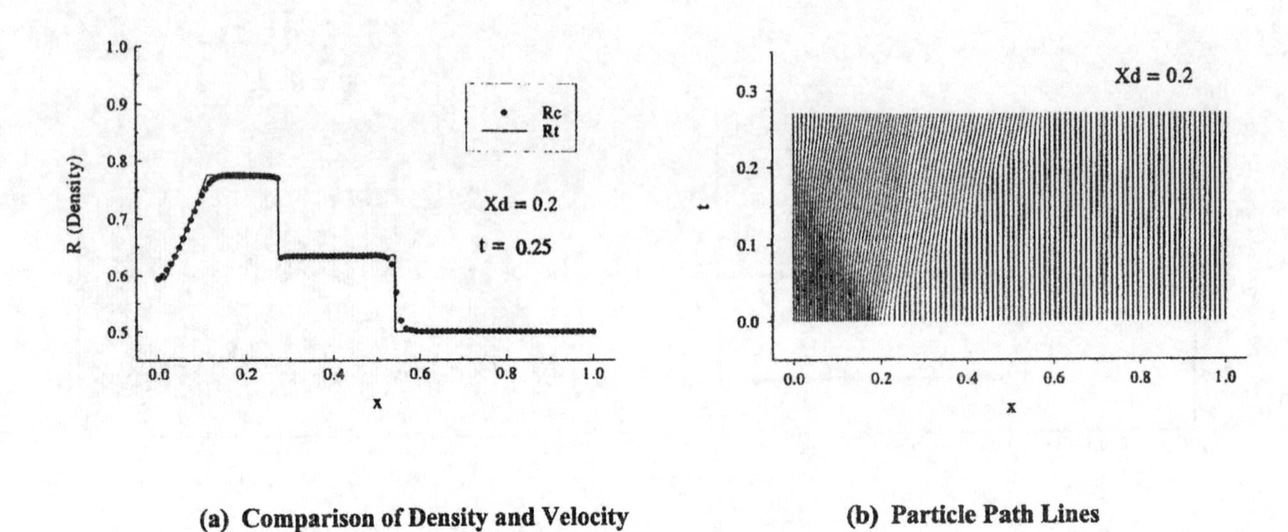

(a) Comparison of Density and Velocity

(b) Particle Path Lines

Figure 7. Reflection of Rarefaction Wave from the Wall

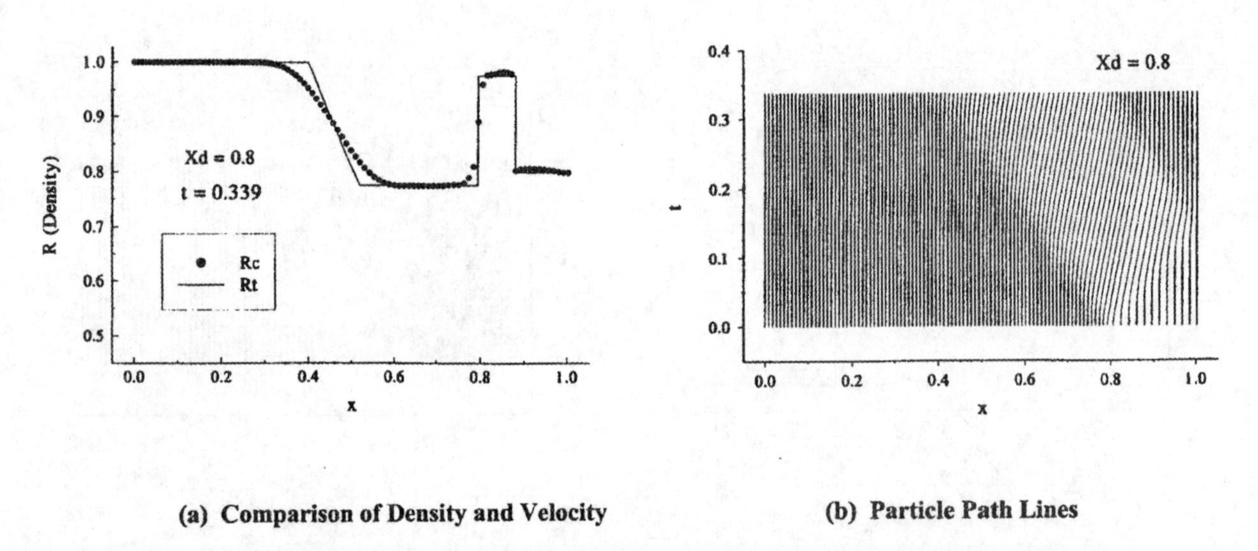

(a) Comparison of Density and Velocity

(b) Particle Path Lines

Figure 8. Shock Wave Reflection and Interaction with Contact Discontinuity

第Ⅲ部分

受聘于加拿大帝国石油公司时期

Changfa An
Ottawa
June 2, 1996

Fourth Annual Conference of the CFD Society of Canada

La quatrième conférence annuelle de la société canadienne de CFD

Proceedings / Compte rendu

CFD96

Société canadienne de
CFD
Society of Canada

Ottawa, Ontario
June 2-6 juin 1996

CFD SIMULATION OF OIL BOOM FAILURE

C.-F. An and E.J. Clavelle
Department of Mechanical Engineering
The University of Calgary
Calgary, Alberta T2N 1N4

H.M. Brown
Imperial Oil Resources Limited
Calgary, Alberta T2L 2K8

R.M. Barron
Fluid Dynamics Research Institute
University of Windsor
Windsor, Ontario N9B 3P4

Abstract

In this paper, two dimensional unsteady flows of two immiscible fluids (oil-water) are calculated using a commercial CFD software, Fluent, to simulate oil containment by booms in flowing water. The test computations show that Fluent is able to meet this need and give reasonable results for both successful oil containment and failure cases. The unsteady evolution process of an oil spill on the surface of flowing water can be simulated by tracking the interface between oil and water. If the boom draft is sufficient and the velocity of the current is not very large, a stable oil slick can be formed and oil containment will be successful. With the same boom draft, if the velocity of the current exceeds a critical value, boom failure by critical accumulation will occur and all of the oil will be lost eventually. The calculated result of the critical velocity is in a good agreement with the experimental result.

Introduction

Petroleum companies in Canada explore, produce and refine crude oils for both the domestic and foreign markets. In the process large quantities of crude oil and products are shipped over large distances. For example, about 45 million m^3 of oil per year is transported through Canada's western coastal waters, primarily through the Strait of Georgia and the international waters of the Strait of Juan de Fuca and northern Puget Sound [1]. Large quantities of oil are also shipped by pipeline. The Trans Mountain Pipeline system between Alberta and Vancouver crosses approximately 450 streams and rivers in its 1000 km length. The possibility of oil spills during transport, while remote, is nevertheless of concern both to the public and to the companies involved. Millions of dollars are spent annually on measures to prevent oil spills and to organize co-operatives which can respond in the event of a spill. Millions more are spent to provide these organizations with the specialized response equipment needed to clean up an oil spill.

If oil is spilled on water, the preferred response is to contain the oil with booms. Booms are barriers that float on the surface of the water and collect oil. The collected oil behind booms can be recovered with skimming devices. However, containment barriers or booms will reliably hold oil only under restricted conditions. If booms are towed by ships too rapidly or if they are deployed in fast moving streams and rivers or if they are used to contain highly viscous oils, failure will occur due to hydrodynamic forces. Typical booms are observed to fail when the relative boom to water velocity is greater than about 0.4 m/sec. A number of failure mechanisms identified as "drainage failure", "droplet entrainment" and "critical accumulation" have been studied but these have not been accurately modelled and the physical processes involved are not well understood.

Research on oil containment and boom failure began in the 1970's and includes experimental work and theoretical analysis [2]-[7]. Only in recent years has the CFD tool been applied to this area [8]-[11]. The primary hydrodynamic consideration in oil containment and boom design is to get a thorough understanding of the effects of various parameters, such as relative oil-water velocity, physical properties of oil and water, water depth and boom draft, sectional geometry and planar deployment of booms, etc., on the flow fields around booms. To reach this goal, computational simulations of the steady or unsteady, two or three dimensional, laminar or turbulent flow of viscous incompressible immiscible two-phase fluids (oil-water) with unknown free interface must be performed. Nowadays, a considerable number of CFD commercial packages are available in the software market dealing with a wide variety of fluid dynamics applications. Among these commercial packages, Fluent[12] is most suitable to our application because a new model VOF (Volume of Fluid), originally developed at Los Alamos Laboratories[13], has been introduced in the most recent version 4.3 to treat two-phase flow with unknown free interface.

During recent years, several experimental projects on the boom failure characteristics at current speed in excess of the critical value have been conducted in the wave basin of Imperial Oil Resources Limited [14, 15, 16]. Some novel boom designs have been tested during these experiments. The present project, CFD simulation of oil boom failure, is the continuation and extension of these studies.

In this paper, the results of exploratory test computations using Fluent are presented. The flows of oil containment by a boom, i.e. two dimensional unsteady flows of two immiscible viscous incompressible fluids (oil-water) around a boom, are simulated numerically. For all test cases, the flows are turbulent and the standard $k-\epsilon$ turbulence model is applied. In section 2, a brief description of oil containment and boom failure is made. In section 3, an outline of Fluent is given. In section 4, several test computations of oil containment are discussed in detail. In section 5, concluding remarks are summarized and further efforts are suggested.

Oil Boom Failure

A typical oil slick contained by a floating boom in moving water is shown in Figure 1(a). Wilkinson[3],[4] divided an oil slick into two zones: a frontal zone where dynamic forces dominate the form of the slick and a viscous zone where viscous shear forces dominate the form of the slick. He analyzed oil slick dynamics and obtained formulas to calculate oil slick shapes for each zone. In the frontal zone, an algebraic equation was derived to determine the slick thickness:

$$F^2 = H(2-H)[\frac{2H}{1-H} + \frac{1}{1-\Delta}]^{-1} \quad (1)$$

where $H = h/d_0$ is the ratio of the oil slick thickness h to the water depth upstream of the slick d_0, $F = U/(\Delta g d_0)^{1/2}$ is densimetric Froude number, U is the velocity of the water current upstream of the slick, g is gravitational acceleration and $\Delta = 1 - \rho_o/\rho_w$ is the ratio of the oil-water density difference to the density of the water, in which ρ is density and subscripts o and w denote oil and water, respectively. In the viscous zone, an ordinary differential equation was derived to determine the slick thickness:

$$8H\frac{dH}{dX}\{1-[H+(\frac{F}{1-H})^2]\} = (\frac{F}{1-H})^2 \times(\frac{f_i}{1-\Delta}\{1-\Delta[H+(\frac{F}{1-H})^2]\}+f_bH) \quad (2)$$

where $X = x/d_0$, x is the coordinate in the current direction from the front of the slick, f_i and f_b are friction coefficients at the oil-water interface and the water bottom, respectively. On the basis of the analysis, Wilkinson concluded that oil containment is impossible if the densimetric Froude number is greater than 0.5.

Lau and Moir[5] confirmed Wilkinson's criterion for the densimetric Froude number from their experiments and further proposed a new criterion for the successful oil containment by a boom with the minimum draft:

$$D > \frac{1-\Delta}{\Delta}\frac{U^2}{2g}. \quad (3)$$

Delvigne[6] classified three important failure mechanisms and gave the following definitions. Drainage failure: The equilibrium between the

安长发博士科研论文集

friction and spreading forces leads to increasing slick thickness h with increasing relative current velocity U. Failure occurs if h exceeds the barrier draft D, cf. Figure 1(b). Droplet entrainment: A high relative oil-water velocity U may cause oil droplets to be torn from the oil-water interface. The droplets may be entrained in the flow underneath the barrier, cf. Figure 1(c). Most importantly, Delvigne discovered the third failure mechanism, critical accumulation, described as the unstable reduction of the slick to an infinitely small length when a layer of highly viscous oil is exposed to a relative oil-water flow velocity exceeding a certain critical velocity U_{ca}. The reduction to an infinitely small slick length causes all the oil to pass underneath the barrier independent of the barrier draft, cf. Figure 1(d).

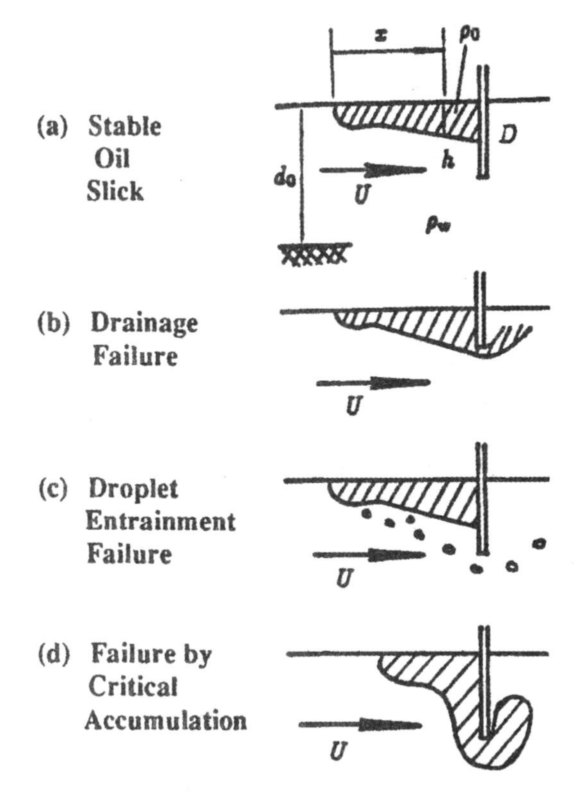

Figure 1. Oil Boom Failure

Johnston et al.[7] also explored the boom failure mechanisms of highly viscous oil in their experimental laboratory study. They found that a surging type of boom failure occurs just before the critical accumulation mode of failure develops and suggested that due to the nature of the failure mechanism, the term "oil creep" is a better description of the behavior than "critical accumulation".

Fluent – A CFD Package

Fluent[12] is a general purpose program for modeling fluid flow problems. It is a finite volume method based code and a complete package consisting of three parts: grid generator, flow solver and graphical visualizer. It is able to simulate various fluid flow problems.

In Fluent, the mean (time-averaged) Navier-Stokes equations and the equations for a selected turbulence model are solved numerically. For the most popular $k-\epsilon$ model, these equations are summarized below. Continuity and momentum equations are:

$$\frac{\partial \rho}{\partial t}+\frac{\partial}{\partial x_i}(\rho U_i)=0, \tag{4}$$

$$\frac{\partial}{\partial t}(\rho U_i)+\frac{\partial}{\partial x_j}(\rho U_i U_j)=-\frac{\partial P}{\partial x_i} +\frac{\partial}{\partial x_j}[\mu_{eff}(\frac{\partial U_i}{\partial x_j}+\frac{\partial U_j}{\partial x_i})]+\rho g_i \tag{5}$$

where ρ is density, U_i is mean velocity, P is mean pressure, g_i is gravitational acceleration, t is time, x_i is Cartesian coordinates and μ_{eff} is effective viscosity defined as the sum of the molecular viscosity μ and the turbulent viscosity μ_t:

$$\mu_{eff}=\mu+\mu_t. \tag{6}$$

In the $k-\epsilon$ model, the turbulent viscosity is determined by

$$\mu_t=\rho C_\mu\frac{k^2}{\epsilon} \tag{7}$$

where k is turbulent kinetic energy per unit mass, ϵ is its dissipation rate and C_μ is an empirical constant. The values of k and ϵ are obtained by solving their conservation equations:

$$\frac{\partial}{\partial t}(\rho k)+\frac{\partial}{\partial x_i}(\rho U_i k)=\frac{\partial}{\partial x_i}(\frac{\mu_t}{\sigma_k}\frac{\partial k}{\partial x_i}) +\mu_t(\frac{\partial U_i}{\partial x_j}+\frac{\partial U_j}{\partial x_i})\frac{\partial U_i}{\partial x_j}-\rho\epsilon, \tag{8}$$

$$\frac{\partial}{\partial t}(\rho \epsilon)+\frac{\partial}{\partial x_i}(\rho U_i \epsilon)=\frac{\partial}{\partial x_i}(\frac{\mu_t}{\sigma_\epsilon}\frac{\partial \epsilon}{\partial x_i}) +C_{1\epsilon}(\frac{\epsilon}{k})\mu_t(\frac{\partial U_i}{\partial x_j}+\frac{\partial U_j}{\partial x_i})\frac{\partial U_i}{\partial x_j}-C_{2\epsilon}\rho(\frac{\epsilon^2}{k}) \tag{9}$$

where $C_{1\epsilon}, C_{2\epsilon}, \sigma_k$ and σ_ϵ are empirical constants. The values of these constants are:

$$C_\mu=0.09,\quad C_{1\epsilon}=1.44,\quad C_{2\epsilon}=1.92,\quad \sigma_k=1.0,\quad \sigma_\epsilon=1.3. \tag{10}$$

Equations (4), (5), (8) and (9) are discretized in space using the control volume method and discretized in time using an implicit scheme. The discretized algebraic equations are solved numerically using various algorithms.

The VOF model is designed for two or more immiscible fluid phases in which the position of the interface between the fluids is of interest. It is suitable to the present oil boom failure problems because oil and water can be considered as two immiscible fluids. Unlike most other multi-phase flow models, the VOF model solves a single set of equations which is shared by the two phases. To distinguish the two phases, a concept of "volume fraction" in a cell, F, is introduced. If $F = 1$, the cell is full of water; if $F = 0$, the cell is empty of water but full of oil; if $0 < F < 1$, the cell contains both water and oil, i.e. the cell is crossed by the interface between the two fluids. Generally, F is a function of space x_i and time t and governed by the conservation equation:

$$\frac{\partial F}{\partial t} + U_i \frac{\partial F}{\partial x_i} = 0. \tag{11}$$

Equation (11) is solved using an explicit time marching scheme after velocity field U_i is obtained. In the progress of time marching, Courant-Fridrich-Lewy (CFL) number is checked and adjusted automatically. On the basis of the local values of F, the interface between oil and water can be tracked. In the VOF model, the momentum equations are solved throughout the domain and the resulting velocity field is shared by the two phases. The values of density and viscosity in the momentum equations depend on the values of F:

$$\begin{aligned} \rho &= F\rho_w + (1-F)\rho_o, \\ \mu &= F\mu_w + (1-F)\mu_o. \end{aligned} \tag{12}$$

Obviously, if $F = 1$, $\rho = \rho_w$ and $\mu = \mu_w$; if $F = 0$, $\rho = \rho_o$ and $\mu = \mu_o$; if $0 < F < 1$, ρ and μ have average values with weight factor F.

Test Computations

Several test computations have been undertaken. In order to compare the calculated results with experimental data, the same boundary and initial conditions as those in the experiments[7] have been used. The flow channel is two dimensional and has length $L = 8m$ and depth $d_0 = 0.23m$. The boom has draft $D = 0.07m$ and thickness $T = 0.01m$ and is placed at $2m$ distance from the inlet. The grid is 281 × 45 and clustered so that the grid is finer in the top region near the boom while it is coarser in the left (inlet), right (outlet) and bottom regions. The finest mesh has the scale $\Delta x = 0.01m$ and $\Delta y = 0.003m$ in the neighborhood of the boom. The oil to be tested is the highly viscous Shell Valvata 1000 oil. The density $\rho_o = 915kg/m^3$ and the dynamic viscosity $\mu_o = 3.294kg/m \cdot s$ (kinematic viscosity $\nu_o = 3600cSt$) at 15°C. To simulate the conditions of Johnston's experiments [7], the computation strategy is designed as follows.

- First, the computation is carried out for a water-only flow field until the steady state is established.
- Then, the oil is carefully "injected" from the top surface of the inlet and forms a thin layer of oil floating on the surface and flowing with water towards the boom. The oil injection does not stop until a specified amount of oil is introduced.
- After the oil injection is stopped, a definite amount of oil flows to the boom, hits it and accumulates against it. The computation runs continually until the steady state is reached. This means that a stable oil slick is formed and the oil containment under this condition is successful.
- On the basis of the above steady state flow, the current velocity is increased by a step and the computation restarts. If another steady state is established, the oil containment is also successful under this current velocity.
- In this way, the current velocity is increased step by step until a considerable amount of oil begins to pass underneath the boom. This means that the boom failure occurs under this current velocity.

In our test computations, the amount of the introduced oil, measured in volume per unit width of the boom, is $Q = 0.005m^3/m$. The current velocity in the water-only flow, in the oil-injection flow and in the flow after stopping the oil-injection and before the first steady state is reached is $U = 0.10m/s$ and then a velocity step $\Delta U = 0.05m/s$ is added to each successive run. The time step from the beginning of the computations is $\Delta t = 0.05sec$. When oil loss begins to occur, the computation oscillates violently, so the time step has to be decreased. The smallest time step is $\Delta t = 0.001sec$. The Reynolds number based on the water density and viscosity, the water depth and the lowest current velocity is about $Re = 20000$ so that the flow must be turbulent. The standard $k - \epsilon$ model is utilized in the computations.

The typical computed flow fields are shown in Figures 2–4. Figure 2 corresponds to the case of the current velocity $U = 0.10m/s$. Figures (a), (b), (c) and (d) show the velocity vectors and the interfaces between oil and water at time $t = 25.3, 35.3, 50.2$ and $52.7sec$, respectively. At $t = 25.3sec$, the oil is flowing to the boom. At $t = 35.3sec$, the oil accumulates against the boom. Comparing the oil slick shapes at $t = 50.2$ and $52.7sec$, the oil slick changes little and the velocity within the oil slick is very small. This means that a stable oil slick is formed and the oil containment under this condition is successful.

Figure 3 corresponds to the case of the current velocity $U = 0.15m/s$. Figures (a), (b), (c) and (d) show the velocity vectors and the interfaces between oil and water at $t = 53.71, 55.71, 59.74$ and $60.24sec$, respectively. At $t = 53.71sec$, the oil slick is shortened and thickened. At $t = 55.71sec$, the oil slick gets shorter and thicker and becomes an "oil lump". Comparing the oil lumps at $t = 59.74$ and $60.24sec$, the shape undergoes little change and the velocity within the oil lump is very small. This means that the oil containment is also successful under this water current.

Figure 4 corresponds to the case of the current velocity $U = 0.20m/s$. Figures (a), (b), (c) and (d) show the velocity vectors and the interfaces between oil and water at $t = 60.26, 60.30, 60.36$ and $60.46sec$, respectively. At $t = 60.26sec$, the oil lump begins to creep underneath the boom and the velocity within the oil lump is not small indicating that the oil is still creeping. At $t = 60.30sec$, more oil has creeped and is entrained into a large vortex downstream of the boom. At $t = 60.36sec$, a large amount of oil has creeped underneath the boom and is totally entrained into the vortex. At $t = 60.46sec$, all of the oil has passed the boom and flows far away downstream. The boom completely fails to contain the oil under this current velocity. According to the Delvigne's definition [6], the failure mechanism is critical accumulation.

The computational results are in a good agreement with Johnston's experiments [7] in which the failure velocity of critical accumulation is $U_{ca} = 0.20m/s$. The whole process of the "oil creep" which was observed in the experiments has been successfully simulated in the present computations. However, the surging phenomenon which was also observed in the experiments has not been captured in the present computations.

Concluding Remarks

After performing these test computations, the following conclusions can be drawn.

- The CFD commercial software, Fluent, is confirmed to be able to simulate the flow field of oil containment by a boom. The calculated critical velocity of failure is in a good agreement with the related experiments.
- The whole unsteady evolution process of oil containment and boom failure, from the oil spill on the surface of flowing water to the total oil loss, can be simulated and demonstrated dynamically using Fluent. The interface between oil and water can be tracked by the VOF technique as time progresses.
- When the current velocity is below the critical value, a stable oil slick can be formed and a steady state flow can be reached.
- When the current velocity is beyond the critical value, the boom failure occurs. For a highly viscous oil ($\nu_o > 3000cSt$), the phenomenon of "oil creep" observed in the experiments can be computationally simulated with satisfactory accuracy.
- Further efforts can be suggested as follows: refine the computational grid; choose smaller time steps; utilize heavier and lighter oils, more and less viscous oils; adjust boom alignment and boom shapes; involve the surface tension between oil and water, and so on.

Acknowledgements

The first author is grateful to NSERC for an Industrial Research Fellowship and to Imperial Oil Resources Limited for the financial support on this project. He also wishes to express his sincere gratitude to Dr.R.H. Goodman for the supervision of this project. Special acknowledgements are due Professor R.D. Rowe of the University of Calgary for offering the computing facilities and the software.

References

[1] D. Dickens, "The double hull issue and oil spill risk on the Pacific West Coast", Report for the Ministry of the Environment, Lands and Parks, BC, Canada, 1995.

[2] M. Wicks III, "Fluid dynamics of floating oil containment by mechanical barrier in the presence of water currents", Proc. of API/FWPCA Joint Conference on the Prevention and Control of Oil Spills, 1969, New York, pp.55-106.

[3] D.L. Wilkinson, "Dynamics of contained oil slicks", J. of the Hydraulics Division, HY6, June 1972, pp.1013-1030.

[4] D.L. Wilkinson, "Limitations to length of contained oil slicks", J. of the Hydraulics Division, HY5, May 1973, pp.701-712.

[5] Y.L. Lau and J. Moir, "Booms used for oil slick control", J. of the Environmental Engineering Division, EE2, April 1979, pp.369-382.

[6] G.A.L. Delvigne, "Barrier failure by critical accumulation of viscous oil", Proc. of 1989 Oil Spill Conference, Feb. 13-16, 1989, San Antonio, TX, USA, pp.143-148.

[7] A.J. Johnston, M.R. Fitzmaurice and R.G.M. Watt, "Oil spill containment: viscous oils", Proc. of 1993 International Oil Spill Conference, March 29-April 1, 1993, Tampa, FL, USA, pp.89-94.

[8] E.J. Clavelle and R.D. Rowe, "Numerical simulation of oil boom failure by critical accumulation", Proc. of 16th Arctic and Marine Oil Spill Program Technical Seminar, June 7-9, 1993, Calgary, AB, Canada, pp.409-418.

[9] E.J. Clavelle and R.D. Rowe, "Simulation of environmental flows with a density interface", Proc. of 3rd Annual Conference of the CFD Society of Canada, June 25-27, 1995, Banff, AB, Canada, pp.441-447.

[10] E.J. Clavelle and R.D. Rowe, "A numerical study of oil boom failure by critical accumulation", Proc. of 9th International Conference on Numerical Methods in Laminar and Turbulent Flow, 1995, Atlanta, GA, USA.

[11] E.J. Clavelle and R.D. Rowe, "Preliminary investigation of oil boom failure by critical accumulation", ASME Energy Sources and Technology Conference, Jan. 30, 1996, Houston, TX, USA.

[12] Fluent User's Guide, version 4.3, Fluent Inc., Lebanon, NH, USA, Jan. 1995.

[13] C.W. Hirt and B.D. Nichols, "Volume of fluid (VOF) method for the dynamics of free boundaries", J. Comp. Phys., Vol. 39, 1981, pp.201-225.

[14] H.M. Brown, R.H. Goodman and P. Nicholson, "The containment of heavy oil in flowing water", Proc. of 15th Arctic and Marine Oil Spill Program Technical Seminar, June 10-12, 1992, Edmonton, AB, Canada, pp.457-465.

[15] H.M. Brown, P. Nicholson and R.H. Goodman, "Novel concepts for the containment of oil in flowing water", Proc. of 16th Arctic and Marine Oil Spill Program Technical Seminar, June 7-9, 1993, Calgary, AB, Canada, pp.485-496.

[16] H.M. Brown, R.H. Goodman and P. Nicholson, "Using vortices to trap oil in flowing water", Proc. of 17th Arctic and Marine Oil Spill Program Technical Seminar, June 8-10, 1994, Vanvouver, BC, Canada, pp.505-512.

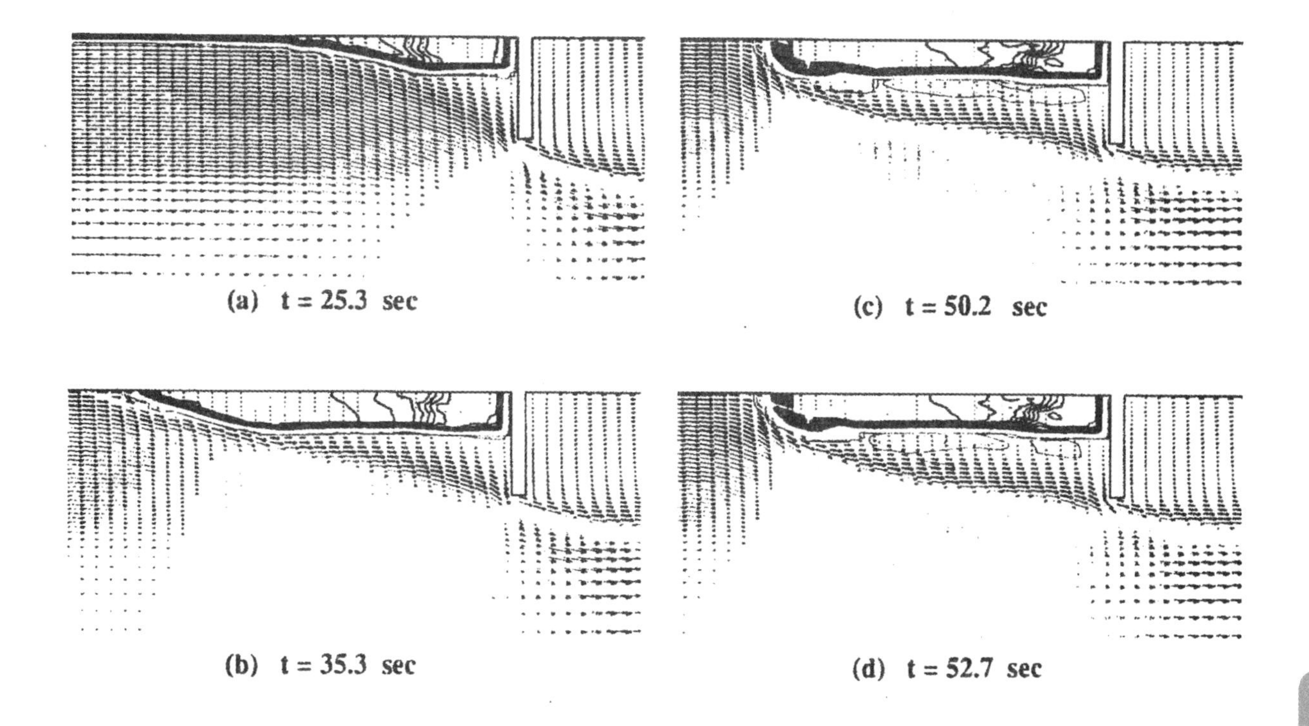

(a) t = 25.3 sec (c) t = 50.2 sec

(b) t = 35.3 sec (d) t = 52.7 sec

Figure 2. Flow Fields for U = 0.10 m/s

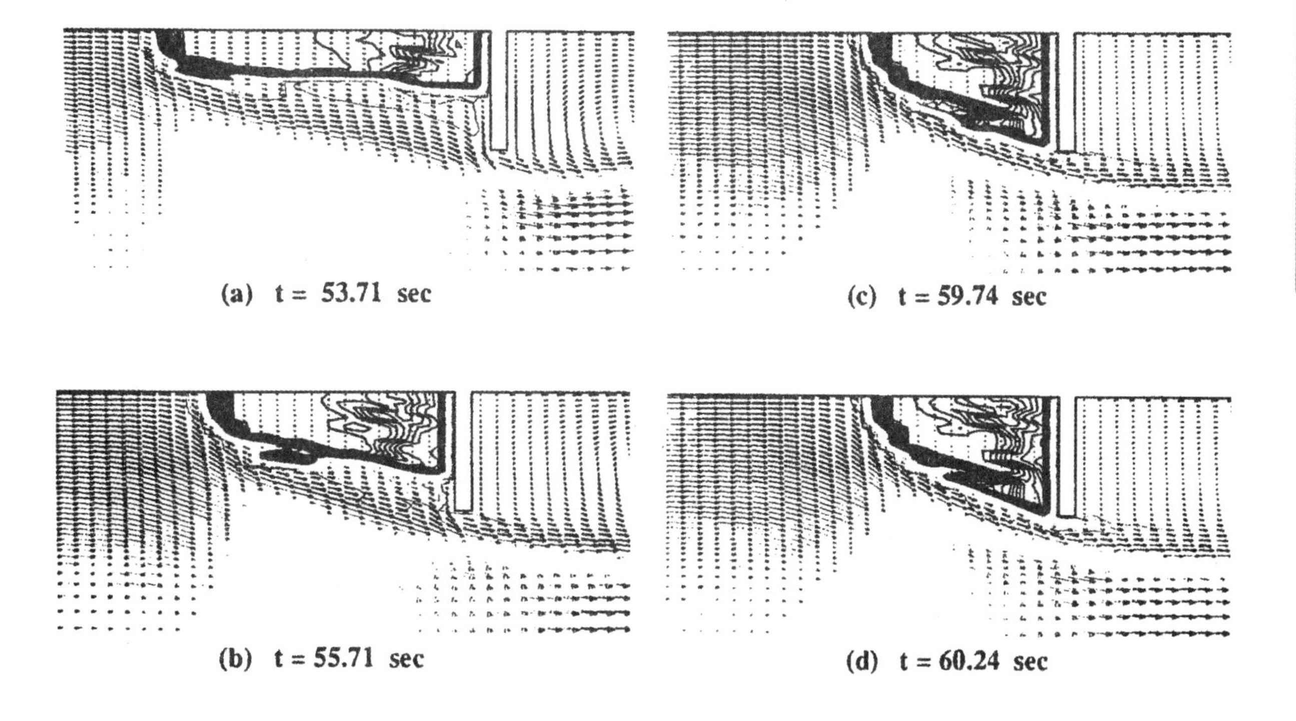

(a) t = 53.71 sec (c) t = 59.74 sec

(b) t = 55.71 sec (d) t = 60.24 sec

Figure 3. Flow Fields for U = 0.15 m/s

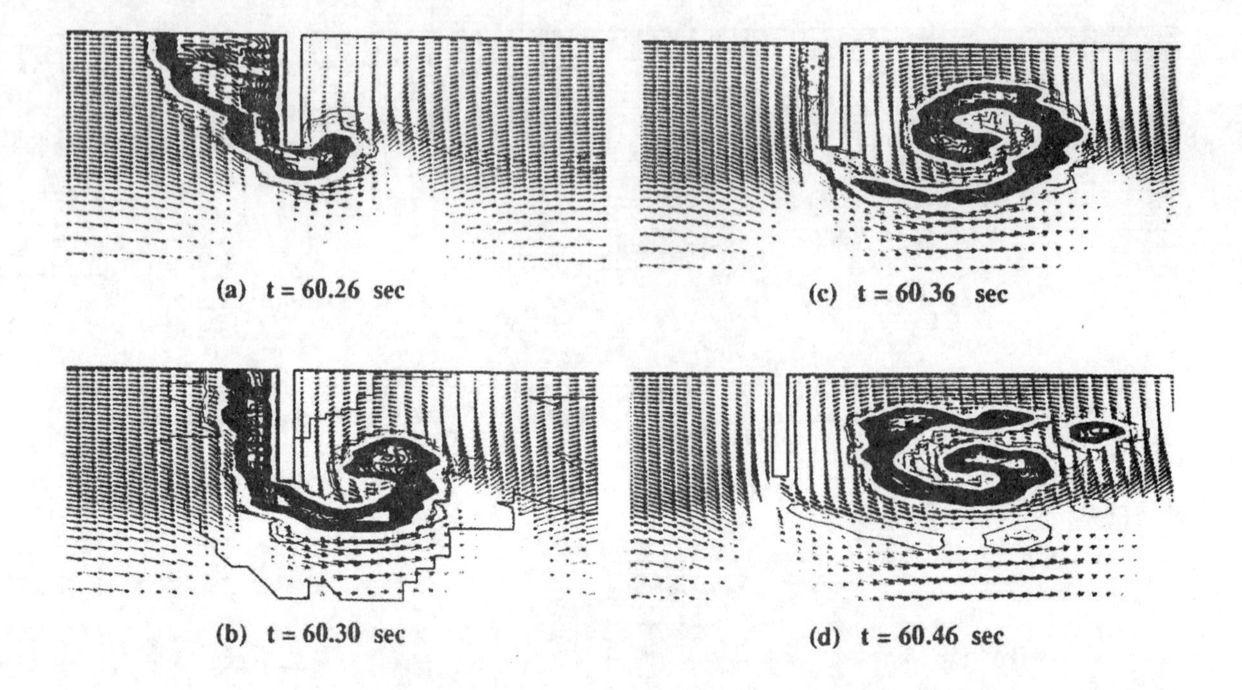

Figure 4. Flow Fields for U = 0.20 m/s

Spill Science & Technology Bulletin, Vol. 3, No. 4, pp. 213–216, 1996

1353-2561/96 $15.00+0.00

PII: S1353-2561(97)00015-7

RESEARCH

Dynamic Modelling of Oil Boom Failure Using Computational Fluid Dynamics

R. H. GOODMAN,*‡ H. M. BROWN,* CHANG-FA AN* & RICHARD D. ROWE†
*Imperial Oil Resources Limited, Calgary, Alberta, Canada T2L 2K8 (Tel: 403 284 7489; Fax: 403 284 7595; e-mail: goodmanr@cia.com)
†Department of Mechanical Engineering, The University of Calgary, Calgary, Alberta, Canada T2N 1N4 (Tel: 403 220 5788; Fax: 403 282 8406)

The common response to an oil spill on water is to contain the oil with booms and recover it with skimming devices. In some situations, however, the booms cannot hold the oil and the oil will escape underneath the boom due to hydrodynamic forces. Computational fluid dynamics (CFD) is a powerful modelling tool combining fluid dynamics and computer technology. We have utilized a commercial CFD program, Fluent, to simulate the oil–water flow around a boom. The studies accurately model channel experiments conducted in recent years. The studies show that the flow patterns around booms are modified by the presence of oil and, therefore, suggest that towing and wave-conformity tests of booms will not be meaningful unless they are undertaken with the presence of oil. © 1997 Elsevier Science Ltd

Keywords: **Numerical modelling; oil spill; boom failure; computational fluid dynamics.**

If oil is accidentally spilled on water, the most common response is to use barriers or booms to collect the spilled oil. The collected oil can be recovered using skimming devices. However, due to hydrodynamic forces, oil containment will be successful only under a limited range of oil characteristics and flow velocity. A number of boom failure mechanisms have been identified and studied, but these have not been accurately modelled. In this study, a commercial computational fluid dynamics (CFD) software, Fluent v4.4 (Fluent Inc, Lebanon, NH, U.S.A.), has been utilized to simulate the oil–water flow around a barrier. Fluent is a general purpose program for modelling fluid flows. In this program, the Reynolds averaged Navier–Stokes equations and the turbulence modelling equations are solved numerically. The volume of fluid (VOF) model is used to distinguish the two immiscible fluids (oil and water) and to track the interface between them. The study has accurately modelled a number of channel experiments reported in the literature. Boom failure mechanisms, such as 'drainage failure', 'droplet entrainment' and 'critical accumulation', have been successfully simulated. The modelling parameters which were used are summarized in Table 1.

The first modelling example was based on a flowing channel experiment conducted in the Imperial Oil test basin (Brown *et al.*, 1997). In that experiment, substantial oil loss was observed at a flow velocity of 0.20 m/s and the mechanism was identified as drainage failure. This result has been successfully simulated using the parameters shown in Table 1 and the model output is illustrated in Fig. 1. The figure shows the colour-coded flow vectors for a longitudinal cross-section of the channel at a sequence of times during the simulation. The red vectors have the highest magnitude, the blue vectors the lowest. Oil encounters the barrier from the left, is briefly stopped by it, and then drains under the barrier until most is lost. When

‡Author to whom correspondence should be addressed.

第Ⅲ部分　受聘于加拿大帝国石油公司时期

Table 1 Modelling parameters

Modelling parameters	(1) Drainage failure	(2) Droplet entrainment	(3) Critical accumulation
Spilled oil	Weathered Federated	Bunker B	Shell Valvata 1000
Density (kg/m^3)	870	888	915
Viscosity (cSt)	241	70	3600
Barrier draft (m)	0.01	0.07	0.07
Velocity (m/s)	0.20	0.09, 0.15, 0.20, 0.24	0.10, 0.15, 0.20
Reynolds No.	200,000	27,000	20,000
Domain (m^2)	8×1	4×0.3	8×0.23
Grid	241×61	281×41	281×45

the oil encounters a deeper barrier and no failure occurs, a 'surging phenomenon' is observed. This is shown in the slick profiles of Fig. 2, in which the oil slick is initially reflected from the barrier and then is re-compressed to a stable length. This modelled result accurately simulated the phenomenon observed in the channel experiment.

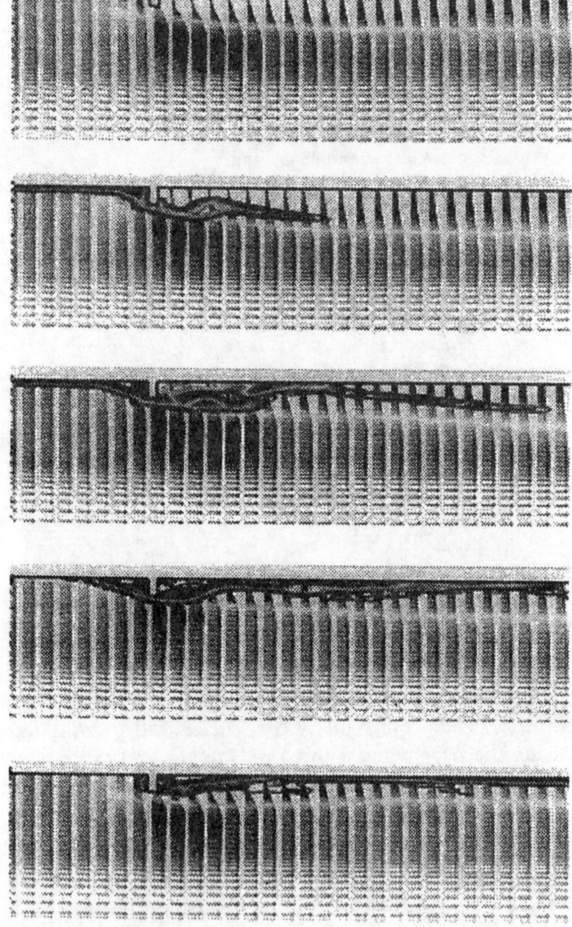

Fig. 1 Drainage failure.

In the second example, the modelling is based on an experiment of Delvigne (1989) in which oil loss occurred at a flow of 0.24 m/s due to droplet entrainment. This experiment has been successfully modelled using the parameters in the table and is illustrated as a time sequence in Fig. 3. In the figure, oil has accumulated behind the barrier, but because of visible instabilities in the oil–water interface, is continually lost under the barrier in the form of oil drops.

The third modelling example is based on an experiment of Johnston *et al.* (1993), in which a viscous oil is lost by critical accumulation or 'oil creep'

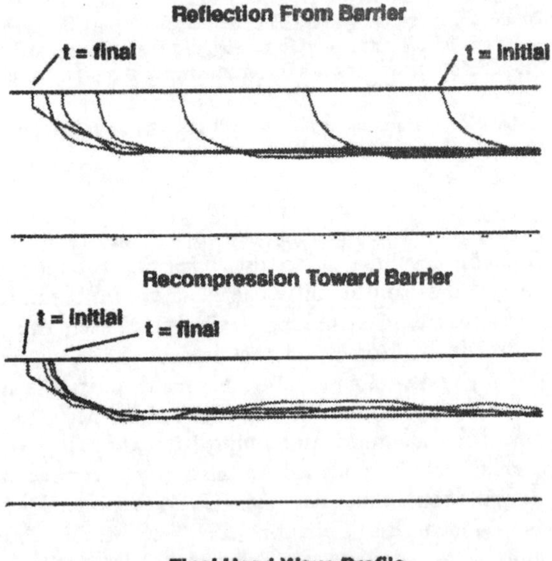

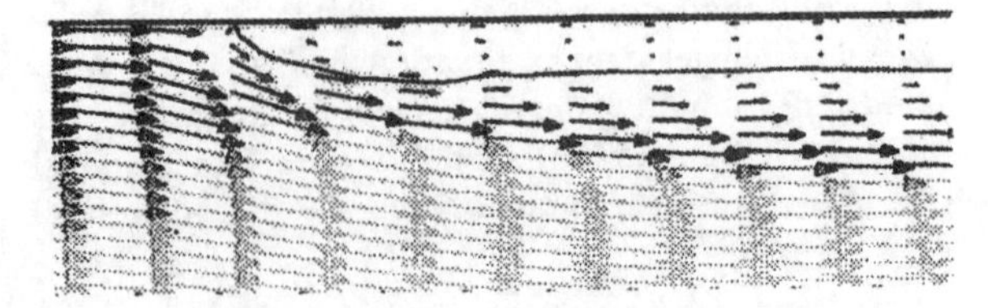

Fig. 2 Surging phenomenon and head wave.

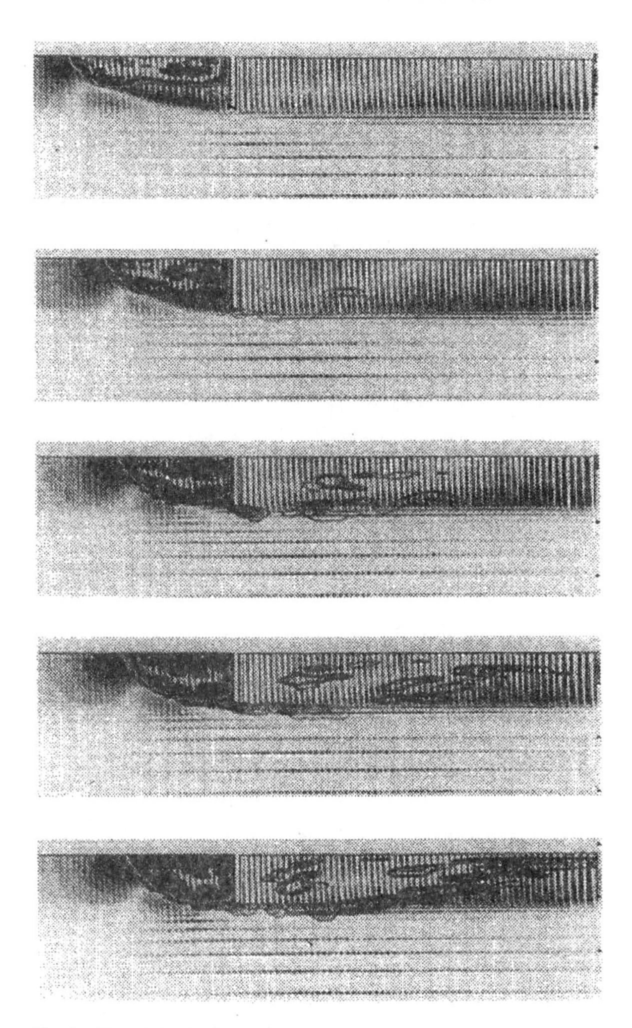

Fig. 3 Droplet entrainment.

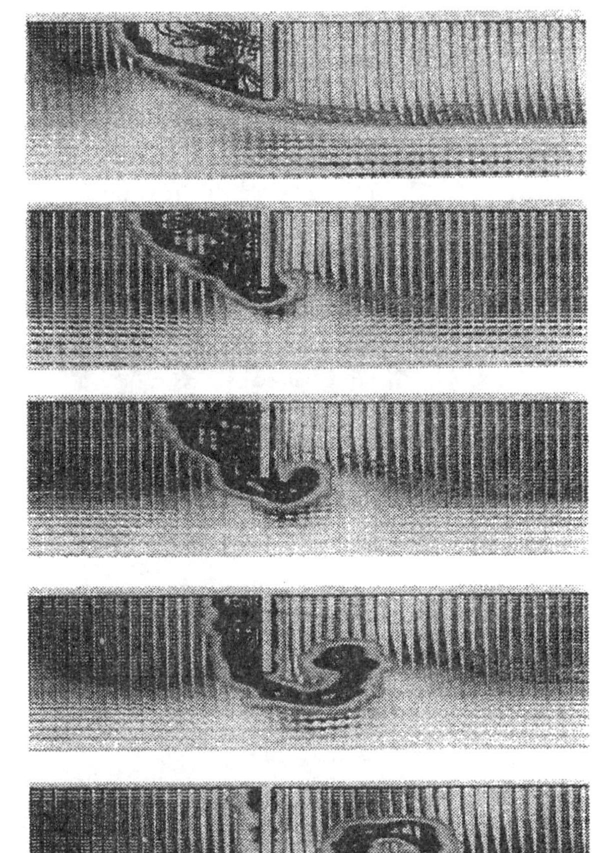

Fig. 4 Critical accumulation.

at a flow velocity of 0.20 m/s. As illustrated in the computer output of Fig. 4, the failure process continues until all of the oil has passed under the barrier in a nearly continuous mass. This model result closely simulates the experimental observations of Johnston *et al.* (1993) and observations of similar experiments conducted by Brown *et al.* (1997). The model output from these simulations also illustrate that the presence of oil on the water surface behind a barrier significantly modifies flow characteristics. Figures 5 and 6 show the velocity vector field, the static pressure contours, velocity magnitude contours, turbulent kinetic energy and turbulent dissipation rate, and stream function for a simulation both with and without the presence of oil. It is clear that there are significant differences in the two flow fields.

Finally it should be noted that each of these three failure processes is quite unique and will occur with particular oil properties and channel flow velocities. In drainage failure, the oil slick is compressed against the barrier until the slick is thick enough to leak under it. During failure due to droplet entrainment, the oil–water interface is unstable and waves propagate along it. Oil droplets may be torn from the interface and swept by the current under the barrier. When boom failure by critical accumulation occurs, the oil remains a single mass due to the high viscosity and moves more readily through the water because its high density.

Acknowledgements—The third author is grateful to the Natural Sciences and Engineering Research Council of Canada for an Industrial Research Fellowship and to Imperial Oil for financial support.

References

Brown, H. M., Goodman, R. H., An, C.-F. and Bittner, J. (1997) Boom failure mechanisms: comparison of channel experiments with computer modelling results. In *20th AMOP*, pp. 457–467, Vancouver, BC, Canada.

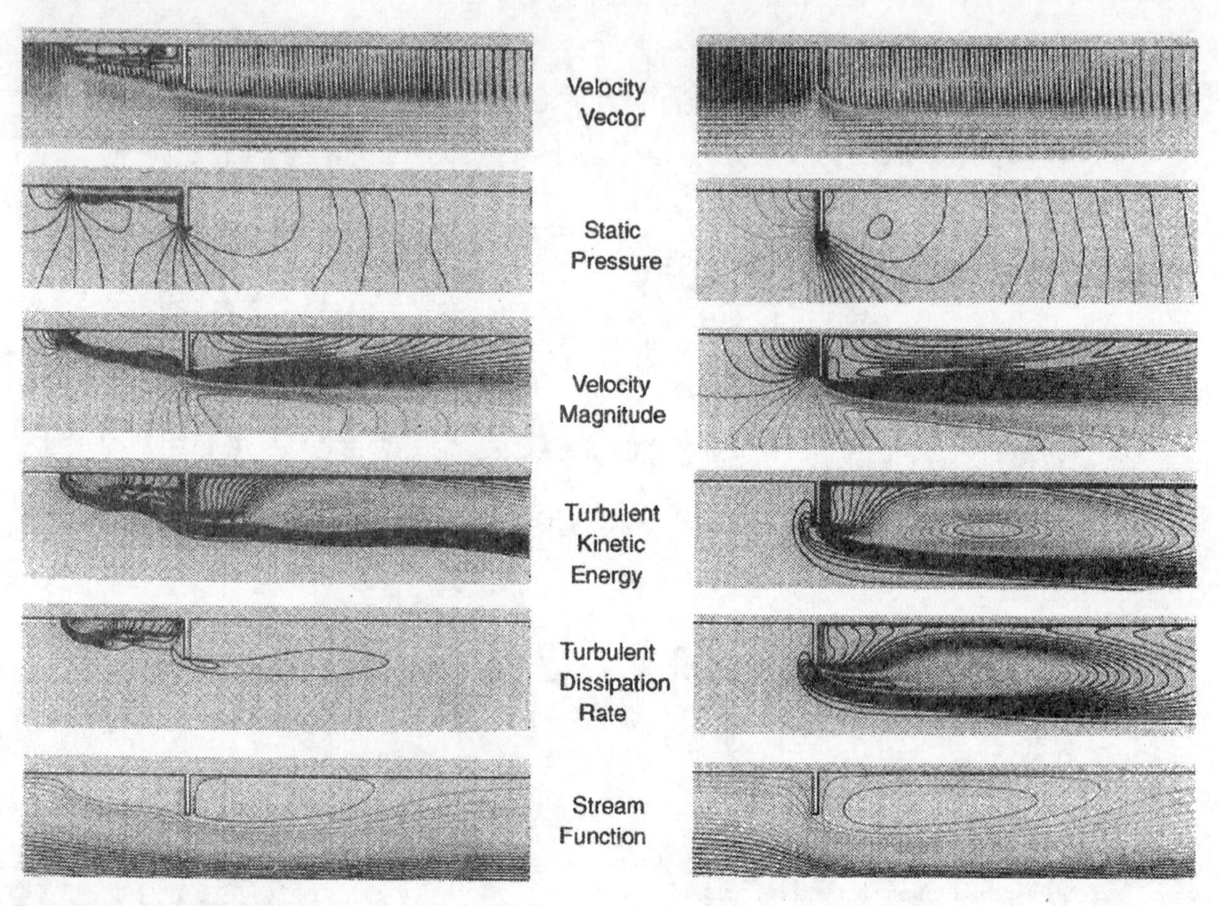

Fig. 5 Flow with stable oil slick.
Fig. 6 Flow with no oil presence.

Delvigne, G. A. L. (1989) Barrier failure by critical accumulation of viscous oil. In *1989 Oil Spill Conf.*, pp. 143–148, San Antonio, TX, U.S.A.

Johnston, A. J., Fitzmaurice, M. R. and Watt, R. G. M. (1993) Oil spill containment: viscous oils. In *1993 Oil Spill Conf.*, pp. 89–94, Tampa, FL, U.S.A.

Pergamon

PII: S1353–2561(97)00017–0

Spill Science & Technology Bulletin, Vol. 3, No. 4, pp. 221–224, 1996

1353–2561/96 $15.00+0.00

RESEARCH

Animation of Boom Failure Processes

CHANG-FA AN*, H. M. BROWN*, R. H. GOODMAN* & ERIC CLAVELLE†
**Imperial Oil Resources Limited, Calgary, Alberta, Canada T2L 2K8 (Tel: 403 284 7541; Fax: 403 284 7595; e-mail: cfan@acs.ucalgary.ca)*
†Department of Mechanical Engineering, The University of Calgary, Calgary, Alberta, Canada T2N 1N4 (Tel: 403 220 4171; Fax: 403 282 8406)

第三部分 受聘于加拿大帝国石油公司时期

Computer animation is a useful tool for demonstrating time-dependent physical processes. In this paper, computer animations of oil–water flow around a boom and boom failure processes are presented. The animations are performed by playing a sequence of images at a speed of 15 frames/s (fps). The images have been obtained from the graphical output of numerical modelling using a computational fluid dynamics (CFD) program, Fluent. The modelling is based on boom failure experiments carried out in flowing water channels. The animations are in audio video interleave (AVI) format and can be viewed easily on personal computers using the Media Player program within Windows '95. Features of the boom failure processes are more evident in the dynamic animation sequences than can be discerned from individual static images.

Keywords: **Animation; oil spill; boom failure; computational fluid dynamics.**

Animation, or dynamic visualization, is an expressive tool for demonstrating time-dependent physical processes. In this paper, the use of computer animation to demonstrate oil boom failure mechanisms is described. The animation is performed by playing a sequence of images at a speed of 15 frames/s (fps). The images were obtained from the graphical output of a numerical model which uses a computational fluid dynamics (CFD) program, Fluent (1996). The model simulated a series of boom failure experiments which have been conducted in flowing water channels (An *et al.*, 1996). Features of boom failure processes are more evident from the entire dynamic sequences than from the individual static images.

A readily available software program, VidEdit (1993) was chosen to create the animation in audio video interleave (AVI) format. The graphical output from the model simulations was obtained in tagged image file format (TIFF) and converted to device independent bitmap (DIB) format for use in VidEdit. Three animated examples of common boom failure mechanisms have been generated using this process and can be suitably displayed on a Pentium PC with the Media Player program supplied with Windows '95.

The first example of animation was based on a model run simulating a channel experiment in which drainage failure was observed (Brown *et al.*, 1997). Several typical frames of the animation are demonstrated in Fig. 1. These show the flow velocity vectors in a cross-section of the channel. The red vectors are the highest magnitude, the blue the lowest. Oil flows in from the left, hits the barrier and drains rapidly under it. In viewing the animation, salient features were the rapidity with which the oil flowed under the barrier, the depth to which it sank and the trapping of oil in the vortex behind the barrier. The animation showed that oil reached the surface behind the barrier and flowed downstream in patches rather than in a continuous slick, and that the leading edge of the oil continued well below the surface for some time. Thick patches of oil remained both in front of and behind the barrier at the end of the sequence.

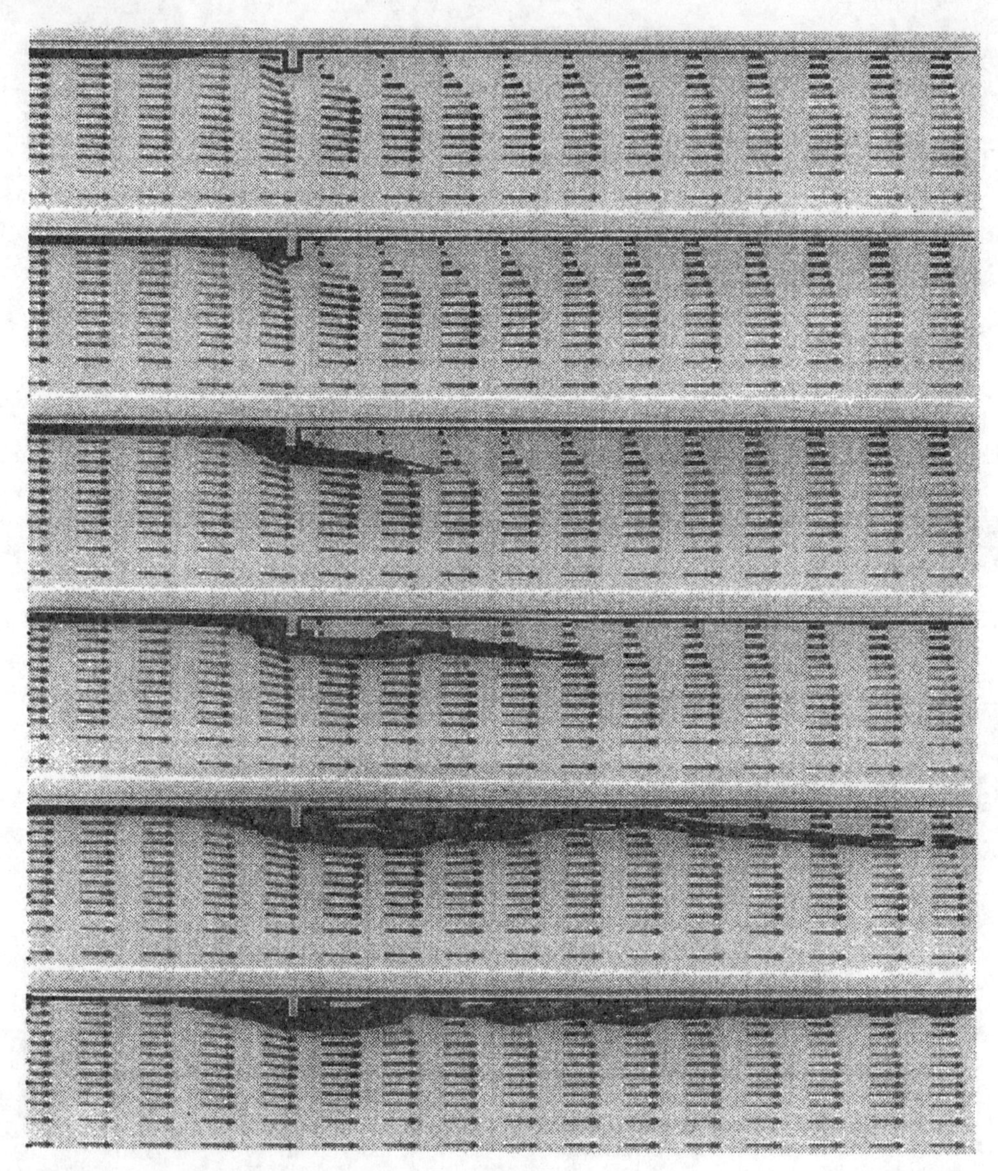

Fig. 1 Animation frames from simulation of drainage failure.

A second example of animation was based on a model run simulating a channel experiment in which the failure mechanism of droplet entrainment was observed (Delvigne, 1989). On viewing the animation several features were immediately apparent. The oil–water interface beneath the accumulated oil was unstable and waves continually propagated along it. Oil continually bled as droplets under the barrier from the interface. The lost oil flowed a considerable distance downstream of the barrier before resurfacing and then flowed back along the water surface to the barrier as it was caught in the large trailing vortex. Circulation patterns in the accumulated oil upstream of the barrier were clearly visible. Several of these features can be seen in the typical frames of the animation shown in Fig. 2.

The third example of animation was based on a model run simulating a channel experiment in which failure by critical accumulation (or more descriptively, 'oil creep') was observed (Johnston *et al.*, 1993). When the animation was viewed, the salient features of oil creep were obvious. A continuous stream of thick oil was observed to creep under the barrier. This was not a thin stream bled from the oil–water interface as was observed in the droplet entrainment example, but a continuous thick mass of oil. The loss continued until all of the accumulated oil had passed under the barrier and occurred despite the relatively deep barrier draft.

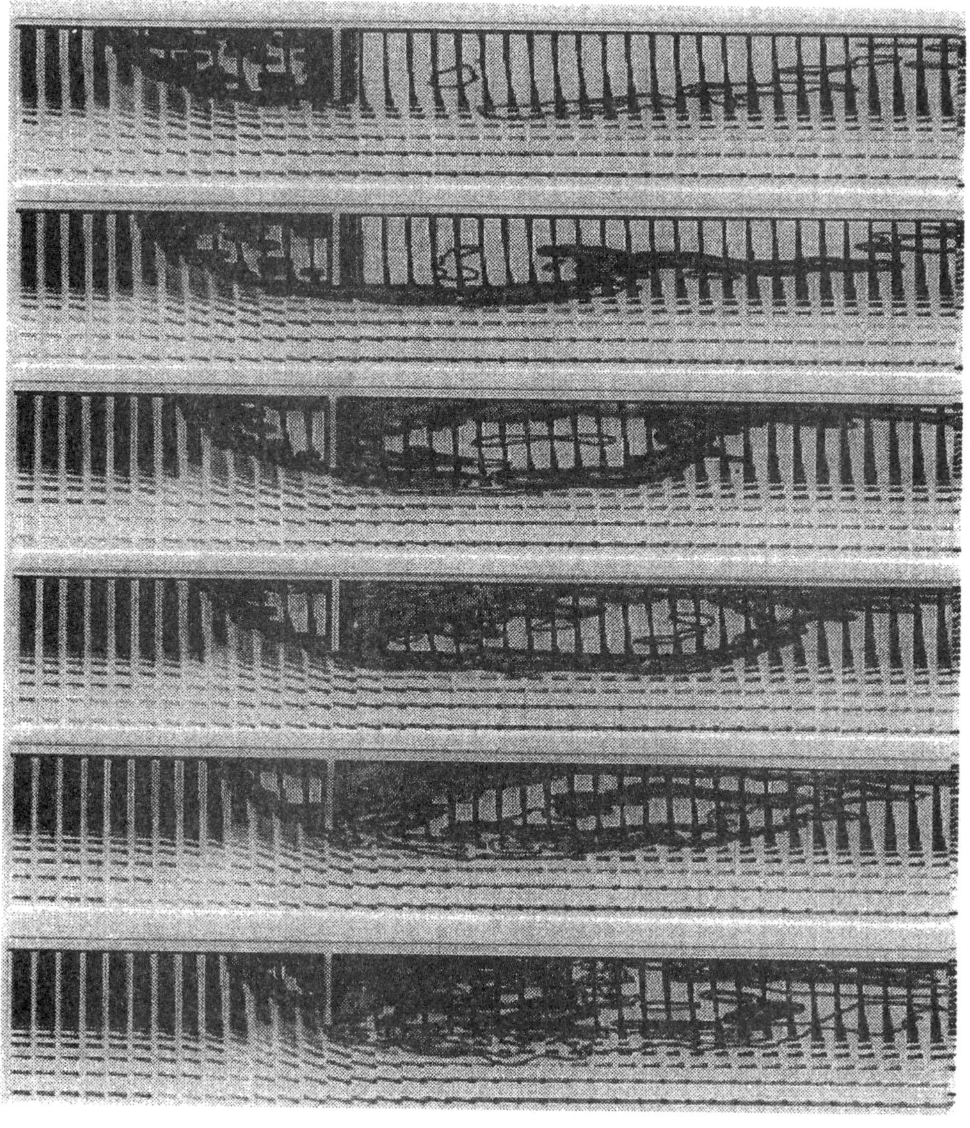

Fig. 2 Animation frames from simulation of droplet entrainment failure.

No visible circulation within the accumulated oil was observed. Once the oil had passed under the barrier it remained caught in the trailing vortex as a single mass of oil. Throughout the simulation the oil behaved as a single mass, which appeared to be squeezed under the barrier by the flowing water. Several typical frames of the animation which show this process are presented in Fig. 3.

Video animation of these dynamic processes using the output of computer models has been very useful for the understanding of the failure mechanisms observed in channel experiments.

Acknowledgements—The first author would like to acknowledge the Natural Sciences and Engineering Research Council of Canada for an Industrial Research Fellowship and Imperial Oil Resources Limited for financial support. He wishes to express sincere gratitude to Prof. R. D. Rowe of the University of Calgary for the sponsorship of a Research Associateship which promotes the current project. He also wishes to recognize Mr. Hank Li of Wisemind Enterprise Co. for his patient help in the computer animation.

References

An, C.-F., Clavelle, E. J., Brown, H. M. and Barron, R. M. (1996). CFD simulation of oil boom failure. In *CFD'96*, pp. 401–408. Ottawa, ON, Canada.

Brown, H. M., Goodman, R. H., An, C.-F. and Bittner, J. (1997). Boom failure mechanisms: comparison of channel experiments with computer modelling results. In *20th AMOP*. Vancourver, BC, Canada.

安长发博士科研论文集

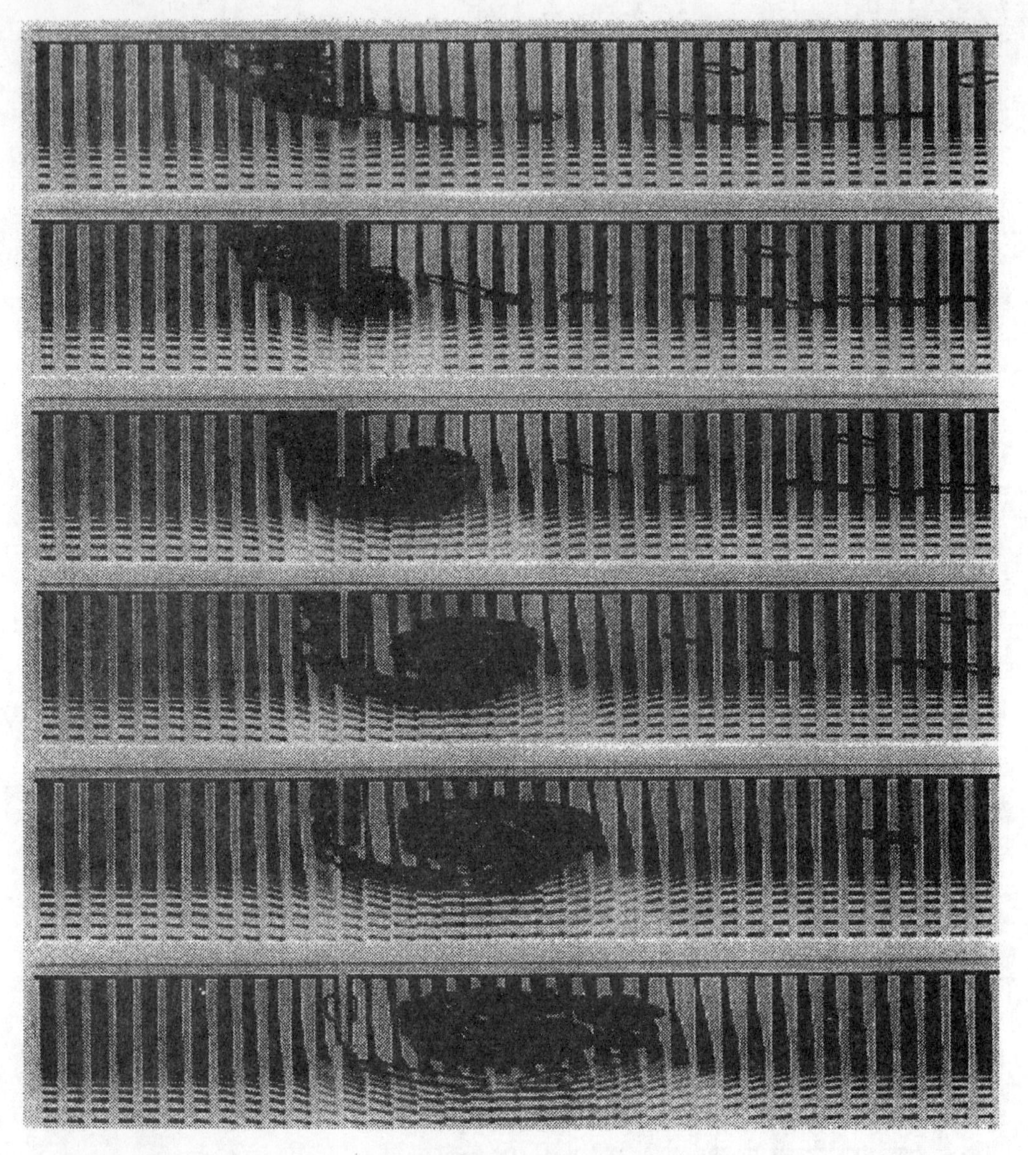

Fig. 3 Animation of frames from simulation of critical accumulation failure.

Delvigne, G. A. L. (1989). Barrier failure by critical accumulation of viscous oil. In *1989 Oil Spill Conf.*, pp. 143–148. San Antonio, TX, U.S.A.
Fluent User's Guide v4.4 (1996). Fluent Inc., Lebanon, NH, U.S.A.
Johnston, A. J., Fitzmaurice, M. R. and Watt, R. G. M. (1993). Oil spill containment: viscous oils. In *1993 Oil Spill Conf.*, pp. 89–94. Tampa, FL, U.S.A.
VidEdit (1993). Microsoft Corporation, Roselle, IL, U.S.A.

Changxa An
Victoria, BC
May 25, 1997

CFD 97

May 25-27, 1997
Du 25 au 27 mai 1997

Proceedings
Comptes rendus

Editors/Editeurs: N. Djilali and S. Dunlop

The Fifth Annual Conference of the Computational Fluid Dynamics Society of Canada

Cinquième Conferénce de la Société Canadienne de CFD

University of Victoria
Victoria, British Columbia

HYDRODYNAMIC BEHAVIOR OF CONTAINED OIL SLICK ON FLOWING WATER

C.-F. An, H.M. Brown & R.H. Goodman
Imperial Oil Resources Limited
Calgary, Alberta, Canada
cfan@acs.ucalgary.ca

E.J. Clavelle & R.D. Rowe
The University of Calgary
Calgary, Alberta, Canada
clavelle@enme.ucalgary.ca

R.M. Barron
University of Windsor
Windsor, Ontario, Canada
az3@uwindsor.ca

ABSTRACT

In this paper, a comparative computational/experimental study of hydrodynamic behavior of an oil slick contained by a barrier on flowing water has been presented. The computational results are obtained using a CFD code, Fluent. The experimental results are from the Imperial Oil wave basin test in which a heavy viscous oil, weathered Hondo, was spilled on the surface of flowing water and contained by a barrier. When the velocity was below a critical value, 0.20 m/s, a stable oil slick was attained and the oil containment was successful. When the velocity reached the critical value, substantial oil loss was observed and the mechanism of barrier failure was believed to be critical accumulation, or oil creep. These experimental results have been accurately modeled in the computational study. For the case of successful oil containment, the simulated profiles of the oil slick are in a good agreement with the experimental measurements. For the case of barrier failure, the unsteady process of oil creep has been simulated numerically.

1. INTRODUCTION

The common response to an oil spill on the surface of flowing water is to contain the oil with barriers and then recover it with skimming devices. However, successful oil containment by a barrier can be attained only under certain conditions. If these conditions are not met, the oil will escape under the barrier due to hydrodynamic forces. Since the early 1970's, considerable effort has been made to investigate the hydrodynamic behavior of the oil slick contained by a barrier on flowing water. The representative work includes theoretical analyses and experimental investigations [1-8]. Most theoretical work was done under a one-dimensional assumption and most experimental work was done for light oils with low viscosity in small-size indoor flumes. Very little computational work can be found in the literature until the 1990's [9-12]. In order to obtain experimental data from a medium-size outdoor channel for various types of oil, an experimental project for oil containment and barrier failure mechanisms has been undertaken in the Imperial Oil wave basin [13].

In the present paper, a comparative computational/experimental study of the hydrodynamic behavior of an oil slick contained by a barrier on flowing water has been conducted. The experimental results are from the wave basin test and the computational results are obtained using a CFD code, Fluent [14]. In the wave basin experiment, the weathered Hondo oil was spilled on the surface of the water channel. The barrier draft was set at 0.05m and the flow velocity ranged from 0.10 to 0.20 m/s. In the computational simulation, the unsteady Navier-Stokes equations with the standard $k-\epsilon$ turbulence model were numerically solved using finite volume method for space and implicit marching for time. The built-in VOF (volume of fluid) multiphase model was invoked to track the interface between the two immiscible liquid phases, oil and water.

In the next section, the mathematical description for the present problem is given. Then, the simulated results of the oil slick behavior are presented and compared with the corresponding experimental measurements and observations. Conclusions are drawn in the last section.

2. MATHEMATICAL DESCRIPTION

In general, the movement of an oil slick on the surface of flowing water and restricted by a barrier is a three dimensional unsteady turbulent flow of multi-component fluids. There exist three fluid components (oil, water and air) and three interfaces (oil-water, oil-air and water-air). If the weather is calm, then the effects of wind and waves do not interfere with the flow significantly. Thus, the oil-air and water-air interfaces can be treated as a single flat plane. Furthermore, the friction effect on the oil-air and water-air interfaces can be neglected compared with the oil-water interface. As a result, the third component (air) can be removed from the system and the oil-air and water-air interfaces can be assumed as a frictionless flat plane (a slip-wall boundary). Finally, the system is reduced to a flow of two liquid phases (oil and water). The compressibility of oil and water are small so that both of them can be treated as incompressible fluids. The mutual solubility between oil and water is also small and the emulsification of oil in water is not significant in a short period of time after a spill. Therefore, the two liquids (oil and water) can be treated as immiscible fluids, for which a simple model, VOF, is suitable with adequate accuracy. The minimum Reynolds number for the present application is about 100,000 based on the minimum flow velocity (0.10m/s), the channel depth (1m) and the kinematic viscosity of water (1cSt). Although the flow within the oil slick may be laminar, turbulent flow must be considered throughout the whole region. In addition, the flow is in the range of low speed and the turbulence is homogeneous so that the standard $k-\epsilon$ model is suitable. Strictly speaking, the flow of an oil slick contained by a barrier on the surface of flowing water is always three dimensional. Nevertheless, the problem can be reduced to a two dimensional one. The ideal planar shape of a deployed boom on flowing water is a catenary curve. If the flow is uniform, the maximum normal velocity to the boom must occur in the middle section, i.e. the apex of the catenary. In other words, the most critical part with threat of oil loss is the middle section which is a two dimensional plane. Therefore, a two dimensional model is appropriate for the present problem.

For a two dimensional unsteady turbulent flow of a viscous incompressible fluid, the Reynolds-averaged Navier-Stokes equations are

$$\frac{\partial U}{\partial x}+\frac{\partial V}{\partial y}=0, \tag{1}$$

$$\rho[\frac{\partial U}{\partial t}+\frac{\partial(U^2)}{\partial x}+\frac{\partial(UV)}{\partial y}]=-\frac{\partial P}{\partial x}+\mu_{eff}[2\frac{\partial^2 U}{\partial x^2}+\frac{\partial}{\partial y}(\frac{\partial U}{\partial y}+\frac{\partial V}{\partial x})], \tag{2}$$

$$\rho[\frac{\partial V}{\partial t}+\frac{\partial(UV)}{\partial x}+\frac{\partial(V^2)}{\partial y}]=-\frac{\partial P}{\partial y}+\mu_{eff}[\frac{\partial}{\partial x}(\frac{\partial V}{\partial x}+\frac{\partial U}{\partial y})+2\frac{\partial^2 V}{\partial y^2}]+\rho g \tag{3}$$

where x and y are Cartesian coordinates, t is the time, ρ is the density, P is the mean pressure, U and V are the mean velocity components in x and y directions, respectively, g is the gravitational acceleration and μ_{eff} is the effective viscosity which is the sum of the molecular viscosity μ and the turbulent viscosity μ_t, i.e. $\mu_{eff}=\mu+\mu_t$.

In the standard $k-\epsilon$ turbulence model, the turbulent viscosity is determined by $\mu_t=\rho C_\mu k^2/\epsilon$ where C_μ is an empirical constant, k is the turbulent kinetic energy per unit mass, defined as $k=(\overline{u'^2}+\overline{v'^2})/2$, and ϵ is the dissipation rate, defined as $\epsilon=\nu[\overline{(\partial u'/\partial x)^2}+\overline{(\partial u'/\partial y)^2}+\overline{(\partial v'/\partial x)^2}+\overline{(\partial v'/\partial y)^2}]$, where ν is the kinematic viscosity. k and ϵ can be solved from the transport equations [15]:

$$\rho[\frac{\partial k}{\partial t}+\frac{\partial(Uk)}{\partial x}+\frac{\partial(Vk)}{\partial y}]=\frac{\partial}{\partial x}(\frac{\mu_t}{\sigma_k}\frac{\partial k}{\partial x})+\frac{\partial}{\partial y}(\frac{\mu_t}{\sigma_k}\frac{\partial k}{\partial y})+G-\rho\epsilon, \tag{4}$$

$$\rho[\frac{\partial \epsilon}{\partial t}+\frac{\partial(U\epsilon)}{\partial x}+\frac{\partial(V\epsilon)}{\partial y}]=\frac{\partial}{\partial x}(\frac{\mu_t}{\sigma_\epsilon}\frac{\partial \epsilon}{\partial x})+\frac{\partial}{\partial y}(\frac{\mu_t}{\sigma_\epsilon}\frac{\partial \epsilon}{\partial y})+C_{1\epsilon}\frac{\epsilon}{k}G-C_{2\epsilon}\rho\frac{\epsilon^2}{k} \tag{5}$$

where $G=\mu_t[2(\partial U/\partial x)^2+(\partial U/\partial y+\partial V/\partial x)^2+2(\partial V/\partial y)^2]$ and the empirical constants are $C_\mu=0.09$, $C_{1\epsilon}=1.44$, $C_{2\epsilon}=1.92$, $\sigma_k=1.0$ and $\sigma_\epsilon=1.3$.

In order to handle the immiscible multi-component fluids, the VOF model is applied. Unlike other multiphase fluid models, the VOF model solves a single set of equations that are shared by all phases. To distinguish the phases, a new scalar variable F, the volume fraction of water in a computational cell, is defined as the ratio of the volume of water in a cell to the total volume of the cell. Obviously, if $F = 0$, the cell is full of oil, if $F = 1$, the cell is full of water, and if $0 < F < 1$, the cell contains both oil and water, i.e. the cell must be crossed by the oil-water interface. In general, F is a function of time and space and governed by the conservation law

$$\frac{\partial F}{\partial t} + U\frac{\partial F}{\partial x} + V\frac{\partial F}{\partial y} = 0. \qquad (6)$$

In the Navier-Stokes equations (1) – (3) and the turbulence equations (4) and (5), the velocity field (U, V) is shared by the two phases or their mixture while the physical properties (ρ, μ) are distinguished by F in an individual cell, $\rho = F\rho_w + (1-F)\rho_o$ and $\mu = F\mu_w + (1-F)\mu_o$, where the subscripts "$o$" and "$w$" represent oil and water, respectively.

3. COMPUTATION vs EXPERIMENT

The water channel which was used to conduct the experiment is 30m long, 1m wide and 1m deep. In the middle of the channel length, a test section was installed consisting of a flat plate barrier as a boom, a flowmeter, an oil slick thickness gauge and a glass window for observation. In this experiment, 10 liters of a heavy viscous oil, weathered Hondo (ρ=952.6kg/m^3, ν=5616cSt), was spilled on the water surface and was contained by the barrier with draft 0.05m. Three flow velocities were taken, i.e. 0.10, 0.15, 0.20m/s. When the flow velocity was 0.10 and 0.15m/s, the flow was able to stabilize and the oil slick profiles were measured. When the flow velocity was increased to 0.20m/s, substantial oil loss occurred and the barrier failure mechanism was observed to be critical accumulation, or oil creep.

These experimental observations and measurements have been accurately modeled in the computational study. The computational domain was a rectangle of 8m × 1m representing the water channel. The barrier was set at 3m distance from the inlet and the draft was set at 0.05m consistent with the experiment. The grid was 241 × 61 and clustered so that the mesh was finer in the neighborhood of the barrier and coarser elsewhere. The standard $k-\epsilon$ turbulence model was applied. The computation was time-dependent and the maximum time step was set to 0.05sec at the beginning and could be reduced afterwards according to the requirements of convergence and stability. The maximum number of iterations within a time step was set to 20 and the maximum residual of the equations was set to 10^{-3}. The VOF model was invoked to track the oil-water interface. On the barrier surface and the bottom of the channel, no-slip walls were assumed. On the oil-air and water-air interfaces, a slip wall was specified. At the inlet, the flow velocity began with 0.10m/s, the turbulence intensity was set to 2% and the characteristic length was 1m, i.e. the channel depth. The initial guess of the flow field was made so that the velocity was uniform eveywhere and equal to that at the inlet. At the beginning of the computation, the oil was injected from the inlet on the water surface. The oil injection was stopped when the exact amount of oil was introduced.

Figure 1(a) shows the evolution of the oil slick at a velocity of 0.10m/s during the period of simulation time t=45.50sec to t=78.05sec. The flow reached a steady state and the oil slick became stable by t=78.05sec. Figure 2(a) presents the comparison of the stable oil slick profiles. The agreement between the computation and the experiment is quite satisfactory.

On the basis of the above steady flow, the velocity was increased to 0.15m/s. Figure 1(b) shows the evolution of the oil slick deformation during the period of t=78.05sec to t=90.19sec. The flow again reached a steady state and the oil slick became stable by t=90.19sec. Figure 2(b) presents the comparison of the stable oil slick profiles. Similar agreement is obtained between the computation and the experiment.

The flow was then accelerated to 0.20m/s. Figure 1(c) shows the evolution of the oil slick behavior during the period of t=90.19sec to t=96.16sec. Figure 3 shows more details of the successive transient flow fields from t=90.19sec to t=98.67sec. At t=91.22sec, the oil slick became shorter and thicker. At t=92.72sec, the oil began to creep under the barrier. At t=94.22sec, more oil crept under the barrier. At t=98.67sec, a considerable amount of oil has crept under the barrier and floats on the water surface. It can be expected that more and more oil will continue to creep under the barrier as time increases further. This process of oil creep was consistent with the observations in the wave basin experiment.

CONCLUSIONS

(1) The comparative study shows that the hydrodynamic behavior of an oil slick contained by a barrier, observed and measured in the wave basin experiment, can be accurately modeled.

(2) For the case of successful oil containment, the predicted oil slick profiles are in a good agreement with the measured ones in the wave basin experiment.

(3) For the case of barrier failure, the predicted critical velocity and the oil slick behavior afterwards are consistent with the observation in the experiment.

(4) It can be concluded from this comparative study that the CFD code, Fluent, is an efficient and accurate tool to guide the new design of oil containment barriers.

ACKNOWLEDGEMENTS

The first author acknowledges the NSERC of Canada for an Industrial Research Fellowship and to Imperial Oil for the financial support on this project.

REFERENCES

1. R.H. Cross and D.P. Hoult, "Collection of oil slicks", J. Waterways, Harbors and Coastal Eng. Div., Vol.97, No.WW2, pp.313-322, 1971

2. D.L. Wilkinson, "Dynamics of contained oil slicks", J. Hydraulics Div., Vol.98, No.HY6, pp.1013-1030, 1972

3. D.L. Wilkinson, "Limitations to length of contained oil slicks", J. Hydraulics Div., Vol.99, No.HY5, pp.701-712, 1973

4. J.H. Milgram and R.J. Van Houlten, "Mechanics of a restrained layer of floating oil above a water current", J. Hydronautics, Vol.12, No.3, pp.93-108, 1978

5. Y.L. Lau and J. Moir, "Booms used for oil slick control", J. Environ. Eng. Div., Vol.105, No.EE2, pp.369-382, 1979

6. G.A.L. Delvigne, "Barrier failure by critical accumulation of viscous oil", Proc. of 1989 Oil Spill Conf., pp.143-148, San Antonio, TX, USA, Feb. 13-16, 1989

7. A.J. Johnston, M.R. Fitzmaurice and R.G.M. Watt, "Oil spill containment: viscous oils", Proc. of 1993 Int'l Oil Spill Conf., pp.89-94, Tampa, FL, USA, Mar.29-Apr.1, 1993

8. H.M. Brown, P. Nicholson, R.H. Goodman, B.A. Berry and B.R. Hughes, "Novel concepts for the containment of oil in flowing water", Proc. of 16th Arctic & Marine Oilspill Program Tech. Seminar, pp.485-496, Calgary, AB, Canada, June 7-9, 1993

9. E.J. Clavelle and R.D. Rowe, "Simulation of environmental flows with a density interface", Proc. of 3rd Ann. Conf. of the CFD Society of Canada, Vol.I, pp.441-447, Banff, AB, Canada, June 25-27, 1995

10. C.-F. An, E.J. Clavelle, H.M. Brown and R.M. Barron, "CFD simulation of oil boom failure", Proc. of 4th Ann. Conf. of the CFD Society of Canada, pp.401-408, Ottawa, ON, Canada, June 2-6, 1996

11. S.T. Grilli, Z. Hu and M.L. Spaulding, "Numerical modeling of oil containment by a boom", Proc. of 19th Arctic & Marine Oilspill Program Tech. Seminar, pp.343-376, Calgary, AB, Canada, June 12-14, 1996

12. R.H. Goodman, H.M. Brown, C.-F. An and R.D. Rowe, "Dynamic modeling of oil boom failure using computational fluid dynamics", Proc. of 20th Arctic & Marine Oilspill Program Tech. Seminar, Vancourver, BC, Canada, June 11-13, 1997

13. H.M. Brown, R.H. Goodman, C.-F. An and J. Bittner, "Boom failure mechanisms: comparison of channel experiments with computer modeling results", Proc. of 20th Arctic & Marine Oilspill Program Tech. Seminar, Vancourver, BC, Canada, June 11-13, 1997

14. Fluent User's Guide, v4.4, Fluent Inc., Lebanon, NH, USA, 1996

15. D.C. Wilcox, "Turbulence modeling for CFD", DCW Industries, Inc., La Canada, CA, USA, 1994

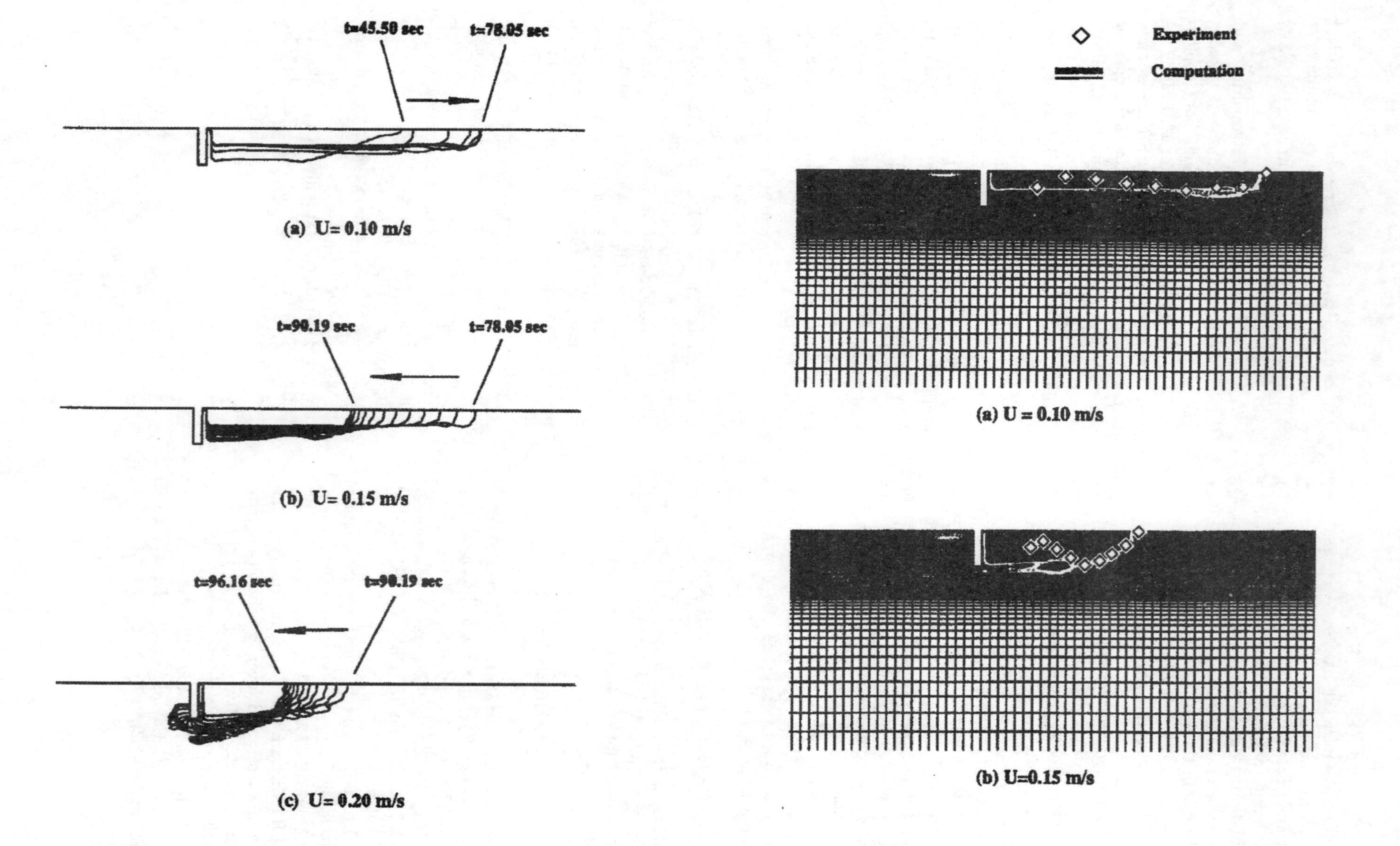

Figure 1 Evolution of Oil Slick

Figure 2 Comparisons of Oil Slick Profiles

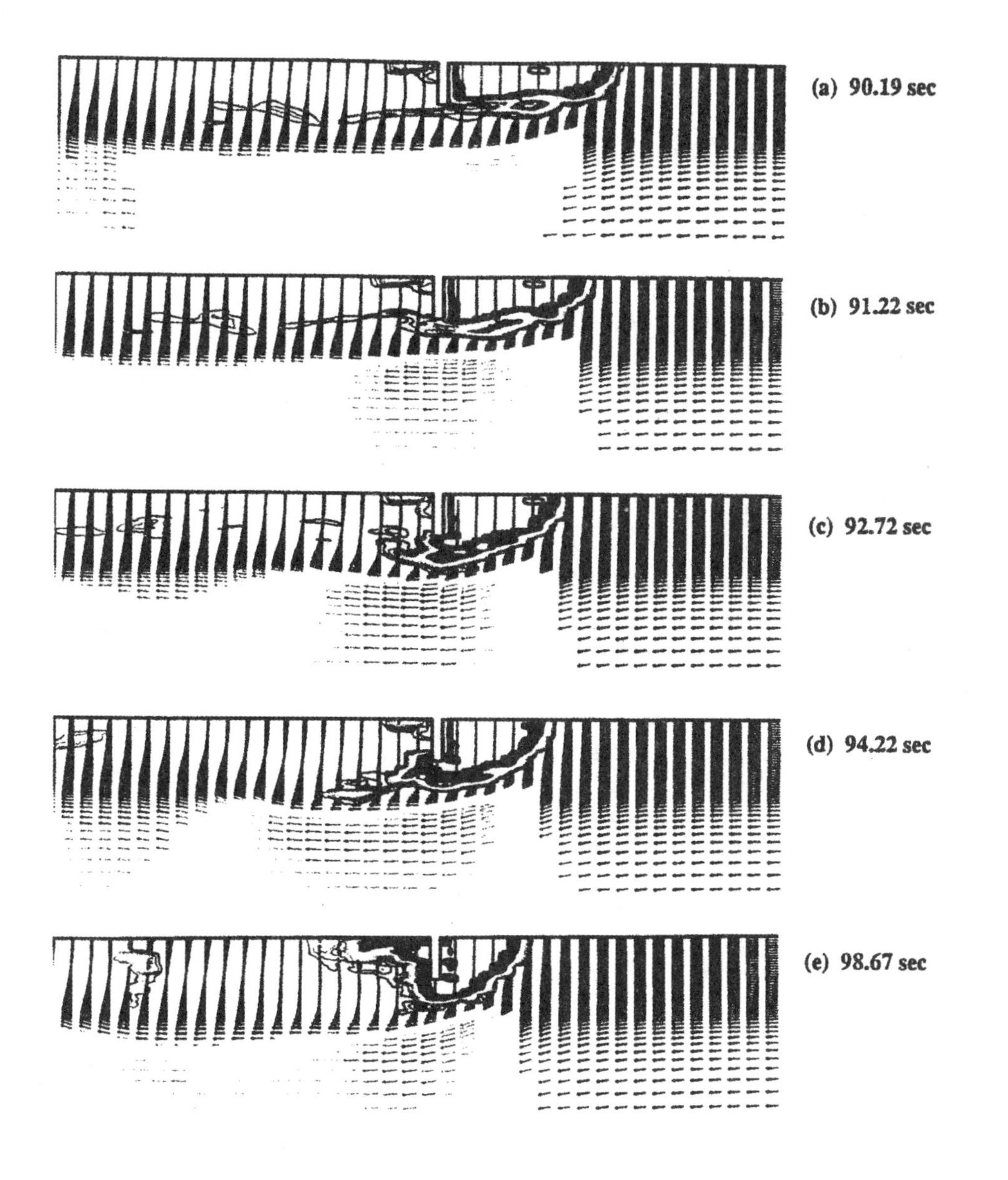

Figure 3 Process of Oil Creep Under a Barrier

COMPUTER ANIMATION OF OIL BOOM DRAINAGE FAILURE

C.-F. An, R.H. Goodman & H.M. Brown
Imperial Oil Resources Limited
Calgary, Alberta, Canada
cfan@acs.ucalgary.ca

E.J. Clavelle
The University of Calgary
Calgary, Alberta, Canada
clavelle@enme.ucalgery.ca

ABSTRACT

In this paper, computer animation of the unsteady process of oil boom drainage failure is presented. The animation is performed by playing a sequence of images at a speed of 15 fps. The individual images are in DIB format that is converted from TIFF format. The TIFF images are exported from the numerical simulation of oil boom failure using the commercial CFD code – *Fluent*. The numerical simulation is based on the corresponding flowing channel experiment. The animation is in AVI format and created by the software *VidEdit* that requests a sequence of DIB images. The conversion from TIFF format to DIB format is conducted by the software *Paint Shop Pro*. The animation example of oil boom drainage failure can be demonstrated on a PC platform under the operating system Windows'95.

INTRODUCTION

Animation is an intuitive way to demonstrate unsteady physical processes[1]. The technique of computer animation developed quickly by taking advantage of two astonishing achievements of computer technology: computer graphics and the Internet. There are articles in the Internet discussing computer graphics and animation. For example, Ladic[2] describes the process for making animation on different platforms using various software packages.

Oil spills may cause pollution to the environment, especially to water resources. If oil is accidentally spilled on water, the usual response is to contain the oil by a barrier, or boom, and then recover it by a skimming device. However, successful oil containment by a boom is not always attainable. If the relative oil-water velocity exceeds a critical value, boom failure will occur and the oil will escape underneath the boom due to hydrodynamic forces. Several mechnisms of oil boom failure are recognized: drainage failure, droplet entrainment and critical accumulation. Drainage failure occurs readily if the boom draft is not deep enough to contain the oil. A computational study was performed to simulate oil containment and boom failure mechanisms[3].

In this paper, a preliminary attempt for generating computer animation of oil boom drainage failure has been made. The source material of the animation is the graphical output of the numerical simulation of oil containment and boom failure using the commercial CFD software – *Fluent* [4]. The numerical simulation is based on the corresponding experiment by Delvigne [5]. The animation format is AVI (audio video interleave). The software that is used to create the animation is a freeware version of *VidEdit* [6] by Microsoft, while the software that is used to convert the image format is a shareware version of *Paint Shop Pro* [7] by Jasc. The computer platform is a PC with a Pentium CPU and 64 MB RAM under the operating system Windows'95.

In the next section, major concepts about making computer animation are explained. Then, the technique of making animation is described in the third section. The animation examples of successful oil containment and oil boom drainage failure are presented in the fourth section and conclusions are drawn in the last section.

CONCEPTS ON ANIMATION

The frequently used **formats** for computer animation are MPEG, QuickTime and AVI. The MPEG is a compression standard for video and audio data and is used on both PC and workstations. QuickTime is used on the Macintosh platform. The AVI format was developed by Microsoft and is used on the PC platforms. The **source material** for animation is a sequence of images in a certain format. The images should be arranged in such a way that the names of all images can be sorted correctly. The most common way to do this is to append a digital identifier in the file names, such as 001, 002, 003, etc. **Conversion** is a necessary step if the source images have a different format from the object images. **Compression** is used in the process of making animation to overcome problems associated with storage space and the speed of computers. The compression standard consists of the corresponding compression/decompression algorithm, referred to as **codec**, that can create and playback the compressed animations. **Encoding** refers to the process of creating animations in a special compressed format from a sequence of images. The most desirable encoders are those that can obtain a high degree of compression while maintaining good playback fidelity. Animation encoders are usually referred to as animation editors. **Decoding** refers to the process of deciphering encoded animations. This is a necessary step to display animations. Most animation decoders are referred to as viewers or players.

MAKING AN ANIMATION

Preparation of Images:

In order to make an animation, a set of sequentially numbered images must be prepared. In this work, the source images are the graphical output of the numerical simulation of oil boom failure using Fluent. Fluent is a general-purpose CFD program for modeling fluid flows. In the flow solver, the unsteady Navier-Stokes equations with proper turbulence modeling are solved numerically using implicit scheme for time and control volume method for space. The turbulence modeling includes several options. The standard $k-\epsilon$ turbulence model is appropriate for the present work. The built-in VOF (volume of fluid) multiphase model is applied to handle the two immiscible liquid phases, oil and water. The graphical output of the simulated results can be in any of seven formats. In the present work, the TIFF (tagged image file format) images are exported from Fluent. While VidEdit can load images one frame at a time in many formats, it can load a sequence of images only in DIB (device independent bitmap) format. Therefore, a batch-conversion of images from TIFF format to DIB format is necessary and was done using Paint Shop Pro.

Creation of an Animation

After the images are converted from TIFF format to DIB format, VidEdit can be run to create the animation. Depending on the size, number of images, codec, compression settings, the CPU speed and system configurations, the process of creating an AVI animation may take several minutes or hours. Some features of editing are frequently used to reduce the size of the resulting AVI file, such as cropping, resizing and color reduction.

Playback of an Animation

There are a number of viewers of AVI animations, but *Media Player* is a simple one to use. Media Player is an executable program called *mplayer.exe* that is included in the Microsoft Windows operating system. To display an AVI animation using Media Player under Windows'95, one should successivelly click on *Start, Program, Accessories, Multimedia* and *Media Player.* A VCR-like icon appears. Open the AVI file that is to be displayed and double click on the *Play* button. The animation can be viewed.

ANIMATION OF BOOM FAILURE

The animation example is based on Run No.15 of the experiment by Delvigne [5] and the corresponding simulation. In the experiment, the channel is $0.5m$ wide and $0.3m$ deep. The boom draft is $0.07m$ and the flow velocity ranges from 0.09 to $0.24m/s$. The spilled oil is an emulsion of a mixture of fuel oil (55%), gasoline (18%) and water (27%) with a density of $924kg/m^3$ and a kinematic viscosity of $1480cSt$. The amount of spilled oil is $0.005m^3/m$. The Reynolds number based on flow velocity, channel depth and water properties is $27,000$. The experiment claims that oil containment is successful until flow velocity reaches a critical value, $0.24m/s$, at which point oil begins to pass under the boom. The oil loss mechanism is drainage failure. The processes of successful oil containment and drainage failure have been accurately modeled in the simulation. The animation is made from these simulated results corresponding to the experiment.

Figure 1 shows typical frames of oil slick movement at a velocity of $0.09m/s$. At simulation time 30.43sec, the oil reaches the boom. At 38.43sec, the oil accumulates against the boom. At 42.43sec, the oil slick is close to the stable shape. Comparing the graphs at 44.45sec and 46.45sec, the oil slick shape has changed little. This means that the flow has reached a steady state, the oil slick has become stable and the oil containment is successful at the velocity of $0.09m/s$. The animation of the oil slick movement is contained in an AVI file.

Figure 2 shows typical frames of the oil slick deformation when the velocity is increased to 0.17m/s. As the simulation time increases, the oil slick becomes thicker and shorter. Comparing the graphs at 52.49sec and 53.49sec, one can see that deformation of the oil slick almost stops. This means that the flow has reached another steady state, the oil slick has become stable again and the oil containment is also successful at the velocity of $0.17m/s$. The animation of this oil slick deformation is contained in another AVI file.

Figure 3 shows typical frames of the oil slick behavior when the velocity is increased to $0.24m/s$. At 54.30sec, the slick thickness exceeds the boom draft and the oil begins to drain. At 55.00sec, the oil continues to drain and the drained oil is broken into pieces. At 56.50sec, the oil pieces are dispersed into the water. At 58.50sec, most of the oil has been drained, dispersed and entrained into the turbulent water . The animation of this oil slick behavior is contained in the third AVI file.

CONCLUSIONS

Computer animation is an intuitive way to demonstrate unsteady physical processes, such as the transient process of oil containment and boom failure.

The source images of the animation are obtained from the numerical simulation by Fluent and the numerical simulation is based on the corresponding flowing channel experiment.

The animation is created by VidEdit and the image format is converted by Paint Shop Pro. The final animation examples are in AVI files and can be viewed by Media Player under Windows'95.

Further improvement can be attempted by using other AVI encoders, other animation formats, other computer platforms and other hardware devices.

ACKNOLEDGEMENTS

The first author is grateful to the NSERC of Canada for an Industrial Research Fellowship and to Imperial Oil Resources Limited for the financial support on this project. He wishes to express his sincere gratitude to Dr. R.D. Rowe of the University of Calgary for offering the usage of both hardware and software. He also would like to thank Mr. Hank Li of Wisemind Enterprice Co. for his patient training of computer animation.

REFERENCES

1. R. Savoie, Y. Gagnon and Y. Mercadier, "Animation of the starting flow down a step", Proc. of the 4th Annual Conference of the CFD Society of Canada, pp.279-284, Ottawa, Ontaio, Canada, June 2-6, 1996

2. L. Ladic, "Making animations for the Web", Internet Article in 1997 at: *http://www.cs.ubc.ca/spider/ladic/anim_unix.html*

3. C.-F. An, E.J. Clavelle, H.M. Brown and R.M. Barron, "CFD simulation of oil boom failure", Proc. of the 4th Annual Conference of the CFD Society of Canada, pp.401-408, Ottawa, Ontario, Canada, June 2-6, 1996

4. Fluent User's Guide, version 4.4, Fluent Inc., Lebanon, New Hampshire, USA, August 1996.

5. G.A.L. Delvigne, "Barrier failure by critical accumulation of viscous oil", Proc. of the 1989 Oil Spill Conference, pp.143-148, San Antonio, Texas, USA, February 13-16, 1989

6. VidEdit, freeware version 1.1, Microsoft Corp., Roselle, Illinois, USA, 1993

7. Paint Shop Pro, shareware version 4.12, Jasc Inc., Eden Prairie, Minnisota, USA, 1996

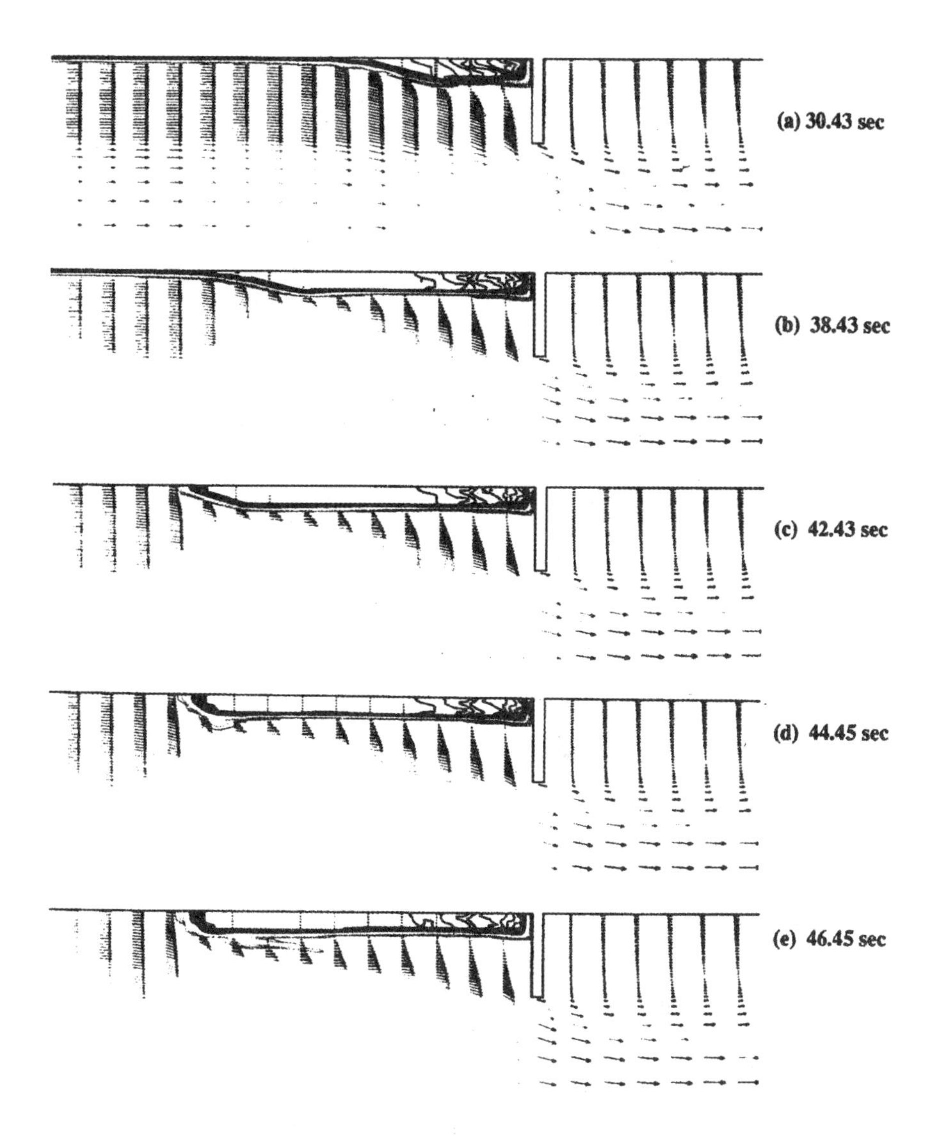

Figure 1 Successful Oil Containment at 0.09m/s

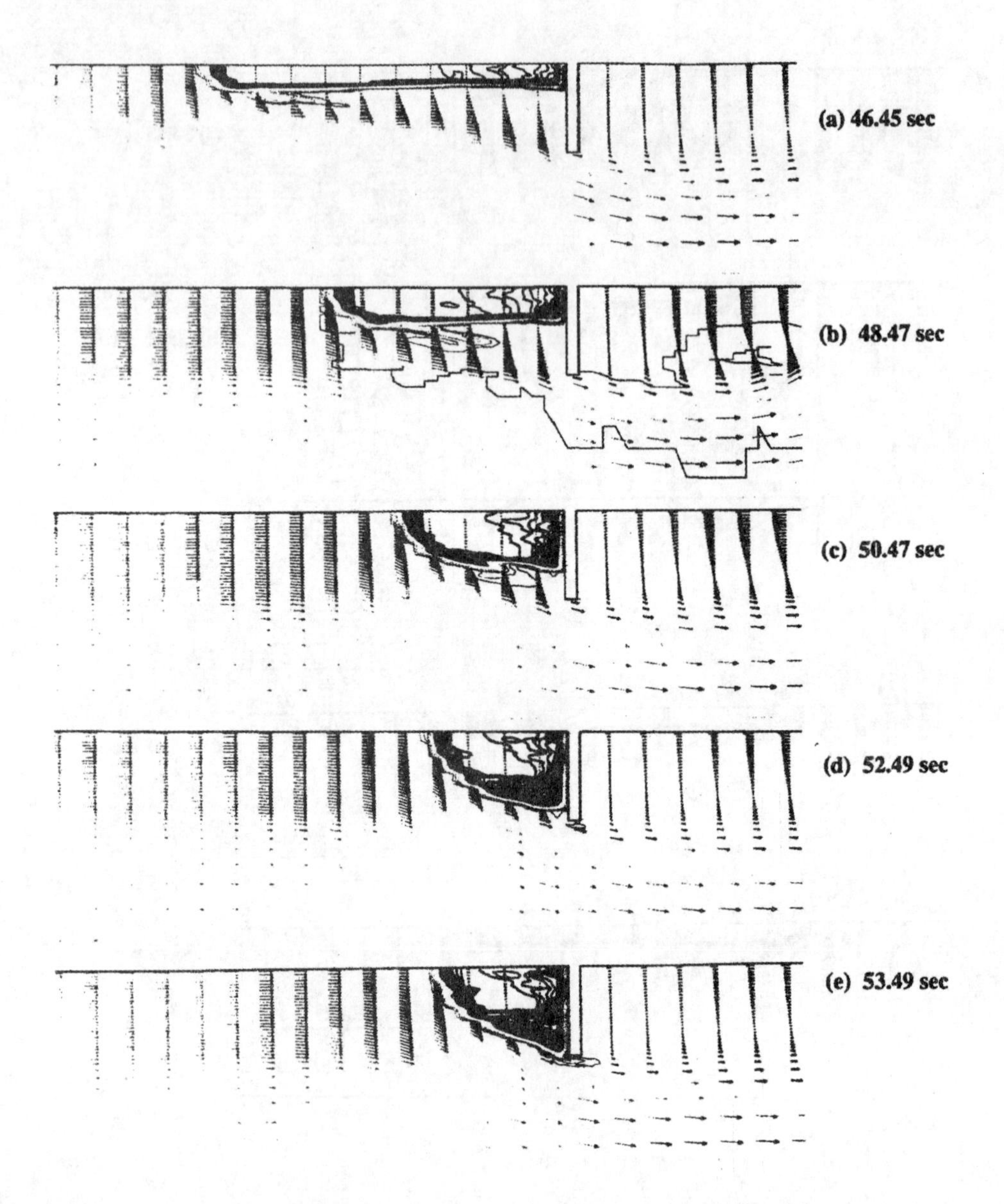

Figure 2 Successful Oil Containment at 0.17m/s

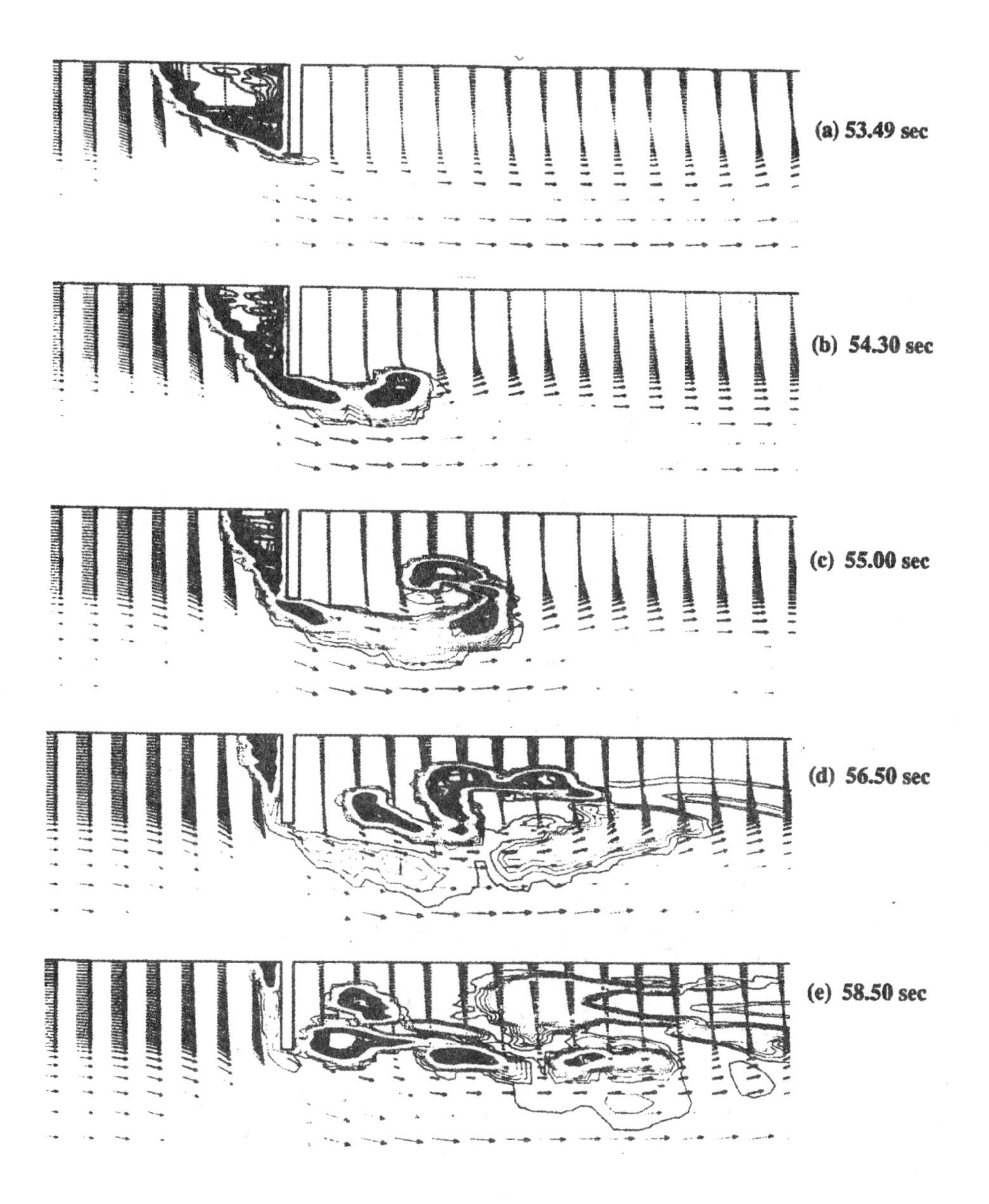

Figure 3 Oil Boom Drainage Failure at 0.24m/s

Chang-fa An
June 10, 1997
Vancouver

VOLUME 1

PROCEEDINGS OF THE TWENTIETH ARCTIC AND MARINE OILSPILL PROGRAM (AMOP) TECHNICAL SEMINAR

COMPTE RENDU: 20° COLLOQUE TECHNIQUE DU PROGRAMME DE LUTTE CONTRE LES DÉVERSEMENTS D'HYDROCARBURES EN MER ET DANS L'ARCTIQUE (AMOP)

JUNE 11 TO 13, 1997
VANCOUVER, BRITISH COLUMBIA
CANADA

DU 11 AU 13 JUIN, 1997
VANCOUVER, COLUMBIE-BRITANNIQUE
CANADA

Environment Canada Environnement Canada

Canada

安长发博士科研论文集

Dynamic Modeling of Oil Boom Failure Using Computational Fluid Dynamics

R.H. Goodman, H.M. Brown and C.-F. An
Imperial Oil Resources Limited
Calgary, Alberta, Canada

R.D. Rowe
The University of Calgary
Calgary, Alberta, Canada

Abstract

The common response to an oil spill on water is to contain the oil with booms and recover it with skimming devices. However, there is no guarantee that the booms will reliably trap and hold oil. In some situations, the oil will escape underneath the boom due to hydrodynamic forces. A number of failure mechanisms have been identified and studied in laboratory experiments, but these have not been accurately modeled. Computational Fluid Dynamics (CFD) is a powerful modeling tool combining fluid dynamics and numerical mathematics with high speed computer technology. CFD packages are now available which are capable of modeling complex hydrodynamic processes. We have utilized a commercial CFD program, ***Fluent***, to simulate the oil-water flow around a barrier. The studies accurately model channel experiments conducted in recent years, and demonstrate boom failure mechanisms such as "drainage failure", "droplet entrainment" and "critical accumulation". The studies show that the flow patterns around barriers are modified by the presence of oil and, therefore, suggest that towing and wave-conformity tests of booms will not be meaningful unless they are undertaken with the presence of oil.

1.0 Introduction

If oil is accidentally spilled on water, the most common response method is to use barriers, or booms, to collect the spilled oil. The collected oil then can be recovered using skimming devices. However, successful oil containment by booms can be attained only under certain conditions. These conditions depend upon the oil-water relative velocity, the physical properties of the oil and water, the depth of the water, the shape and draft of the boom and so on. If particular conditions are not met, boom failure will occur and the oil will escape underneath the boom due to hydrodynamic forces.

Previous research on oil containment and boom failure has included both theoretical studies and experimental investigations. Wilkinson(1972, 1973) proposed a two-zone oil slick model that assumed a frontal zone where dynamic forces dominate the oil slick and a viscous zone where viscous shear forces dominate the oil slick. He analyzed oil slick dynamics, derived formulas to calculate the oil slick shape in each zone, verified these formulas in his experiment and concluded that successful oil containment is impossible if the densimetric Froude number is greater than 0.5. Cross and Hoult(1972) derived a formula in which the maximum length of the contained oil slick is proportional to the square of the barrier draft. Milgram and Van

Houlten(1978) measured the shape of the interface between oil and water, analyzed oil slick dynamics and concluded that droplet entrainment failure is the result of breaking of the Kelvin-Helmholtz waves on the oil-water interface. Lau and Moir(1979) confirmed Wilkinson's criterion for the densimetric Froude number and further proposed a criterion for minimum boom draft to prevent drainage failure. These early studies only considered one dimensional analyses and only light oil with low viscosity was used in the experimental studies. Delvigne(1989) was the first investigator to study the slick behavior of heavy oil with high viscosity. He observed a failure mechanism in which viscous oil continuously leaks under a barrier independent of its draft. He classified this behavior as critical accumulation failure. Johnston et al(1993) explored this boom failure mechanism with highly viscous oil. They observed a surging phenomenon of the oil slick before the failure by critical accumulation developed. They suggested that the term "oil creep" is a better description for the failure mechanism with highly viscous oil than the term "critical accumulation".

Beginning in the 1990's, computational fluid dynamics has been used to investigate the fundamentals of boom failure mechanisms. Clavelle and Rowe(1993, 1995, 1996) have applied the method of volume of fluid in CFD to predict boom failure by critical accumulation. An et al (1996) have explored oil containment and boom failure using computational fluid dynamics. Grilli et al(1996) have applied the concept of the vortex-sheet and boundary element method to model the Kelvin-Helmholtz instability of the oil-water interface.

Computational Fluid Dynamics (CFD), originally developed in the aeronautical and space industry, has become a powerful research tool in mamy engineering fields. Currently, there are many commercial CFD packages dealing with a wide variety of applications in fluid flow, heat and mass transfer. In this paper, the commercial CFD software, *Fluent* (1996), is used to simulate the flow of oil and water around barriers.

In the following, an outline of the CFD software Fluent is given, several modeling examples are described in detail, and the simulated results are compared with the corresponding experiments.

2.0 A CFD Software - Fluent

Fluent is a general-purpose program for modeling fluid flow problems. It uses a finite volume based algorithm and consists of a pre-processor or grid generator, a main processor or flow solver and a post-processor or graphical visualizer. In Fluent, the Reynolds-averaged Navier-Stokes equations and the turbulence modeling equations are solved numerically. For the commonly used k-ε turbulence model, the equations to be solved in Fluent are summarized below. The mean continuity and momentum equations in Cartesian coordinates are:

$$\frac{\partial \rho}{\partial t} + \frac{\partial}{\partial x_i}(\rho U_i) = 0 \tag{1}$$

$$\frac{\partial}{\partial t}(\rho U_i) + \frac{\partial}{\partial x_j}(\rho U_i U_j) = -\frac{\partial P}{\partial x_i} + \frac{\partial}{\partial x_j}[\mu_{eff}(\frac{\partial U_i}{\partial x_j} + \frac{\partial U_j}{\partial x_i})] + \rho g_i \tag{2}$$

where ρ is the density, U_i is the mean velocity, P is the mean pressure, g_i is the gravitational acceleration, t is the time, x_i is the Cartesian coordinate and μ_{eff} is the effective viscosity, defined as the sum of the molecular viscosity μ and the turbulent viscosity μ_t

$$\mu_{eff} = \mu + \mu_t \tag{3}$$

In the standard k-ε turbulence model, the turbulent viscosity is determined by:

$$\mu_t = \rho C_\mu \frac{k^2}{\epsilon} \tag{4}$$

where k is the turbulent kinetic energy per unit mass, ε is its dissipation rate and C_μ is an empirical constant. The values of k and ε are obtained by solving their transport equations:

$$\frac{\partial}{\partial t}(\rho k) + \frac{\partial}{\partial x_i}(\rho U_i k) = \frac{\partial}{\partial x_i}(\frac{\mu_t}{\sigma_k}\frac{\partial k}{\partial x_i}) + \mu_t(\frac{\partial U_i}{\partial x_j} + \frac{\partial U_j}{\partial x_i})\frac{\partial U_i}{\partial x_j} - \rho\epsilon \tag{5}$$

$$\frac{\partial}{\partial t}(\rho\epsilon) + \frac{\partial}{\partial x_i}(\rho U_i \epsilon) = \frac{\partial}{\partial x_i}(\frac{\mu_t}{\sigma_\epsilon}\frac{\partial \epsilon}{\partial x_i}) + C_{1\epsilon}\frac{\epsilon}{k}\mu_t(\frac{\partial U_i}{\partial x_j} + \frac{\partial U_j}{\partial x_i})\frac{\partial U_i}{\partial x_j} - C_{2\epsilon}\rho\frac{\epsilon^2}{k} \tag{6}$$

where $C_{1\varepsilon}$, $C_{2\varepsilon}$, σ_k and σ_ε are empirical constants. The usually accepted values of these constants are:

$$C_\mu = 0.09, \quad C_{1\epsilon} = 1.44, \quad C_{2\epsilon} = 1.92, \quad \sigma_k = 1.0, \quad \sigma_\epsilon = 1.3 \tag{7}$$

Equations (1), (2), (5) and (6) are discretized in space using the control volume method and discretized in time using an implicit scheme. The discretized algebraic equations are solved numerically using various algorithms. In Fluent, the volume of fluid (VOF) model is used to solve two or more immiscible fluid phases where the position and shape of the interface between the phases are of interest. This model is suitable for the present problem because oil and water can be considered as two immiscible fluids and the oil-water interface is of great interest. Unlike other multi-phase fluid models, the VOF model solves a single set of equations (1), (2), (5) and (6) which are shared by all phases. To distinguish between phases, a variable F, the volume fraction in a cell , is introduced. If F=0, the cell is full of oil; if F=1, the cell is full of water; if $0<F<1$, the cell contains both oil and water. Generally, F is a function of space and time and is governed by the conservation equation:

$$\frac{\partial F}{\partial t} + U_i\frac{\partial F}{\partial x_i} = 0 \tag{8}$$

This equation is solved using an explicit time marching scheme after the velocity field U_i is obtained. In the time marching progress, the Courant-Fridrich-Lewy (CFL) number must be less than 1 to guarantee the stability of the computation. The CFL number is checked and adjusted automatically. The governing equations (1), (2), (5) and (6) are solved throughout the domain. The resulting velocity field is shared by the fluid phases or their mixture, but the values of density ρ and molecular viscosity μ in the governing equations (1), (2), (5) and (6) depend on the local value of *F*:

$$\rho = F\rho_w + (1 - F)\rho_o \tag{9}$$

$$\mu = F\mu_w + (1 - F)\mu_o \tag{10}$$

where the subscripts "*o*" and "*w*" represent oil and water, respectively. On the basis of the local value of *F*, the interface between oil and water can be tracked easily.

After the equations are solved, the flow field can be viewed in a number of ways, including by a vector plot, a contour plot, or a profile plot. This graphical feature provides a good opportunity to analyze the flow field quantitatively.

3.0 Modeling Examples

To verify the validity and applicability of the modeling method, several test examples have been modeled and the simulated results have been compared with the corresponding experiments. Three types of oils were chosen to conduct the simulations. For simplicity, only a flat plate barrier was considered as a boom. For all computations, the flows are turbulent based on the corresponding Reynolds numbers and only standard *k*-ε turbulence model is applied. In the CFD terminology, the tasks are the numerical solutions of two dimensional unsteady turbulent flows of two immiscible viscous incompressible fluids, i.e. oil and water, around a simple barrier. The modeling parameters for the simulations are summarized in the table below.

Table 1 Simulation Parameters

Simulation Parameters	Drainage Failure	Droplet Entrainment	Critical Accumulation
Oil Density (kg/m^3)	870	888	915
Oil Viscosity (cSt)	241	70	3600
Reynolds Number	200,000	27,000	20,000
Computational Domain (m x m)	8 x 1	4 x 0.3	8 x 0.23
Barrier Position (m)	3	2	2
Barrier Draft (m)	0.01	0.07	0.07
Computational Grid	241 x 61	281 x 41	281 x 45
Flow Velocity (m/s)	0.20	0.09, 0.15, 0.20, 0.24	0.10, 0.15, 0.20

The computational domain refers to the length and depth in meters of the channel being modeled and the barrier position is measured from the upstream entrance to the channel. The computational grid refers to the number of grid elements used in the computer simulation.

3.1 Modeling of Drainage Failure

The computation in this example is based on an experiment in the Imperial Oil wave basin (Brown et al, 1997) using a water channel 30*m* long, 1*m* wide and 1*m* deep. The barrier was installed at a distance of 14*m* from the channel entrance and the boom draft was adjustable up to $0.5m$. In the test run of the experiment for which the modeling is conducted, the barrier draft was $0.01m$ and the flow velocity was $0.20m/s$. Five liters of a weathered Federated crude oil with a density of $870kg/m^3$ and a kinematic viscosity of $241cSt$ was used in the experiment. The water density and viscosity were $1000kg/m^3$ and $1cSt$, respectively. In the wave basin experiment, oil loss was observed and the oil loss mechanism was believed to be drainage failure.

This boom failure process has been successfully modeled using the parameters shown in Table 1. The computational grid of 241x61 is clustered in such a way that the mesh is finer in the neighborhood of the barrier while it is coarser elsewhere. At the beginning of the simulation, water flows into the channel from the right inlet with velocity $0.20m/s$. At the same time, oil is injected into the channel in a thin layer on the water surface with the same velocity. The Reynolds number based on water properties, flow velocity and channel depth is about 200,000. Hence, the flow must be turbulent and the standard k-ε turbulence model is invoked. On the bottom of the channel and on the surface of the barrier, a no-slip wall is assumed. On the top of the channel, a slip wall is assumed to simulate the free surface. This is called the "rigid-lid approximation" which is commonly used in channel flow problems.

Figure 1 demonstrates the transient flow fields at a sequence of simulation times from t=13.70 to 25.25*sec*. At t=13.70*sec*, the oil layer reaches the barrier and by t=15.75*sec*, has started to pass under the barrier. At t=19.25*sec*, the lost oil is entrained into a downstream vortex and by t=22.50*sec*, has refloated to the water surface. At t=25.25*sec*, most oil has passed under the barrier. These graphs successfully model the drainage failure observed in the wave basin experiment.

3.2 Modeling of Droplet Entrainment

In this example, the computation is based on an experiment of Delvigne(1989). The simulation parameters are shown in Table 1. The grid of 281×41 is clustered so that the mesh is finer near the barrier and the water surface while it is coarser elsewhere. Initially, the flow velocity is set at $0.09m/s$ and is then increased until substantial oil loss appears. With a Reynolds number of about 27,000 the flow must be turbulent and the standard k-ε turbulence model is invoked.

Figure 2 shows the transient flow fields from t=54.70 to 104.8*sec* at a velocity of $0.09m/s$. From t=54.70*sec* to t=64.70*sec*, the oil floats on the surface of the water and accumulates against the barrier. At t=74.75*sec*, the oil is shown reflecting back from the barrier. This reflection continues until t=94.80*sec*. After this time, the oil slick is re-compressed until t=104.8*sec* when the flow reaches a steady state and the barrier successfully contains the oil. The phenomenon of oil slick reflection and re-compression was first observed by Johnston et al (1993) and called the "surging motion". This surging phenomenon has been captured in the computation and can be seen more clearly in the evolution history in Figure 3 (a) and (b). Figure 3(c) gives a close-up of the stable oil slick at t=104.8*sec*. A typical head wave and recirculation pattern within the oil slick can be seen clearly.

On the basis of the above results, the flow velocity was increased to $0.15m/s$. Figure 4 shows the succeeding transient flow fields from t=108.9 to 124.9*sec*. As time increases, the oil slick becomes shorter and thicker. After t=120.9*sec*, the shape of the oil slick changes little as a stable oil slick is formed and oil containment is successful.

At this point of the simulation, the flow velocity was increased to 0.20*m/s*. Figure 5 shows the resultant transient flow fields from t=124.9 to 137.0*sec*. Clearly, the oil slick becomes shorter and thicker. Again, a stable oil slick is formed and oil containment is successful.

When the flow velocity is further increased to 0.24*m/s*, the flow pattern is changed significantly. Figure 6 shows the transient flow fields from t=137.0 to 145.1*sec*. At t=137.0sec, the flow is the same as that in Figure 5(e). By t=139.1*sec*, the oil slick has become shorter and a slight oil loss has begun. At t=141.1*sec*, an obvious interfacial wave can be seen on the oil-water interface. This kind of wave was observed by Delvigne(1989) in a low viscous oil experiment (see Figure 2(a) in his paper). Between t=143.1 and 145.1*sec*, the oil loss becomes more and more severe. The oil loss mechanism is believed to be droplet entrainment because the barrier draft is still deep enough to prevent oil loss from drainage failure.

In Delvigne's experiment (1989), a substantial loss of oil occurred at a critical velocity of 0.24*m/s*. The failure mechanism was due to droplet entrainment. This experimental observation is exactly modeled in the present computation.

3.3 Modeling of Critical Accumulation

This example is based on an experiment by Johnston et al (1993) and again uses the parameters shown in Table 1. The grid of 281×45 is clustered similarly to the previous examples. The oil is highly viscous Shell Valvata 1000 with a density of 915kg/m^3 and a kinematic viscosity of 3600*cSt*. The simulation process was similar to that used in previous examples.

Initially, the simulation was carried out for a water-only flow with an inlet velocity of 0.10*m/s*. Then, the oil was carefully "injected" from the top of the inlet. After oil injection was stopped, the computation was run continually until a steady state was reached. At this point, the velocity was increased by a step of 0.05*m/s* and the computation was restarted until another steady state was established. The flow velocity was increased step by step until substantial oil loss was observed.

Figure 7 depicts the transient flow fields from t=20.20 to 52.70*sec* during which time the oil slick flows to the barrier and accumulates against it. The flow reaches a steady state and oil containment is successful.

When the velocity is subsequently increased to 0.15*m/s*, the oil slick gets thicker and shorter and again reaches a steady state. This is illustrated in Figure 8.

After the velocity is increased to 0.2*m/s*, progressive oil loss is observed. For time t=60.24, 60.25, 60.27, 60.30 and 60.35sec, the transient flow fields are shown in Figures 9. At t=60.25sec, the oil begins to "creep" underneath the barrier and by t=60.30sec, the lost oil is entrained into the downstream vortex. By t=60.35sec, almost all of the oil has passed under the barrier.

In this example, the barrier completely fails to contain the oil at a velocity of 0.20*m/s*. This exactly coincides with the experimental observation by Johnston et al (1993). According to the classification by Delvigne, the failure mechanism is critical accumulation but is called "oil creep" by Johnston et al (1993). The whole process of oil creep observed in that experiment has been dynamically modeled in the present computation.

To demonstrate the change in the flow pattern due to the presence of oil, two flow fields are shown in Figures 10 and 11. Figure 10 shows the steady state water-only flow field at a velocity of 0.10*m/s* while Figure 11 shows the steady state flow during successful oil containment by a barrier at the same velocity. From these plots one can see that the flow patterns have been considerably modified by the presence of oil. On comparing the flow fields of Figures 10 and 11, one can see that the flow patterns downstream of the barrier are similar, but the flow patterns upstream of the barrier are significantly altered. In front of the barrier when oil is present, the velocity is very low, cf. 11(a) and 11(c); the pressure is very high, cf. 11(b); the turbulence quantities attain a maximum at the tip of the oil slick, cf. 11(d) and 11(e), and the streamlines are quite different from those in water-only flow field, cf. 10(f) and 11(f).

It should be pointed out that the flow characteristics for different failure mechanisms are quite different. In the drainage failure process, the oil slick is compressed against the barrier until the slick is deep enough to leak under the barrier (Figure 1). During boom failure due to droplet entrainment, the oil-water interface of the oil slick is wavy and unstable. The oil droplets may be torn from the interface and swept underneath the barrier by the water stream (Figure 6). During boom failure by critical accumulation, the oil remains a single mass due to the high viscosity and moves under the barrier more readily because of its near neutral density (Figure 9).

The boom failure processes of these three mechanisms are dynamically displayed in an animation paper at the poster session of this conference (An et al, 1997).

4.0 Concluding Remarks

After modeling these examples, the following concluding remarks can be drawn:

(1) The commercial CFD software, Fluent, is suitable for modeling boom failure problems. The entire unsteady process of oil containment and boom failure can be modelled numerically and expressed graphically.

(2) When the flow velocity is below a critical value, a steady state flow is reached and a stable oil slick is formed. The "surging phenomenon" observed experimentally can be simulated.

(3) When the current velocity is beyond a critical value, boom failure occurs. For light or medium oils, with low or moderate viscosity, drainage failure and droplet entrainment may occur and the oil loss processes can be modeled. For a heavy oil with a high viscosity, failure by critical accumulation may occur and the process of "oil creep" observed in the experiment can be modeled.

(4) The evolution of the oil-water interface can be tracked as time progresses. The flow features and tendencies after boom failure occurs can be predicted computationally and demonstrated graphically.

(5) The simulated results show that the flow pattern is significantly modified by the presence of oil. This suggests that boom towing and wave-conformity tests will not be meaningful unless these are conducted with the presence of oil.

5.0 Acknowledgment

The authors would like to thank Imperial Oil Resources Limited for the financial support on this project. The third author is grateful to NSERC of Canada for an Industrial Research Fellowship. He also wishes to express his sincere gratitude to the University of Calgary for the use of its computer facility for this project.

6.0 References

An, C.-F., E.J. Clavelle, H.M. Brown and R.M. Barron, "CFD Simulation of Oil Boom Failure", in *Proc. of 4th Ann. Conf. of CFD Soc. of Canada*, Ottawa, ON, Canada, June 2-4, 1996

An, C.-F., H.M. Brown and R.H. Goodman and E.J. Clavelle, "Animation of Boom Failure Processes", in *Proc. of 20th AMOP Tech. Seminar*, Vancouver, BC, Canada, June 11-13, 1997

Brown, H.M., R.H. Goodman, C.-F. An and J. Bittner, "Boom Failure Mechanisms: Comparison of Channel Experiments with Computer Modeling Results", in *Proc. of 20th AMOP Tech. Seminar*, Vancouver, BC, Canada, June 11-13, 1997

Clavelle, E.J. and R.D. Rowe, "Numerical Simulation of Oil Boom Failure by Critical Accumulation", in *Proc. of 16th AMOP Tech. Seminar*, Calgary, AB, Canada, pp.409-418, June 7-9, 1993

Clavelle, E.J. and R.D. Rowe, "Simulation of Environmental Flows with a Density Interface", in *Proc. of 3rd Ann. Conf. of CFD Soc. of Canada*, Banff, AB, Canada, pp.441-447, June 25-27, 1995

Clavelle, E.J. and R.D. Rowe, "Simulation of Oil Boom Failure by Critical Accumulation", in *ASME Energy Week'96*, Vol. 2, pp.191-195, Houston, TX, USA, Jan. 30, 1996

Cross, R.H. and D.P. Hoult, "Oil Booms in Tidal Currents", *J. Waterways, Harbors & Coastal Eng. Division*, Vol. 98, No. WW1, pp.25-34, 1972

Delvigne, G.A.L., "Barrier Failure by Critical Accumulation of Viscous Oil", in *Proc. of 1989 Oil Spill Conf.*, San Antonio, TX, USA, pp.143-148, Feb. 13-16, 1989

Fluent User's Guide, version 4.4, Fluent, Inc., Lebanon, NH, USA, 1996

Grilli, S.T., Z. Hu and M.L. Spaulding, "Numerical Modeling of Oil Containment by a Boom", in *Proc. of 19th AMOP Tech. Seminar*, Calgary, AB, Canada, pp.343-376, June 12-14, 1996

Johnston, A.J., M.R. Fitzmaurice and R.G.M. Watt, "Oil Spill Containment: Viscous Oils", in *Proc. of 1993 Oil Spill Conf.*, Tampa, FL, USA, pp.89-94, Mar. 29-Apr. 1, 1993.

Lau, Y.L. and J. Moir, "Booms Used for Oil Slick Control", *J. Environ. Eng. Division*, Vol. 105, No. EE2, pp.369-382, 1979

Milgram, J.H. and R.J. Van Houlten, "Mechanics of a Restrained Layer of Floating Oil above a Water Current", *J. Hydronautics*, Vol. 12, No. 3, pp.93-108, 1978

Wilkinson, D.L., "Dynamics of Contained Oil Slicks", *J. Hydraulics Division*, Vol. 98, No. HY6, pp.1013-1030, 1972

Wilkinson, D.L., "Limitations to Length of Contained Oil Slicks", *J Hydraulics Division*, Vol. 99, No. HY5, pp.701-712, 1973

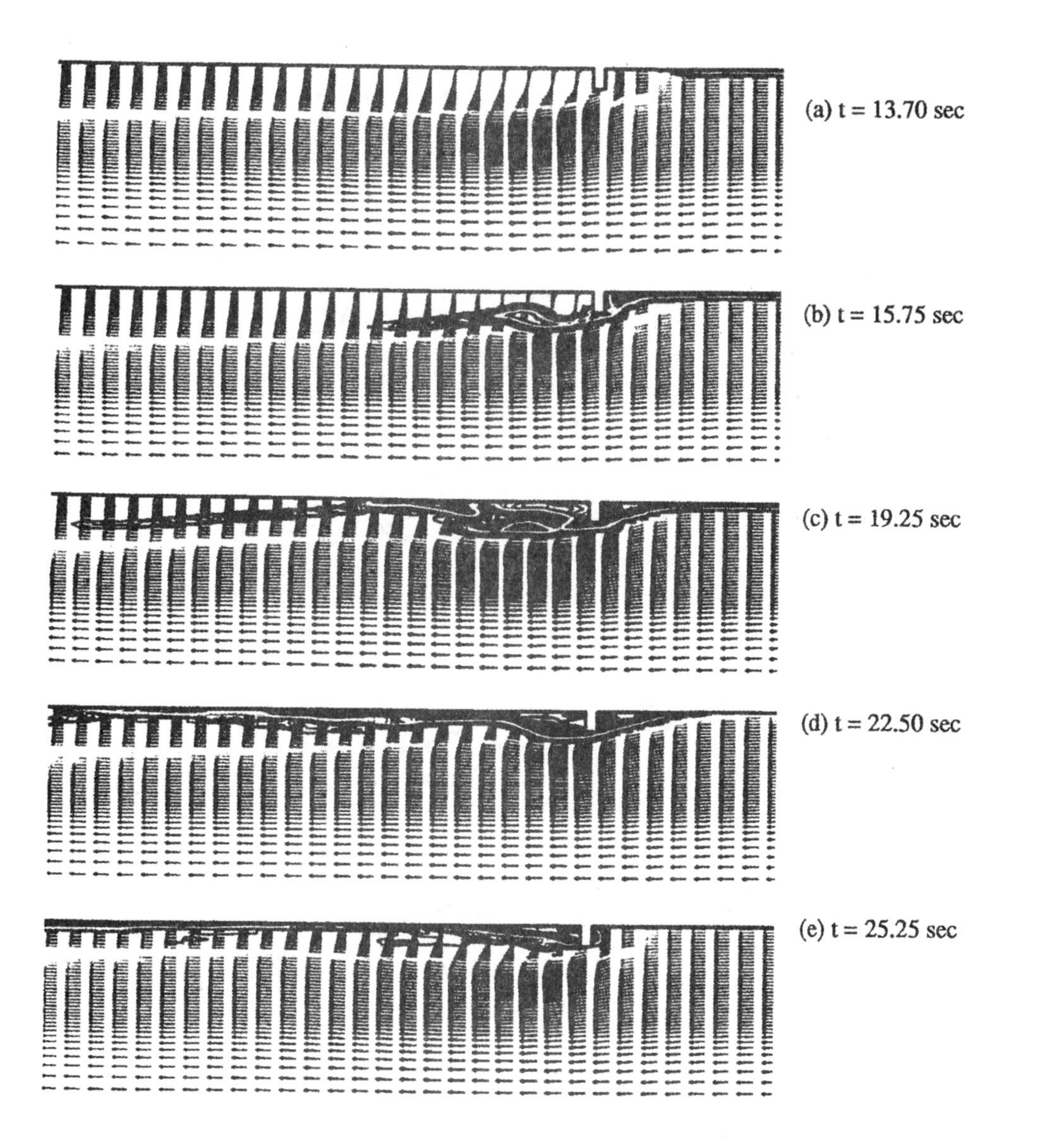

Figure 1 Drainage Failure at Velocity 0.20m/s

Based on: Imperial Oil Wave Basin Experiment
Oil: Weathered Federated, Boom Draft: 0.01m

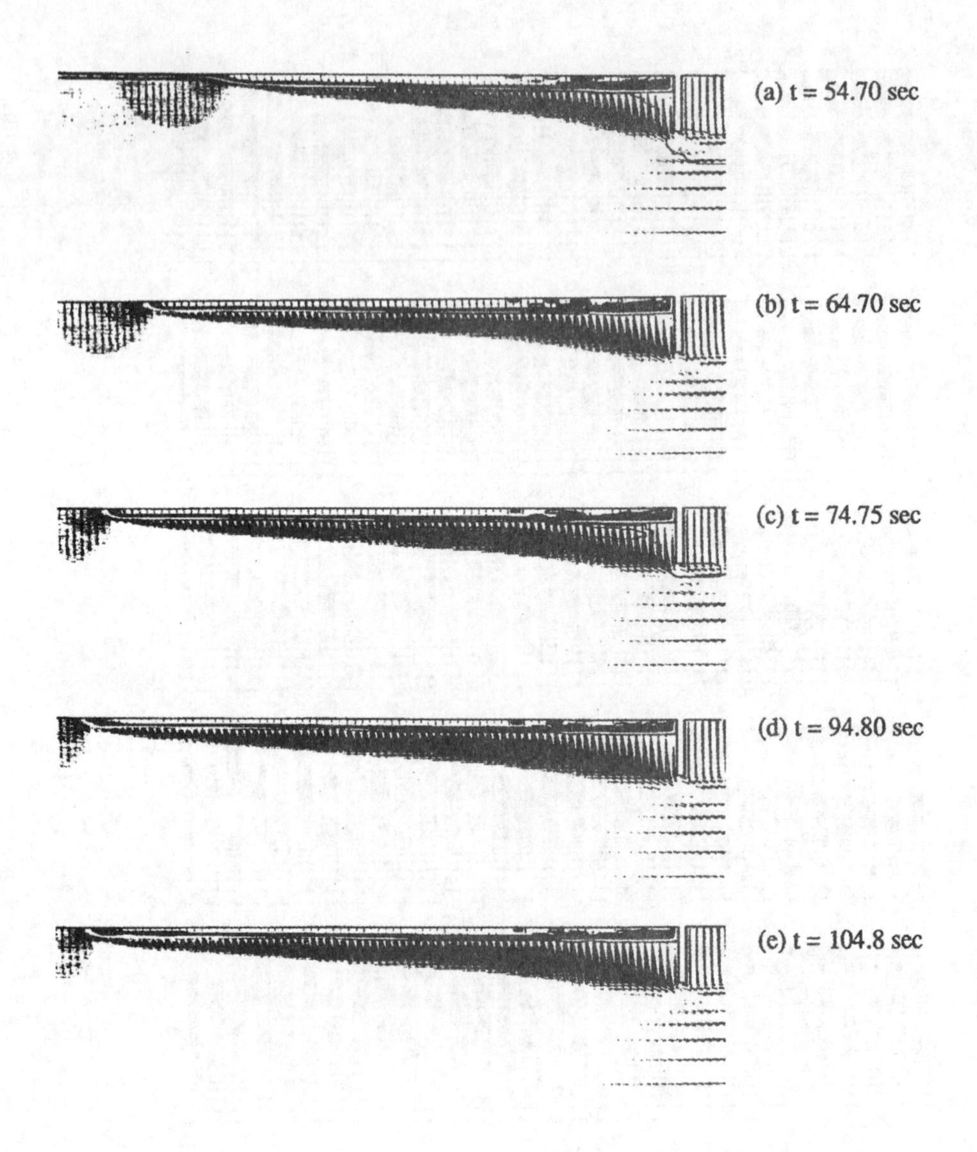

Figure 2 Successful Oil Containment at Velocity 0.09m/s

Based on: Delvigne's Experiment Run No. 3
Oil: Bunker B, Boom Draft: 0.07m

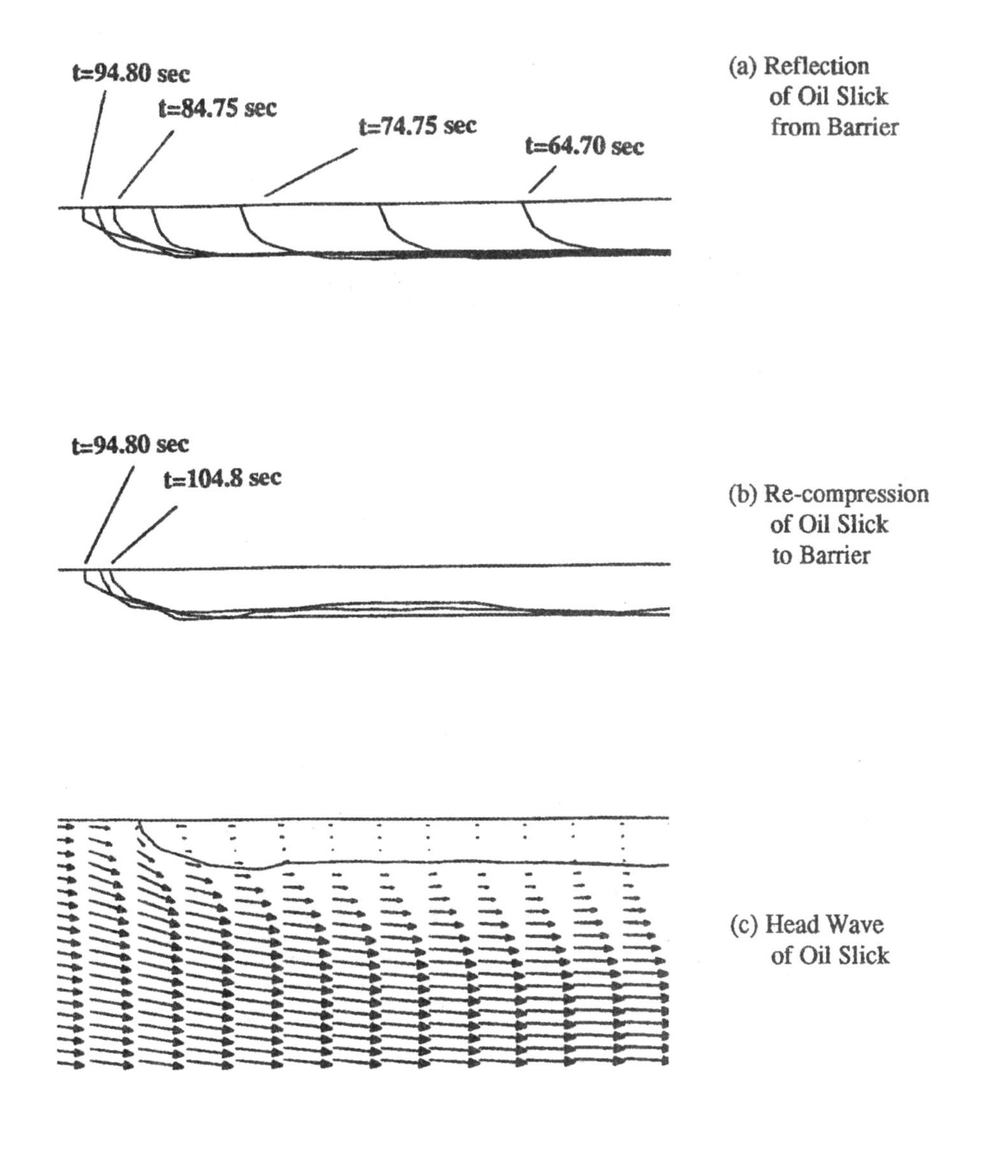

Figure 3 Surging Phynomenon and Head Wave

Based on: Delvigne's Experiment Run No. 3
Oil: Bunker B, Boom Draft: 0.07m, Velocity: 0.09m/s

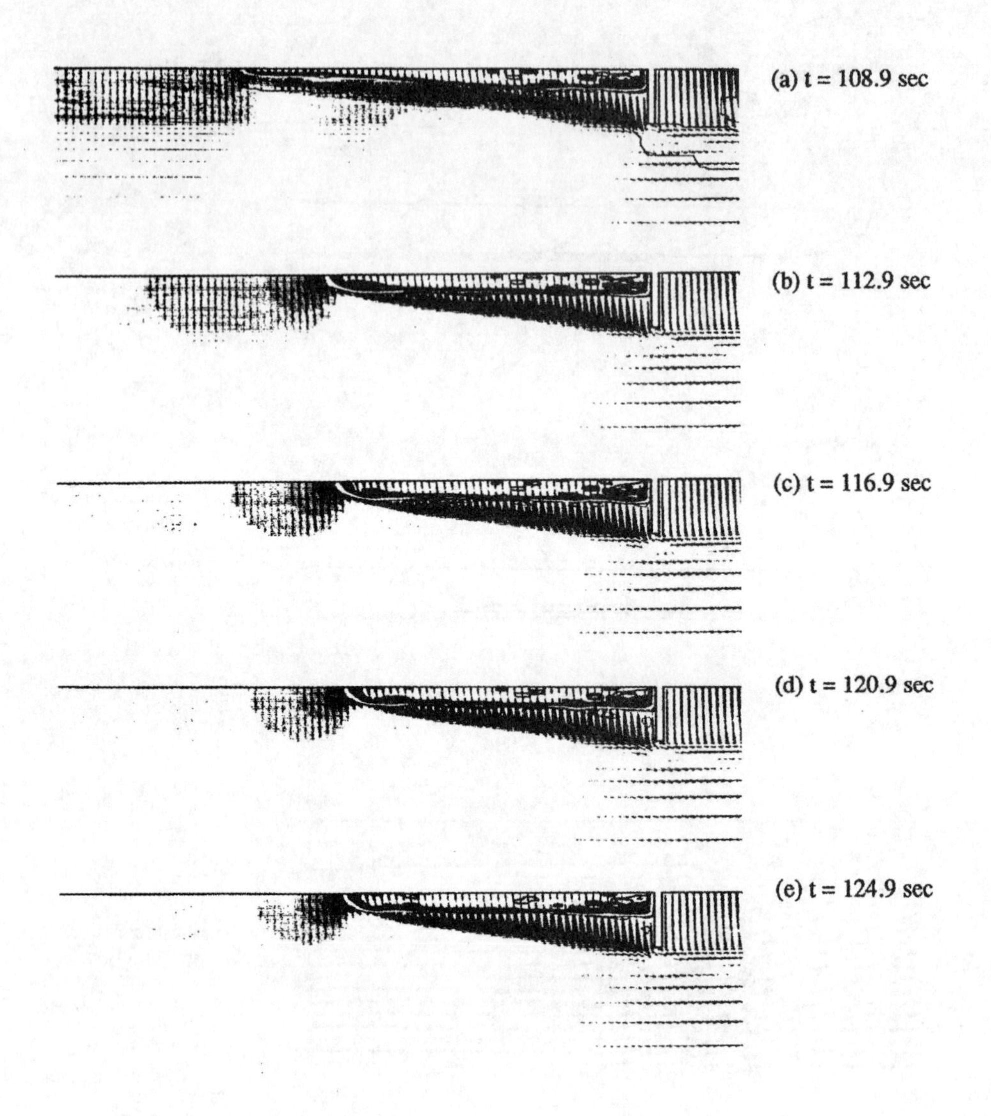

Figure 4 Successful Oil Containment at Velocity 0.15m/s

Based on: Delvigne's Experiment Run No. 3
Oil: Bunker B, Boom Draft: 0.07m

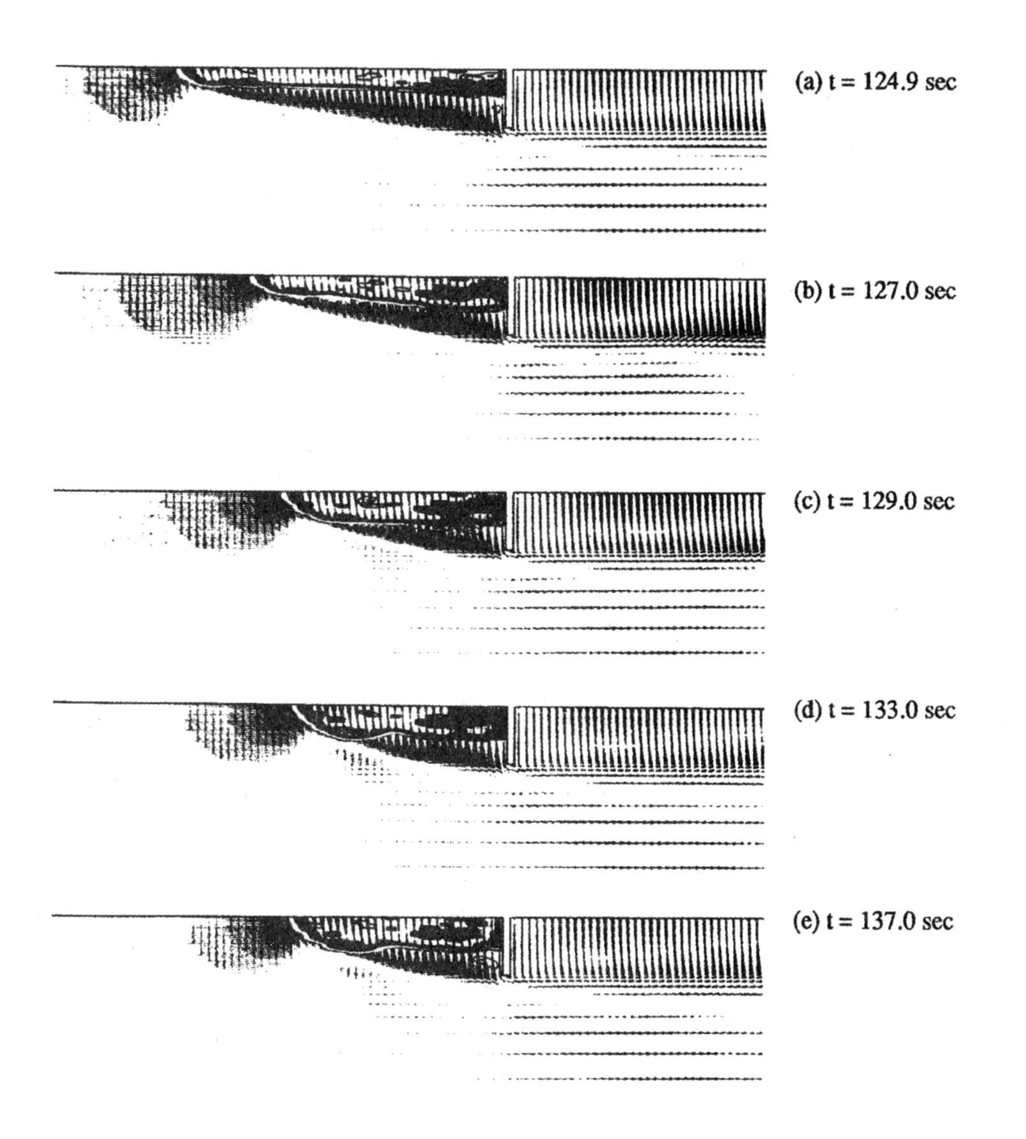

Figure 5　Successful Oil Containment at Velocity 0.20m/s

Based on: Delvigne's Experiment Run No. 3
Oil: Bunker B, Boom Draft: 0.07m

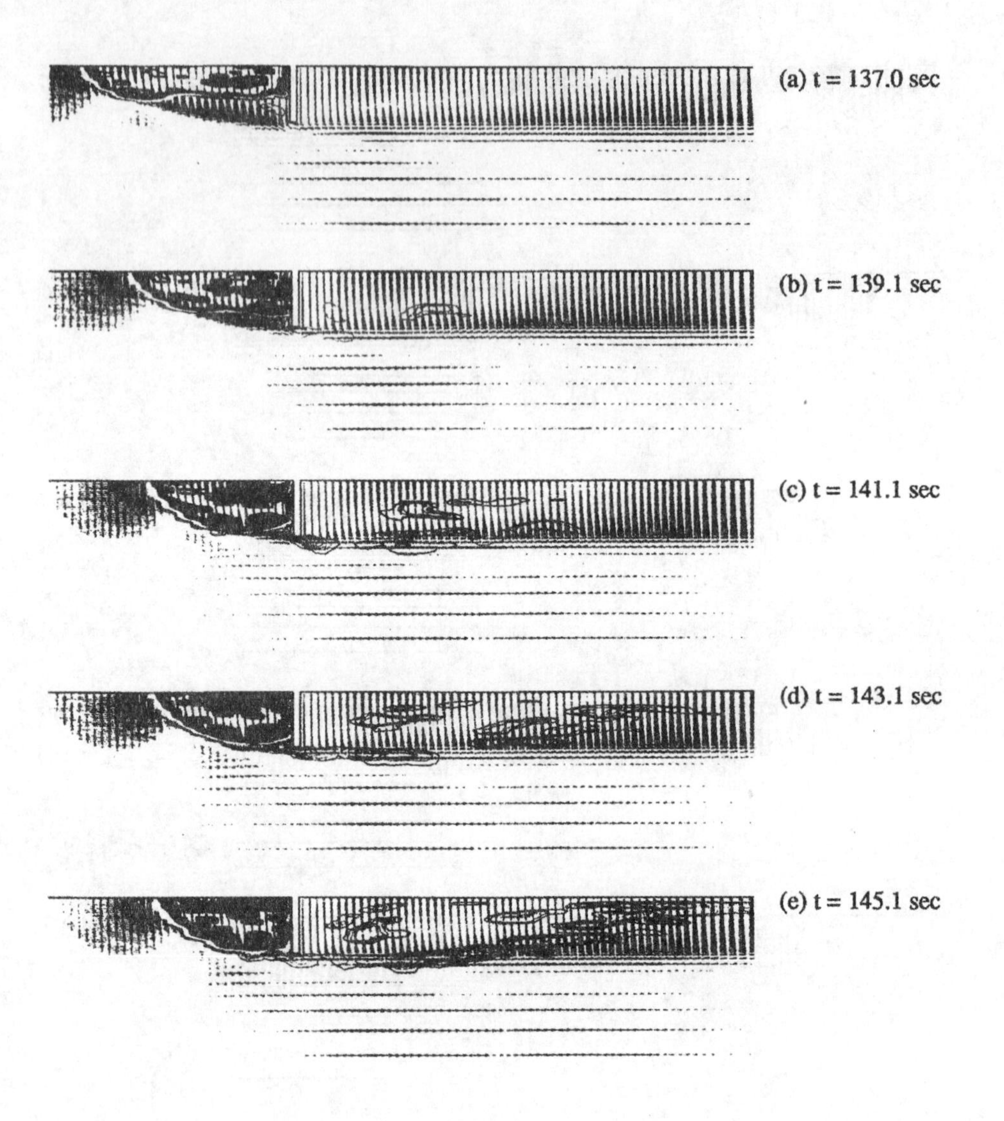

Figure 6　Failure Due to Droplet Entrainment at Velocity 0.24m/s

Based on: Delvigne's Experiment Run No. 3
Oil: Bunker B, Boom Draft: 0.07m

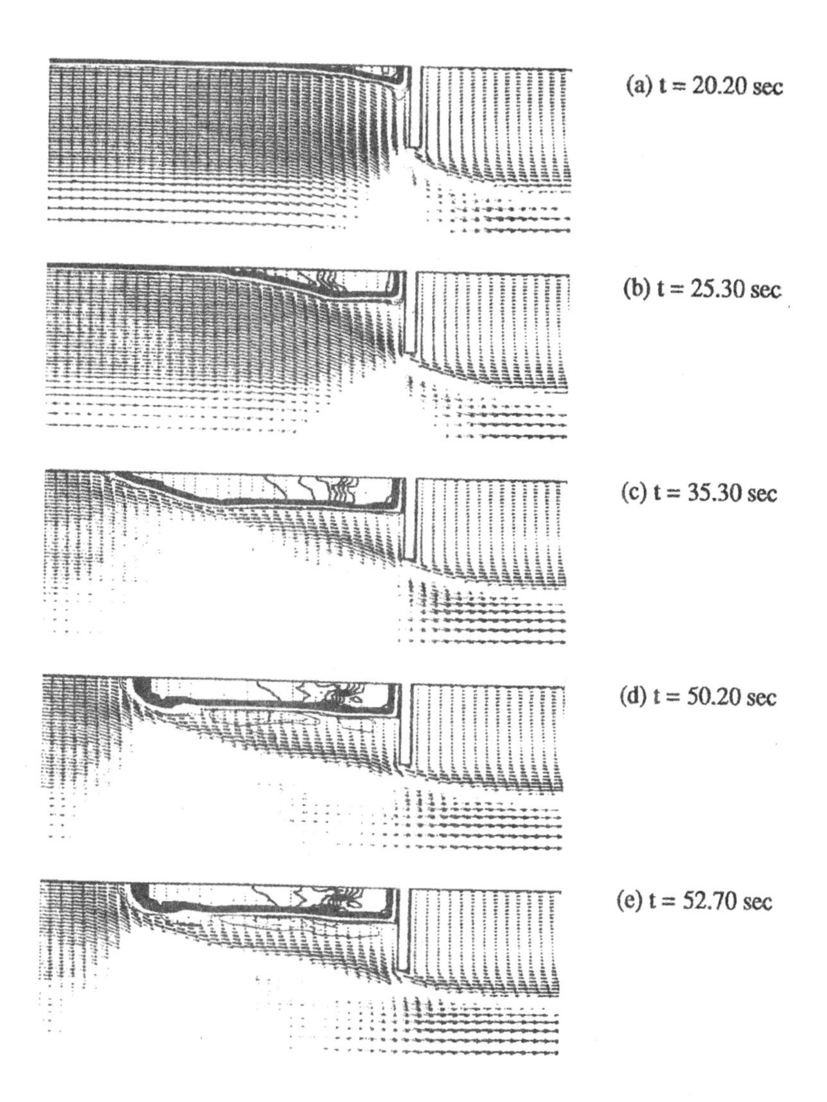

Figure 7 Successful Oil Containment at Velocity 0.10m/s

Based on: Johnston's Experiment Run No. 4
Oil: Shell Valvata 1000, Boom Draft: 0.07m

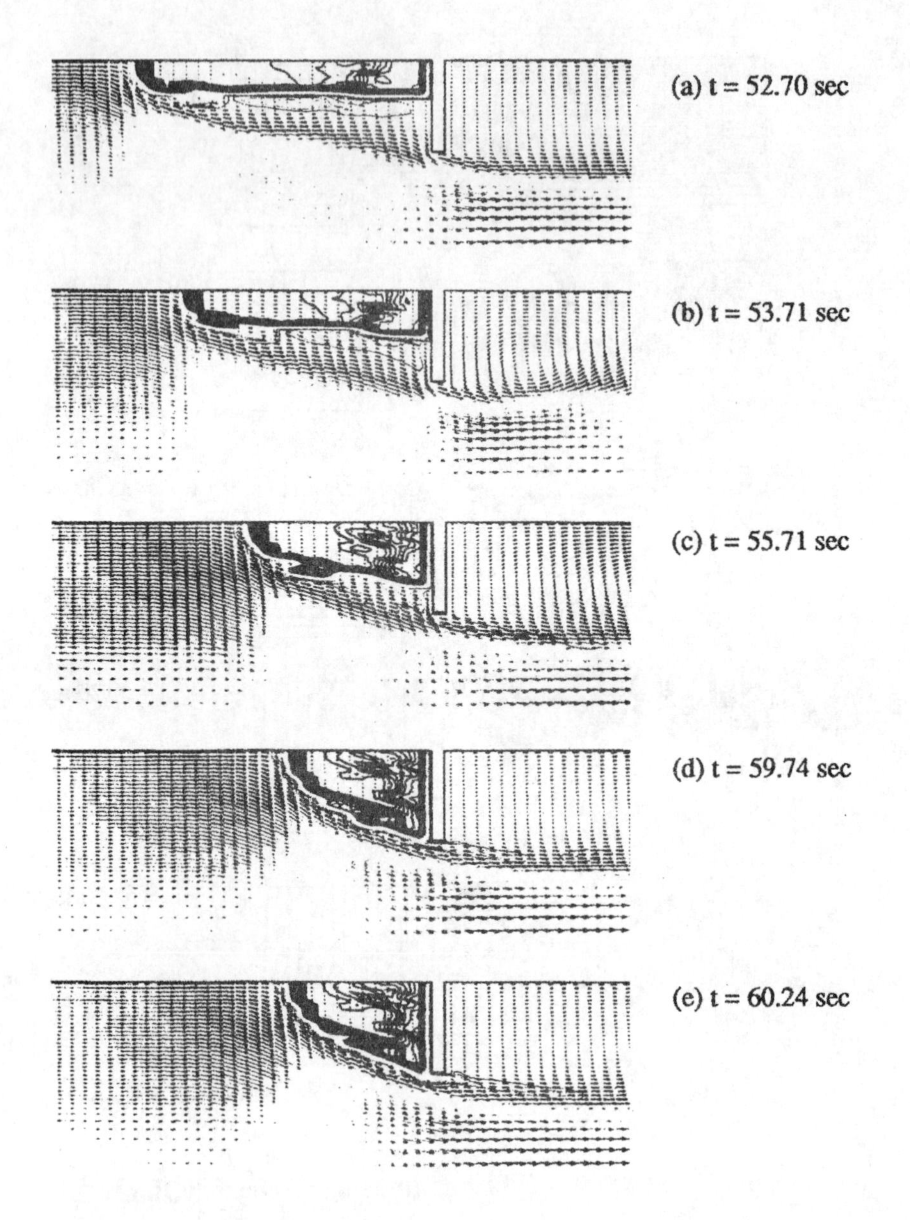

Figure 8　Successful Oil Containment at Velocity 0.15m/s

Based on: Johnston's Experiment Run No. 4
Oil: Shell Valvata 1000,　Boom Draft: 0.07m

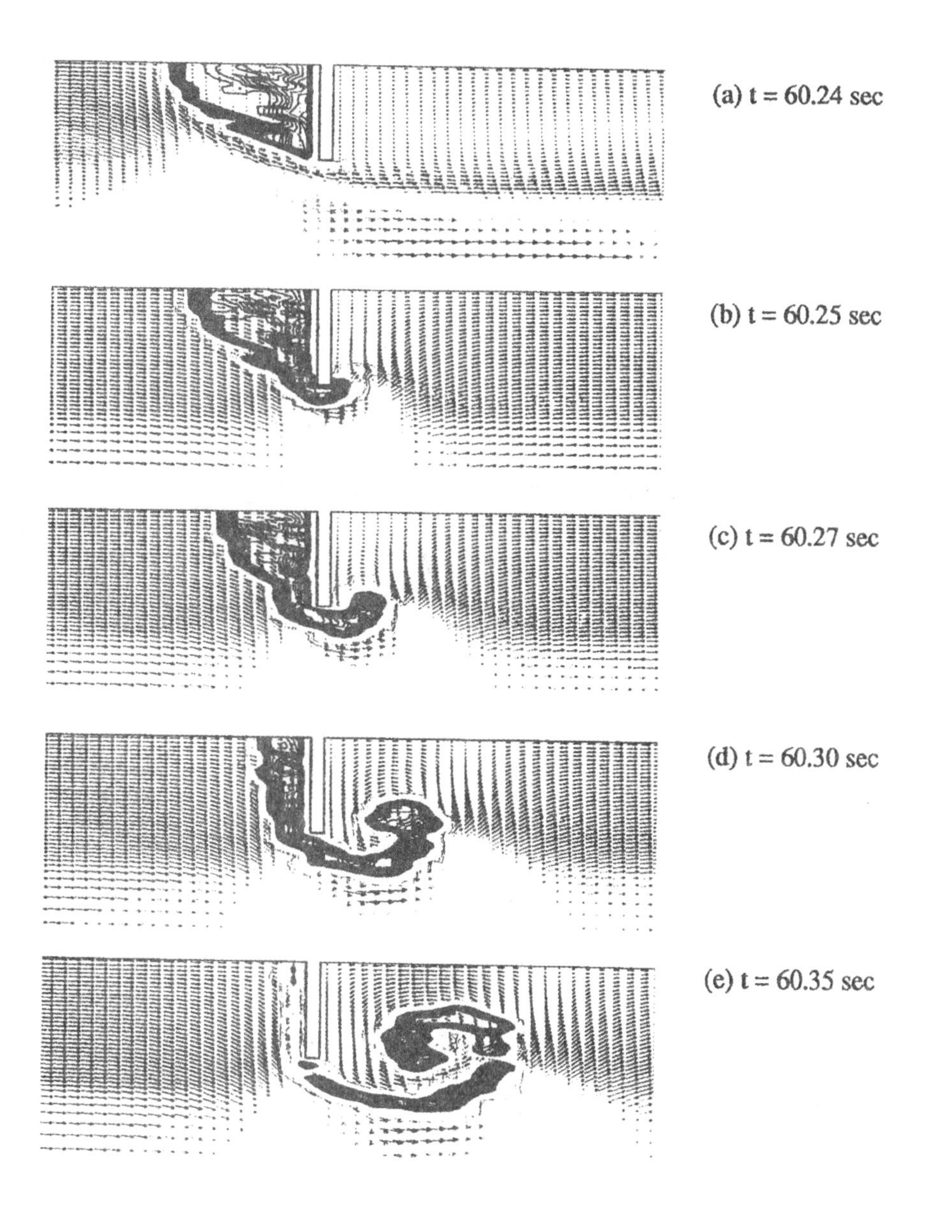

Figure 9　Failure by Critical Accumulation at Velocity 0.20m/s

Based on: Johnston's Experiment Run No. 4
Oil: Shell Valvata 1000,　Boom Draft: 0.07m

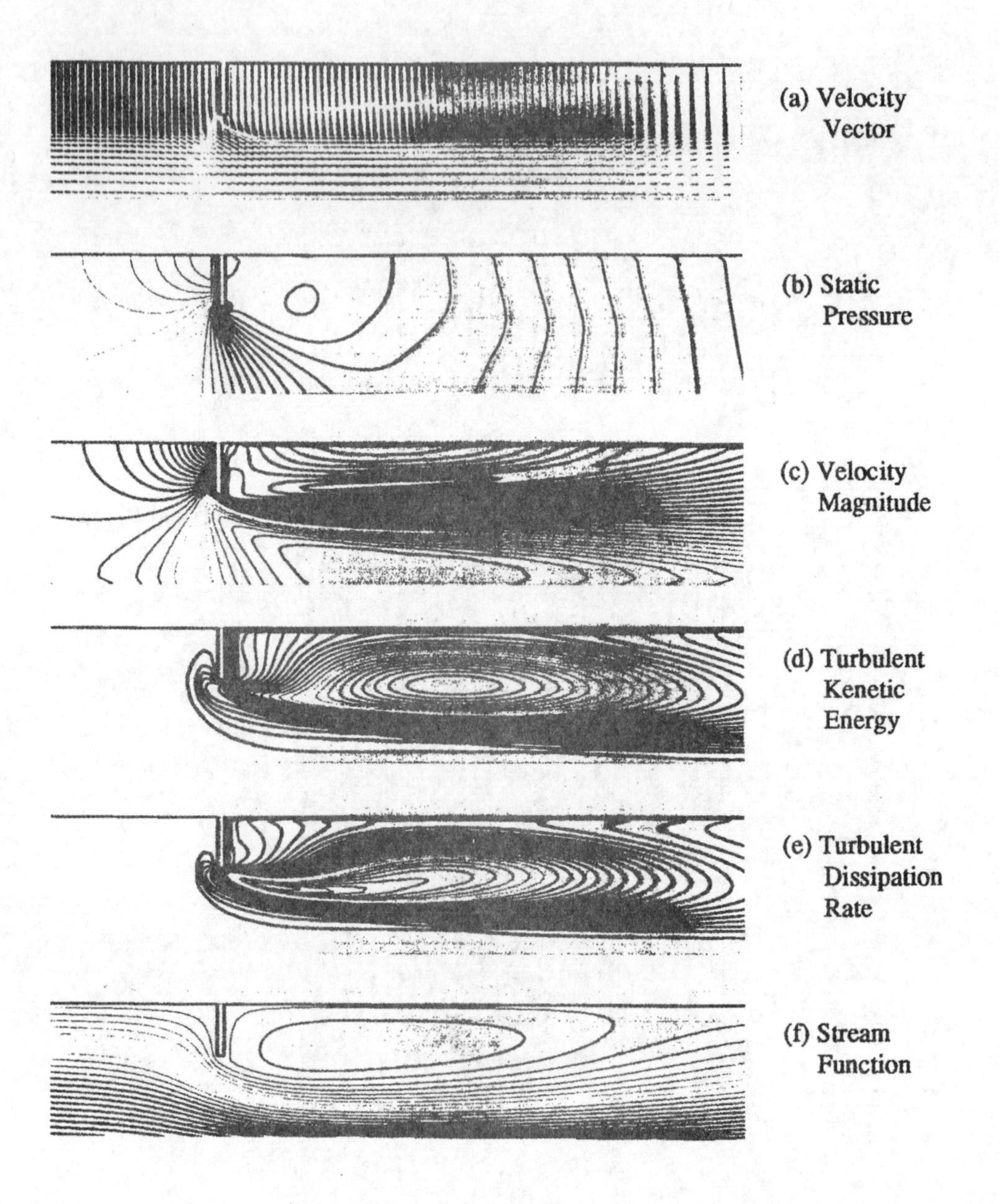

Figure 10 Steady State Flow of Water-Only Channel

Based on: Johnston's Experiment Run No. 4
Oil: Shell Valvata 1000, Velocity: 0.10m/s

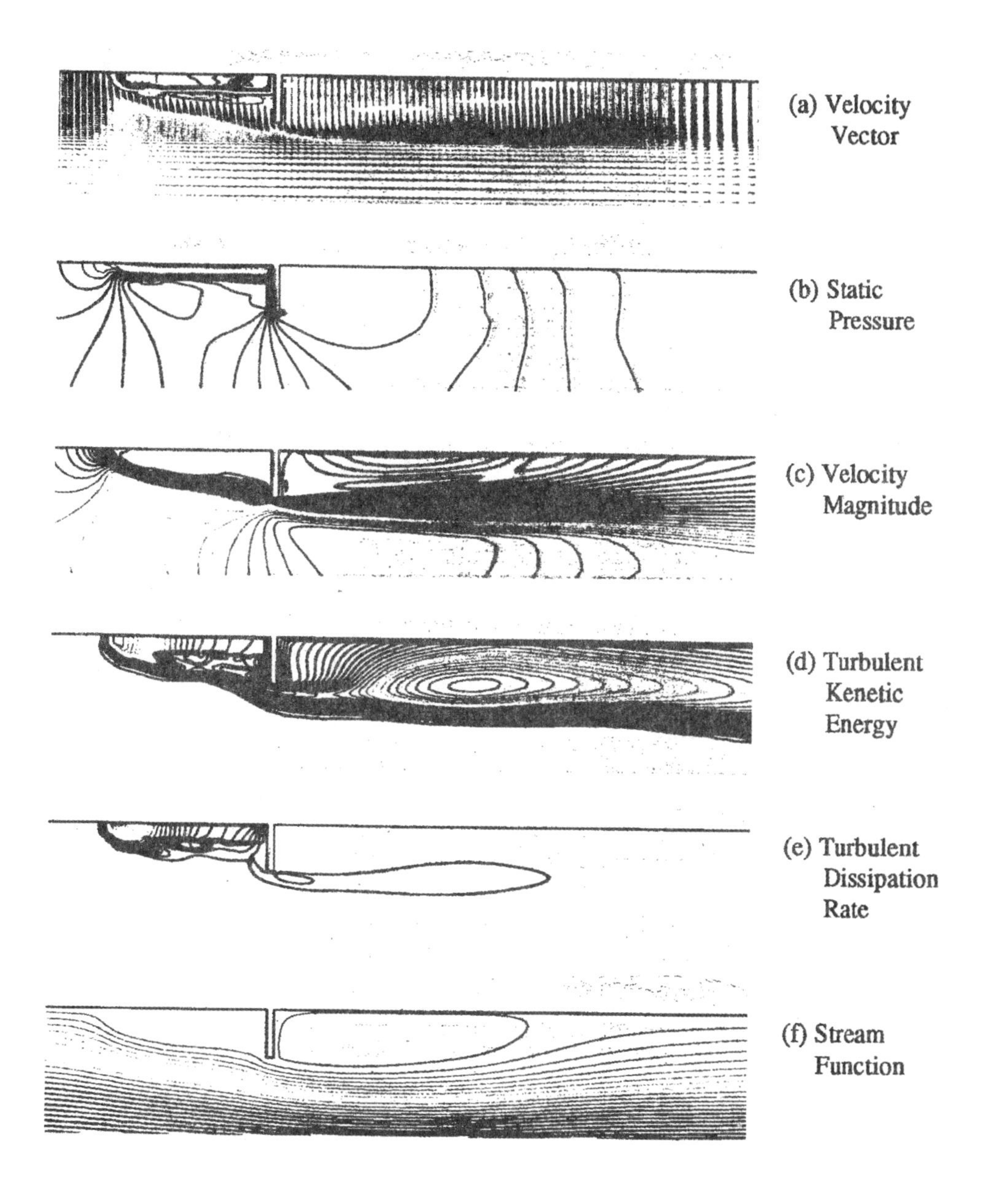

Figure 11 Steady State Flow of Successful Oil Containment

Based on: Johnston's Experiment Run No. 4
Oil: Shell Valvata 1000, Velocity: 0.10m/s, Draft: 0.07m

Boom Failure Mechanisms: Comparison of Channel Experiments with Computer Modelling Results

H.M. Brown, R.H. Goodman, C.-F. An, and J. Bittner
Imperial Oil Resources Limited
Calgary, Alberta, Canada

Abstract

A large outdoor flowing water channel has been used to obtain experimental data on boom failure mechanisms. Oil containment and failure around a simple barrier has been observed for oil viscosities from 10cSt to 5600cSt at relatively low flow velocities of 0.10m/s to 0.20m/s. The centre-line profiles of stable contained slicks have been measured and underwater videos of escaping oil have been made when the barrier failed. These experiments have been duplicated with a computational fluid dynamics model of the channel and barrier, and satisfactory agreement between the simulated and experimental measurements has been obtained. The study indicates that computer simulations of these complex processes can be used to obtain data about failure mechanisms that would be difficult to measure experimentally.

1.0 Introduction

The preferred response to an oil spill on water is to sweep the oil into towed booms where it can be recovered by skimming devices. Unfortunately, booms are effective only under very limited towing conditions and sea states. For spill responders, this poses a serious problem because the rate at which oil can be swept into booms and effectively trapped there is the limiting rate in overall oil recovery. For example: if a boom with a swath width of 100m is towed at 1kt through a continuous slick 1mm thick, oil will be trapped at a rate of 180m^3/hr. This optimistic trapping rate is within the recovery capacity of a large skimmer. In reality, the highest encounter rate ever experienced during the Valdez spill was only about 18m^3/hr and it was typically about 8m^3/hr in the first days of the recovery operation (Watkins, 1995). This low recovery rate arises because of the patchy nature of oil slicks and the low effectiveness of booms when they are not towed at very low speeds in relatively calm seas. Poor recovery rates with booms dramatically impact cleanup costs, particularly if the oil reaches a shoreline. Historical recovery costs have been about US$6500/m^3 for oil recovered on water but between US$130,000 and US$230,000/m^3 for oil cleaned from shore (Oil Spill U.S. Law Report, 1994). Thus any improvement in boom holding capabilities will have a direct impact on spill cleanup and recovery.

A more effective boom design will require a clearer understanding of the hydrodynamics of oil and water flow around a boom. Although a number of studies have been made on boom deployment and handling characteristics and on wave conformance, these have generally been done in the absence of oil. Channel or flume experiments to study the fluid dynamics of oil and water around a boom are few and many of the earlier studies are not well documented. Some of the findings of these studies are summarized in Table 1. As can be seen from the table, each study investigated one aspect of slick stability or dynamics over a small

Table 1. Channel Studies of Oil Containment and Boom Failure

Reference	Flume Length (m)	Flume Width (m)	Flume Depth (m)	Boom Draft (m)	Flow Velocity (m/s)	Oils Used	Comment
Cross et al. (1971, 1972)	12.2	0.76	0.30	N/A	N/A	Soybean Oil, #2 Fuel Oil	Calculated slick length
Wilkinson (1972, 1973)	2.3	0.38	0.22	0.038	0.158	Peanut Oil, Kerosene	Defined minimum Froude number
Milgram and Van Houlten (1978)	6.25	0.46	0.53	0.175	0.16-0.50	Mineral Heavy & Light, Arzew, Diesel #2	Demonstrated slick head & interfacial wave
Lau and Moir (1979)	15.0	0.60	0.06-0.20	N/A	N/A	Synthetic	Calculated boom minimum draft
Delvigne (1989)	N/A	0.50, 2.0	0.30, 0.70	0.07-0.13	0.09-0.26	Arabian Heavy & Light, Bunker B, Heavy Fuel, Diesel	Demonstrated critical accumulation (oil creep)
Johnston et al. (1993)	24.0	0.40	0.40	0.05-0.08	0.10-0.25	Range of refined oils	Surging before oil creep phenomena

range of oil characteristics, often with small channel depths and artificial oils. Hence, the relative water to boom speeds at which a particular oil will escape, or the nature of the escape mechanisms in terms of oil and oil/water interface characteristics are not well understood.

Experimental studies of the flow of oil and water around booms are difficult to conduct. If real booms are used, the cost becomes prohibitive even if permission to spill oil could be obtained. If model booms are used, scaling of the channel dimensions, boom shapes, and fluids may be necessary. In addition, the measurement of fluid flow and interfacial properties in a dynamic experiment becomes very difficult or impossible. In this study, we have used measurements of water flow around barriers and oil slick profiles in a meso-scale channel to parameterize and verify the graphical results of barrier failure obtained from computer simulations. The confidence gained in computer simulations of flow around simple barriers will allow the study of more complex three dimensional booms by computational fluid dynamics.

2.0 Computer Modelling

Computer simulations of complex fluid dynamics problems have only been possible within the last 10 years with the development of both very large, fast computers and sophisticated software programs. There are now a number of general computational fluid dynamics (CFD) programs available for solving problems in fluid flow, heat and mass transfer. The general problem is the solution of the Navier-Stokes equations with a set of boundary conditions which specify the particular scenario. In this study, the problem is the numerical solution of two dimensional unsteady turbulent flow of two immiscible viscous incompressible fluids, oil and water, around a simple barrier. The simulation requires the oil characteristics, the channel and barrier dimensions, the water flow and assumptions about the channel walls and free surface. Output from the simulation consists of cross-sectional graphs of the channel, oil and barrier at various times during the simulation. The graphs may show flow velocity vectors or velocity contours, pressure contours, turbulent kinetic energy, turbulent dissipation, or stream function. A number of channel experiments equivalent to the computer simulations have been conducted to verify the computer output.

3.0 Experimental Procedures

A channel 30m long, 1m wide, and 1m deep was constructed within an existing wave basin. A four horsepower variable speed underwater mixer with a 40cm propeller was situated at the exit end of the channel to move water through the channel and back into the main area of the basin. A flow straightener consisting of 2m lengths of stacked 10cm I.D. PVC piping was placed at the channel entrance. An underwater viewing box with a 1m by 0.80m plexiglass window was placed midway along the channel. A straight barrier, which could be moved vertically to vary the draft, acted as a boom and was placed across the channel at the midpoint of the viewing window. From the viewing box it was possible to see and photograph both the upstream and downstream sides of the barrier from below the water surface. This channel provided sufficiently uniform flow to study boom or barrier characteristics up to a flow velocity of 0.22m/s (slightly less than 0.5kt). Typical velocity profiles were used to standardize the computer model simulations. For these flow velocities and the barrier drafts used in this study, the Reynolds number was typically about 100,000.

Three oils of varying densities and viscosities were used in the experiments. These are listed in Table 2. For each oil, the slick behaviour was studied for four boom drafts with flow velocities varying from 0.10 to 0.20m/s. The barrier draft or flow velocity was always adjusted so that ultimately the barrier failed and the failure mechanism could be photographed or video taped.

Table 2. Properties of the Oils Used

Oil Type	Density (kg/m^3)	Viscosity (cSt)	Amount Spilled (l)	Boom Draft (m)
Federated Crude	834	10	5	0.015 - 0.15
Weathered Federated	870	240	5	0.010 - 0.15
Weathered Hondo	953	5600	10	0.050 - 0.15

Channel flow velocity profiles were measured with a Marsh-McBirney Model 2000 flowmeter. This device utilizes a Faraday-law electromagnetic induction sensor. A rigid positioning device which spanned the channel, was built to hold the sensor so that it could be accurately and reproducibly placed in both cross-channel and depth below-the-surface positions. The sensor was satisfactory for measuring channel flow profiles but, because of its size (5cm in diameter by 10cm long), could not be used to measure flow velocities near the oil-water interface or adjacent to the barrier.

When oil properties and flow velocities resulted in the formation of stable slicks behind the barrier, the slick profiles along the channel centre-line were measured using one of two different techniques. A resistance probe was constructed consisting of a digital vernier micrometer screw with a needle tip attached to a travelling carriage but electrically insulated from it. The carriage slid along a rigid bar which was attached to the channel sides but positioned over the centre-line of the slick. The micrometer screw tip was connected in series to an ohmmeter and to a copper plate submerged in the water of the channel. The carriage was moved to a position along the length of the slick and the micrometer screwed downward through the slick until the needle penetrated the oil and touched the water. When this occurred, the resistance of the ohmmeter circuit dropped dramatically and the screw position was recorded.

An acoustic device was also used to measure slick thicknesses. A Krautkramer Branson Model CL304 ultrasonic thickness gauge with a 10Mhz submersion probe was used successfully to measure static oil thicknesses as thin as 0.5mm on a 50cm column of water. The device measures the delay between acoustic pulses reflecting off the oil-water interface and the oil-air interface. It appeared to be most reliable when the oil-water interface was clean with little surface turbulence.

Experiments were conducted by carefully spilling either 5 or 10 litres of the selected oil onto the channel surface 3m upstream of the barrier after the initial channel flow had stabilized. Once the slick had been stopped by the barrier and had stabilized, the centre-line slick profile was measured. The flow velocity was then increased and if the slick remained behind the barrier, the profile was again measured. If a stable slick was retained behind the barrier at the highest flow velocity (~0.20m/s), the series of experiments was repeated with a fresh aliquot of oil but with a smaller barrier draft. When a barrier draft and flow velocity combination was reached in which the oil began to escape, video and still photographs of the failure process were taken through the underwater viewing window.

4.0 Comparison of Experimental Results with Computer Simulations

A representative sample of the centre-line slick profiles measured in the flowing channel is shown in Figure 1. Profiles at flow velocities of 0.10m/s, 0.15m/s, and 0.20m/s are shown for each of the three oils in order of increasing viscosities. In each case the barrier draft was 0.15m and the water depth was 1m - the top of the slick is shown as that elevation. The barrier is located at the origin and channel flow proceeds from the right. These profiles were measured after the oil had encountered the barrier and had time to rebound and stabilize. As expected, increasing the flow velocity shortens and thickens the slick as does increasing the oil viscosity for a given flow velocity. Note that a head wave has developed in the 10cSt oil at a velocity of 0.20m/s and it is likely that a small increase in flow velocity would cause entrainment failure to begin. The head wave was observed to be stable at this velocity, however and no loss of oil occurred. No distinct head wave is apparent in the profiles of the 240cSt viscosity oil, although the slicks are shorter and thicker. Significant shortening occurs at a flow of 0.20m/s in the 5600cSt oil slick and an increase in flow velocity would likely precipitate oil creep under the barrier. A similar set of profiles was measured for this viscous oil but with a shallower barrier. In this case the slick had a similar profile to that shown with U = 0.20m/s but at a flow of 0.15m/s. Oil creep failure was observed when the flow was increased to 0.20m/s.

These flowing channel experiments were duplicated using the computational fluid dynamics software and procedures described in a companion paper (Goodman et al. 1997). In the computer simulations, a two dimensional numerical solution of the turbulent flow along the channel centre-line for the two immiscible fluids, oil and water, is obtained. A portion of the channel 8m long by 1m deep and containing the straight uniform barrier at the 3m mark is modelled. This cross-sectional area is divided into a nonuniform grid of 241 by 61 within which the Navier-Stokes equations and the κ-ε turbulence modelling equations are solved for appropriate boundary conditions. Three examples of computed slick profiles (one for each of the oil viscosities used in the channel experiments) are shown in Figure 2 and compared to the measured thicknesses. There is satisfactory agreement for both the slick lengths and general thickness profiles. As in the previous figure, the slick upper surface is shown as the 1m water depth and flow proceeds from the right of the figure. All of the simulation results duplicated satisfactorily the measured slick profiles as either the water flow or oil viscosity was increased or the barrier draft was altered. In addition, the initial surging motion, observed in the lighter oil slick as the barrier was encountered and stability was established, was accurately modelled.

Two failure mechanisms were observed in the channel experiments and recorded with video or photographs from below the water surface. With a viscous oil, failure can occur even with a very large draft barrier if the flow velocity is sufficient. An example of this 'critical accumulation' failure, or more descriptively, 'oil creep' (Johnston et al., 1993) with the 5600cSt oil at a flow of 0.20m/s is shown in the photographs of Figure 3. The views show the bottom of the oil slick which was initially pooled to the right of the barrier in the right background. Oil can be seen creeping under the barrier, which is diagonal across the view and rising back up to the water surface on the downstream side of the barrier. The two photographs were taken about 15 seconds apart and the progression of the thick oil mass is evident. This failure mechanism continued over a period of several minutes until all of the trapped oil had escaped. A computer simulation of this experiment was conducted and the slick profiles as a function of time are shown in Figure 4. These show a very similar failure mechanism in progress which eventually resulted in the loss of all of the trapped oil.

With both of the lighter oils, no failure was observed at the highest flow velocity of 0.20m/s as long as the barrier was of sufficient draft. However, drainage failure could be initiated if the barrier was shallow enough or sufficient oil was added to the slick. Underwater photographs of this failure phenomena were duplicated by computational fluid dynamics modelling.

5.0 Conclusion

These comparisons demonstrate that the computer simulations mimic the experimental studies satisfactorily and imply that the processes occurring in the channel can be studied using computational fluid dynamics. There are a number of advantages to this study method. Channel experiments are expensive to conduct. More importantly, the simulations provide much information that can not currently be measured with any accuracy. This includes vector flow values at the oil-water interface, velocity magnitude profiles around the barrier, static pressure contours, contours of turbulent energy and dissipation, and stream functions. We hope to use this information to fully understand barrier failure processes with a view to designing better oil containment devices.

6.0 Acknowledgements

The authors wish to acknowledge the support of Imperial Oil Resources and the cooperation of Prof. R.D. Rowe of the Mechanical Engineering Department of the University of Calgary.

7.0 References

Cross, R.H. and D.P. Hoult, "Collection of Oil Slicks", *J. Waterways, Harbors & Coastal Eng. Div.*, Vol. 97, WW2, pp.313-322, 1971

Cross, R.H. and D.P. Hoult, "Oil Booms in Tidal Currents", *J. Waterways, Harbors & Coastal Eng. Div.*, Vol. 98, WW1, pp. 25-34, 1972

Delvigne, G.A.L., "Barrier Failure by Critical Accumulation of Viscous Oil", in *Proc. of 1989 API Oil Spill Conf.*, San Antonio, TX., USA, pp. 143-148, 1989

Goodman R.H., H.M. Brown, C.-F. An, and R.D. Rowe, "Dynamic Modelling of Oil Boom Failure Using Computational Fluid Dynamics", in *Proc. of 20th Arctic & Marine Oilspill Prog. Tech. Sem.*, Vancouver, Canada, 1997

Johnston, A.J., M.R. Fitzmaurice, and R.G.M. Watt, "Oil Spill Containment: Viscous Oils", in *Proc. of 1993 API Oil Spill Conf.*, Tampa, FL.,USA,1993

Lau, Y.L. and J. Moir, "Booms Used for Oil Slick Control", *J. Environ. Eng. Div.*, Vol. 105, No. EE2, pp. 369-382, 1979

Migram, J.H. and R.J. Van Houlten, "Mechanics of a Restrained Layer of Floating Oil Above a Water Current", *J. Hydronautics*, Vol. 12, No.3, pp. 93-108, 1978

Oil Spill U.S. Law Report, Cutter Information Corp., March 1994

Watkins, R.L., "Effectiveness of Oil Recovery Systems", in *Proc. of 1995 API Oil Spill Conf.*, Long Beach, CA., USA, pp. 985-987, 1995

Wilkinson, D.L., "Dynamics of Contained Oil Slicks", *J. Hydraulics Div.*, Vol. 98, No. HY6, pp. 1013-1030, 1972

Wilkinson, D.L., "Limitations to Length of Contained Oil Slicks", *J. Hydraulics Div.*, Vol. 99, No. HY5, pp. 701-712, 1973

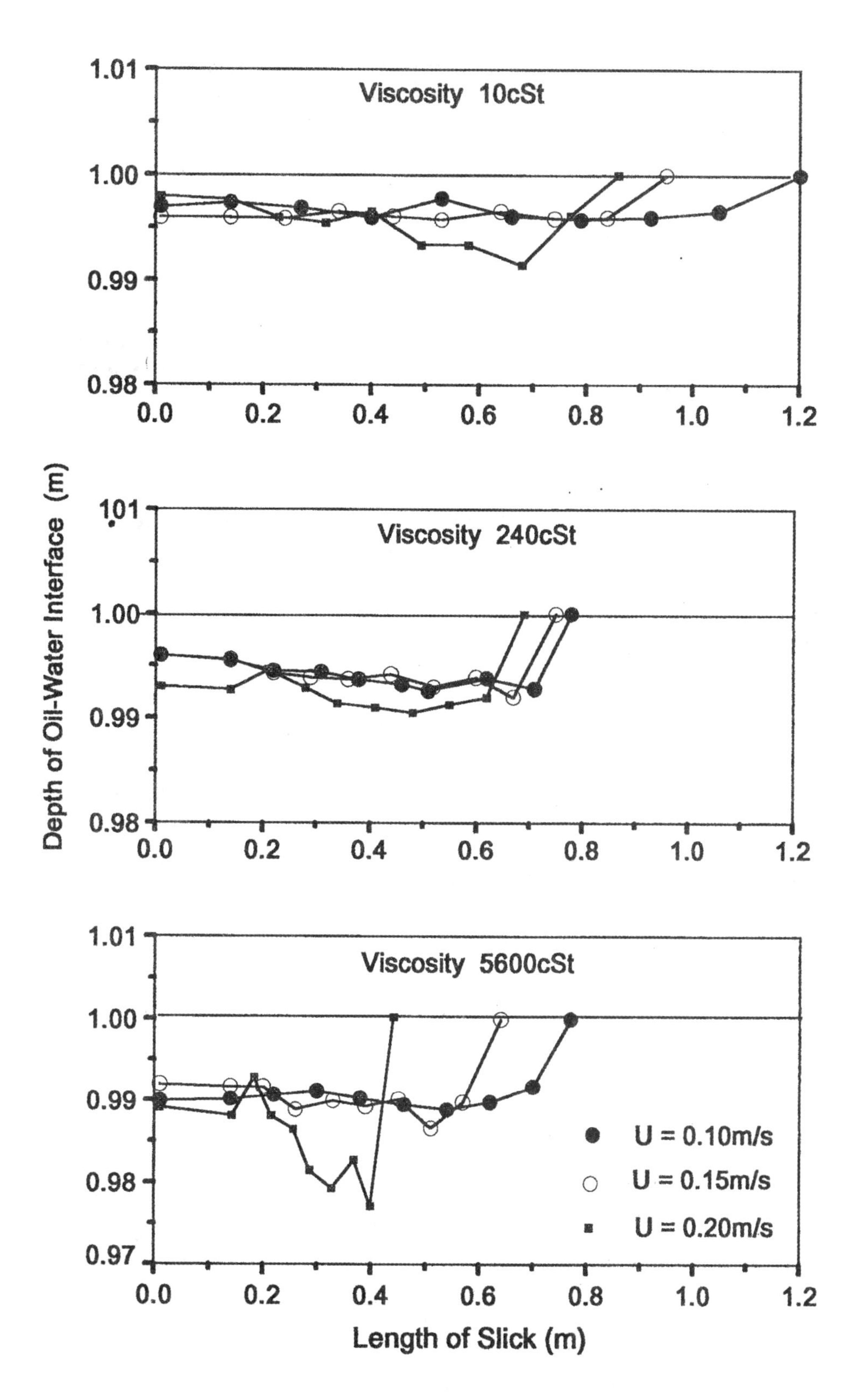

Figure 1 Oil Slick Profiles - 0.15m Barrier Draft

第Ⅲ部分 受聘于加拿大帝国石油公司时期

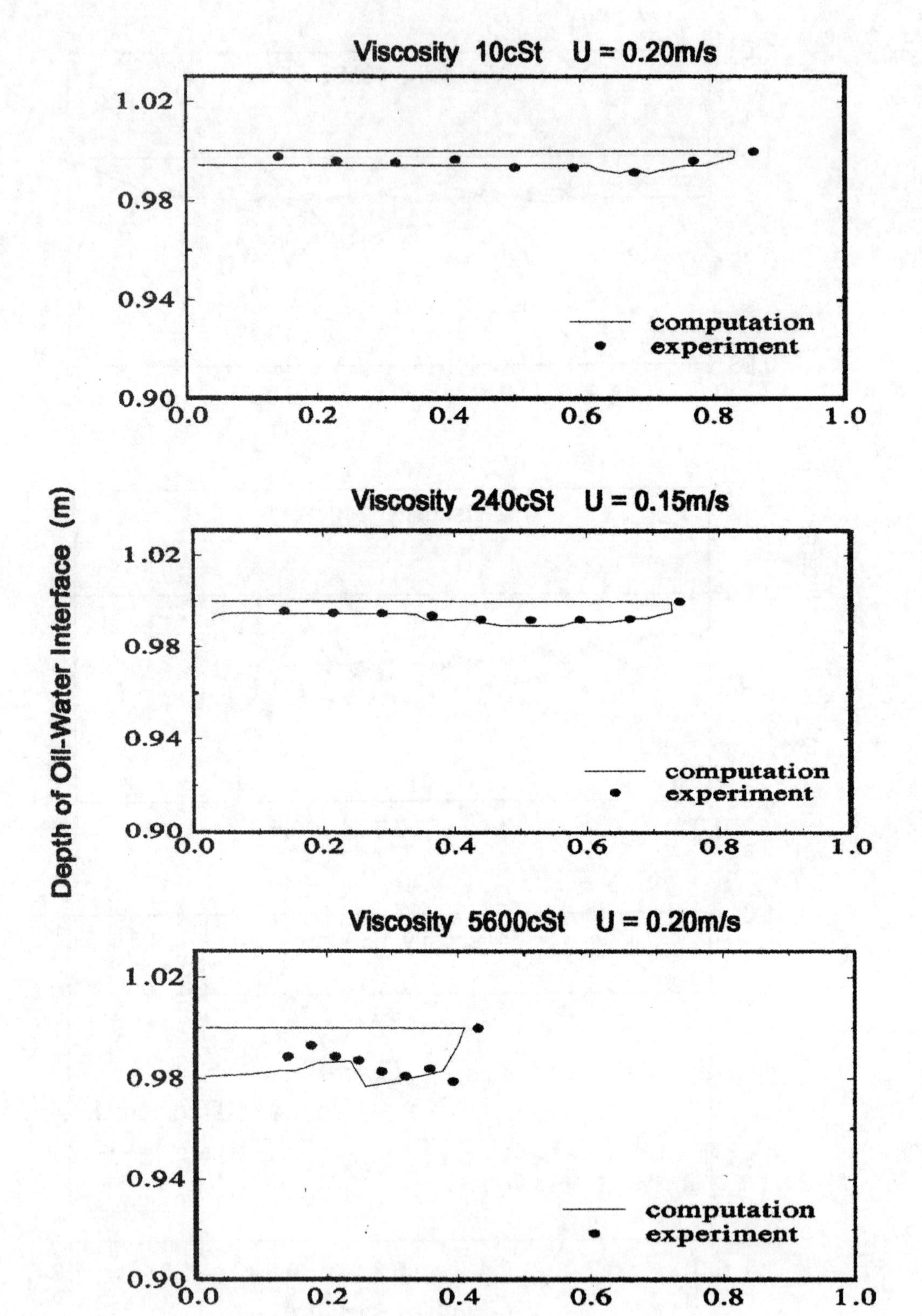

Viscosity 10cSt U = 0.20m/s
1.02
0.98
0.94
0.90
0.0
0.2
0.4
0.6
0.8
1.0
computation
experiment
Viscosity 240cSt U = 0.15m/s
Depth of Oil-Water Interface (m)
Viscosity 5600cSt U = 0.20m/s
Length of Slick (m)

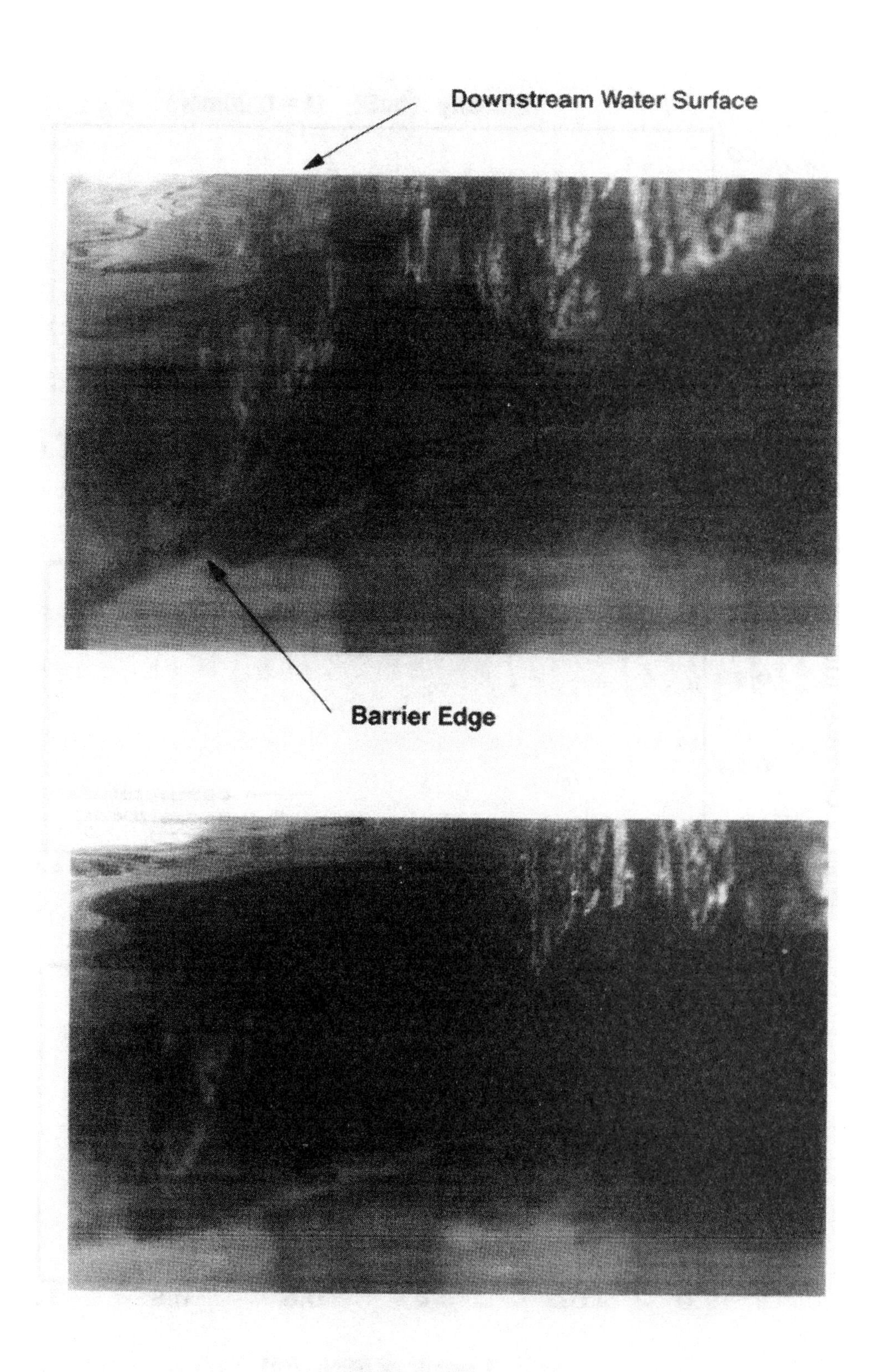

Figure 3 Viscous Oil Creep Under Barrier

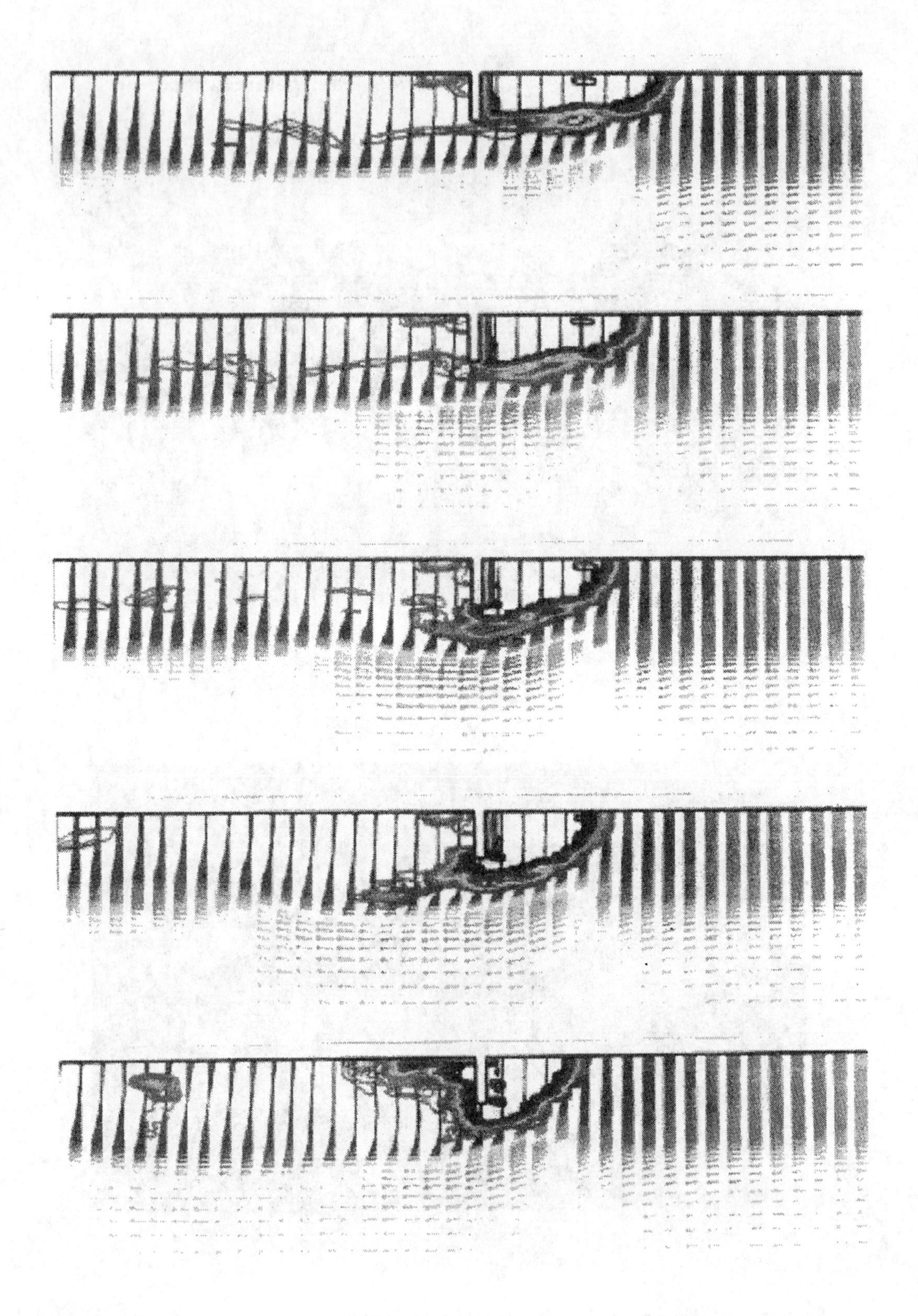

Figure 4 Computed Slick Profiles

Chang-Fa An
June 10, 1997
Vancouver

VOLUME 2

PROCEEDINGS OF THE TWENTIETH ARCTIC AND MARINE OILSPILL PROGRAM (AMOP) TECHNICAL SEMINAR

COMPTE RENDU: 20°COLLOQUE TECHNIQUE DU PROGRAMME DE LUTTE CONTRE LES DÉVERSEMENTS D'HYDROCARBURES EN MER ET DANS L'ARCTIQUE (AMOP)

JUNE 11 TO 13, 1997
VANCOUVER, BRITISH COLUMBIA
CANADA

DU 11 AU 13 JUIN, 1997
VANCOUVER, COLUMBIE-BRITANNIQUE
CANADA

 Environment Canada Environnement Canada

Canada

第三部分 受聘于加拿大帝国石油公司时期

Animation of Boom Failure Processes

C.-F. An, H.M. Brown and R.H. Goodman
Imperial Oil Resources Limited
Calgary, Alberta, Canada

E.J. Clavelle
The University of Calgary
Calgary, Alberta, Canada

Abstract

Computer animation is a very useful tool for demonstrating time-dependent physical processes, such as the flow of oil and water around a barrier. In this paper, computer animations of oil boom failure mechanisms are presented. The animations are performed by playing a sequence of images which have been obtained from the graphical output of a computational fluid dynamics (CFD) program, *Fluent*. The modeling is based on boom failure experiments which have been carried out in flowing water channels. The animations are in Audio Video Interleave (AVI) format and can be viewed easily on personal computers using the Windows'95 operating system. Three animation examples of boom failure processes, i.e. drainage failure, droplet entrainment and critical accumulation, are presented. Features of the boom failure mechanisms are more evident in the animation sequences than can be discerned from individual images.

1.0 Introduction

Animation, or dynamic visualization, is an expressive tool for demonstrating any time-dependent process in science or engineering research. It depends on techniques developed for computer graphics, a subject area which has evolved rapidly in recent years. The topic is widely discussed on the Internet and a large volume of software for computer animation is available there (Ladic, 1997, Walker, 1997, Woillez , 1997).

During the past two years, studies of oil boom failure mechanisms have been carried out by Imperial Oil Resources Limited. These studies have included computational fluid dynamic (CFD) simulations of oil boom failure (An et al, 1996, Goodman et al, 1997) and the comparison of flowing channel experiments with computer modeling (Brown et al, 1997). The graphical output of the computational fluid dynamic models shows the time-dependent processes of oil boom failure mechanisms and provides suitable source material for animation. For this work the output was generated by a commercial CFD program - *Fluent* (1996).

Below, the technique of making computer animation is described briefly and three animation examples created from the modeling results of oil boom failure simulations are presented. One animation of each of the three main mechanisms of boom failure is included, i.e. drainage failure, droplet entrainment and critical accumulation. These animated examples of boom failure processes can be demonstrated on any personal computer of reasonable speed.

2.0 Animation Techniques

There are a number of software programs available for animating time-dependent processes. For this work the program *VidEdit* was chosen as this software was available with the Windows operating system. *VidEdit* uses the Audio Video Interleave (AVI) format which is commonly used on personal computers. In order to make an animation, a set of sequentially arranged images must be prepared. These were obtained as the graphical output from the numerical modeling of oil boom failure using the computational fluid dynamics modeling program *Fluent*. *Fluent* (1996) is a general-purpose CFD program for modeling fluid flow problems. Graphical output from the modeling program was obtained at set time intervals as a series of Tagged Image Format Files (TIFF). These were given sequential file names prior to incorporation in the animation program. The *VidEdit* program, which was used to create the animation from the file sequences, requires that the source files be in the Device Independent Bitmap (DIB) format and so it was necessary to convert the file sequence from TIFF to DIB format before animation. This was done using *Paint Shop Pro*, a program which can also be used to correct the color balance and contrast of individual images if necessary.

The *VidEdit* program compresses and stores the image sequence in AVI format as a single file. It features editing and compression options which can be used to crop or resize the image and chose the level of compression. The compression level affects the graininess or pixellation of the resulting video presentation - a sharper image requiring more storage space and possibly a faster computer. Full motion video requires 30 frames per second (fps) which would necessitate the storage of about 7 megabytes of information per second if each image was 320 x 240 pixels. In order to store and handle this large information flow, one must employ a fast computer and an efficient compression algorithm or accept a lower quality video image. Satisfactory animation of the three boom failure sequences, each of which lasts about 10 seconds, can be achieved with data stored on three 1.4 MB floppy disks. A PC with a Pentium CPU operating under Windows'95 has been used to generate the animation. The sequences can be viewed with the *Media Player* program which is included with this operating system.

3.0 Animation Examples

3.1 Model Simulation of Drainage Failure

The first animation sequence is based on a model run in which 5 liters of Federated Crude oil were spilled onto a channel of water flowing at 0.20 m/s. The oil had a density of 834 kg/m^3 and a viscosity of 10 cSt. It encountered a very shallow barrier of 0.015m draft in a channel 1m deep. This modeling experiment simulated a meso-scale channel experiment in which drainage failure was observed (Brown et al, 1997).

The sequence was made from the graphical output of the model taken in time steps of 0.05sec from time t=13.50sec to time t=21.00sec of the simulation. Thus 151 source images were used to make the animation. After compression, the final animation file occupied 0.85 MB of disk space. Satisfactory video resolution and motion were obtained at a playback speed of 15 fps for a sequence which lasts 10 seconds.

Several typical frames of the animation are demonstrated in Figure 1. These show the oil flowing from the right, hitting the barrier and rapidly draining under it. In viewing the motion in the animation, salient features are the rapidity with which the oil flows under the barrier, the depth to which it sinks, and the trapping of oil in the vortex behind the barrier. The motion also shows that oil reaches the surface behind the barrier and flows downstream in patches rather than in a continuous slick and that the leading edge of the oil continues well below the surface for some time. The submerged leading edge probably consists of fine oil droplets. Thick patches of oil remain both in front of and behind the barrier at the end of the sequence. Some of these features can be seen in Figure 1.

3.2 Model Simulation of Droplet Entrainment

The example of droplet entrainment is based on a model run in which 5 liters of Bunker B oil with a density of 888 kg/m^3 and a viscosity of 70 cSt was spilled onto a channel flowing at 0.24m/s. The oil encountered a barrier with a shallow draft of 0.07m in a channel 0.3m deep. This model run simulated a channel experiment in which droplet entrainment had been observed (Delvigne 1989).

The animation was made from the graphical output of the modeling results by taking images at 0.02sec intervals from time t=145.1sec to time t=177.1sec of the simulation. This resulted in 161 images for the animation which after compression required 1.25 MB of disk space. As in the previous example, satisfactory motion and resolution were obtained at a playback speed of 15 fps. This gave an animation length of 9.4 seconds.

On viewing the animation several features are immediately apparent. The oil-water interface beneath the accumulated oil is unstable and waves continually move along it. Oil continually bleeds under the barrier from this interface, although it should be noted that the animation shows this as a bleeding oil-water mixture (i.e. oil droplets), not pure oil. The lost oil droplets flow a considerable distance downstream of the barrier before resurfacing and then flow back along the water surface to the barrier as they are caught in the large trailing vortex. Circulation patterns in the accumulated oil upstream of the barrier are clearly visible. Several of these features can be seen in the typical frames of the animation shown in Figure 2 and all are similar to features observed in various channel experiments.

3.3 Model Simulation of Critical Accumulation

This example was based on a model run of the CFD program *Fluent* in which 2 liters of the heavy oil, Shell Valvata 1000, was spilled onto a flowing water channel 0.23m deep. This oil had a density of 915kg/m^3 and a viscosity of 3600 cSt. The oil encountered a barrier of 0.07m draft in the water channel. The model run simulated a channel experiment in which the critical accumulation failure mechanism (or more descriptively, "oil creep") had been observed (Johnston et al., 1993).

For this animation, the images produced from *Fluent* were taken at intervals of 0.001sec from time t=60.24sec to time t=60.46sec of the simulation. This involved converting and compressing 220 images and required 1.15 MB of disk space. Once again the motion and resolution were found satisfactory when the video frames were played back at 15 fps. At this speed the animation lasts 8.25 seconds.

When the animation is viewed, the salient features of oil creep or critical accumulation failure are very apparent. A continuous stream of thick oil is observed to flow under the barrier. This is not a thin stream bled from the oil-water interface as was observed in the droplet entrainment example but a continuous thick mass of oil. The loss continues until all of the accumulated oil has passed under the barrier and occurs despite the relatively large barrier draft. No circulation within the accumulated oil is observed before loss and once the oil has passed under the barrier it remains caught in the trailing vortex as a single mass of oil. Throughout the simulation the oil behaves as a single mass which appears to be squeezed under the barrier by the flowing water.

Several typical frames of the animation are presented in Figure 3 and some of the features described above can be seen. All of these observations are consistent with those described in channel experiments conducted by ourselves or others.

4.0 Conclusions

(1) Computer animation is a very useful tool to demonstrate the transient processes of oil boom failure.

(2) Readily available software can be used to create satisfactory animations.

(3) Details of the three boom failure mechanisms are more apparent in the animation sequences than can be discerned with static output graphs from the models.

5.0 Acknowledgment

The authors would like to thank Imperial Oil Resources Limited for the support on this project. The first author wishes to acknowledge the Natural Science and Engineering Research Council of Canada for an Industrial Research Fellowship. He also wishes to express his sincere gratitude to Prof. R.D. Rowe of the University of Calgary for his kind sponsorship of a Research Associateship which allows the author to use both hardware and software of the university to conduct this project. He also would like to recognize Mr. Hank Li of Wisemind Enterprise Co. for his patient help in the computer animation.

6.0 References

An, C.-F., E.J. Clavelle, H.M. Brown and R.M. Barron, "CFD Simulation of Oil Boom Failure", in *Proc. of 4th Ann. Conf. of CFD Soc. of Canada*, Ottawa, ON, Canada, June 2-4, 1996

Brown, H.M., R.H. Goodman, C.-F. An and J. Bittner, "Boom Failure Mechanisms: Comparison of Channel Experiments with Computer Modeling Results", in *Proc. of 20th AMOP Tech. Seminar*, Vancouver, BC, Canada, June 11-13, 1997

Delvigne, G.A.L., "Barrier Failure by Critical Accumulation of Viscous Oil", in *Proc. of 1989 Oil Spill Conf.,* San Antonio, TX, USA, pp.143-148, Feb. 13-16, 1989

Fluent User's Guide, version 4.4, Fluent, Inc., Lebanon, NH, USA, 1996

Goodman, R.H., H.M. Brown, C.-F. An and R.D. Rowe, "Dynamic Modeling of Oil Boom Failure Using Computational Fluid Dynamics", in *Proc. of 20th AMOP Tech. Seminar*, Vancourver, BC, Canada, June 11-13, 1997

Johnston, A.J., M.R. Fitzmaurice and R.G.M. Watt, "Oil Spill Containment: Viscous Oils", in *Proc. of 1993 Oil Spill Conf.*, Tampa, FL, USA, pp.89-94, Mar. 29-Apr.1, 1993.

Ladic, L., "Making Animations for the Web", *Internet Article* in 1997 at: *http://www.cs.ubc.ca/spider/ladic/anim_unix.html*

Paint Shop Pro, shareware version 4.12, Jasc, Inc., Eden Prairie, MN, USA, 1996

VidEdit, freeware version 1.1, Microsoft Corporation, Roselle, IL, USA, 1993

Walker, J., "Graphics Viewers, Editors, Utilities and Info", *Internet Article* in 1997 at: *http://www.bae.ncsu.edu/bae/people/faculty/walker/hotlist/graphics.html*

Woillez, S., "Multimedia Utilities", *Internet Article* in 1997 at: *http://www.prism.uvsq.fr/public/wos/multimedia*

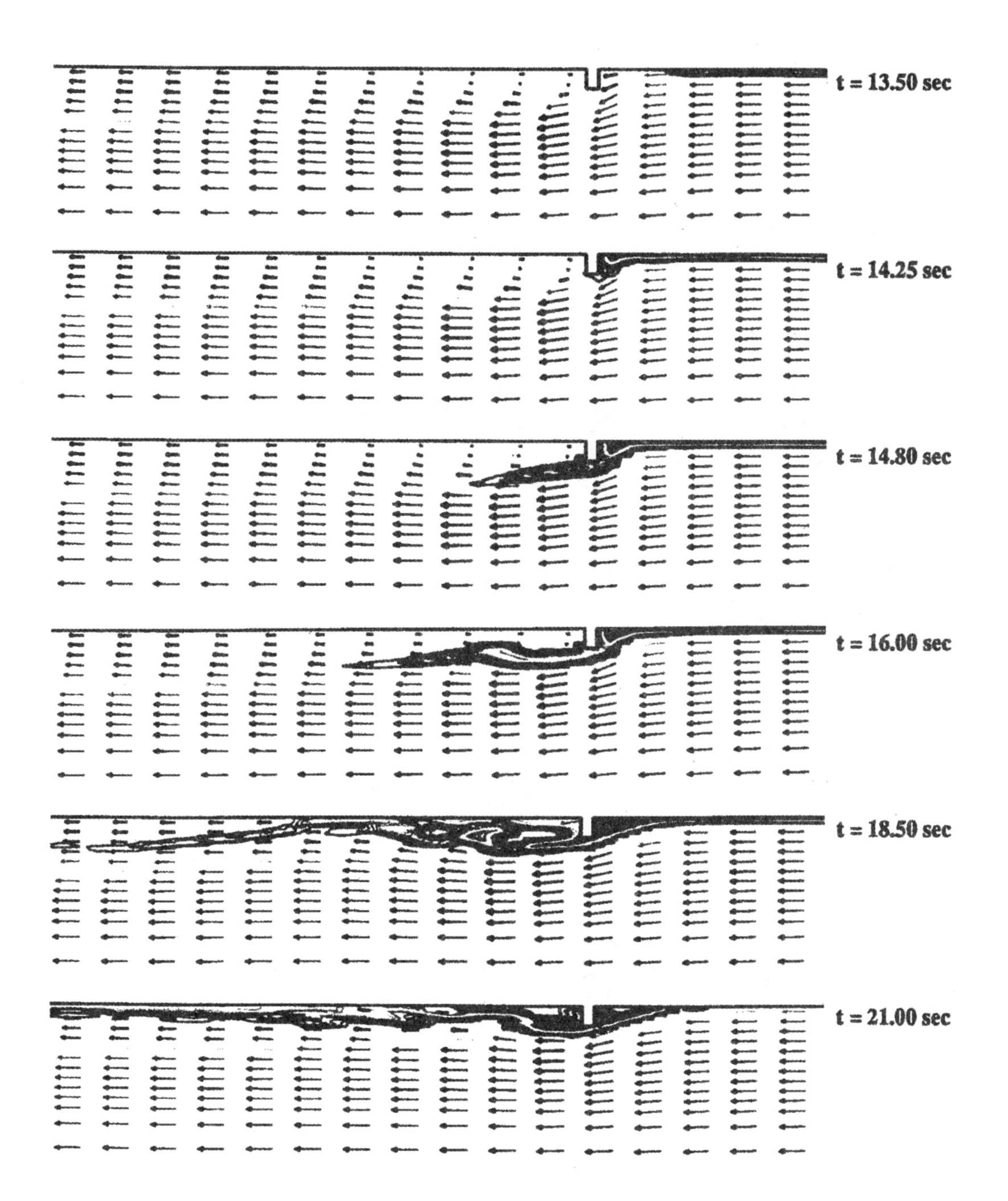

Figure 1 Animation of Drainage Failure

Based on Wave Basin Test and Modeling
Oil: Federated Crude, Amount: 5 Liters
Draft: 0.015m, Velocity: 0.20m/s

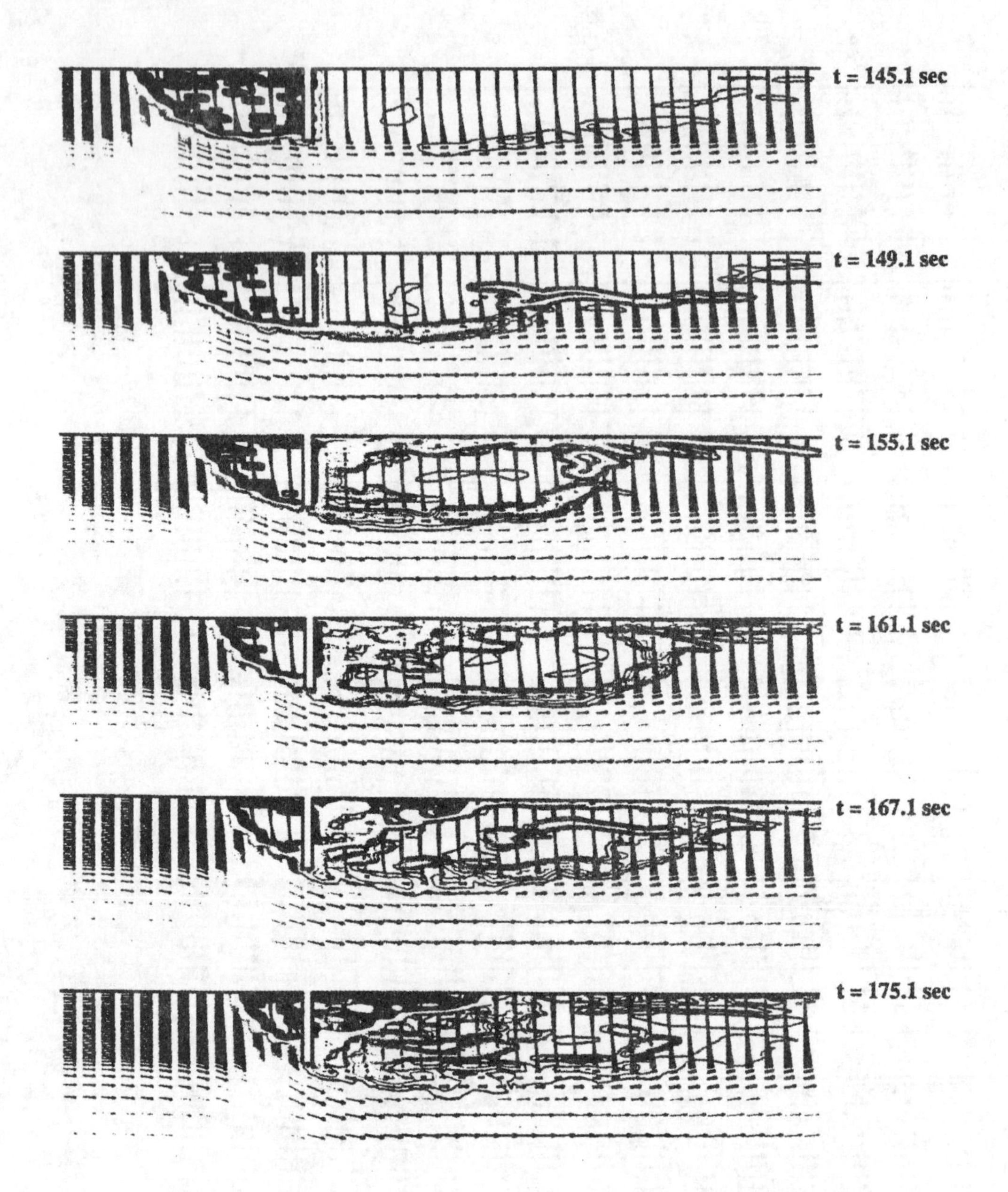

Figure 2 Animation of Droplet Entrainment

Based on Delvigne's Test and Modeling
Oil: Bunker B, Amount: 5 Liters
Draft: 0.07m, Velocity: 0.24m/s

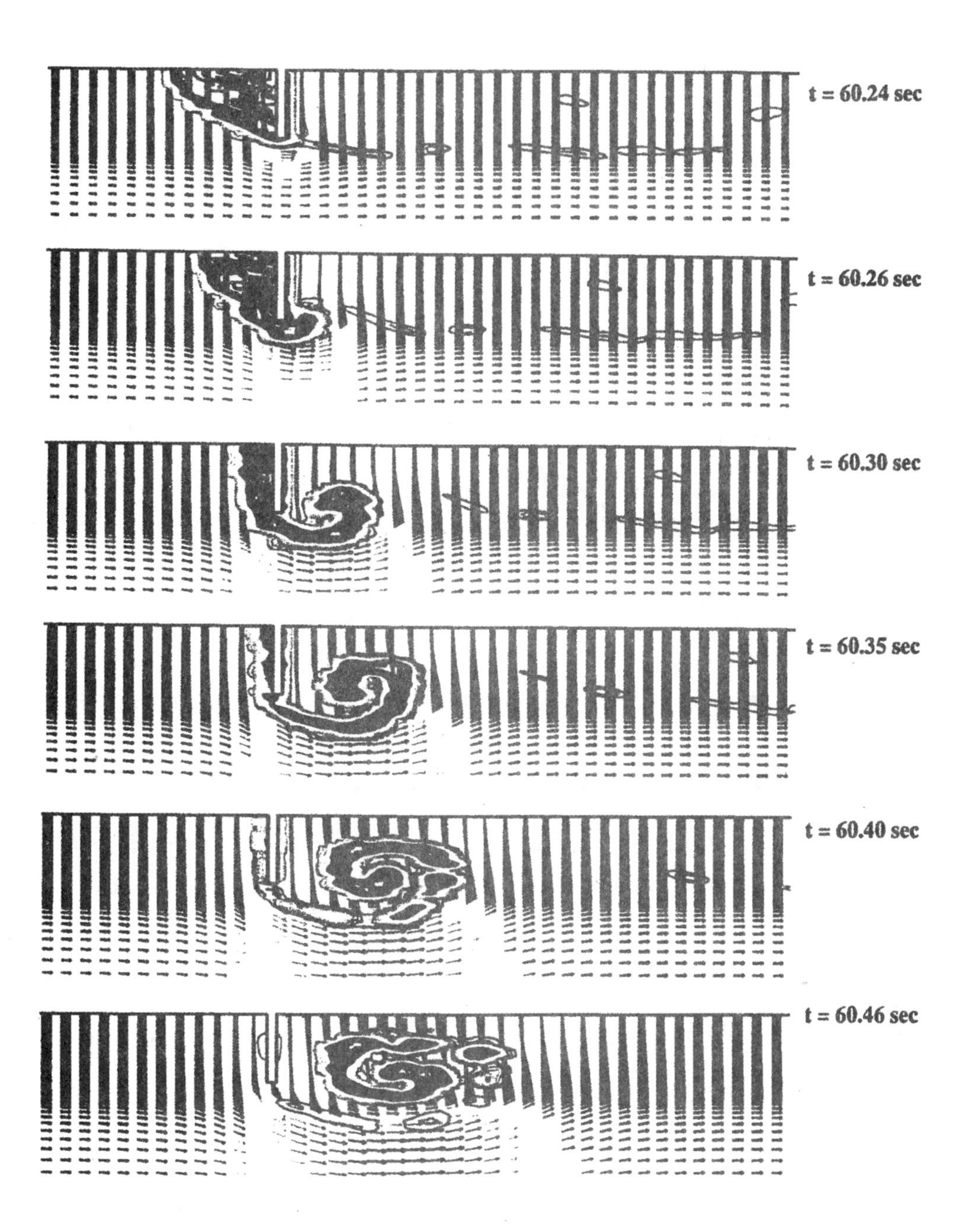

Figure 3 Animation of Critical Accumulation

Based on Johnston's Test and Modeling
Oil: Shell Valvata 1000, 2 Liters

应用VOF方法对水流中拦油栅失效进行数值模拟尝试

程石勇　张政*
（北京化工大学 化学工程学院，100029 北京）
安长发
（加拿大帝国石油公司）

摘要

作为对流体体积分数法（VOF方法）的应用尝试，本文通过自编程序用这种方法模拟计算了两种拦油失效的情形：低粘度的油类在水流速度超过某一临界值时发生了油滴夹带失效；粘度很大的油类在水流速度超过临界值时发生了临界累积失效。作为第一步，文中将流体运动简化为层流。模拟结果在定性上与已有的实验结果和C.-F An博士的模拟计算相一致。但计算出的临界水流速度还存在偏差，这将在以后的工作中通过加入湍流模型进行进一步的探讨。

1 概述

由于开采，运输过程中甚至人为造成的海洋，河流上的石油泄漏造成的水污染已成为世界各国关注的环境问题。目前，处理此类污染最常用的方法是用水面拦油栅[1]拦集，然后用撇油器回收或直接焚烧。拦油栅即漂浮在水面上用以收集或存放油类的障碍物，或称挡板。拦油栅拦集石油的示意图见于下图：

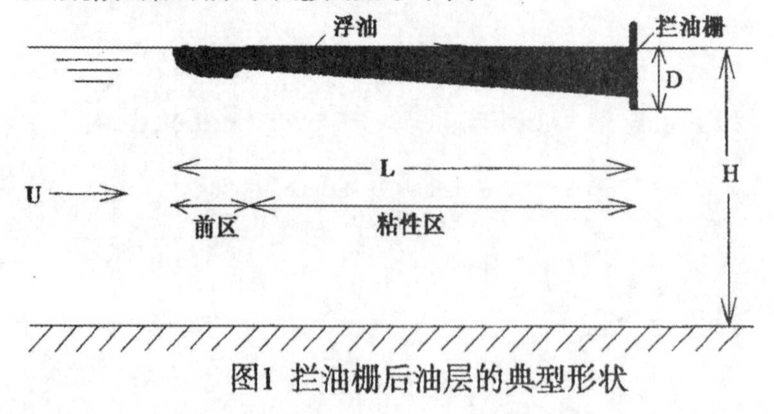

图1 拦油栅后油层的典型形状

* 通讯联系人

实践表明，采用栏油栅拦集石油的效果并不理想。广阔海面上，由于风浪作用或固定不合理经常造成栏油栅被破坏而发生失效。即使排除风浪影响因素，由于油层下水流作用的影响，经常造成油从栅板下被水流带走而造成拦油失效。

有关拦油失效的实验研究[2,3,4]表明主要存在三种失效形式[3]：a.油滴夹带失效，b.油层流失，c.临界累积失效。油滴夹带失效一般发生于低粘度油层，当油与水间相对速度足够大时，油滴从油－水界面处被撕下来，如果没有足够的时间使油滴在浮力作用下重返油层主体，它将被水流从栅板下带走，发生失效。对于中等粘度的油，当油水相对速度增大时，界面应力增大，从而使油层变短变厚，当它超过栅板深度以后，就会发生油层的流失。一般说来，对于上述二种情况，只要栅板深度足够，油层总是可以拦住的。临界累积失效发生于石油运动粘度$\upsilon > 3000cs$和水流速度$u > 0.15m/s$时，油－水界面剪切力超过了油层内部循环流可以抵销的范围，使油层变短变厚，直到整个油层从栅板下逃逸出去。需要指出的是，这种失效几乎与栅板深度无关，而且，由于风化作用，石油的粘度很容易会超过3000*cs*,油水相对速度大于0.15*m/s*的情况也很普遍。因此，临界累积失效是很重要的一种失效形式。

有关拦油动力学的机理，早期的研究[5,6]建立在简化模型的基础上，水流被视为一维均匀流，对象是运动粘度$\upsilon < 3500cs$的油类，但这些研究结果均无法说明拦油失效的机理。近年来，随着计算机技术的发展，应用数值模拟的方法研究拦油失效成为一种有效的手段。

E.J.Clavelle 和 R.D.Rowe[7]采用了SOLA－VOF程序对拦油的累积失效问题进行了数值模拟。其研究对象是运动粘度$\upsilon = 2300cs$的重油。模拟结果表明，水流速度$v = 0.14m/s$时，即发生油层完全从栅板下流走的所谓临界失效状态，该结果与Delvigine[3]的实验结果相吻。

目前，加拿大帝国石油公司的安长发（C.-F.An）博士等人[8,9,10]利用商用软件FLUENT 成功地模拟了二维情况下拦油失效的三种形式。计算采用了 VOF 方法，并应用$k-\varepsilon$模型计算水流的湍流粘度。根据模拟结果：对于一定体积的低粘度轻油，水流速度低于临界值时，油层最终会达到稳态，拦油是成功的；当水流速度增大至超过临界值时，发生了油滴夹带失效。其模拟计算得到的前区油层厚度和临界水流速度分别与 Wilkson 的理论值和 Delvigine 的实验结果相吻合。对于高粘度的重油，当水流速度小于临界值时，油层能达到稳定状态，在挡板前形成既短且厚的“油堆”；水流速度超过临界值时，发生了临界累积失效，油层从挡板下“爬走”，最终油全部绕过挡板流失掉了。模拟计算的临界值与 Delvigine 的实验相符合。

帝国石油公司拟与我们合作开发，应用VOF方法研究多相流，尤其是拦油失效问题的应用程序。本文目的就是在 An 博士等人工作的基础上，自己动手编制程序，探讨用VOF方法获得与 C.-F.An 应用 FLUENT 软件计算的结果。为进一步深入研究拦油失效打下基础。

2 计算方法

实际上，排除风浪影响，拦油栅拦油是一个三维的多相流问题。由于空气－水、空气－油界面上的摩擦力与油－水界面上的摩擦力相比很小，计算中忽略气相的影响，研究的问题可以简化为油－水两相流动问题。并且，水面上拦油时，展开的拦油栅呈一弧形，中心处水流速度最高。考虑到水流速度是拦油失效的重要影响因素，拦油栅的中心区是计算模拟的重点。因而，可以把问题进一步简化为二维，本文实际研究的是二维条件下油－水两相非定常流动问题。

因为油水两相不互溶，存在明显的相界面，数值模拟的关键部分是跟踪两相界面。解决这一问题的一种有效方法是流体体积分数法[11,12]（VOF 方法），它采用体积分数函数跟踪两相界面，通过求解同一控制方程计算两相的流场分布。VOF 方法定义了体积分数函数 $F(i,j,t)$，它表示单位体积内特定流体所占据的体积分数。若网格被一种特定流体完全充满，$F=1.0$；若网格中全部为另一种流体，$F=0.0$；F 取 0～1之间值时，说明网格中存在两相界面。

2.1 基本控制方程

过程中不存在热量和质量传递，并且流体被认为是不可压缩的。根据简化的模型，控制方程为二维 Navier－Stokes 方程：

$$\frac{\partial \mathbf{u}}{\partial t}+(\nabla\cdot\mathbf{u})\mathbf{u}=\frac{1}{\rho}\nabla\cdot p+\upsilon(\Delta\mathbf{u})+g \quad (\text{I})$$

流体满足连续性方程：

$$\nabla\cdot\mathbf{u}=0 \quad (\text{II})$$

为了区分油水两相，引入体积分数函数 $F(i,j,t)$，F函数满足：

$$\frac{\partial F}{\partial t}+(\mathbf{u}\cdot\nabla)F=0 \quad (\text{III})$$

2.2 VOF 方法求解要点

（1）动量方程方程的离散化：

方程（Ⅰ）的离散采用“半隐式”格式。其中，压力为隐式压力 p^{n+1},扩散项和对流项中的速度为显式速度 $\mathbf{u}^n$，从而得到了 n+1 时刻速度 $\mathbf{u}^{n+1}$ 关于压力 p^{n+1} 和n 时刻速度 $\mathbf{u}^n$ 的关系式。为了计算 n+1 时刻的压力 p^{n+1}，将方程（Ⅰ）用显式格式离散，得到估计速度 $\mathbf{u}^*$ 关于显式压力 p^n 与速度 $\mathbf{u}^n$ 的关系式，两离散化方程相减，得到：

$$\mathbf{u}^{n+1}=\mathbf{u}^*-\frac{\Delta t}{\rho}\nabla\delta p \quad (\text{IV})$$

其中，$\delta p=p^{n+1}-p^n$ （Ⅴ）

把（Ⅳ）代入连续性方程（Ⅱ）中，得到压力方程：

$$\frac{\Delta t}{\rho}\Delta\delta p=\nabla\cdot\mathbf{u}^{*}\quad(\text{VI})$$

(2) 求解压力方程:

对压力方程的求解采用牛顿迭代法，计算修正压力δp。将δp代入式（Ⅳ）和式（Ⅴ），验算新的速度场是否满足连续性方程，若不满足则进一步迭代，直到$\nabla\cdot\mathbf{u}^{*}=0$。经过压力修正和速度修正，得到了n+1时刻的速度场。将此速度场代入方程（Ⅲ）中，计算体积分数函数F。

（3）F函数的求解

由于F函数是一阶梯形函数，采用标准的插值计算将引起F函数的抹平效应。为了保证F函数的明确意义,计算采用了给体－受体方法（DA方法）[13]。根据上游和下游的F信息确定自由表面的形状,然后根据该形状计算网格面上F通量.

根据求出的流体体积分数和速度分布，得到新时刻的流场，并以此为初始条件进行下一时间步的迭代计算。

（4）边界条件的处理

流体的上边界是与空气接触的自由表面，考虑到气相作用的体系影响较小，且实验观察到油、水上表面高度差别细微，因此上边界被简化为无摩擦的平面板，边界上y方向速度为零；底部边界视为固定无滑脱表面，此边界上x方向的速度为零；流体从左边界流入，给定入口速度u_i ;右边界为连续性边界。y方向重力加速度g=-9.81N。含有油、水两相的网格中，密度和粘度取算术平均值。

3 计算试例与结果比较

本文分别针对两种不同粘度的油类进行了模拟计算，为了和Delvigine等人的实验结果相比较，油类的物性参数与已有的实验值相同。计算域的选取以及挡板尺寸等几何参数根据实验确定，并与C.-F An模拟计算中的取值相同。采用不均匀网格划分，在挡板附近和计算域上部加密网格（如图2），以保证油－水界面和挡板附近的计算精度。

3.1 低粘度油的计算模拟

低粘度油类的模拟计算基于Delvigine的实验。计算域为一矩形通道，长（x方向)4.0m,深（y方向）0.30m。挡板置于距入口2.0m处，其深度为0.07m（y方向），厚（x方向)0.01m。网格数目142×31，最小网格尺寸取$\Delta x=0.02m$, $\Delta y=0.005m$。油的密度$\rho_{O1}=888\ kg/m^3$，其运动粘度$\upsilon_{O1}=70.0cs$，水的密度取20℃时值$\rho_W=998.2kg/m^3$，运动粘度$\upsilon_W=1.004cs$。

未加入油之前，水从左边界以$u=0.09m/s$的速度给入计算域，经过30.0s后得到一稳态流场。以此为初始条件，油从左边界最上方网格给入，当单位宽度上的油量达到$Q=0.01m^3/m$时停止给油。计算水流速度为0.09m/s，0.15m/s，0.20m/s，0.24m/s时及0.28m/s时油层在挡板附近的动力学行为，其结果见于图3中。

$t=0.0$时刻，油开始从流场左边界给入，22.2s后停止进油，图3a 给出了此时挡板前的油层侧形和流场速度分布：油层的前锋已到达挡板并在挡板前集聚，其主体仍在向挡板流动；继续进水，油层不断变短，$t=40.2s$时（图3b），油层内出现了负向u速度，此后油层开始从挡板“反射”回去，变长变薄，直到$t=85.0s$时（如图3c），油层又有所收缩并达到稳态。比较图 3b和3c可以发现，后者反而更长。此结果与实验现象相符，Johnston 称之谓“波动”。

增大入口水流速度至$u=0.15m/s$，油层又开始向挡板方向收缩，$t=110s$时（如图3d）达到稳态，油层上游部分有离开主体的油滴存在，没有油流失于挡板之后。在此水速下，油层被成功地拦住了。

继续增大入口水流速度至$u=0.20m/s$（如图3e），油层变得更短更厚，其上游油－水相对速度较大，油从主体被“剪切”下来形成油滴，挡板附近油层较稳定，说明油滴又重返油层主体，没有发生失效。

水流速度增大至$u=0.24m/s$,油层变得更短，上游处更不稳定，但并没有出现油的流失（如图3f）。这时，给定入口水流速度为$u=0.28m/s$, 2.5秒后（如图3g），油层的下缘已接近挡板底部，油滴已开始流失到下游，$t=143s$时（图3h），更多的油流失于挡板之后。流失的油呈分散状，体积较小；而油层的主体被拦在挡板之前，因此判断发生了油滴夹带失效。

3.2 高粘度油的计算模拟

对高粘度油类的计算模拟基于 Johnston 的实验。计算域长（x 方向)4.3m,深（y方向）0.23m。挡板距入口 2.0m，深0.07m（y 方向），厚（x 方向)0.01m。网格数目220×27，最小网格尺寸取$\Delta x=0.01m$, $\Delta y=0.005m$。油的密度$\rho_{O2}=915kg/m^3$，其运动粘度$\upsilon_{O2}=3600cs$，水的密度取15℃时值$\rho_W=999.1kg/m^3$，运动粘度$\upsilon_W=1.138cs$。

以入口水流速度为$u=0.1m/s$的水相稳态流场为初始条件，油从左边界最上方网格给入。当油量达到$Q=0.005m^3/m$时停止进油，计算水流速度分别为0.1m/s，0.15m/s，0.20m/s，0.25m/s时油层在挡板附近的动力学行为，其结果见于图4中。

$t=0.0s$时，油开始从流场左边界给入，10.0秒后停止进油，水继续以$u=0.10m/s$的速度进入流场。$t=20.0s$时刻（如图4a），油层正在向挡板流动，其前锋已到达挡板并在此集聚，油层内u速度为正。随着时间的增加，油层不断变厚。图4b 和4c 分别给出了$t=65.0s$和$t=70.0s$时刻挡板前的油层形状和流场分布：油层变得较厚并且出现了负的u速度,即油层内出现了循环流。比较两图，5秒内油层形状和流场分布几乎没有变化，油层到达了稳态，没有发生失效。与低粘度油类不同，由于油水间摩擦力较大，过程中没有出现油层的“反射”现象。

在$t=70.0s$时刻，增大入口水流速度至$u=0.15m/s$，经过11秒，油层在挡板前重新到达平衡。由图4d可以看到：$t=81.0s$时，油层继续变短变厚，堆积在挡板前。油－水界面呈波形，且其上游油层较不稳定，有较大的油团与油层主体分离。此时刻没有油层的流失，拦油是成功的。

继续增大入口水流速度至$u=0.20m/s$，根据图4e，$t=83.5s$时，油层变得既短且厚并超过了挡板，油层开始绕过挡板并开始流失于挡板之后。但经过2.5秒后发现油层的流失速度减慢。在$t=86.0s$时增大水流入口速度至$u=0.25m/s$。图4f给出了$t=89.0s$时刻的油层侧形和流场分布：油一团团地流失于挡板之后，挡板前只有薄薄的一层油并继续流向挡板下游。由于挡板前已没有达到平衡状态而存在的主体油层，故在此水流速度下发生的是临界累积失效。

3.3 结果比较

根据 Delvigine 和Johnston 的实验结果：对于低粘度油类，发生油滴夹带失效的临界水流速度为$u=0.24m/s$左右；而高粘度（$\upsilon>3000cs$）的油类，在$u>0.20m/s$时会发生临界累积失效。An 博士利用 FLUENT 软件计算出的两种失效的临界值分别为$0.24m/s$和$0.20m/s$。而根据本模拟计算的结果，油滴夹带失效发生于水速为$0.28m/s$时，而高粘度油在水速为$0.25m/s$发生了临界累积失效，比实验值和An博士的计算结果偏高。并且，以低粘度油为例，根据An博士的计算结果，当水流速度分别为0.09*m/s*,0.15*m/s*,0.20*m/s*时，达到平衡后挡板前油层厚度约为0.016*m*,0.03*m*,0.04*m*。而本模拟计算的结果分别为0.008*m*,0.016*m*,0.025*m*.与 An 博士的计算结果相比，在相同的入口水流速度下，油层的累积程度较低，相对较薄。

根据入口处的水流速度、流道几何尺寸计算得出雷诺数$R_e>20,000$,属于湍流流动，而计算中未加入湍流模型，所以考虑模拟计算出现偏差可能由此造成。

4 结论

模拟计算的结果表明：较低水流速度下，拦油栅拦油是成功的。在较高水流速度下（$u=0.28m/s$），低粘度油类发生了油滴夹带失效，油层的主体被拦在挡板之前，但少量的油以油滴的形式流失于挡板之后；对于高粘度的油类，随着水速的增加，油层在挡板前累积并变得很短很厚，当水流速度超过临界值（$u=0.25m/s$）时，整个油层都流失到挡板之后，发生了临界累积失效。

与Delvigine 等人的实验结果相比，模拟计算的结果在定性上是与之相符的，模拟出了油滴夹带失效和临界累积失效两种失效现象，以及在低粘度油类发生的“反射”现象；但在定量上，模拟计算得出的水流速度临界值比实验值偏高。与An博士的计算结果相比，相同水流速度下，油层累积程度较低。考虑可能是由于计算中未加入湍流模型所致，在今后的工作中将引入湍流模型，作更深一步的研究。

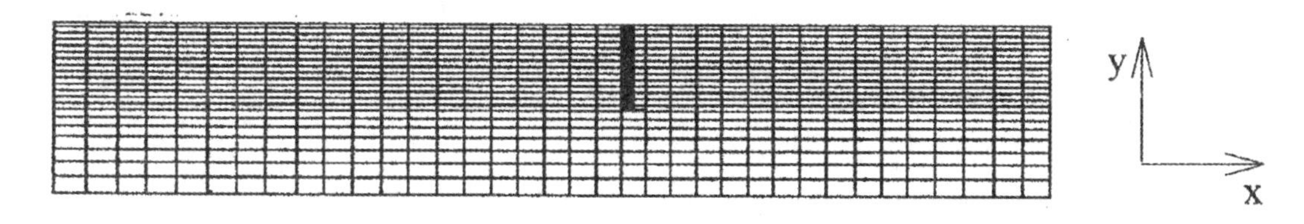

图2 网格划分（局部）

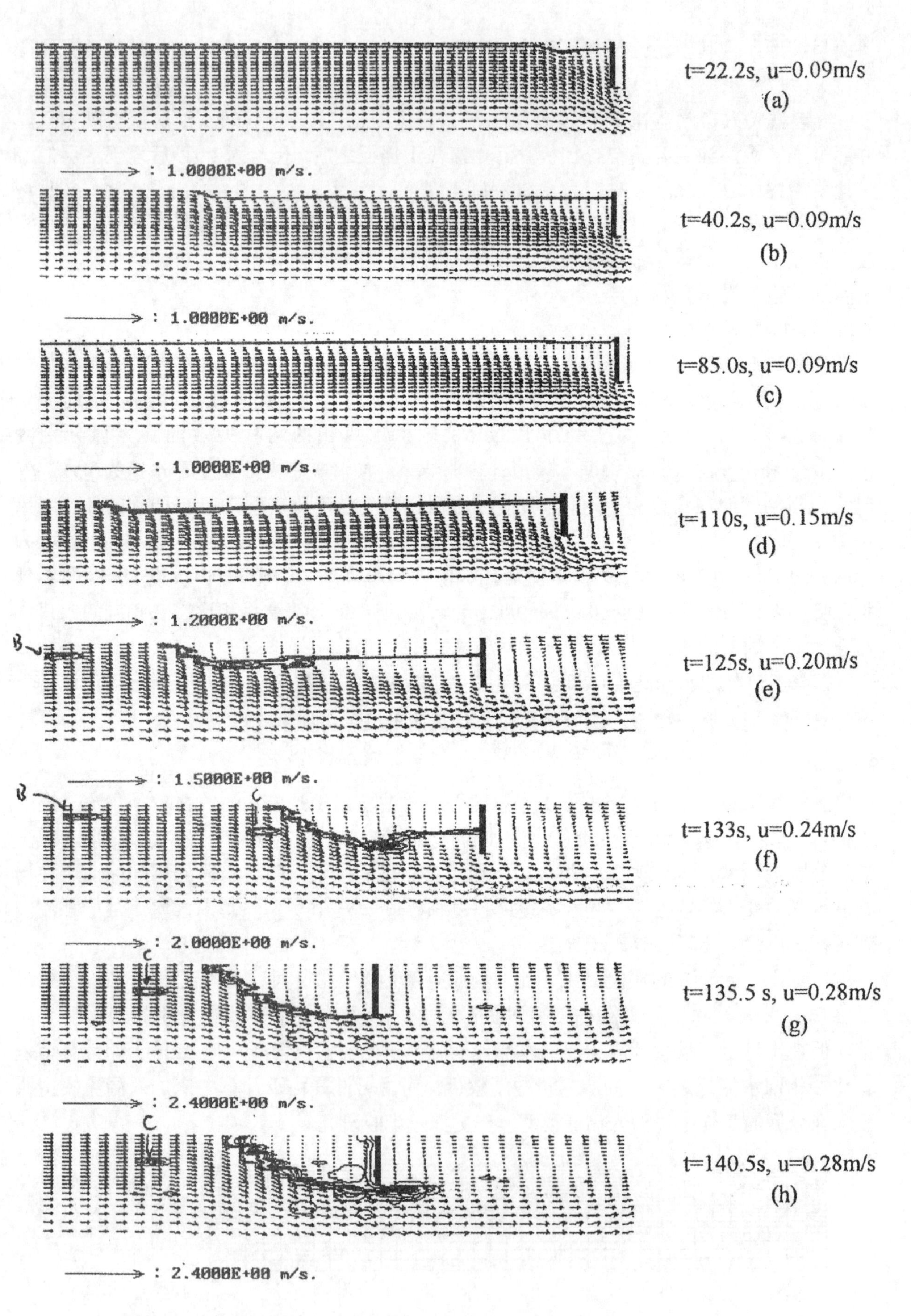

图3 不同水流速度下低粘度油在挡板前的动力学行为

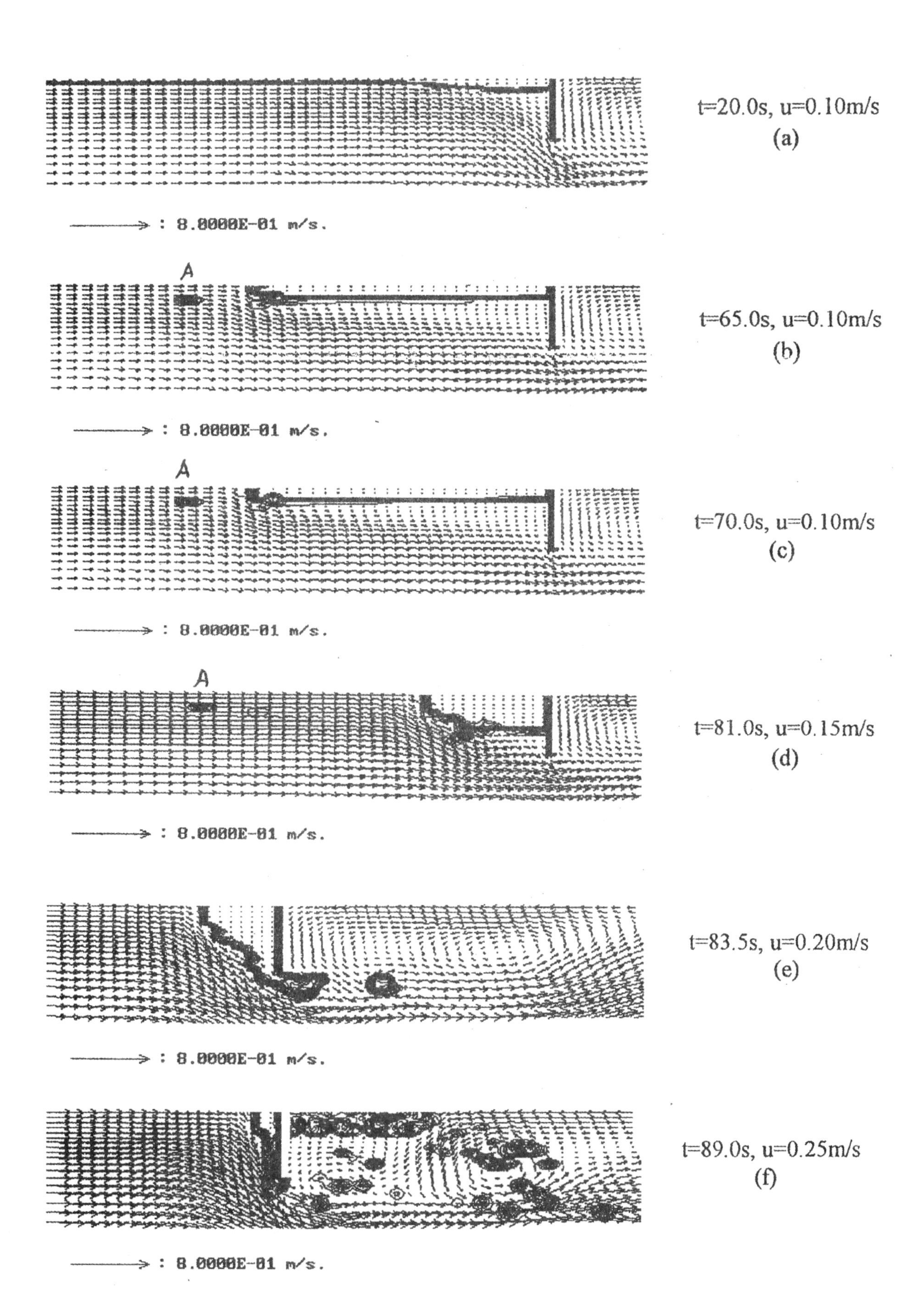

图4 不同水流速度下高粘度油在挡板前的动力学行为

参考文献:

1 · David P. Hoult, "Containment And Collection devices For Oil Slicks", Oil In Water,65-79,1965

2 · Delvigine, G.A.L.," Laboratory Experiments on Oil Spill Protection of a Water Intake:, Monograph: Oil in Freshwater, 446-458 (1985)

3 · Delvigine, G.A.L., " Barrier Failure by Critical Accumulation of Viscous Oil ", Proceedings Oil Spill Conference, San Antonio, 143-148 (1989)

4 · Johnston, A.J., Fitzmaurice, M.R. and Watt, R.G.M,, Oil spill containment: viscous oils, Proc. of 1993 International Oil Spill Conference, March 29 1993, Tampa, FL, USA, 89-94.

5 · Wilkinson, D., " Dynamics of Contained Oil Slicks " , Journal of the Hydraulics Division , ASCE, 98, 1014-1030 (1972)

6 · Wikinson, D., " Limitations to the Length of Contained Oil Slicks:, Journal of the Hydraulics Division, ASCE, 99, 701-712 (1973)

~~USA (1995)~~

7 · E.J. Clavelle and R.D. Rowe, "A Numerical Study of Oil Boom Failure by Critical Accumulation, " Proc. of 9th Conf. on Num. Methods in Laminar and Turbulent Flow", Atlanta, USA (1995)

8 · Chang-Fa An, "CFD Simulation of Oil Boom Failure vol(i)", Technical Report,1996

9 · Chang-Fa An, "CFD Simulation of Oil Boom Failure vol(ii)", Technical Report,1996

10 · C.-F. An, H.M. Brown, R.G. Goodman, R.D. Rowe, Z. Zhang, "Numerical Modeling of The Dynamics of An Oil Slick Spill On Flowing Water", 1997

11 · C.W.Hirt and B.D.Niichols, "Volume of Fluid (VOF) Method for the Dynamics of Free Boundaries", Journal of Computational Physics 39, 201-225 (1981)

12 · 范维澄，万跃硼．流动及燃烧的模型与计算，中国科学技术大学出版社

13 · W.E. Johnson, "Development and Application of Computer Programs Related to Hypervelocity Impact", Systems Science, an Software report 3SR-353, 1970.

MULTIPHASE FLUID, NON-NEWTONIAN FLUID AND PHYSICO-CHEMICAL FLUID FLOWS

Proceedings of the International Symposium on Multiphase Fluid, Non-Newtonian Fluid and Physico-Chemical Fluid Flows

(ISMNP '97)

October 7-10, 1997 Beijing, China

Edited by: ZHOU Lixing
LI Xiangfang

Sponsored by

Chinese Society of Theoretical and Applied Mechanics
Chinese Society of Petroleum Engineering
Chinese Society of Engineering Thermophysics
Chemical Industry and Engineering Society of China
National Natural Science Foundation of China

Organized by

University of Petroleum (Beijing)

Changfa. An
Beijing, China
Oct. 7, 1997

IAP International Academic Publishers

NUMERICAL MODELING OF THE DYNAMICS OF AN OIL SLICK SPILLED ON FLOWING WATER

C.-F. An, H. M. Brown, R. H. Goodman
Imperial Oil Resources Limited
3535 Research Road, N.W.
Calgary, Alberta, Canada T2L 2K8
Tel (403)284-7541 Fax (403)284-7595
E-mail: cfan@acs.ucalgary.ca

R. D. Rowe
Department of Mechanical Engineering
The University of Calgary
Calgary, Alberta, Canada T2N 1N4

Z. Zhang
Department of Chemical Engineering
Beijing University of Chemical Technology
Beijing, China 100029

ABSTRACT

Numerical modeling of oil slick dynamics behind a containment boom has been conducted based on experiment in a flowing channel. The modeling results are obtained using a commercial computational fluid dynamics (CFD) software, Fluent, and are in a good agreement with the channel experiment. The oil slick movement around the boom observed in the channel experiment has been accurately modeled. For the case of successful oil containment, the predicted and measured oil slick profiles are consistent with each other. For the case of oil loss, the predicted and measured critical velocities are identical. Therefore, the numerical modeling is well verified and validated by the channel experiment and can be used to guide the design of new booms.

I. INDUSTRIAL BACKGROUND

Oil plays an important role in our modern life, but oil pollution due to spill may jeopardize the environment, especially water resources. If oil is spilled on the surface of water, the most common response is to contain the oil by a boom and recover it by a skimmer. However, successful oil control can be attained only under certain conditions. If these conditions are not met, boom failure will occur and the oil will escape underneath the boom due to hydrodynamic forces. A number of boom failure mechanisms, identified as drainage failure, droplet entrainment and critical accumulation, have been studied during the past quarter of century, but the hydrodynamic mechanisms of boom failure are not well understood. In order to observe physical phenomena of oil contained behind a boom and to obtain first-hand experimental data, an experimental study was conducted in the Imperial Oil wave basin (Brown et al, 1997). Valuable observations and measurements have been acquired from this experiment. Computational studies of oil boom failure mechanisms have also been undertaken using computational fluid dynamics (CFD) by the authors (An et al, 1996 and Goodman et al, 1997). These efforts provide a possibility to model the dynamic characteristics of oil spilled on flowing water and controlled by a boom.

In this paper, numerical modeling of the dynamics of oil behind a boom on flowing water is reported. The modeling is carried out using a commercial CFD software, Fluent (1996), and based on the channel experiment. The unsteady Navier-Stokes equations with the standard k-ε turbulence modeling equations are solved using a finite volume scheme for space and implicit marching for time. In order to track the oil-water interface, the volume of fluid (VOF) model which is suitable for immiscible multi-phase flow is invoked. The modeling results are in a good agreement with the channel experiment. The evolution process of the oil slick movement has been successfully simulated. On the basis of the modeling results, computer animation has been made to demonstrate the oil slick movement.

II. PHYSICAL DESCRIPTION AND MATHEMATICAL FORMULATION

In general, the movement of an oil slick on the surface of flowing water is a complicated three dimensional unsteady turbulent flow of multi-phase fluids. Three fluids are involved: oil, water and air. If the weather is calm, then the wind and waves do not affect the flow significantly. The difference between the elevations of oil and water is negligibly small according to experimental observation. Thus, the oil and water surfaces can be treated as a single flat plane. Furthermore, the friction effects on the oil-air and water-air interfaces are small and can be neglected compared to that on the oil-water interface. As a result, the third phase (air) can be removed from the system and the oil-air and water-air interfaces can be assumed as a flat frictionless plane, i.e. a slip wall. The system, therefore, is reduced from a three-phase flow to a two-phase flow or, to be precise, a stratified flow of two liquids with different physical properties. In addition, the compressibility of oil and water is small so that both of them can be treated as incompressible fluids. The mutual solubility between oil and water is also small and the emulsification of oil in water is not significant during a short period of time after a spill. Hence, the two liquids can be considered as two immiscible fluids for which a simple model with adequate accuracy is available, i.e. volume of fluid (VOF) model. For the case of a water channel flow in the wave basin experiment, the minimum Reynolds number is 100,000 based on the water depth, flow velocity and water viscosity. Obviously, the flow is turbulent at least within the water region. Although the flow within the oil region may be laminar, turbulent flow must be considered for the whole region of the flow. As the flow can be treated as homogeneously turbulent, the standard k-ε model is suitable. In order to trace the oil slick movement, the flow should be considered unsteady. Strictly speaking, the flow of an oil slick controlled by a deployed boom on flowing water is always three dimensional. Compared to the two dimensional case, three dimensional computation requires more space and a faster computer. However, the present problem can be reduced to a two dimensional one. The ideal planar shape of the deployed boom is a catenary curve. If the water current is uniform, the maximum normal velocity to the boom is in the middle part of the boom, i.e. at the apex of the catenary. In other words, the critical section of the boom with threat of oil loss is the middle section which can be considered as a two dimensional plane. Making a compromise between accuracy and economy, a two dimensional model is relevant for the problem of oil slick movement around a boom.

For a two dimensional unsteady flow of a viscous incompressible fluid without heat and mass transfer and chemical reaction, the governing equations are the Navier-Stokes equations:

$$\frac{\partial u}{\partial x}+\frac{\partial v}{\partial y}=0 \tag{2.1}$$

$$\rho[\frac{\partial u}{\partial t}+\frac{\partial(u^2)}{\partial x}+\frac{\partial(uv)}{\partial y}]=-\frac{\partial p}{\partial x}+\mu[2\frac{\partial^2 u}{\partial x^2}+\frac{\partial}{\partial y}(\frac{\partial u}{\partial y}+\frac{\partial v}{\partial x})] \tag{2.2}$$

$$\rho[\frac{\partial v}{\partial t}+\frac{\partial(uv)}{\partial x}+\frac{\partial(v^2)}{\partial y}]=-\frac{\partial p}{\partial y}+\mu[\frac{\partial}{\partial x}(\frac{\partial v}{\partial x}+\frac{\partial u}{\partial y})+2\frac{\partial^2 v}{\partial y^2}]+\rho g \tag{2.3}$$

where x and y are Cartesian coordinates, t is time, ρ is density, p is static pressure, u and v are velocity components in x and y directions, respectively, g is the gravitational acceleration and μ is the molecular viscosity which is assumed to be a constant.

Theoretically, the Navier-Stokes equations are valid for both laminar and turbulent flows. In practice, however, direct numerical simulation (DNS) of turbulent flow is currently studied at the preliminary research stage (Wilcox, 1994). The most common technique to treat turbulent flow is to apply an appropriate turbulence model. Among numerous turbulence models, the standard k-ε model is suitable for the present problem. After Reynolds averaging operations, equations (2.1)-(2.3) become:

$$\frac{\partial U}{\partial x}+\frac{\partial V}{\partial y}=0 \tag{2.4}$$

$$\rho[\frac{\partial U}{\partial t}+\frac{\partial(U^2)}{\partial x}+\frac{\partial(UV)}{\partial y}]=-\frac{\partial P}{\partial x}+\mu_{eff}[2\frac{\partial^2 U}{\partial x^2}+\frac{\partial}{\partial y}(\frac{\partial U}{\partial y}+\frac{\partial V}{\partial x})] \tag{2.5}$$

$$\rho[\frac{\partial V}{\partial t}+\frac{\partial(UV)}{\partial x}+\frac{\partial(V^2)}{\partial y}]=-\frac{\partial P}{\partial y}+\mu_{eff}[\frac{\partial}{\partial x}(\frac{\partial V}{\partial x}+\frac{\partial U}{\partial y})+2\frac{\partial^2 V}{\partial y^2}]+\rho g \tag{2.6}$$

where the capital letters represent the mean flow variables and the effective viscosity is defined as the sum of the molecular viscosity and turbulent viscosity:

$$\mu_{eff}=\mu+\mu_t \tag{2.7}$$

Equations (2.4)-(2.6) have the same format as equation (2.1)-(2.3) except that the molecular viscosity is replaced by the effective viscosity. In the standard k-ε model, the turbulent viscosity is determined from

$$\mu_t=\rho C_\mu\frac{k^2}{\epsilon} \tag{2.8}$$

where k is the turbulent kinetic energy per unit mass

$$k=\frac{1}{2}(\overline{u'^2}+\overline{v'^2}) \tag{2.9}$$

and ε is the dissipation rate of turbulence

$$\epsilon=\nu[\overline{(\frac{\partial u'}{\partial x})^2}+\overline{(\frac{\partial u'}{\partial y})^2}+\overline{(\frac{\partial v'}{\partial x})^2}+\overline{(\frac{\partial v'}{\partial y})^2}] \tag{2.10}$$

The values of k and ε can be solved from their transport equations (Wilcox, 1994):

$$\rho[\frac{\partial k}{\partial t}+\frac{\partial(Uk)}{\partial x}+\frac{\partial(Vk)}{\partial y}]=\frac{\partial}{\partial x}(\frac{\mu_t}{\sigma_k}\frac{\partial k}{\partial x})+\frac{\partial}{\partial y}(\frac{\mu_t}{\sigma_k}\frac{\partial k}{\partial y})+G-\rho\epsilon \tag{2.11}$$

$$\rho[\frac{\partial \epsilon}{\partial t}+\frac{\partial(U\epsilon)}{\partial x}+\frac{\partial(V\epsilon)}{\partial y}]=\frac{\partial}{\partial x}(\frac{\mu_t}{\sigma_\epsilon}\frac{\partial \epsilon}{\partial x})+\frac{\partial}{\partial y}(\frac{\mu_t}{\sigma_\epsilon}\frac{\partial \epsilon}{\partial y})+C_{1\epsilon}\frac{\epsilon}{k}G-C_{2\epsilon}\rho\frac{\epsilon^2}{k} \tag{2.12}$$

where

$$G=\mu_t[2(\frac{\partial U}{\partial x})^2+(\frac{\partial U}{\partial y}+\frac{\partial V}{\partial x})^2+2(\frac{\partial V}{\partial y})^2] \tag{2.13}$$

and the empirical constants are chosen to be

$$C_\mu=0.09,\ C_{1\epsilon}=1.44,\ C_{2\epsilon}=1.92,\ \sigma_k=1.0,\ \sigma_\epsilon=1.3 \tag{2.14}$$

To handle a multi-phase fluid flow, the volume of fluid (VOF) model can be used for the two immiscible fluids (Hirt and Nichols, 1981). Unlike other multi-phase fluid models, the VOF model solves a single set of equations which is shared by all phases. To distinguish different phases, a scalar variable F (the volume fraction of water in a computational cell) is introduced and defined as the ratio of water volume in a cell to the total volume of the cell. In general, F is a function of time and space and governed by the conservation law:

$$\frac{\partial F}{\partial t}+U\frac{\partial F}{\partial x}+V\frac{\partial F}{\partial y}=0 \qquad (2.15)$$

In the Navier-Stokes equations (2.4)-(2.6) and the turbulence modeling equations (2.11)-(2.13), the velocity components U and V, the static pressure P and the turbulent quantities k and ε are shared by oil, water and their mixture in a cell. However, the physical properties of the material in the cell are distinguished through F:

$$\rho=F\rho_w+(1-F)\rho_o \qquad (2.16)$$

$$\mu=F\mu_w+(1-F)\mu_o \qquad (2.17)$$

where the subscripts "w" and "o" represent water and oil, respectively. Based on the local values of F, the interface between oil and water can be traced easily.

III. MODELING RESULTS COMPARED WITH CHANNEL EXPERIMENT

A water channel of 30m long, 1m wide and 1m deep was used to conduct the oil control and boom failure experiment in the wave basin. In the middle of the channel, a test section was constructed consisting of a flat plate barrier (as a boom), a flow velocity measurement device, an oil slick thickness measurement device and an observation window. The boom draft can be adjusted up to 0.5m deep below the water surface. Three types of oil were spilled onto the channel surface. For each type of oil, four boom drafts were set. For each boom draft, three flow velocities were taken. Therefore, a total of 36 runs of oil slick profile measurements and boom failure observations were made. Some of the experimental results have been reported by Brown et al (1997).

On the basis of the wave basin experiment, numerical modeling has been executed using Fluent. The computational domain is a two dimensional rectangle 8m long and 1m deep to simulate the water channel. The boom is put at 3m distance from the inlet. The computational grid is 241x61 and clustered so that the mesh is finer in the neighborhood of the boom while it is coarser elsewhere. All computations are carried out in the category of turbulence and the standard k-ε model is applied. To record the history of the oil slick movement, all computations are performed on the time-dependent basis. At the beginning of the computation, the time step is set to be 0.05sec and the maximum number of iterations is set to be 20. After some iterations, the computation could be oscillatory, especially when the oil reaches the boom. If this occurs, the time step is decreased accordingly. The maximum normalized residual of the governing equations is set to be 10^{-3}. In order to simulate the two phases, the VOF model is enabled. The upper boundary of the channel is assumed to be a slip wall on which the tangent velocity component is permitted but the normal velocity component is prohibited. On the boom surface and the channel bottom, no-slip walls are assumed. At the inlet, the water flow is started at 0.10m/s and the oil is floated on the water surface. It flows to the boom with the same velocity as the water until a certain amount of oil has been introduced. After that moment, the oil introduction is stopped and only water flows into the channel. Also at the inlet, the turbulence intensity is set to be 2% and the characteristic length is taken to be the channel depth. The initial guess of the flow field within the channel is made so that the velocity is uniform everywhere and equal to that at the inlet.

Deep Boom Case

Three types of oil were used in the channel experiment, but only the heavy viscous oil, weathered Hondo, was used in the numerical modeling. The density of this oil is 952.6 kg/m^3 and the kinematic viscosity is 5616 cSt. The amount of spilled oil is 10 liters in the experiment and 0.01m^3/m in the modeling. The boom draft is relatively deep (0.15m) and the flow velocity begins with 0.10m/s. Figure 1 shows the transient flow fields at simulation times of 15.20sec - 60.26sec. At 15.20sec, the oil layer drifts towards the boom. At 32.20sec, the oil reaches the boom and accumulates against the boom. At 46.20sec, the oil continues to accumulate. Comparing the flow fields at 58.26sec and 60.26sec, the oil slick shape has changed little. This means that the oil slick has become stable and the flow field has attained its steady state. The comparison of the oil slick profiles between experiment and computation is given in Figure 4(a). In this figure, the horizontal line at 1m height represents the channel surface. One can see from this figure that the modeling result is accurate compared to the channel experiment for this case.

After a steady state flow at 0.10m/s is achieved, the flow is accelerated to 0.15m/s. Figure 2 shows the evolution of the oil slick movement from simulation time 60.26sec to 77.86sec. It can be seen that the oil slick is compressed continually. Some oil (possibly oil droplets) is torn from the head wave, but there is no significant oil loss. Figure 4(b) gives the comparison of the oil slick profiles between the experiment and the computation. Again, the accuracy of the simulation in this case is satisfactory.

Next, the flow velocity is increased to 0.20m/s from the previous flow field. The transient flow fields during the period of 77.86sec to 95.00sec are shown in Figure 3. From these graphs one can see that the oil-water interface, especially the head wave, is not calm due to strong compression. Nevertheless, there is no apparent oil loss in this case. Figure 4(c) presents the comparison of the oil slick profiles between the experiment and the computation. It can be seen that the predicted result has the same order as the experimental result.

Shallow Boom Case

In this case, the heavy viscous oil, weathered Hondo, was also spilled but the boom draft was relatively shallow

(0.05m) and the flow velocity began at 0.10m/s. Figure 5 shows the evolution process of the oil slick movement from 32.80sec to 78.05sec. At 32.80sec, the oil layer approaches the boom. At 42.80sec, the oil accumulates against the boom. At 50.50sec, the oil slick begins to reflect and the oil slick gets longer after this moment. Comparing the flow fields at 68.50sec and 78.05sec, the oil slick becomes stable and the flow attains its steady state. A small amount of oil loss at the moment 42.80sec is probably due to dynamic effects and there is no oil loss afterwards. Figure 8(a) presents the comparison of the oil slick profiles between the experiment and the computation. The predicted result is of the same order as the measured result in the experiment.

After the steady state flow is established at 0.10m/s, the flow velocity is accelerated to 0.15m/s. Figure 6 shows the process of oil slick compression during the period from 78,05sec to 90.19sec. The oil slick is compressed continually and, eventually, it becomes stable and the flow reaches its steady state (cf. graphs at 89.19sec and 90.19sec). Although some oil escapes under the boom, there is no substantial oil loss in this case. Figure 8(b) shows the comparison of the oil slick profiles between the experiment and the computation. The accuracy of the simulation is acceptable for this case compared with the channel experiment.

Next, the flow velocity is increased to 0.20m/s. Figure 7 shows the successive transient flow fields at a time sequence of 90.19sec to 98.67sec. The oil slick cannot become stable and the flow cannot reach a steady state anymore for this case. At 91.22sec, the oil slick gets shorter and thicker. At 92.72sec, the oil begins to creep underneath the boom. At 94.22sec, more oil is creeping underneath the boom. At 98.67sec, a substantial amount of oil has crept under the boom and has floated on the surface of the water. It can be expected that more and more oil will pass the boom from then on. This modeled result exactly coincides with the process of boom failure observed in the channel experiment. The boom failure mechanism for this case in the experiment was identified as "critical accumulation" according to the definition of Delvigne (1989) or, more descriptively, "oil creep" according to the suggestion of Johnston et al (1993). The dynamic process of oil slick movement in the experiment has been recorded on a video tape (Brown et al, 1997) and has also been animated based on the corresponding modeling (An et al, 1997).

IV. CONCLUSIONS

The numerical modeling of oil slick dynamics shows that the modeling results are in a good agreement with the channel experiment.

The evolution process of oil slick movement around a boom observed in the channel experiment has been accurately modeled. This process includes: oil spill on the surface of flowing water; oil drifting towards the boom; formation of a stable oil slick if oil control is successful and oil loss if boom failure occurs.

For the case of successful oil control, the predicted oil slick profiles are consistent with the measured ones along the channel centerline.

For the case of boom failure, the predicted critical velocities from the simulation are identical to the experimental observations.

It can be concluded that the CFD software Fluent proves to be an accurate and efficient tool to guide the design of new booms.

Keywords: Multi-phase Flow, Petroleum Engineering, Environmental Engineering, CFD

Acknowledgments

The authors would like to acknowledge Imperial Oil Resources Limited for the financial support on this project. The first author is especially grateful to the Natural Sciences and Engineering Research Council of Canada for an Industrial Research Fellowship.

References

An, C.-F., Brown, H.M., Goodman, R.H. and Clavelle, E.J., 1997, "Animation of Boom Failure Processes," *Proc. of 20th Arctic & Marine Oilspill Program (AMOP) Technical Seminar*, Vancouver, BC, Canada, Vol. 2, pp. 1181-1188

An, C.-F., Clavelle, E.J., Brown, H.M. and Barron, R.M., 1996, "CFD Simulation of Oil Boom Failure," *Proc. of 4th Annual Conference of the CFD Society of Canada*, Ottawa, ON, Canada, pp. 401-408

Brown, H.M., Goodman, R.H., An, C.-F. and Bittner, J., 1997, "Boom Failure Mechanisms: Comparison of Channel Experiments with Computer Modeling Results," *Proc. of 20th Arctic & Marine Oilspill Program (AMOP) Technical Seminar*, Vancouver, BC, Canada, Vol. 1, pp. 457-467

Delvigne, G.A.L., 1989, "Barrier Failure by Critical Accumulation of Viscous Oil," *Proc. of 1989 Oil Spill Conference*, San Antonio, TX, USA, pp. 143-148

Fluent User's Guide, version 4.4, 1996, Fluent, Inc., Lebanon, NH, USA

Goodman, R.H., Brown, H.M., An, C.-F. and Rowe, R.D., 1997, "Dynamic Modeling of Oil Boom Failure Using Computational Fluid Dynamics," *Proc. of 20th Arctic & Marine Oilspill Program (AMOP) Technical. Seminar*, Vancouver, BC, Canada, Vol. 1, pp. 437-455

Hirt, C.W. and Nichols, B.D., 1981, "Volume of Fluid (VOF) Method for the Dynamics of Free Boundaries," *J. Comp. Physics*, Vol. 39, pp. 201-225

Johnston, A.J., Fitzmaurice, M.R. and Watt, R.G.M., 1993, "Oil Spill Containment: Viscous Oils," *Proc. of 1993 International Oil Spill Conference*, Tampa, FL, USA, pp. 89-94

Wilcox, D.C., 1994, "Turbulence Modeling for CFD," DCW Industries, Inc., La Canada, CA, USA

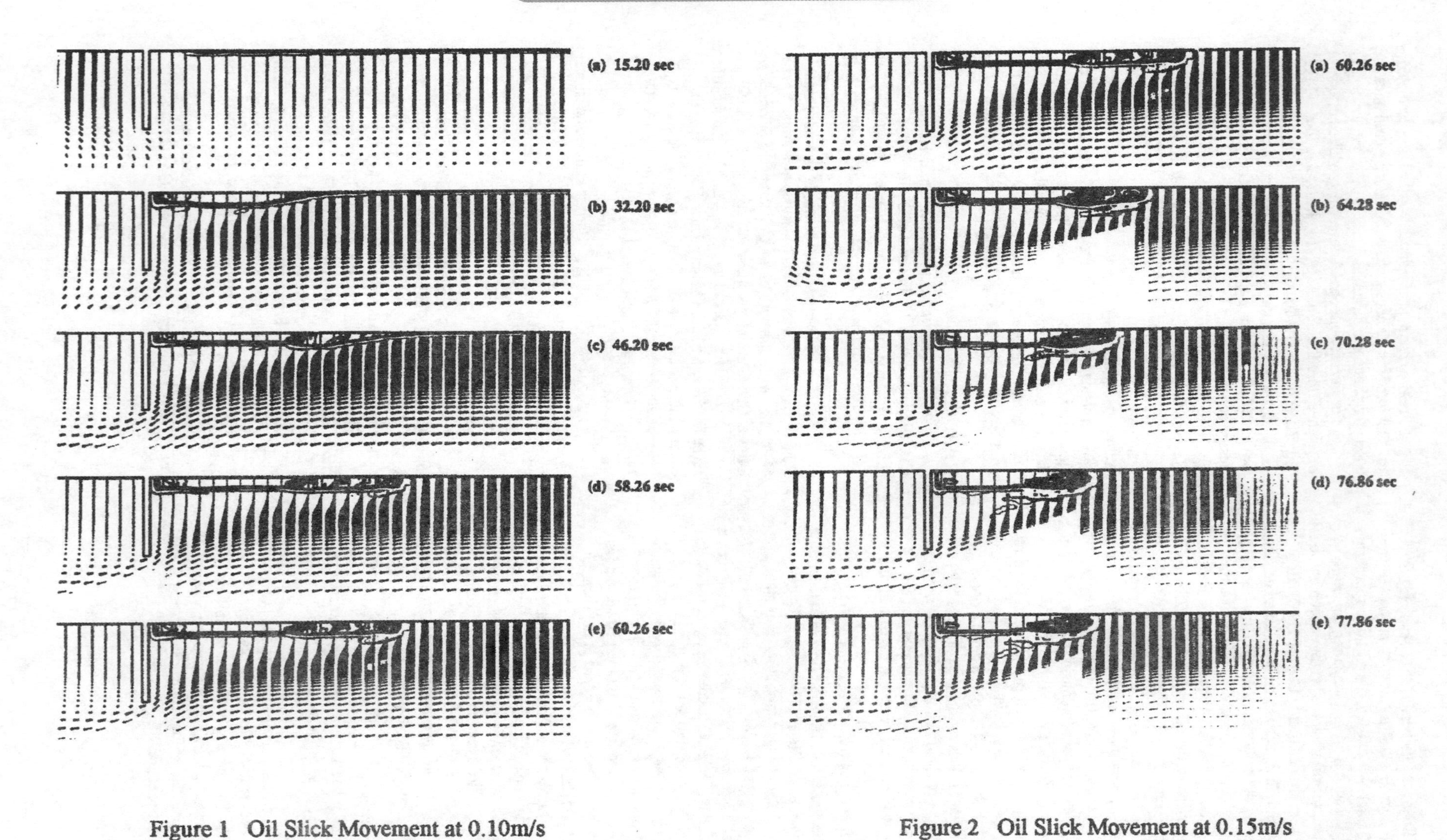

Figure 1 Oil Slick Movement at 0.10m/s

Boom Draft: 0.15m

Figure 2 Oil Slick Movement at 0.15m/s

Boom Draft: 0.15m

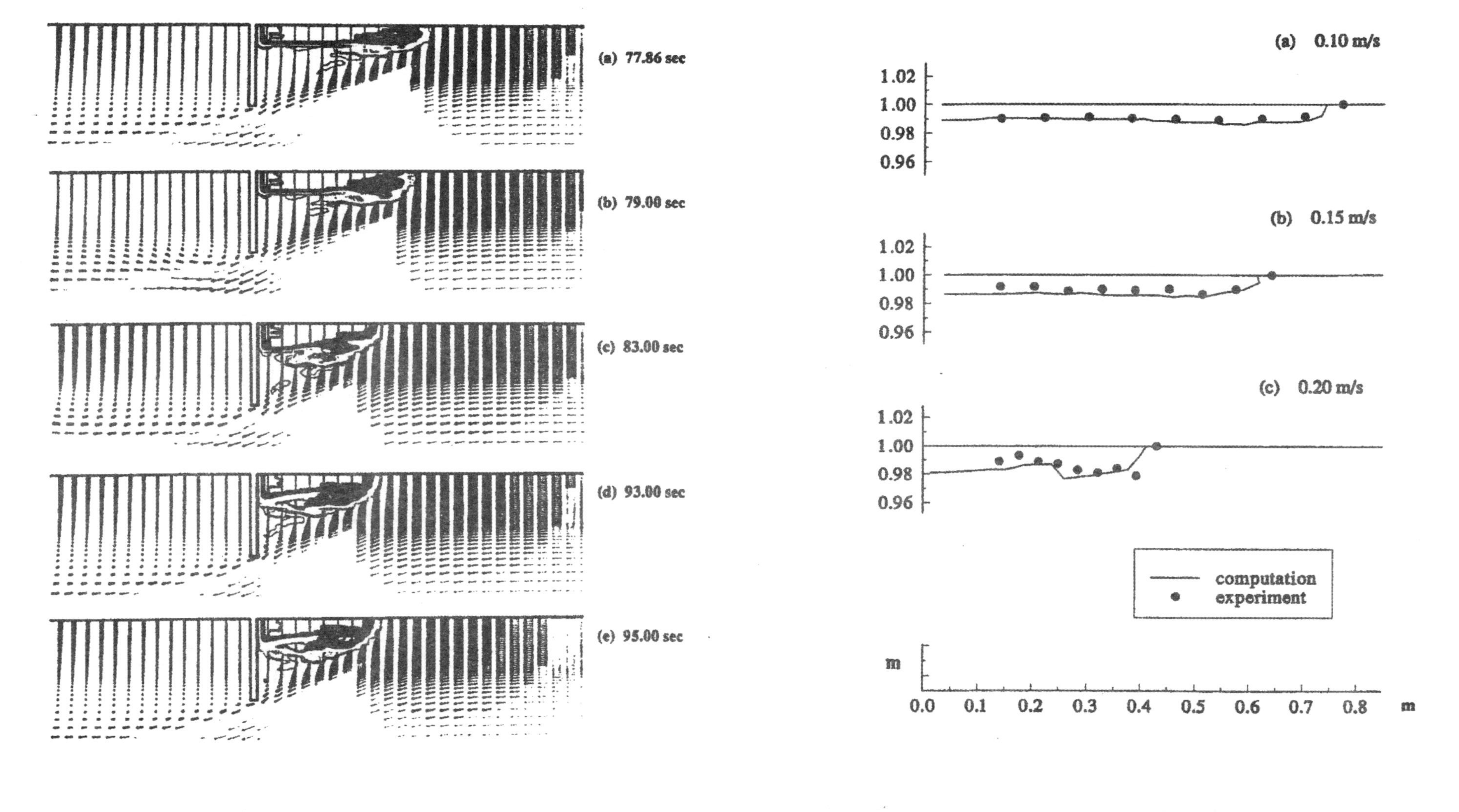

Figure 3 Oil Slick Movement at 0.20m/s

Boom Draft: 0.15m

Figure 4 Comprison of Oil Slick Profiles

Boom Draft: 0.15m

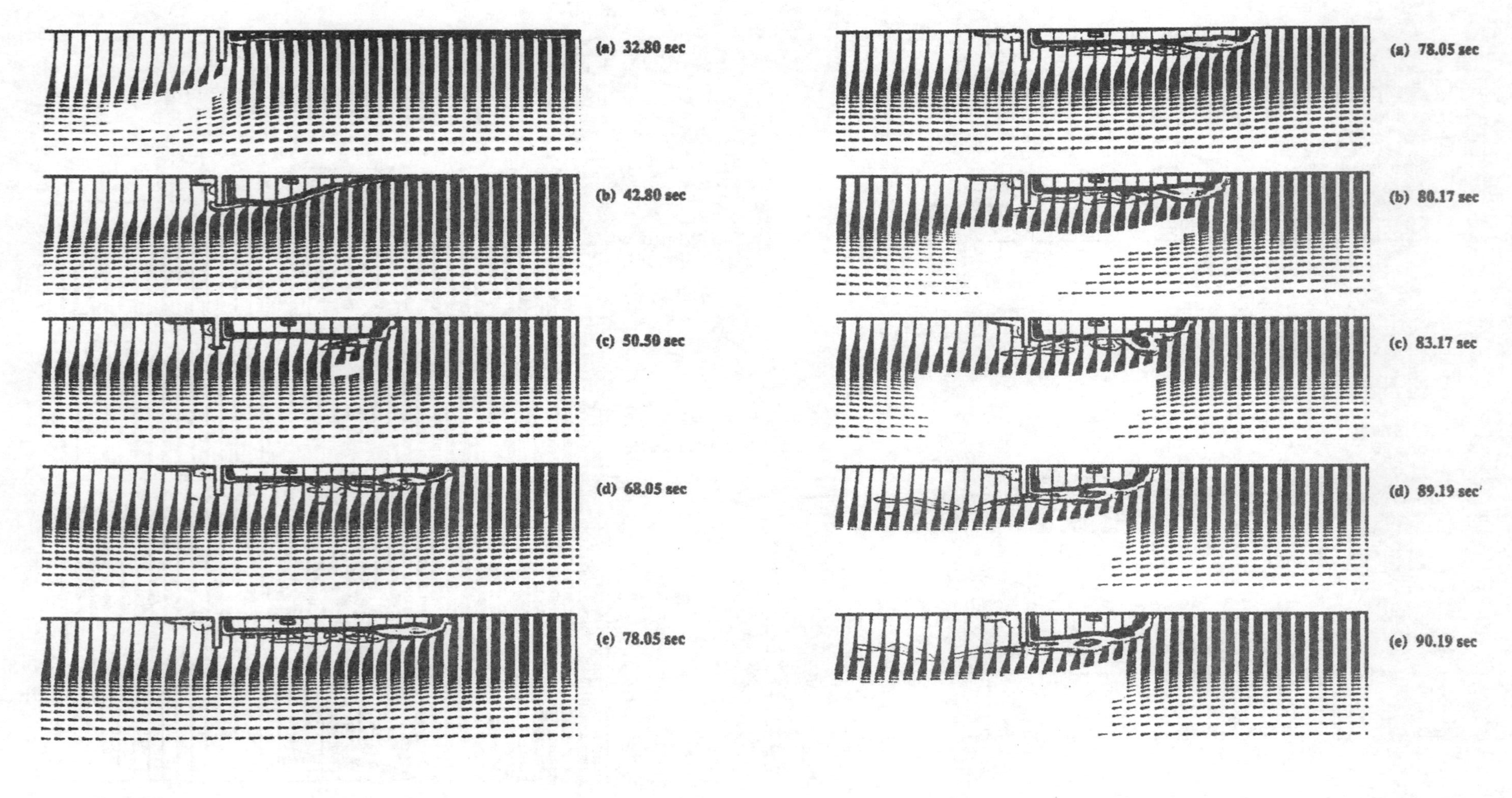

Figure 5 Oil Slick Movement at 0.10m/s

Boom Draft: 0.05m

Figure 6 Oil Slick Movement at 0.15m/s

Boom Draft: 0.05m

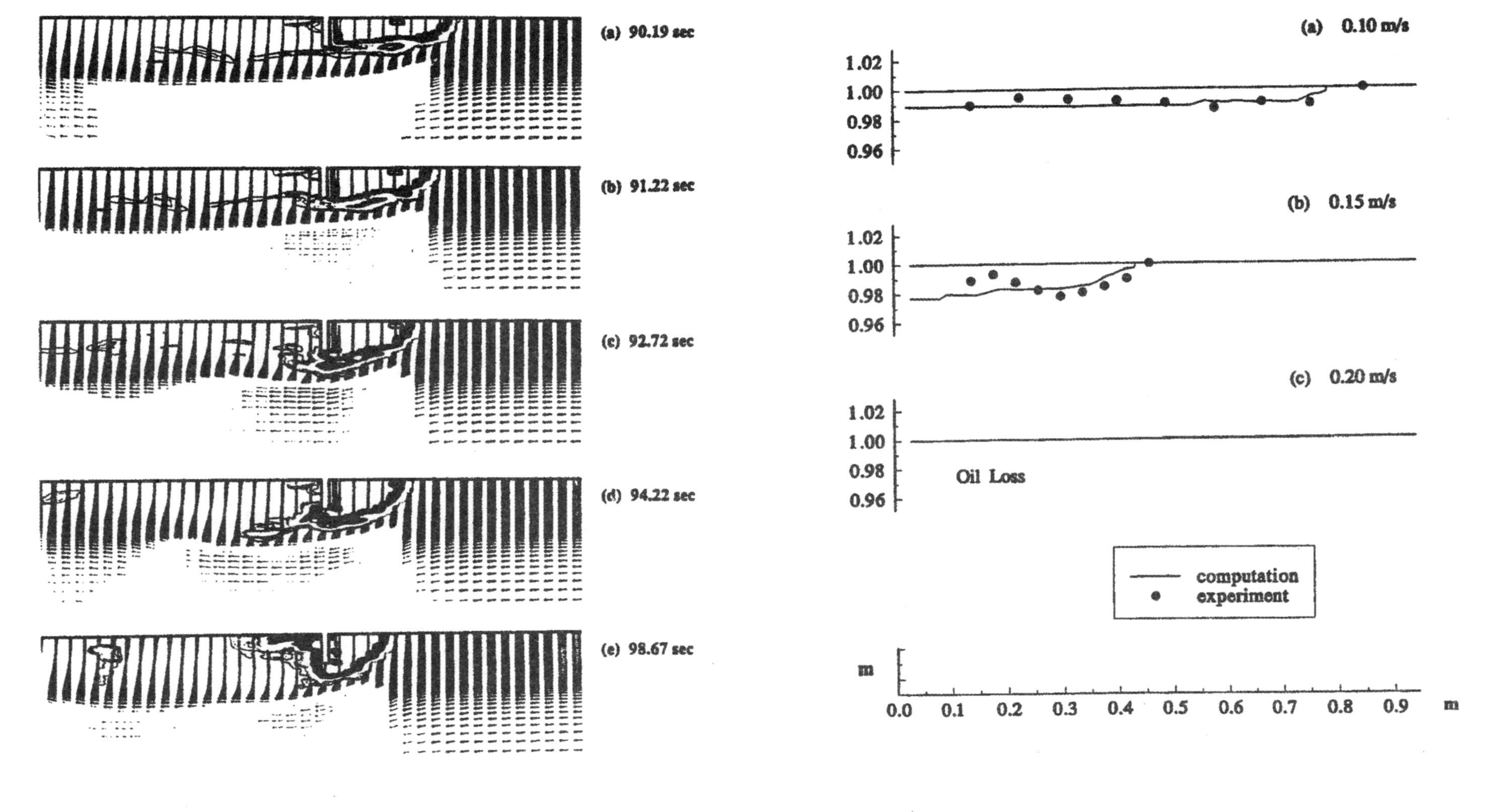

Figure 7 Oil Slick Movement at 0.20m/s

Boom Draft: 0.05m

Figure 8 Comprison of Oil Slick Profiles

Boom Draft: 0.05m

Chang-Fa A.
Dec. 1, 1997
Kyoto, Japan

Transport Phenomena
in
Thermal Science
and
Process Engineering

Volume 1

Plenary Session
Sessions 1-8
Poster Session 1

November 30 - December 3, 1997
Kyoto Research Park, Kyoto, Japan

OIL-WATER INTERFACIAL PHENOMENA BEHIND A BOOM ON FLOWING WATER

C.-F. An, R.H. Goodman & H.M. Brown
Imperial Oil Resources Limited
3535 Research Road, N.W.
Calgary, Alberta, T2L 2K8
Canada

E.J. Clavelle
Dept. of Mechanical Engineering
The University of Calgary
Calgary, Alberta, T2N 1N4
Canada

R.M. Barron
Fluid Dynamics Research Institute
University of Windsor
Windsor, Ontario, N9B 3P4
Canada

ABSTRACT

In this paper, a computational study of oil-water interfacial phenomena of an oil slick contained by a boom on flowing water is presented. The emphasis is put on the Kelvin-Helmholtz (K-H) instability, and its consequence, droplet entrainment boom failure. The K-H instability is analyzed using a small disturbance theory and simulated using a computational fluid dynamics software Fluent. The results of analysis and simulation for an example of K-H instability are consistent with each other. A numerical modeling of droplet entrainment based on an experiment is carried out and the agreement between computational prediction and experimental observation is quite satisfactory.

1. INTRODUCTION

Oil spills occur from time to time during the process of oil production, storage and transportation. The pollution due to oil spills severely jeopardizes the environment, especially the water resources. For example, a recent oil spill in Japan threatened the whole country and a national emergency was called by the Prime Minister (Calgary Herald, 1997). In this accident, the supertanker "Diamond Grace" ran aground and released about 1300 tonnes of crude oil into Tokyo Bay, one of the most populated areas in the world. Therefore, a prompt and efficient response to an oil spill and a proper method for cleanup are of great importance. If oil is spilled on the surface of water, the usual response is to contain the oil with booms and then recover it with skimming devices. However, successful oil containment can be attained only when the relative boom-water (oil-water) velocity is below a certain critical value. If the relative velocity is in excess of the critical value, boom failure will occur and the oil will escape underneath the boom due to hydrodynamic forces. During the past three decades, a number of boom failure mechanisms have been identified and studied, including drainage failure, droplet entrainment, splash-over and critical accumulation. The hydrodynamic processes for different boom failure mechanisms are quite different. The physical phenomena at the oil-water interface are very complex when boom failure occurs, especially for the case of droplet entrainment.

The phenomenon of droplet entrainment was recognized 28 years ago (Wicks, 1969) and was described comprehensively by Delvigne (1989). The onset of droplet formation from the oil-water interface was interpreted to be the breaking of interfacial waves which result from the amplifying of Kelvin-Helmholtz waves (Leibovich, 1976). In other words, the hydrodynamic mechanism of droplet entrainment failure is due to Kelvin-Helmholtz (K-H) instability.

The analysis of K-H instability can be traced

to the pioneering work of Kelvin and Helmholtz in the last century. Lamb(1932) first formulated the mathematical description of K-H instability in his well-known book. Several scholars showed interest in this topic in their monographs: Birkhoff and Zarantonello(1957), Chandrasekhar(1961), Yih(1965), Drazin and Howard(1966) and Pai(1977), while others studied K-H instability analytically, experimentally or numerically in their papers: Squire(1933), Thorpe(1969), Drazin(1970), Patnaik et al(1976) and Zalosh(1976). The following investigations of oil boom failure relating to K-H instability have been conducted: Jones(1972), Zalosh and Jensen(1975), Kordyban(1990), Rangel and Sirignano(1991), Ertekin and Sundararaghvan(1995) and Grilli et al(1996). In spite of these efforts on studying the phenomena of K-H instability and droplet entrainment, very little work has been done using computational fluid dynamics (CFD).

In this paper, a computational study of oil-water interfacial phenomena behind a boom on flowing water is presented. The emphasis is put on the mechanism of droplet entrainment failure and related K-H instability. In section 2, K-H instability is analyzed theoretically and simulated numerically. In section 3, a case of droplet entrainment failure is modeled and compared to the corresponding experiment. In section 4, concluding remarks are drawn and future studies are suggested.

2. KELVIN-HELMHOLTZ INSTABILITY

The Kelvin-Helmholtz instability arises from the characteristics of the equilibrium of a stratified heterogeneous fluid flow when different layers are in relative motion. Following Lamb(1932) and Yih(1965), we will analyze the critical condition at which K-H instability occurs.

Suppose that we have two layers of fluids with densities ρ and ρ', one (ρ) beneath the other (ρ'), moving parallel to the horizontal x direction with velocities U and U', respectively. The interface between the two fluids (when undisturbed) is plane and horizontal chosen as $y = 0$. Based on the steady motion of fluids, a small perturbation wave in the vertical y direction is applied. Assuming the fluids are incompressible and inviscid, and the flow is irrotational, velocity potentials $\phi(x, y, t)$ and $\phi'(x, y, t)$ can be defined as the sums of the steady potentials Ux and $U'x$ plus the perturbation potentials $\phi_1(x, y, t)$ and $\phi_1'(x, y, t)$, i.e.

$$\phi = Ux + \phi_1, \quad \phi' = U'x + \phi_1' \tag{1}$$

where ϕ_1 and ϕ_1' are small by hypothesis. The velocity of either fluid at the interface may be regarded as made up of the velocity of this surface itself, and the velocity of the fluid relative to it. Letting $\eta(x, t)$ be the y coordinate of the displaced surface, assumed to be small, the following kinematic conditions for vertical velocity components hold at $y = 0$ in the lower fluid,

$$\frac{\partial \eta}{\partial t} + U\frac{\partial \eta}{\partial x} = \frac{\partial \phi}{\partial y} \tag{2}$$

and in the upper fluid,

$$\frac{\partial \eta}{\partial t} + U'\frac{\partial \eta}{\partial x} = \frac{\partial \phi'}{\partial y} \tag{3}$$

where quadratic terms in η, ϕ_1 and ϕ_1' are neglected. Since the flow is assumed to be irrotational, the Bernoulli quation can be used to evaluate pressure in the lower fluid,

$$p = -\rho(\frac{\partial \phi_1}{\partial t} + U\frac{\partial \phi_1}{\partial x} + gy) \tag{4}$$

and in the upper fluid,

$$p' = -\rho'(\frac{\partial \phi_1'}{\partial t} + U'\frac{\partial \phi_1'}{\partial x} + gy). \tag{5}$$

At the interface $y = \eta$, pressure is continuous, i.e.

$$\rho(\frac{\partial \phi_1}{\partial t} + U\frac{\partial \phi_1}{\partial x} + g\eta) = \rho'(\frac{\partial \phi_1'}{\partial t} + U'\frac{\partial \phi_1'}{\partial x} + g\eta). \tag{6}$$

Assuming both fluids to be of infinite depth, the perturbation potentials can be written as

$$\phi_1 = Ce^{ky+i(\sigma t - kx)}, \quad \phi_1' = C'e^{-ky+i(\sigma t - kx)} \tag{7}$$

and the displaced interface,

$$\eta = ae^{i(\sigma t - kx)} \tag{8}$$

where k is wave number which is related to wave length λ by $k = 2\pi/\lambda$ and σ is the circular frequency of the perturbation wave. It is obvious that the perturbation potentials ϕ_1 and ϕ_1' in (7) satisfiy the boundary conditions at infinity: $\phi_1 \to 0$ as $y \to -\infty$ and $\phi_1' \to 0$ as $y \to +\infty$. From equation (8) one can see that if, for some reason, σ becomes a complex number with a negative imaginary part, a factor of η will be e^{Rt}

where R is a positive real number. As a result, the amplitude of the perturbation wave ae^{Rt} will increase without limit. That is, the interfacial wave will become unstable and K-H instability will occur. Substituting (7) and (8) into (2) and (3), one gets

$$ia(\sigma - kU) = kC, \quad ia(\sigma - kU') = -kC'. \quad (9)$$

Substituting (7) and (8) into (6) gives

$$\rho[i(\sigma - kU)C + ga] = \rho'[i(\sigma - kU')C' + ga]. \quad (10)$$

Solving (9) for C and C', substituting them into (10) and eliminating a, we have

$$\rho(\sigma - kU)^2 + \rho'(\sigma - kU')^2 = gk(\rho - \rho'). \quad (11)$$

Solving the quadratic equation (11) for σ, we obtain

$$\frac{\sigma}{k} = \frac{\rho U + \rho' U'}{\rho + \rho'} \pm \sqrt{\frac{g}{k}\frac{\rho - \rho'}{\rho + \rho'} - \frac{\rho\rho'(U - U')^2}{(\rho + \rho')^2}}. \quad (12)$$

The first term on the right hand side may be called the mean velocity of the two fluids. Relative to this there are waves travelling with velocities

$$c = \pm\sqrt{c_0^2 - \frac{\rho\rho'(U - U')^2}{(\rho + \rho')^2}} \quad (13)$$

where

$$c_0 = \pm\sqrt{\frac{g}{k}\frac{\rho - \rho'}{\rho + \rho'}} \quad (14)$$

is the wave-velocity in the absence of relative motion between the two fluids, i.e. $U - U' = 0$. It is easily noticed that the values of σ given by (12) are imaginary if

$$(U - U')^2 > \frac{g}{k}\frac{\rho^2 - \rho'^2}{\rho\rho'}. \quad (15)$$

This is the condition under which the interface becomes unstable and K-H instability occurs.

In order to investigate the developing process of K-H instability, a numerical simulation has been performed using a computational fluid dynamics (CFD) code Fluent (Fluent User's Guide, 1996). The specific description of Fluent for the current problem can be found in An et al (1996). The set up of K-H instability is explained below. Figure 1(a) shows the initial flow field at time $t = 0$. The upper fluid (darker) is a type of oil with density $\rho' = 834kg/m^3$ at rest, i.e. velocity $U' = 0$, while the lower fluid (lighter) is water with density $\rho = 1000kg/m^3$ and velocity $U = 0.25m/s$. An initial sinusoidal perturbation wave

$$\eta = \epsilon sin(2\pi\frac{x}{\lambda}), \quad -\frac{\lambda}{2} \le x \le \frac{\lambda}{2} \quad (16)$$

is imposed upon the oil-water interface. The wave length $\lambda = 0.03m$ and the amplitude of the wave $\epsilon = \lambda/40 = 0.00075m$. The computational domain is a rectangle of length λ and height $\lambda/2$. The grid is 121×71 and clustered so that it is finer in the center part while it is coarser elsewhere. The Euler model is chosen for the simulation and the VOF model is invoked to track the interface. At the inlet (left boundary), $U' = 0$ and $F = 0$ for the upper half while $U = 0.25m/s$ and $F = 1$ for the lower half where F denotes volume fraction of water. Both upper and lower outer boundaries are assumed to be slip-walls and the outlet (right boundary) is left free without any specification. The time step is chosen $\Delta t = 0.001sec$.

Figures 1(b)-(e) show the transient flow fields at times $t = 0.05, 0.10, 0.15, 0.20sec$. From these graphs one can see that the amplitude of the interfacial wave is continuously increased and a vortex is generated. As the vortex becomes stronger and stronger with time increasing, it can be expected that the vortex will break eventually and oil droplets will be formed and dispersed into the water. This simulation provides a description of droplet formation and shedding which are considered as the mechanism of droplet entrainment boom failure. From the previous analysis, the critical velocity difference can be obtained from equation (15),

$$\Delta U_{cr} = \sqrt{\frac{g\lambda}{2\pi}\frac{\rho^2 - \rho'^2}{\rho\rho'}} \quad (17)$$

which gives $\Delta U_{cr} = 0.12m/s$ for the present simulation. The velocity difference in this simulation is $\Delta U = U - U' = 0.25m/s$, greater than ΔU_{cr}. Therefore, the numerical prediction is consistent with the theoretical analysis.

3. DROPLET ENTRAINMENT

In order to simulate the process of boom failure due to droplet entrainment, a numerical simulation has been undertaken based on an experiment by Delvigne(1989). In his experiment, a channel $0.5m$ wide and $0.3m$ deep was used to provide a uniform water flow. A flat plate barrier was installed as a boom and the boom draft was set at $0.07m$. The oil that was spilled onto the water surface and contained by the boom was Bunker B with density $888kg/m^3$ and kinematic viscosity $70cSt$. The amount of spilled oil was $0.01m^3/m$. The experiment claims that boom failure occurs at a critical velocity of $0.24m/s$ and that the failure mechanism is droplet entrainment. This experimental result has been accurately simulated by the authors and presented recently (Goodman et al, 1997). The major result will be used here to interpret the process of droplet entrainment.

In the boom failure simulation, the computational grid is 281×41 and clustered so that the mesh is finer near the boom while it is coarser elsewhere. Initially, the flow velocity is set at $0.09m/s$ and then increased until substantial oil loss appears. With the Reynolds number of 27,000 the flow must be turbulent and the standard $k-\epsilon$ turbulence model is used. To track the oil-water interface, the VOF model is invoked. When the flow velocities are $0.09, 0.15, 0.20m/s$, the flow can reach a steady state, the oil slick can be stabilized and no obvious oil loss can be seen. This means that the oil containment is successful at these velocities. When the flow velocity is increased to $0.24m/s$, the flow pattern changes significantly. Figures 2(a)-(e) show the transient flow fields at times $t = 137.0, 139.1, 141.1, 143.1, 145.1sec$. From these graphs one can see that the flow has the following features:

i) The boom draft is still deep enough to prevent the oil from draining;

ii) The oil-water interface is wavy and unstable;

iii) More and more oil is torn from the interface and swept downstream by the main flow;

iv) The lost oil is dispersed into the water downstream of the boom.

Delvigne(1989) described the features of droplet entrainment based on his experimental observations. For low-viscosity oil at high relative oil-water velocity, the oil slick length decreases with increasing relative velocity and fast-moving interfacial waves with increasing wave height appear. These gravity shear waves become unstable due to Kelvin-Helmholtz instability at a certain critical velocity and the oil droplets are torn from the wave tops. Depending on droplet buoyancy, current velocity and boom draft, a droplet may pass underneath the boom causing droplet entrainment failure, or it may re-attach to the oil slick.

Comparing the flow features in Figure 2 with Delvigne's statement, we are confident that the mechanism of boom failure in this case is due to droplet entrainment and the onset of the droplet entrainment is due to K-H instability even though the computational grid is not fine enough to capture the details of oil droplets. The mechanism of droplet entrainment can be evidenced more confidently by watching the animation (An et al, 1997). The fast-moving interfacial waves, the oil shedding from the interface and the oil escape underneath the boom are clearly visible in the animation.

4. CONCLUSIONS

(1) The oil-water interfacial phenomena of an oil slick behind a boom on flowing water can be investigated using the computational fluid dynamics (CFD) tool, Fluent.

(2) The evolution of Kelvin-Helmholtz instability on an oil-water interface can be numerically simulated and dynamically demonstrated. The critical condition of K-H instability evaluated by the theoretical analysis is confirmed by the simulation.

(3) The process of oil boom failure due to droplet entrainment can be simulated and the simulated flow features are consistent with the corresponding experimental observations.

(4) More efforts are needed for a thorough understanding of the relationship between K-H instability and droplet entrainment. These efforts include various aspects: experimental, theoretical and computational.

KEYWORDS

Kelvin-Helmholtz instability, Droplet entrainment, Interfacial phenomena, Multiphase flow, CFD

ACKNOWLEDGEMENTS

The first author acknowledges the Natural Sciences and Engineering Research Council of Canada for an Industrial Research Fellowship and Professor R.D. Rowe of the University of Calgary for his assistance with respect to the computing facilities and the simulation software.

REFERENCES

An, C.-F., Brown, H.M., Goodman, R.H. and Clavelle, E.J., 1997, "Animation of boom failure processes", *Proc. of the 20th Arctic & Marine Oilspill Program Tech. Seminar*, Vol.2, pp.1181-1188, Vancouver, British Columbia, Canada

An, C.-F., Clavelle, E.J., Brown, H.M. and Barron, R.M., 1996, "CFD simulation of oil boom failure", *Proc. of the 4th Ann. Conf. of the CFD Society of Canada*, ed. by D. Jones, pp.401-408, Ottawa, Ontario, Canada

Birkhoff, G. and Zarantonello, E.H., 1957, *Jets, Wakes, and Cavities*, pp.254-255, Academic Press, Inc., New York, USA

Calgary Herald, *Newspaper* of Calgary, Alberta, Canada, A7: World, Thursday, July 3, 1997

Chandrasekhar, S., 1961, *Hydrodynamic and Hydromagnetic Stability*, pp.481-498, Dover Publications, Inc., New York, USA

Delvigne, G.A.L., 1989, "Barrier failure by critical accumulation of viscous oil", *Proc. of 1989 Oil Spill Conf.*, pp.143-148, San Antonio, Texas, USA

Drazin, P.G., 1970, "Kelvin-Helmholtz instability of finite amplitude", *J. Fluid Mech.*, Vol.42, Part 2, pp.321-335

Drazin, P.G. and Howard, L.N., 1966, "Hydrodynamic stability of parallel flow of inviscid fluid", *Advances in Applied Mechanics*, Vol.9, pp.1-89, ed. by G.G. Chernyi et al, Academic Press, New York, USA

Ertekin, R.C. and Sundararaghavan, H., 1995, "The calculation of the instability criterion for a uniform viscous flow past an oil boom", *J. Offshore Mechanics & Arctic Eng.*, Vol.117, pp.24-29

Fluent User's Guide, version 4.4, 1996, Fluent Inc., Lebanon, New Hampshire, USA

Goodman, R.H., Brown, H.M., An, C.-F. and Rowe, R.D., 1997, "Dynamic modeling of oil boom failure using computational fluid dynamics", *Proc. of the 20th Arctic & Marine Oilspill Program Tech. Seminar*, Vol.1, pp.437-455, Vancouver, British Columbia, Canada

Grilli, S.T., Hu, Z. and Spaulding, M.L., 1996, "Numerical modeling of oil containment by a boom", *Proc. of the 19th Arctic & Marine Oilspill Program Tech. Seminar*, pp.343-376, Calgary, Alberta, Canada

Jones, W.T., 1972, "Instability at an interface between oil and flowing water", *J. Basic Eng.*, pp.874-878

Kordyban, E., 1990, "The behavior of the oil-water interface at a planar boom", *J. Energy Resources Technology*, Vol.112, pp.90-95

Lamb, H., 1932, *Hydrodynamics*, 6th Ed., pp.373-374, Cambridge University Press, Cambridge, UK

Leibovich, S., 1976, "Oil slick instability and the entrainment failure of oil containment booms", *J. Fluids Eng.*, Vol.98, pp.98-105

Pai, S.-I., 1977, *Two-Phase Flows*, pp.108-111. Wieweg & Sohn Verlagsgesellschaft, Braunschweig, Germany

Patnaik, P.C., Sherman, F.S. and Corcos, G.M., 1976, "A numerical simulation of Kelvin-Helmholtz waves of finite amplitude", *J. Fluid Mech.*, Vol.73, Part 2, pp.215-240

Rangel, R.H. and Sirignano, W.A., 1991, "The linear and nonlinear shear instability of a fluid sheet", *Phys. Fluids A*, Vol.3, No.10, pp.2392-2400

Squire, H.B., 1933, "On the stability for three-dimensional disturbances of viscous fluid flow between parallel walls", *Proc. of Royal Society, A*, Vol.142, pp.621-628

Thorpe, S.A., 1969, "Experiments on the instability of stratified shear flows: immiscible fluids", *J. Fluid Mech.*, Vol.39, Part 1, pp.25-48

Yih, C.-S., 1965, *Dynamics of Nonhomogeneous Fluids*, pp.159-163, MacMillan Company, New York, USA

Zalosh, R.G., 1976, "Discretized simulation of vortex sheet evolution with buoyancy and surface tension effects", *AIAA J*, Vol.14, No.11, pp.1517-1523

Zalosh, R.G. and Jensen, D.S., 1975, "A numerical model of droplet entrainment from a contained oil slick", *Fluid Mechanics in the Petroleum Industry*, ed. by C. Dalton and E. Denison, pp.17-27, American Society of Mech. Eng., New York, USA

Wicks, M. III, 1969, "Fluid dynamics of floating oil containment by mechanical barriers in the presence of water currents", *Proc. of API/FWPCA Joint Conf. on Prevention & Control of Oil Spills*, pp.55-106, New York, USA

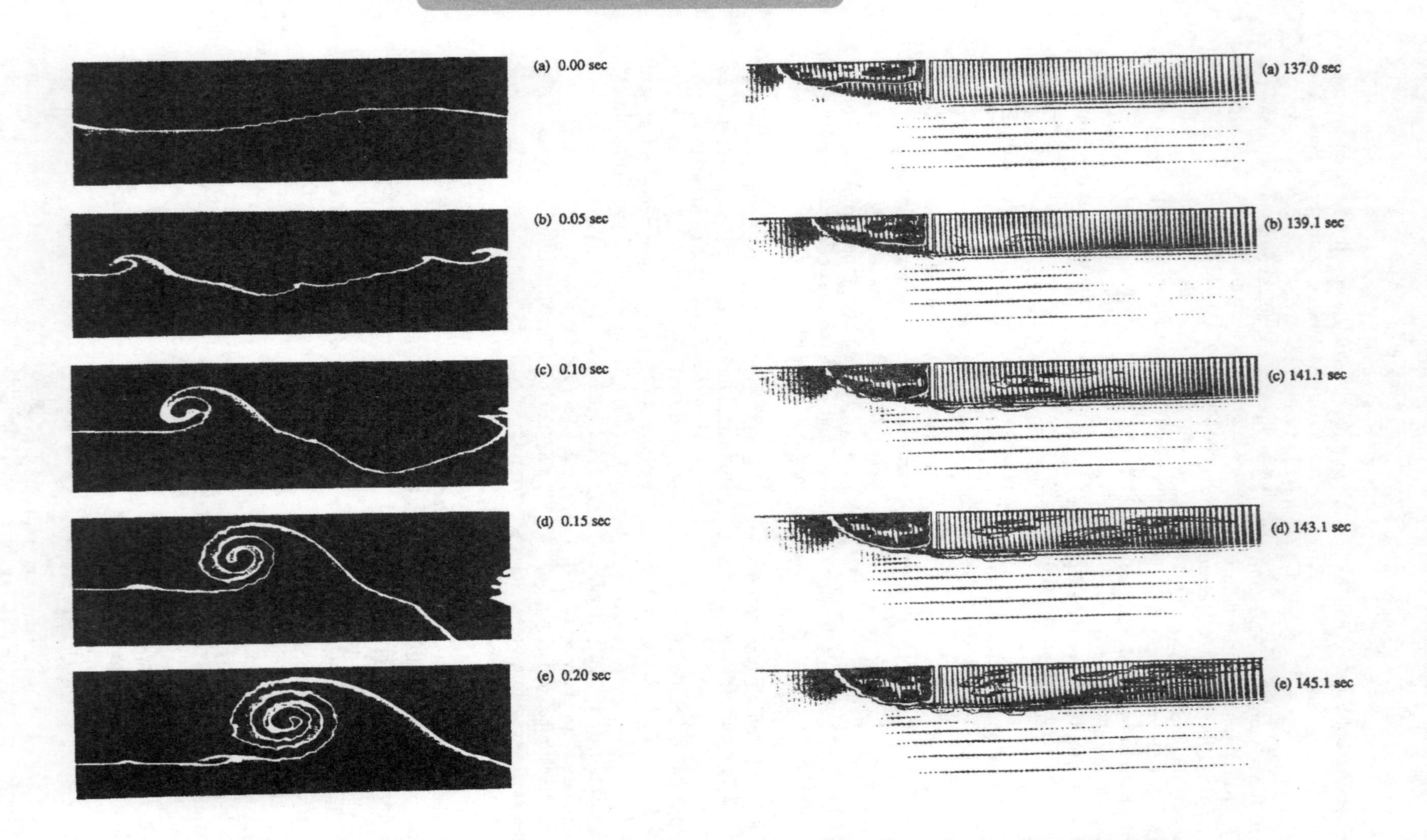

Figure 1. Kelvin-Helmholtz Instability

$U=0.25m/s$, $U'=0$
$\rho=1000kg/m^3$, $\rho'=834kg/m^3$
$\lambda=0.03m$, $\varepsilon=0.00075m$

Figure 2. Droplet Entrainment Failure

Boom Draft: $D=0.07m$
Flow Velocity: $U=0.24m/s$
Oil: Bunker B, $Q=0.01m^3/m$

The Proceedings of
the Eighth (1998) International
OFFSHORE AND POLAR ENGINEERING
CONFERENCE

Montréal, Canada

VOLUME II, 1998

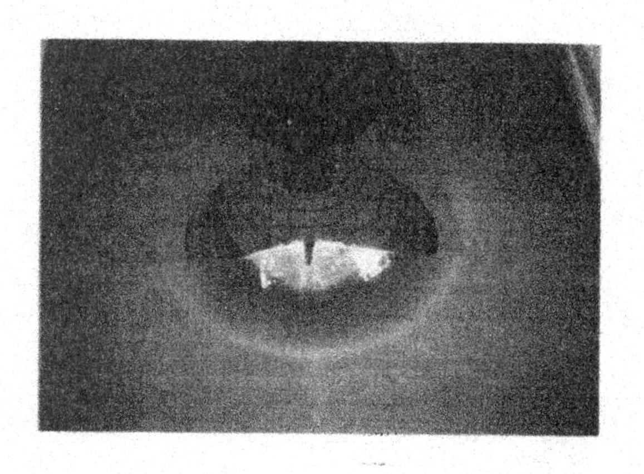

ISOPE International Society of
Offshore and Polar Engineers

Proceedings of the Eighth (1998) International Offshore and Polar Engineering Conference
Montreal, Canada, May 24-29, 1998

ISBN I-880653-34-6 (Set); ISBN I-880653-36-2 (Vol. II)

Droplet Entrainment Boom Failure and Kelvin-Helmholtz Instability

C.-F. An & R.M. Barron
University of Windsor,
Windsor, Canada

H.M. Brown & R.H. Goodman
Imperial Oil Resources Limited
Calgary, Canada

ABSTRACT

In this paper, a comparative study of the relationship between droplet entrainment and Kelvin-Helmholtz (K-H) instability is presented. In a channel experiment, oil droplets were formed on the oil-water interface and entrained by the water current causing droplet entrainment when the velocity was about 0.25m/s for a flat plate boom with 0.05m draft in a channel of 1m width and 1m depth. This experimental result has been successfully simulated using the computational fluid dynamics (CFD) software Fluent. The onset of droplet entrainment is due to K-H instability, which has been modeled by Fluent. Based on numerical computations, the process of droplet entrainment and the evolution of K-H instability are animated.

KEY WORDS: Marine environment, Oil boom, Droplet entrainment, Kelvin-Helmholtz instability, Hydrodynamics, CFD

1. INTRODUCTION

Oil spills on water occurring when shipping crude oil or its products through marine waterways may pollute the environment. The performance of oil recovery systems is an important factor in mitigating the effects of oil spilled on water. If oil is spilled on the surface of water, the most common response is to contain the oil using a boom towed by boats and, then, to recover it using a skimmer. A boom is a barrier that floats on the surface of the water and contains the spilled oil. A typical boom is composed of a cylindrical float with a skirt. Experiments and experience have shown that when the towing speed is about 0.5kt, the boom often fails to contain the oil and the oil escapes underneath the boom due to hydrodynamic forces. This severely limits the capability of oil containment booms. The oil spill response community has a serious concern about how to improve the capability of current booms and how to design more efficient new booms. Boom characteristics have been studied since the late 1960's. A number of boom failure mechanisms have been identified, such as drainage failure, droplet entrainment and critical accumulation. Nevertheless, most boom studies are experimental, with at most a simple one-dimensional analysis. Very little computer modeling work can be found in the literature. The hydrodynamic details of oil-water flow around a boom are not well understood.

During recent years, Imperial Oil has conducted investigations on boom failure mechanisms, including experimental studies (Brown et al, 1997) and computer modeling (An et al, 1997, Goodman et al, 1997). On the basis of these investigations, a joint research project was established in the summer of 1997 under which channel experiments were undertaken at the Imperial Oil Wave Basin and, concurrently, computer simulations were carried out at the Fluid Dynamics Research Institute, University of Windsor.

In this paper, the boom failure mechanism referred to as droplet entrainment is of particular interest and a comparative study of the relationship between droplet entrainment boom failure and Kelvin-Helmholtz (K-H) instability is presented. In section 2, the experimental results from the Imperial Oil Wave Basin tests on a flat plate boom failure due to droplet entrainment are reported. In section 3, the results of computer modeling of droplet entrainment and K-H instability are presented. Conclusions are given in section 4.

2. CHANNEL EXPERIMENTS

Figure 1 shows a sketch of the water channel (top view) used to conduct oil boom experiments in the Imperial Oil Wave Basin. The channel is 30m long, 1m wide and 1m deep, stretching along the south bank of the basin. Water was moved by a 15HP pump (ITT Flygt model 4660 mixer) controlled by a Toshiba variable frequency drive. At each end of the channel, a flow straightener was installed consisting of a rack of closely packed pipes. A flat plate barrier of transparent plexiglass, as a boom, was installed at about mid-length of the channel. The draft of the boom can be adjusted up to 0.5m deep from the water surface. Two underwater observation windows were installed in large viewing boxes (one upstream and one downstream of the boom location) so that an observer, a 35mm camera and a camcorder could be held in each of the two boxes. A flow meter was installed at the reference point (indicated in the figure) and two types of oil slick thickness measurement devices were installed.

After the water channel was calibrated, several tests were performed. In this paper, only the tests of a flat plate boom with 0.05m draft resulting in droplet entrainment boom failure is presented. Five liters of light oil, Federated crude, with density 834kg/m^3 and kinematic viscosity 10cSt were poured onto the channel surface upstream of the boom. When the flow velocity was low, the oil was contained by the boom and a stable oil slick was formed. When the flow velocity was increased, the oil slick became shorter and thicker. At a velocity of about 0.25m/s, the oil-water interface became wavy and unstable. Figures 2 and 3 show photographs of oil slick behavior against the boom when droplet entrainment occurred. After the flow velocity reached the critical velocity, 0.25m/s, oil droplets were formed at the oil-water interface (Figure 2) and some of them were entrained by the main stream of the water current causing substantial oil loss (Figure 3). The dark region on the right part of the photograph is the shadow of the oil slick. The ripples on the left part of the photograph are the images of the waves on the water surface. The water flows from the right to the left.

The phenomenon of droplet entrainment boom failure was recognized at an early stage of boom studies (Wicks, 1969), but the hydrodynamic rationale of droplet entrainment was not well understood. Leibovich (1976) analyzed the process of droplet entrainment theoretically and concluded that the onset of droplet formation from an oil-water interface is the amplifying and breaking of the interfacial waves, i.e. Kelvin-Helmholtz waves. This means that the hydrodynamic mechanism of droplet entrainment boom failure is due to interfacial K-H instability. Delvigne (1989) comprehensively described the characteristics of droplet entrainment boom failure from his experimental observations as follows. When the relative oil-water velocity increases, the oil slick length decreases and fast-moving interfacial waves appear with increasing amplitude. These waves become unstable at a certain critical velocity and oil droplets are torn from the wave tops. Depending on droplet buoyancy, current velocity and boom draft, a droplet may re-attach to the oil slick, or pass underneath the boom causing droplet entrainment boom failure. Most of these features were observed in the present channel experiment.

3. COMPUTER MODELING

In order to investigate the process of boom failure due to droplet entrainment and the evolution of interfacial K-H instability, computer modeling has been performed using a computational fluid dynamics (CFD) package Fluent. The specific description of the software can be found in the manual (Fluent User's Guide v4.4, 1996).

3.1 Droplet Entrainment

In the wave basin channel experiment, the oil can be successfully contained by a flat plate boom at low flow velocity. At a critical velocity of about 0.25m/s, boom failure occurred and the mechanism was droplet entrainment. This experimental result has been qualitatively modeled using Fluent.

The computational mesh is 251x61 and clustered so that the mesh is finer near the boom while it is coarser elsewhere. Initially, the oil is introduced from the left on the surface at low velocity and, then, the flow velocity is increased until obvious oil loss appears. With the Reynolds number of about 10^6 based on the flow velocity, channel depth and kinematic viscosity of the water, the flow must be turbulent and the standard k-ε turbulence model is used. To track the oil-water interface, the volume of fluid (VOF) model is applied. When the flow velocity is increased to 0.25m/s, the oil loss appears. Figures 4(a) – (e) show the transient flow fields at simulation times t = 37, 47, 57, 67 and 77sec. At t = 37sec, some oil detaches from the oil-water interface. At t = 47sec, the detached oil is entrained by the main flow underneath of boom. At t = 57sec, the lost oil is swept away downstream. At t = 67sec, more oil is detached from the interface. At t = 77sec, this detached oil is also entrained by the main flow underneath the boom. From these graphs one can see that the oil-water flow has the following features:

i) The boom draft is deep enough to contain the oil and to avoid drainage failure;
ii) The oil-water interface is unstable and the oil detaches from it from time to time;
iii) More and more oil is torn from the interface and entrained by the main flow as time increases.

Comparing the flow features in Figure 4 with Delvigne's description cited in section 2, we are confident that the mechanism of boom failure in this case is due to droplet entrainment, even though the computational mesh is not fine enough to capture the details of individual oil droplets. The mechanism of droplet entrainment can be further confirmed by watching the animation. The fast-moving interfacial waves, the oil shedding from the interface, the circulating flow pattern within the oil slick and the oil escaping underneath the boom are clearly visible in the animation.

3.2 K-H Instability

The setup of the K-H instability modeling is shown in Figure 5(a). The upper fluid is oil with density ρ_2 = 834kg/m^3 and velocity U_2 = 0. The lower fluid is water with density ρ_1 = 1000kg/m^3 and velocity U_1 = 0.25m/s. An initial small disturbance wave

$$\eta = \varepsilon \sin(2\pi \frac{x}{\lambda}), \qquad -\frac{\lambda}{2} \le x \le \frac{\lambda}{2}$$

is imposed on the oil-water interface at time t = 0. The wavelength λ is 0.03m and the wave amplitude ε is λ/40.

The computational domain is a rectangle of λxλ/2. The mesh is 121x71 and packed at the center of the region. The Euler solver is selected for the modeling and the VOF model is invoked to track the oil-water interface. At the left inlet boundary, U_2 = 0 for the upper half boundary and U_1 = 0.25m/s for the lower half boundary. Both upper and lower far field boundaries are assumed to be slip-walls on which a horizontal velocity component is permitted, but a vertical velocity component is prohibited. The right outlet boundary is free without any specification. The time step is chosen as Δt = 0.001sec and the residual tolerance of the governing equations is set to be 10^{-3}.

Figure 5(b) – (e) show the modeled transient flow fields at times t = 0.06, 0.12, 0.18 and 0.24sec. These graphs show that the wave amplitude is continuously increased and that the wave shape is severely distorted so that a concentrated vortex is generated. As the vortex becomes stronger and stronger with increasing time, it will be broken eventually. Therefore, oil droplets will be formed and may be swept away by the water flow. The vortex has already started to break at t = 0.24sec. This simulation models the process of K-H instability, which is considered as the mechanism of droplet entrainment boom failure. From the linear analysis of K-H instability based on the assumption of potential flow (Lamb, 1932), the critical velocity difference can be estimated by

$$\Delta U_{cr} = \sqrt{\frac{g\lambda}{2\pi}\frac{\rho_1^2-\rho_2^2}{\rho_1\rho_2}}$$

where g is the acceleration due to gravity. This equation gives $\Delta U_{cr} = 0.12$m/s for this case. The actual velocity difference is $\Delta U = U_1 - U_2 = 0.25$m/s, much greater than ΔU_{cr}. Therefore, the oil-water interface is expected to be unstable and the K-H instability occurs. The computer modeling verifies the observations from the experiments, but the theoretical prediction from the linear analysis underestimates the critical velocity.

4. CONCLUSIONS

(1) The phenomenon of droplet entrainment boom failure has been observed in a wave basin experiment at flow velocity 0.25m/s for a flat plate boom with 0.05m draft in a channel of 1m width and 1m depth.

(2) The process of boom failure due to droplet entrainment can be simulated using CFD software Fluent and the simulated flow patterns are consistent with the observations in the wave basin channel experiment.

(3) The evolution of K-H instability on an oil-water interface can be modeled using Fluent and the modeling result agrees with experimental observations.

(4) Further research is needed to gain insight into the relationship between droplet entrainment boom failure and interfacial K-H instability. This will include extensive theoretical, computational and experimental studies.

安长发博士科研论文集

ACKOWLEDGEMENT

The authors acknowledge Imperial Oil Resources Limited for the financial support on this project.

REFERENCES

An, CF, Brown, HM, Goodman, RH, Clavelle, EJ, Rowe, RD, and Barron, RM (1997). "Hydrodynamic Behavior of Contained Oil Slick on Flowing Water," *Proc. of 5th Ann. Conf. of CFD Society of Canada (CFD97)*, Victoria, BC, Canada, pp 6:31–36

Brown, HM, Goodman, RH, An, CF, and Bittner, J (1997). "Boom Failure Mechanisms: Comparison of Channel Experiment with Computer Modeling Results," *Proc. of 20th Arctic & Marine Oilspill Program (AMOP) Technical Seminar*, Vancouver, BC, Canada, Vol. 1, pp 457–467

Delvigne, GAL (1989). "Barrier Failure by Critical Accumulation of Viscous Oil," *Proc. of 1989 Oil Spill Conf.*, San Antonio, TX, USA, pp 143-148

Fluent User's Guide v4.4 (1996). Fluent Inc., Lebanon, NH, USA

Goodman, RH, Brown, HM, An, CF, and Rowe, RD (1997). "Dynamic Modeling of Oil Boom Failure Using Computational Fluid Dynamics," *Proc. of 20th Arctic & Marine Oilspill Program (AMOP) Technical Seminar*, Vancouver, BC, Canada, Vol. 1, pp 437-455

Lamb, H (1932). *Hydrodynamics*, Cambridge University Press, Cambridge, UK, pp 373-374

Leibovich, S (1976). "Oil Slick Instability and the Entrainment Failure of Oil Containment Booms," *J. Fluids Eng.*, Vol. 98, pp 98-105

Wicks, M III (1969). "Fluid Dynamics of Floating Oil Containment by Mechanical Barriers in the Presence of Water Currents," *Proc. of API/FWPCA Joint Conf. on Prevention & Control of Oil Spills*, New York, USA, pp 55-106

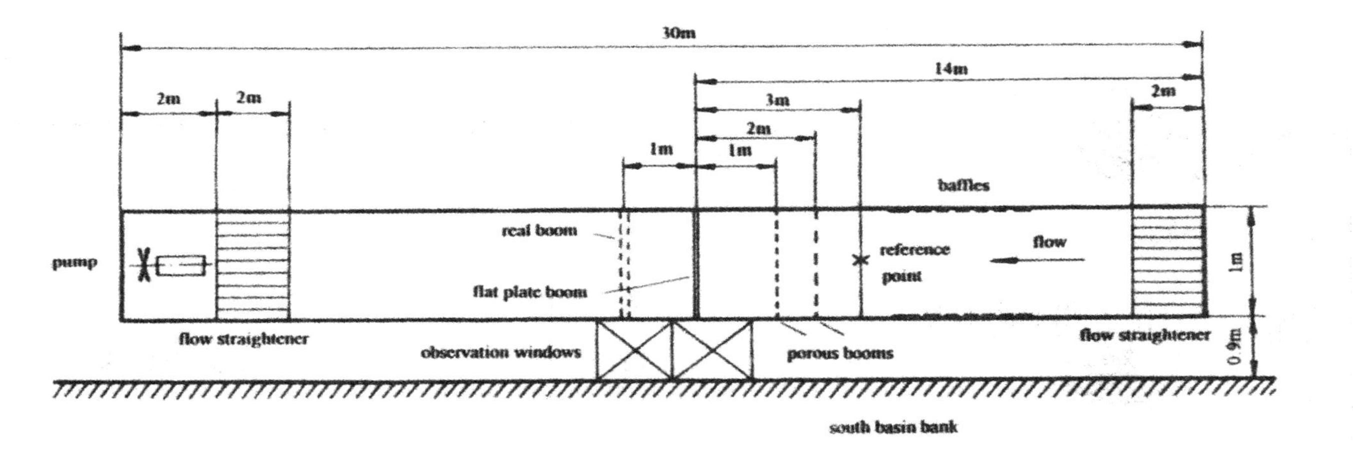

Figure 1. Sketch of Water Channel in Wave Basin

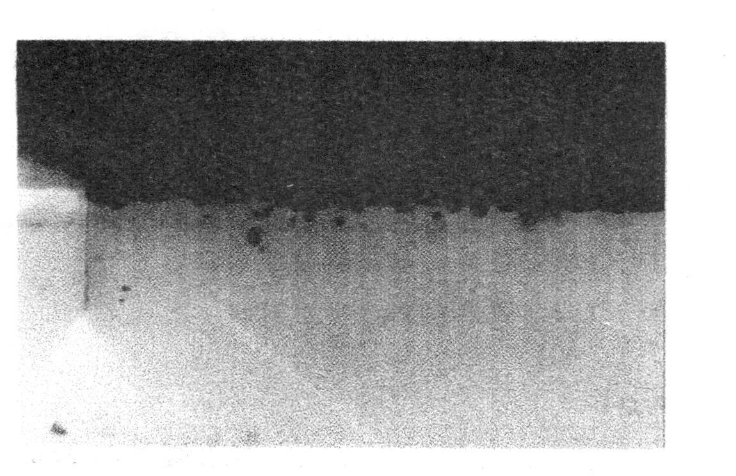

Figure 2. Droplet Formation

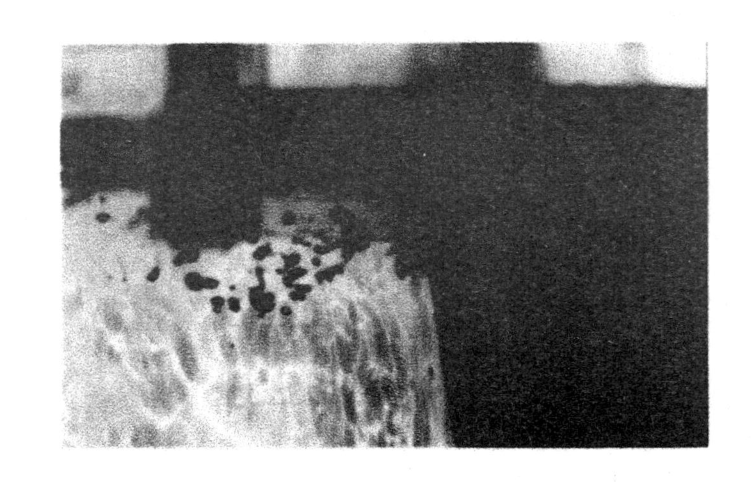

Figure 3. Droplet Entrainment

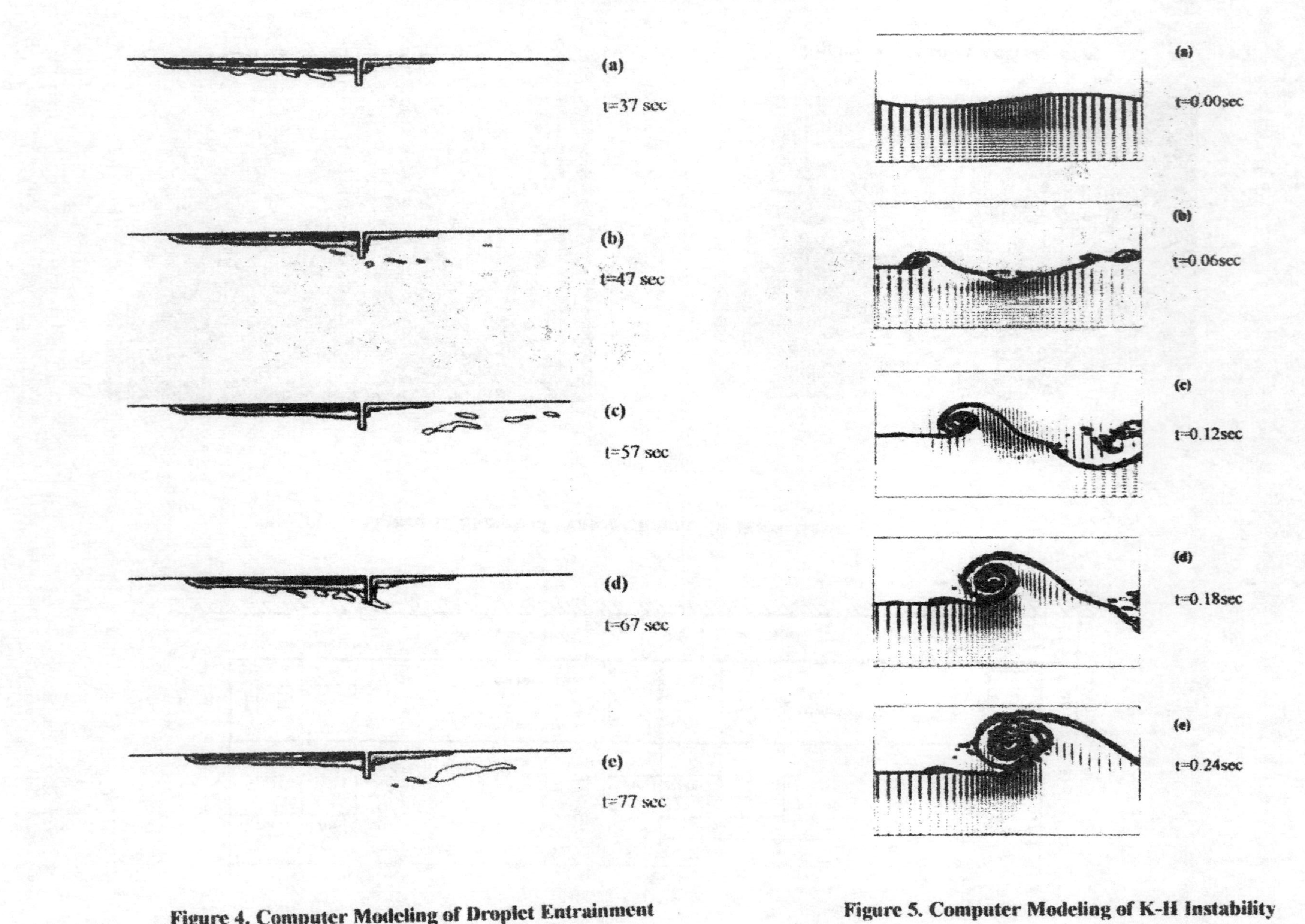

Figure 4. Computer Modeling of Droplet Entrainment

Figure 5. Computer Modeling of K-H Instability

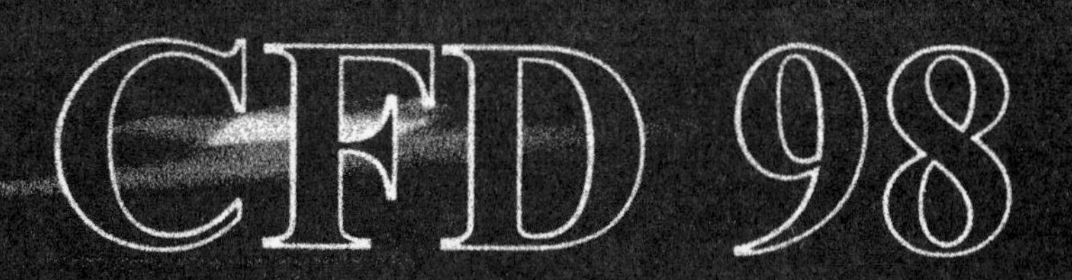

COMPTE-RENDU
PROCEEDINGS

7 - 9 juin 1998

La sixième conférence annuelle de la Société canadienne de CFD

CRDV / DREV

DÉPARTEMENT DE GÉNIE MÉCANIQUE
FACULTÉ DES SCIENCES ET DE GÉNIE

June 7-9, 1998

The Sixth Annual Conference of the Computational Fluid Dynamics Society of Canada

COMPARATIVE STUDY OF MODELS FOR OIL BOOM SIMULATION

C.-F. An & R.M. Barron
Fluid Dynamics Research Institute
University of Windsor
Windsor, Ontario N9B 3P

Abstract: Experimental and computational investigations of oil boom failure typically replace the boom with a simple flat plate. A computational study has been conducted to compare the flows around a flat plate barrier and a real boom configuration with the same draft. The simulated results show that the flow patterns for the two cases at low velocity are almost identical except for the boom tip and oil slick regions where minor differences in velocity and pressure can be discriminated. For a high velocity, boom failure due to droplet entrainment begins to occur for both cases. The study enhances our confidence that a simple flat plate barrier can be used to represent a real boom configuration with the same draft in studying boom failure due to droplet entrainment.

1. INTRODUCTION

Accidental oil spills may severely pollute water resources. For example, the Russian tanker Nakhodka broke in two during a strong storm in the Sea of Japan near Mikuni on January 2, 1997. More than 1.5 million gallons of heavy fuel oil were released along 460 kilometers of Japanese coastline [1]. If oil is spilled on the surface of water, one of the most common responses is to use a boom to contain the oil and a skimmer to recover it. In order to increase the efficiency of oil recovery and prevent the oil from spreading, the speed of oil herding should be as high as possible. However, if the speed of oil herding exceeds a critical value, boom failure may occur and the oil may escape underneath the boom due to hydrodynamic forces. The mechanisms of oil loss have been studied since the early 1970's and categorized as the following three modes: drainage failure, droplet entrainment and critical accumulation [2]. The critical velocity, at which oil loss begins to occur, depends on a number of factors, such as water depth, boom shape and draft, oil type and properties, etc. It is expensive to cover all the factors in an experimental project due to the limitations of experimental facilities, manpower costs and time.

Computational fluid dynamics (CFD) is a powerful tool for simulating a variety of fluid flows and can be much less expensive compared with experimental investigations. Since 1993, CFD technology has been introduced to simulate oil-water flow around an oil containment boom [3] and, since then, it has been utilizing extensively [4-6]. The boom configuration typically consisting of a cylindrical float that is used to provide buoyancy and a flexible or rigid skirt that is used to contain the oil, has been represented in computational studies by a flat plate barrier with the same draft. This same simplified configuration is normally used in experimental investigations as well. Some questions arise from this representation. Is it reasonable to make such a simplification? Are the two flow patterns similar to each other when the oil is successfully contained? Are the critical velocities the same or close to each other when the boom fails to contain the oil? If these velocities are close, what percentage of deviation can be expected? Many of these questions were posed by oil spill professionals, who are non-hydrodynamicists, after presentation at a recent oil spill conference [6].

This paper attempts to answer these questions and provides evidence and confidence for the CFD simulation of oil boom failure with the above simplified model. A comparative study is conducted using the commercial CFD code Fluent [7] to simulate oil-water flows around a flat plate barrier and a real boom configuration that is composed of a cylindrical float and a rigid skirt. The drafts of the flat plate barrier and the real boom configuration are the same and all the other geometrical and physical factors are kept the same. The final simulated results show that there are very minor differences between the two flows and the significantly simplified flat plate barrier can be used to represent the real boom configuration when CFD modeling of oil boom failure is performed. Section 2 describes the computational grids for the two physical flow regions. Section 3 compares the simulated results between the two flows for the case of successful oil containment at low velocity and the case of boom failure at high velocity. Section 4 gives concluding remarks.

2. COMPUTATIONAL GRIDS

The computational grid for the region of the flow around the real boom configuration was generated using GeoMesh [8]. GeoMesh is a CAD program for geometry creation and grid generation. It has three important components, DDN, P-Cube and Leo. DNN is used to create the geometry and P-Cube is used to generate quadrilateral (2D) or hexahedral (3D) grids. Leo is used to check grid quality, such as Jacobian, skewness, aspect ratio and so on. The grid files generated by GeoMesh can be directly imported into Fluent. The topological association of the physical region with the computational domain for the present real boom configuration is shown in Figure 1. The lower half of the cylindrical float submerged under the water surface is represented by a semi-circle and the rigid skirt is represented by a vertical segment. This geometry was created in DDN based on the hierarchy of the geometry entities: points, curves (including lines and arcs) and surfaces. In order to generate a grid system in P-Cube, this hierarchy of entities should be associated with the hierarchy of the grid entities: vertices, edges and faces. Uniform grid point distribution is used in the vertical direction for regions 1 and 5 and in the horizontal direction for region 3. Exponential grid point distribution is used for the rest of the vertical and horizontal lines. Smooth transition is maintained across the borders of the blocks. The float and skirt in the physical region are represented by dead cells in the computational domain. The final computational grid for the real boom configuration is shown in Figure 2. This is a 209x59 grid covering a 4mx1m flow region. The diameter of the float is 0.08m and the thickness of the skirt is 0.02m. The draft of the boom is 0.22m. The finest cell near the boom and the surface has a dimension of Δx=0.01m and Δy=0.005m. The generated grid can be checked for its quality using Leo. The worst Jacobian for this grid is 99.5 (ideal: 100) and appears below the float. The worst skewness is 1.056 (ideal: 1) and also appears below the float. The worst aspect ratio is 7.66 and appears near the channel bottom and close to the upper corners of the inlet and outlet boundaries. In general, the quality of this computational grid is quite good. The computational grid for the flow around the flat plate barrier was generated within Fluent and is shown in Figure 3. It is also a 209x59 grid covering a 4mx1m flow region. A group of dead cells represents the barrier with the draft of 0.22m and the thickness of 0.02m. The finest cell near the barrier also has a dimension of Δx=0.01m and Δy=0.005m. At the left inlet boundary for both the flat plate case and the real boom case, the upper 4 cells are used to introduce the oil at the beginning of the computation. The flow surface is assumed to be a slip-wall on which horizontal velocity component is allowed but vertical velocity component is not. The channel floor and the surfaces of the booms are specified as no-slip walls. The right outlet boundary of the flow region is free of specification. The flow is turbulent and the standard k-ε model is used to close the Navier-Stokes equations. The volume of fluid (VOF) model is used to distinguish the oil and water phases. Other solver settings are similar to those in [4].

3. SIMULATED RESULTS

Figure 4 shows the comparison of the steady state flow patterns between the real boom configuration and the flat plate barrier at a velocity of 0.15m/s. In Figure 4(a), velocity profiles at several channel stations are shown for both cases. From these profiles one can see that the two flow patterns are quite similar. The recirculation region downstream of the booms is apparent. The maximum velocity in the main stream region is 0.260m/s for the real boom case and 0.252m/s for the flat plate case. By carefully checking the graphs, one can identify the thin oil slick against the boom for the two cases. The length of the oil slick is 0.80m for the real boom case and 0.79m for the flat plate case. The maximum oil slick thickness is 0.0015m for both cases. Figure 4(b) shows the comparison of streamlines. The recirculation regions are more visible. In Figure 4(c), the contours of static pressure are compared between the two cases. Pressure is from -43.1 to 27.5Pa for the real boom case and from -42.3 to 27.5Pa for the flat plate case. The highest pressure is in the oil slick region and the lowest pressure is in the recirculation region for both cases. In Figures 4(d) and 4(e), turbulent kinetic energy and its dissipation rate are shown, respectively. Comparing the corresponding graphs, one can easily conclude that the flow patterns for the two cases are almost identical, at least from a macroscopic point of view.

In order to distinguish any minor differences between the two flows, more comparisons have been made at a local level as shown in Figures 5-8. Figures 5 and 6 show the velocity and pressure profiles, respectively, at channel stations x=1.95m (a), 2.00m (b) and 2.05m (c) for both the real boom and flat plate cases. The boom is located at x=2.00m. It is clear that velocity and pressure profiles for both cases are very close except near the boom tip where a small deviation can be identified, cf. Figures 5(b) and 6(b) at y=0.8m. For pressure profiles, a small difference also can be found in the recirculation region at x=2.05m, y>0.8m in Figure 6 (c). Both velocity and pressure profiles are almost identical at other stations upstream and downstream of the boom (not shown here).

In Figures 7 and 8, velocity and pressure distributions are shown, respectively, for both cases at channel heights y=0.50m (a), 0.75m (b) and 0.99m (c). The boom tip is at y=0.78m. It can be seen that velocity and pressure distributions for the two flows are almost identical in the main stream region at y=0.50m (a), are slightly deviated near the boom tip at y=0.75m (b) and are distinguishably different in the oil slick region at y= 0.99m (c). The maximum difference between the real boom case and the flat plate case is about 0.02m/s for velocity and about 3Pa for pressure in the oil slick region, cf. Figures 7(c) and 8(c). When the flow velocity is increased to 0.20m/s and 0.25m/s, the comparisons between the two flows are similar to those at velocity 0.15m/s. When the flow velocity is accelerated to 0.30m/s, substantial oil loss begins to occur for both cases. Figure 9 shows the simulated oil loss for the real boom case (a) and the flat plate case (b), respectively, at the velocity of 0.30m/s. The mechanism of oil loss for both cases is considered as droplet entrainment.

An experimental study [9] has confirmed the simulated results as shown in Figure 10. The photo in Figure 10(a) shows the oil loss in a discrete form and the photo in Figure 10(b) shows the oil loss in a concentrated form. This flow was around a real boom configuration with the draft of 0.22m at a velocity of 0.31m/s. The flow direction in this experiment was from right to left.

4. CONCLUDING REMARKS

The flow pattern around a flat plate barrier is almost identical from a macroscopic point of view to that around a real boom configuration consisting of a cylindrical float and a rigid skirt at low flow velocity, i.e. 0.15m/s, for which a steady state can be reached.

Comparing the variables for the two flows, only minor differences in velocity and pressure can be discriminated at the boom tip and in the oil slick regions.

The difference between the two flows does not seem to have much effect on the boom failure mechanism. For both flows, when the velocity reaches 0.30m/s, oil loss begins to occur and the mechanism is considered as droplet entrainment.

This work increases our confidence that a real boom configuration can be represented by a simplified flat plate barrier with the same draft when oil boom failure due to droplet entrainment is to be studied.

Acknowledgement: The authors acknowledge Drs. R.H. Goodman and H.M. Brown for their enlightening discussions on this topic.

Keywords: Multi-phase flow, environmental flow, oil boom, CFD

References:

1 Oil Spill Intelligence Report, 1997 in Review: Fewer Spills, Smaller Spills Worldwide, *International Newsletter*, Vol.XXI, No.1, Cutter Information Corp., Jan. 1, 1998

2 Delvigne, G.A.L., Barrier failure by critical accumulation of viscous oil, *Proc. 1989 Oil Spill Conference*, pp.143-148, San Antonio, Texas, USA, Feb. 13-16, 1989

3 Clavelle, E.J. and Rowe, R.D., Numerical simulation of oil boom failure by critical accumulation, Proc. of *16th Arctic & Marine Oilspill Program (AMOP) Tech. Sem.*, pp.409-418, Calgary, Alberta, June 7-9, 1993

4 An, C.-F., Clavelle, E.J., Brown, H.M. and Barron, R.M., CFD simulation of oil boom failure, Proc. of *4th Ann. Conf. of CFD Society of Canada (CFD96)*, pp.401-408, Ottawa, Ontario, June 2-4, 1996

5 An, C.-F., Brown, H.M., Goodman, R.H., Clavelle, E.J., Rowe, R.D. and Barron, R.M., Hydrodynamic behavior of contained oil slick on flowing water, Proc. of *5th Ann. Conf. of CFD Society of Canada (CFD97)*, pp.6:31-36, Victoria, BC, May 25-27, 1997

6 Goodman, R.H., Brown, H.M., An, C.-F. and Rowe, R.D., Dynamic modeling of oil boom failure using computational fluid dynamics, Proc. of *20th Arctic & Marine Oilspill Program (AMOP) Tech. Sem.*, Vol.1, pp.437-455, Vancouver, BC, June 11-13, 1997

7 Fluent User's Guide version 4.3, Fluent Inc., Lebanon, NH, USA, 1996

8 GeoMesh User's Guide version 3, Fluent Inc., Lebanon, NH, USA, 1996

9 Brown, H.M., Goodman, R.H. and An, C.-F., Flow around oil containment barriers, *21st Arctic & Marine Oilspill Program (AMOP) Tech. Sem.*, Edmonton, Alberta, June 10-12, 1998

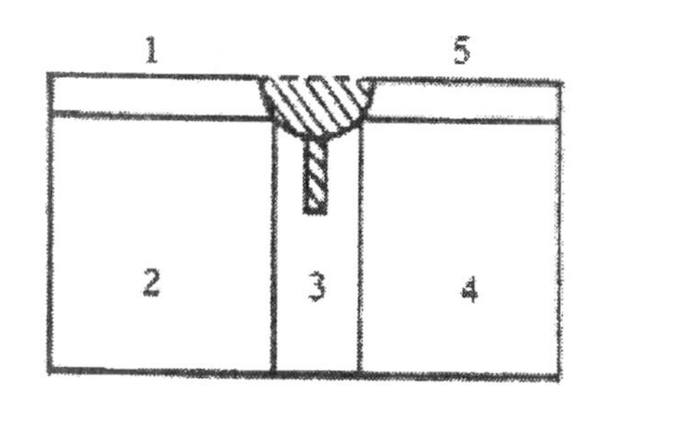

(a) Physical Region

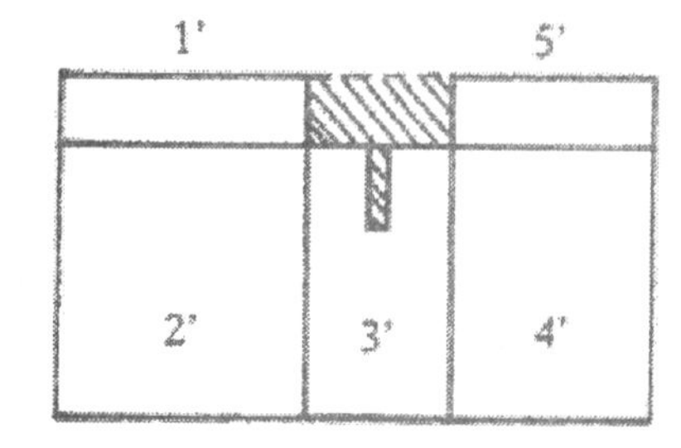

(b) Computational Domain

Figure 1. Topological Association of Real Boom Configuration

Figure 2. Computational Grid for Real Boom Configuration

Figure 3. Computational Grid for Flat Plate Barrier

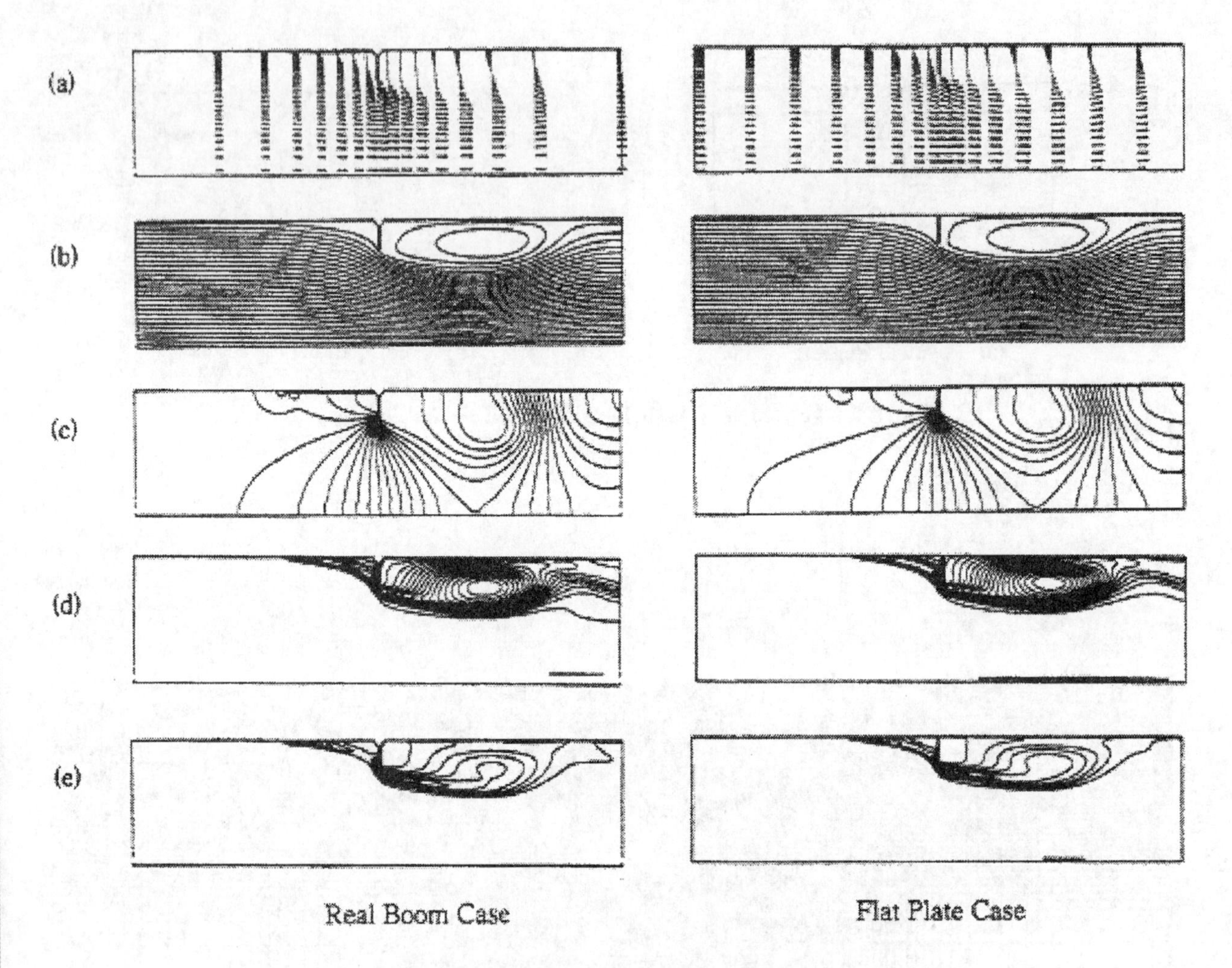

Figure 4. Steady State Flow Patterns at 0.15m/s

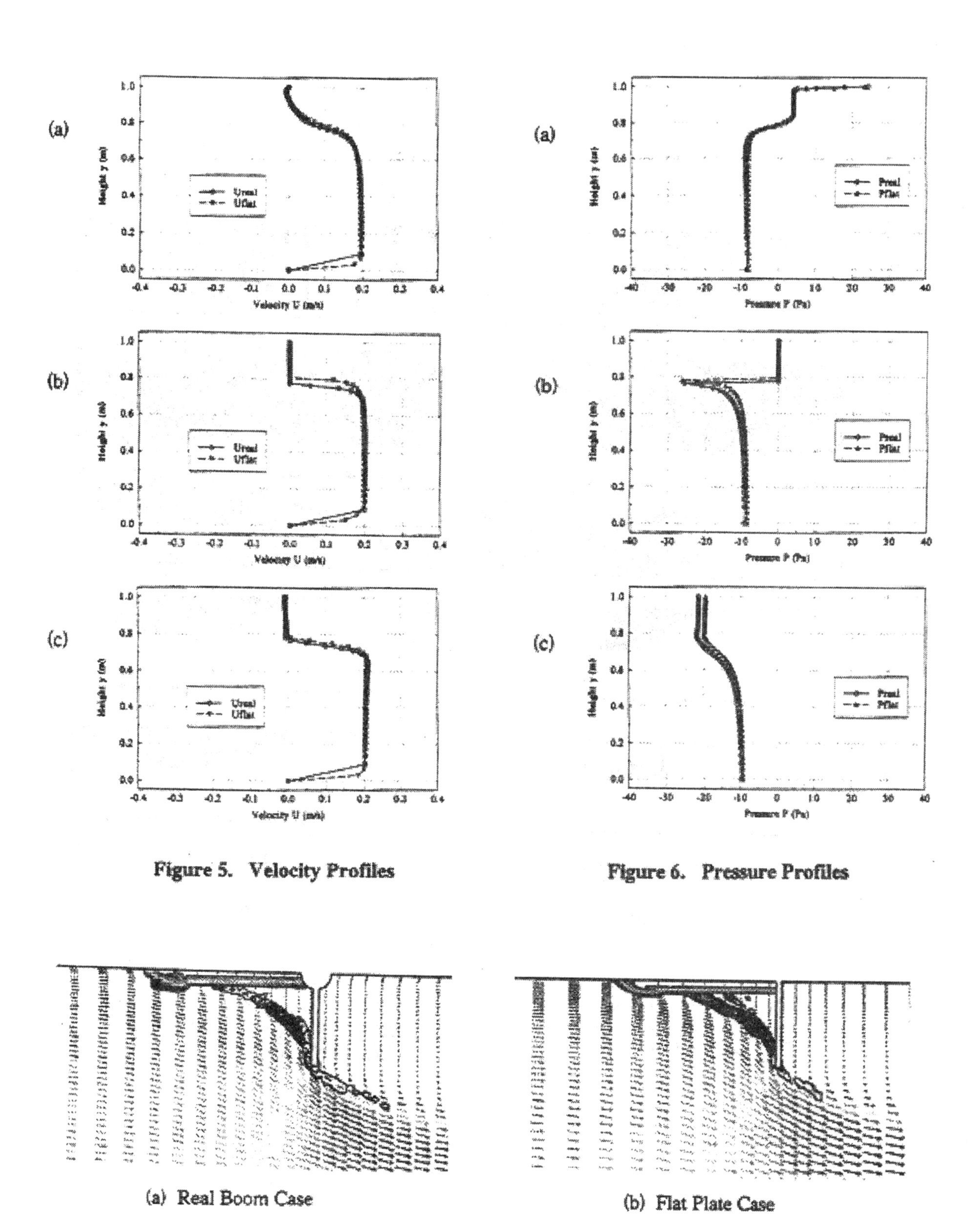

Figure 5. Velocity Profiles

Figure 6. Pressure Profiles

(a) Real Boom Case

(b) Flat Plate Case

Figure 9. Simulated Oil Loss at 0.30m/s

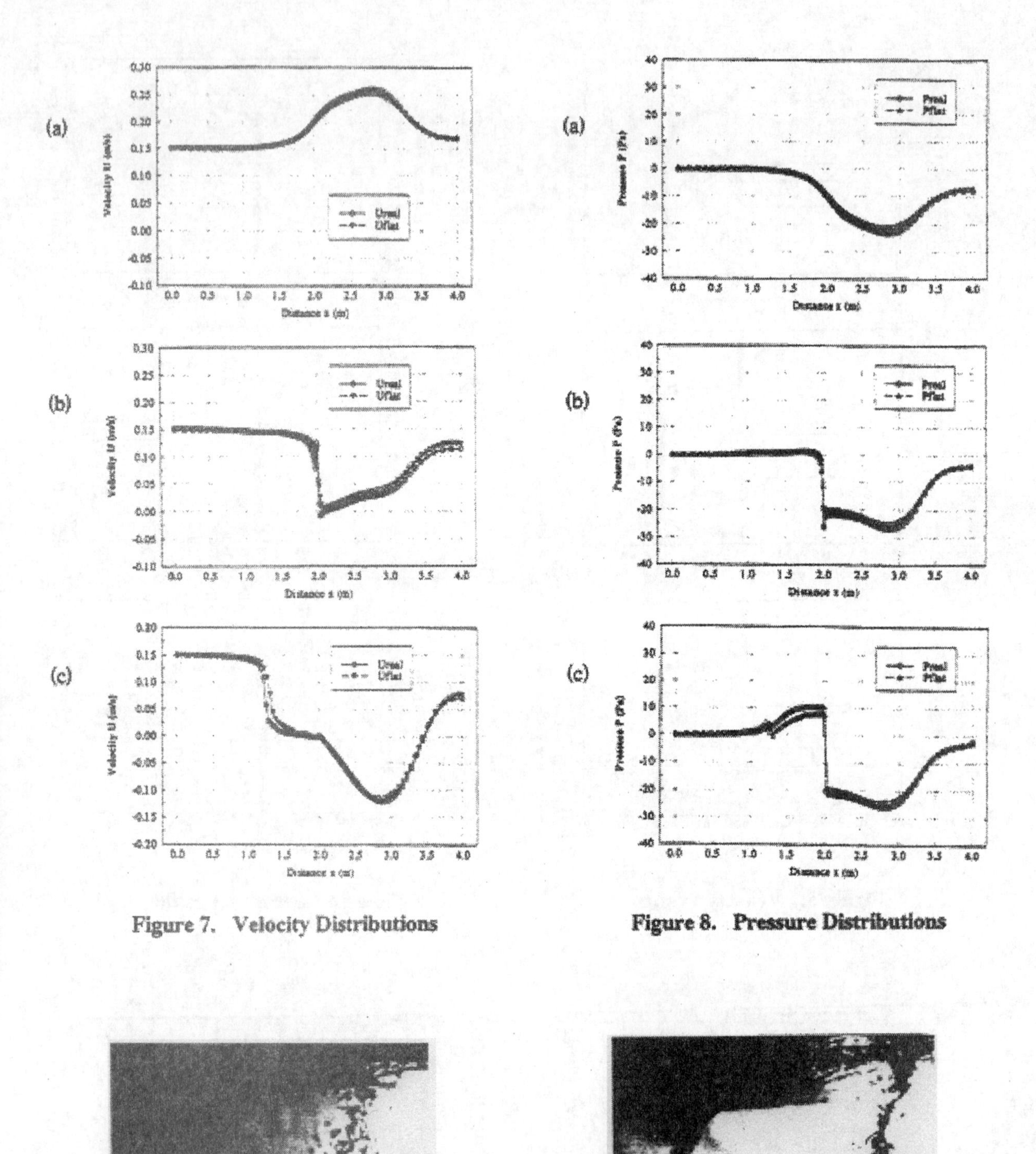

Figure 7. Velocity Distributions

Figure 8. Pressure Distributions

(a) In Discrete Form

(b) In Concentrated Form

Figure 10. Observed Oil Loss at 0.31m/s

VOLUME 1

Chang-Fan

PROCEEDINGS OF THE TWENTY-FIRST ARCTIC AND MARINE OILSPILL PROGRAM (AMOP) TECHNICAL SEMINAR

COMPTE RENDU: 21ᵉ COLLOQUE TECHNIQUE DU PROGRAMME DE LUTTE CONTRE LES DÉVERSEMENTS D'HYDROCARBURES EN MER ET DANS L'ARCTIQUE (AMOP)

JUNE 10 TO 12, 1998
EDMONTON, ALBERTA
CANADA

DU 10 AU 12 JUIN, 1998
EDMONTON, ALBERTA
CANADA

Environment Canada Environnement Canada

Canada

345

Flow Around Oil Containment Barriers

H.M. Brown and R.H. Goodman
Imperial Oil Limited, Calgary
Canada

C.-F. An
University of Windsor, Windsor,
Canada

Abstract

Measurements in a large flowing water channel of contained oil slick thickness profiles and the critical water current velocities at which oil loss begins, confirm the predictions of computational fluid dynamics (CFD) simulations of flow around a simple barrier. Over the past two years, CFD simulations with oils whose viscosity varies from 9 cSt to 5600 cSt and whose density varies from 834 kg/m^3 to 953 kg/m^3 have been made of flow around a simple barrier. All containment failure mechanisms observed along the boom centre-line with these oils have been accurately simulated by the two-dimensional CFD models. Studies of the containment capability of a more realistic flexible boom configuration show that its behaviour is not significantly different from that of a simple flat boom. Measurements around a porous boom system show that it is capable of oil containment at higher flow velocities than a single boom and have been used to validate a CFD model of the system.

1.0 Introduction

When oil spills occur on water and shorelines are threatened, the preferred response is to contain the oil and recover it. At present, recovery equipment can generally pick up oil from the water surface faster than it can be reliably contained by booms, even when seas are calm and currents are low (Schulze, 1993). The problem of containment failure at low current speeds was recognised in the early 1970's and a number of novel containment solutions were proposed (Graebel and Phelps, 1975). Despite some experimental work on new boom designs sponsored by the U.S. Coast Guard and others, no significant improvement in containment capability ensued. Renewed effort, however, as described by Coyne et al. (1998) has resulted in oil barriers which can contain heavy oils at speeds higher than standard booms.

In 1995 Imperial Oil Limited began a collaboration with the University of Calgary and the University of Windsor to apply computer modelling techniques to the problem of containment boom failure as a complement to flume studies. At that time computing power had become inexpensive enough that computational fluid dynamics (CFD) could be applied to this problem. If it could be demonstrated that CFD simulations correctly predicted the results of experimental measurements, then computer generated data could be used to understand boom failure processes and on this basis to develop better containment devices. The process of developing a validated CFD model involved a number of steps as follows:

- Choice of Appropriate Code: A commercial CFD code known as FLUENT was chosen as a possible simulator for flow around a simple flat plate barrier. The computational task is to find the solution of the Navier-Stokes equations for two-phase unsteady, incompressible flow in a two-dimensional grid. To the solution of the flow equations, a turbulence model must be added because the Reynolds number for typical flow parameters around a boom indicates a turbulent flow regime. Finally, the oil / water interface must be delineated during the simulation. This can be done by activating a volume of fluid (VOF) model in the code. This effort differs from that of Grilli et al. (1996) in which new code is being developed from first principles. The completed code, as chosen, was tested in a series of computations which insured that the flow field appeared to mimic experimental observations in a flowing channel, that the gridding was fine enough and correctly oriented to track the oil / water interface, and that the computations converged in a reasonable number of iterations.

- Verification of the Code: This step insured that, with appropriate input parameters, the model could simulate the results of documented experiments. Two flume experiments, described in the literature, were simulated with the final software code configuration. The flume experiments of Delvigne (1989) and Johnston (1993) which demonstrated droplet entrainment failure and creep (critical accumulation) failure respectively, were accurately reproduced by the CFD technique. Critical failure velocities were predicted and the graphical output of the simulations mimicked the visual observations and photographs reported in the literature.
- Extension of Code Verification: A final verification step before the completed code could be used in a predictive mode, was to show that the CFD code could be used to simulate experiments conducted with commonly transported crude oils and typical barrier configurations. In previously reported work, the computer simulations were compared to experimental measurements conducted in a large flowing channel for three oils of increasing density and viscosity (Brown *et al.*, 1997). Contained slick profiles, critical failure velocities, and qualitative flow behaviour along the channel centre-line were accurately simulated by the 2-D model for both drainage failure and creep failure. This paper reports experiments to confirm the model applicability to droplet entrainment failure and to study the behaviour of a more complex flexible boom configuration. Measurements to validate a porous boom model are reported.

2.0 Experimental Procedures

Experiments were conducted in a flowing channel 30m long by 1m wide and 1m deep which was built within an existing wave basin. Channel flow was driven by a 15hp ITT Flygt underwater mixer (Model 4660) which was controlled by a Toshiba variable frequency drive. Sets of two metre long flow straighteners consisting of bundles of 10cm I.D. PVC piping were placed at each end of the channel. This combination of drive and flow straightener provided useable flow velocities of up to 0.4m/s. Two large underwater viewing boxes were built into the channel walls adjacent to the boom position.

In laboratory experiments it was shown that an acoustic probe could be used in the water beneath an oil layer to measure the floating oil thickness. This technique was applied in the water channel to measure the thicknesses of slicks contained by booms. A Krautkramer model USD 15 ultrasonic flaw detector was used to generate pulses of 10MHz acoustic waves directed at the underside of floating slicks. The acoustic probe transmitted the pulses and detected their reflections from both the oil / water interface and the oil / air interface and measured the time delay between them. From this the oil thickness could be determined with an accuracy of $\pm$0.3mm. The probe was mounted on a track fixed to the channel floor and could be rapidly positioned anywhere along the centre-line of the channel under a slick by means of a screw mechanism.

The CFD model provides an estimate of the pressure field around a barrier. In order to confirm the model predictions, a method for measuring pressure drops within the water flow was devised. As the pressure variations were anticipated to be in the order of 50 Pascals (approximately 0.007 psi), a very sensitive pressure device was required. A Data Instruments Sursense model DCAL4 ultra low pressure transducer was used. This transducer is intended for air flow monitoring and is capable of measuring differential pressures over a range of 0 to 250 Pa. The sensor was mounted on a rigid frame above the water channel and attached to two stainless steel tube probes which extended into the water. The pressure difference between the two probe positions was recorded with a Campbell Scientific Model 21X micro logger and was estimated to be accurate to $\pm$2.5 Pa.

All other procedures used in the experiments were similar to those reported previously (Brown *et al.*, 1997).

3.0 Results and Observations

Three interrelated experiments were conducted to complete the validation of the CFD model. The results from these are described below.

3.1 Droplet Entrainment Failure

Droplet entrainment failure was observed for several boom drafts with Federated crude oil, which has a density of 834 kg/m^3 and a viscosity of 10 cSt at 15°C. Typical results for a boom draft of 10cm are shown in Figures 1 and 2. Figure 1 shows the measured slick profiles as water flow velocities are increased up to the point where the slick becomes unstable and droplet entrainment begins. As the flow velocity (indicated in the legend in m/s) increases, the slick shortens and the development of a head wave is apparent. The formation and entrainment of droplets is illustrated in the two photographs of Figure 2 taken through the viewing window of the channel and showing the underside of the slick. Oil loss due to droplet entrainment begins at a channel flow of about 0.30m/s as measured in mid-channel 3m upstream from the flat boom. The CFD model, with Federated oil properties and these channel dimensions, predicted a boom failure at a critical channel flow velocity of 0.29m/s. Qualitative observations of the onset and progression of droplet entrainment for this experiment were accurately portrayed in the model and have been reported elsewhere (An *et al.*, 1998a). It has been speculated that droplet entrainment is initiated by Kelvin-Helmholtz instabilities in the oil / water interface (Leibovich, 1976) and this has been confirmed by computer simulations as well.

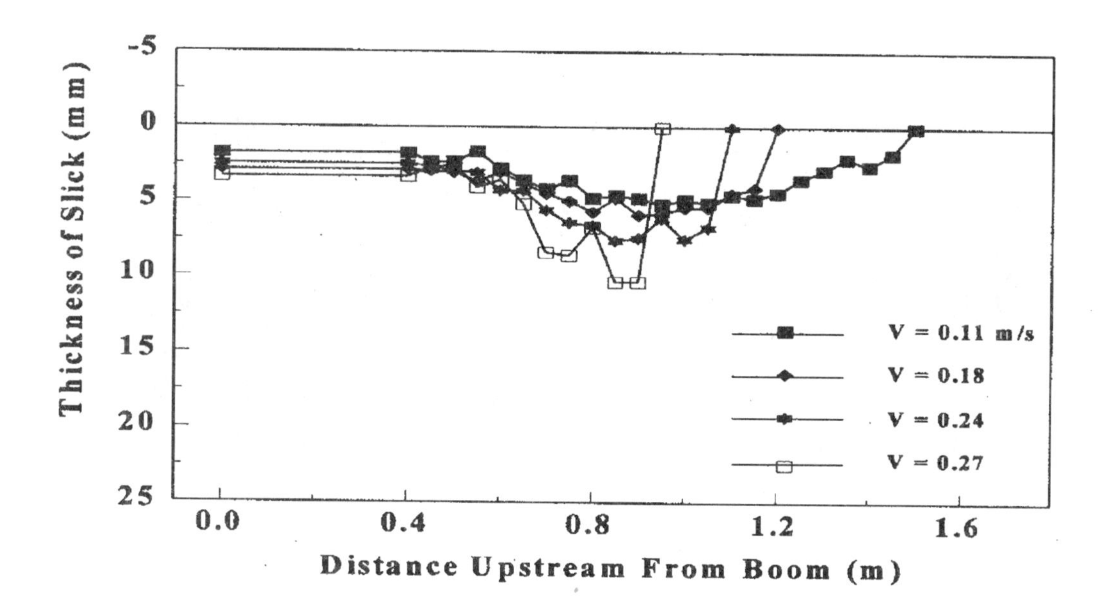

Figure 1 Slick Profile Prior to Droplet Entrainment

To date, CFD simulations of channel experiments have been two-dimensional and accurately describe failure mechanisms occurring along the channel centre-line. In a large channel it is difficult to prevent some flow velocity drop-off along the channel walls which results in a velocity gradient away from the walls. This leads to three-dimensional effects which become more pronounced as the boom draft increases. We have observed the formation, movement, and disappearance of vortices near the channel walls. These were unstable and often extended underneath the barrier entraining oil with them. This process competed as an oil loss mechanism with droplet entrainment for the low viscosity Federated oil. Measured critical velocities for these failure mechanisms are shown below as a function of boom draft.

Table 1 Effect of Boom Draft on Failure Mechanism

Draft (m)	Entrainment Loss (m/s)	Vortex Loss (m/s)
0.05	0.29	---
0.10	0.31	0.31
0.15	---	0.29
0.23	0.33	0.23

As the boom draft increases, oil loss by vortex entrainment becomes the dominant mechanism. For true two-dimensional behaviour at the channel centre-line, it appears that the boom draft can be no deeper than 10% of channel depth at maximum. Three-dimensional CFD modelling will be required to predict oil loss where velocity gradients perpendicular to the general flow direction occur.

Figure 2a Formation of Oil Drops on Slick Surface near the Critical Velocity

Figure 2b Loss of Oil Drops Under Barrier at Critical Velocity

3.2 Behaviour of Flexible Booms

Previous comparisons between the CFD model predictions and channel measurements have utilised simple flat plate booms. In this experiment, contained slick profiles and critical failure velocities were measured for a flexible skirt boom. The intent was to determine if there were significant differences between this more realistic boom configuration and the flat plate boom and to determine if the CFD modelling was capable of detecting any differences. The flexible boom consisted of an 8.5cm diameter cylindrical float which supported an 18cm deep skirt. The boom was anchored in position by attaching a stiff rod through the bottom of the skirt to the channel side walls but otherwise was allowed float freely. Figure 3 shows the shapes taken by the boom as flow velocities were increased.

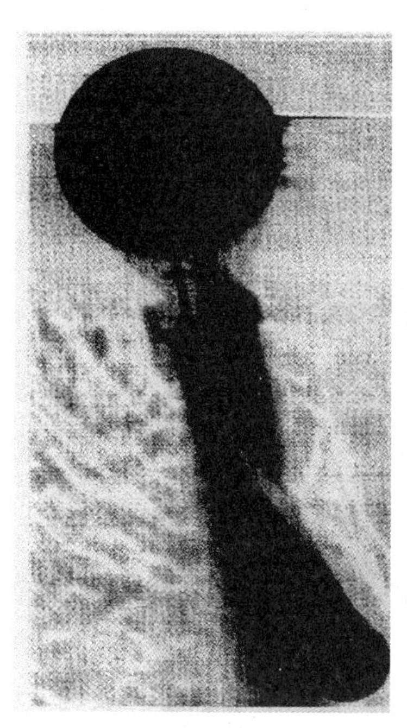

Figure 3 Crosssection of Flexible Boom for Increasing Flows

Slick thicknesses and lengths were measured and found to be similar to those for a flat plate barrier of similar draft. Because this boom had a very deep draft, vortex entrainment was the dominant loss mechanism and this failure mode began at 0.23m/s. However, droplet entrainment was also observed when the flow velocity reached 0.33m/s. A CFD simulation of this experiment, which models only the two-dimensional effects occurring along the channel centre-line, was carried out and showed droplet entrainment loss at a flow velocity of about 0.30m/s, in reasonable agreement with the measured value. This simulation required an advanced gridding procedure to model the cylindrical float and curved skirt (An *et al.*, 1998b). Both the CFD modelling results and the channel measurements show that there was no significant difference in the behaviour of the more realistically shaped flexible boom and that of the simple flat plate boom.

3.3 Porous Booms

In the past, experiments have been conducted with one or more porous booms preceding a solid boom. The intent of this configuration is to slow the floating oil and upper water layers to speeds below those at which the solid boom will fail (Norton and Rand, 1975). An experiment was carried out with a porous boom system of 0.10m draft consisting of one simple flat barrier with a 50% porosity and a second flat barrier with 30% porosity. These were positioned 2m and 1m respectively upstream of a flat solid barrier. The porous barriers consisted of flat metal plates with horizontal rows of circular holes. Figure 4 shows the measured velocity profiles along a vertical line at the mid-point between the solid boom and the second porous barrier. It is evident that the velocity in the upper layer of the channel is decreased

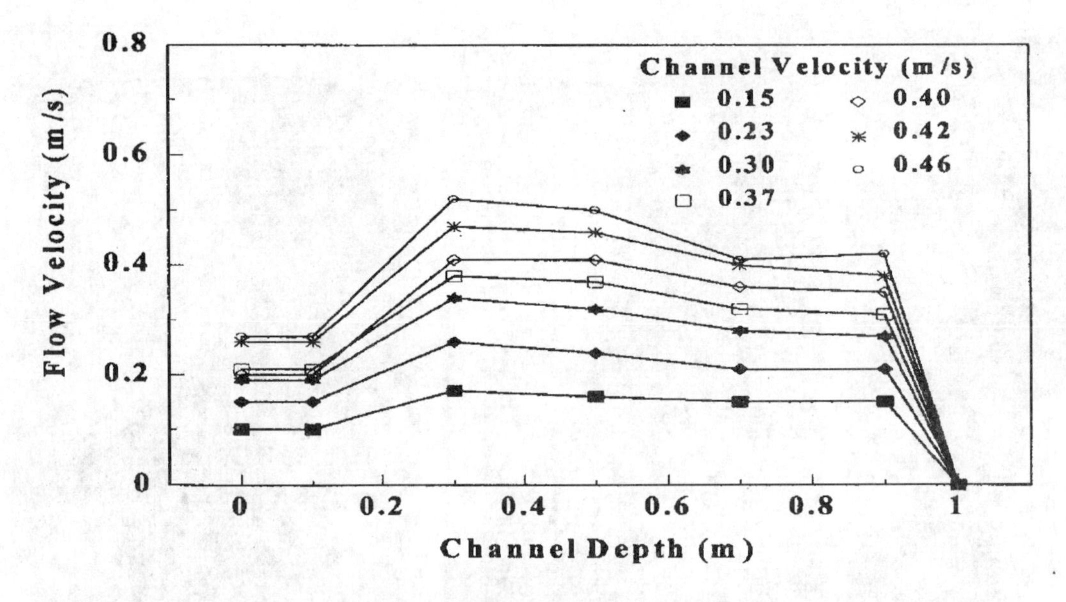

Figure 4 Velocity Depth Profile with Two Porous Booms

significantly for all reference velocities. When Federated oil was spilled in the channel upstream of this set of booms, the critical channel velocity at which the solid boom failed was found to be 0.39m/s, an improvement over the critical failure velocity of about 0.29m/s for a single flat boom. Measurements were made of the pressure drop along the channel centre-line of the three boom system using the pressure probe device described above. Figure 5 shows the results of these measurements at a depth of 0.10m below the water surface. Each point on the graph is the pressure difference between two probes, the sensing probe 15cm downstream of the graphed position and a reference probe placed 15cm upstream of the graphed position. Thus the difference in pressure across the solid boom at the 0m channel position is about -25Pa, that is, the pressure downstream of the boom is 25Pa lower than the upstream reference. There is a similar pressure drop across the 30% porous boom and a lesser pressure drop across the 50% boom. At intermediate channel positions between the solid boom and 30% porous boom, the pressure increases as one approaches the solid boom. The measurements are consistent for all velocities.

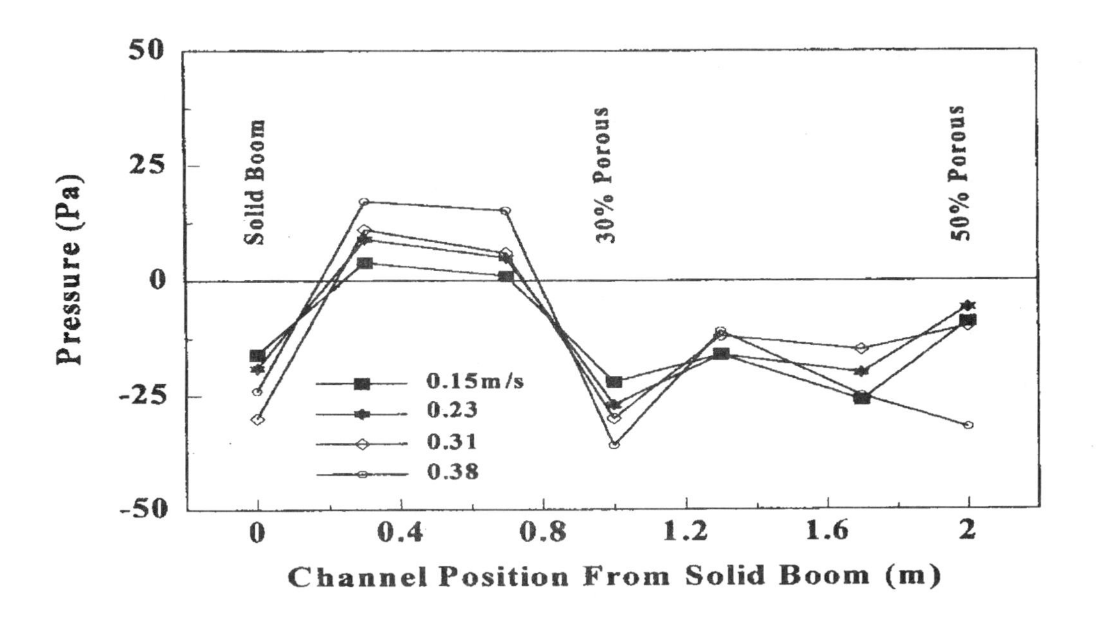

Figure 5 Pressure Drop Across Multiple Porous Booms

A pressure profile derived from a CFD simulation of this boom system is shown in Figure 6. While the figure above shows only pressure measurements for discrete points along the boom system axis, Figure 6 is a continuous profile. There are obvious pressure drops across the upper (50%) porous boom and the solid boom. However, the magnitude of these and the relative drops for each boom differ from those measured. This indicates that some of the model parameters are incorrect and demonstrates the need for validation of a model with physical experiments before reliable predictions from the model can be made. When the validation has been completed, it will be far easier to model boom parameters to achieve optimum oil retention than to attempt to do this physically in a flowing channel.

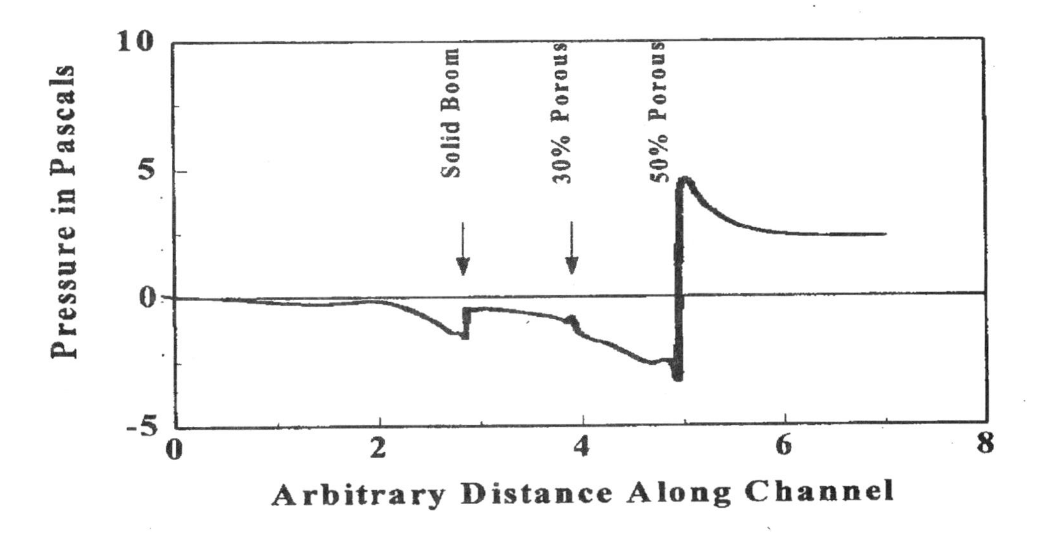

Figure 6 Pressure Profile From Porous Boom Simulation

4.0 Conclusions

Computational fluid dynamics modelling has been validated by reproducing experiments on boom failure reported in the literature and by comparison with measurements in a flowing channel. All three failure mechanisms which occur at the boom centre-line have been accurately simulated by two-dimensional CFD modelling. Velocity gradients near the channel walls create vortices which will require three-dimensional CFD modelling to correctly describe this loss mechanism. Studies of boom failure using a conventional boom shape show that CFD predictions using a flat plate boom provide good approximations for this case. Measurements around a porous boom were used to improve a CFD model of these boom systems.

5.0 References

An, C-F., R.M. Barron, H.M. Brown, and R.H. Goodman, "Droplet Entrainment Boom Failure and Kelvin-Helmholtz Instability", paper # 98-Mi-01, accepted for publication in *Proceedings of the 8th International Offshore and Polar Engineering Conference*, Montreal, Canada, 1998a.

An, C-F. and R.M. Barron, "A Comparative Study of Oil Boom Failure by Critical Accumulation", accepted for publication in *Proceedings of the 6th Annual Conference of the Computational Fluid Dynamics Society of Canada (CFD 98)*, Quebec City, Canada, 1998b.

Brown, H.M., R.H. Goodman, C-F. An, and J. Bittner, "Boom Failure Mechanisms: Comparison of Channel Experiments with Computer Modelling Results", *Proceedings of the Twentieth Arctic and Marine Oil Spill Program (AMOP) Technical Seminar*, Environment Canada, Ottawa, Canada, pp. 457-467, 1997.

Coyne, P.M., D.S. DeVitis, and W.T. Schmidt, "1997 Ohmsett Test Tank Utilization", *Proceedings of the Twenty-first Arctic and Marine Oil Spill Program (AMOP) Technical Seminar*, Environment Canada, Ottawa, Canada, 1998.

Delvigne, G.A.L., "Barrier Failure by Critical Accumulation of Viscous Oil", *Proceedings of Oil Spill Conference*, API, Washington, D.C.,USA, pp.143-148, 1989.

Graebel, W.P., and V.A. Phelps, "Fast Current Oil Response Systems, Tests on Phase I Concepts", Report # Cg-D-20-76, U.S. Coast Guard, Washington, D.C., 1975.

Grilli, S.T., Z. Hu and M.L. Spaulding, "Numerical Modeling of Oil Containment by a Boom", *Proceedings of the Nineteenth Arctic and Marine Oil Spill Program (AMOP) Technical Seminar*, Environment Canada, Ottawa, Canada, pp. 343-376, 1996.

Johnston, A.J., M.R. Fitzmaurice, and R.G.M. Watt, "Oil Spill Containment: Viscous Oils", *Proceedings of the 1993 Oil Spill Conference*, API / USCG, Washington, D.C., USA, pp. 89-94, 1993.

Leibovich, S., "Oil Slick Instability and the Entrainment Failure of Oil Containment Booms", Journal of Fluids Engineering, Vol .98, pp. 98-105, 1976.

Norton, D.J., and J.L. Rand, "The Hydrodynamics of Porous Barrier Oil Recovery Systems", Report # CG-D-146-75, U.S. Coast Guard, Washington D.C., 1979.

Schultze, R., "Oil Spill Encounter Rate: A Means of Estimating Advancing Skimmer Performance", *Proceedings of the Sixteenth Arctic and Marine Oil Spill Program (AMOP) Seminar*, Environment Canada, Ottawa, Canada, pp.419-433, 1993.

PROCEEDINGS OF THE THIRD INTERNATIONAL CONFERENCE ON FLUID MECHANICS

BEIJING, 1998

Chang fa An
Beijing, China
July 7, 1998

Beijing Institute of Technology Press

安长发博士科研论文集

UNSTABLE INTERFACE BETWEEN OIL AND WATER WITH FAST RELATIVE MOTION

R. M. Barron & C.-F. An
Fluid Dynamics Research Institute, Univ. of Windsor, Windsor, Ontario, Canada N9B 3P4
Z. Zhang
Dept. of Chemical Engineering, Beijing Univ. of Chemical Technology, Beijing, China 100029

Abstract In this paper, a study of the unstable interface between oil and water with fast relative motion is presented. The objective is to develop a clear understanding of Kelvin-Helmholtz (K-H) instability on oil-water interface. This instability is considered to be the mechanism of droplet entrainment in oil boom failure. It has been observed that at the critical relative velocity 0.33m/s, droplet entrainment failure occurred for a realistic boom configuration. This experimental result has been simulated using the computational fluid dynamics (CFD) code Fluent. Computation of the K-H instability on the oil-water interface has further confirmed the experimental and computational results.

1. INTRODUCTION

The motivation of this study is to provide a scientific basis for the strategy of dealing with accidental oil spills on water to environmental agencies, coast guards and oil companies. Oil spills occur every year and may severely pollute water resources, for example, Exxon Valdez spill in Alaska, USA, in 1989 [1] and Diamond Grace spill in Tokyo Bay, Japan, in 1997 [2]. If oil is spilled on the surface of water, the immediate response is to use booms to contain the oil and skimmers to remove it. Booms are barriers that float on the surface of water and contain the spilled oil. However, oil containment by a boom is successful only under slow relative velocity between oil (with boom) and water. If the relative velocity exceeds a certain value, even in a calm water area without effects of wind and waves, boom failure may occur and the oil may escape underneath the boom due to hydrodynamic forces. This fact severely limits the capability of booms and there is a need within the oil spill response community to understand the hydrodynamic mechanisms of boom failure.

Boom failure mechanisms, or modes, have been studied since 1970's and a number of boom failure mechanisms have been recognized, such as drainage failure, droplet entrainment and critical accumulation [3]. Drainage failure usually occurs for a shallow boom draft when the thickness of the oil slick becomes larger than the draft. This failure can be avoided by deepening the boom. Critical accumulation usually occurs for heavy and highly viscous oils. For frequently encountered light and less viscous oils, droplet entrainment is the most threatening failure mode. It usually occurs at a relative velocity of 0.3-0.4m/s, which is a common speed in the cleanup process of an oil spill.

The droplet entrainment mode of boom failure was noticed at the beginning of boom studies, but the hydrodynamic mechanism was not clear. Leibovich [4] analyzed the mechanism of droplet entrainment and posed an explanation that droplet formation on the oil-water interface is the result of an instability of Kelvin-Helmholtz (K-H) type. The onset of droplet formation from contained oil slicks is due to the breaking of finite amplitude interfacial waves. He concluded that the minimum velocity is about $2.2U_{cr}$ below which no droplets can be formed. When the velocity is between $2.2U_{cr}$ and $3.1U_{cr}$, droplets can be formed but only from the head wave. When the velocity is beyond $3.1U_{cr}$, droplets can be formed from any part of the interface. Here U_{cr} is the critical velocity derived from linear K-H theory. Other researchers have shown interest in K-H instability [5-7] and the relation between K-H instability and droplet entrainment [8-10]. However, very little work has been done using computational fluid dynamics (CFD), especially with turbulent flow conditions.

In this paper, a study of unstable interfacial phenomena between oil and water with fast relative motion is presented. The study is directly dedicated to the understanding of the relationship between droplet entrainment boom failure and the instability of oil-water interface. Section 2 reports the experimental results regarding droplet entrainment failure of a realistic boom configuration. Section 3 presents the computational results based on the flow conditions in the above channel experiments using the commercial CFD code Fluent. The K-H instability on the oil-water interface is also simulated for a simple model using Fluent. Section 4 gives some comments, remarks and conclusions. Videotapes on the experimental observations and the animations based on the CFD simulations help to achieve a clearer understanding of this flow.

2. EXPERIMENTAL STUDY

Measurements and observations of the interface between oil and water with relative motion were conducted in a 30m long, 1m wide and 1m deep channel [11]. The water flow was driven by an underwater pump, which was manipulated by a variable frequency controller. Two sets of flow straighteners consisting of bundles of plastic pipes were placed at each end of the channel. A realistic boom configuration was installed about 1.5m downstream of the channel mid-length. An underwater window box was fixed on the channel wall in the vicinity of the boom. A flow meter was mounted at a reference point in the channel centerline, 3m upstream from the flat plate barrier and 0.5m below from the surface. The channel was able to provide flow velocity of up to 0.4m/s, enough to make the oil-water interface unstable and reach the state of droplet entrainment failure.

The realistic boom configuration consists of a cylindrical float and a flexible skirt with a rod in its bottom hem. The float is used to provide buoyancy and the skirt is used to contain the oil. In the field condition, the boom may be towed or held in place by cables connecting either the float or the skirt hem. In the present experiment, the boom had an 8.5cm diameter cylindrical float and an 18cm deep skirt. The boom was anchored in position by attaching the rod through the bottom hem of the skirt to the channel sidewalls. The float and the skirt can move freely in the water current. Figure 1 shows the boom shapes at different current velocities.

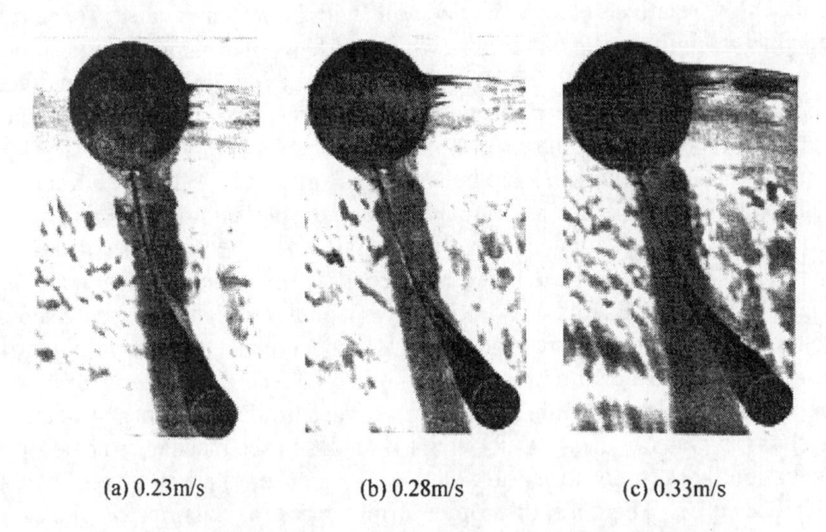

(a) 0.23m/s (b) 0.28m/s (c) 0.33m/s

Figure 1. Boom Shapes at Different Current Velocities

Five liters of Federated crude oil with density 834kg/m^3 and viscosity 10cSt were poured onto the channel surface upstream of the boom. When the flow velocity was low, a stable oil slick was formed. When the flow velocity was increased to 0.33m/s, droplet entrainment failure occurred. Figure 2 is a photograph taken from the underwater window and shows the close up of the wavy and unstable oil-water interface and the strong droplet entrainment at velocity 0.38m/s. Because this boom has a very deep draft of 22cm, i.e. the ratio of boom draft to water depth is 22%, three-dimensional effects were very strong. As a result, a pair of vortices was formed, moved around and disappeared intermittently in front of the boom. These vortices usually dived underneath the boom bringing some oil with them. At times, oil loss due to this "vortex diving" was as strong as that due to droplet entrainment (Figure 3).

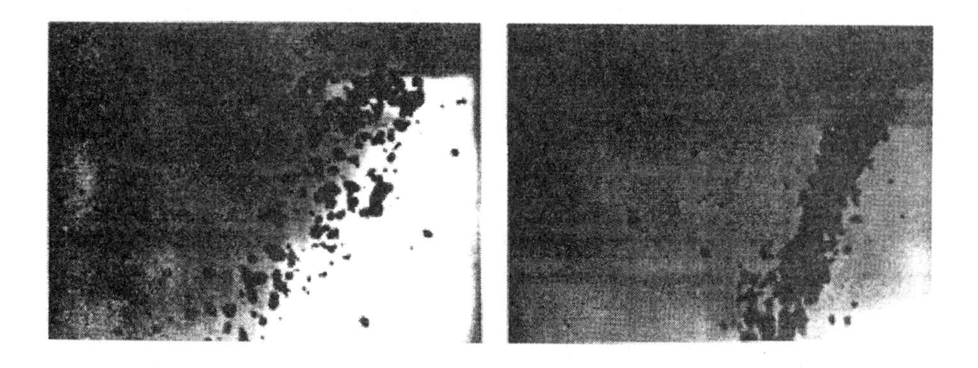

Figure 2. Droplet Entrainment (0.38m/s) **Figure 3. Vortex Diving** (0.38m/s)

3. COMPUTATIONAL STUDY

3.1 Droplet Entrainment

Measurements and observations from the channel experiment show that the realistic boom configuration is able to successfully contain 5 liters of light oil up to the flow velocity 0.33m/s. Beyond this velocity, the oil-water interface becomes wavy and unstable and substantial oil loss begins due to droplet entrainment. These results have been simulated using the commercial CFD code Fluent [12] by the assumption that the flow in the channel centerline can be considered as two-dimensional.

The computational grid was generated by Geomesh that is available with the Fluent package and is shown in Figure 4. The mesh is 207x54 and clustered so that the cells are denser near the boom and the upstream surface and sparser elsewhere. The float is approximated by a semi-circle and the skirt is approximated by a straight segment at a small angle to the vertical direction.

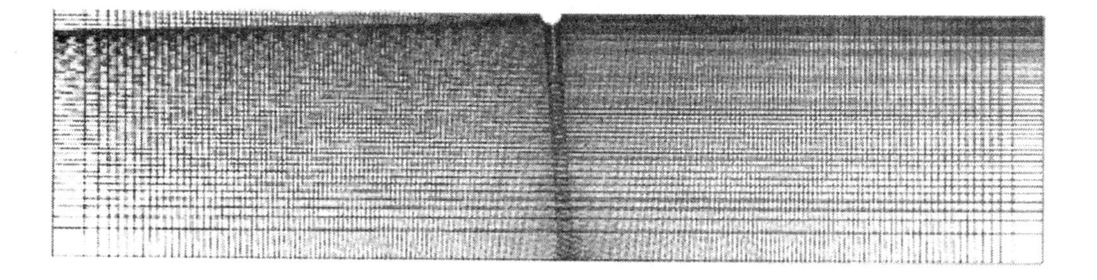

Figure 4. Computational Grid

The right boundary is taken as the inlet, the left boundary is an outlet, the bottom of the channel and the surface of the boom are solid walls and the free surface of the channel is a slip-wall on which tangent velocity component is permitted but normal velocity component is prohibited. Under the experimental condition, the Reynolds number based on the current velocity, channel depth and water viscosity is about 10^4 so that the flow is generally turbulent. The standard k-ε turbulence model is used to close the Navier-Stokes equations in the solver. Because the flow is essentially unsteady, especially when oil loss occurs, the time dependent solver is activated and the time step is chosen to be Δt=0.05sec at the beginning of the computation and decreased afterwards if necessary. In order to track the oil-water interface, the VOF (volume of fluid) model is invoked. After the water-only flow attains its steady state at velocity 0.10m/s, a thin layer of oil is introduced from the top of the inlet with the same velocity. When the exact amount of oil has been spilled into the flow region, oil introduction is stopped and the computation continues until a steady state is reached. Then, the flow velocity is increased at increments of 0.05m/s and the computation continues until another steady state is reached. In this way, the computation continues until substantial oil loss begins. Figure 5 shows a steady state flow field at flow velocity 0.20m/s. The graph is a contour plot of the volume fraction of oil overlapped on the background of the velocity vector plot. The oil slick becomes stable against the boom and no obvious oil loss can be found. Figure 6 shows an instantaneous flow field at velocity 0.30m/s at simulation time 162.5sec. From this figure one can see that some oil is detached from the oil-water interface and swept underneath the boom by the water current. It is not possible to capture the details of oil droplets that are entrained by the main flow because the dimension of the droplets is about 3-5mm but the dimension of the finest cell is about 10mm. In spite of this, the oil loss mechanism can be identified as droplet entrainment based on the following facts:

(i) The draft of the boom skirt is deep enough to contain oil and to avoid drainage failure;
(ii) The oil is not heavy and highly viscous so that it is not critical accumulation failure;
(iii) The head wave can be seen as wavy and unstable;
(iv) The oil is intermittently detached from the interface and entrained by the main flow.

It should be pointed out that the channel flow velocity was measured at the reference point. The average velocity is usually smaller than that value. For the critical velocity 0.33m/s at which droplet entrainment occurred, an average velocity of 0.30m/s (10% deduction) should be reasonable. Therefore, the computational prediction has attained good agreement with the experimental measurements.

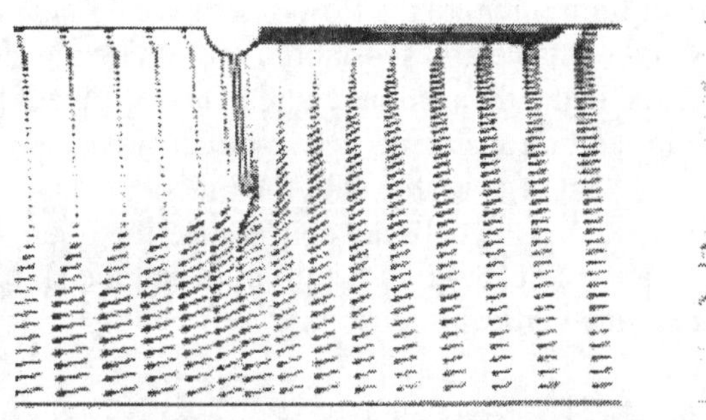

Figure 5. Stable Oil Slick (0.20m/s)

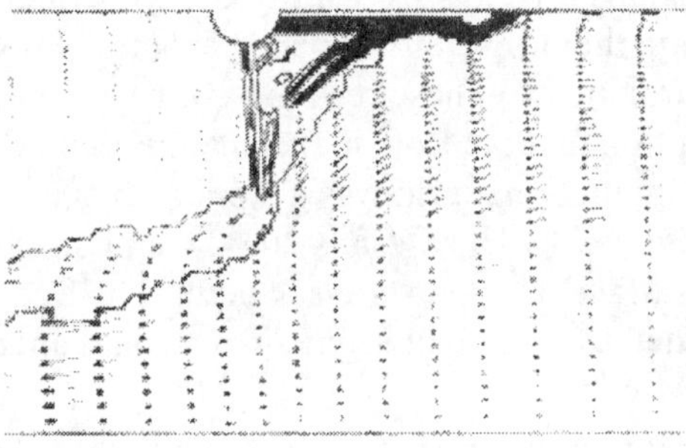

Figure 6. Droplet Entrainment (0.30m/s)

3.2 Kelvin-Helmholtz Instability

In order to investigate the details of the unstable interface between oil and water, a simple model has been simulated using Fluent. Suppose we have two layers of fluids with different densities and velocities. The upper fluid is oil with density ρ_o= 834kg/m^3 and velocity U_o = 0. The lower fluid is water with density ρ_w = 1000kg/m^3 and velocity U_w = 0.30m/s. At time t=0, an initial small disturbance wave (Figure 7a)

$$\eta = -\varepsilon \sin(2\pi x/\lambda), \quad 0 \le x \le \lambda$$

is imposed on the oil-water interface. Here, λ=0.03m is the length of the disturbance wave and $\varepsilon=\lambda/40$ is the amplitude of the disturbance wave. The computational domain is chosen as a rectangle of $4\lambda x \lambda/2$ and the grid is 101x71 clustered near the interface. The time-dependent Euler solver is used for the computation and the VOF model is activated to trace the oil-water interface. At the left inlet boundary, U_o=0 for the upper part and U_w=0.30m/s for the lower part. The right outlet boundary is free of specification. The upper and lower horizontal boundaries are assumed to be slip-walls. Figure 7 shows the shapes of the oil-water interface at different simulation times.

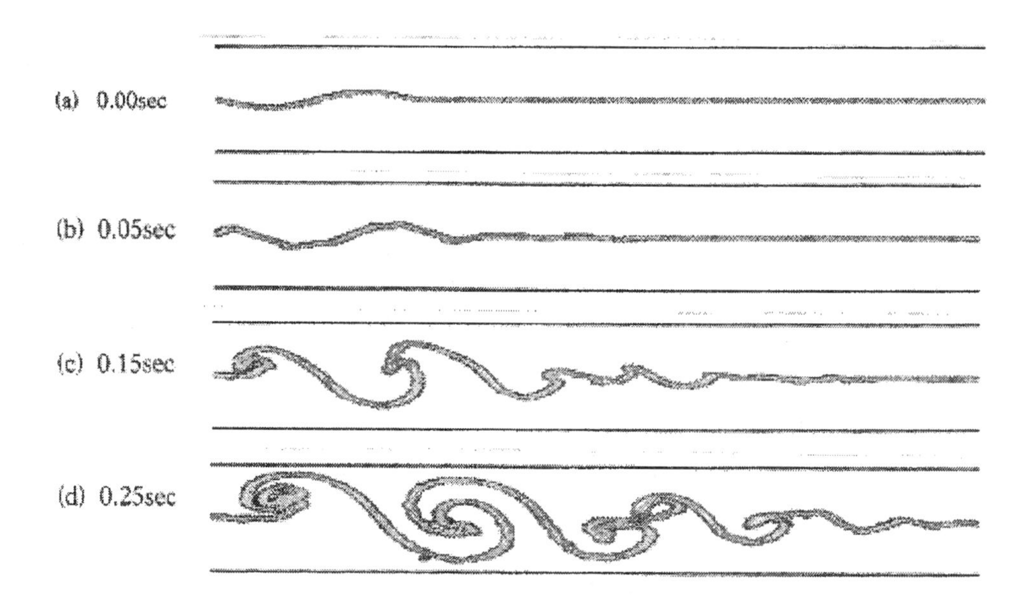

Figure 7. Kelvin-Helmholtz Instability

From these graphs it can be seen that the amplitude of the interfacial waves is amplified considerably and the shape is severely distorted. Concentrated vortices are generated (Figure 7c) and strengthened significantly (Figure 7d). It can be expected that the interfacial waves, or the concentrated vortices, will be broken eventually as time increases. The broken oil mass will become droplets and may be swept by the underlying water current causing droplet entrainment. From linear analysis [13], the critical relative velocity in this case is $U_{cr} = 0.12$m/s. According to the analysis of Leibovich [4], droplets should be formed from the head wave when the relative velocity is between $2.2U_{cr}$ and $3.1U_{cr}$, i.e. between 0.264m/s and 0.372m/s. The relative velocity U_w-U_o = 0.30m/s falls in this range. This result is consistent with the channel experiments reported in section 2 where the critical velocity is 0.33m/s and with the computational simulation presented in section 3.1 where the critical velocity is 0.30m/s.

4. CONCLUSIONS

The instability on the interface between oil and water with fast relative motion and its consequence, droplet entrainment failure of oil containment booms, have been studied experimentally and computationally. The experimental study has provided a scenario in which droplet entrainment occurred at 0.33m/s for a realistic boom configuration with 0.22m draft in a 1m wide and 1m deep water channel. The computational study has simulated the experimental scenario with an acceptable accuracy using the commercial CFD code Fluent. A simple model simulation of K-H instability has further confirmed the experimental and computational studies of the unstable interface between oil and water with fast relative motion.

Keywords
Oil spill, Droplet entrainment, Interfacial phenomena, Kelvin-Helmholtz instability, CFD

REFERENCES

[1] Wells, P.G., Butler, J.N. and Hughes, J.S., eds., Exxon Valdez oil spill: fate and effects in Alaskan waters, Proc. of 3rd Symp. on Environ. Toxicology & Risk Assessment, Atlanta, GA, USA, Apr. 26-28, 1993

[2] An, C.-F., Goodman, R.H., Brown, H.M., Clavelle, E.J. and Barron, R.M., Oil-water interfacial phenomena behind a boom on flowing water, Proc. of 10th Int'l Symp. on Transport Phenomena (ISTP-10), Vol.1, pp.13-18, Kyoto, Japan, Nov.30-Dec.3, 1997

[3] Delvigne, G.A.L., Barrier failure by critical accumulation of viscous oil, Proc. of 1989 Oil Spill Conf., pp. 143-148, San Antonio, TX, USA, Feb. 13-16, 1989

[4] Leibovich, S., Oil slick instability and the entrainment failure of oil containment booms, J. Fluids Eng., Vol. 98, pp. 98-105, 1976

[5] Yih, C.-S., Dynamics of Nonhomogeneous Fluids, pp.159-163, MacMillan Co., New York, 1965

[6] Drazin, P.G., Kelvin-Helmholtz instability of finite amplitude, J. Fluid Mech., Vol.42, Part 1, pp.321-335, 1970

[7] Pai, S.-I., Two-Phase Flows, pp.108-111, Weiweg & Sons, Braunschweig, Germany, 1977

[8] Thorpe, S.A., Experiments on the instability of stratified flows: immiscible fluids, J. Fluid Mech., Vol.39, Part 1, pp.25-48, 1969

[9] Jones, W.T., Instability at an interface between oil and flowing water, J. Basic Eng., pp.874-878, 1972

[10] Kordyban, E., The behavior of the oil-water interface at a planar boom, J. Energy Resources Tech., Vol.112, pp.90-95, 1990

[11] Brown, H.M. Goodman, R.H. and An, C.-F., Flow around oil containment barriers, 21st Arctic & Marine Oilspill Program (AMOP) Tech. Sem., Edmonton, AB, Canada, June 10-12, 1998

[12] Fluent User's Guide v4.4, Fluent Inc., Lebanon, NH, USA, 1996

[13] Lamb, H., Hydrodynamics, 6th edition, pp. 373-375, Cambridge Univ. Press, Cambridge, UK, 1932

VOLUME 2

Chang-fa An

PROCEEDINGS OF THE TWENTY-SECOND ARCTIC AND MARINE OILSPILL PROGRAM (AMOP) TECHNICAL SEMINAR

COMPTE RENDU: 22ᵉ COLLOQUE TECHNIQUE DU PROGRAMME DE LUTTE CONTRE LES DÉVERSEMENTS D'HYDROCARBURES EN MER ET DANS L'ARCTIQUE (AMOP)

JUNE 2 TO 4, 1999
CALGARY, ALBERTA
CANADA

DU 2 AU 4 JUIN, 1999
CALGARY, ALBERTA
CANADA

Environnement
Canada

Canadä

Development of Containment Booms for Oil Spills in Fast Flowing Water

H.M. Brown and R.H. Goodman
Imperial Oil Limited, Calgary, Canada
and
C.-F. An
University of Windsor, Windsor, Canada

Abstract

Two double-boom systems designed to contain oil at high water currents have been developed and prototype sections of these tested in a flowing channel. Computational fluid dynamics (CFD) modelling was used to determine the feasibility of these designs. Measurements of the current velocity profiles along the stream centre-line of the booms were then made to confirm that the designs would successfully trap oil. Tests with three oils, whose viscosities varied from 10 cSt to 10,420 cSt and whose densities varied from 834 kg/m^3 to 964 kg/m^3, showed that the designs or modifications to them were successful in containing the oils. These oils were contained at twice the current speeds at which a simple flat barrier failed.

1.0 Introduction

The weakest component in the rapid recovery of spilled oil on water is the containment of the oil. Containment booms generally do not hold oil reliably when towed on open water at speeds above about 0.5 m/s (1 knot). Boom models tested in flowing channels usually fail at 0.25 m/s (0.5 knot) or less depending on the Froude number of the particular experiment (Wicks, 1969; Delvigne, 1989; Johnston *et al.*, 1993). Most of the relevant theoretical and experimental work on improved boom designs or other oil collection devices was carried out in the 1970's (e.g.: Hoult *et al.*, 1970; Hale *et al.*, 1974; Norton and Rand, 1975). The majority of this work is reported in about 25 papers, including those which developed a mathematical description of oil spill containment (e.g.: Wicks, 1969; Cross and Hoult, 1971; Wilkinson, 1972, 1973; Milgram and Van Houten, 1978). This latter work considered the forces acting on the oil / water interface to devise relations for the critical failure speeds in terms of Froude and Weber numbers and in a few cases, to describe mathematically the slick profile. The general features so described were confirmed by laboratory measurements in small flumes using relatively light oils. All theoretical work assumed a two - dimensional system and none tackled the dynamic problem of solving the Navier-Stokes equations.

On the basis of this theoretical and experimental work, a number of oil collection and recovery devices were proposed. Prototypes of these were tested on behalf of the U.S. Coast Guard at both Texas A&M and the University of Michigan (Graebel and Phelps, 1975). The devices generally fell into three categories: devices which used some energy dissipative mechanism to slow the oil and water to below the critical loss velocity before encountering the boom, shields which separated the contained oil slick from the faster underlying water, or devices for the rapid removal of oil from the water surface without first containing it. Seven prototype devices, including at least one from each of these categories, were eventually tested and while most succeeded in containing the oil or recovering it at water currents of up to 1.5 m/s (3 knots), all had significant design or handling problems at that development stage. It does not appear that any of these designs reached commercial production. Recently, several groups have tested improved designs at OHMSETT, although the details of these have not been reported (Coyne *et al.*, 1998).

Imperial Oil has collaborated with the University of Calgary and the University of Windsor to investigate boom failure mechanisms using computational fluid dynamics modelling (An *et al.*, 1998; An and Barron, 1998). The present boom designs were devised on the basis of this work and on previous studies using a flowing channel (Brown *et al.*, 1997; Brown *et al.*, 1998).

2.0 Experimental Procedures

Prototype sections of the multiple boom designs were tested in a flowing channel constructed within an existing wave basin in Calgary, Canada. The channel was approximately 30 m long and had a cross-section of 1.25 m in width by 0.80 m in depth. Computational fluid dynamics (CFD) modelling of flow around a simple solid barrier has shown that significant vortex generation occurs along the channel walls if the channel width to depth ratio is less than 1.5. With low width to depth ratios some vortices may be formed even at the channel centre-line (Zhang *et al.*, 1999). The channel cross-section was chosen in order to minimize these effects so that true entrainment losses could be observed. Channel flow was driven by a 15 hp ITT Flyte underwater mixer (Model 4660) which was controlled by a 20 hp Toshiba variable frequency drive. Sets of flow straighteners consisting of two metre long bundles of 10 cm I.D. PVC piping were placed at each end of the channel. This combination of drive power and straighteners provided uniform flow at velocities of up to 0.5 m/s or approximately twice the flow velocities at which containment failure is observed in a channel around a simple solid boom. Observations of the oil and water behavior around the multiple booms could be made from two adjacent underwater viewing boxes built into the channel walls at the mounting position of the booms. Each box was large enough to hold an observer and their camera equipment.

Measurements of the flow velocity along the channel centre-line were made at several depths for each boom configuration. These were repeated for each of four nominal channel flow rates. In addition, depth-velocity profiles were made at a position 20 cm upstream of the solid boom component of multiple boom systems (see Figure 1). Nominal channel flow rates were established by measuring the flow as a function of mixer input power at 24 points on a channel cross-sectional grid 3 m upstream of the solid boom position. The average of these grid points for a particular power input was taken as the nominal flow rate. Nominal channel flow rates were determined both with and without a simple solid boom in place.

A Faraday law electromagnetic flow meter (Marsh-McBirney Flo-Mate Model 2000) was used for flow measurements. Usually, the probe of the flow meter was attached to a stand on a track-mounted dolly on the channel floor, which could be moved rapidly along the channel centre-line by means of a screw mechanism. The latter was operated by a hand crank which protruded out of the water downstream of the boom position. The depth of the probe below the water surface could be reproducibly set at 5 cm intervals by means of a détente on the dolly stand.

After the flow profiles had been measured as a function of channel input power for each boom configuration, 5 litres of oil were spilled 3 m upstream of the boom and observations of the oil / water flow around the boom were made from the underwater viewing boxes. Oil was typically spilled at the lowest nominal channel flow rate and if successfully contained by the boom configuration, the flow rate was increased to the next nominal channel flow. This scheme was repeated for each of three oil types for each boom configuration. The oils used and their properties at 15°C are shown in Table 1. After each test, the booms were cleaned with an oil solvent or were replaced with new material.

Table 1. Properties of Oils Used in the Experiments

Oil Type	Density (kg/m^3)	Viscosity (cSt)
Federated Crude	834	10
Hondo Crude	939	1443
Weathered Hondo Crude	964	10420

Straight sections of model booms were constructed and fixed in position across the channel in front of the viewing boxes. This provided observers with an above and below waterline cross-sectional view of water and oil movement through and around the multiple boom systems. The two boom configurations, which were designed in conjunction with CFD simulations, are shown in cross-section in Figure 1. These consist of a simple solid boom represented here by a plexiglas plate, preceded upstream by a cylindrical float boom with an angled skirt. In one design, the skirt is porous and has attached to it a mop-like mat of oleophilic fibres. In a second design, the float is held slightly below the water surface so that most of the oil and some water would pass over it. In both designs, variations in the fibre density, skirt porosity, and bottom net were tested as shown in Table 2.

Table 2. Design Configurations Tested in the Flowing Channel

Boom Design	Fibres	Skirt Porosity	Bottom Type
A	single density	50 %	net
	double density	60 %	no net
	no fibres		
B	single density	50 %	net
	no fibres	0 %	no net
Plate	no fibres	no skirt	no net

Computer simulations suggested that designs of this type should trap and hold oils at current velocities above those at which a simple boom would fail. The channel

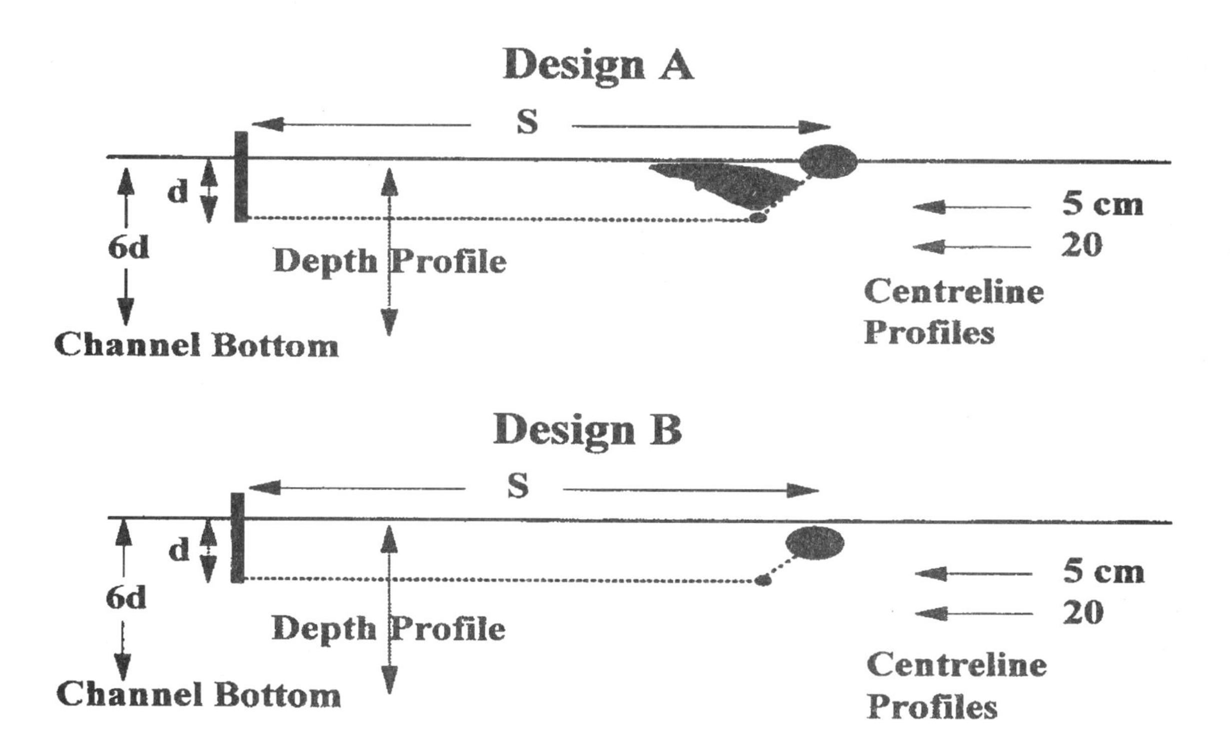

Figure 1. Cross-sections of Model Booms

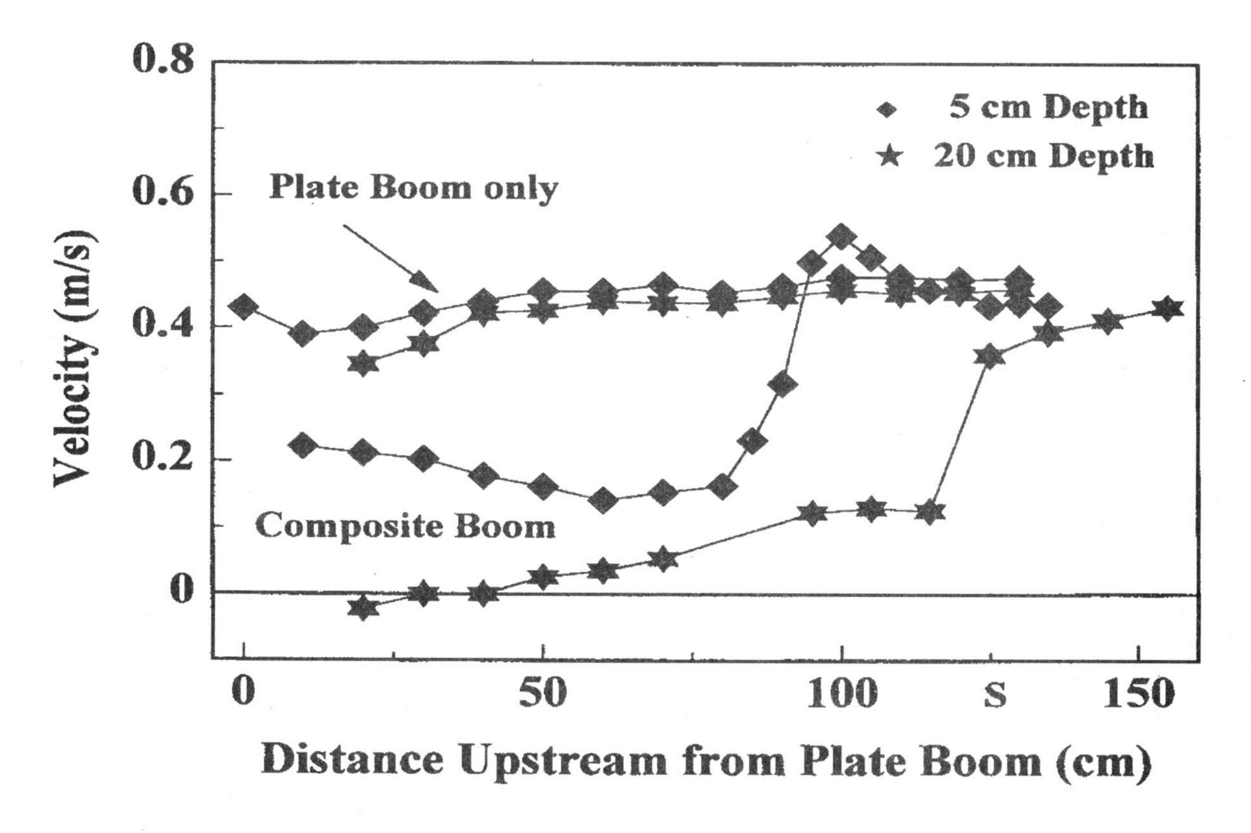

Figure 2. Centre-line Velocity Profile of Design A Boom

experiments were carried out to confirm these predictions and to examine the effect on failure velocities of variations in the construction materials and configuration.

3.0 Results and Discussion

3.1 Flow Velocities Around Complex Booms

An example of the effect on water flow profiles of a multiple boom design as compared to those measured around a simple plate boom is shown in Figure 2. This compares the centre-line flow velocities of Design A with single density oleophilic fibres, a 50% porosity skirt, and bottom net and velocities of the plate barrier alone. Velocities at two depths are shown: one depth just below the water surface and the second just lower than the boom draft. These data were measured at a nominal channel velocity of 0.5 m/s. The figure shows that the surface velocity between the porous skirt with the fibre mop and the downstream solid barrier is reduced to zero whereas there is practically no reduction in surface velocity with the solid barrier alone. There is also a significant reduction in the velocity of the water passing just beneath the skirt and bottom net but no reduction at this depth beneath the simple barrier. Figure 3 compares the vertical velocity profile at a position 20 cm in front of the solid boom for both the simple boom and complex boom cases. For the simple boom case, there is a small reduction in velocity near the water surface but for the complex boom case, the water velocity is zero between the water surface and the boom net bottom. These measurements show that oil which penetrates the upstream fibre skirt of the complex boom should pool and remain trapped indefinitely within the boom in the quiescent area upstream of the solid barrier. Similar results were obtained for lower nominal channel velocities.

The effect on flow patterns of the oleophilic fibres attached to the porous skirt of Design A is shown in Figure 4. These serve to reduce the flow near the water surface within the boom system compared to the flow when only the porous skirt is used. They also reduce the flow velocity directly beneath the bottom net, possibly because the skirt appears less porous and diverts a faster stream of water further down into the channel below the upstream boom. This streamlining may create a more quiescent zone close to the bottom net. The same effect is observed if a more porous skirt is used to support the fibre strands. Figure 5 shows the centre-line velocities at the same two depths but compares a skirt with a 60% porosity to that with a 50% porosity. When a larger proportion of the flow passes through the skirt, the velocity close to, but below the bottom net, also increases. A similar phenomenon is observed if the bottom net between the fibrous skirt and the solid boom is removed. The effect on flow velocities is shown in Figure 6. With no connecting bottom net, a reverse surface current is set up within the complex boom due to currents just below the boom system impinging the bottom of the downstream plate. The current velocity just below the boom draft is increased. These measurements suggest that the presence of the bottom net serves to separate the faster moving layers of water beneath the boom system from those within the boom.

In Design B the upstream boom acts as a weir and serves to skim off the top layer of water and oil while diverting most of the water beneath the boom system. That it should be effective in trapping oil is indicated by the centre-line velocities

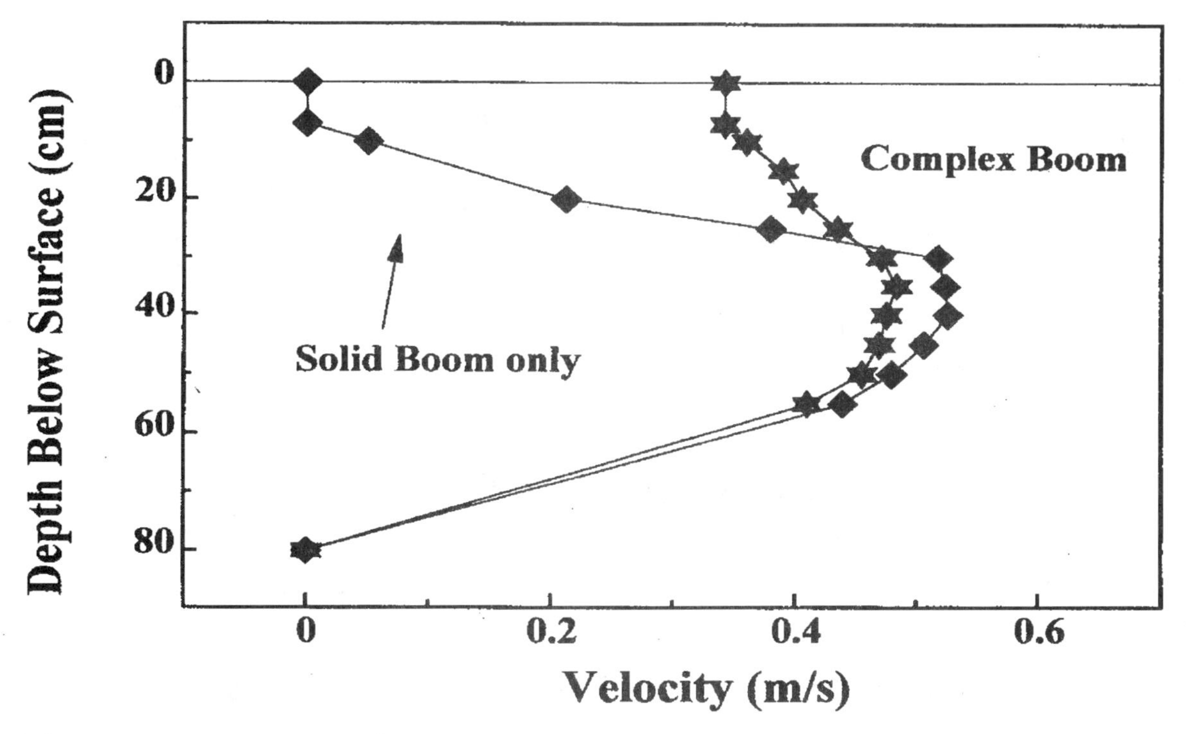

Figure 3. Comparison of Depth Velocity Profiles

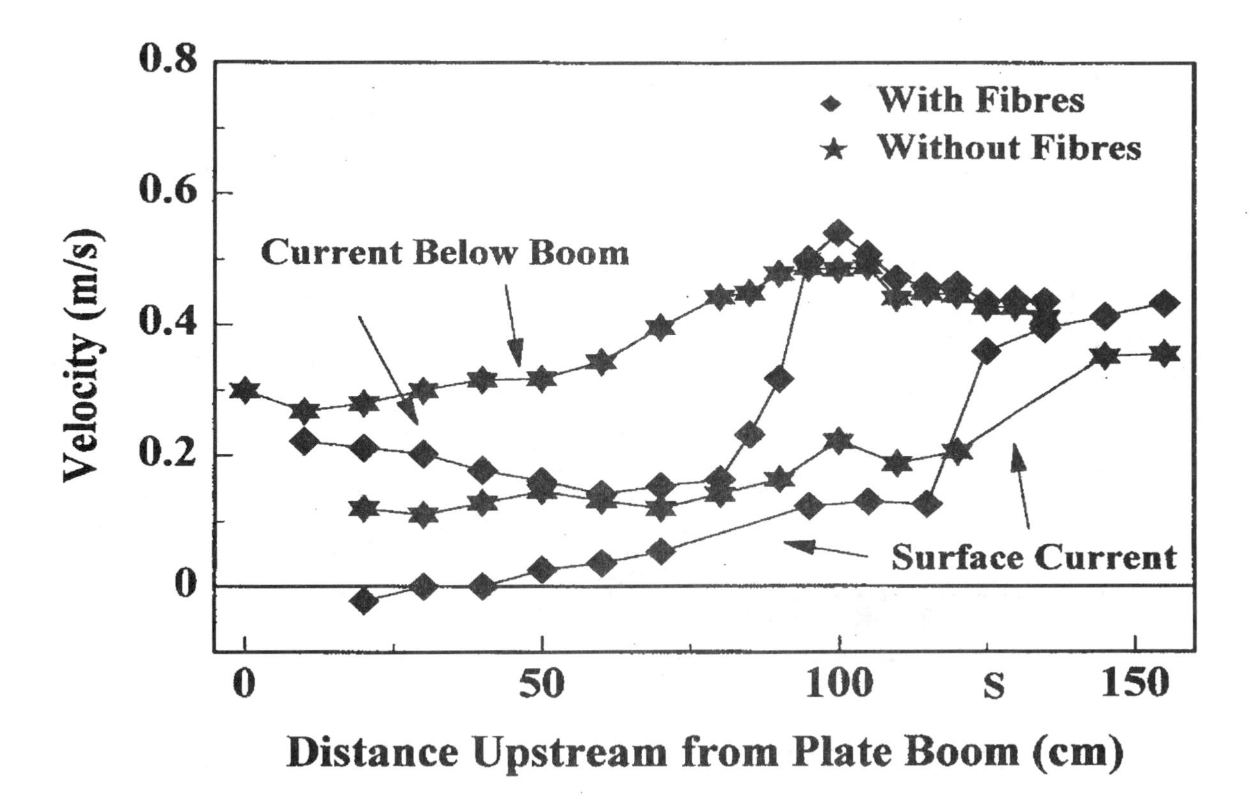

Figure 4. Comparison of Velocity Profiles with and without Fibres

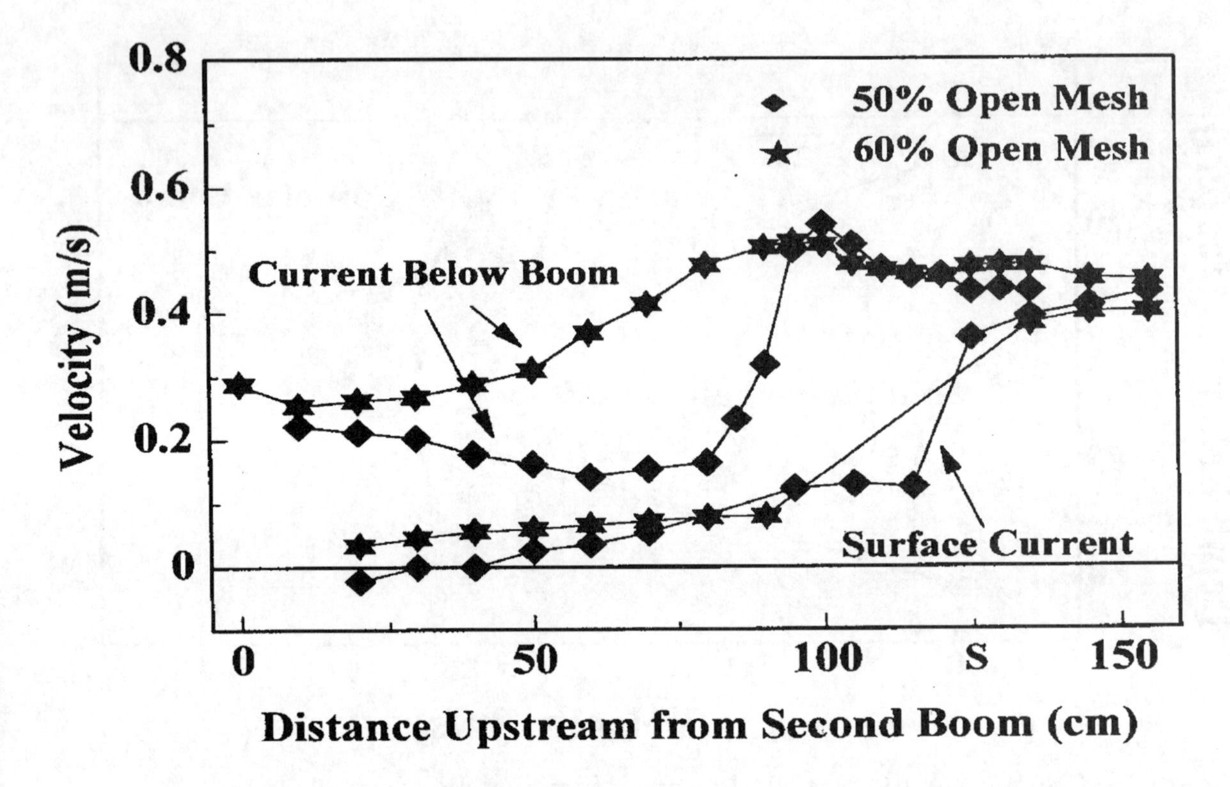

Figure 5. Centre-line Velocity Profiles for Different Mesh Sizes

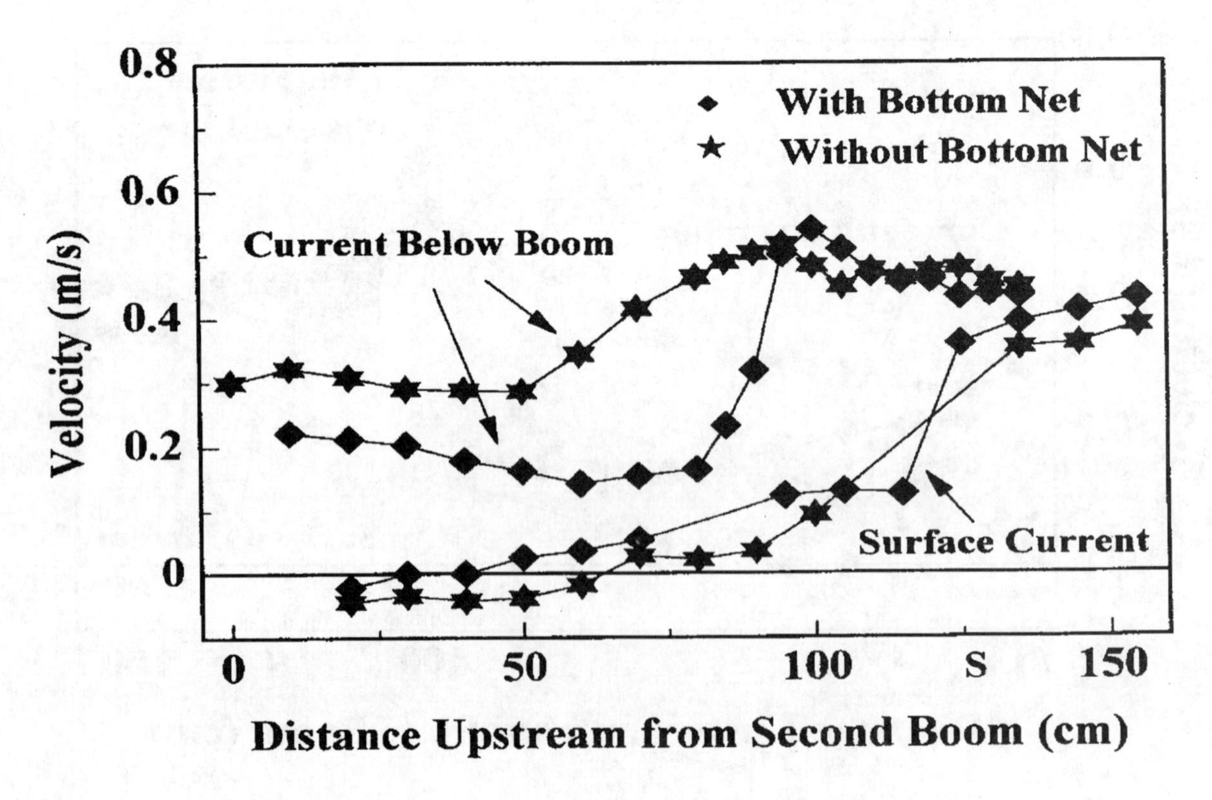

Figure 6. Comparison of Velocity Profile with and without Net

shown in Figure 7. Again the surface velocities within the boom system are seen to be very low compared to those of the simple solid boom. However, any oil which is not skimmed by the submerged upstream boom may be lost in the high currents which flow immediately beneath the boom bottom net.

3.2 Oil Interaction with the Boom Systems

The measurements of current velocities around the boom systems suggest that oil which penetrates the upstream barrier should be trapped in the quiescent area of these multiple booms. Observations of oil movement around the booms were made for the range of oils and their properties noted above, in order to confirm that oil would be trapped. Thirty-five mm photographs were taken through the channel underwater viewing windows to document these observations.

Design A performed well with the light Federated oil. Five litres of oil were spilled upstream of the boom, encountered the first cylindrical barrier, were carried under it and penetrated the mesh skirt to pool in the area between the upstream cylinder and the solid plate barrier. It appeared that this oil would remain trapped for an indefinite time at a channel velocity of 0.5 m/s. When the heavier and more viscous Hondo oil was spilled, it encountered the cylindrical boom and crept slowly down the mesh skirt. A very small fraction of this oil did not extrude through the skirt and continued to creep along the bottom net. A few drops of this creeping oil were torn away and lost under the plate barrier at 0.5 m/s. This very minor loss was due to the long rise time associated with this dense oil and the slow extrusion through the relatively fine mesh bottom net.

Design A was modified by replacing the 50% porosity net skirt and bottom with a more open mesh material. This material had a nominal open area to fibre area of 0.6 (60% porosity), but consisted of fewer, wider fibres and larger openings. It was clear from the experiments that specifying the mesh porosity alone was insufficient to predict correctly the extrusion rate of viscous oils. With this modification the boom system was again tested with Hondo oil and trapped all the spilled oil at the highest attainable channel velocity of 0.5 m/s. This test was repeated with an even more viscous weathered Hondo oil ($\rho = 964\ kg/m^3$ and $\gamma = 10420$ cSt) and again trapped all the oil at the highest channel velocity. In both cases, oil which crept down the mesh skirt was extruded into the fibre mop and ultimately into the quiescent zone of the system.

A test was conducted to determine the effectiveness of the mop-like fibres attached to the mesh skirt. When these were removed and the light Federated oil was spilled upstream, oil passed rapidly under the cylindrical boom and through the mesh skirt. At high channel velocities there was sufficient turbulence at the mesh skirt that some water-in-oil-in-water bubbles were formed. As these were neutrally buoyant, it was possible for a few to float through the quiescent zone where there was a low current (see Figure 4) and pass under the solid plate boom. Although this was a very minor oil loss, a thin sheen was formed downstream of the boom system. With higher viscosity oils, some drops were formed in the turbulent zone and these were generally trapped between the booms.

A similar loss in system effectiveness was observed when the bottom net connecting the upstream boom skirt and the solid plate boom was removed. This

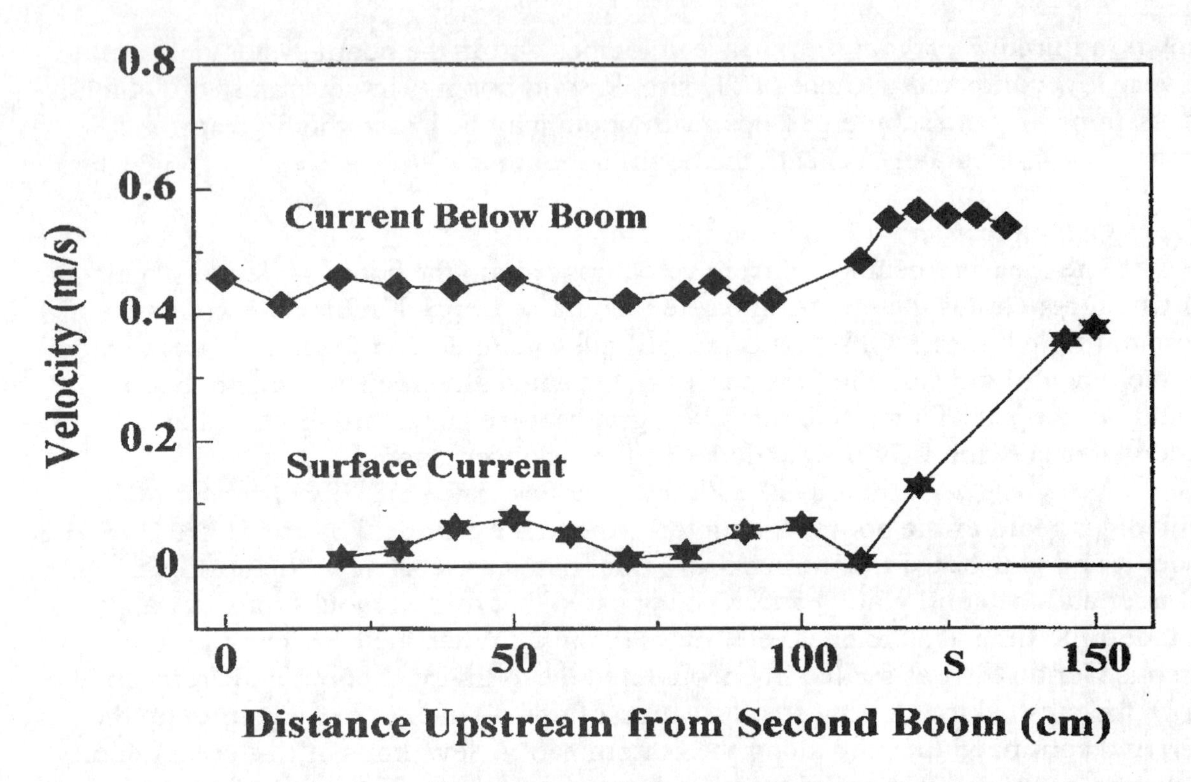

Figure 7. Centre-line Velocity Profile of Boom Design B

appeared to be due to the observed increase in current velocity down the skirt when the bottom net was removed (see Figure 6). Any oil which did not penetrate the skirt mesh was entrained in the high speed current passing beneath the boom system. Although this was not a significant loss mechanism for the achievable channel velocities, it could become a problem at higher velocities. Most oil did reach and pool in the quiescent zone.

Whereas Design A created an oil pooling zone by using a porous mesh and oleophilic fibre barrier to slow the water and separate out the oil, Design B used a weir and solid skirt to separate the oil and create a calm collection zone. When Federated oil was spilled upstream of this barrier, a very small amount of oil was lost through drop formation at the weir. As the oil and water flows over the shallow weir, a shear zone is set up which creates a large number of oil drops, particularly of the light oils. A few of these were lost through the bottom net, although most resurfaced and were trapped behind the solid boom. No oil was lost when the boom was used to trap the Hondo oil. However, with the heavier, more viscous weathered Hondo oil, some oil was lost. This occurred when some of the oil adhered to the cylindrical weir and crept down the solid-skirt slope instead of passing over the weir. This oil extruded into long strands which did not rise quickly enough to pass through the bottom net. In a modification of the weir design, the solid skirt was replaced with the oleophilic fibre and 50% mesh skirt of Design A. This design dampened the shear zone behind the weir and decreased the formation of oil drops but did not solve the problem of viscous oil creep down the outside of the skirt. The use of a larger mesh size skirt and bottom net should prevent this loss.

4.0 Conclusions

Measurements of the water velocities along the centre-line of two multiple booms suggested that these should successfully trap and hold oils over a wide range of viscosities. Observations of spilled oil encountering these booms confirmed that the designs were effective. The presence of oleophilic fibres attached to the upstream boom skirt increased the effectiveness of the multiple booms by slowing the water entering the oil trapping region and by providing an oil-attracting pathway into the trapping area. Bottom nets between the booms decreased the water velocity within the boom system and improved oil trapping. Viscous oils of up to 10,000 cSt were successfully trapped if a more open mesh was used in the upstream boom skirt together with oleophilic fibres. A weir boom was less successful in trapping these very viscous oils. Water velocities of twice those at which simple flat barriers have been shown to fail were used in the tests.

5.0 References

An, C.-F., R.H. Goodman, H.M. Brown and R.M. Barron, "Droplet Entrainment Boom Failure and Kelvin-Helmholtz Instability", *Proceedings of the Eighth International Offshore and Polar Engineering Conference*, Montreal, Vol.II, pp.322-325, 1998.

An, C.-F. And R.M. Barron, "Comparative Study of Models for Oil Boom Simulation", *Proceedings of the Sixth Annual Conference of the Computational Fluid Dynamics Society of Canada*, Quebec, pp.IV-37-42, 1998.

Brown, H.M., R.H. Goodman, C.-F. An, and J. Bittner, "Boom Failure Mechanisms: Comparison of Channel Experiments with Computer Modelling Results", *Proceedings of the Twentieth Arctic and Marine Oil Spill Program (AMOP) Technical Seminar*, Environment Canada, Ottawa, Canada, pp.457-468, 1997.

Brown, H.M., R.H. Goodman, and C.-F. An, "Flow Around Oil Containment Barriers", *Proceedings of the Twenty-first Arctic and Marine Oil Spill Program (AMOP) Technical Seminar*, Environment Canada, Ottawa, Canada, pp.345-354, 1998.

Coyne, P.M., D.S. DeVitis, and W.T. Schmidt, "From High Speed Skimmers to Sorbents: 1997 Ohmsett Test Tank Utilization", *Proceedings of the Twenty-first Arctic and Marine Oil Spill Program (AMOP) Technical Seminar*, Environment Canada, Ottawa, Canada, pp.949-956, 1998.

Cross, R.H. and D.P. Hoult, "Collection of Oil Slicks", *Journal of the Waterways, Harbors, and Coastal Engineering Division*, WW2, pp.313-322, 1971.

Delvigne, G.A.L., "Barrier Failure by Critical Accumulation of Viscous Oil", *Proceedings of the 1989 Oil Spill Conference*, API, San Antonio, Texas, pp.143-148, 1989.

Graebel, W.P. and V.A. Phelps, *Fast Current Oil Response Systems*, U.S. Coast Guard report CG-D-20-76, Washington D.C., 1976.

Hale, L.A., D.J. Norton and C.A. Rodenberger, *The Effects of Currents and Waves on an Oil Slick Retained by a Barrier*, U.S. Coast Guard report CG-D-53-75, Washington D.C., 1974.

Hoult, D.P., R.H. Cross, J.H. Milgram, E.G. Pollak and H.J. Reynolds, *Concept Development of a Prototype Lightweight Oil Containment System for Use of the High Seas*, U.S. Coast Guard report 714102/A/003, Washington D.C., 1970.

Johnston, A.J., M.R. Fitzmaurice, and R.G.M. Watt, "Oil Spill Containment: Viscous Oils", *Proceedings of the 1993 Oil Spill Conference*, API, Tampa, Florida, pp.89-94, 1993.

Milgram, J.H. and R.J. Van Houten, "Mechanics of a Restrained Layer of Floating Oil above a Water Current", *Journal of Hydronautics*, Vol.12, pp.93-108, 1978.

Norton, D.J. and J.L. Rand, *The Hydrodynamics of Porous Barrier Oil Recovery Systems*, U.S. Coast Guard report CG-D-146-75, Washington D.C., 1975.

Wicks, Moye, "Fluid Dynamics of Floating Oil Containment by Mechanical Barriers in the Presence of Water Currents", *Proceedings of the Joint Conference on the Prevention and Control of Oil Spills*, API - FWPCA, New York, N.Y., pp.55-106, 1969.

Wilkinson, D.L., "Dynamics of Contained Oil Slicks", *Journal of the Hydraulics Division, Proc. of A.S.C.E.*, pp.1013-1030, 1972.

Wilkinson, D.L., "Limitations to Length of Contained Oil Slicks", *Journal of the Hydraulics Division, Proc. of A.S.C.E.*, pp.701-712, 1973.

Zhang, Z., C.-F. An, and R.M. Barron, "A Numerical Study on Porous Boom Systems ", *Proceedings of the Twenty-second Arctic and Marine Oil Spill Program (AMOP) Technical Seminar*, Environment Canada, Ottawa, Canada, 1999.

Numerical Study on (Porous) Net-Boom Systems-
Front Net Inclined Angle Effect

Z. Zhang, C.-F. An & R.M. Barron
Fluid Dynamics Research Institute
University of Windsor
Windsor, Ontario, Canada
zzhang@uwindsor.ca &/or zhangz@buct.edu.cn

H.M. Brown & R.H. Goodman
Imperial Oil Resources Limited
Calgary, Alberta, Canada

Abstract

In this paper, a commercial CFD software, Fluent, is used to simulate the flow of oil and water against a family of net-boom structures (a solid rear boom with front and bottom nets) in which the angle of inclination of the front net is varied. From a plot of inclination angle against critical velocity, it is found that the highest critical velocity is achieved when the front net is nearly vertical or slightly inclined forward. Detailed analyses of the flow fields for these structures, at the same upstream flow velocities and at the critical velocity of each case, lead to a clearer understanding of the physical mechanisms which cause oil to escape under the rear boom. The simulations compare well to the available experimental data.

1.0 Introduction

An oil boom is one of the major tools to deal with oil spill on the sea surface. It is known that the spilled oil can be well contained only under certain restricted conditions. If those conditions are not satisfied, the oil will escape underneath the boom by hydrodynamic forces and the oil boom containment will fail. The boom failure can be studied by both experimental work and theoretical analysis (An, 1996*a*). There are a number of papers studying the boom failure problem by using numerical simulation, the computational fluid dynamics (CFD) method, in the past years. Clavelle and Rowe (1993, 1995*a*, 1995*b*) employed the volume of fluid (VOF) method to predict the boom failure critical accumulation and the results were consistent with Delvigne's observation (Delvigne, 1989) for a highly viscous oil. Grilli *et al* (1996) used a numerical method to study oil containment by a boom. An (1996*b*) numerically simulated three failure mechanisms by using a commercial CFD software, Fluent. The numerical results coincided well with the experiments done by Delvigne (1989) and Johnston *et al* (1993), both qualitatively and quantitatively. The computational results captured the "surging" phenomenon and "oil creep" phenomenon for a heavy oil observed in Johnston's experiment. Also, for a light and low viscous oil, droplet entrainment failure was reached as that in Delvigne's observations. Recently, oil spill stability problems have been of interest (An *et al*, 1998, An and Barron, 1998) to explore the mechanism of droplet entrainment and other instability features in the oil spill by using the CFD method. All of these show that the CFD method has strong potential of studying oil spill behavior and predicting critical velocity. However, the studies above are concentrated on discussing the situation only for a single boom structure. There are many factors, such as boom structure, boom draft, water depth, oil and water properties, water velocities, etc., that contributes to the success or failure of oil containment. Among them, boom structure may be one of the most important factors. Obviously, a good boom structure should be able to collect a greater amount of oil as quickly as possible when an oil spill occurs. Recently, efforts have been made to invent new types of oil booms with better performance. Ueda and Yamanouchi (1998) reported some experiments using several types of net structure. Unfortunately, they were only concerned with obtaining a dimensionless relationship between parameters and no critical velocity was reported. In the meantime, some experiments have been done at Imperial Oil using two separate perforated plates with different porosity located at some distance upstream from the solid boom (An *et al*, 1997). It was found that the perforated plates can improve the oil containment by increasing the critical velocity from 0.34 m/s to 0.42-0.46 m/s for Federated Crude oil (ρ=834 kg/m^3 and ν = 10 cs). Some numerical work was also done by Cheng *et al* (1997), who simulated the flow field and predicted critical velocity for the same porous boom system tested at Imperial Oil, using their own developed code based on VOF method, finite volume scheme and k-ε turbulence model. The predicted critical velocity and velocity profiles coincide well with the experimental results at Imperial Oil. Furthermore, Cheng *et al* (1997) found that if the front porous boom had a small modification with about 1cm solid length on the upper part of the perforated plate, the critical velocity could be further increased up to 1.0 m/s, double the value for a fully-holed perforated plate. The reason for this improvement is that the upper solid part of the perforated plate makes the water flow near the surface much more quiescent. This not only indicates a promising direction for boom structure improvement, but also confirms that numerical simulations have the capability of such prediction and structure improvement. Of course, the prediction needs to be verified by the corresponding experiments.

Recent experiments in the Imperial Oil Test Basin (Brown *et al*, 1999) have shown that the oil containment capability of a solid boom can also be improved by adding some nets in front of it. Based on these (porous) net-boom prototype designs, some numerical simulation of oil-water flow around various net-boom structures has been reported (Zhang *et al*, 1999). It has been found that the angle of inclination of the front net has a significant effect, and this effect will be discussed in this paper.

2.0 Boom Structure Configuration

The net-boom structure under consideration is a solid boom with front and bottom nets. The bottom net is horizontal and the front net is inclined from the vertical at an angle α as shown in Figure 1. The basic case, **A-3**, has an angle α of 67° in accordance with the prototype design **A$_3$** tested in the Imperial Oil Test Basin (Brown *et al*, 1999).

In order to study the front net inclined angle effect, several variants of the **A-3** net-boom structure with the front net at different angles are considered, labeled **H-B, Q-F, H-F, F-F** and **VER**. Here, **H-B** refers to **Half-Backward** inclined front net compared to the fully-backward inclined front net in **A-3. Q-F, H-F** and **F-F** are variants of **A-3** with **Q**uarter-, **H**alf- and **F**ully-**F**orward inclined front nets, respectively. The angles for **Q-F, H-F** and **F-F** are equal to 30°, 49° and 67° respectively, but inclined forward. **VER** means that the front net is standing vertical and α=0. Figure 2 illustrates **A-3** and its variations.

The computational domain and physical dimensions are schematically shown in Figure 3. The dimensions coincide with the experimental settings at the Imperial Oil Test Basin. The channel depth is 0.8 m, the solid boom draft is 0.15 m, the distance between the solid boom and front net is about 1.35 m-1.40 m. For the **A-3** case, the front net is inclined backwards about 0.25 m and the inclined angle is about 67° from the vertical. In the experiments, the channel width is 1.25 m and the amount of oil spilled on the surface is five litres. This means that the oil volume per meter of channel width is 4.0 litres/m.

For computational convenience, some simplifications are made as described below. The computational domain length is set at 14.4m. The length after the solid boom is taken to be 3.0 m. The semicircular float of the front net is replaced by a rectangle of the same area, as illustrated in Figure 4. A zigzag "chain" of porous cells is used to represent the inclined front net, as shown in Figure 5. This makes the problem setup and calculation easier because it allows use of Cartesian coordinates rather than more complicated body-fitted curvilinear coordinates.

3. 0 Description of Modeling Methods

The simulation of oil-water flow around a porous boom is conducted using the commercial CFD software, Fluent. For solving oil boom problems, the major modeling strategies used, such as k-ε turbulence modeling, finite volume scheme and the VOF technique, have already been described by An (1996*a*). Additional details about the code can be found in Fluent User's Guide (1994). This information will not be repeated here. However, the net-boom structure is being numerically studied for the first time in this work. It is necessary to describe the flow model for the net. Since the net can be taken as a porous medium with a certain of porosity, a proper porous medium model needs to be adopted.

In Fluent, porous media are modeled as an additional momentum sink in the governing momentum equations. The added momentum sink can be split into two terms, a permeability term, or Darcy term, and an inertial resistance term, which can be written as:

$$\frac{\mu}{\beta}\mathbf{v} + C_2\left(\frac{1}{2}\rho\mathbf{v}|\mathbf{v}|\right) \tag{1}$$

where β is the permeability in the direction of the velocity vector $\mathbf{v}$, C_2 is the inertial resistance factor, μ is viscosity and ρ is density. The sink terms contribute to the pressure gradient in the porous cell. Generally speaking, for laminar flows or low Reynolds number flows, the first term is dominant and the inertial resistance may be neglected. On the contrary, if the flow is turbulent and the Reynolds number is high, the second term is dominant and β should be set as a large real number to eliminate the effect of the first term. The manual gives the calculation method for these two empirical constants for the flow passing through a perforated plate. Since the flow passing through a porous net is similar to this case, modeling flow through a perforated plate is of particular interest to us. In that case, the first term is eliminated and the second term in (1) creates a pressure drop Δp in the flow direction given by

$$\Delta p \cong C_2 \cdot \Delta n \cdot \frac{1}{2}\rho v|v| \tag{2}$$

where Δn is the thickness of the perforated plate in the velocity direction and Δp is the pressure drop when the flow passes through the perforated plate. The pressure drop in each coordinate direction (for 2-D modeling) can be written as:

$$\Delta p_x \cong C_{2x} \cdot \Delta n_x \cdot \frac{1}{2}\rho v_x|v_x| \tag{3}$$

$$\Delta p_y \cong C_{2y} \cdot \Delta n_y \cdot \frac{1}{2}\rho v_y|v_y| \tag{4}$$

Since $v_x^2+v_y^2=v^2$ and the pressure drop in the flow direction is always positive, Equations. (2), (3) and (4) can be combined to give

$$\frac{\Delta p}{C_2\Delta n} = \frac{|\Delta p_x|}{C_{2x}\Delta n_x} + \frac{|\Delta p_y|}{C_{2y}\Delta n_y} \tag{5}$$

For an inclined net with an angle α as shown in Figure 6, it is apparent that

$$\Delta n_y = \Delta n \cdot \sin\alpha \qquad and \qquad \Delta n_x = \Delta n \cdot \cos\alpha \tag{6}$$

Using (6), Equation. (5) can be rewritten as

$$\frac{\Delta p}{C_2} = \frac{\Delta p_x}{C_{2x}\cos\alpha} + \frac{\Delta p_y}{C_{2y}\sin\alpha} \tag{7}$$

It is also seen from Figure 6 that $\Delta p=p_1-p_3$, $\Delta p_x=p_1-p_2$, and $\Delta p_y=p_2-p_3$. Therefore, Δp is always equal to $|\Delta p_x|+|\Delta p_y|$, no matter what value α is. This implies that C_2 must be equal to $C_{2x}\cos\alpha$ and $C_{2y}\sin\alpha$, so that the input data for Fluent, C_{2x} $(C_2/\cos\alpha)$ and C_{2y} $(C_2/\sin\alpha)$ can be calculated from the resistance coefficient of the net, C_2, and the inclined angle α.

To calculate C_2, Smith and Van Winkle's correlation recommended by Perry and Chilton (1974) is adopted. The original formula is more complicated, but after some transformations, C_2 can be calculated from a relatively simple equation:

$$C_2 = \frac{1}{C\Delta n}\left(\left(\frac{A_p}{A_f}\right)^2 - 1\right) \tag{8}$$

Here A_p is the total cross-section area of the perforated plate, A_f is the total free area of holes, Δn is the plate thickness and C is a dimensionless orifice coefficient depending on hole Reynolds number Re, plate thickness-diameter ratio (Δn/D) and hole pitch-diameter ratio (P/D). For a wide range of Re, P/D and Δn/D, C varies between 0.75 and 1. For smaller Δn/D, C ranges from 0.75 - 0.9. Therefore, 0.8 is adopted in our calculations.

A_f/A_p in Equation (8) is actually equal to ϕ, the porosity in our case, and the net thickness is taken to be 4.3 x 10^{-3}m, which is equal to the local grid size in the y direction. Therefore, Equation (8) can be rewritten as:

$$C_2 = \frac{1}{3.44\times10^{-3}}(\frac{1}{\phi^2}-1) \quad (9)$$

For all calculations reported here, a 142×57 grid is used. Non-uniform grid spacing has been adopted to allow more cells clustered near the boom, net and free surface to secure higher accuracy in calculation for a limited number of grid points. A time-dependent procedure is also adopted to obtain the evolution of the oil slick and real unsteady water-oil flow. Since the time step is small enough and the underrelaxation factor is optimized in each case, in most cases only one iteration is needed to make the actual maximum normalized residuals below 10^{-4} for each time step. At least 10^4 time steps are calculated to ensure a relatively steady state has been reached.

Based on this, the oil-water flow velocity distributions, its flow pattern and the oil volume of fluid (VOF) of the six variants of inclined front net structure are calculated from relatively smaller entry water velocity up to their critical points.

4.0 Main Results-Effect of Front Net Inclined Angle

4.1 Comparison Based on Same Velocity

The calculated oil slick VOF contours and flow velocity vectors in the boom-net area for different front net angles are shown in Figure 7. The results for all cases are based on the same entry water velocity (1.0 m/s), the same net porosity (0.5) and Federated Crude oil. Like the solid upper part of the front perforated plate used in the numerical simulation by Cheng *et al* (1997), the solid rectangular float in this study may play a similar role. It can be seen that at this relatively high water velocity, 1.0 m/s, oil can be contained inside the net-boom area in most cases, except for **A-3** and **F-F.**

For case **A-3**, it can be seen from oil VOF contours in Figure 7 that most of the oil has been swept away by the water flow. This can be explained from the velocity distribution shown in the same figure. The highly backward-inclined front net allows more water to enter the net-boom area and sweep onto the water surface. This causes the water velocity in the net-boom area, especially on the water surface, to be much higher. This stresses and pushes the oil slick out through the bottom net near the solid boom. The figure indicates that the critical velocity in case **A-3** must be less than 1.0 m/s.

In case **F-F,** the large angle of the forward-inclined front net allows the front net to draw more water into the net-boom region than the other cases. A large amount of the entering water cannot escape through the front part of the bottom net. Therefore, more water is forced to turn upwards and move to the surface. The water velocity near the surface and the solid boom in case **F-F** is higher than all the other cases except **A-3.** However, the situation in case **F-F** is not as severe as case **A-3.** Its critical velocity should be higher than 1.0 m/s.

Among the other four cases, **H-B** clearly does not perform as well as the other three. In **H-B,** the oil slick is thicker and shorter. The highest VOF value of the oil contained in front of the solid boom is about 0.65 for the **H-B** case. This means that it contains only about 65% or less oil. This is smaller than the VOF value (about 0.9) of the trapped oil slick in the other three cases. Also, only a small amount of oil is left behind the net float. It appears that this velocity (1.0 m/s) is close to the critical value. Since most oil still remains inside the region, however, the critical velocity is actually slightly higher for this case. Similar to case **A-3,** the backward-inclined front net draws water to the water surface, but not as strongly as in case **A-3.** Therefore, the **H-B** configuration has better performance than **A-3.**

There is very little difference in performance among the remaining three cases. The length of the oil slick in front of the solid boom is almost the same. The VOF value for case **VER** is slightly higher and more oil is trapped behind the net float. Also, it seems that the velocity vectors are smoother in case **VER.** The flow entering the region through the vertical front net is straighter and more horizontal, and evenly spreads to both sides. Furthermore, the water velocity near the surface is smaller than the other two cases. All these features mean that case **VER** may be the best case. Figure 8 gives the comparison of streamlines for the above-mentioned cases. Since the contour levels have been set to be identical at the same stream function difference (40 streamlines/4×10^{-4} m^2s^{-1}) for all cases, the flowrate between the neighbouring streamlines must be the same (10^{-3} m^2s^{-1}). Therefore, the distance between neighbouring lines provides relative information about local velocity. For example, a wider distance means a smaller local velocity. Additionally, the streamlines themselves give the flow pattern. From these streamline plots, the following information and ideas may be obtained:

(1) For case **A-3,** more than 80% of the flowrate coming through the front net leaves the region through the downstream part of the bottom net close to the solid boom. This flowrate causes more water to sweep along the water surface and solid boom in this case.

(2) About 60-70% of the incoming flowrate leaves the region through the downstream part of the bottom net in case **H-B.** This means that case **H-B** has better performance to contain the oil than case **A-3.**

(3) Only 30-40% of the fluid coming in through the front net leaves the region through the downstream part of the bottom net for the three cases **VER, Q-F** and **H-F.** This means that compared to cases **A-3** and **H-B,** a smaller amount of flowrate passes through the net-boom area and escapes in front of the solid boom. Therefore, more oil is contained in these cases, than in cases **A-3** and **H-B.**

(4) By counting the number of streamlines over a fixed distance, it is obvious that about twice as many streamlines pass through the net-boom area in case **A-3** than cases **VER, Q-F**, and **H-F**. This means that about twice the water flowrate passes through the net-boom area in case **A-3** than the above-mentioned three cases, at the same entry water flow velocity of 1.0 m/s. This implies that for these three cases, the critical velocity must be higher than case **A-3**. The critical velocity for these three cases appears to be around 2.0 m/s.

(5) For case **F-F**, about twice the flowrate enters the net-boom area compared to case **VER.** Although some leaves the area through the upstream part of the bottom net, a

large amount of the flowrate passes through the area near the solid boom. This amount is close to that observed in case **H-B**, but slightly less than case **A-3**.

4.2 Comparison Based on Critical Velocity

Comparison of the oil slick contours and velocity vectors for different front net angles at the corresponding critical velocity for each case is shown in Figure 9. The other parameters are kept the same as those in Figure 7. As expected from the analysis of Sub-section 4.1, cases **VER, H-F** and **Q-F** may have critical velocities two times higher than case **A-3.** In case **A-3,** a substantial amount of oil gets lost at an entry water velocity of 0.8 m/s. This implies that the critical velocity for this case is about 0.8 m/s. Substantial oil loss has occurred when the entry water velocity reaches 1.2 m/s for both cases of **H-B** and **F-F,** so that the critical velocity for these cases may be set at 1.2 m/s. The remaining three cases can keep the oil well contained until entry water velocity up to 1.8 m/s, which is more than double the critical velocity for case **A-3.** The critical velocities for the last three cases are very close. It appears that case **VER** may be the best among the three, as expected in the previous subsection.

It should be pointed out that for each of these six cases, the water surface velocity in the net-boom region, especially near the solid boom, is relatively close when the entry water flow velocity reaches the critical value. This implies that the water velocity near the surface may be the most important factor influencing the oil loss. A good boom structure should have the capacity to reduce the flow velocity near the surface and solid boom. Figure 10 confirms that for all six cases, the flowrates entering inside the net-boom region are very close at their critical velocities. Nearly the same numbers of streamlines pass through the region, except that case **F-F** has one more streamline. Figure 11 summarizes the critical velocities for all six cases studied in this section, among which both **VER** and **Q-F** seem to be the highest.

5.0 Comparison with Experimental Measurements

Although the available experimental results are limited, it is still helpful to use those to validate the predictions from the numerical simulations. Brown et al (1999) have reported on experiments conducted using Federated Crude oil ($\rho = 834$ kg/m^3, $\mu = 10$ cs) and a net-boom structure corresponding to our case A-3. The experimental results for the velocity distribution are used here to compare to the numerical results. The horizontal velocity profiles vs the distance upstream from the solid boom, at 20 cm and 10 cm depths from the surface, for case A-3, are shown in Figures. 12 and 13 respectively. These figures show that the experiment and computation are in reasonable agreement for both lines above and under the bottom net, especially when the entry water velocity is lower (0.3 m/s and 0.4 m/s). There are some differences at the higher velocity (0.5 m/s). These comparisons show that the numerical results are acceptable, and that CFD can be used as a reliable predictive tool for oil boom analysis and design.

6.0 Conclusions

From the simulations and analysis, some important conclusions may be drawn:

(1) The flow velocity near the water surface and the solid boom, and the flowrate inside the net-boom region, are important parameters affecting the critical velocity.
(2) Among the six cases, **VER** appears to be the best. Furthermore, a front net slightly inclined forward may not cause any real difference.
(3) There is excellent potential to raise the critical velocity by deliberately designing net- boom structures. For a single solid boom, the critical velocity is found to be about 0.3- 0.4 m/s. A solid boom with front perforated plates may raise the critical velocity to 0.5 m/s. A solid boom with a partially perforated front plate, which has a solid upper part, will raise the critical velocity to 1.0 m/s. From the present investigation it can be seen that a properly designed net-boom structure can raise the critical velocity to about 2.0 m/s. There is also a realistic potential to raise the critical velocity even higher. In order to achieve this goal, further experimental and numerical studies are needed.
(4) The numerical velocity profiles are in good agreement with experimental data and show that the numerical results are acceptable.

7.0 Acknowledgement

The authors acknowledge Imperial Oil Resources Limited, Calgary for the financial support for this research.

8.0 References

An, C-F., *CFD Simulation of Oil Boom Failure (1) Explaining Test Computations*, Technical Report No. IPRCC.OM.96.13, Imperial Oil Resources Limited, Calgary, AB, Canada, 1996*a*.

An, C-F., *CFD Simulation of Oil Boom Failure (2) Comparisons with Theory and Experiments,* Technical Report No. IPRCC.OM.96.14, Imperial Oil Resources Limited, Calgary, AB, Canada, 1996*b*.

An, C-F. and H.M. Brown, *Oil Containment in Flowing Water, Studies of Several Concepts,* Technical Report No. IPRCC.OM.97.19, Imperial Oil Resources Limited, Calgary, AB, Canada, 1997.

An, C-F., R.M. Barron and Z. Zhang, "Unstable Interface Between Oil and Water with Fast Relative Motion", *Proceedings of the Third International Conference of Fluid Mechanics*, Beijing, China, pp.247-252, 1998.

An, C-F. and R.M. Barron, *Numerical Study of Oil Boom Models for Droplet Entrainment and Related K-H Instability,* FDRI-TR-98-01, FDRI, University of Windsor, Windsor, ON, Canada, April 1998.

Brown, H.M., R.H. Goodman and C-F. An, "Development of Containment Booms For Oil Spills in Fast Flowing Water", to be presented at the 22nd AMOP Conference, Calgary, AB, Canada, June 1999.

Cheng, S.Y., Z. Zhang and C.-F. An, "Preliminary Numerical Simulation of Boom Failure with VOF Model", *Proceedings of the 7th National Symposium on Numerical Heat Transfer*, p.189, Beijing, China, Nov. 1997 (in Chinese).

Clavelle, E.J. and R.D. Rowe, "Numerical Simulation of Oil Boom Failure by Critical Accumulation", *Proceedings of 16th Arctic and Marine Oil Spill Program Technical Seminar*, Calgary, AB, Canada, pp.409-418, June 1993.

Clavelle, E.J. and R.D. Rowe, "Simulation of Environmental Flow with a Density Interface", *Proceedings of Third Annual Conference of the CFD Society of Canada*, Banff, AB, Canada, pp. 441-447, 1995*a*.

Clavelle, E.J. and R.D. Rowe, "A Numerical Study of Oil Boom Failure by Critical Accumulation", *Proceedings of 9th International Conference on Numerical Methods in Laminar and Turbulent Flow*, Atlanta, GA, USA, 1995*b*.

Delvigne, G.A.L., "Barrier Failure by Critical Accumulation of Viscous Oil", *Proceedings of 1989 International Oil Spill Conference*, San Antonio, TX, USA, pp.143-148, 1989.

Grilli, S.T., Z. Hu, and M.L. Spaulding, "Numerical Modelling of Oil Containment by a Boom", *Proceedings of 19th AMOP Technical Seminar*, Calgary, AB, Canada, pp. 343-376, June 1996.

Johnston, A.J., M.R. Fitzmaurice and R.G.M. Watt, "Oil Spill Containment: Viscous Oils", *Proceedings of 1993 International Oil Spill Conference*, Tampa, FL, USA, pp.89-94, 1993.

FLUENT, *User's Guide*, Version 4.4, Fluent Inc., Lebanon, NH, USA, 1994.
Perry, R.H. and C.H. Chilton, *Chemical Engineers' Handbook*, 5th Edition, p. 5.34-35, McGraw-Hill Book Co., NY, USA, 1974.

Ueda, K. and H.Yamanouchi, "Oil Containment on Water Currents and a Method of Presentation of Escaping Oil with Net", Presented at the Speaker's Corner of the 21st AMOP Technical Seminar, Edmonton, AB, Canada, June 1998.

Zhang, Z., C.-F. An and R.M. Barron, *Modeling of Oil-Water Flow around Porous Boom Structures*, FDRI-TR-99-01, FDRI, University of Windsor, Windsor, ON, Canada, March, 1999.

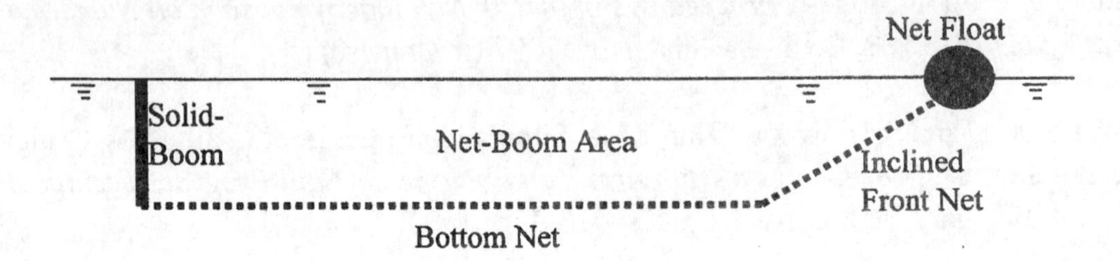

Figure 1 Basic Net-Boom Structure

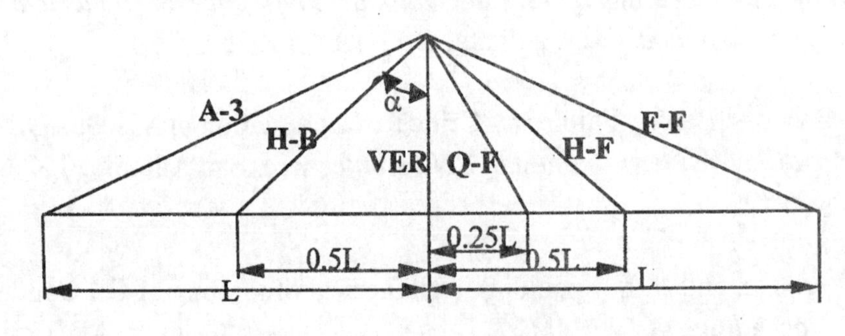

Figure 2 Inclined Front Net Cases

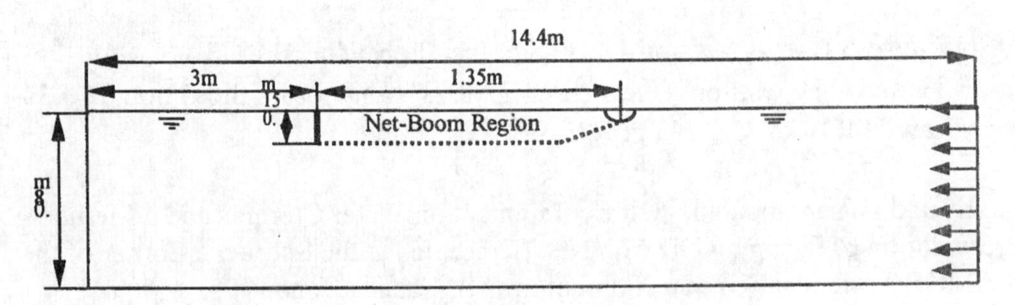

Figure 3. Physical Dimensions and Flow Region

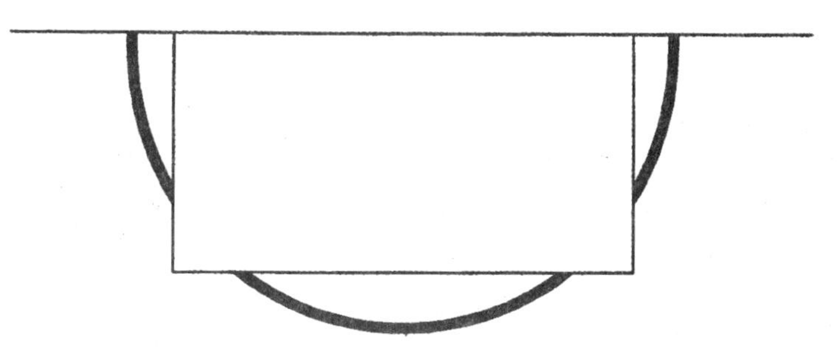

Figure 4 Replacement of Float Geometry

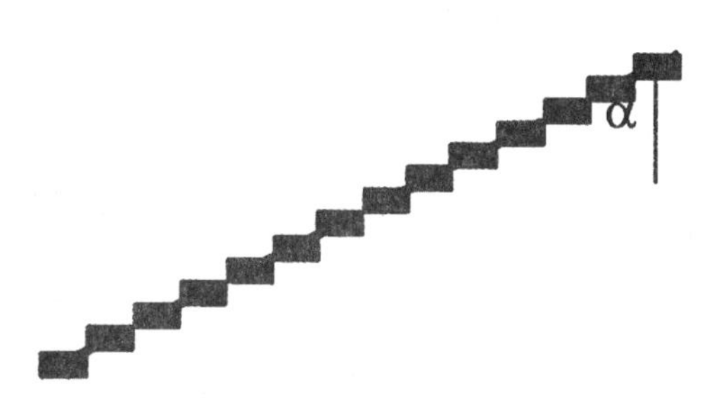

Figure 5 Zigzag - Chain Net

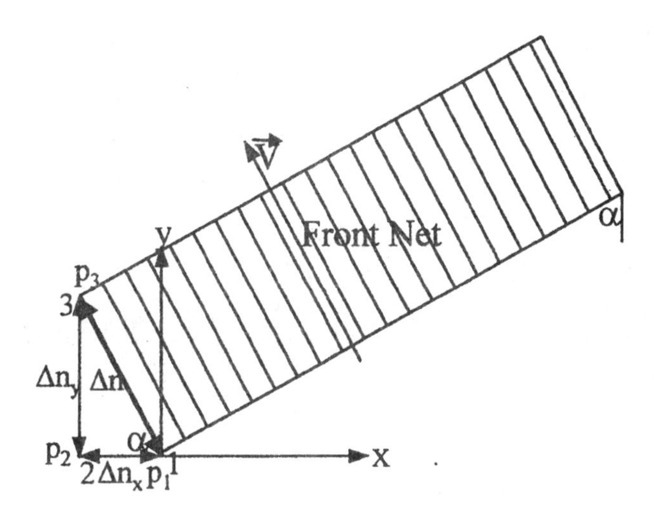

Figure 6 Pressure Drop & Thickness of the Front Net

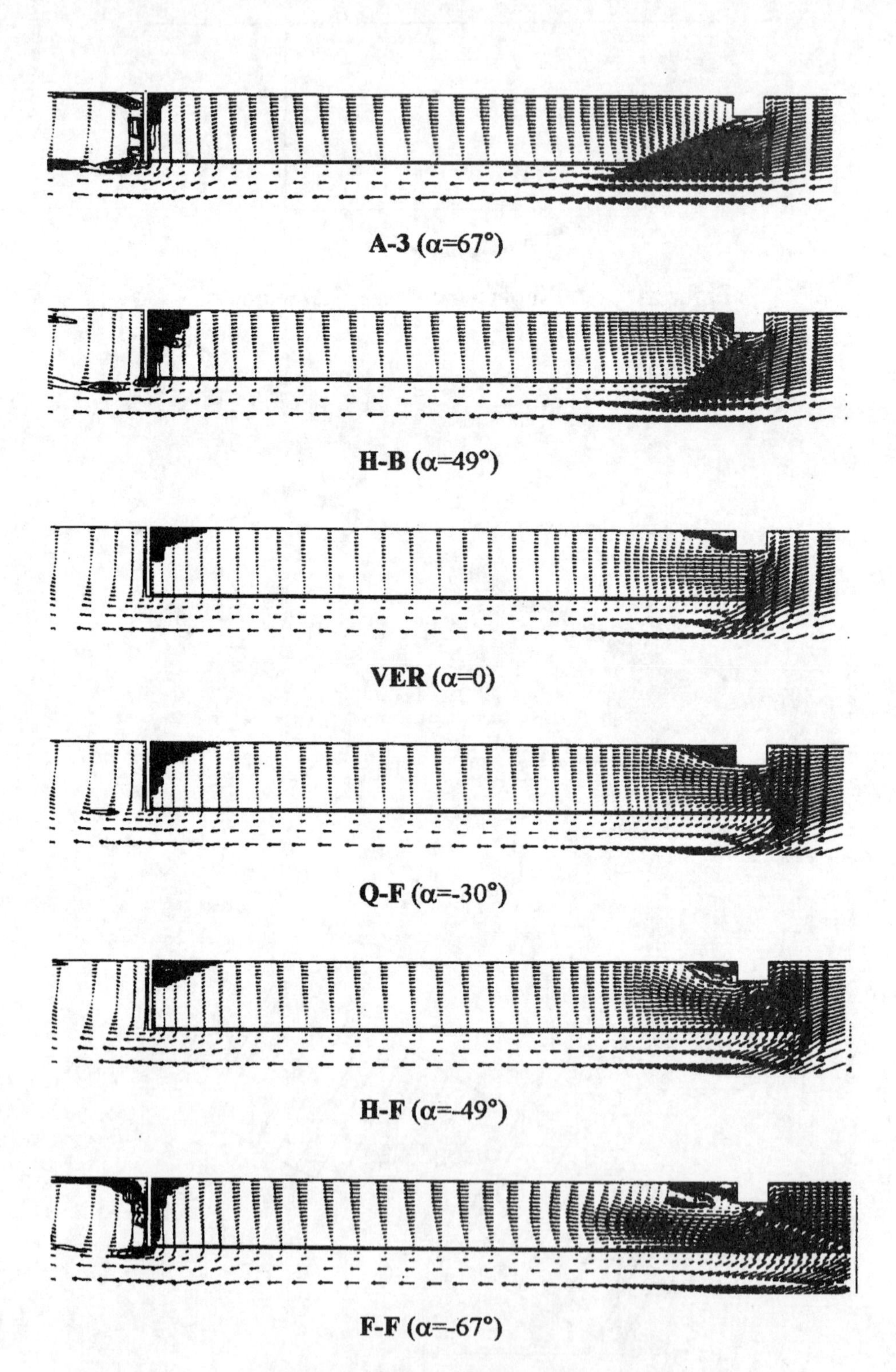

Figure 7 Comparison of Oil VOF and Flow Velocity Vectors Based on Same Entry Water Velocity, 1.0m/s

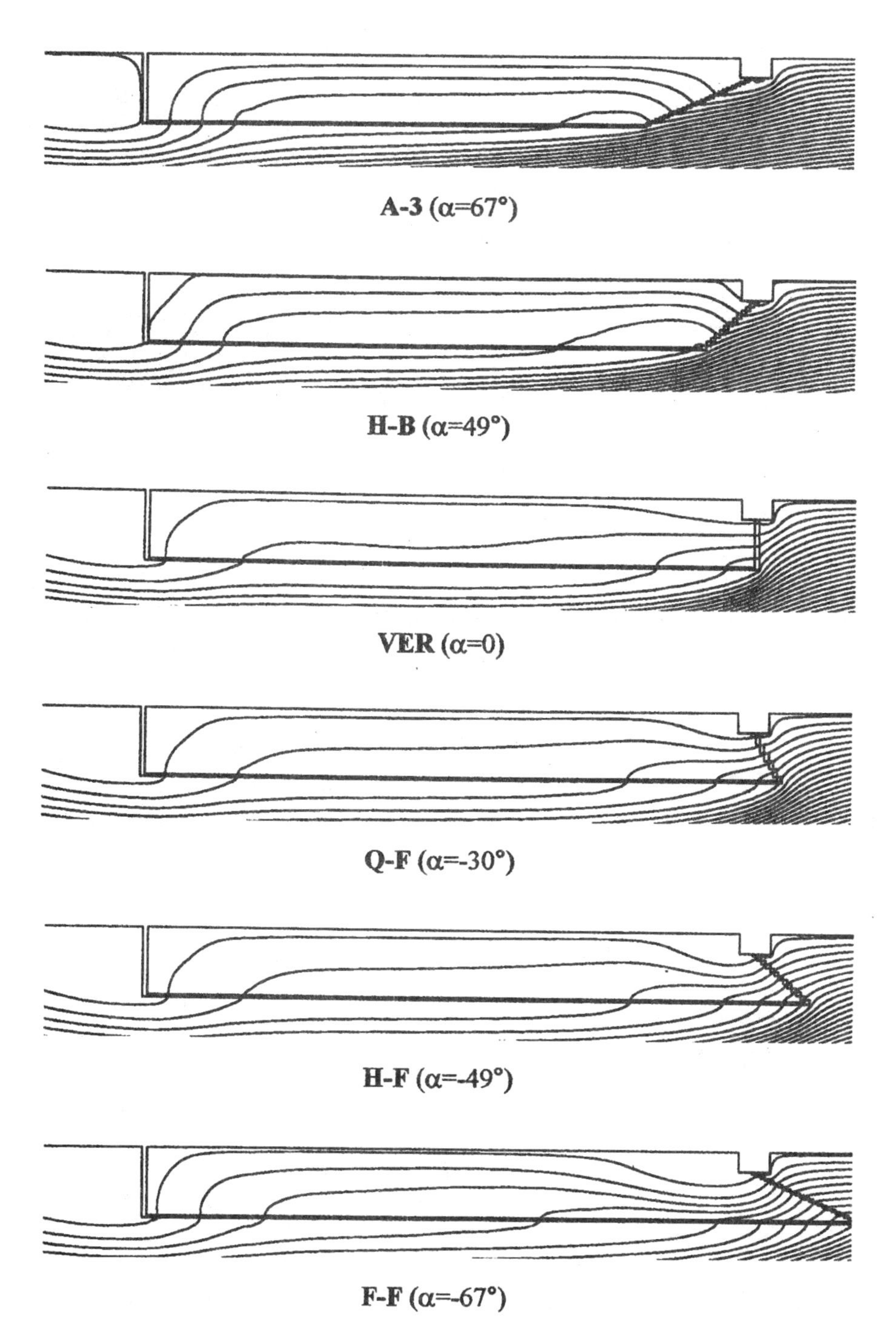

Figure 8 Comparison of Stream Function Contours Based on Same Entry Water Velocity, 1.0m/s

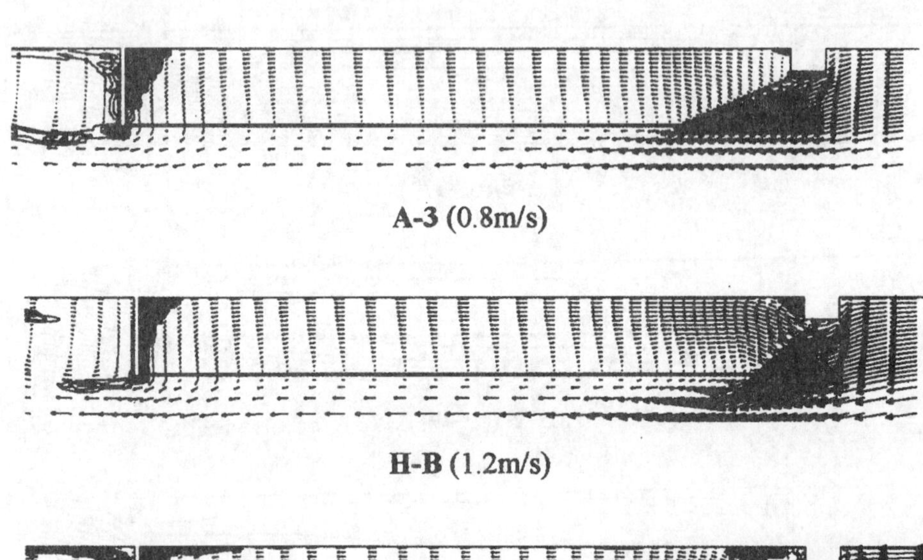

A-3 (0.8m/s)

H-B (1.2m/s)

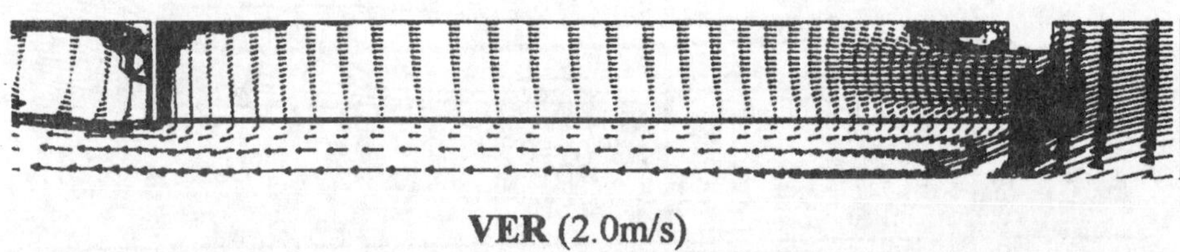

VER (2.0m/s)

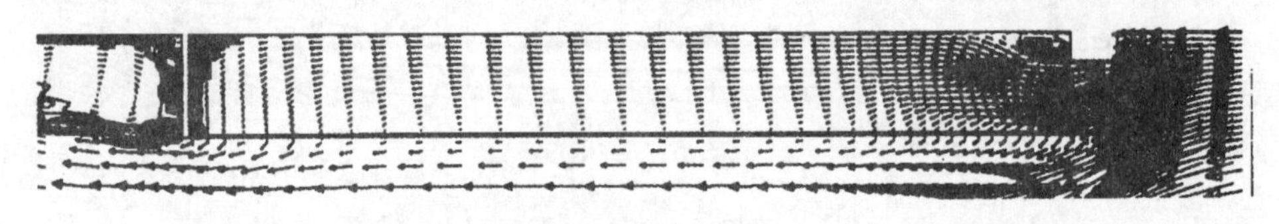

Q-F (2,0m/s)

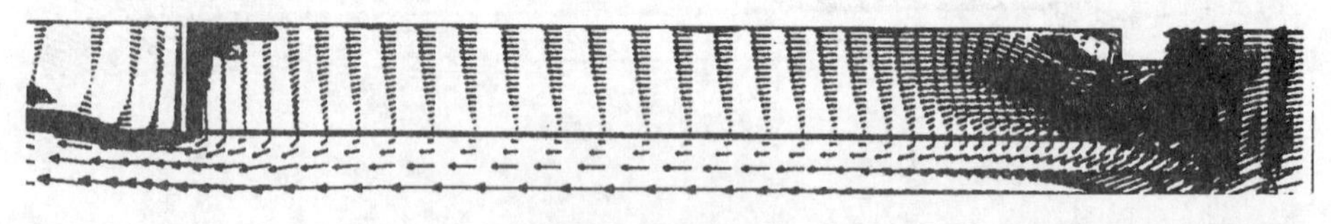

H-F (1.8m/s)

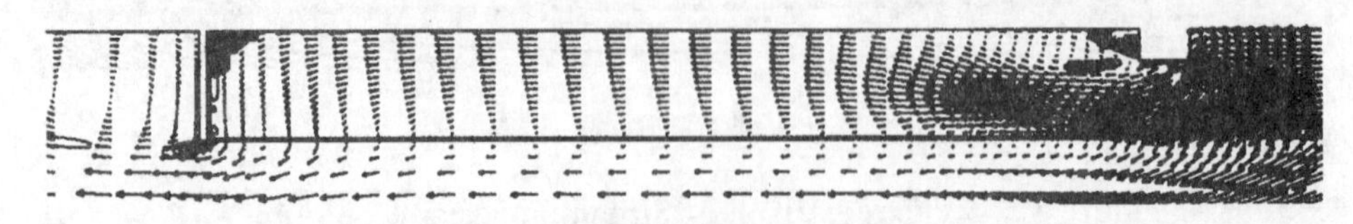

F-F (1.2m/s)

Figure 9 Comparison of Oil VOF and Flow Velocity Vectors at Critical Water Velocity for Each Case

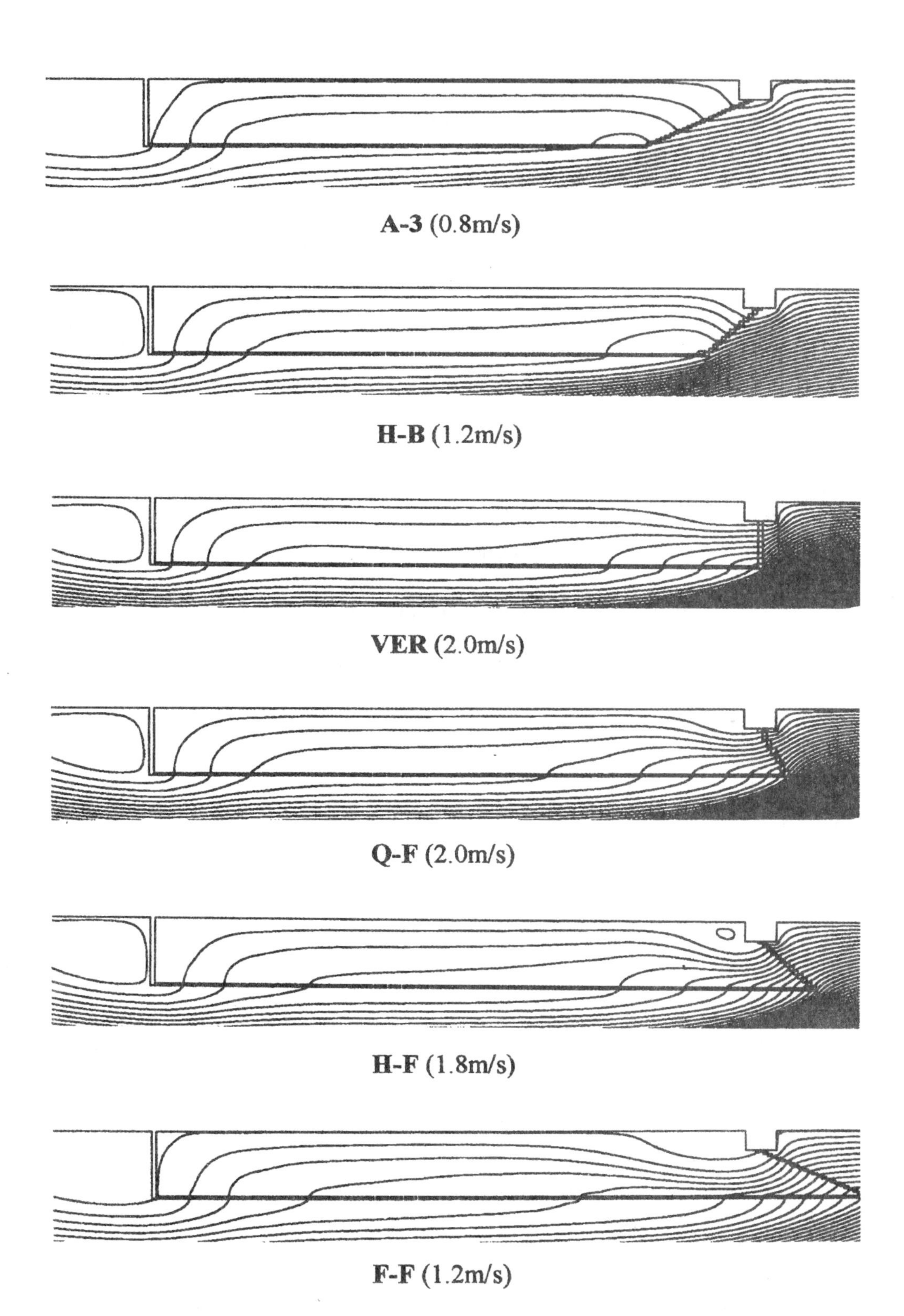

Figure 10 Comparison of Flow Stream Function Contours at Critical Water Velocity for Each Case

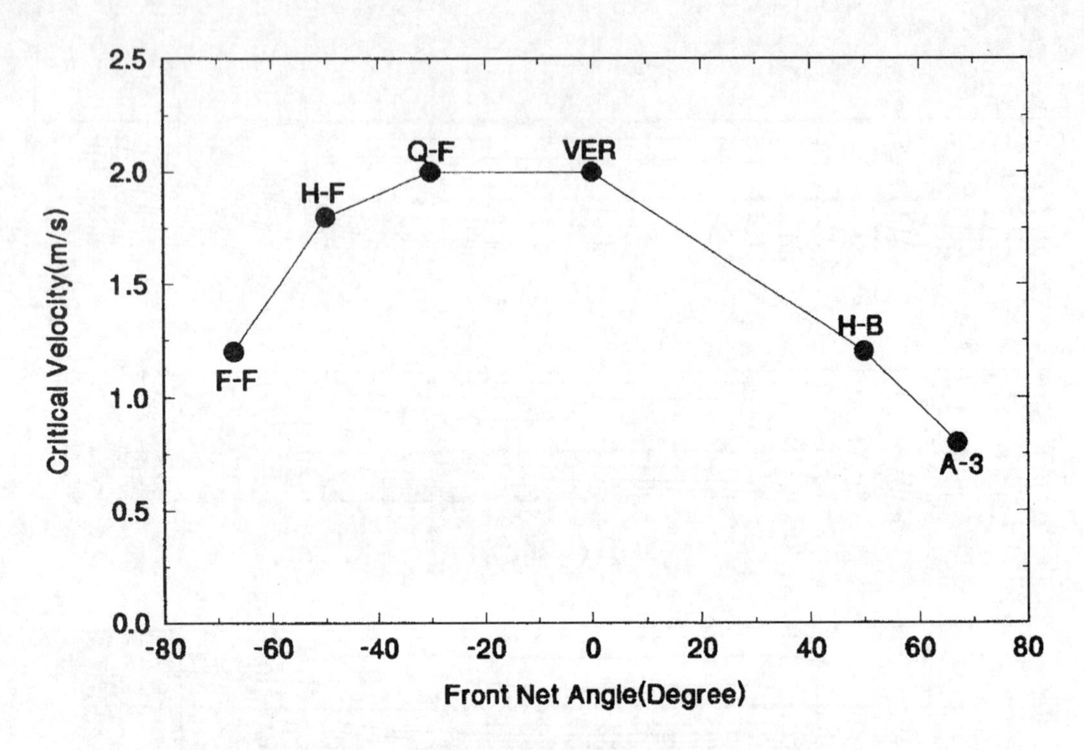

Figure 11 Critical Velocity Vs Front Net Angle

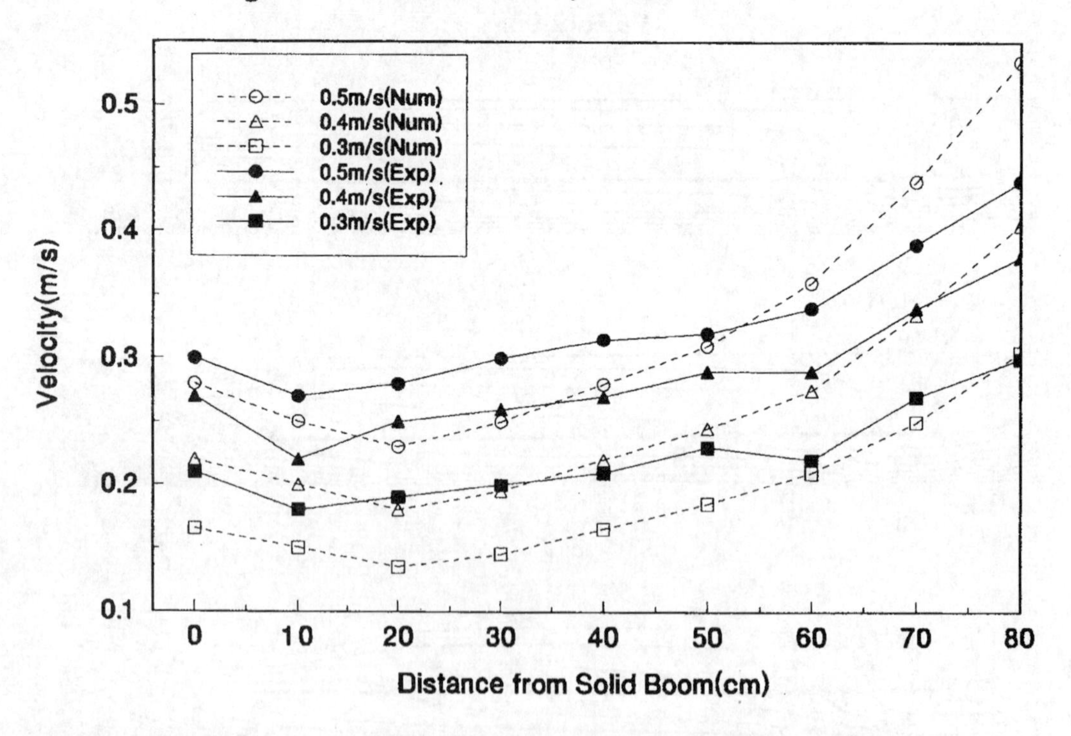

Figure 12 Velocity Profile on a line 20 cm below Surface

安长发博士科研论文集

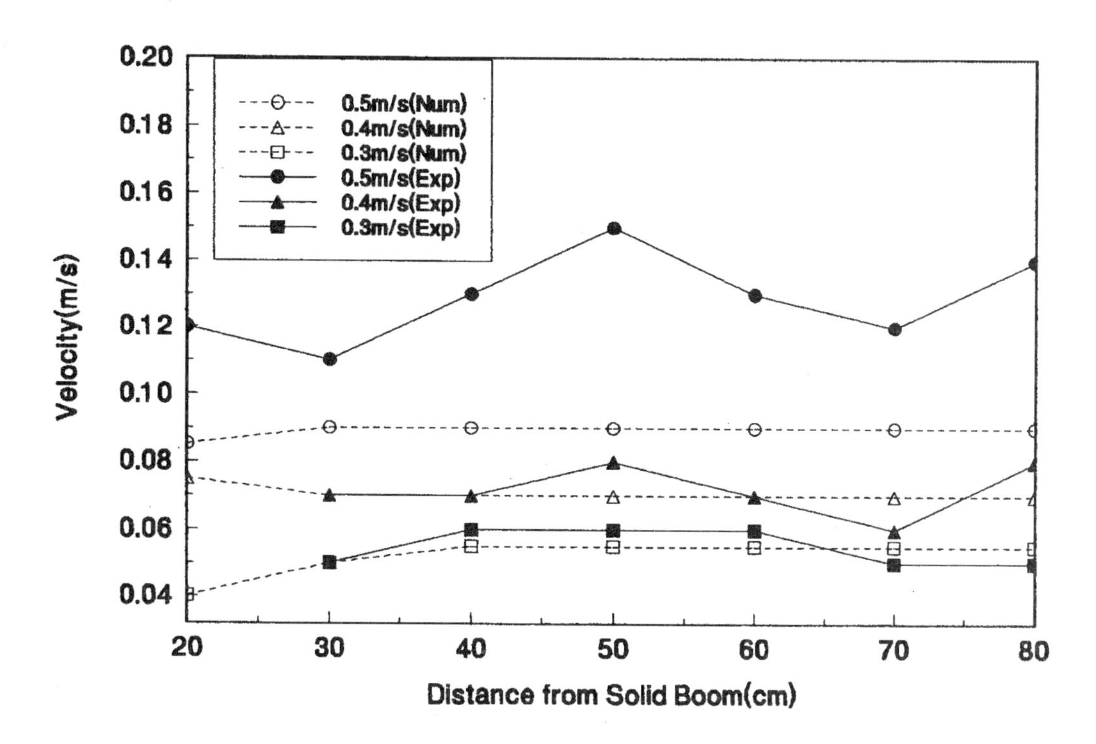

Figure 13　　Velocity Profile on a line 10 cm below Surface

PROCEEDINGS OF THE EIGHTH ASIAN CONGRESS OF FLUID MECHANICS

December 6 - 10, 1999 Shenzhen, China

Editor in Chief: Erjie Cui

International Academic Publishers

Proceedings of the Eighth Asian Congress of Fluid Mechanics
December 6-10, 1999, Shenzhen, China

NUMERICAL STUDY ON NEW NET-BOOM STRUCTURES

Z. Zhang*
(College of Chem. Eng., Beijing University of Chemical Technology, Beijing 100029, China)
R.M. Barron & C.-F. An
(Fluid Dynamics Research Institute, Univ. of Windsor, Windsor, Ontario, N9B 3P4 Canada)

I. INTRODUCTION

An oil boom is one of the major tools to deal with an oil spill on the sea surface. The oil will escape underneath the boom by hydrodynamic forces and the oil boom containment will fail if certain fluid-dynamic conditions are not satisfied. The boom failure involves interesting liquid-liquid interfacial phenomena and can be studied by both experimental work and theoretical analysis (An, 1996*a*). There are also a number of papers studying the boom failure problem using computational fluid dynamics (CFD) method. Clavelle and Rowe (1993) employed the volume of fluid (VOF) method to predict the boom failure critical accumulation and the results were consistent with Delvigne's observation (Delvigne, 1989) for a highly viscous oil. Grilli et al (1996) used a numerical method to study oil containment by a boom. An (1996*b*) numerically simulated three failure mechanisms using a commercial CFD software, Fluent. The numerical results coincided well with the experiments done by Delvigne (1989) and Johnston et al (1993), both qualitatively and quantitatively. However, the studies above are concentrated on discussing the situation only for a single boom structure. There are many factors, such as boom structure, boom draft, water depth, oil and water properties, water velocities, etc., that contribute to the success or failure of oil containment. Among them, boom structure may be one of the most important factors. Cheng et al (1997) simulated the flow field and predicted critical velocity for the same porous boom system tested at Imperial Oil (An and Brown, 1997) by using their own code based on VOF method, finite volume scheme and k-ε turbulence model. The predicted flow field and critical velocity coincide well with the experimental results. Furthermore, they found that a small modification of the front porous boom with about 1cm solid length on the upper part of the perforated plate could increase the critical velocity up to 1.0m/s, double the value for a fully-holed perforated plate. This shows that improvements to the net-boom structure can significantly increase the boom system performance.

Based on the net-boom prototype designs recently tested at the Imperial Oil Test Basin (Brown et al, 1999), some numerical simulations of oil-water flow around various net-boom structures has been reported (Zhang et al, 1999*a*). The study of the front net inclined angle effect (Zhang et al, 1999*b*) indicates that the best angle is close to 0° (case **VER**) The critical velocity is more than twice the value of that obtained for a backward-inclined front net structure used in the tests at Imperial Oil. This paper will further discuss and modify the bottom net based on the best structure in the previous study.

II. MODELING METHODS

The simulation of oil-water flow around a porous boom is conducted using the

---

* Currently working at Fluid Dynamics Research Institute, University of Windsor, Windsor, On., Canada

commercial CFD software, FLUENT. For solving oil boom problems, the major modeling strategies, such as k-ε turbulence modeling, finite volume scheme and the VOF technique are directly adopted from FLUENT' s code (FLUENT User' s Guide, 1994). The calculation of the net porosity has been described in detail by Zhang et al. (1999*b*).

III. IMPROVED BOOM STRUCTURE CONFIGULATION AND STUDY

From the previous study, it is found that the case **VER** is the best one of those studied (Zhang et al., 1999*b*). The case **VER** is a net-boom structure of a solid boom with vertical front net and bottom net. The analysis suggested that the design could be improved by creating a quiet region near the upper part of the solid boom where the oil can be trapped. Based on the **VER** configuration, a new structure is proposed, in which a section of the bottom net immediately upstream of the solid boom, about 1/5 of the total bottom net length, is replaced by a net with smaller porosity as shown in Fig. 1.

3.1 Partly impervious bottom net

With porosity equal to zero, the short section of bottom net is impervious and no fluid can pass through it, referred to as case **D-4.** This should make the area near the solid boom inside the net-boom region very quiet, i.e., with essentially no flow. It is expected that the oil will get trapped once it enters this area. However, our simulations have shown that this dead zone idea does not work. It is obvious that the oil cannot enter this dead region and is swept away by the oncoming water flow. This result is reasonable, because the dead zone is totally stagnant. Since the water inside this zone does not move out, there is no room for the oil to move in. The water and oil turn downward in front of this dead zone and escapes through the normal section of bottom net. In fact, this structure is similar to making the net-boom region shorter, which results in poorer oil containment capability. Therefore, the critical velocity will be even smaller. Even though this structure is not successful, the concept that reducing the flow velocity near the corner of the water surface and the solid boom may increase the critical velocity is still worthwhile. The problem with this case **(D-4)** is that the structure does not allow the water to leave and thereby create empty space for the oil to enter. This has led us to consider a new structure with some holes in the short section of the bottom net, but at a lower porosity that the main part of the net. This is referred to as a partly denser bottom net structure.

3.2 Partly denser bottom net

Based on analysis in the previous subsection, a new net-boom structure, referred to as the **D**-Series, has been designed. As described above, this new structure has a denser section of bottom net near the solid boom. In the present investigation, three cases of this new structure, **D-1**, **D-2**, and **D-3**, are studied. The short section of net is considered to be a porous medium with porosities 0.2, 0. 1, and 0.05 respectively. The rest of the bottom net has porosity 0.5. The calculation of the oil contours and the flow field for these three cases, along with **D-4** and **VER** are based on a high entry water velocity of 2.0m/s with the other parameters remaining the same as in the previous study (Zhang et al., 1999). The results show that D-4 cannot contain the oil slick at this water speed. Case **VER** is around its critical condition. Case **D-1** is close to its critical condition and some oil has begun to escape under the solid boom, but the containment of the oil slick is better than case **VER.** In this case, most of the oil can still be kept in the net-boom region. Case **D-2** is better, in which the oil slick is longer and contains more oil with higher VOF value. It seems that 2.0m/s is below its critical condition. Finally, the case **D-3** appears to be the best case of the five. Its oil slick is thinner and longer than in the other cases. This implies that the critical velocity for case **D-3** may be well above 2.0m/s.

Figure 2 gives the comparison of the streamlines for all five cases mentioned above. The streamlines clearly show that the smallest amount of flowrate goes into the area near the solid boom for the **D-3** case, followed by case **D-2**, **D-1** and **VER**.

IV. CONCLUSIONS

(1) A good design of net-boom structure should satisfy the following conditions:

a. The oil slick should be able to enter the net-boom region before it escapes under the net. The front net should not have too large a resistance to the oil, but it should have some resistance to the water; and

b. The captured oil should be able to move to a relatively calm area, e.g. the water surface and the solid boom corner, where the oil can be contained.

(2) A new net-boom structure with a composite bottom net concept has been proposed and studied. The front part of the bottom net has a small resistance so that the water in the region can easily escape. The back part of the bottom net has a large resistance so that a calm back area around the solid boom is created.

(3) Additional experiments are needed to confirm the new findings and predictions of the computational model.

ACKNOWLEDGEMENT

The authors acknowledge Imperial Oil Resources Limited for the financial support for this research. They are particularly grateful to Dr. Ron Goodman and Dr. Hugh Brown for their enthusiastic support for the Oil Boom Research Project.

REFERENCES

[1] An, C.-F., Technical Report No. IPRCC.OM.96.13, Imperial Oil Resources Limited, Calgary, AB, Canada, 1996*a*.

[2] An, C.-F., Technical Report No. IPRCC.OM.96.14, Imperial Oil Resources Limited, Calgary, AB, Canada, 1996*b*.

[3] An, C.-F. and H.M. Brown, Technical Report No. IPRCC.OM.97.19, Imperial Oil Resources Limited, Calgary, AB, Canada, 1997.

[4] Brown, H.M., R.H. Goodman and C.-F. An, to be presented at the 22nd AMOP Technical Seminar, Calgary, AB, Canada, June 1999.

[5] Cheng, S.Y., Z. Zhang and C.-F. An, In: *Proceedings of the 7th National Symposium on Numerical Heat Transfer*, p.189, Beijing, China, Nov. 1997 (in Chinese).

[6] Clavelle, E.J. and R.D. Rowe, In: *Proceedings of 16th Arctic and Marine Oil Spill Program Technical Seminar*, Calgary, AB, Canada, pp.409-418, June 1993.

[7] Delvigne, G.A.L., In: *Proceedings of 1989 International Oil Spill Conference*, San Antonio, TX, USA, pp.143-148, 1989.

[8] Grilli, S.T., Z. Hu, and M.L. Spaulding, In: *Proceedings of 19th AMOP Technical Seminar*, Calgary, AB, Canada, pp. 343-376, June 1996.

[9] Johnston, A.J., M.R. Fitzmaurice and R.G.M. Watt, In: *Proceedings of 1993 International Oil Spill Conference*, Tampa, FL, USA, pp.89-94, 1993.
[10] FLUENT, *User's Guide*, Version 4.4, Fluent Inc., Lebanon, NH, USA, 1994.
[11] Ueda, K. and H.Yamanouchi, Presented at the Speaker's Corner of the 21st AMOP Technical Seminar, Edmonton, AB, Canada, June 1998.
[12] Zhang, Z., C.-F. An and R.M. Barron, FDRI-TR-99-01, FDRI, University of Windsor, Windsor, ON, Canada, March 1999*a*.
[13] Zhang, Z., C.-F. An, R.M. Barron, H.M. Brown, R.H. Goodman, to Proceedings of the 22nd AMOP Technical Seminar, Calgary, AB, Canada, June 1999*b*.

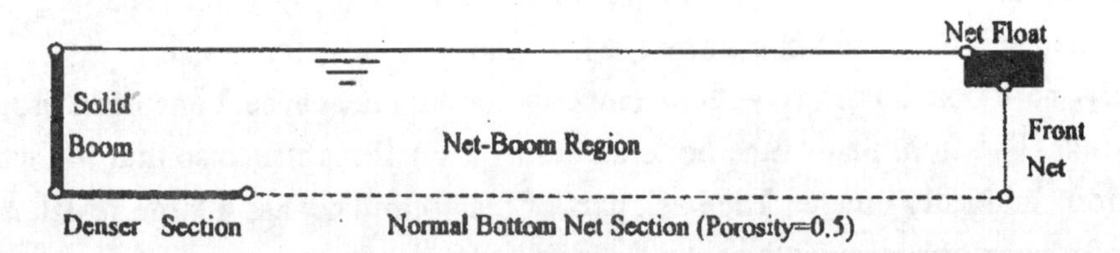

Fig.1 Vertical front net and modified bottom net structure

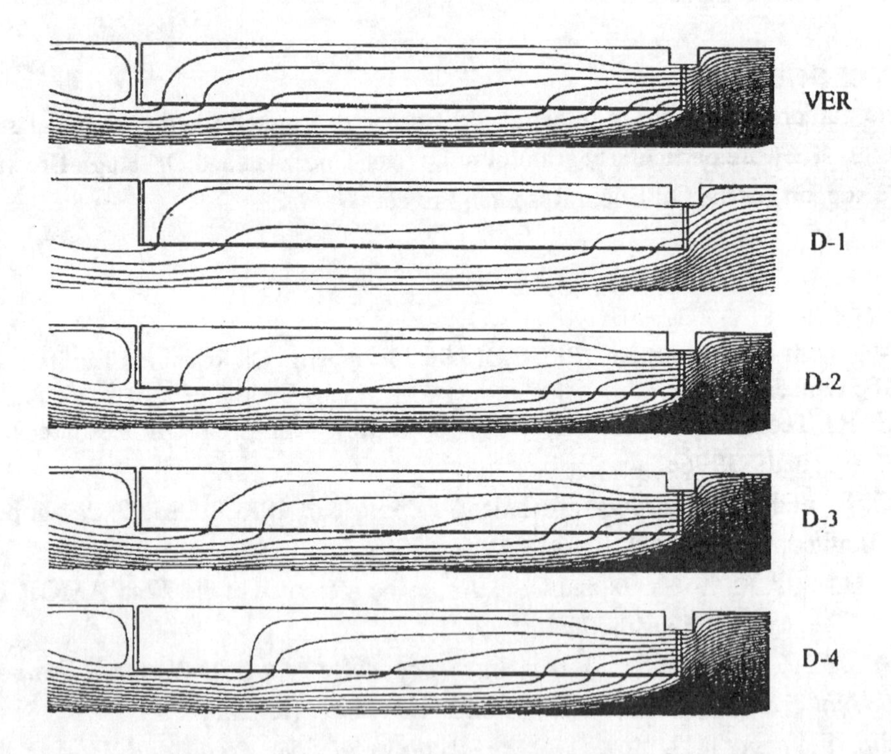

Fig.2 Comparison of stream function contours based on same entry water velocity, 2.0m/s

Proceedings of
The Fourth Asian Computational Fluid Dynamics Conference

September 18-22, 2000 Mianyang, China

Edited by ZHANG Hanxin

Changfa An
Mianyang, China
Sept. 18, 2000

University of Electronic Science and Technology of China Press

NUMERICAL INVESTIGATION OF *3*-D EFFECT OF WATER CHANNEL IN OIL BOOM EXPERIMENTS

Z. Zhang[=]
College of Chemical Engineering
Beijing University of Chemical Technology
Beijing, 100029, China

R.M. Barron & C.-F. An
Fluid Dynamics Research Institute
University of Windsor
Windsor, Ontario, Canada

Abstract

In order to investigate the performance of oil containment booms experimentally, a test facility must be properly designed. The three-dimensional (*3*-D) effect for a water channel is usually strong so that a pair of vortices appears in front of the boom near the channel walls and dives underneath the boom. As a result, when the spilled oil is contained by boom, the oil may escape underneath the boom together with the diving vortices. This phenomenon causes oil loss in the laboratory condition and may hide the real mechanisms of boom failure in the field condition. Therefore, the *3*-D effect of the water channel and the related oil loss due to vortex diving should be reduced to a minimum level. In this paper, computational fluid dynamics (CFD) has been used to simulate water flows in channels with different ratio of width to depth. Compared to the square-section channel, the wide channel reduces the *3*-D effect and weakens the vortices and, therefore, is recommended for oil boom experiments. In contrast, the narrow channel enhances the *3*-D effect and strengthens the vortices and, therefore, should be avoided for such experiments.

Keywords: Channel Flow, Oil Boom, Vortex Diving, *3*-D Effect, CFD

1. Introduction

Oil spills may occur on the surface of seas and oceans due to accidents of ships and tankers, such as collision, grounding, storm and so on. If oil is spilled on the surface of water, the most common response is to use a boom to contain the oil and a skimmer to recover it. Booms are barriers that float on the surface of water and are towed by boats to collect the spilled oil. To increase the efficiency of oil collection, the towing speed should be as high as possible. However, if the towing speed exceeds a critical value, boom failure may occur and the oil may escape underneath the boom due to hydrodynamic forces. The study of oil boom failure is of interest to the petroleum industry, environmental agencies and ocean engineering departments. In order to investigate boom failure mechanisms experimentally, the relative movement between boom and water is necessary and this movement can be implemented either by towing the boom on calm water or by driving water through a fixed channel with a fixed boom.

In the past years, Imperial Oil conducted several channel experiments on boom failure mechanisms in its wave basin. Some of the results have been reported in ISOPE'98 [1] and the details can be found in a technical report [2]. One of the problems observed in these experiments was that when the water flow velocity reached a moderate value, a pair of vortices appeared near the channel walls in front of the boom and subsequently passed underneath the boom. The phenomenon of vortex formation and vortex diving is believed to be the combined effect of horizontal velocity gradient near the channel walls and the vertical blunt obstacle (boom), i.e. the *3*-D effect. If oil is contained with the boom under this circumstance, the oil will escape underneath the boom together with the diving vortices. This mechanism of oil loss was observed in the laboratory channel condition, but may not occur during the boom towing process in the field condition. In order to match the laboratory condition with the field condition, the *3*-D effect and the resulting vortex diving should be restricted to a minimum level. Furthermore, to thoroughly understand the *3*-D effect of the channel flow near the boom, the details of the channel flow in the vicinity of the boom are of great importance.

In order to provide a guideline for new water channel design, the computational fluid dynamics (CFD) technology is used in this paper to study the *3*-D effect of water channel for oil boom experiments. In Section 2, the CFD methodology is described and the CFD simulation results for three typical channels are reported in Section 3. Section 4 gives some concluding remarks.

[=] Currently working at Fluid Dynamics Research Institute, University of Windsor, Windsor, ON, CA

2. CFD Methodology

According to Chow [3], the ratio of width to depth of a channel is the major factor affecting the *3*-D effect of the channel. Other parameters, such as flow velocity, boom draft, channel depth and so on, may also introduce *3*-D effect. As a preliminary study, however, it is considered that the influence of the ratio of width to depth of the channel is the primary factor and the other parameters are kept the same as in the experimental condition. Three channel flows are simulated in this study: a square-section channel ($1m \times 1m$), a wide channel ($1.25m \times 0.8m$) and a narrow channel ($0.8m \times 1.25m$). To reduce the computer space requirement, only *4m* length is chosen as the computational channel length and the boom is installed in the middle of the channel, i.e. *2m* from the inlet. The draft of the boom is *0.15m* and the flow velocity is *0.3m/s*. Only water flow is simulated in the present study. The computational grid is generated using ICEM CFD [4] and the fluid flow is solved and post-processed using FLUENT [5]. The main functionality of ICEM CFD and FLUENT are described in the following subsections.

2.1 Grid Generator - ICEM CFD

ICEM CFD is a commercial code that is able to generate a computational grid on geometrically complex configurations. ICEM CFD embodies full CAD tools for creating geometry and importing geometry from various CAD systems. ICEM CFD can generate both structured (HEXA) and unstructured (TETRA) grids. In this study, ICEM CFD HEXA is used to generate structured grids for the channel flow region.

ICEM CFD HEXA has three major modules: DDN, Mesher Interface and HEXA. DDN is used to create or import geometry. The geometry is defined based on the hierarchy of geometry entities: point, curve (including line and arc), surface (including plane) and volume. The Mesher Interface is used for tagging information on the geometry entities, such as curves and surfaces. HEXA reads the geometry file that is created by Mesher Interface. To create a grid system, the hierarchy of geometry entities created in DDN, i.e. point, curve, surface and volume, should be associated with the hierarchy of grid entities, i.e. vertex, edge, face and block. HEXA also provides sophisticated surface smoothing and volume relaxation algorithms for grid computation and a set of tools for grid quality checking. Using the grid quality checking option in HEXA, cells with undesirable skewness or angles can be displayed on the screen and the corresponding block edges can be repaired interactively to obtain a better volume grid. After a satisfactory mesh is generated, the grid file can be exported in several forms to match different flow solvers. In this study the CFD software FLUENT is used to solve and post-process the flow.

2.2 Flow Solver - FLUENT

FLUENT is a general-purpose commercial CFD code for modeling fluid flow problems. It is a complete package consisting of three parts: pre-processor, flow solver and post-processor. It is able to simulate various fluid flow and heat transfer problems. In FLUENT, the mean (time-averaged) Navier-Stokes equations and some additional equations for a selected turbulence model are solved numerically. For the frequently used k- ε turbulence model, these equations are summarized below. The continuity and momentum equations are

$$\frac{\partial \rho}{\partial t}+\frac{\partial}{\partial x_i}(\rho U_i)=0 \tag{1}$$

$$\frac{\partial}{\partial t}(\rho U_i)+\frac{\partial}{\partial x_j}(\rho U_i U_j)=-\frac{\partial p}{\partial x_i}+\frac{\partial}{\partial x_j}[\mu_{eff}(\frac{\partial U_i}{\partial x_j}+\frac{\partial U_j}{\partial x_i})]+\rho g_i \tag{2}$$

for $i=1,2,3$, where ρ is density, U_i is mean velocity, p is static pressure, g_i is gravitational acceleration, t is time, x_i is Cartesian coordinates and μ_{eff} is effective viscosity defined as the sum of the molecular viscosity of the fluid, μ, and the turbulent viscosity of the flow, μ_t ($\mu_{eff}=\mu+\mu_t$). In the standard k-ε turbulence model, the turbulent viscosity is determined by $\mu_t=\rho C_\mu k^2/\varepsilon$, in which k is turbulent kinetic energy per unit

第三部分 受聘于加拿大帝国石油公司时期

mass, ε is dissipation rate and C_μ is an empirical constant. The quantities k and ε are solved from their transport equations

$$\frac{\partial}{\partial t}(\rho k)+\frac{\partial}{\partial x_j}(\rho U_j k)=\frac{\partial}{\partial x_j}(\frac{\mu_t}{\sigma_k}\frac{\partial k}{\partial x_j})+\mu_t(\frac{\partial U_i}{\partial x_j}+\frac{\partial U_j}{\partial x_i})\frac{\partial U_j}{\partial x_i}-\rho\varepsilon \tag{3}$$

$$\frac{\partial}{\partial t}(\rho\varepsilon)+\frac{\partial}{\partial x_j}(\rho U_j\varepsilon)=\frac{\partial}{\partial x_j}(\frac{\mu_t}{\sigma_\varepsilon}\frac{\partial\varepsilon}{\partial x_j})+C_{1\varepsilon}(\frac{\varepsilon}{k})\mu_t(\frac{\partial U_i}{\partial x_j}+\frac{\partial U_j}{\partial x_i})\frac{\partial U_j}{\partial x_i}-C_{2\varepsilon}\rho(\frac{\varepsilon^2}{k}) \tag{4}$$

where σ_k, σ_ε , $C_{1\varepsilon}$ and $C_{2\varepsilon}$ are also empirical constants. Their default values in FLUENT are:

$$C_\mu=0.09,\quad C_{1\varepsilon}=1.44,\quad C_{2\varepsilon}=1.92,\quad \sigma_k=1.0,\quad \sigma_\varepsilon=1.3 \tag{5}$$

Equations (1-4) are discretized in space using the finite volume method and discretized in time using an implicit scheme. The discretized algebraic equations can be solved numerically using various algorithms.

FLUENT has excellent post-processing capabilities including vector plot, contour plot, *x-y* plot, streamline plot and animation. In this study, the streamline plot is used to demonstrate the process of vortex formation and vortex diving.

3. Simulated Results

First, the water flow in a square-section channel (*1m×1m*) is simulated. Due to symmetry of the channel flow, only half width of the channel is considered. As mentioned before, the geometry and the computational grid are generated using ICEM CFD HEXA. The dimension of the "half width channel" is *4m* long, *0.5m* wide and *1m* deep. In the channel length direction, *150* grid points are exponentially clustered so that the finest cell is *0.002m* long near the boom. In the channel width direction, *41* grid points are clustered exponentially so that the smallest cell is *0.002m* wide near the channel wall. In the vertical direction, the height is divided into two portions. The upper portion is uniformly divided into *28* grid points and each cell is *0.01m* thick. The lower portion contains 31 geometrically distributed points. The grid distribution across the border between the two portions is continuous. The total number of grid points is *356,700*. The computational grid is shown in Figure 1.

The generated grid is imported into FLUENT for simulation. The density of water is $1000kg/m^3$ and the kinetic viscosity is *1.5cSt*. The flow velocity is *0.3m/s*. The Reynolds number based on the channel depth, flow velocity and water viscosity is about *200,000* so that the flow is turbulent. The standard *k-ε* turbulence model is applied. The entry turbulent intensity is set to be *0.05* and the characteristic length is chosen to be the channel depth, *1m*. The flow is considered steady and the tolerance of the normalized residual is set to be *0.001*. The boundary conditions are as follows. The inlet velocity is *0.3m/s* and the outlet is entirely outflow. No-slip wall condition is set on the sidewall, the bottom floor and the surface of the boom. Symmetry boundary conditions are set on the symmetry plane. On the surface of the channel, "rigid lid approximation" is applied so that the surface is treated as a slip wall.

The computation converges at iteration number *57* which means that the steady state flow is obtained. Figures 2(a), (b) and (c) show the streamlines at the heights *0.80, 0.97* and *1.00m*, respectively. At height *0.80m*, i.e. *0.20m* below the water surface, the streamlines look like a well-ordered laminar layer. At height *0.97m*, i.e. *0.03m* below the water surface, the streamlines near the wall are distorted and the cross flow can easily be seen. At height *1.00m*, i.e. on the water surface, the streamline distortion becomes more severe. A pair of vortices is visible at the corners of the channel walls in front of the boom and the streamlines are tightened to the symmetry plane and the walls. To capture the details of the vortices, only a few streamlines are chosen at the corners as shown in Figure 3. Here, the pair of vortices can be seen more clearly. About two and a half to three complete circles are made for each vortex before it dives underneath the boom.

For the wide channel, the dimension is *4m* long, *1.25m* wide and *0.8m* deep. The ratio of width to depth is *1.5625*. The physical model, the boundary conditions and the solver settings are the same as those in the square-section channel. The computation converges after *53* iterations. The streamlines at the height *0.70, 0.79* and *0.80m* are very similar to those in Figures 2(a), (b) and (c), but the degree of distortion is weaker. The concentrated vortices are shown in Figure 4. It is obvious that the strength of the vortex pair is weaker than that in the case of the square-section channel. For each of the vortices, only about one and a half to two complete circles are made for each vortex before it dives underneath the boom. Since the vortex strength is weaker, the wide channel is better to reduce the 3-D effect compared to the square-section channel.

The dimension of the narrow channel is *4m* long, *0.8m* wide and *1.25m* deep. The ratio of width to depth is *0.64*. The physical model, the boundary conditions and the solver settings are the same as before. The computation converges after *31* iterations. The streamlines at the heights *1.00, 1.20* and *1.25m* are similar to those in Figures 2(a), (b) and (c), but the degree of distortion is stronger. Figure 5 shows the concentrated vortices and about three and a half to four complete circles for each vortex can be identified before the vortex dives underneath the boom. It is noted that the narrow channel is worse compared to the square-section channel because it enhances the 3-D effect.

From the above-simulated results, we can see that compared to the square-section channel, the wide channel decreases the 3-D effect while the narrow channel increases it. Therefore, a wider water channel is recommended for the oil boom experiments to reduce the 3-D effect and, in turn, to suppress the oil loss due to vortex diving.

4. Concluding Remarks

(1) Water channel is a useful laboratory facility to conduct experiments for oil containment with booms and boom failure mechanisms.

(2) The formation of a pair of vortices in front of the boom near the channel walls and vortex diving underneath the boom are phenomena observed only in the laboratory condition and, therefore, should be suppressed. In fact, they may hide the real boom failure mechanisms in the field condition of oil containment by towing booms.

(3) The phenomenon of vortex formation and vortex diving is attributed to the 3-D effect of the water channel. Among a number of factors, the ratio of width to depth of the channel plays the most important role.

(4) CFD technology can be used to simulate the 3-D effect of the water channel flow and to capture the concentrated vortices.

(5) The simulated results of three typical channels can provide a guideline for the channel design. A wider channel can reduce the 3-D effect and weaken the vortex strength and, therefore, is recommended for oil boom experiments. In contrast, a narrower channel will enhance the 3-D effect and strengthen the vortices and, therefore, should be avoided.

(6) Some questions remain open. The effects of other factors, such as the boom draft, the channel depth, the flow velocity and so on, need investigation before thorough understanding of the 3-D effect of the water channel can be obtained.

Acknowledgement

The authors are grateful to Drs. Hugh Brown and Ron Goodman for the results of channel experiments and to Imperial Oil Resources Limited for the financial support.

References

[1] An, CF, Barron, RM, Brown, HM and Goodman, RH (1998). Droplet Entrainment Boom Failure and Kelvin-Helmholtz Instability, *Proc. of 8th Int'l Offshore & Polar Eng. Conference*, Vol.II, pp.322-326.

[2] An, CF and Brown, HM (1997). Oil containment in flowing water – studies of several concepts, *Tech. Report, IPRCC.OM.97.19*, Imperial Oil Resources Limited, Calgary, Alberta, Canada.

[3] Chow, VT (1959) *Open-Channel Hydraulics*, McGraw-Hill Book Co., New York, pp.26-27.

[4] ICEM CFD (1997) *ICEM CFD HEXA Manual*. ICEM CFD Engineering Inc., Livonia, MI, USA.

[5] FLUENT (1996) *Fluent User's Guide v4.4*. Fluent, Inc., Lebanon, NH, USA.

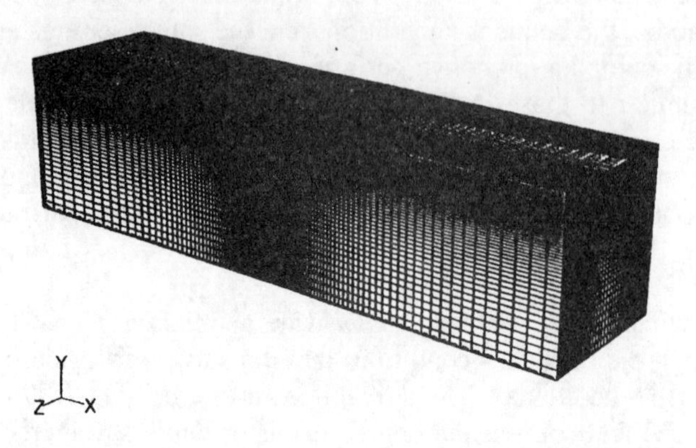

Figure 1. Computational Mesh for Square-Section Channel

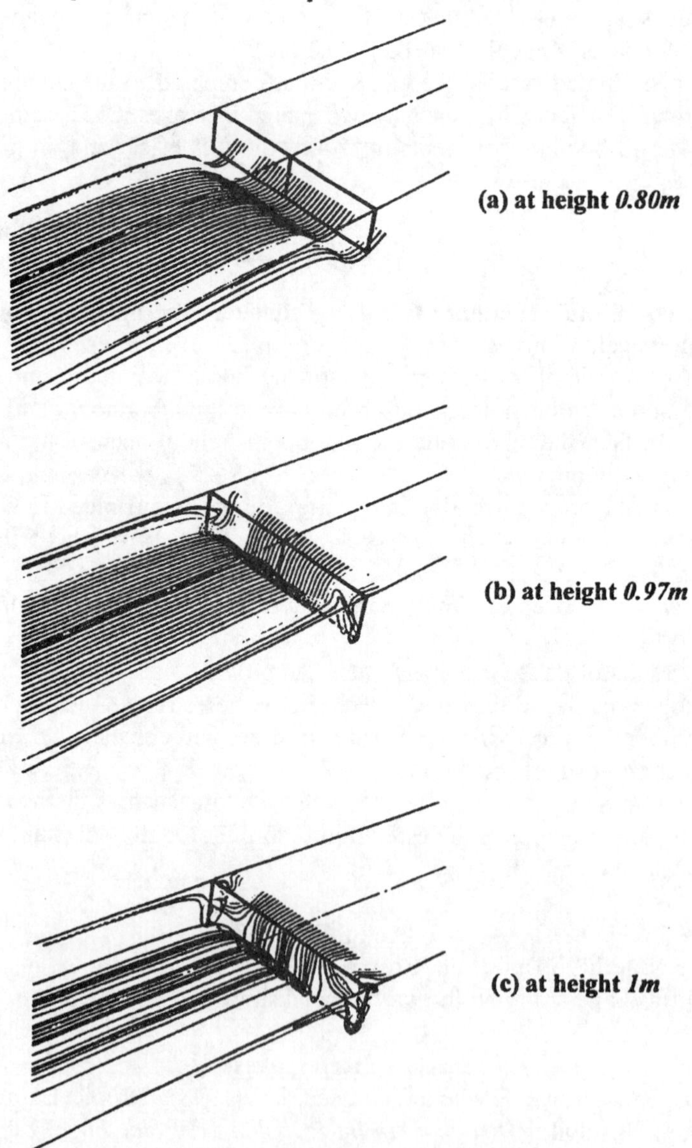

Figure 2. Streamlines in Square-Section Channel Flow

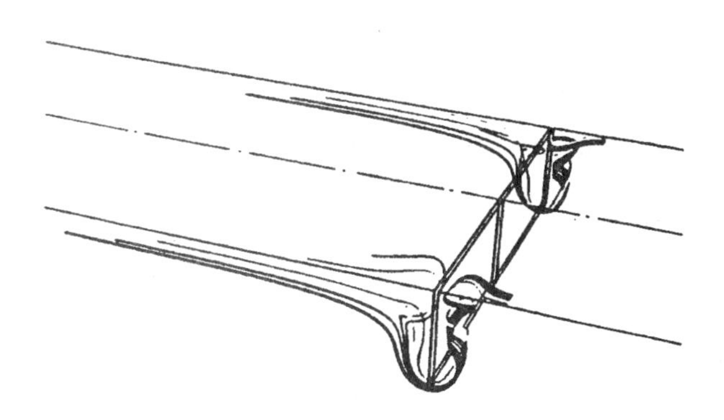

Figure 3. A Pair of Vortices in Square-section Channel Flow

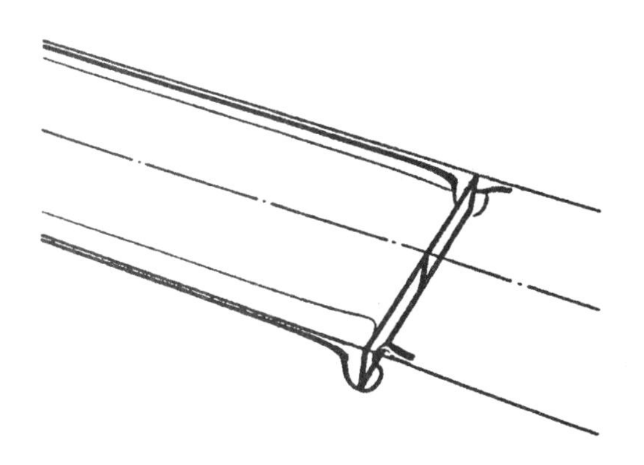

Figure 4. A Pair of Vortices in Wide Channel Flow

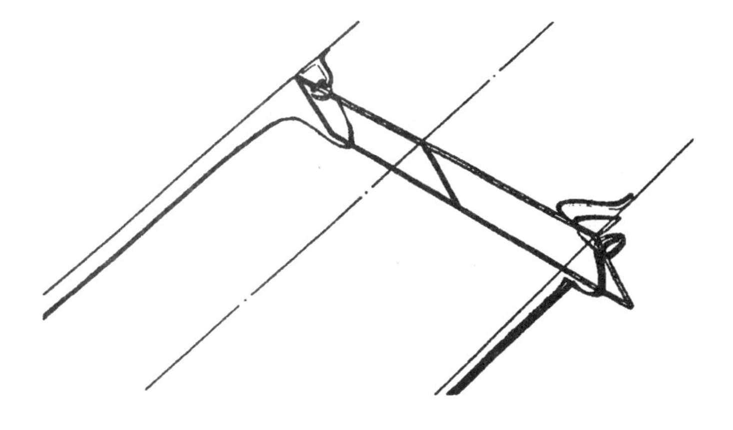

Figure 5. A Pair of Vortices in Narrow Channel Flow

Conference Proceedings

ACFD 2000 Beijing

International Conference on Applied Computational Fluid Dynamics

October 17–20, 2000 Beijing China

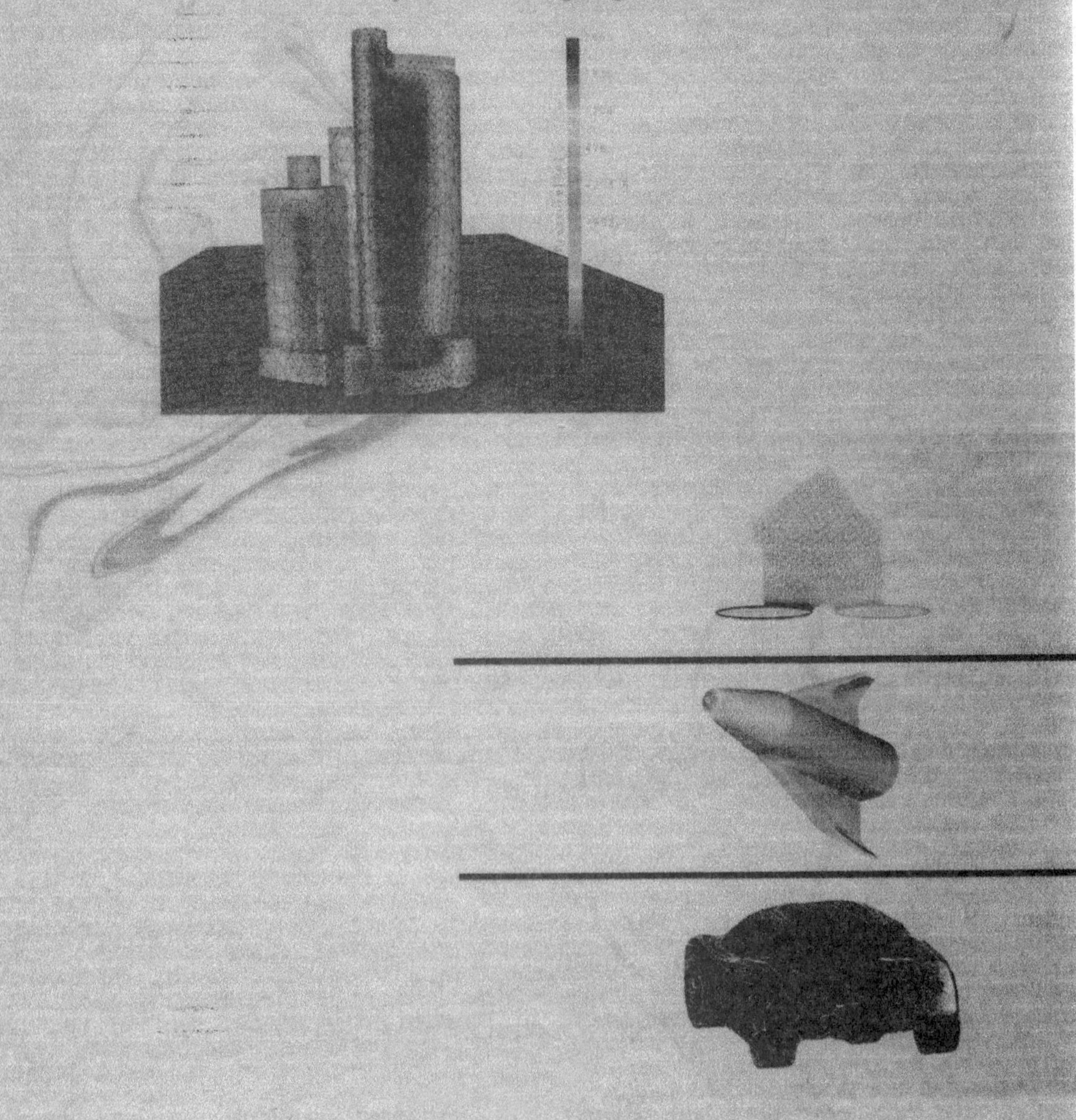

EFFECTS OF FLOAT SHAPE AND SIMULATION COORDINATES FOR THE NUMERICAL STUDY OF OIL NET-BOOM STRUCTURES

BARRON Ronald M. & AN Chang-Fa
Fluid Dynamics Research Institute, University of Windsor, Windsor, ON, N9B 3P4, Canada
Az3@uwindsor.ca and ca56@daimlerchrysler.com

ZHANG Zheng
College of Chemical Engineering, Beijing University of Chemical Technology, Beijing, 100029, China
Zhangz@buct.edu.cn and zzhang@uwindsor.ca

ABSTRACT

This paper emphasizes the numerical study of the effects of float shape and simulation coordinates for some net-boom structures. Firstly, the modeling methodology in a body-fitted coordinate (BFC) system is presented, the grid generation for the particular geometry is briefly described and the porous medium resistance formula for the inclined net is derived. Secondly, the flow patterns and oil losses in three net-boom structures with a circular float are numerically studied. Thirdly, the results for net-boom structures with a circular float are compared to the results for the rectangular float with the same section area. Fourthly, a numerical study for some net-boom structures with a "stepped round" float using a Cartesian grid is conducted and compared with the previous two studies. It is found that the circular float smoothens the flow pattern to bring more water into the net-boom region under the same entry water velocity and may easily cut off the oil into small pieces. This causes a significant decrease of the oil volume of fraction in the oil lump. Therefore, the critical velocity for the circular float case is reduced. It is also found that the effect of float shape is more important, and the particular grid system used for the simulation has little effect.

KEYWORDS Oil Spill, Net-boom Structure, Float Shape, Mesh Effect, CFD

INTRODUCTION

An oil boom is one of the major tools used to deal with oil spills on the water surface. It is known that the spilled oil can be well contained only under a certain critical boom-towing velocity. If the boom-towing velocity is above the critical one, the oil will escape underneath the boom due to hydrodynamic forces and the oil containment will fail. A single plate barrier is a typical boom structure and has been used for many years. In recent years, considerable attention has been given to investigate new types of oil boom structures to improve the capability of oil containment. Ueda and Yamanouchi (1998) reported some experiments using several types of net structures. Some experiments have been done at Imperial Oil where two separate perforated plates with different porosities were located upstream from the solid boom (An and Brown, 1997). It was found that the perforated plates improved the oil containment by increasing the critical velocity from 0.34 to 0.42-0.46m/s for Federated Crude oil. Cheng et al (1997) numerically simulated the flow field and predicted critical velocity for the same porous boom system tested at Imperial Oil. The predicted critical velocity coincides with the experimental results at Imperial Oil and it was also found that a small modification might significantly raise the critical velocity.

Recent experiments at Imperial Oil (Brown et al, 1999) have shown that the oil containment capability of a solid boom can also be improved by adding some nets in front of it. Based on these (porous) net-boom prototype designs, Zhang et al (1999a) reported some numerical simulations of oil-water flow around various net-boom structures, from which it was concluded that a net-boom structure offered great advantage to raise the critical velocity of boom failure compared to a single solid boom system. This value is much higher than the upper limit of the experiment capability in the Imperial Oil Test Basin (Brown et al, 1999), where some net-boom structures have been tested. It was found that the angle of inclination of the front net has a significant effect (Zhang et al, 1999b). Furthermore, some new net-boom structures were proposed (Zhang et al, 1999c) based on the numerical simulation. In some cases, the critical velocity for Federated Crude oil may be raised to as high as 2.0m/s, which is four times higher than a single plate boom. In the experimental condition, the incoming water velocity can only reach 0.5-0.6m/s. Under the best working conditions and boom structures used in the experiment, boom failure did not seem to occur at this velocity. Nevertheless, even in that case, some droplets could be still formed and escape in an actual tested net-boom system. It seems that the actual net-boom systems used in the experiments do not support the results obtained in our previous study (Zhang et al 1999b, c). In those simulations, the float that holds the front net has a rectangular shape with the same area as the original circular float. However, a small change in the net-boom structure in a flow-sensitive area may cause a significant change in flow pattern and, in turn, a significant change in critical velocity. In the front net area, water brings oil into the net-boom region through the front net. It is a highly sensitive region of the flow. Since the circular float may cause a different flow pattern from the rectangular float, it is necessary to explore the float shape effect for these types of net-boom structures. In this paper, the effects of float shape and the related coordinates for several net-boom structures have been studied. The study will be carried out in two coordinate systems, BFC system and Cartesian coordinate system.

MODELING THE NET-BOOM STRUCTURE

Similar to the previous study (Zhang et al, 1999b), the commercial CFD software, FLUENT, is also adopted to simulate the oil-water flow around and inside a net-boom structure. FLUENT includes all the conventional modeling strategies, such as k-ε turbulence modeling, finite volume scheme and the volume of fluid (*VOF*) technique. The details are described in the FLUENT guide (1994). Zhang et al (1999a, b) have successfully modeled the net as a porous medium. In this model, the net is considered as an additional momentum sink in the governing momentum equations. The studied basic net-boom structure is sketched in Fig. 1. It includes a basic vertical solid boom, a bottom net, an inclined front net and its supporter, a circular float. Normally, the float is partially under and partially above the water surface. To simplify the treatment without losing generality, it is assumed that the float is half-under and half-above the water surface as shown in Fig. 1. Since the flow passing through a porous net is similar to that passing through a perforated plate, the Darcy term in the additional momentum sink is neglected and only the inertial resistance term is considered. This term creates a pressure drop Δp in the flow direction given by

$$\Delta p \cong C_2 . \Delta n \frac{1}{2} \rho |\mathbf{v}|^2 \tag{1}$$

where C_2 is the inertial resistance factor, Δn is the thickness of the perforated plate and **v** is the superficial flow velocity passing through the plate. Accordingly, the pressure drop in x and y directions, Δp_x and Δp_y, can be written as

$$|\Delta p_x| \cong C_{2x} . \Delta n_x . \frac{1}{2} \rho |v_x|^2 \tag{2}$$

and

$$|\Delta p_y| \cong C_{2y} . \Delta n_y . \frac{1}{2} \rho |v_y|^2 \tag{3}$$

where C_{2x} and C_{2y} are inertial resistance factor components, Δn_x and Δn_y are Δn components, and v_x and v_y are flow velocity components in x and y direction, respectively. From $v_x^2 + v_y^2 = v^2$, Eqs. (1)-(3) can be combined to give the following relation:

$$\frac{\Delta p}{C_2 \Delta n} = \frac{|\Delta p_x|}{C_{2x} \Delta n_x} + \frac{|\Delta p_y|}{C_{2y} \Delta n_y} \tag{4}$$

For an inclined net with an angle α as shown in Fig. 2, it is apparent that

$$\Delta n_x = \Delta n \cdot Cos\alpha \qquad and \qquad \Delta n_y = \Delta n \cdot Sin\alpha \tag{5}$$

It is also seen from Fig.2 that $\Delta p = p_1 - p_3 = (p_1 - p_2) + (p_2 - p_3) = |\Delta p_x| + |\Delta p_y|$. Therefore, no matter what value α is, Eq. (4) and Eq. (5) imply that C_2 Δn must equal to $C_{2x}\Delta n_x$ and $C_y \Delta n_y$, or $C_2 = C_{2x}$ $Cos\alpha = C_{2y} Sin\alpha$. Therefore, C_{2x} and C_{2y} can be calculated from C_2. Adopting Smith & Van's correlation recommended by Perry & Chilton (1974) and making a small modification, C_2 can be calculated from

$$C_2 = \frac{1}{C \Delta n}((\frac{1}{\Phi})^2 - 1) \tag{6}$$

where C is a dimensionless orifice coefficient depending on the hole Reynolds number *Re*, the plate thickness-diameter ratio (Δn/D) and the hole pitch-diameter ratio (*P/D*). Φ is the ratio of total free hole area to cross section area of the perforated plate or the porosity of the porous medium. Since the value of C does not change significantly (0.75 to 0.9 in our conditions), a value of 0.8 was taken for all cases in the previous papers (Zhang et al, 1999a, b). In the present paper, all of these parameters are kept the same except that the grids are different in treating the circular float and inclined front net.

There are two kinds of grids, BFC and Cartesian, used in this paper. To create body-fitted coordinates, a software package called Geomesh (1996) is used. The generated BFC grids in the front net and float area for three cases considered in this paper, **A-3**, **VER** and **Q-F**, are shown in Fig. 3 (a)-(c), respectively. Here, **A-3** is a basic case in accordance with the prototype design **A-3** tested at Imperial Oil (Brown et al, 1999). For this case, the backward inclined angle α is about 67°. **Q-F** is a variant of **A-3** with a quarter-forward inclined front net. The angle α is about 30°. **VER** means that the front net is vertical and α is 0°. The grids in Cartesian coordinates use a zig-zag technique to approximate the circular float and the inclined front net. As an example, the grids for case **Q-F** is shown in Fig. 4. The calculation domain and the dimensions in this study are the same as those in the previous study (Zhang et al, 1999b) and coincide with the experimental settings at

Imperial Oil (Brown et al, 1999). The channel depth is 0.8m, the solid boom draft is 0.15m and the distance between solid boom and front net is about 1.35-1.4m. The oil volume per unit channel width is 4.0 litters/m. The details can be found in Zhang et al (1999a).

MAIN RESULTS FOR THE CIRCULAR FLOAT

1. BASIC CASE A-3

In case **A-3**, the front net is highly back-inclined (67°). When the water flows towards the net-boom region, most of the water is diverted and flows outside the front net instead of entering the net-boom region as shown in Fig. 5(a)-(c). From these figures, it can be seen that only 17 streamlines penetrate the front net and enter the net-boom region out of a total of 300 streamlines in the whole channel depth for all three water velocities from 0.5 to 1.0m/s. This means that the ratio of flow rate of entering water to flow rate of the oncoming water is only 5.7%. This is quite small compared to the ratio of boom draft to channel depth of 18.75%. Figure 5 also shows that the water flow outside of the front net is almost parallel to it. This may cause a situation where the oil lump, which attempts to enter the net-boom region through the front net due to the buoyancy force, is broken into small droplets as shown in Fig. 6(a)-(c). This drastically reduces the *VOF* value of the oil slick and reduces the density difference between oil and water. The color scale is shown on Fig.7. The number in this figure gives the value of oil *VOF*. 1 means full of oil without water and 0 means full of water without oil. Therefore, the buoyancy forces acting on the oil slick are reduced and the oil slick becomes easier to be swept by the water flow. Furthermore, Fig. 5 also shows that the water entering the net-boom region tends to flow towards the water surface. The entering water flow near the float changes direction right behind it and drives the oil slick along the water surface towards the solid boom. Therefore, no oil can stay right behind the float, as shown in Fig. 8(a)-(c). As a result, the maximum water velocity along the water surface inside the net-boom region, u_{smax}, for all water velocities of 0.5, 0.75 and 1.0m/s, is still larger than that in the other two cases, as shown in Table 1. Also the vertical velocity near the solid boom, v_{out}, is another important velocity to drive the oil slick out of the net-boom region. Table 1 shows that this velocity is also larger than that in the other two cases. Both u_{smax} and v_{out} were defined as the characteristic velocities in the net-boom region to determine the critical velocity, cf. Zhang et al (1999b). All of the above-mentioned facts make the oil slick difficult to be contained by this boom system. Figure 8(a) shows that for the case of water velocity 1.0m/s, all of the oil escapes in less than 20 seconds, except some small pieces of oil or droplets circulating behind the solid boom with a strong water vortex. Figure 8(b) shows a similar situation for the case of water velocity 0.75m/s after longer time. There is no doubt that boom failure occurs for both 1.0m/s and 0.75m/s cases. Figure 8(c) shows that almost all of the oil (with *VOF* value 0.7) is kept in front of the solid boom in the case of water velocity 0.5m/s. There are only very few droplets occasionally escaping from the net-boom region. It can be concluded that boom failure does not occur for this case. This means that the critical velocity for case **A-3** is between 0.5 and 0.75 m/s. This result is close to the experimental results at Imperial Oil (Brown et al, 1999).

2. VERTICAL FRONT NET CASE, VER

In the vertical front net case, **VER**, when the water flow reaches the net-boom region, the incoming water flow is almost perpendicular to the front net, except for a slightly downward flow at the net bottom as shown in Fig. 9(a)-(c). There are 24 streamlines out of a total of 300 streamlines entering the net-boom region through the front net for all the water velocities 0.5, 0.75 and 1.0m/s. This means that about 8% of the total flow in the channel enters the net-boom region in this case. This ratio is larger than that for case **A-3** under the same flow condition. Since the water flow coming to the front net is almost perpendicular to the front net, the shear force which can cut the oil lump into pieces is much smaller than that in case **A-3**. Therefore, the oil lump entering the net-boom region adheres together and not much is broken into small pieces, as shown in Fig. 10(a)-(c) (see Fig. 6 for comparison). In this case, the value of *VOF* can be kept very high (close to 1). For the case of water velocity 0.5m/s, the oil lump entering the net-boom region turns around the circular float with small vortex behind the float as shown in Fig. 10(a). This is consistent with the streamlines shown in Fig. 9(a). For larger water velocities, 0.75 and 1.0m/s, the circulations behind the float are larger, cf. Fig. 9(b) and 9(c). The oil lump entering the net-boom region leaves from the float to the vortex behind it, cf. Fig. 10(b) and 10(c). As shown in Fig. 9(a)-(c), the water flow entering the net-boom region is kept almost horizontal (cf. Fig. 5) and more entering water turns down and passes through the bottom net than that for case **A-3**. As a result, the water velocity along the surface is smaller than that for case **A-3**. This also reduces the downward velocity in front of the solid boom and thus significantly delays the occurrence of boom failure. The maximum surface water velocity, u_{smax}, and the downward velocity in front of the solid boom, v_{out}, are shown in Table 1. It is obvious that both velocities for case **VER** are much smaller than those for case **A-3**. Figs. 11(a)-(c) show the final oil slick status of case **VER** for three water velocities, 0.5, 0.75 and 1.0m/s, respectively. It can be seen that the oil can be kept inside the net-boom region with *VOF* value as high as close to 1, cf. Fig. 8(c), for all three velocities. As shown in Fig. 11(a), the oil slick remains highly stable in front of the solid boom with a thin and long lump shape for the small velocity 0.5m/s. Almost no oil is left behind the float as the vortex there is weak when water velocity is low. At higher velocities, 0.75m/s and 1.0m/s, oil can also be kept in front of the solid boom, but some oil may remain behind the float as shown in Fig. 11(b) and (c). It seems that more oil is left behind the float in the case of 1.0m/s. However, the oil lumps kept in front of the solid boom for higher velocities are not as stable as that for the case of 0.5m/s. They are thicker and shorter with lower *VOF* value in them. Nevertheless, for all three velocities, boom failure does not occur. Furthermore, it is found that when the velocity is increased to 1.5m/s, the oil lump can not be kept inside the net-boom region at all. Therefore, the critical velocity for case **VER** is likely between 1 and 1.5m/s.

3. QUARTER-FORWARD INCLINED FRONT NET CASE, Q-F

In the case of quarter-forward inclined front net, the front net is slightly inclined forward a quarter of the distance compared to the basic case **A-3**. In this case, the incoming water flows perpendicularly to the front net as seen in Figs. 12(a)-(c), for incoming velocities 0.5, 0.75 and 1.0m/s, respectively. This causes the oil lumps to pass through the front net very smoothly without breaking as shown in Fig. 13. In this case, the front net can easily accept the incoming water and even more water can enter the net-boom region. The ratio of the entering water flow rate to the total flow rate of the channel is about 8.7% in this case. This value is higher than that for **VER** case. It may be also seen from Fig. 12 that a large portion of entering water flows directly out from the net-boom region through the bottom net. As a result, both the water surface velocity and the downward velocity in front of the solid boom are relatively smaller than their counterparts in **VER** case as shown in Table 1. Since there is only a very weak vortex behind the float for the small velocity case, 0.5m/s, as shown in Fig. 12(a), the entering oil lump almost clings underneath the float before it reaches the water surface as shown in Fig. 13(a). For the medium velocity case 0.75m/s, the oil lump leaves the float before it touches the water surface and then turns around under the circulating forces by a vortex behind the float as shown in Fig. 12(b) and Fig. 13(b). Figure 13(c) shows that the oil lump entering the net-boom region is largely turned around behind the float when the water velocity is 1.0m/s. This is because there is a strong flow vortex behind the float as shown in Fig. 12(c). Figure 14(a) shows that most oil is finally kept in front of the solid boom except that a small amount of oil remains behind the float for water velocity 0.5m/s case. In this case, the contained oil lump is thin, long and stable with high value of oil *VOF*. For two larger velocity cases, 0.75 and 1.0m/s, most oil remains in the recirculation area behind the float as shown in Fig. 14(b) and (c). Especially for the case of 1.0m/s, almost all of the oil is kept in the large recirculation zone. From these figures, neither 0.75m/s nor 1.0m/s can be considered as a critical status. Therefore, the critical velocity in this case should be larger than 1.0m/s and at least as high as that in case **VER**. Again in Table 1, case **Q-F** has the smallest u_{smax} and v_{out} among the three cases.

COMPARISON BETWEEN TWO STUDIES

The results of the circular float and inclined net with body-fitted coordinates (BFCs) are compared to the results of the previous study (Zhang et al, 1999b) for a rectangular float with Cartesian coordinates. The differences are as follows. For the case of **A-3**, since the BFCs in the present study make the front net straight, the shear force causing breakage of the entering oil lump is stronger than that for a rectangular float with a zig-zag net used in the previous study. Therefore, the entering oil lump is more unstable and easy to be broken into small pieces. Moreover, the *VOF* value of the oil slick in this study (~0.6) is smaller than the *VOF* values (0.6-0.8) in the previous study (Zhang et al, 1999b). Since the circular float has less resistance to the oncoming water flow, it allows more water to enter the net-boom region. This may cause the result that both the surface water velocity and the downward flow velocity in front of the solid boom become larger than their counterparts in the rectangular float studied previously. For comparison, the characteristic velocities (u_{smax} and v_{out}) for the rectangular float case in the previous study (Zhang et al, 1999b) are also shown inside the brackets in Table 1. This table shows that: (1) the surface water velocity is increased about 40% for **A-3** case and about 80-100% for the cases of **VER** and **Q-F**, and (2) the downward flow velocity in front of the solid boom is increased even higher, over 1.5 times for **A-3** case and about 3-4 times for the other two cases. Based on the above analysis, the critical velocities for all the studied cases are significantly reduced. In case **A-3**, the critical velocity is reduced from 0.8-1.0m/s in the previous study to about 0.65m/s in this study, which is about 20-35% reduction. For the cases of **VER** and **Q-F**, the critical velocities are reduced from 1.8-2.0m/s in the previous study to 1.2-1.3m/s in this study, about 30-40% reduction.

CARTESIAN COORDINATES STUDY

There are two major differences between this study and the previous one (Zhang et al, 1999b), i.e. float shape and coordinates. The float shape effect is discussed in the previous section. In order to investigate the effect of the coordinates, we used Cartesian coordinates with zig-zag circular float and inclined front net to simulate the oil-water flow and critical velocities for all three configurations, **A-3, VER, Q-F**, and all three velocities, 0.5, 0.75 and 1.0m/s. It was found that the differences between the results with Cartesian coordinates and the results with BFCs were very small. The BFC grids for all three cases in this study have been shown in Figs. 3(a)-(c). As an example in this study, only the results of the case **Q-F** with 1.0m/s are shown here. The Cartesian grids for this case are illustrated in Fig. 4. Figures 15(a) and (b) show that at an early stage the oil smoothly enters the net-boom region and recirculates with the vortex behind the front net float. This flow feature is quite similar to its counterpart for the BFCs as shown in Fig. 13(c). Figure 16 shows the final oil slick shape. Almost all the oil has high *VOF* value (~1.0) and turns around in a recirculation zone behind the float as observed in the case of BFCs, cf. Fig.14(c). Finally, it can be concluded that the results from zig-zag type objects in Cartesian coordinates are very close to the results from the BFCs as long as the grids are fine enough.

CONCLUDING REMARKS

From the results of the numerical study for the three cases of a net-boom structure with a circular float and an inclined front net in BFCs, the following conclusions can be drawn. (1) The inclined angle of the front net has a significant effect on the flow pattern and the flow rate through the net-boom region, especially the water surface velocity and the downward velocity in front of the solid boom. (2) The inclined angle significantly affects the pattern of oil lump entering the net-boom region through the front net and, in turn, affects the final status of oil slick. (3) Similar to the results for the case of a rectangular float and a zig-zag front net in Cartesian coordinates, both case **VER** and case **Q-F** are much better than case **A-3** to increase critical velocity. The critical velocities for case **VER** and case **Q-F** are about 1.0-1.5m/s, while the critical velocity is about 0.5-0.75 for case **A-3**. The latter is quite close to the experimental results at Imperial Oil (Brown et al, 1999).

The float shape has a significant effect on the flow pattern and the ability to keep the oil inside the net-boom region. Comparing to the rectangular float, the circular float can smoothen the incoming water flow to enter the net-boom region and can push the entering oil lump to move faster and escape from the region more easily. Therefore, the circular configuration has lower critical velocity compared to the rectangular float configuration.

The results of a zig-zag circular float and an inclined front net with a fine grid in Cartesian coordinates are close to the results in the body-fitted coordinates. This suggests that the reason for the differences in the critical velocities between this study and the previous study (Zhang et al, 1999a) are primarily due to the difference in the float shape, rather than the difference in the coordinates.

ACKNOWLEDGEMENT

The authors acknowledge Imperial Oil Resources Limited for financial support of this study. They are especially grateful to Drs. Ron Goodman and Hugh Brown for their encouragement and discussions on this project.

REFERENCES

AN C.F. and BROWN H.M., (1997), Oil Containment in Flowing Water - Studies of Several Concepts, *Technical Report No. IPRCC.OM.97.19*, Imperial Oil Resources Limited, Calgary, AB, Canada.

BROWN H.M., GOODMAN R.H. and AN C.F., (1999), Development of Containment Booms for Oil Spills in Fast Flowing Water, *Proceedings of the 22nd Arctic & Marine Oilspill Program (AMOP) Technical Seminar*, Vol.2, pp.813-823, Calgary, AB, Canada.

CHENG S.Y., ZHANG Z. and AN C.F., (1997), Preliminary Numerical Simulation of Boom Failure with VOF Model, *Proceedings of the 7th National Symposium on Numerical Heat Transfer*, pp. 189-193, Beijing, China (in Chinese).

FLUENT, (1994), *Fluent User's Guide*, Fluent Inc., Lebanon, NH, USA

GEOMESH, (1996), *Geomesh User's Guide*, Fluent Inc., Lebanon, NH, USA

PERRY R.H. and CHILTON, C.H., (1974), *Chemical Engineers Handbook, 5th* Edition, p.5.34-35, McGraw-Hill Book Co., NY, USA.

UEDA, K. and YAMANOUCHI H., (1998), Oil Containment on Water Currents and a Method of Presentation of Escaping Oil with Net, Presented at the Speaker's Corner of *The 21st Arctic & Marine Oilspill Program (AMOP) Technical Seminar*, Edmonton, AB, Canada.

ZHANG Z., AN C. F. and BARRON R.M., (1999a), Modeling of Oil-Water Flow around Porous Boom Structures, *FDRI-TR-99-01*, Fluid Dynamics Research Institute, University of Windsor, Windsor, ON, Canada.

ZHANG Z., AN C.F., BARRON R.M., BROWN H.M. and GOODMAN R.H. (1999b), Numerical Study on (Porous) Net-Boom Systems - Front Net Inclined Angle Effect, *Proceedings of the 22nd Arctic & Marine Oilspill Program (AMOP) Technical Seminar*, Vol.2, pp. 903-919, Calgary, AB, Canada.

ZHANG Z., BARRON R.M. and AN C.F., (1999c), Numerical Study on New Net-Boom Structures, *Proceedings of the Eighth Asian Congress of Fluid Mechanics*, pp.806-809, Shenzhen, China.

Table 1 Some important velocities in different cases for circular float
(Data in brackets are for rectangular float)

Case	Incoming Water Velocity (m/s)	u_{smax}* (m/s)	v_{out}** (m/s)
A-3	1.00	0.62(0.45)	0.34(0.135)
	0.75	0.46(0.34)	0.26(0.098)
	0.50	0.31(0.22)	0.15(0.067)
VER	1.00	0.45(0.17)	0.27(0.056)
	0.75	0.34(0.13)	0.20(0.040)
	0.50	0.23(0.08)	0.13(0.028)
Q-F	1.00	0.40(0.22)	0.25(0.053)
	0.75	0.31(0.17)	0.17(0.040)
	0.50	0.21(0.11)	0.12(0.030)

* u_{smax} : maximum water velocity along the surface
** v_{out} : downward velocity at the bottom of the solid boom

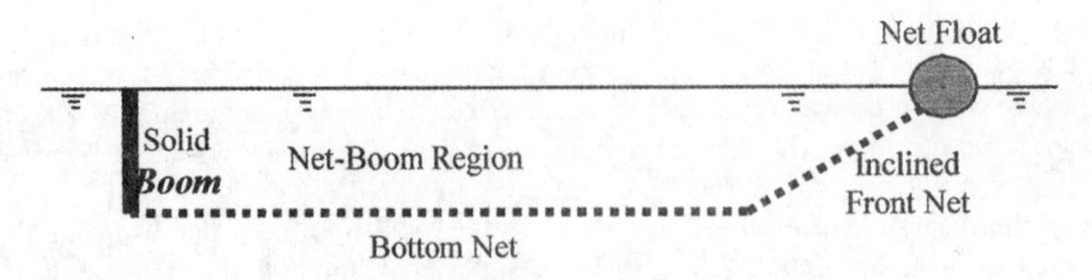

Fig. 1 Net-Boom Structure

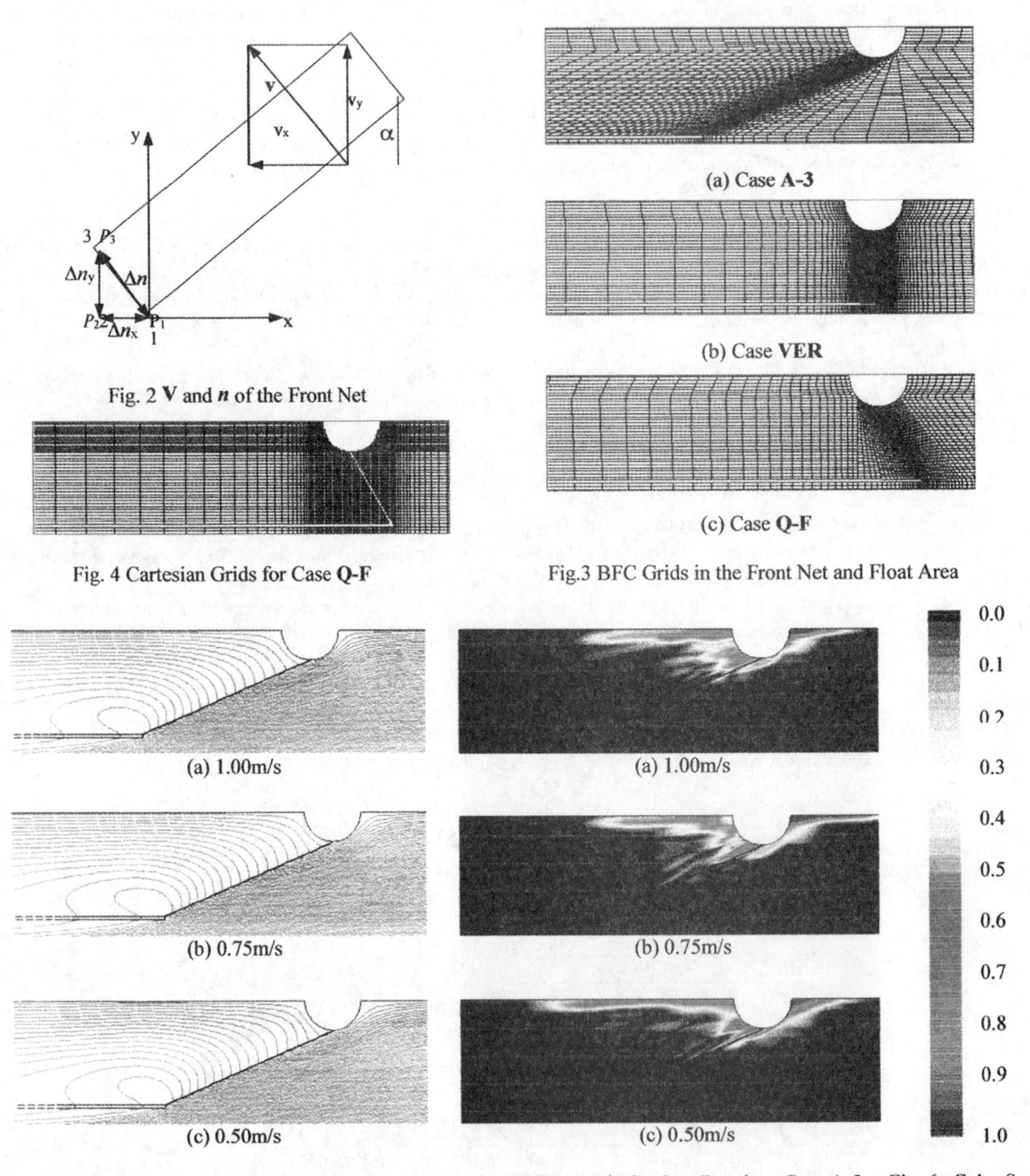

Fig. 2 **V** and ***n*** of the Front Net

(a) Case **A-3**

(b) Case **VER**

(c) Case **Q-F**

Fig.3 BFC Grids in the Front Net and Float Area

Fig. 4 Cartesian Grids for Case **Q-F**

(a) 1.00m/s

(b) 0.75m/s

(c) 0.50m/s

Fig. 5 Streamlines through the Front Net, Case **A-3**

(a) 1.00m/s

(b) 0.75m/s

(c) 0.50m/s

Fig. 6 Oil Lump is Cut into Droplets, Case **A-3**

Fig. 6a Color Scale

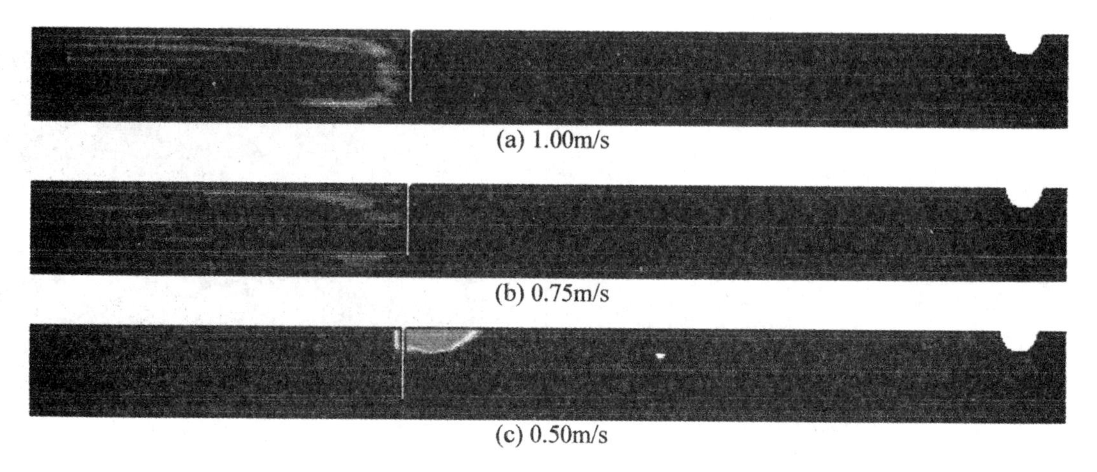
(a) 1.00m/s

(b) 0.75m/s

(c) 0.50m/s

Fig. 7 Final Oil Slick Status, Case **A-3**

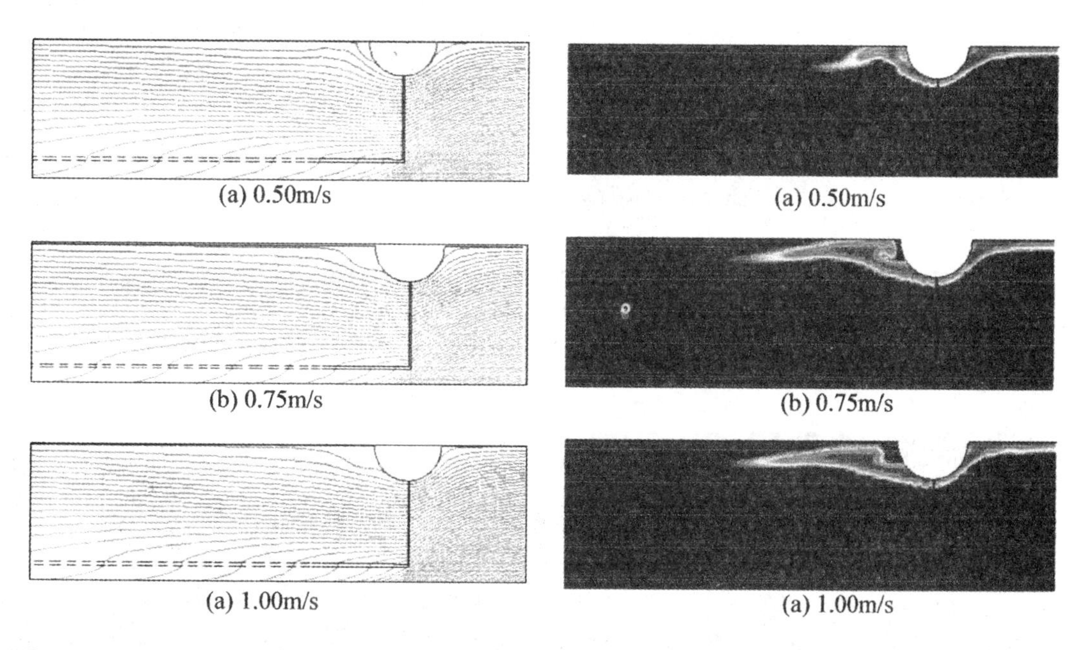
(a) 0.50m/s (a) 0.50m/s

(b) 0.75m/s (b) 0.75m/s

(a) 1.00m/s (a) 1.00m/s

Fig. 8 Streamlines through the Front Net, Case **VER**

Fig. 9 Oil Lump through the Front Net, Case **VER**

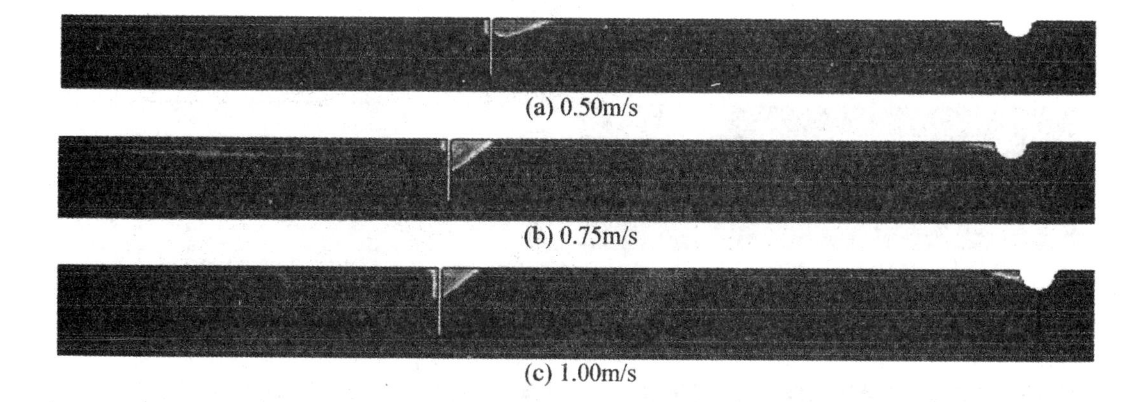
(a) 0.50m/s

(b) 0.75m/s

(c) 1.00m/s

Fig.10 Final Oil Slick Status, Case **VER**

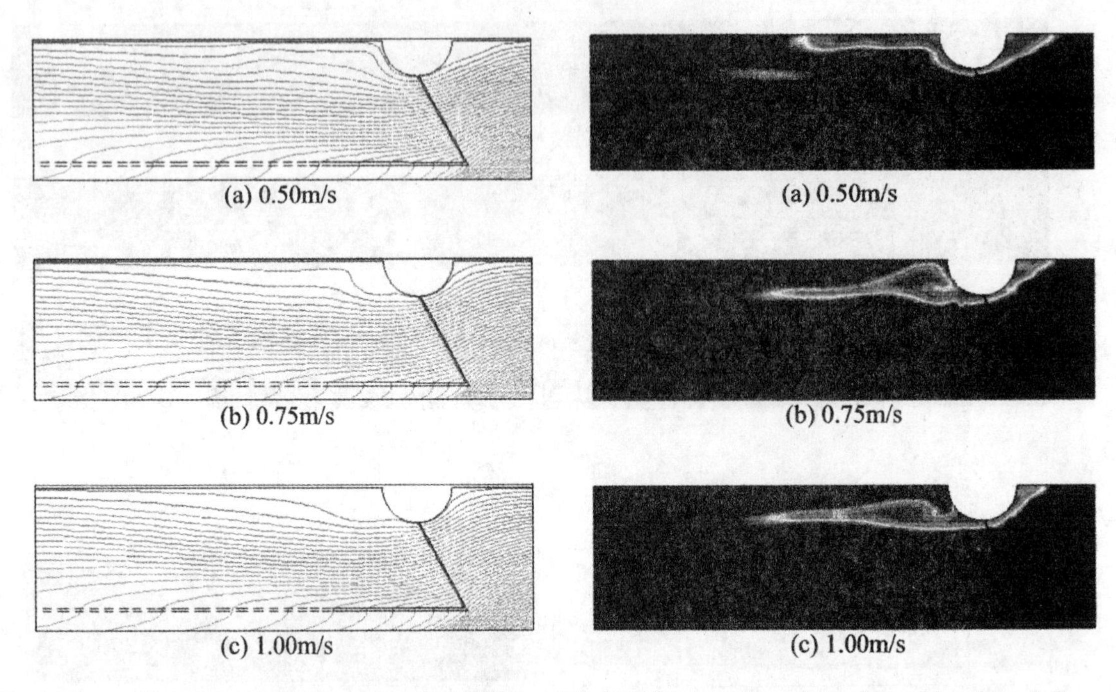

(a) 0.50m/s (a) 0.50m/s

(b) 0.75m/s (b) 0.75m/s

(c) 1.00m/s (c) 1.00m/s

Fig. 11 Streamlines through the Front Net, Case **Q-F**

Fig. 12 Oil Lump through the Front Net, Case **Q-F**

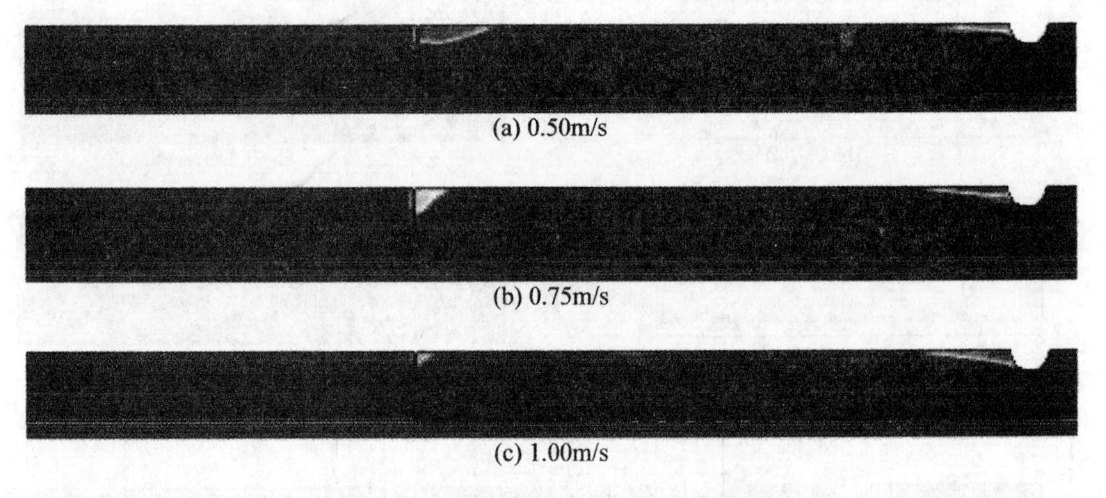

(a) 0.50m/s

(b) 0.75m/s

(c) 1.00m/s

Fig. 13 Final Oil Slick Status, Case **Q-F**

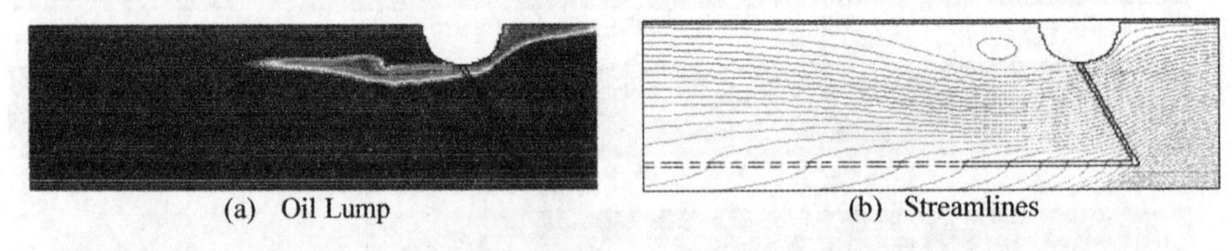

(a) Oil Lump (b) Streamlines

Fig. 14 Cartesian Coordinates, Case **Q-F** (1.00m/s)

Fig. 15 Final Oil Slick Status, Case **Q-F**
(Cartesian Coordinates, 1.00m/s)

June, 2001 • Volume 34 • Number 3

Richard Rimbach, Sr. and Richard Rimbach, Jr., Founding Publishers

Norberta Rimbach, Publisher, Editorial Director
Dick Bronner, Associate Publisher
Raquel Rimbach, Assistant Publisher

David Lavender, Managing Editor
Lisa Cinna, Product Editor
Amy Peindl, Graphic/Production Editor

Karen Galante, Circulation Manager
Renee Rimbach, Advertising Manager
Jeff Bronner, Regional Editor

Pollution Equipment News (USPS 349-750, ISSN 0032-3659) is published 7 times per year, February, April, June, August, October, November and December by Rimbach Publishing Inc. There is a shipping/handling fee of $25.00 per year for subscriptions outside of the United States. March and September issues continue to be mailed internationally at no charge. Periodicals postage paid at Pittsburgh, PA and at additional offices. POSTMASTER: send change of address to *Pollution Equipment News*, 8650 Babcock Blvd., Pittsburgh, PA 15237-5010.

BPA INTERNATIONAL

Printed in USA

8650 Babcock Blvd.
Pittsburgh, PA 15237-5821
E-mail: info@rimbach.com
1-800-245-3182 • FAX: 412-369-9720
Website: http://www.rimbach.com

Sales Representatives

George Allison
8650 Babcock Blvd., Pittsburgh, PA 15237
412/364-5366, 800/245-3182
FAX: 412/369-9720
PA, WV, NC, SC, GA, AL, MS, IN, IA, WI, IL, LA

Jeff Bronner
783 Ranch Rd., Tarpon Springs, FL 34689
412/364-5366, 800/245-3182
FAX: 412/369-9720
Florida and all companies outside the U.S.

Jim Campbell
8650 Babcock Blvd., Pittsburgh, PA 15237
412/364-5366, 800/245-3182
FAX:412/369-9720
NY, PA, TN, KY, OH, MI, MN, SD, ND, MO, KS, NE, AR, OK, TX

Don McCann
180 Tuckerton Rd., Ste 6, Medford, NJ 08055
412/364-5366, 800/245-3182
FAX: 412/369-9720

Kyle McCann
180 Tuckerton Rd., Ste 6, Medford, NJ 08055
412/364-5366, 800/245-3182
FAX: 412/369-9720
MA, NH, ME, VT, NJ, NY, PA, RI, CT, DE, DC, MD, VA

Ken Stoutenburg
701 High St., Ste. 227, Auburn, CA 95603
530/823-8160, FAX: 530/823-8161
AZ, NM, Southern CA

Phil Stoutenburg
701 High St., Ste. 227, Auburn, CA 95603
530/823-8160, FAX: 530/823-8161
MT, CO, WY, ID, UT, NV, HI, OR, WA, AK, Northern CA

TABLE of CONTENTS

Articles

4 CFD Analysis Helps Develop Oil Containment Booms
34 Cut Gas Sensor Maintenance Costs with Auto Calibration System
34 Hydrothermal Oxidation Unit Processes Sludge Economically and Produces By-product Income
36 Air & Waste Management Show Exhibition List
37 Controlling CO_2: Is There a Handle?
38 Optimize Sludge Dewatering

Charts

50 Gas Detectors
54 Air Cleaning Equipment
56 Dataloggers
58 Continuous Emission Monitors

Product Emphasis

44 Environmental Software
46 Laboratory Instruments
46 Spill Control
47 Air Quality Monitors
53 Air Cleaning Equipment

Sections

42 Literature Showcase
40 Industry News
58 Trade Shows
59 Advertiser Index
59 Searchlite Section

Rimbach Publishing Co., publisher of *Pollution Equipment News*, is a member of the following associations: Water Environment Federation, ISA, American Chemical Society, Spill Control Association of America, American Water Works Association, Fluid Controls Institute, and American Society of Mechanical Engineers.

Questions or comments about *Pollution Equipment News*, call

(888) RIMBACH or (412) 364-5366

CFD Analysis Helps Develop Faster Oil Containment Boom

By Chang-Fa An

Researchers using computational fluid dynamics (CFD) analysis have improved the velocity at which oil containment booms can operate successfully, increasing the current level of under 0.4 m/sec, where traditional booms fail, to a theoretically possible 2.0 m/sec. The boom velocity is critical because oil spill cleanup is limited by how fast the booms can be towed before hydrodynamic forces cause them to fail.

To develop a boom that operates at faster velocities, oil companies have been evaluating a number of new designs, including angled booms and porous nets that operate below and upstream of traditional booms.

Previously, researchers performed water channel tests to evaluate new designs, but this added significantly to development costs. Now, CFD simulation is used instead.

This alternative allows researchers to simulate many design alternatives quickly and inexpensively. Recently, researchers have also begun to animate CFD results to enhance their understanding of boom failure mechanisms.

When oil is accidentally spilled on water, the most common response is to use barriers or booms to collect the oil. A typical boom floats on the surface of the water and is composed of a cylindrical float with a skirt.

In open water, it is towed behind a boat. In rivers and streams, it is placed in the current. In either situation, as oil collects behind the boom, it is recovered with skimming devices. Booms reliably hold oil only under certain conditions. If they are towed too rapidly by ships or are deployed in fast moving streams or rivers, failure occurs due to hydrodynamic forces.

Traditional booms have been observed to fail when the water velocity is greater than about 0.4 m/sec. In recent years, oil companies and others have attempted to find new boom designs that can work in greater water velocity. This would permit faster cleanup operations in open water because it would allow the boats to travel faster. It would also make booms more effective in a greater number of rivers and streams.

Difficult Testing

As with any new product, part of designing a new boom involves evaluating its performance under typical operating conditions. Oil boom testing is normally done in a tank or channel that is at least 30 m long by 15 m wide by 2 m deep. It must be this large because this type of hydrodynamic study is affected by scale. The closer the scale is to real world conditions, the more accurate the results are.

A boom is placed in the middle of the channel where there is an underwater observation window. The tank is instrumented with a flow velocity device. As water flows through the channel, oil is poured onto the water. From the observation window, researchers watch what happens to the oil when it reaches the boom. The speed of the water can be varied, as can the depth of the boom placement.

What makes this type of testing expensive is the cost of the power needed to accelerate that volume of water through the channel. Also, once the experiment is over, the channel must be thoroughly cleaned to remove the oil.

Researchers have turned to CFD analysis as a less expensive way of predicting boom performance. A CFD simulation provides fluid velocity, pressure, and species concentration values throughout the solution domain for problems with complex geometries and boundary conditions.

As part of the analysis, a researcher may change the geometry of the system or the boundary conditions such as inlet velocity for example, and view the effect on fluid flow patterns or concentration distributions.

CFD also can provide detailed parametric studies that can significantly reduce the amount of experimentation necessary to develop a device and thus reduce design cycle times and costs.

In simulating a water channel test with CFD, the goal is to get a thorough understanding of the effects of various parameters on the flow. For example, effects such as relative oil-water velocity, physical properties of oil and water, water depth, and boom draft can be investigated. Although most CFD programs can simulate a variety of fluid dynamics situations, an oil boom simulation adds a complicating factor that many programs cannot handle. The code must be able to distinguish two immiscible fluids, oil and water, and track the interface between them. The software, FLUENT, from Fluent Inc. has a volume of fluid (VOF) model that is capable of modeling this situation accurately.

Preparing the Analysis Model

The first step in preparing a water channel simulation is the creation of a grid to conform to the problem geometry. GeoMesh, from Fluent Inc. is used for this purpose.

GeoMesh has three components: a CAD system called DDN, an interactive computational grid generation tool called P-Cube, and mesh visualization and manipulation software called Leo.

DDN is used to create the geometry representing the water channel and the boom. The lower half of the cylindrical float, the submerged part, is represented by a semi-circle. The rigid skirt of the boom is represented by a vertical segment. P-Cube generates quadrilateral (2D) or hexahedral (3D) grids from this geometry. Leo is used to check grid quality and display the mesh.

Next, the grid files generated by GeoMesh are imported into FLUENT where boundary conditions and other information about the problem are specified. For example, an inlet boundary is defined at the left of the model, and the region where the oil enters is specified. The flow surface is designated as a slip-wall on which a horizontal velocity component is allowed, but a vertical velocity component is not. The channel floor and the surfaces of the boom are specified as no-slip walls. The right outlet boundary of the flow region is free of specification. The flow is turbulent and the standard $k-\varepsilon$ model is used to close the Navier-Stokes equations. The VOF model is used to track the flow of the oil and water phases.

Evaluating Boom Designs

This model has been validated through comparisons to channel testing and researchers can now quickly evaluate the performance of new oil boom designs. To test a new design, engineers modify the boom geometry with GeoMesh, regenerate the grid, import the new grid into FLUENT, and run a new analysis.

One of the recently studied designs called a net-boom structure, consists of a solid boom with front and bottom porous nets. The bottom net is horizontal. The front net can be vertical or positioned at an angle in relation to the boom. Because it is so easy to evaluate different design alternatives, researchers looked at numerous positions for the front net, including $0°$, $30°$, $49°$ and $67°$ as measured from vertical. To adapt the analysis model to the new design with nets, engineers only needed to add porous regions (to represent the nets) as an additional momentum sink in the governing momentum equations.

After this series of analyses was completed, the researchers created a plot showing inclination angle against critical velocity. It was determined that the highest critical velocity was achieved when the front net was nearly vertical or inclined slightly forward. More importantly, this study proved that there is excellent potential to raise the critical velocity by putting a porous net structure in front of the boom. For a single, solid boom (the traditional design), critical velocity is found to be 0.3 to 0.4 m/sec. A solid boom with a partially perforated net will raise the critical velocity to 1.0 m/sec. The researchers believe that with further modifications to the net-boom structure design, they can raise the critical velocity to 2.0 m/sec and perhaps higher.

Animating the Results

Until recently, researchers relied on static images of CFD results to understand oil boom failure and evaluate new designs. They have now started to combine these images into animated sequences that offer even greater insight into this time-dependent physical process.

The animation is created by playing a sequence of FLUENT images at a speed of 15 frames/sec. The graphical output from the simulations was obtained in TIFF format and converted to device independent bitmap (DIB) format for use in VidEdit from Microsoft Corp. This program is used to create the animations in audio video interleave (AVI) format. The animations can be viewed on a Pentium PC with the Media Player program supplied with Windows 95. The researchers involved in this work have found the animations to be particularly useful in understanding the failure mechanisms observed in channel experiments.

Using software to simulate the interaction between oil, water, and boom offers a less expensive alternative to channel testing, thereby reducing the cost of designing improved oil booms. Once they have a validated CFD model, researchers can quickly change the boom geometry and evaluate design alternatives.

Using CFD, researchers have now identified a type of boom that has the potential to increase critical velocity by more than a factor of four. When implemented, the new net-boom structure will decrease the time needed for oil spill cleanup, protecting the environment from the effects of the oil.

For more information, contact Fluent Inc., 10 Cavendish Court, Centerra Resource Park, Lebanon, NH, 03766, 603/643-2600, Fax: 603/643-3967, web:www.fluent.com or circle 314 on the reader service card. .

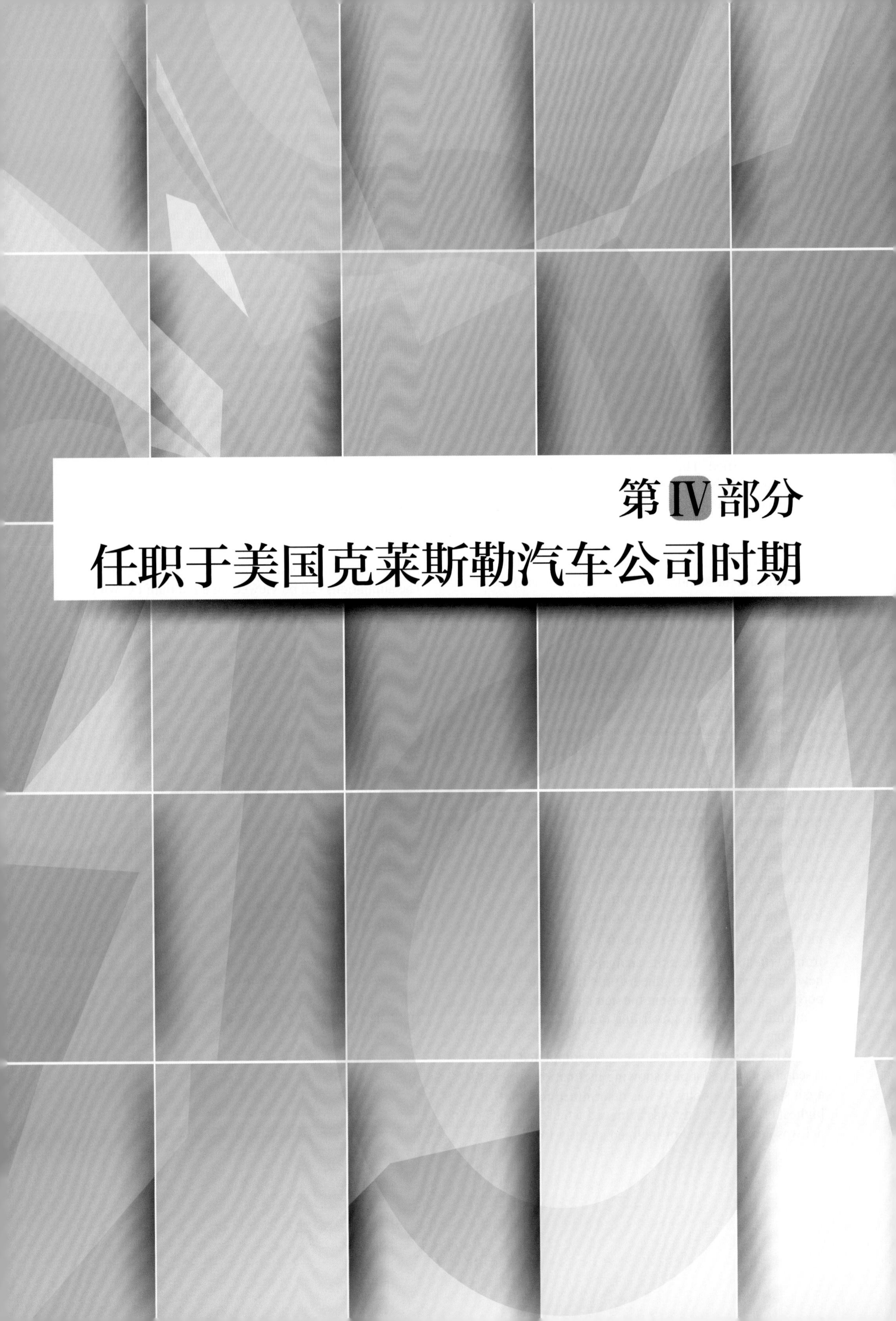

第Ⅳ部分
任职于美国克莱斯勒汽车公司时期

8e Conférence annuelle de la
8th Annual Conference of the

Comptes rendus
Proceedings

Volume 2

Montréal
11 au 13 juin 2000
June 11-13, 2000

Organisée par / Organized by :

cerca CENTRE DE RECHERCHE EN CALCUL APPLIQUÉ

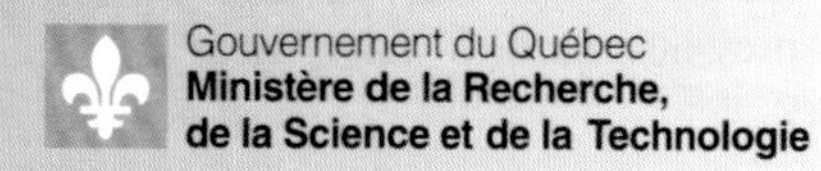

NUMERICAL SIMULATION FOR AN AUTOMOTIVE ENGINE COOLING FAN

R.M. Barron, H. Yang, A. El Saheli, C.F. An and G.W. Rankin
Fluid Dynamics Research Institute
University of Windsor
Windsor, ON., N9B 3P4
Email (Barron): az3@uwindsor.ca

1. INTRODUCTION

The numerical simulation of the underhood flow in an automobile is a major challenge to both the computer and CFD technologies. The underhood environment has many components with complex geometry concentrated in a narrow space, and very complicated fluid flow and heat transfer phenomena take place under the hood. One practical and feasible way to simulate the underhood flow is to develop various numerical models for different components, and then to couple those models together to resolve the whole flowfield. Among all the underhood components, the engine cooling fan is one of the most difficult objects to be numerically simulated due to the characteristics of its flowfield. Therefore, a reasonable modeling of the engine cooling fan is very important to the complete underhood flow simulation.

Recently we have conducted a series of numerical tests to establish a robust and reasonable "fan model". This paper presents numerical simulations of the three-dimensional viscous flow through an axial cooling fan with complex geometry. Both design and typical off-design operating conditions of the fan have been considered, and the calculated results are compared to available experimental data. The multiblock structured mesh topology has been adopted for this fan, and a high quality grid is generated with good orthogonality near solid walls and with high resolution in the boundary layers and wake. A state-of-the-art 3-D Navier-Stokes solver using the SIMPLE algorithm and the $k-\varepsilon$ turbulence model is used. Visualization of the computed results reveals the complicated features of the flow through the fan.

Additionally, comparative studies have been undertaken to determine the effects of the meshing strategy and the degree of numerical diffusion related to the order of the differencing for the convective terms in the transport equations. The effect of modeling the hub is also investigated since many previous studies have not considered the hub geometry.

2. CONFIGURATION AND MESH

The axial fan modeled in this investigation has five blades that are forward swept and radially twisted in space. Furthermore, the inclusion of the outer ring and the hub in the numerical modeling introduces significant difficulties for the generation of a block-structured mesh with high quality. Taking advantage of the periodicity of the fan configuration, only one-fifth of the fan is used to construct the solution domain. Thus, the solution domain contains only one blade connecting with one-fifth of the hub and ring, and accordingly a periodic boundary pair must be added to the solution domain (see Fig. 1). The inlet and outlet boundaries are placed two and a half and four axial chord lengths upstream and downstream of the fan, respectively. As shown in Fig. 1, an "axis boundary" in the vicinity of the rotating axis is created to connect the inlet with the hub in the inflow region while three cylindrical boundary surfaces are produced to link the ring with the inlet and outlet, and to extend the hub downstream to the outlet, respectively. As a result, the simulation on the above domain is somewhat similar to classical turbomachinery computation in the sense that the fan is placed into a duct with an inlet and an outlet, and therefore it might be referred to as a "design calculation" in order to distinguish it from the unconfined flow simulation for the fan. Since there is no such duct in the real fan testing configuration, we impose slip wall boundary conditions on the surface of this duct. To improve the mesh quality, we create 12 control surfaces around the blade profile, and then divide the solution domain into 25 blocks to build the block-structured mesh. Figure 2 shows the surface grids on the hub at the intersection with the fan blade for two different meshes used in this study. These meshes differ mainly in the distribution of nodes in the trailing edge and wake regions. More attention is given to the mesh quality in the first mesh (a), while the second mesh (b) is easier to construct but has more skewness in the trailing edge region.

3. COMPUTATIONAL MODEL AND N-S SOLVER

Using a rotating frame of reference fixed on the fan, the airflow in the fan is assumed to be steady, adiabatic, incompressible and turbulent. The finite volume based Navier-Stokes solver FLUENT [1] is applied in this work, with the SIMPLE algorithm implemented to deal with pressure-velocity coupling, and the standard $k-\varepsilon$ two equation model with wall functions selected to provide turbulence closure for the solution of the N-S equations. The convective terms are discretized using either first or second order accurate upwind schemes while the diffusive terms are discretized by second order central differencing. For the boundary conditions, velocity boundary conditions are specified at the inlet boundary, and the solver computes the mass flow rate and momentum fluxes into the domain through the inlet. The pressure outlet boundary condition is applied at the outlet. All the rotating solid walls are designated with moving-wall boundary conditions, namely the no-slip boundary conditions in the rotating reference frame. Slip-wall boundary conditions are applied on the outer cylindrical boundaries. Periodic boundary conditions are used for the periodic boundary pair, and the axis boundary conditions are employed on the axis boundary.

4. NUMERICAL SIMULATION AND ANALYSIS

The numerical simulations have been carried out at both design and typical off-design operating conditions of the fan. The convergence criteria are set as 10^{-4} for continuity and 10^{-3} for other equations, and the specified convergence state is reached after about 1200 iterations in most cases. The calculated characteristics of the fan are compared to the available experimental data in Fig. 3. Even though the second mesh (Fig. 2b) provides good results at the higher flow rates, convergence cannot be achieved at the low flow rates. As discussed later, the flow is much more complicated at the low flow rate. Comparing the results from first and second order differencing of the convective terms, we see that the first order scheme gives a slight improvement in the predicted pressure rise. However, velocity contour maps on planes immediately downstream of the fan show a large amount of numerical diffusion in the wake region for the first order scheme (cf. Fig. 4).

The effects of the hub are seen to be most significant at the higher flow rates. For this simulation, the hub is replaced by a long cylinder extending upstream to the inlet boundary. The primary advantage of this configuration is that the meshing is greatly simplified. Many previous fan simulations have incorporated this assumption. This simplified configuration yields a reasonable prediction of the fan pressure rise and downstream velocity magnitude and axial velocity. For example, the flow pattern at 20 mm behind the fan is similar to Fig. 4, with only some minor differences in the passage region between the blades. Overall, one may conclude, at least for this fan, that the hub does not significantly alter the flow through the fan and therefore could be replaced by the extended cylinder.

The following discussions are based on calculations with the second order scheme on the mesh in Fig. 2(a). Figure 3 demonstrates that for this simulation the calculated pressure rise of the fan is generally in good agreement with the experimental data over the entire range of operating conditions. However, it was found that after the design point, if the flow rate is increased, the discrepancy between the calculated and the measured pressure rise increases. This implies that the justification of the above numerical model with confined domain decreases with the increase of flow rate after the design point.

To gain a deeper insight into the flow mechanism of the fan, its 3-D flowfield has been visualized and analyzed under three flow rates, i.e., Q_{design}, $0.2Q_{design}$ and $1.8Q_{design}$. Firstly, Figs. 4, 5, 6 and 7 are used to illustrate the flowfield under the design operating condition. Figure 4(b) shows the velocity magnitude contours on a plane referred to as *S_3-20*, which is located 20 mm downstream of the fan. From this figure, one can easily see the wake region corresponding to high gradients of velocity. Figures 5 and 6 present the details of the airflow on a meridian plane cutting at about 2/3 of the span and on a revolution surface (S_1) close to the hub, respectively. In Fig. 5, a passage vortex near the blade tip has been found behind the fan, while in Fig. 6, one can see a separation bubble triggered from the suction surface in the vicinity of the leading edge. Figure 7 shows pathlines distributed around the suction side in the relative reference frame, from which one can see that the strong migration of particles in both radial and circumferential directions has resulted in a passage vortex near the suction surface. These visualizations demonstrate that the present numerical modeling is able to capture the important viscous phenomena associated with this flow.

Figure 8 shows the vorticity contours on plane *S_3-20* for two off-design operating conditions, $0.2Q_{design}$ and $1.8Q_{design}$. These figures illustrate that with a reduction of flow rate, the vorticity distribution becomes less uniform and the width of the wake is broader. Thus, the induced mixing losses will be larger after the fan if the flow rate is decreased. Figure 9 presents oil flow patterns on the suction surface under the three flow rates Q_{design}, $0.2Q_{design}$ and $1.8Q_{design}$. This figure indicates that even at the design flow rate, the fan has strong secondary flow, since the low energy fluid particles display an obvious tendency of radial migration around the blade suction surface, and this tendency will be strengthened if the flow rate is decreased. In contrast with lower flow rate cases, at higher flow rates such as $1.8Q_{design}$, the flow is smoother passing over the suction surface.

When the flow rate is small, say $0.2Q_{design}$, the state of the flow is found to be more complicated. Figures 8 and 9 have already provided images where, at $0.2Q_{design}$, the wake region is larger than the one at the higher flow rate, and the secondary flow is intensified. In Fig. 9(b), one can see that there is a large boundary layer separation area near the tip on the suction surface. Actually, at this small flow rate, the positive incidence is so large in the tip area that it triggers the boundary layer separation inducing a stall along the suction surface (see Fig. 10). In addition, for the $0.2Q_{design}$ case, the pathlines in Fig. 11 reveal a large reverse flow originating from the upper part of the fan and going back towards its lower part.

The above 3-D visualizations and analyses have manifested the complicated mechanism of the airflow in this fan. It is also important to note that the strong secondary flow in this fan, especially the radial and circumferential migration of the low energy fluid particles, accounts for most of the energy losses within the flowfield.

5. CONCLUDING REMARKS

The flow mechanisms of an automotive engine cooling fan has been investigated with three-dimensional Navier-Stokes simulations under both design and typical off-design operating conditions. Due to the complex geometry of the fan, a high quality multiblock structured grid is developed. The first order accurate upwind scheme for convective fluxes is found to introduce non-trivial numerical diffusion. The standard $k-\varepsilon$ two-equation turbulence model has proven to be robust and acceptable for all operating conditions. The calculated characteristics of the fan are in good agreement with the experimental data.

The visualization and analysis of the computed results have captured many important viscous phenomena in the flowfield, and thus the flow mechanisms of the fan can be revealed with the aid of the present numerical modeling. Comparison of the computed results at different operating conditions has demonstrated that different flow rates can trigger widely different flowfields in terms of the low energy particle migration, boundary layer separation, stall occurrence, vortex movement, etc. Deep insight into the flow mechanisms of the fan is very valuable to the development of a "fan model" for the whole underhood flow simulation. In addition, the accuracy of the "confined duct" model is found to decrease with the increase of flow rate after the design point, and therefore for large flow rate, it is suggested that the numerical model allow the airflow to expand freely after the fan.

REFERENCES

[1] FLUENT Inc., "Fluent 5, User's Guide", July 1998.

ACKNOWLEDGEMENTS

The authors are very grateful to Dr. Richard Sun and Mr. John Billington (*DaimlerChrysler*) for their support during the course of this research. The authors are also thankful to Dr. Evangelos Hytopoulos (*SGI*) for his helpful discussion and suggestions and Mr. Jun Sun for carrying out some of the computation.

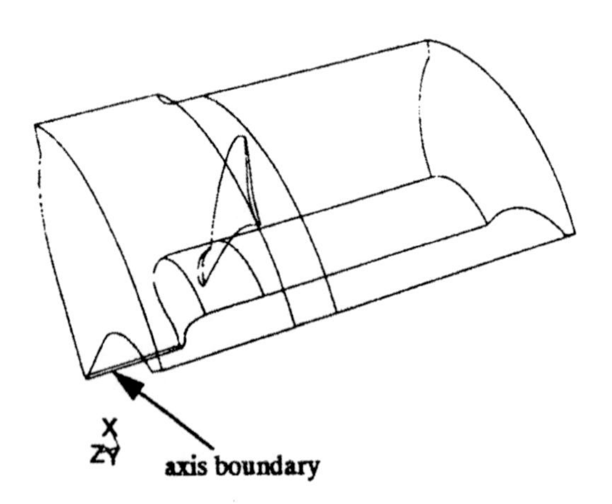

Figure 1 Solution domain

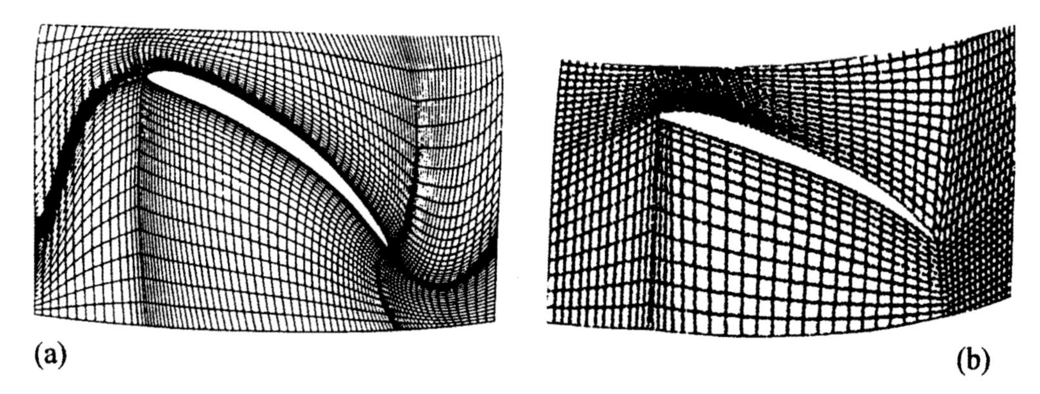

Figure 2 Surface meshes on the hub

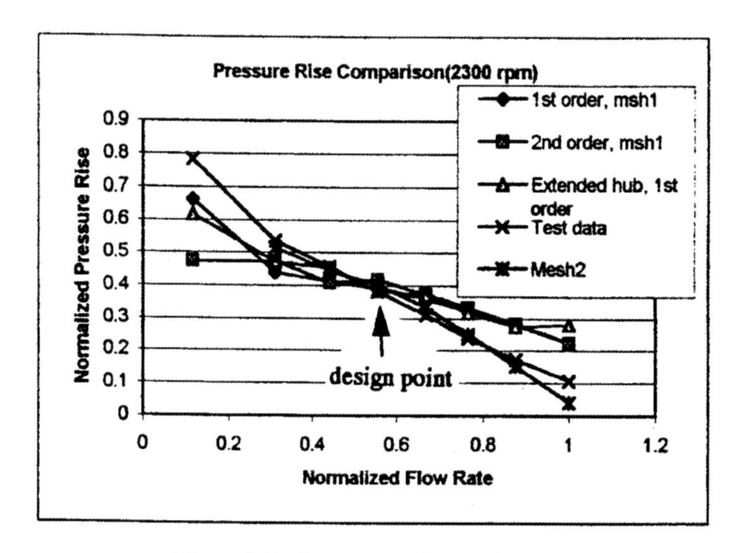

Figure 3 Static pressure rise vs. flow rate

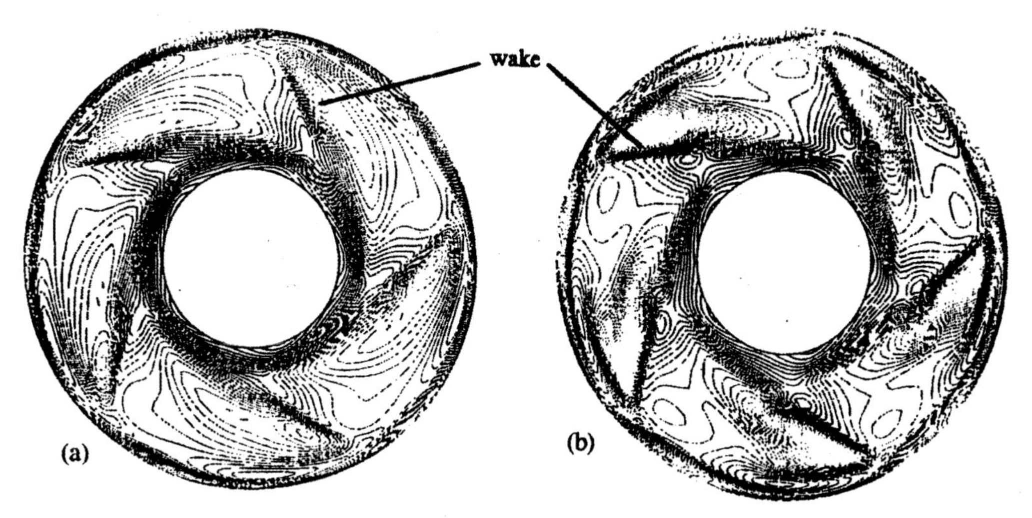

Figure 4 Velocity magnitude contours on plane *S3-20*
(a) first order scheme; (b) second order scheme

第IV部分 任职于美国克莱斯勒汽车公司时期

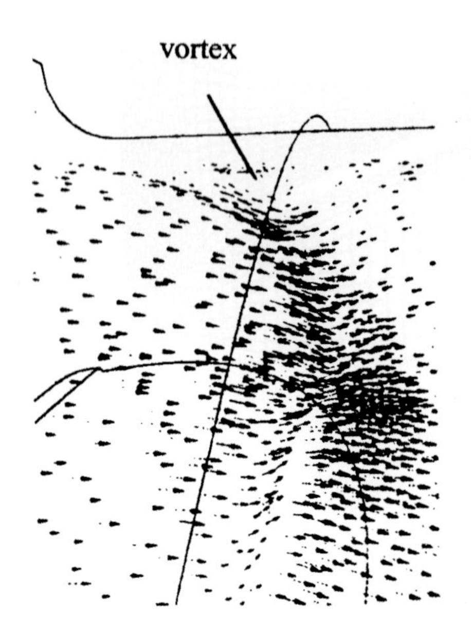

around the tip

Figure 5 Rel. velocity vectors on a meridian plane

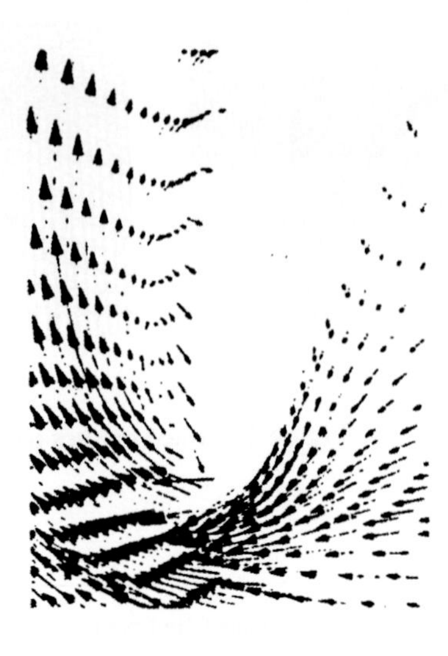

on S_1 surface

Figure 6 Rel. velocity vectors close to hub (around leading edge)

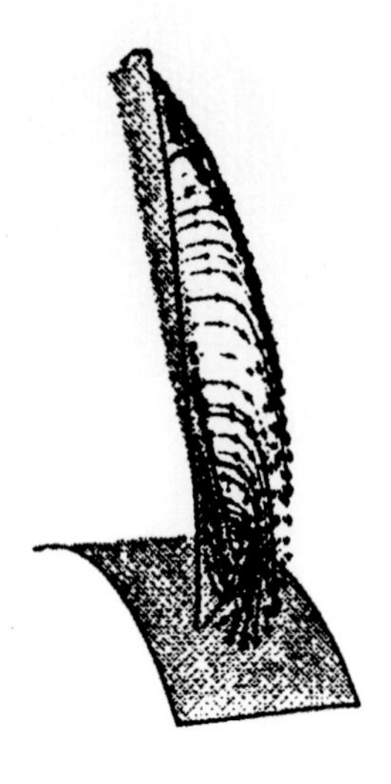

rear view

Figure 7 Pathlines around suction surface

安长发博士科研论文集

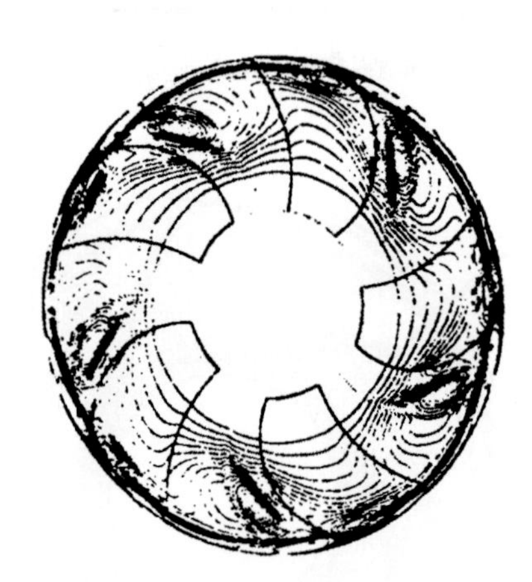

(a) 0.2 Q_{design}

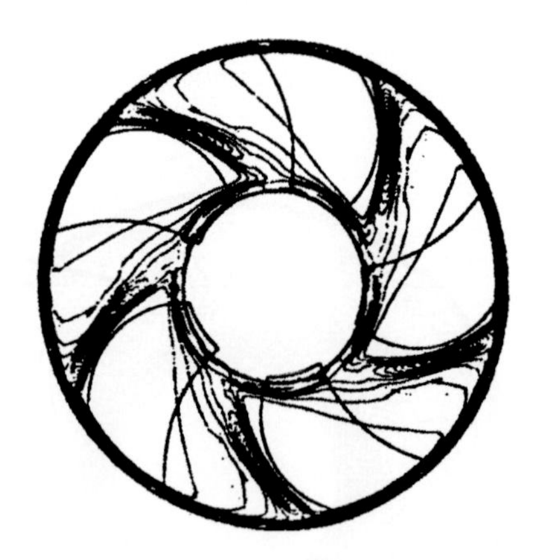

(b) 1.8 Q_{design}

Figure 8 Vorticity magnitude contours on plane S_3-20

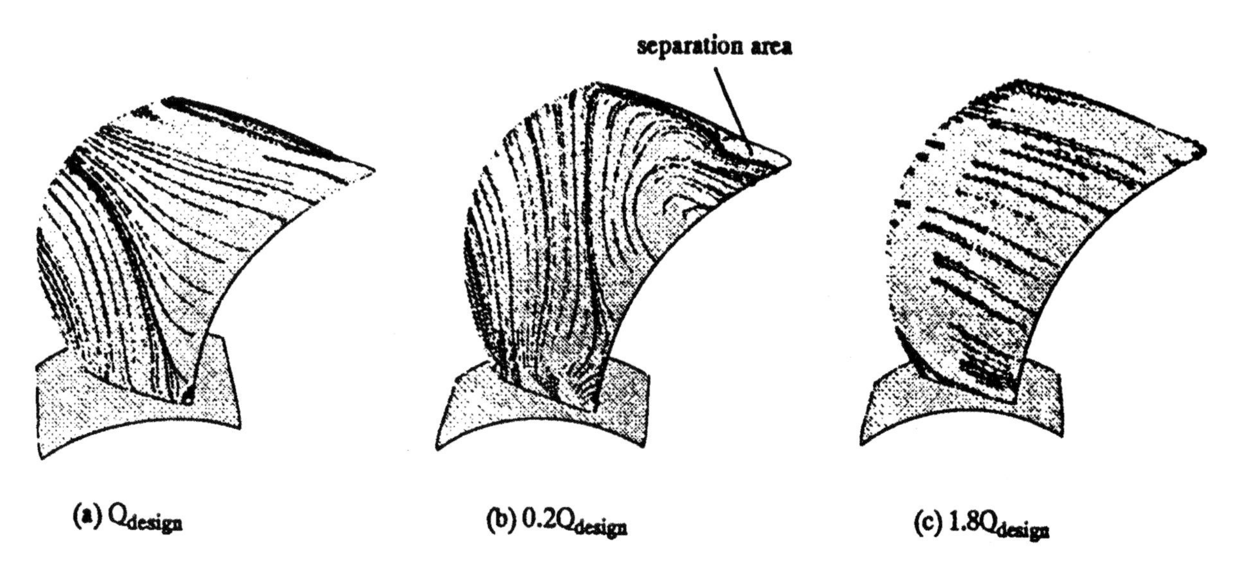

(a) Q_{design} (b) $0.2Q_{design}$ (c) $1.8Q_{design}$

Figure 9 Oil flow on suction surface (in relative reference frame)

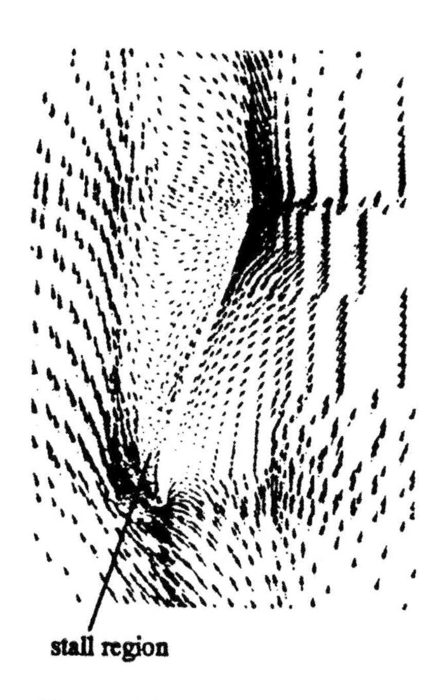

Figure 10 Rel. velocity vectors close to the ring ($0.2Q_{design}$)

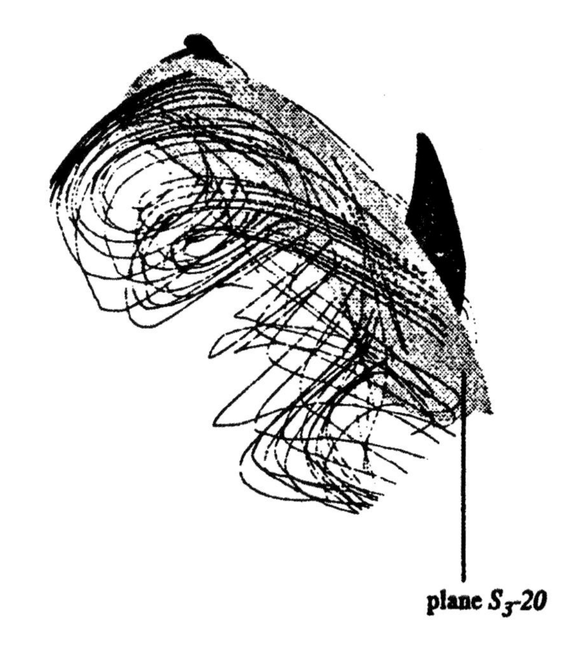

Figure 11 Pathlines traced back to plane S_3-20 in absolute reference frame ($0.2Q_{design}$)

OCTOBER 20–22, 2003

COMPUTATIONAL AEROACOUSTICS (CAA) WORKSHOP ON BENCHMARK PROBLEMS

安长发博士科研论文集

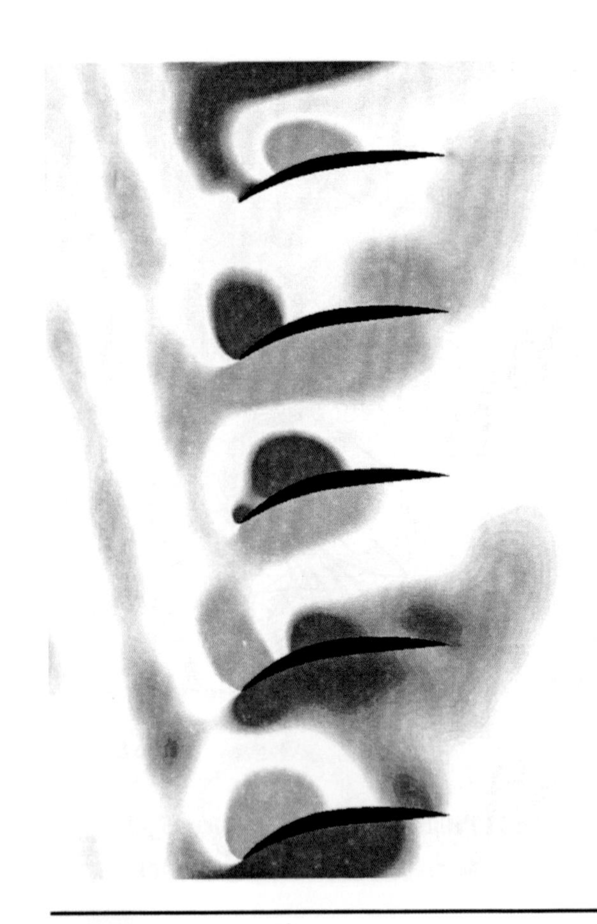

Changfa An

Cleveland, OH, USA
Oct. 20, 2003

Z. Zhang, R. Barron, and C.-F. An
University of Windsor
401 Sunset Avenue, Windsor
Ontario, Canada N9B 3P4

ABSTRACT

Noise and vibration may be induced when air passes over a cavity. This phenomenon is currently of significant practical interest in several industries, especially aeronautical and automotive, and it is the motivation for one of the benchmark problems suggested by the CAA4 workshop organizers. In the present work, power spectrum analysis of airflow passing over a cavity is studied numerically using the commercial Computational Fluid Dynamics (CFD) software, Fluent. The solution domain includes the cavity with a rectangular above it. The rectangle is 0.6m long (0.3m upstream and 0.3m downstream from the cavity) and 0.3m high. The lower edge of the rectangle coincides with the upper edge of the cavity. The flow velocity on the left side of the rectangle (inlet boundary) is uniform and equal to 50 m/s. The top of the computational domain is set to be a slip wall, or a symmetry line. The right side of the rectangle (outlet boundary) is set to be a pressure outlet. In general, he results from a CFD simulation may depend on many parameters in the problem settings, such as grid size, time step size, relaxation factors, turbulent models, order of time and space discretization, initial flow field, etc. The major findings of the present study are summarized below.

(1) The initial flow field does not have much effect on the power spectrum as long as the computations are converged at each time step. As shown in Table 1, the peak frequency (F) and peak sound pressure level (SPL) are the same for all four cases studied.

(2) Relatively large relaxation factors can be used for both continuity (0.5) and momentum (0.7) equations. Since turbulence residuals (for κ and ε) may fluctuate violently during computation, smaller relaxation factors (0.15-0.2) should be used. It was found that very few computations diverged with this combination of factors.

(3) The particular turbulence model used in the simulation is an important consideration. For a specific case (see Table 2), the results for $\kappa-\varepsilon$ type turbulence models (including RSM) are very close to each other. However, the results for $\kappa-\omega$ type turbulence models show lower peak frequencies and higher SPL compared to the $\kappa-\varepsilon$ models. Furthermore, the results using first order upwind space discretization for the convective terms are quite close to those of second order, and the SPL predicted by a laminar flow model simulation is close to the prediction by $\kappa-\varepsilon$ model.

(4) It is found that the power spectrum for the three particular sensor locations, specified by the benchmark coordinator, is essentially the same. Only average static pressure differs at different sensor location (see Table 3).

(5) The effect of time step size on the result is always significant. The results for four different grids (10x10, 15x15, 20x20 and 30x30 in the cavity neck) shows great variation for both peak frequency and peak SPL. Generally, the larger the time step is, the smaller the peak frequency and SPL are (see Table 4). Therefore, it is extremely important to choose a proper time step in the CFD simulation for power spectrum analysis of airflow over a cavity.

Table 1 Initial Velocity Field Effect*

Initial velocity field	F_{max} (Hz)	PYY_{max} (Pa^2/Hz)	P_{max} (Pa)	SPL_{max} (db)
A data file of $\delta t=5\times10^{-5}$s	120	13.78	2.171	100.7
All velocity=0	120	10.154	2.178	100.7
U=20 m/s	120	10.148	2.177	100.7
A data file of $\delta t=2\times10^{-4}$s	120	10.196	2.182	100.8

Table 2 Turbulence Model and Solution Scheme Effect at 20×20 grid

Model & Scheme	F_{max} (Hz)	PYY_{max} (Pa^2/Hz)	P_{max} (Pa)	SPL_{max} (db)
Standard k-ε	111	18.13	2.53	102.0
k-ε-RNG	112.5	11.81	1.82	99.2
Reynolds stress	111.1	11.2	2.59	102.3
k-ω	66.3	435.12	12.36	115.8
k-ω-sst-c	80.6	209.4	8.58	112.7
k-ω-sst-t	79.6	217.2	8.75	112.8
1st-order space	112.4	19.25	2.53	102.1
Laminar flow	127.6	8.56	1.95	99.8

Table 3 Sensor Location Effect at $\delta t=10^{-3}$ s and 2×10^{-5} s (bracketed)

Sensor location	F_{max} (Hz)	PYY_{max} (Pa^2/Hz)	P_{max} (Pa)	SPL_{max} (db)	Static P (Pa)
Cav. floor	45.3 (129.1)	3.81 (65.01)	0.652 (5.77)	90.3 (109.2)	17-21(11-27)
Cav. front wall	45.3 (129.1)	3.8 (64.71)	0.651 (5.75)	90.2 (109.2)	8-12 (2-18)
Cav. rear wall	45.3 (129.1)	3.79 (64.55)	0.650 (5.74)	90.2 (109.2)	6-10 (0-16)

Table 4 Time Step Effect at Different Mesh Fineness

Mesh	δt (s)	F_{max} (Hz)	PYY_{max} (Pa^2/Hz)	P_{max} (Pa)	SPL_{max} (db)
10×10	5e-5	125.2	119.1	6.36	110.1
	1e-4	104.6	41.15	3.77	105.5
	2e-4	83.6	0.13	0.19	79.3
15×15	2e-5	129.1	65.01	5.77	109.2
	5e-5	120.0	13.78	2.17	100.7
	2e-4	88.9	3.36	0.94	93.4
	5e-4	62.6	4.95	1.01	94.1
	1e-3	45.3	3.81	0.65	90.3
20×20	2e-5	125	113.6	6.3	110.0
	5e-5	112.4	19.25	2.53	102.1
	1e-4	98.5	5.72	1.18	95.4
30×30	5e-5	84.7	1032.5	19.09	119.6
	1e-4	67.7	2053	23.26	121.3

* (Basic working condition: 50 m/s, 15×15 grid, sensor located at cavity floor, $\delta t=5\times10^{-5}$ s, k-ε-RNG model, first order in time, second order for convective term)

第IV部分　任职于美国克莱斯勒汽车公司时期

Vehicle Aerodynamics 2004

SP-1874

Changfa An
Detroit, MI, USA
March 8, 2004

SAE International™

Published by:
Society of Automotive Engineers, Inc.
400 Commonwealth Drive
Warrendale, PA 15096-0001
USA
Phone: (724) 776-4841
Fax: (724) 776-5760
March 2004

2004-01-0230

Side Window Buffeting Characteristics of an SUV

Chang-Fa An
Belcan Corporation, Novi, MI
Seyed Mehdi Alaie
Optimal CAE, Inc., Novi, MI
Sandeep D. Sovani
Fluent Incorporated, Ann Arbor, MI
Michael S. Scislowicz
Truck Aero-Thermal Development
DaimlerChrysler Corporation, Detroit, MI
Kanwerdip Singh
Center of Competency, CFD Core Group
DaimlerChrysler Corporation, Auburn Hills, MI

ABSTRACT

Buffeting is a wind noise of high intensity and low frequency in a moving vehicle when a window or sunroof is open and this noise makes people in the passenger compartment very uncomfortable. In this paper, side window buffeting was simulated for a typical SUV using the commercial CFD software Fluent 6.0. Buffeting frequency and intensity were predicted in the simulations and compared with the corresponding experimental wind tunnel measurement. Furthermore, the effects of several parameters on buffeting frequency and intensity were also studied. These parameters include vehicle speed, yaw angle, sensor location and volume of the passenger compartment. Various configurations of side window opening were considered. The effects of mesh size and air compressibility on buffeting were also evaluated. The simulation results for some baseline configurations match the corresponding experimental data fairly well. This gains the confidence that the current method can be used to predict, improve and optimize SUV buffeting characteristics.

INTRODUCTION

In a recent survey, J.D. Power and Associates found that wind noise is one the topmost crucial complaints of new vehicle buyers. Buffeting is an important constituent of wind noise. From an aerodynamic point of view buffeting refers to temporally varying force exerted on passengers by pressure fluctuations of high intensity and low frequency. Such pressure fluctuations arise primarily from vortices created in the flow field by free shear structures such as jets, wakes, mixing layers and so on.

The investigation of buffeting noise was initiated as early as mid-1960's and it was called "wind throb" at that time [1-2]. Some researchers accept the terminology [3], while others call this phenomenon "booming" [4] or simply "resonance" [5]. However, most researchers in recent years use the term "buffeting" to express this type of wind noise [6-10] and this terminology is followed in this paper.

Buffeting can be considered as cavity noise since the entire passenger compartment acts as a cavity when the sunroof or a window is open. Cavity noise is caused by the unstable shear layer established at the upstream edge of the cavity. Vortices shed from the front edge of the opening are convected downstream along the flow. When they impinge onto the rear edge of the opening, vortices break down and a pressure wave is generated and propagates inside as well as outside the cavity. When the new wave reaches the front edge of the opening, it triggers another set of vortex shedding. This process occurs periodically with a specific frequency. If this frequency coincides with the natural frequency of the cavity, as a Helmholtz resonator, resonance will occur. The resonance frequency depends on the flow speed, geometry of the opening, volume of the cavity and so on. For automobiles, this resonance phenomenon is buffeting and the frequency is usually very low (<20 Hz, infrasonic), but the intensity is very strong (>100 dB). Although human ears cannot "hear" such a low frequency, it can be "felt" as a pulsating wind force that may be very annoying and fatiguing. Therefore, it is important to consider aerodynamic buffeting at an early stage of automobile design to assure passenger comfort.

Traditionally, most studies on automobile buffeting were based on wind tunnel experiments or road tests. However, robust CFD software has been developed and computers with sufficiently high speed and large memory have become available in recent years. This provides the possibility of computational simulation of buffeting. Ota et al [5] reported an early CFD simulation of sunroof resonance (buffeting) where a 2D flow field on the symmetry plane of a passenger car was considered. A structured mesh was used and flow field was solved

第IV部分 任职于美国克莱斯勒汽车公司时期

using a finite volume based CFD code (Golde). Free stream velocities ranging from 11.0 - 18.5 m/s were simulated to find the peak of resonance intensity. The highest level observed at the driver's ear was 115 dB at speed of 14 m/s. This result was compared well with the corresponding experimental measurement that made during a road test. However, the resonance frequency was over-predicted compared to the theoretical estimation and road test as well. In addition, a wind deflector study was also done to determine its adequate angle to suppress resonance.

With rapid development of high speed computers, it became possible to consider larger model sizes. An early 3D study of sunroof wind throb (buffeting) was reported by Ukita et al [3]. They modeled the flow field in and around a simplified passenger car body using a structured mesh with a finite difference code NAGARE. Simulation results were compared well with experimental measurements from a water tunnel with equivalent Reynolds number. They also studied sunroof buffeting with and without a wind deflector.

In recent years, more detailed CFD models have been studied. Karbon et al [6-8] reported sunroof buffeting with complex vehicle models that included detailed representation of the wind deflector and passenger compartment. An unstructured 3D mesh was used with the finite element based software PAM-FLOW. A vehicle speed of 50 kph was simulated for the cases with two noise control mechanisms: wind deflector and sunroof glass comfort positions. Simulation results for both cases compared well with the corresponding wind tunnel experimental measurements.

Recently Hendriana and Sovani [9-10] conducted an extensive study of side window buffeting for a passenger car side window buffeting using the finite volume based CFD software Fluent. They showed that transient CFD simulations are able to predict peak buffeting frequency within one Hz of wind tunnel measurements. Likewise, sound pressure level (SPL) at the buffeting frequency can be captured within 4 decibels of experimental data. Furthermore, their buffeting simulations replicated the experimentally observed trends of change in peak frequency and SPL caused by variations in vehicle speed, yaw angle and mirror housing shape. They also found that all passengers inside the vehicle experience the same peak buffeting level and frequency.

This paper focuses on the side window buffeting of an SUV vehicle. Four vehicle configurations are considered: (a) front right window open, (b) rear right window open, (c) front left window open and (d) rear left window open (see Figure 1). The external geometry, open windows, mirrors and dummies can be recognized clearly. The major parts inside the cabin, such as seats, quarter trims, pillars, etc., are also well considered.

(a) Front right window open

(b) Rear right window open

(c) Front left window open

(d) Rear left window open

Figure 1. Different window opening configurations

Table 1. Study Parameters

Parameters
1. Vehicle speed (30, 40, 50, 60 and 70 mph)
2. Yaw angle (-10, -5, 0, 5 and 10 degrees)
3. Sensor location (left & right ears of dummies)
4. Cabin volume (0 - 7 dummies)
5. Mesh size (coarse, medium and fine)
6. Air Compressibility (incompressible & ideal gas)

In this paper first the methodology used for CFD simulation is described including meshing, solver set-up and solution procedure. Then, the simulation results are reported and compared with wind tunnel measurements. After validation of the simulation, the effects of various parameters on buffeting frequency and peak SPL are studied using the validated procedure. The study parameters are listed in Table 1.

METHODOLOGY

MESHING

In this study, a completely tetrahedral mesh is chosen as it can handle more complex geometry with relative ease. Typical medium mesh size of each model has about 2.5 million tetrahedral cells. In order to find a reasonable compromise between mesh size and solution accuracy, two more mesh sizes are also considered. One is a coarser mesh of 1.5 million cells and the other is a finer mesh of 4.5 million cells.

The vehicle is placed inside a virtual wind tunnel, see Figure 2. The vehicle geometry was created using the CAD software CATIA. The inlet and outlet boundaries are kept at distances of 4 and 5.5 vehicle lengths upstream and downstream, respectively. This ensures that there is enough distance for the flow to develop from the inlet boundary. It also ensures that there is enough distance for the wake to die down and constant free stream properties to re-establish before the flow encounters the outlet boundary. The domain includes the entire passenger compartment that is connected to the external flow domain through the open window. Dummies representing the driver, co-driver and the second row left passenger are also included in the model as shown in Figure 3. The purpose of including dummies in the model is to correctly represent the volume and shape of the passenger compartment as it was at the time of test.

The vehicle surface is modeled to a significant level of detail to capture flow development from the vehicle front end to the window opening. The features modeled to the greatest detail are the A-pillar, side view mirror, window opening and the cabin interior in the neighborhood of the open window. The pre-processor ANSA is used to create surface mesh. For the medium size mesh (2.5 million cells), the cell edge length in the window opening and side view mirror is 8-10 mm and 20–30 mm elsewhere on the vehicle surface. A view of the surface mesh in the area of front window opening, including side view mirror and a dummy representing the driver is shown in Figure 4.

Features far away from the window are modeled to a lesser level of detail. The model does not include underhood and underbody details and the effect of underhood and underbody flows on window buffeting is assumed to be insignificant. For further simplification the vehicle front end and all the HVAC duct openings in the cabin are kept closed. Also, the cabin interior is kept perfectly sealed.

The surface mesh is transferred to the mesh generator TGRID and volume mesh is created there. Four levels of local refinements for volume mesh are utilized to ensure that the volume mesh is sufficiently fine in the regions of predominant flow unsteadiness such as the region of window opening and the wake of side view mirror. Two local refinement levels are applied outside the vehicle (see outer regions 1 and 2 in Figures 5 and 6). The 3rd local refinement level is applied inside vehicle to capture wave propagation in the cabin. The 4th and finest local refinement level is applied in the area of window opening and side mirror wake to capture shear layer shedding and the wake behind the mirror. Local refinement control parameters are detailed in Table 2. Although most of the work was done with 2.5 million cell model, two more models with finer mesh (4.5 million cells) and coarser mesh (1.5 million cells) were also considered to investigate mesh dependency.

SOLVER SET-UP

The solver set-up in Fluent 6.0 is listed in Table 3.

The large eddy simulation (LES) turbulence model is chosen in the simulations. It might be treated as "coarse LES" since the mesh used here is very coarse for well-resolved LES. The implementation of LES in Fluent 6.0 is described in detail in [11]. Some highlights of the

Table 2. Local Refinement Control Parameters

Refinement Region	Region Dimension $X \times Y \times Z$ (m^3)	Cell Count Control
Wind Tunnel	$50 \times 20 \times 10$	1,800E+6
Outer 1	$10.5 \times 3.5 \times 2.75$	500E+3
Outer 2	$7.5 \times 2.6 \times 2.25$	100E+3
Inner	$3.5 \times 1.7 \times 1.34$	17E+3
Window	$1.3 \times 0.7 \times 0.65$	2.5E+3

Note: Directions: X = flow-wise, Y = width-wise, Z = height-wise

Table 3. Solver Set-Up

Function	Setting
Solver	Segregated Unsteady
Time marching	2nd Order Implicit
Pressure discretization	2nd Order
Momentum discretization	2nd Order Upwind
Pressure-velocity coupling	SIMPLE
Energy discretization	2nd Order Upwind
Fluid	Air (Ideal Gas)

model are outlined below. The LES approach uses the RNG sub-grid scale model for effective viscosity that is expressed as

$$\mu_{eff} = \mu + \mu_t = \mu\left[1 + H(x)\right]^{1/3} \quad (1)$$

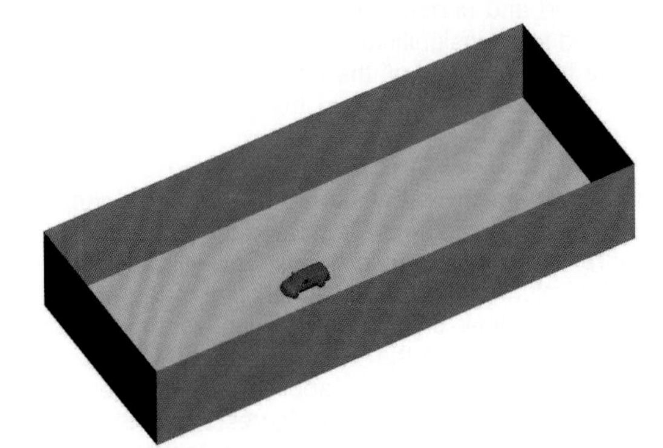

Figure 2. Virtual wind tunnel

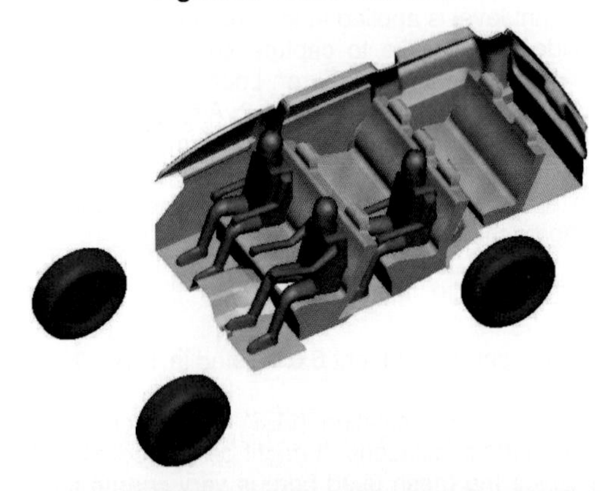

Figure 3. View of the dummies

Figure 4. Surface mesh at front window opening

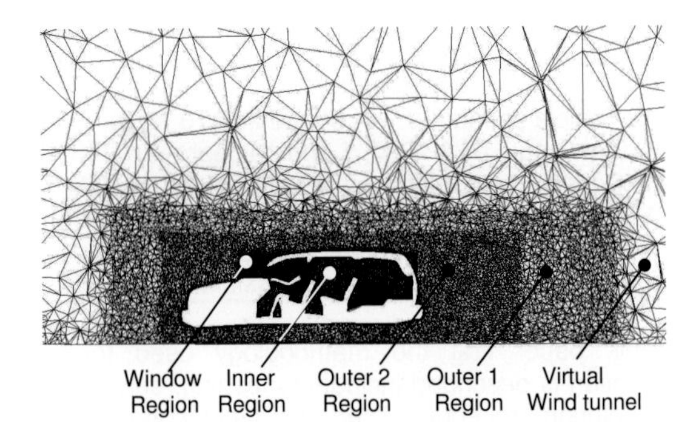

Figure 5. Vertical cross-section of the mesh

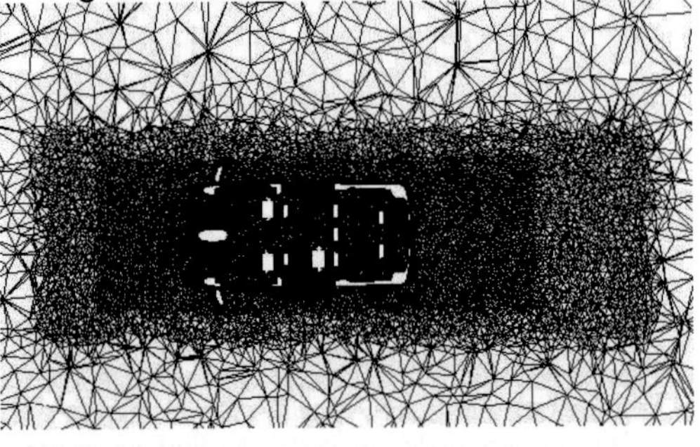

Figure 6. Horizontal cross-section of the mesh

where *H(x)* is the Heaviside function:

$$H(x) = \begin{cases} x, & x > 0 \\ 0, & x \le 0 \end{cases} \quad (2)$$

where

$$x = \frac{\mu_s^2 \mu_{eff}}{\mu^3} - C \quad (3)$$

and

$$\mu_s = \left(C_{RNG} V^{1/3}\right)^2 \sqrt{2\bar{S}_{ij} . \bar{S}_{ij}} \quad (4)$$

where μ is the molecular viscosity, μ_t is the turbulent viscosity, C is a constant with value 100, C_{RNG} is a constant having a value of 0.157 (derived from the RNG theory), V is the volume of the computational cell, and $\bar{S}_{ij}$ is the rate of strain tensor. The LES model automatically uses the following laminar stress-strain relationship to compute wall shear stress if the mesh resolution is found to be fine enough to resolve the laminar sub-layer,

$$\frac{\bar{u}}{u_\tau} = \frac{\rho u_\tau y}{\mu} \quad (5)$$

where $\bar{u}$ is the mean velocity, u_τ is the friction velocity, y is distance from the wall and μ is dynamic viscosity. If the mesh is too coarse to resolve the laminar sub-layer, the law-of-the-wall is automatically applied [12].

The boundary conditions used in the simulation for zero yaw angle case are listed in Table 4. For non-zero yaw angle case the same mesh was used, but boundary conditions at left and right sides of the tunnel are different. If left window is open, for the positive yaw angle case the left side of the tunnel is given a velocity inlet boundary condition with appropriately angled flow just like the inlet boundary. The right side of the tunnel is given a pressure outlet boundary condition just like the outlet boundary. Likewise, for the negative yaw angle case the right side of the tunnel is given a velocity inlet boundary condition with appropriately angled flow just like the inlet boundary. The left side of the tunnel is given a pressure outlet boundary condition just like the outlet boundary. If a right window is open, the settings are opposite to the left window open case.

Table 4. Boundary Conditions For Zero Yaw Angle

Boundary	Condition	Value
Inlet	Constant Velocity	60 mph, etc.
Outlet	Constant Pressure	101325 Pa
Tunnel left side	Symmetry (slip wall)	
Tunnel right side	Symmetry (slip wall)	
Ground	Symmetry (slip wall)	
Tunnel top	Symmetry (slip wall)	
Vehicle surface	No slip wall	

It should be noted that in the simulations, interior surfaces of the passenger compartment were assumed to be a solid wall instead of soft surfaces of fabric or carpeting. Solid surfaces reflect pressure waves more strongly than soft surfaces. This assumption along with the assumption of not including cabin leakage may be a cause of the small difference observed between the predicted buffeting level and the corresponding experimental value. In the simulations, compressible fluid model (ideal gas) was used to capture the wave propagation and resonance effects inside the cabin. To investigate the effect of fluid compressibility, some more runs were also preformed under the assumption of incompressible fluid, i.e. constant air density.

SOLUTION PROCEDURE

Each case is first run in steady state mode for about 100 iterations using the standard k-ε turbulence model. Then the steady state flow field is used to initialize the unsteady flow. A time step of 0.002 second is chosen to run unsteady flow. It is much smaller than the time period of the frequency of interest, ~20 Hz. Within each time step, the number of sub-iterations is set to be 15. The residual convergence criteria are chosen as follows: continuity 1e-4, momentum and turbulence 1e-3 and energy 1e-7. The equations are considered converged when the residuals reach the criteria or drop more than 3 orders of magnitude within each time step. It typically took about 15 sub-iterations for the solution to converge for each time step. Sensors are installed at the driver's and second row left passenger's ears. Static pressure is recorded at each time step. After the initial process stabilizes in about 300-500 time steps, the pressure signal reaches a dynamically stable periodic fluctuation. Subsequently, time history of pressure fluctuation is recorded for the time duration of about 2 seconds.

The pressure fluctuation signals were then acoustically post-processed using the software XMGR. A discrete Fourier transform with a Hanning window is used to convert the recorded signals to the spectral format where the amplitude of fluctuating signal is expressed as a function of frequency. Finally, the amplitude is converted to the format of sound pressure level (SPL) in dB units using the formula:

$$SPL(dB) = 10\log_{10}\left(\frac{p}{p_{ref}}\right)^2 \quad (6)$$

where p is the amplitude of pressure fluctuation in Pa and the reference pressure $p_{ref} = 20 \times 10^{-6}$ Pa.

RESULTS

A typical simulation took about 6-7 days of run time for every 1000 time steps with 15 sub-iterations per time step while running on a SGI Origin 2000 machine with 64 300-MHz CPU and 64 GB memory using the IRIX 6.5 operating system using 6 processors.

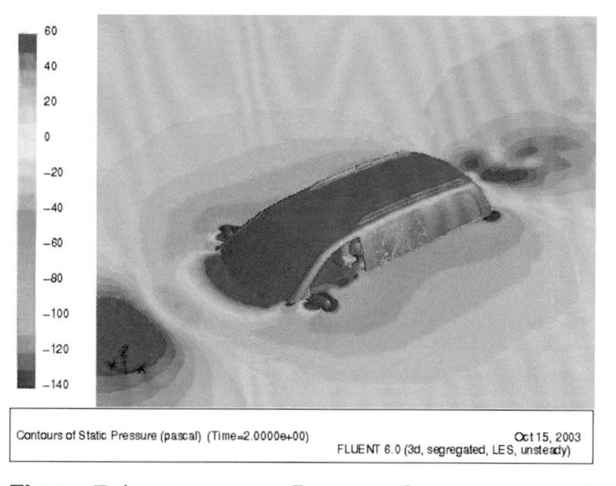

Figure 7. Instantaneous Pressure Contours Around the Vehicle

Figure 7 shows the instantaneous pressure contours for the case of front left window open at 60 mph and 5 degree yaw angle on a horizontal cut plane passing through the mid-point of the side view mirror. The graph gives a generic overview of the flow field outside and inside of the vehicle.

Figures 8 and 9 more closely show the shear layer occurring in the window opening area and its interaction with the side view mirror wake. In Figure 8, an imaginary rectangular plate is put vertically in the front left window opening area and displays pressure contours in the filled format. In the mean time, an unbounded cut plane is put parallel to the rectangular plate and passing through the mid-point of side view mirror. This cut plane displays pressure contours in a transparent manner. The filled pressure contours mainly display the shear layer developments behind the A-pillar while the transparent contours display the wake behind the mirror. The alternating structure of high and low pressure seen on the rectangular plate demonstrates the shear layer shedding at A-pillar. A dynamic animation of these pressure contours clearly shows the undulations of the shear layer. The animation further shows that as the shear layer impinges onto the B-pillar a wave is generated which propagates into the passenger compartment.

In Figure 9, the filled contours display pressure on a horizontal cut plane passing through the mid-point of the side view mirror while the transparent contours display pressure on a horizontal cut plane passing through the driver's ears. The pressure contours on the upper cut plane illustrate the vortex generation and developments at A-pillar while the contours on the lower cut plane display the wake behind the mirror. Pulsating high and low pressure contours are indicative of shear layer at A-pillar and the wake behind the mirror. The animations can show the vortex shedding and wave propagation

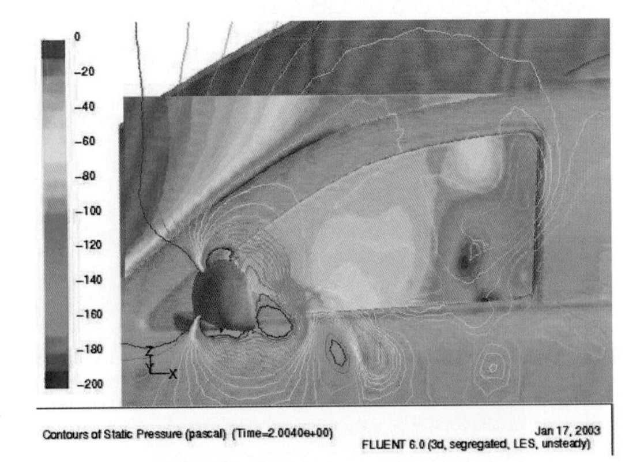

Figure 8. Pressure contours on vertical cut planes around open front window

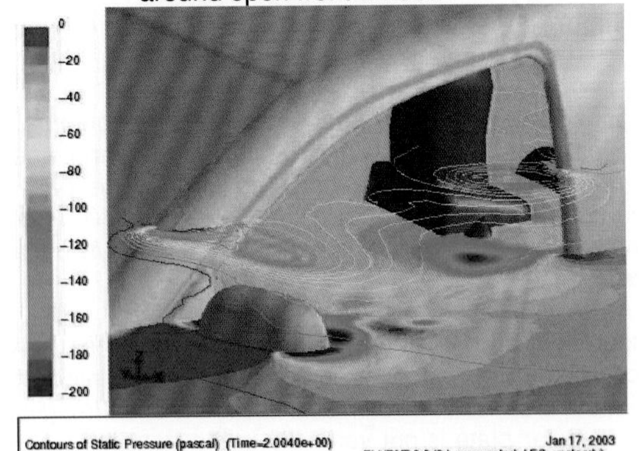

Figure 9. Pressure contours on horizontal cut planes around open front window

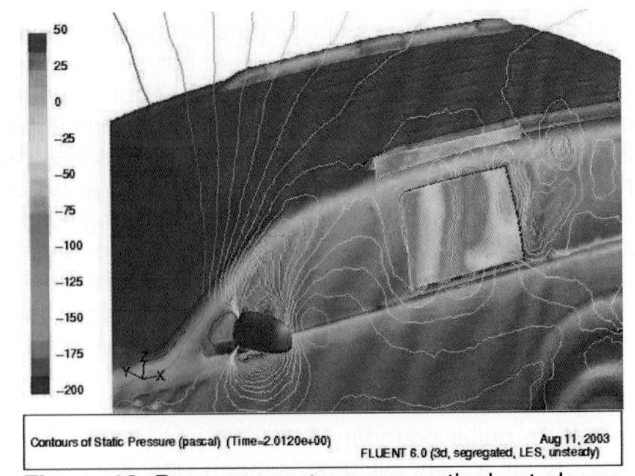

Figure 10. Pressure contours on vertical cut planes around open rear window

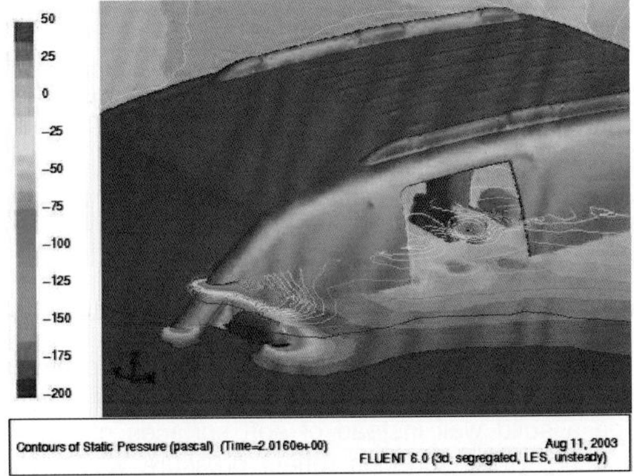

Figure 11. Pressure contours on horizontal cut planes around open rear window

around the window opening. The interactions between the vortices and the mirror wake can be seen in the lower rear corner of the window opening. From these figures it is concluded that both vortex shedding from A-pillar at the window opening and the wake behind the side view mirror play an important role in determining side window buffeting characteristics.

Figures 10 and 11 show the instantaneous pressure contours for the case of rear left window open. In these figures the filled and transparent contours are plotted on the similar cut planes as those in Figures 8 and 9, respectively. Figures 10 and 11 indicate that the flow features causing buffeting for the case of rear window open are similar to those for the case of front window open. Alternating stripes of low and high pressure seen behind B-pillar in Figure 10 indicate the shear layer and vortex shedding. Similar to the case of front window open, the animation clearly shows the movement of shear layer, shed vortices and their impingement onto C-pillar. This observation is in agreement with the buffeting theory described by Hucho [4]. Figure 11 indicates that the vortices behind the side view mirror continue to remain coherent as far downstream as the rear window opening. Therefore, these vortices also play an important role in the rear window buffeting phenomenon.

The animation shows more temporal details of this process.

Figure 12 shows time history of pressure fluctuation recorded at the right ear of the second row left passenger for the case of rear right window open at 60 mph and 0 yaw angle. The absolute values of pressure fluctuations are proprietary and not open for publication at this time. Hence, no numbers are attached to the vertical axis of Figure 12. From the figure, however, we can see that the initial transient process dies out after 0.5 second. We also observe that pressure amplitude increases from 0.5-1.0 second and does not change much after 1.0 second. For this case, we truncate pressure fluctuation signal and choose the duration from 1.0-2.0 seconds to make discrete Fourier transform. After the amplitude of pressure fluctuation is obtained as a function of frequency, equation (6) is then used to convert the amplitude to SPL in dB units. The curve of SPL vs frequency is usually referred to as frequency spectrum, or simply spectrum, of the selected signal. All the other cases are treated in the same manner.

In the following section, we consider two validation cases where buffeting frequency spectrum predicted by CFD simulation is compared with the corresponding wind tunnel data. The effects of various parameters are discussed subsequently.

VALIDATION CASES

The accuracy of CFD simulation, as compared to the wind tunnel experimental data, is assessed using two validation cases. Both cases are for 60 mph vehicle speed and 0 yaw angle. The front right window is open in one case while the rear right window is open in the other, as shown in Figures 1(a) and 1(b), respectively.

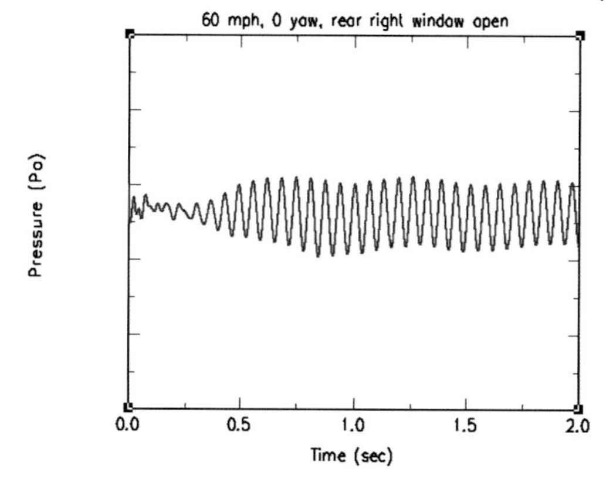

Figure 12. Time history of pressure fluctuation at second row left passenger's right ear for the case of rear right window open at 60 mph and 0 yaw

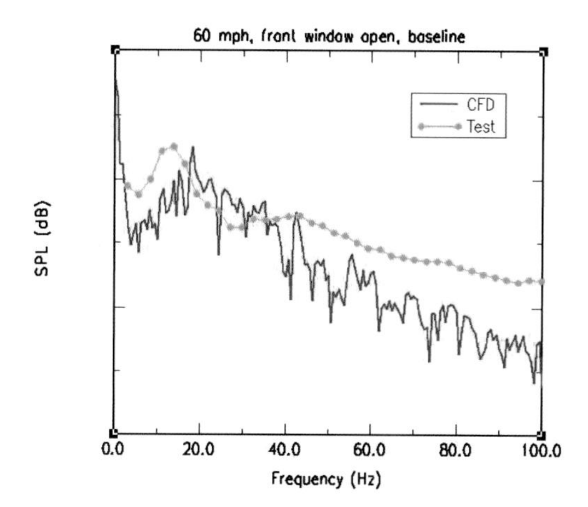

Figure 13. Buffeting spectrum at second row left passenger's right ear for the case of front right window open at 60 mph and 0 yaw

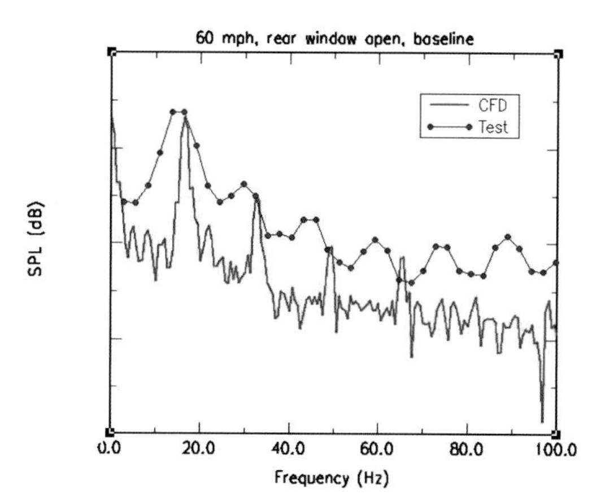

Figure 14. Buffeting spectrum at second row left passenger's right ear for the case of rear right window open at 60 mph and 0 yaw

The `experimental measurements are obtained at the DaimlerChrysler aeroacoustic wind tunnel (AAWT) in Auburn Hills, MI. During the measurement, a person sat in the co-driver's position to experience buffeting and gave orders to the person in the control room for recording. A sensor (microphone) was installed at the right ear of the dummy sitting on the left seat of the second row. Quantitative measurement of the pressure fluctuation was recorded using this microphone.

Figures 13 and 14 show the comparison of the predicted buffeting spectrum with the corresponding wind tunnel measurements for the two validation cases, respectively. Again, the absolute values of the sound pressure level are proprietary and not open for publication at this time. Hence, no numbers are attached to the vertical axis of Figures 13, 14 and all other figures in this paper that contain buffeting sound pressure levels. The difference between any two adjacent markings on the vertical axis is 20 dB. However, the vertical axis may or may not intersect the horizontal axis at 0 dB. Although these figures do not have absolute values of SPL on the vertical axis, they do provide a clear quantitative comparison between the experimental and simulation results.

The first peak in the buffeting frequency spectrum is of primary interest since pressure fluctuation at the first peak is predominantly sensed, or felt, by human. For both front and rear window open cases CFD simulations predict the peak SPL accurately within 1-2 dB of the corresponding experimental measurements. For the front window open case CFD predicts that the first peak occurs at 18 Hz, 4 Hz greater than the experimentally measured value. For the rear window open case the first peak frequency predicted by CFD is 16 Hz, 1Hz more than the experimentally measured value. It should be noticed in Figure 14 that the predicted SPL spectrum also captures some higher harmonic peaks, which are seen in the corresponding experimental spectrum.

For buffeting phenomenon peak SPL is more significant than peak frequency from the point of view that passenger discomfort correlates directly to peak SPL rather than peak frequency. For both the validation cases CFD predicts peak SPL very accurately. Thus, CFD technology can be considered as a useful tool for prediction, improvement and optimization of buffeting characteristics.

EFFECT OF VEHICLE SPEED

Simulations were conducted for five vehicle speeds ranging from 30 to 70 mph at 10 mph increments for the case of rear left window open at zero yaw angle to assess the effect of vehicle speed on buffeting characteristics. The variations of SPL and frequency at the first predominant peak of the buffeting spectrum are shown in Figure 15. Notice that no numbers are provided on the vertical axis of the SPL figure, as they are proprietary. Nevertheless, the difference between any two consecutive marks on the vertical axis is 20 dB for SPL. It can be seen from the figure that as vehicle speed increases from 30 to 70 mph, the peak SPL increases nearly linearly by about 22 dB. Vehicle speed in this range seems not to have a significant effect on the frequency of the first predominant peak which remains at 18 ± 1Hz throughout the vehicle speed range studied. These findings agree with those of Hendriana et al [10] who reported a 5 dB increase in peak SPL with a 10 mph increment in vehicle speed and a frequency constant within 1 Hz.

EFFECT OF YAW ANGLE

The effect of yaw angle on buffeting characteristics was also studied for the case of rear left window open at

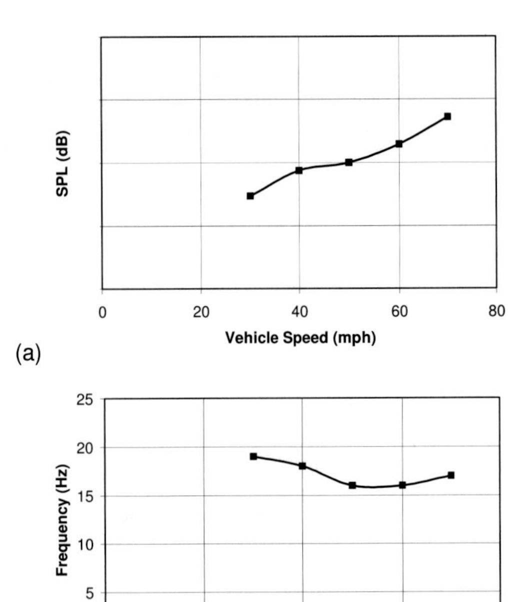

Figure 15. Effect of vehicle speed on (a) peak SPL and (b) frequency

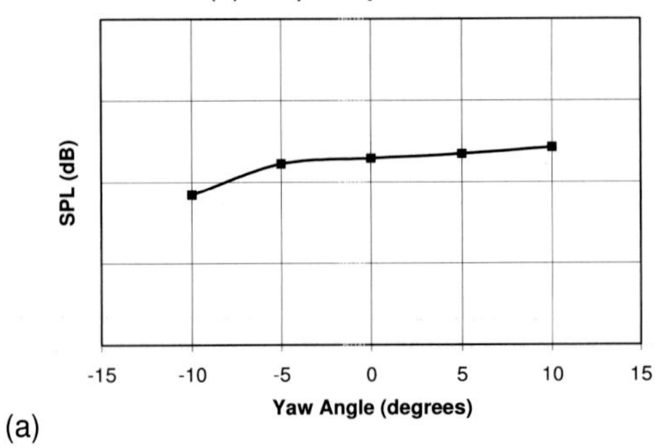

(a)

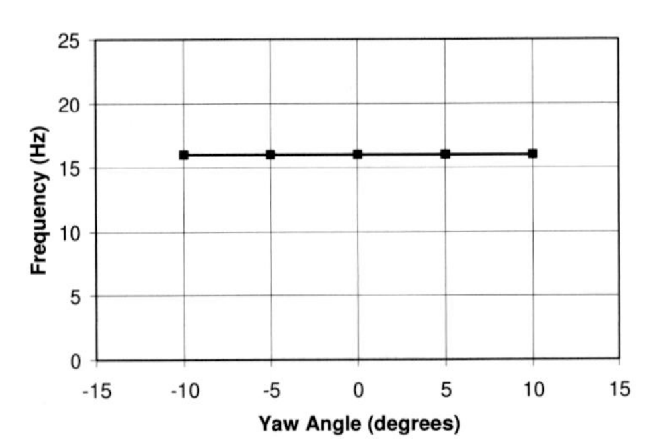

(b)

Figure 16. Effect of yaw angle on (a) peak SPL and (b) frequency

vehicle speed 60 mph. Five yaw angles were considered ranging from –10 to +10 degrees at 5 degree increments. Negative yaw angle indicates the case when the incident free stream has a component from right to left (co-driver side to driver side) and vice versa for positive yaw angle. The variations of SPL and frequency at the first predominant peak of the buffeting frequency spectrum with yaw angle are shown in Figure 16.

From this figure we can see that peak SPL increases by 8 dB as the yaw angle changes from –10 to –5 degrees. Thereafter it linearly increases by 4 dB as the yaw angle increases from –5 to 10 degrees. The frequency of the first peak buffeting seems not to have obvious change and keeps to be about 16 Hz when yaw angle varies from –10 to 10 degrees.

EFFECT OF SENSOR LOCATION INSIDE THE PASSENGER COMPARTMENT

Figure 17 shows the frequency spectra of the signals at the left and right ears of the driver and the second row left passenger for the case of front left window open at 50 mph and 5 degree yaw angle. The figure shows that all of the spectral curves have similar frequency and SPL for the first predominant peak. This indicates that all passengers inside the passenger compartment encounter the same buffeting characteristics. Hendriana et al [10] reported the same finding in their study.

EFFECT OF FLUID COMPRESSIBILITY

In this work, the air is assumed compressible (ideal gas) and good agreement has been attained for the baseline cases (Figures 13 and 14). However, the simulation did not demonstrate a resonance or critical buffeting speed in the range of 30-70 mph, but it showed a relatively linear trend instead (Figure 15a). In order to examine if

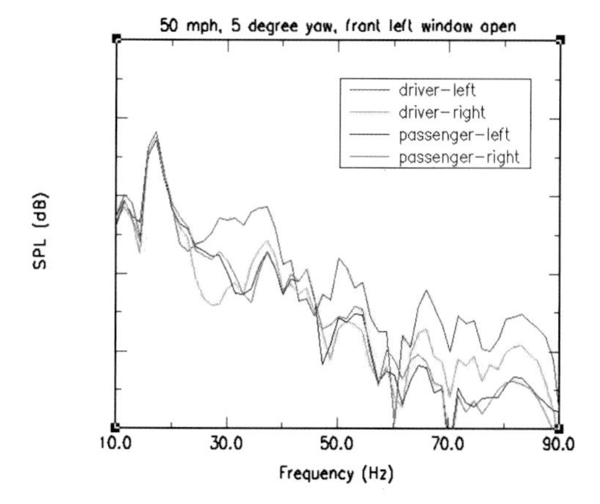

Figure 17. Effect of sensor location on peak SPL and frequency

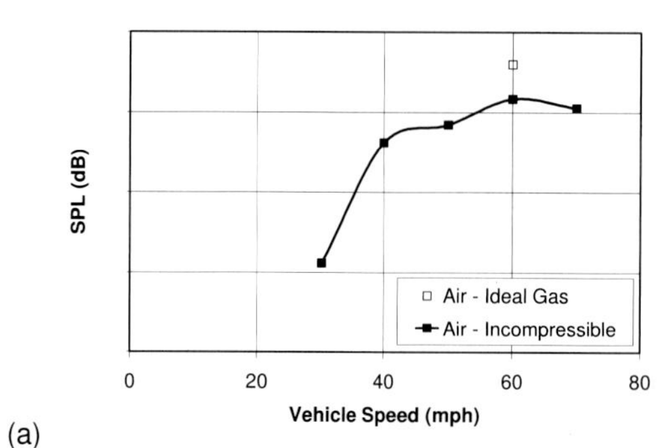

(a)

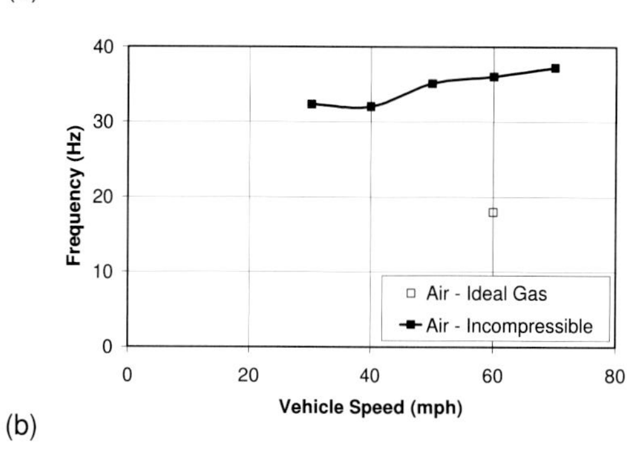

(b)

Figure 18. Effect of air compressibility on (a) peak SPL and (b) frequency

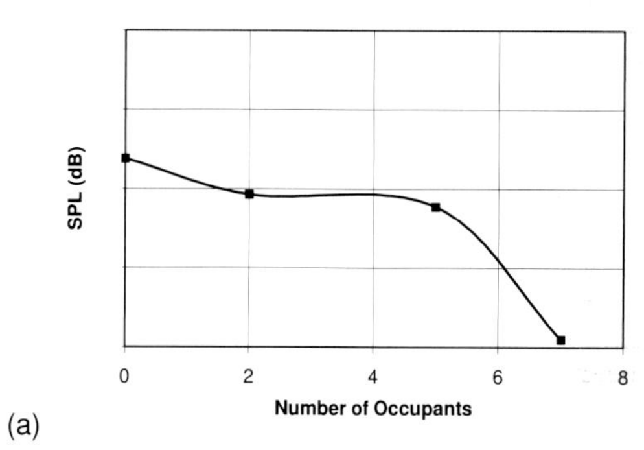

(a)

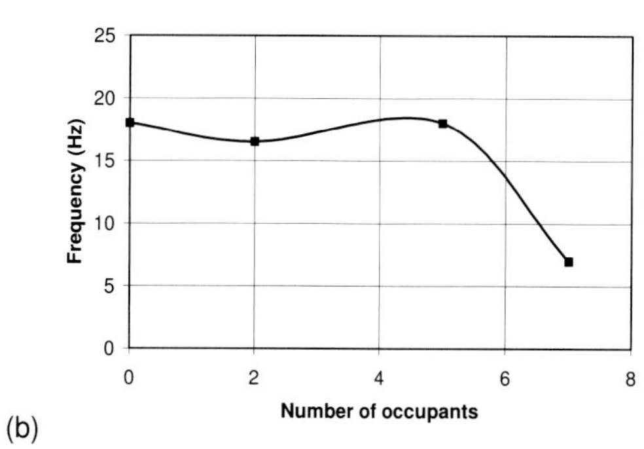

(b)

Figure 19. Effect of passenger compartment volume on (a) peak SPL and (b) frequency

incompressible fluid model can predict such a resonance more runs were performed for the case of front right window open using the incompressible (constant density) air model in the same range of vehicle speed at 0 yaw angle. The results are shown in Figure 18. Although the SPL curve seems to have a critical vehicle speed 60 mph, the value of SPL for 60 mph was significantly under-predicted (more than 10 dB) and the value of buffeting frequency for 60 mph was significantly over-predicted (about double), compared to the wind tunnel measurements (cf. Figure 13). This discrepancy was considered not acceptable. Therefore, the incompressible fluid model was discarded and ideal gas model was employed throughout the study. It is still not clear if the ideal gas model is missing the resonance effect or if this particular vehicle lacks a resonance. More wind tunnel experiments or road tests are needed for this purpose.

EFFECT OF VOLUME OF THE PASSENGER COMPARTMENT

The effect of volume of the passenger compartment on buffeting frequency and peak SPL was studied for the case of front left window open at 50 mph and 0 yaw angle. Figure 19 shows the variations of buffeting frequency and SPL with the number of dummies in the passenger compartment (0-7). The general trend is that with increase in the number of dummies, i.e. with decrease in the passenger compartment volume, the buffeting SPL decreases and no buffeting appears when all 7 seats are occupied by dummies. The buffeting frequency seems not to be sensitive to the number of dummies except for the case of 7 dummies in which the frequency decreases to about half of the other cases. The interior shape of the passenger compartment is so complicated that it is not practical to find its natural frequency through a simple formula. The dummies affect the interior volume and shape of the cabin, where both of the changes have influence on the buffeting characteristics. More work is needed to thoroughly capture the influence.

EFFECT OF MESH SIZE

Buffeting simulation is a new challenge to CFD in the following sense. (1) Both external and internal flows have to be solved simultaneously. (2) Flow is always transient. (3) In the acoustically sensitive region cell dimension should be small enough. (4) Time step should be very small and time duration should be long enough for establishment of periodically pulsating flow. Therefore, such simulations are computationally expensive. As the first step of CFD simulation, it is better to keep mesh size as small as possible under the condition of acceptable accuracy.

At early stage of this work, a study of mesh size effect on buffeting characteristics was done for the case of front window open at 50 mph and 0 yaw angle to find a suitable mesh size. As mentioned previously, three tetrahedral meshes were considered. A coarse mesh

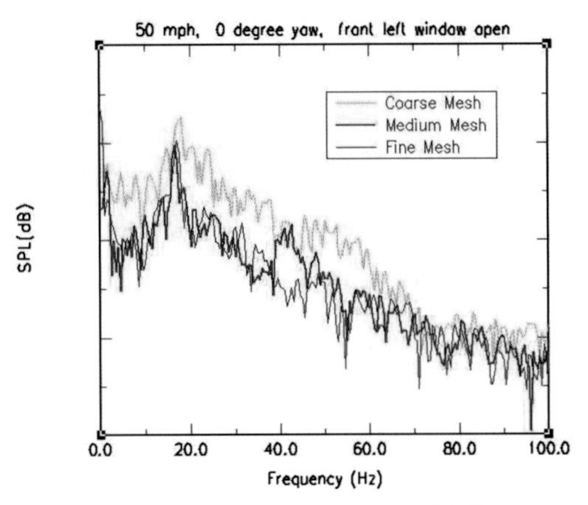

Figure 20. Effect of mesh size on peak SPL and frequency

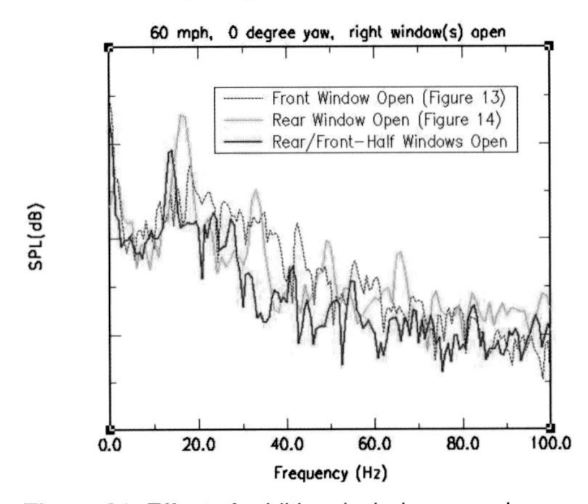

Figure 21. Effect of additional window opening on peak SPL and frequency

has total 1.5 million cells with 15mm minimum cell edge length at the window opening region. A medium mesh has 2.5 million cells with 8-10mm minimum. A fine mesh has 4.5 million cells and 5mm minimum. The results of buffeting spectra for three meshes are shown in Figure 20. The cases of medium and fine meshes give very close peak SPL (difference~2dB) and buffeting frequency (difference~1Hz). The two spectral curves are very similar to each other at low frequency. However, the computational time is quite different. The medium mesh took about 7 days of run time, while the fine mesh took about 21 days. The coarse mesh only took 4 days, but the peak SPL was over-predicted by about 10dB compared to the other two cases. Moreover, the entire spectral curve diverted considerably from the other two curves, especially for frequency < 70Hz. Therefore, compromising between solution accuracy and computational efficiency, the medium mesh of 2.5 million cells was chosen for all simulations.

It should be emphasized at this point that the functionality of local refinement in TGRID, described in the section of "meshing", improves mesh quality significantly. As shown in Figures 5 and 6, the mesh far away from the vehicle is very coarse. Around and inside the vehicle, 4 levels of local refinement establish a fairly good mesh system to guarantee the solution accuracy. Although the mesh size was not very large and no prism layer was used, the simulations are able to provide reasonably accurate solutions. On the other hand, however, prism layer near-wall mesh will be considered for better accuracy in the future work.

EFFECT OF ADDITIONAL WINDOW OPENING

As indicated in Figures 13 and 14, rear window buffeting is worse than front window buffeting and peak SPL of the rear window open case is about 10 dB higher than the front window open case. In addition, in order to examine the "exhaust" effect of the cabin, a 5th CFD model was run in which rear window is fully open and front window is half open. The predicted buffeting spectra for the three models are presented in Figure 21 where the buffeting curves of front and rear window open cases are of the duplicate from Figures 13 and 14 while the 3rd curve is of the 5th model. It is seen from the figure that the 5th model does alleviate buffeting level, which confirms the "exhaust" effect of the cabin.

CONCLUSION

Simulations of side window buffeting characteristics for an SUV vehicle using the commercial CFD code Fluent 6.0 have been presented in this paper. Details of passenger compartment and exterior surface are included in the model. Further simplifications, such as a closed grille and flat underbody, are assumed since the effects of underhood and underbody flows on buffeting characteristics are believed to be insignificant. The simulations are able to predict peak SPL accurately, within 1-2 dB of the corresponding experimental data. The frequency of the first peak in the buffeting spectrum can be predicted to the accuracy of 1 to 4 Hz.

The effects of various parameters on buffeting characteristics were also studied. These parameters include vehicle speed, yaw angle, location of measurement, air compressibility, cabin volume, mesh size and additional window opening. Study of vehicle speed ranging from 30 to 70 mph leads to a roughly 5 dB rise per 10 mph of the peak SPL for the case of left rear window open at 0 yaw angle. When yaw angle increases from –10 to 10 for the case of rear left window open at 60 mph, the peak SPL increases roughly 4-5 dB per every 5 degrees increase. Within the passenger compartment all locations have the same peak SPL and frequency for the case of front left window open at 50 mph and 5 degree yaw. The ideal gas seems to be a proper fluid model since incompressible fluid model significantly under-predicts peak SPL and over-predicts buffeting frequency. Mesh size study indicates that the medium size mesh of 2.5 million cells proves to be a reasonable compromise between solution accuracy and computational efficiency. Cabin volume study shows that buffeting intensity decreases with increase of the number of occupants. At last, simulation of additional window opening confirms the "exhaust" effect of the cabin to alleviate buffeting. In order to apply the CFD technique to the vehicle design in the future, further development and validation are needed via wind tunnel experiments and road tests.

ACKNOWLEDGEMENTS

The authors would like to thank Kenneth M. Albers of DaimlerChrysler Corporation for providing wind tunnel experimental data.

REFERENCES

[1] Bodger, W.K. and Jones, C.M., Aerodynamic wind throb in passenger cars, SAE Transactions, Vol. 73, pp. 195-206 (1965)

[2] Aspinall, D.T., An empirical investigation of low frequency wind noise in motor cars, MIRA report No.1966/2, Warwickshire, UK (1966)

[3] Ukita, T., China, H. and Kanie, K., Analysis of vehicle wind throb using CFD and flow visualization, SAE paper 970407 (1997)

[4] Hucho, W.H., *Aerodynamics of road vehicles*, fourth edition, Society of automotive engineers, Inc., Warrendale, PA (1998)

[5] Ota, D.K., Chakravarthy, S.R., Becker, T. and Sturzenegger, T., Computational study of resonance suppression of open sunroofs, *Journal of Fluids Engineering*, Vol. 116, pp. 877-882 (1994)

[6] Karbon, K. and Kumarasamy, S., Computational aeroacoustics applications in automotive design, First MIT Conference on Computational Fluid and Solid Mechanics, June (2001)

[7] Karbon, K., Kumarasamy, S. and Singh, R., Applications and issues in automotive computational aeroaco-ustics, 10th Annual Conference of the CFD Society of Canada, Windsor, Canada, June 9-11 (2002)

[8] Karbon, K., Singh, R., "Simulation and design of automobile sunroof buffeting noise control", 8th AIAA/CEAS Aeroacoustics Conference & Exhibit, June 17-19 (2002)

[9] Sovani, S.D. and Hendriana, D., Predicting Passenger Car Window Buffeting With Transient External-Aerodynamics Simulations, 10th Annual Conference of the CFD Society of Canada, Windsor, Canada, June 9-11 (2002)

[10] Hendriana, D., Sovani, S.D. and Schiemann, M.K., On Simulating Passenger Car Side Window Buffeting, SAE Paper No. 2003-01-1316 (2003)

[11] Fluent 6.0 User's Guide, Fluent Inc., Lebanon, NH (2001)

[12] Tennekes, H. and Lumley, J.L., *A First Course in Turbulence*, MIT Press, Cambridge, MA (1972)

2004

PVP-Vol. 491-1 ASME/JSME PRESSURE VESSELS AND PIPING CONFERENCE

COMPUTATIONAL TECHNOLOGIES FOR FLUID/THERMAL/STRUCTURAL/CHEMICAL SYSTEMS WITH INDUSTRIAL APPLICATIONS

VOLUME 1

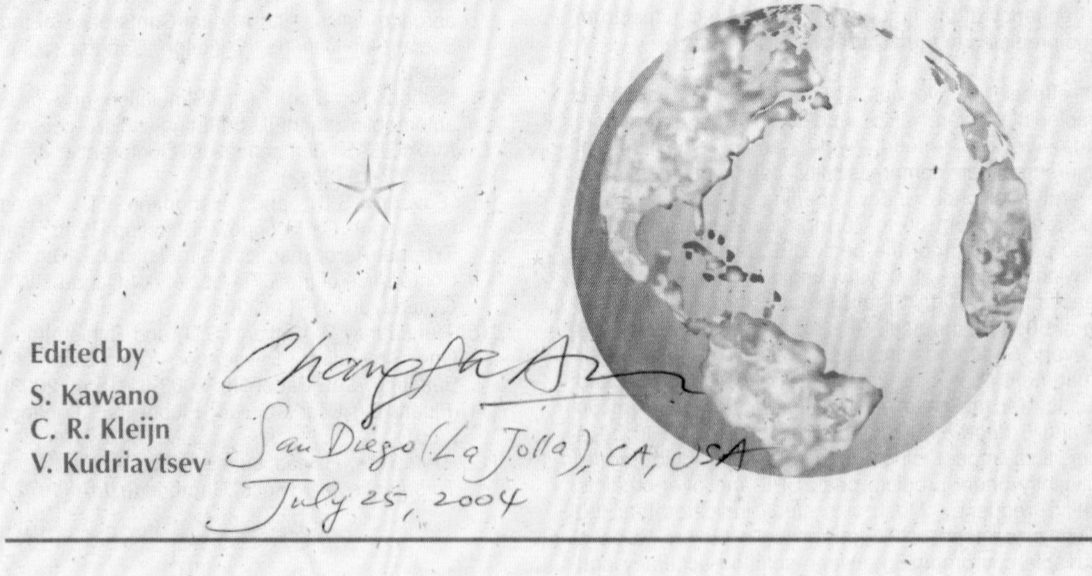

Edited by
S. Kawano
C. R. Kleijn
V. Kudriavtsev

Changfa An
San Diego (La Jolla), CA, USA
July 25, 2004

安长发博士科研论文集

Proceedings of 5th International Bi-Annual ASME/JSME Symposium on Computational Technology
For Fluid / Thermal / Structural / Chemical Systems with Industrial Applications
San Diego / La Jolla, California, USA, July 25-29, 2004

PVP2004-3099

IMPACT OF CAVITY ON SUNROOF BUFFETING - A TWO DIMENSIONAL CFD STUDY

Chang-Fa An
Belcan Corporation
Novi, Michigan 48888 USA
ca56@dcx.com

Seyed Mehdi Alaie
Optimal CAE, Inc.
Novi, Michigan 48888 USA
sma20@dcx.com

Michael S. Scislowicz
DaimlerChrysler Corporation
Detroit, Michigan 48227 USA
mss2@dcx.com

ABSTRACT

Driven by fluid dynamics principles, the concept for buffeting reduction, a cavity installed at the leading edge of the sunroof opening, is analyzed. The cavity provides a room to hold the vortex, shed from upstream, and prevents the vortex from escaping and from directly intruding into the cabin. The concept has been verified by means of a two dimensional simulation for a production SUV using the CFD software – FLUENT. The simulation results show that the impact of the cavity is crucial to reduce buffeting. It is shown that the buffeting level may be reduced by 3 dB by adding a cavity to the sunroof configuration. Therefore, the cavity could be considered as a means of buffeting reduction, in addition to the three currently-known concepts: wind deflector, sunroof glass comfort position and cabin venting. Thorough understanding of the buffeting mechanism helps explain why and how the cavity works to reduce buffeting. Investigation of the buffeting-related physics provides a deep insight into the flow nature and, therefore, a useful hint to geometry modification for buffeting reduction. The buffeting level may be further reduced by about 4 dB or more by cutting the corners of the sunroof opening into smooth ramps, guided by ideas coming from careful examining the physics of flow. More work including three dimensional simulation and wind tunnel experiment should follow in order to develop more confidence in the functionality of the cavity to hopefully promote this idea to the level that it can be utilized in a feasible way to address sunroof buffeting.

Keywords: Buffeting, Cavity, Vortex, and CFD

INTRODUCTION

In the current automobile market wind noise is one of the major customer complaints, especially buffeting. Buffeting is a low frequency and high intensity wind noise, which can occur when a side window or sunroof is open on a moving vehicle. The buffeting frequency is usually low (~20Hz), even lower than the audible range (infrasonic) for which people cannot "hear" but "feel". Therefore, buffeting has an effect of "pumping" or "beating" the people in the passenger compartment. Buffeting could make people in the cabin uncomfortable or even "carsick" if it lasts for a long time. The buffeting frequency and intensity depend on many parameters. These parameters include vehicle speed, yaw angle, vehicle geometry near the opening, cabin volume, size and shape of the opening and some others that are yet to be found.

Early work on the buffeting noise of road vehicles, mainly experimental, started from mid-1960s. Bodger and Jones [1] reported the results of an empirical study on side window "wind throb", the early terminology for buffeting. The prediction for natural frequency was well compared with the experimental data. Aspinall [2] investigated rear window wind throb and suggested some means to reduce the noise, wind deflector for example. Hucho [3] introduced another term "booming" for this type of wind noise in his book of road vehicle aerodynamics. With the developments of high-speed computers and computational fluid dynamics (CFD) technology in 1990s, numerical simulation for buffeting became possible. Ota et al [4] reported an early two dimensional CFD computation for a car model to investigate the sunroof "boom phenomenon" using the CFD code Golde. They also tried to find an adequate angle of wind deflector to suppress the resonance. Ukita et al [5] presented the results of a three dimensional CFD study for sunroof wind throb for a car model using a finite difference code NAGARE. The calculation was well compared with the corresponding water tunnel visualization. They also verified the ability of a wind deflector in reducing the intensity of wind throb. Since the new millennium, more and more CFD studies have been undertaken and more and more realistic road vehicles have been taken under consideration. In the mean time, the phenomenon of such wind noise has been given a name "buffeting". Karbon et al [6-7] reported a study of sunroof buffeting control for a car model using a finite element CFD code PAM-FLOW. The simulated results were in good agreement with wind tunnel experiments for both magnitude and frequency of pressure fluctuations. They also proved that both types of buffeting control mechanisms, wind deflector and the comfort position, work well to reduce buffeting. In 2003, Hendriana et al [8] undertook a CFD study of side window buffeting for a passenger car using a finite volume CFD code FLUENT. The results compared well with wind tunnel test results. Almost at the same time, many other automotive companies also took action towards sunroof buffeting studies under the collaboration with CFD software vendors. This includes Audi's study with EXA's PowerFLOW [9], SAAB's study with adapco's STAR-CD [10] and Volkswagen's study also with PowerFLOW [11]. Recently, An et al [12] presented the results of CFD simulation of side window buffeting for a DaimlerChrysler's production SUV. The simulated results for the baseline settings were well correlated to the wind tunnel measurements. In addition, sensitivity studies were also undertaken, such as vehicle speed, yaw angle, compressibility of fluid, volume of the cabin, mesh size, additional window opening, etc. With the demand of high-quality low-noise road vehicles, more and more people from industry, research institutions and universities will join the team studying road vehicle buffeting.

The mechanism of buffeting is a very complicated aeroacoustic phenomenon. At the leading edge of the opening (either side window or sunroof), there exists a shear layer between the fast sweeping airflow outside the opening and the relatively stationary air inside the cabin. The shear layer is usually unstable in nature and vortices are generated frequently and shed periodically. Accordingly, static pressure fluctuates periodically with the same frequency as vortex shedding. The cabin can be considered as a resonance chamber, which has a certain natural frequency. If the frequency of the pressure fluctuation happens to coincide with the natural frequency of the cabin, then resonance may be excited and buffeting may occur. In this circumstance, the vehicle cabin serves as a Helmholtz resonator.

A Helmholtz resonator is a chamber that is connected to a fluid flow field outside the camber through a narrow neck. Communicating with the outside fluid flow, the air is forced to move-in and -out through the neck. In this way aeroacoustic noise is made with a certain frequency and intensity. Rayleigh [13], Pierce [14] and Kinsler et al [15] deduced the natural frequency of a Helmholtz resonator as $f_n = c/(2\pi)\sqrt{A/(VL)}$ where f_n is natural frequency, c is speed of sound, V is volume of the resonator, L and A are length and cross-area of the neck, respectively. Blevins [16] collected natural frequency for many structural and fluid systems, including a Helmholtz resonator. However, this formula usually over-predicts natural frequency if the length of the neck is short compared to the scale of the cross-area of the neck, as in the case of a vehicle cabin. Alster [17] deduced a modified formula for the natural frequency of a simple Helmholtz resonator, $f_n = c/(2\pi)\sqrt{A/[1.21(V+AL_N)L_1]}$ where L_N and L_1 depend on specific geometry. The modified formula gives more accurate natural frequency of a Helmholtz resonator that has the size of a vehicle.

Another subject that is related to sunroof buffeting is "cavity". As a matter of fact, the cabin of a vehicle, in the sunroof buffeting study, may be imagined as a cavity with a big volume and a small opening. This is because the flow over the sunroof opening has a similar feature to the flow over a cavity. As described later, a cavity is intentionally added to the leading edge of the sunroof opening to reduce buffeting. This makes the sunroof a "double cavity" structure. Therefore, it is of essential importance that the characteristics of flow over a cavity should be thoroughly understood. Most early work on flow over a cavity was done in aerospace industry. For example, Roshko [18] measured pressure and velocity on the walls of a cavity with various ratios of length to depth. Block [19] investigated noise response of cavities to the flow with subsonic speeds. Tam [20] conducted a theoretical study of flow over a rectangular cavity with the interest in aeroacoustic noise. In 1978, Rockwell and Naudascher [21] published a review paper of self-sustaining oscillations of flow past cavities. They grouped unstable flow over a cavity into three categories: fluid-dynamic, fluid-resonant and fluid-elastic. Fluid-dynamic oscillations are attributed to instability of the shear layer and enhanced through a feedback mechanism. Fluid-resonant oscillations are governed by resonance conditions associated with compressible wave phenomena. Fluid-elastic oscillations are primarily controlled by the elastic displacement of a solid boundary. They further categorized fluid-resonant cavity into two types: shallow cavity and deep cavity, depending on whether or not the length is larger than the depth of the cavity. A shallow cavity usually generates a longitudinal wave in the flow direction while a deep cavity usually generates a lateral wave perpendicular to the flow direction. Rockwell and Knisely [22] observed the vortex impingement upon the trailing edge (or corner) of a cavity using laser anemometry and hydrogen bubble technique in a water tunnel. Their visualization of vortex movement showed that an impinging vortex might experience one of the following three events. (a) Complete clipping – the vortex is completely plunging down into the cavity; (b) Partial clipping – which results in vortex breakdown; (c) Escape – involving vortex deformation, where it is swept downstream over the trailing edge (corner). This observation provides a useful hint on how to decrease disturbance to the cavity. In mid-1980s, Blake [23] reviewed the flow over cavities with some complex details of the shear layer impingement and feedback. Some years later, Blevins [24] summarized the state-of-the-art of the studies on the sound excited due to flow over cavities. Since 1990s, the study of noise generation from flow over cavities has been applied to automotive industry. Mongeau et al [25] investigated flow-induced pressure fluctuations in a door gap cavity in a quiet wind tunnel. They found that the primary excitation mechanism of pressure fluctuations was an "edge tone" phenomenon. They also claimed that modifying the geometry of trailing edge could reduce cavity pressure fluctuations and tapering the trailing edge is beneficial since it tends to deflect the flow away from the cavity. This observation can be directly utilized as a useful guide to modify the geometry of the sunroof opening for buffeting control. The automotive application of cavity flow gained so much interest that the door gap cavity flow was adopted as one of the benchmarks in the 3th computational aeroacoustics workshop [26] and some papers were contributed to this benchmark [27-28]. Three years later, the automotive door gap cavity flow was still presented as a benchmark to the 4th workshop [29] and more contributions are expected to be made, Ashcroft et al [30] for example.

Shear layer instability at the leading edge of the cavity is believed to be the source of pressure fluctuations in the flow over a cavity. Early work on shear layer instability for a "step" velocity profile, described in Lamb's book [31], was named Kelvin-Helmholtz instability. In this instance, velocity gradient at the interface between the upper and lower layers is singular and this singularity constitutes a vortex sheet. Traditionally, analysis of flow stability begins with an assumption that a small disturbance is present in the flow and interest is the growth of the disturbance in time or space, whether decays or runs away. Using this procedure, it was known that the step velocity profile is unstable to any disturbance and the vortex sheet develops unboundedly, if the difference of velocities between the two layers exceeds a certain critical value. Applying this theory, An et al [32] conducted the analysis of Kelvin-Helmholtz instability with application to the oil-water interfacial phenomena. Nevertheless, the step velocity profile is less realistic for most circumstances. Linear, curved or boundary layer velocity profiles were used in the subsequent investigations of shear layer instability. Esch [33] examined the shear layer instability for a linear velocity profile between upper and lower uniform streams using an eigenvalue technique and determined the region of instability. Michalke adopted the linear stability theory for the case of temporary growing disturbances to the hyperbolic-tangent velocity [34], analyzed and explained the vortex formation, development and rolling up [35], and finally extended the analysis to the case of spatially growing disturbances [36]. Experimentally, Browand [37] undertook an investigation of shear layer instability in a small wind tunnel. He found that the primary oscillation, predicted by the linear stability theory, grows rapidly with a gradual growth of higher harmonics. This indicates a secondary instability in the lower portion of the shear layer. The shear layer instability for boundary layer velocity profile were extensively investigated and summarized by Lin [38] and Bechov and Criminale [39] in their monographs.

Vortex shedding is an immediate outcome of shear layer instability. If a cavity exists at the downstream of the shear layer, noise excitation due to vortex shedding will follow. Abernathy and Kronauer [40] proposed a vortex formation theory using the concept of point vortices and vortex sheets. The formation of discrete vortices in the wake of a blunt body can be interpreted as non-linear interaction of two vortex sheets with opposite directions (clockwise and counter-clockwise), initially at a distance h in the fluid. They found that the number and strength of the discrete vortices depend strongly on the ratio of h/a where a is the wavelength of initial disturbance. The theory also explained the phenomenon of vortex disappearance and broadening observed in the body wake. Boldman et al. [41] undertook flow visualization of vortex shedding from a blunt trailing edge of a plate in a wind tunnel. They found an interesting fact that strong von Karman vortices develop behind the trailing edge when the velocities on upper and lower surfaces are equal and the vortices tend to disappear when the velocities are unequal. This observation was in good agreement with the theory of vortex formation explained elsewhere [40]. Nakamura and Nakashima [42] conducted an experimental investigation of vortex excitation where two shear layers were separated from a bluff body of various cross-sections, both in a wind tunnel for measurement, and in a water tank for visualization purpose. They observed that, the two unstable shear layers interact with each other when they meet together downstream of the body and that, von Karman vortices appear with the same frequency of the oscillation. They concluded that the impinging-shear-layer is responsible for vortex excitation.

Up to the present, popular concepts of buffeting control includes at least the following three types: wind deflector at leading edge and/or training edge, sunroof glass comfort position at which the noise is minimal and cabin venting by a second opening. The objective of this paper is to analyze a concept for buffeting control, a "cavity" installed at the leading edge of the sunroof opening. In order to understand this concept, a two dimensional simulation for a production SUV is performed using the CFD software – FLUENT [43]. The simulated results for sunroof configurations both with and without a cavity are compared to find out which one produces less noise. Moreover, based on fluid dynamics principles, the features of local flow (mainly vortex dynamics) at the sunroof opening for both cases are examined in more detail to unveil the associated physics causing the cavity functionality. Then guided by those local flow feature investigations, modification of sunroof opening is made and the modified model is simulated again to find out if the modification is in the correct direction. For all sunroof configurations, the simulated flows are visualized using both static (snapshot) and dynamic (animation) tools to explain why and how one is better than the other in the sense of buffeting reduction.

METHODOLOGY

A two dimensional model of a realistic vehicle with a sunroof opening, as shown in the lower part of Figure 1, is chosen for buffeting study. In order to speed up the process without losing the main feature of buffeting characteristics, further simplification is made. The exterior and interior of the vehicle are represented by the external and internal shape contour. For two dimensional case, the seats shown in Figure 1 are simply because seats could partially block the internal air flow to the extent that it could create problems on obtaining converged solutions. For three dimensional case, unlike two dimensional case, internal air flow can pass through the gaps between seats and side panels of the body and between seats. Therefore, the seats were removed to avoid any possible "flow blockage" for two dimensional simulations. Obviously, wheels are not a part of solution domain either. Three cases of sunroof configurations are considered as shown in the upper part of Figure 1. Case A is conventional sunroof geometry without a cavity. In case B, a cavity is intentionally added at the leading edge of the sunroof opening to control the vortex shed from the upstream shear flow. For case C, the cavity is further modified by cutting off the upper and lower corners of the trailing edge and the lower corner of the leading edge to three ramps. The whole flow domain is represented as a rectangle. The inlet (left) boundary of the domain is about 3m upstream from the vehicle's front end and the outlet (right) boundary is about 3m downstream from the vehicle's tail. The upper boundary of the domain is about 4.5m high above the ground. The vehicle dimension is about 5m x 2m. The sunroof opening is 270 mm long and 60 mm high (thickness of the vehicle roof), and the cavity in case B and case C is 75 mm long and 40 mm deep.

The sketches of the three sunroof configurations are shown in the upper part of Figure 1. The objective for studying these three cases is to find out which one produces the lowest buffeting noise. In order to perform aeroacoustic analysis a sensor should be installed somewhere in the cabin. The blue point near the sunroof in the cabin (Figure 1) is used to record the signal of pressure fluctuation.

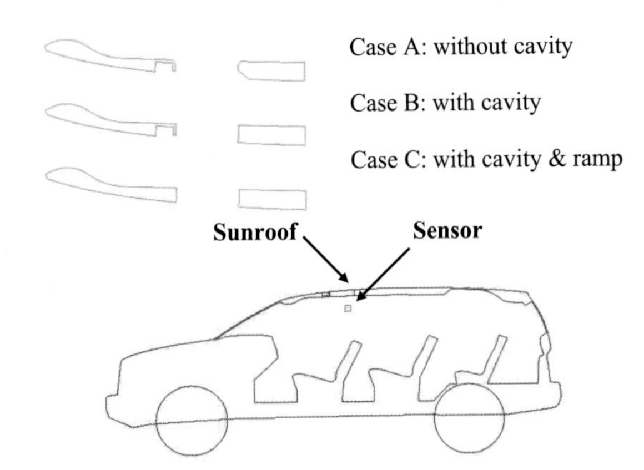

Figure 1. Two Dimensional Model of a Vehicle with Three Sunroof Configurations and a Sensor

For each sunroof configuration, a triangular mesh for the whole flow domain is generated from a CATIA geometry model using a pre-processor ANSA and, then, the mesh is further refined using another pre-processor TGRID. A local refinement option, available in TGRID, is utilized to refine the mesh in the region of the sunroof opening. The final mesh at the cavity and the sunroof opening for case B is shown in Figure 2. Dimensions of cavity in millimeters are also shown for reference. The total number of cells for the whole flow domain is about 360,000 and the minimal dimension of the cells at the cavity and sunroof opening is about 1 mm.

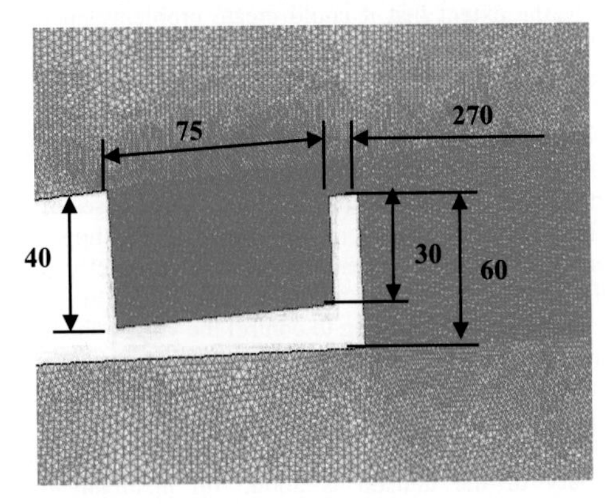

Figure 2. Computational Mesh at Cavity (Case B)

The commercial CFD software FLUENT is used to simulate fluid flow. The solver settings and boundary conditions are listed in Table 1 and Table 2, respectively.

Table 1. Solver Settings

Function	Setting
Solver	Segregated Transient
Fluid	Air (Ideal Gas)
Time Marching	2nd Order Implicit
Space Discretization	2nd Order Upwind
P-V Coupling	2nd Order SIMPLE
Turbulence Modeling	RNG k-ε
Time Step	0.002 Seconds

Table 2. Boundary Conditions

Boundary	Condition
Inlet	Constant Velocity (20m/s)
Outlet	Constant Pressure
Upper & Ground	Frictionless Wall
Vehicle Surface	Frictional Wall

SOLUTION PROCEDURE

For each case of the sunroof configurations, the entire flow field is solved in steady state mode for about 800 iterations using the standard k-ε turbulence model. The relaxation parameters for steady state simulation are set to be pressure 0.1, momentum 0.5, turbulence 0.7 and energy 0.8. The steady state flow is then used as initial field of the transient flow and the RNG k-ε turbulence model is activated for the transient simulation. The relaxation parameters for transient simulation are set to be pressure 0.2, momentum 0.4, turbulence 0.5 and energy 1. A time step of 0.002 seconds is set to run the transient flow. This time step is small enough to capture buffeting feature because the typical frequency of buffeting is around 20 Hz, or equivalently, the typical period of buffeting is around 0.05 seconds. Hence, there are around 25 sampling points within a period to reveal unsteadiness of the flow. Due to the option of implicit scheme in the solver, a certain number of sub-iterations have to be set and this number is 30 for this study. If initial field of the transient flow is "good" enough, the simulation will converge within 30 sub-iterations for each time step.

A sensor (microphone) is installed at point (1.6m, 1.15m), i.e. about the height of human ears in the cabin (see Figure 1), to record pressure fluctuation at each time step. After about 500-800 time steps, the transient simulation usually completes the initiating process and the static pressure reaches a periodic or quasi-periodic oscillation mode. The time duration for signal processing is chosen to be 2 seconds, or 1000 time steps, after the completion of initiating process for transient simulation.

The recorded signal of pressure fluctuation, as a function of time, is then processed using a software package XMGR that has an option of Fourier transform. After performing Fourier transform the recorded signal is converted to the spectral format in which the amplitude of pressure fluctuation is expressed as a function of frequency. Finally, the amplitude of pressure fluctuation is converted to the format of sound pressure level (*SPL*) in dB. The formula $SPL = 10\ log_{10}\ (P/Pref)^2$ [14] is used to make such conversion where *P* is the amplitude of pressure fluctuation in Pa and $Pref = 20 \times 10^{-6}$ Pa. Finally, the simulated transient flow for each case of the sunroof configurations is processed to get buffeting spectrum for comparison. At the same time, either static snapshot or dynamic animation of the simulated flow is used to investigate flow characteristics.

SIMULATED RESULTS AND DISCUSSIONS

The time history of pressure fluctuation signals at the sensor location for the three sunroof configurations are shown in Figures 3 and 4. As mentioned earlier, the signals before t = 2 seconds are in initiating process of the transient simulation and they have not reached a periodic or quasi-periodic oscillation mode. Therefore, the time range of t = 2 ~ 4 seconds is selected to process the signal, unless otherwise mentioned.

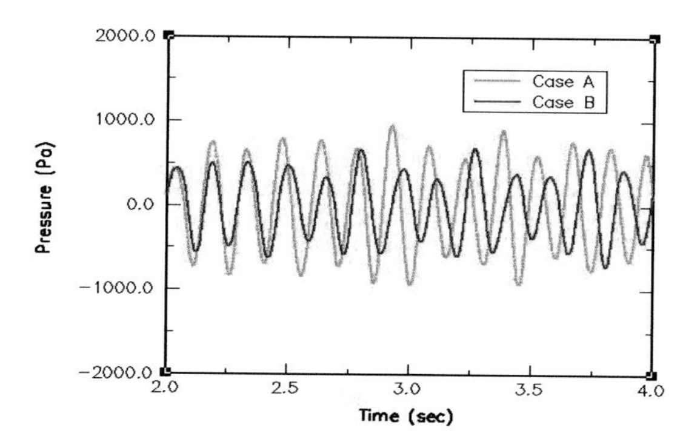

Figure 3. Time History of Signals for Cases A & B

It is clear from Figure 3 that the amplitude of pressure fluctuation for case B (red solid curve) is smaller than that of case A (green dot curve). Figure 4 shows that the amplitude of pressure fluctuation for case C (blue dash curve) is even smaller that that of cases B (red solid curve). These two figures imply that buffeting intensity among the three sunroof configurations should be rated as C < B < A.

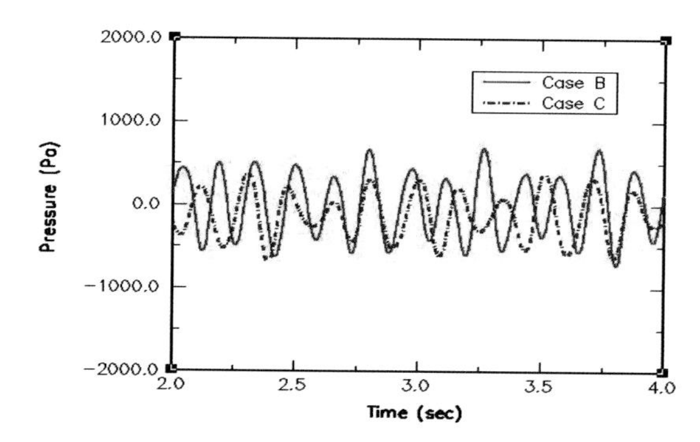

Figure 4. Time History of Signals for Cases B & C

The selected noise signals in Figures 3 & 4 are then processed using XMGR for Fourier transform. Because XMGR is a general purpose plotting software package, it is necessary to validate its ability and accuracy for Fourier transform. For this purpose a simple noise signal is processed and compared with the analytic solution (see Appendix). The final results demonstrate that it is able to give acceptably accurate results by its option of Discrete Fourier Transform (DFT) with a Hanning window.

Figures 5 & 6 depict the corresponding frequency spectra of the noise signals for the three cases in Figures 3 & 4. The frequency spectrum of a signal is a curve of sound pressure level (SPL) in dB as a function of frequency in Hz. The first peak in a curve represents the buffeting frequency in Hz and the buffeting sound pressure level (SPL) in dB.

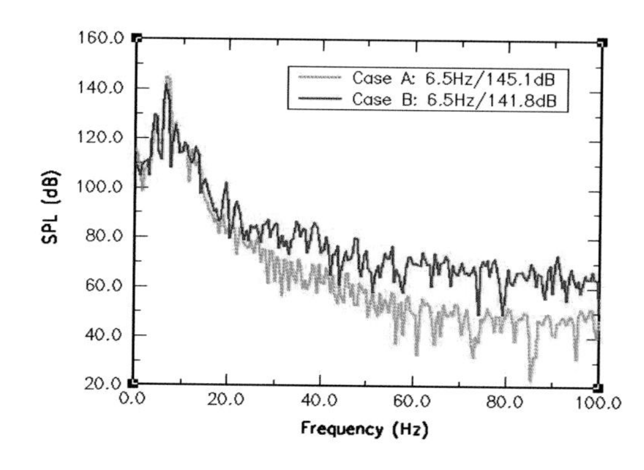

Figure 5. Buffeting Level for Cases A & B

From Figure 5 it is found that case B (red solid curve) has buffeting frequency 6.5Hz and buffeting SPL 141.8dB, while case A has the same buffeting frequency but higher buffeting SPL 145.1dB. From now on, "buffeting intensity", "buffeting SPL", and "buffeting level", or just "level" are going to be used interchangeably. Therefore case B (red solid curve) reduces the buffeting level by about 3dB compared to case A (green dot curve). This result is consistent with the discussion for Figures 3 & 4, which proves that the cavity in the sunroof configuration plays an important role to suppress, or reduce, buffeting level. It is worthwhile to mention that the first peak of SPL at low frequency is the principal index of buffeting. Therefore, the difference of the SPL curves between cases A and B is of no practical concern and is neglected here.

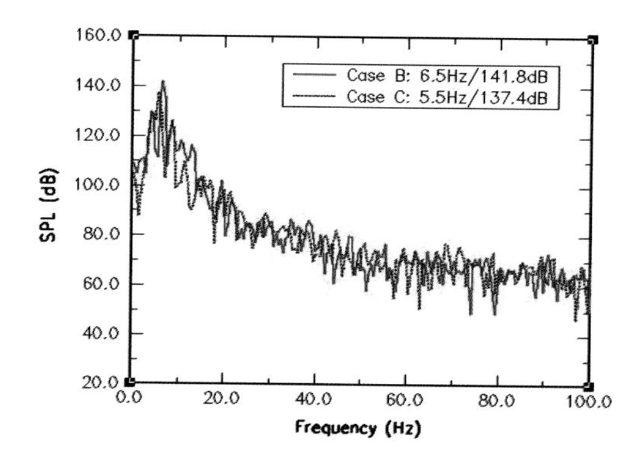

Figure 6. Buffeting Level for Cases B & C

From Figure 6 we can see that case C (blue dot curve) has even lower buffeting level, 137.4dB at 5.5Hz, compared to case B (red solid curve), 141.8dB at 6.5Hz. This means that the ramps at the corners of the leading and trailing edges enhance the effect of buffeting reduction by about 4 more dB.

It should be pointed out here that compared to buffeting intensity, buffeting frequency is less critical because passenger's comfort is dependent on buffeting intensity. It should also be pointed out that, due to inherent simplifications associated with a two dimensional approach, absolute value of buffeting intensity SPL predicted here may not be exactly comparable to that found for the real vehicle. However, the observed relative trend among the three cases of the sunroof configurations should be trustworthy. Therefore, the present study has revealed the fact that the concept of using a cavity has the ability to reduce buffeting.

The primary impact of the cavity is due to superimposing a local cavity-like flow, at the leading edge of the sunroof opening, over the main flow. As a result, the strong vortex, shed from the upstream, gets partially trapped in the cavity and turns into a continuously re-circulating flow there. It is believed that this re-circulating flow pushes the next coming vortex upward away from the cabin. This local flow acts as an absorptive bed to reduce the degree of instability that the coming vortex would otherwise have experienced due to suddenly sweeping over the almost stationary air in the cabin.

As a side-discussion, it should be mentioned here that it is true that a cavity-like structure might have been already used at the leading edge of the sunroof opening in some automobiles. But it has always owed its presence, solely as a side-feature or accessory, for other primary purposes mainly related to water management, wind deflector "hold-in-room", and so on. Therefore, it is fair to mention that the important role of cavity-like structure in buffeting reduction has not yet been emphasized. In fact this role can be viewed as a by-product of the cavity-like structure which needs to be understood and then to be employed in an optimal way. As the major outcome of many buffeting studies done so far, at least three concepts have already been known: wind deflector, glass comfort position and cabin venting. Each of these concepts has proved to be efficient to some extent by studies done through experimental, theoretical, and computational approaches. In a similar way, the cavity-like structure is suggested now to be considered as a concept for buffeting control.

One of the objectives of this paper is to explain the reasons responsible for favorable effect of a cavity on buffeting reduction. In the following sections, based on fluid dynamics principles, the local flow in the region of the sunroof opening for each case will be analyzed. This analysis will address the observations gained from static (snapshot) as well as dynamic (animation) visualizations of the results, such as velocity vector and vorticity magnitude. It is the authors' hope that this analysis paves the way to develop a better understanding of the flow nature and to explain why and how the cavity helps reduce buffeting. Next, additional physical discussions, based on post-processing of the flow patterns in the region of sunroof opening at some typical time moments, will be followed. The specific chosen time moment in a periodic cycle is not crucial here and the time moments for post-processing are not necessarily sequential either. These time moments are selected solely to identify and uncover some special feature within the fluid flow that are believed to be essential to cavity performance.

Figure 7 shows the velocity vector of the flow in the region of the sunroof opening for case A at time moment t=4.624sec from the initiation of transient simulation. More attention should be paid to the following points. (a) A clockwise vortex is shed from the upper corner of the leading edge. This vortex shedding is due to the instability of shear layer on the upper roof surface upstream the opening. (b) The shear layer between high-speed flow outside the vehicle and relatively stationary air inside the cabin is highly curved and unstable. (c) The strong vortex shed from the leading edge travels downstream to the trailing edge of the opening. (d) The clockwise rotational flow (strong vortex) impinges onto the trailing edge of the opening, reflects from there and deeply intrudes into the cabin. (e) Moreover, the clockwise vortex shed from the upper corner of the leading edge is so strong that it rolls over the lower corner of the leading edge and intrudes into the cabin.

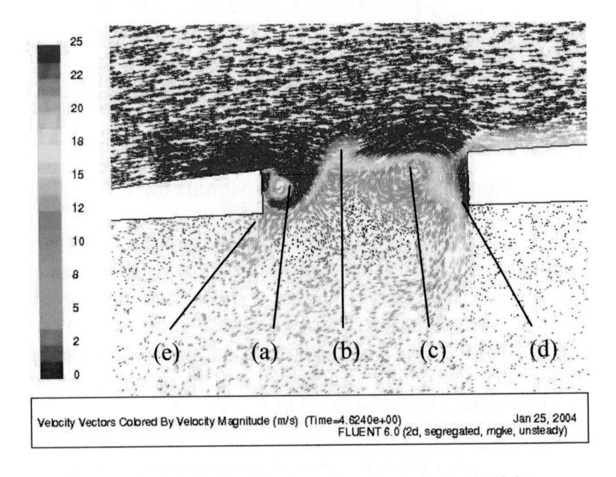

Figure 7. Velocity Vector for Case A at 4.624 sec

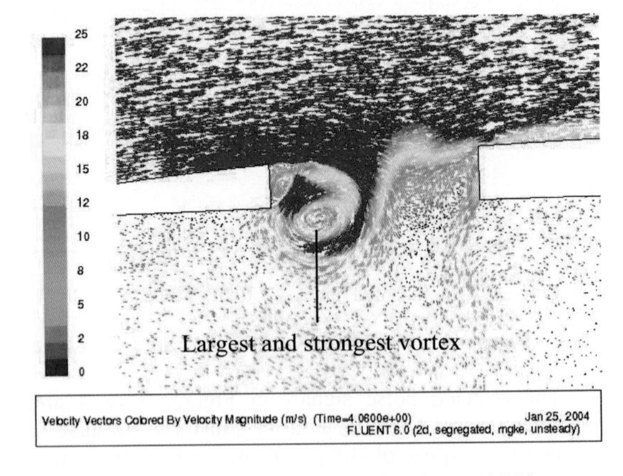

Figure 8. Velocity Vector for Case A at 4.060 sec

Figure 8 shows the largest and strongest clockwise vortex behind the leading edge of the opening in a periodic cycle.

Figure 9 shows the strongest reflection of the flow from the trailing edge of the sunroof opening and the deepest intrusion of the flow into the cabin in a periodic cycle.

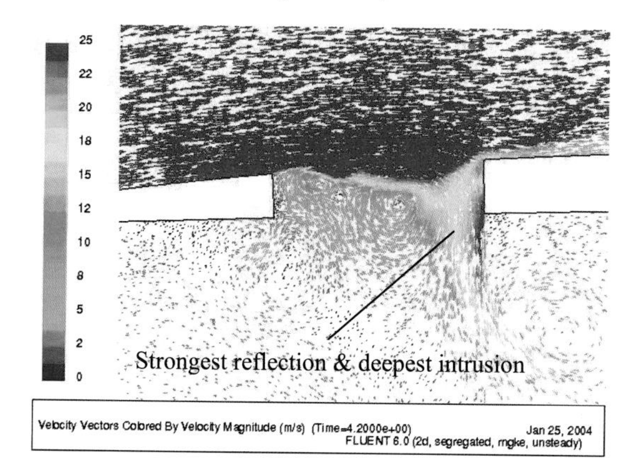

Figure 9. Velocity Vector for Case A at 4.200 sec

Because the nature of the local flow at the sunroof opening is highly rotational, vortex dynamics is more helpful than the conventional visualization (pressure, velocity, turbulence, etc.) to capture the flow characteristics.

In the following context, we will look at the vorticity magnitude contour at some typical time moments. In these contour graphs, the vorticity magnitude is defined by Vx-Uy and the value of vorticity magnitude may be either positive or negative. If the value of vorticity is positive then the vortex is counter-clockwise (red). If the value of vorticity is negative then the vortex is clockwise (blue). If the value of vorticity is near zero then the vortex is relatively weak (green).

Figure 10 shows the vorticity magnitude contour of the flow at sunroof opening for case A at time moment t=4.180sec from the initiation of transient simulation. As mentioned before, the exact moment in time is not crucial but typical flow feature is a major concern. In accordance to the velocity vector graphs in Figure 7, attention should be paid to the following points in this figure. (a) A clockwise vortex is shed from the upper corner of the leading edge and this vortex shedding is due to the instability of shear layer on the upper roof surface. (b) A strong vortex, shed from the leading edge, travels downstream towards the trailing edge. (c) A counter-clockwise vortex is generated at the lower corner of the trailing edge and it deeply intrudes into the cabin. (d) The clockwise vortex at the leading edge is so strong that it rolls over the lower corner of the leading edge and intrudes into the cabin (cf. Figure 7 at point e). It should be pointed out that a counter-clockwise vortex (red) at the lower corner of the leading edge is not too harmful because it rolls outward away from the cabin. However, a clockwise vortex (blue) at the same corner is more harmful because it rolls inward into the cabin. Although the clockwise vortex at this corner is not very strong more attention should be paid to this phenomenon.

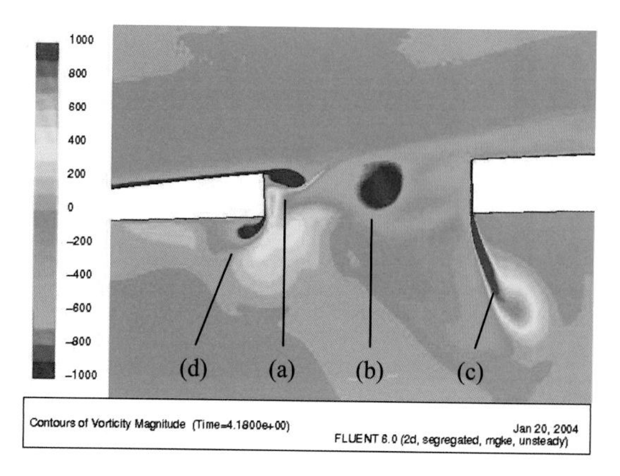

Figure 10. Vorticity Magnitude for Case A at 4.180 sec

Figure 11 shows the vorticity magnitude contour of the flow at the sunroof opening at t=4.500sec. The major features of the flow are as explained as below.

(a) The clockwise vortex, shed from the upper corner of the leading edge, is magnified and strengthened (cf. Figure 10 at point a). (b) The shear layer between the high-speed flow outside the vehicle and the relatively stationary air inside the cabin is highly curved and unstable (cf. Figure 7 at point b). (c) The downstream traveling clockwise vortex (blue) impinges onto the trailing edge and is broken down into two major portions. A smaller portion passes over the opening and a larger portion is reflected from the trailing edge and intrudes into the cabin together with the counter-clockwise vortex (red) generated there.

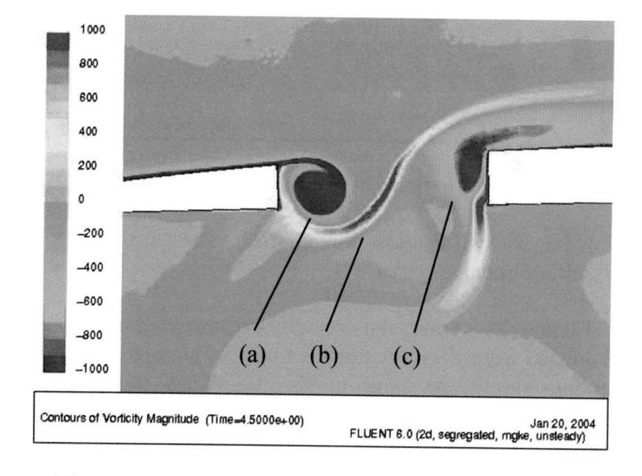

Figure 11. Vorticity Magnitude for Case A at 4.500 sec

Figure 12 shows the largest and strongest clockwise vortex behind the leading edge of the sunroof opening in a periodic cycle (cf. Figure 8).

第IV部分 任职于美国克莱斯勒汽车公司时期

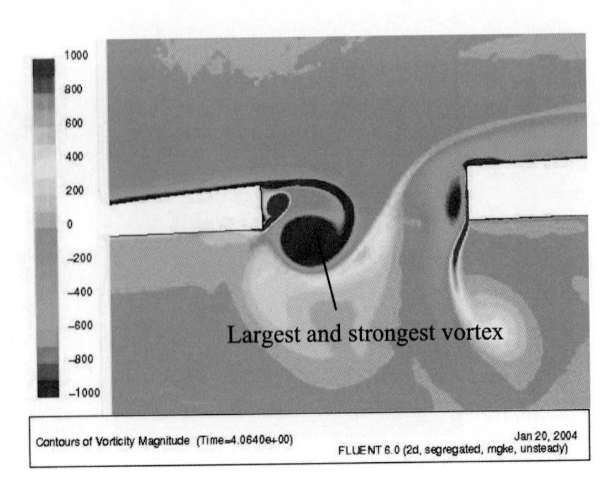

Figure 12. Vorticity Magnitude for Case A at 4.064 sec

Figure 13 shows the strongest flow reflection from the trailing edge and the deepest intrusion into the cabin. Similar to the discussion for Figure 10 at the lower corner of the leading edge (d), for Figure 13 at the lower corner of the trailing edge, a counter-clockwise vortex (red) is more harmful regarding the intrusion of the flow into the cabin.

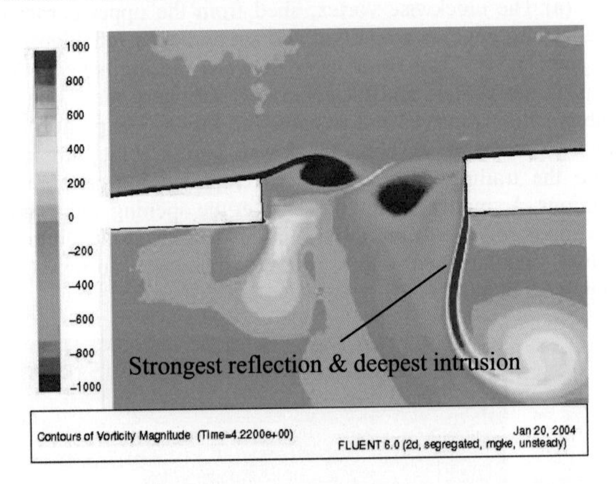

Figure 13. Vorticity Magnitude for Case A at 4.220 sec

Next, we will discuss flow characteristics in the region of the sunroof opening for case B.

Figure 14 shows the velocity vector at time moment t=2.044sec from the initiation of transient simulation. Again, attention should be paid to the following points. (a) A clockwise vortex generated from the upper roof surface in the upstream of the flow is trapped within the cavity. This is the most important feature of the cavity to reduce buffeting. (b) Behind the cavity there is still a vortex but it is not as strong as the one shown for case A (cf. Figure 7 at point a). (c) The shear layer between the fast sweeping outer flow and the relatively stationary air in the cabin is not as curved as the one for case A (cf. Figure 7 at point b). (d) The clockwise vortex, shed from the leading edge of the opening, is traveling downstream. (e) The flow impingement onto the trailing edge and the reflection are not as strong as case A (cf. Figure 7 at point d).

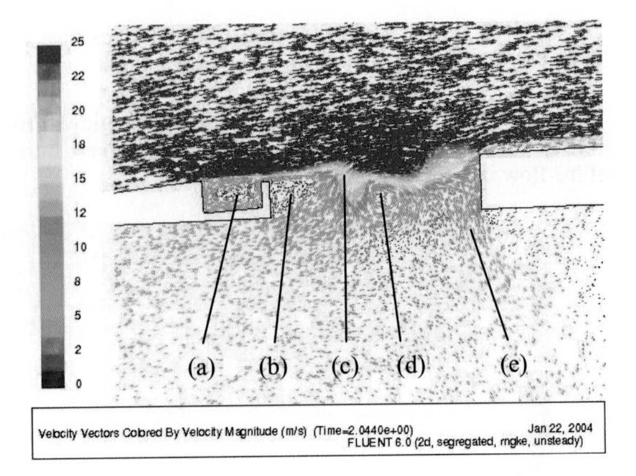

Figure 14. Velocity Vector for Case B at 2.044 sec

As described before, the vorticity magnitude contour can provide more information about the flow feature than the usual post-processing of pressure and velocity. Now, let us look at the vorticity magnitude contour for case B at some typical time moments.

Figure 15 shows the vorticity magnitude at time moment t=2.044sec from the initiation of transient simulation. In accordance to the velocity vector graphs in Figure 14, attention should be paid to the following points. (a) A clockwise vortex (blue) generated from the upstream of the flow is trapped within the cavity. (b) Some clockwise vortex escapes from the cavity but the escaping vortex is not as strong as case A (cf. Figures 10 & 11 at point a and Figure 12). (c) The escaping vortex is traveling downstream. (d) The traveling clockwise vortex (blue) impinges onto the trailing edge and breaks down there. However, the impingement and breakdown are not as strong as case A. As a result, the portion of the vortex that intrudes into the cabin is much smaller than the portion that passes over the opening (cf. Figure 11 at point c and Figure 12). (e) The counter-clockwise vortex (red), shed at the lower corner of the trailing edge, is weaker than case A and the intrusion of the vortex into the cabin is not as deep as case A (cf. Figure 13).

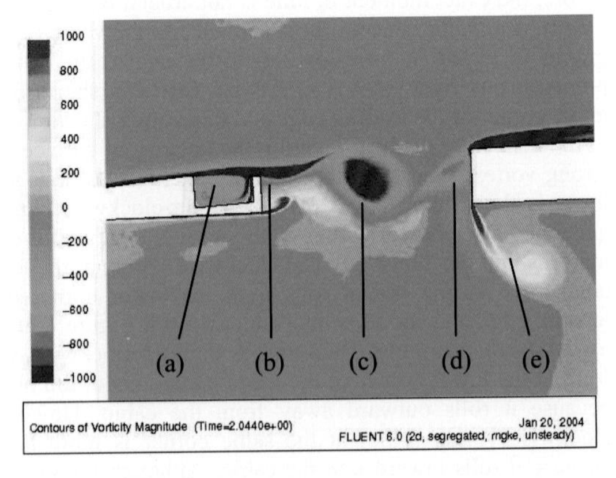

Figure 15. Vorticity Magnitude for Case B at 2.044 sec

安长发博士科研论文集

At some other time moment, the harmful clockwise vortex still exists at the lower corner of the leading edge (see Figure 16) but it is weaker than case A (cf. Figure 10 at point d). At most instances in time, however, a less harmful counter-clockwise vortex (red) usually appears at the lower corner of the leading edge (see Figure 15).

In general, the shear layer between the flow outside the vehicle and the air inside the cabin for case B is flatter than case A. The intrusion of the vortex into the cabin for case B is milder than case A. From this discussion we conclude that the cavity plays a critical role to reduce buffeting. This explains how the cavity works in the sunroof configuration and why the buffeting characteristics behaves like that in Figures 3 and 5 between cases A and B.

In order to enhance the capability of buffeting reduction of the sunroof configuration case B, further modifications are made. Cutting the corners into ramps (see Figure 1 case C) completes these modifications. The ramp at the upper corner of the trailing edge is to control the clockwise vortex (blue) breakdown so that a greater portion of the vortex passes over the opening and smaller portion of the vortex reflects and intrudes into the cabin. The ramp at the lower corner of the trailing edge is to guide the counter-clockwise vortex (red) to move along the inner surface of the roof without strongly intruding into the cabin. The ramp at the lower corner of the leading edge is to remove the clockwise vortex (blue) shown in Figure 16. The counter-clockwise vortex (red) is less harmful than the clockwise vortex (blue) at the lower corner of the leading edge because the former moves outward to the ambient but the latter moves inward to the cabin.

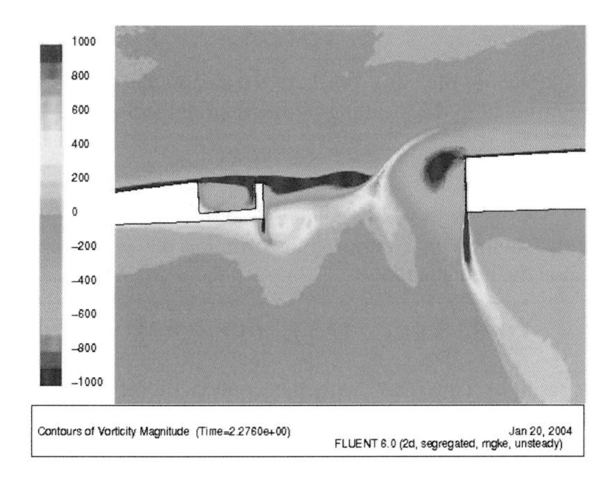

Figure 16. Vorticity Magnitude for Case A at 2.276 sec

Figure 17 shows the velocity vector for case C at time moment t=2.000sec from the initiation of transient simulation. Similar to case B in Figure 14, attention should be paid to the following points. (a) A clockwise vortex generated from the upstream of the flow is trapped within the cavity. (b) Behind the cavity there is still a vortex. (c) The shear layer between the fast sweeping outer flow and the relatively stationary air in the cabin is even flatter than case B (cf. Figure 14 at point c). (d) The flow impinges onto the trailing edge more mildly than case B. The reflection from the trailing edge is not as strong as case B. (e) The counter-clockwise vortex, shed at the lower corner of the trailing edge, moves more horizontally along the lower surface of the roof. Therefore, this vortex has a minimal effect of intrusion into the cabin.

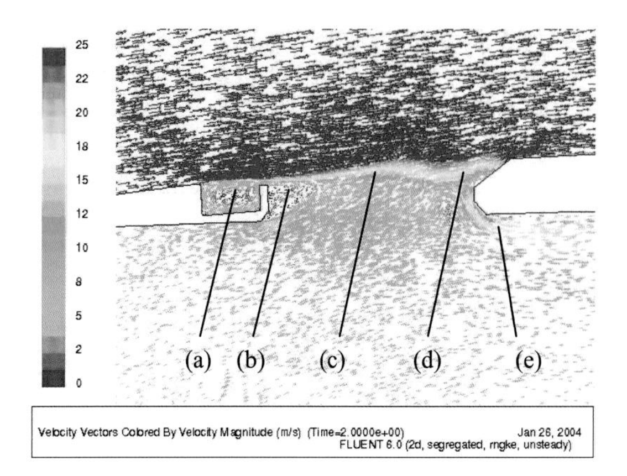

Figure 17. Velocity Vector for Case C at 2.000 sec

Figure 18 shows the vorticity magnitude of the flow for case C at t=2.216sec. It can reveal the flow feature more clearly than the velocity vector graph. The following points should be paid more attention. (a) A clockwise vortex (blue) is trapped within the cavity. (b) A counter-clockwise vortex (red) is shed from the lower corner of the leading edge. (c) The clockwise vortex (blue), escaping from the cavity, is traveling downstream, but its strength is weaker the case B. (d) The ramp at the upper corner of the trailing edge guides the clockwise vortex (blue) to pass over the opening. It also reduces the vortex reflection and intrusion into the cabin. (e) The counter-clockwise vortex (red) is flowing along the inner surface of the roof without deeply intruding into the cabin.

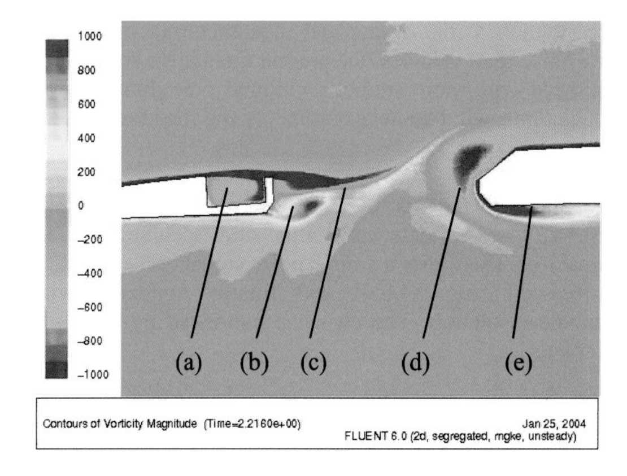

Figure 18. Vorticity Magnitude for Case C at 2.216 sec

It should be pointed out that the clockwise vortex (blue) at the lower corner of the leading edge disappears in case C (cf. Figure 18 at point b and Figure 16). This is a positive effect to reduce the disturbance to the cabin and, therefore, reduce buffeting in case C. All of these simulated flows can be visualized dynamically by animations that will be available at the conference.

CONCLUSIONS

Sunroof buffeting for the two dimensional model of a production SUV can be numerically simulated using the CFD procedure of CATIA/ANSA/TGRID/FLUENT/XMGR.

The impact of a cavity installed at the leading edge of the sunroof opening is crucial to reduce buffeting. The buffeting level for the sunroof with a cavity (case B) may be reduced by 3dB compared to the one without a cavity (case A). Additional intelligent modifications of sunroof configuration with a cavity by cutting the corners into ramps (case C) may further reduce the buffeting level by 4dB or more compared to the one without the ramps (case B).

Thorough understanding of the buffeting mechanism helps explain why and how the cavity decreases the disturbance to the air in the cabin and, hence, reduces buffeting. Investigating the physics, related to the buffeting phenomenon, such as Helmholtz resonance, cavity characteristics, shear layer instability and vortex dynamics (formation, shedding, traveling, reflection, breakdown, etc.), provides a deep insight into the flow nature and, therefore, a guide towards geometry modification for buffeting reduction.

Although a cavity-like structure might have been already integrated within the sunroof configuration of some automobiles for the purpose of water management and/or deflector "hold-in-room", it appears that up to now it has not yet been emphasized that the cavity-like structure plays an important role in reduction of buffeting. The cavity is suggested to be considered as a buffeting reducer, in addition to the currently-known three concepts: wind deflector, the glass comfort position and cabin venting by a second opening.

In principle, the concept of cavity analyzed in this paper for reduction of sunroof buffeting could also be extended to deal with side window buffeting which forms the future continuation of this study. In order to provide a well-founded contribution to the design and manufacturing of high quality low noise road vehicles, the present qualitative study should be followed with more studies including three dimensional CFD simulations, wind tunnel experiments and road tests.

APPENDIX

Since XMGR is a general-purpose plotting tool, it is necessary to validate its ability and accuracy for Fourier transform. Therefore, a simple noise signal is constructed, then processed using XMGR and finally compared with the analytical solution. The signal is expressed by the following function:

$$p=\sum_{k=1}^{4} A_k \sin(2\pi f_k t) \qquad (A-1)$$

where p is static pressure (in Pa), t is time (in sec) and the amplitudes and frequencies of the 4 harmonics are:

A_1=40, A_2=20, A_3=10, A_4=5 (Pa)
f_1=10, f_2=20, f_3=40, f_4=80 (Hz)

It is expected to have 4 peaks in the frequency spectrum of the signal at $f=f_k$, k=1,2,3,4. The time history of the signal is drawn in Figure A1.

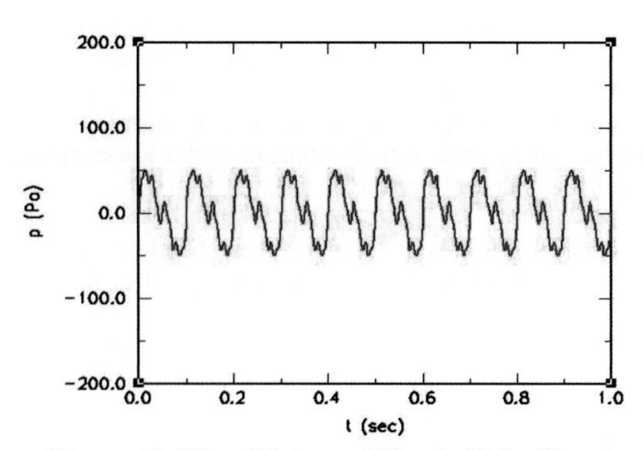

Figure A1. Time History of Simple Noise Signal

Fourier transform is defined as:

$$F=\int_{-\infty}^{\infty} p e^{-i(2\pi f t)} dt \qquad (A-2)$$

where F is Fourier transform of p, f is frequency (in Hz) and $i=\sqrt{-1}$. Applying this definition the Fourier transform for the signal can be derived as:

$$F=\int_{-\infty}^{\infty} \sum_{k=1}^{4} A_k \sin(2\pi f_k t) e^{-i(2\pi f t)} dt$$

$$=-i\sum_{k=1}^{4} A_k \int_{-\infty}^{\infty} \sin(2\pi f_k t)\sin(2\pi f t) dt$$

$$=i\sum_{k=1}^{4}(A_k/2)\{\int_{-\infty}^{\infty}\cos[2\pi(f+f_k)t]dt$$

$$-\int_{-\infty}^{\infty}\cos[2\pi(f-f_k)t]dt\} \qquad (A-3)$$

The integrals in the curly brackets are meaningless unless they are interpreted in the meaning of distribution theory [44]:

$$\int_{-\infty}^{\infty}\cos(2\pi f t) dt=\delta(f) \qquad (A-4)$$

where $\delta(f)$ is the Dirac impulse function. Therefore, the analytic solution of Fourier transform for the signal can be obtained mathematically:

$$F=i\sum_{k=1}^{4}(A_k/2)[\delta(f+f_k)-\delta(f-f_k)] \qquad (A-5)$$

This result is consistent with the context of [45]. However, negative frequency has no physical meaning and, hence, only the second part of equation (A-5) is remained:

$$F=-i\sum_{k=1}^{4}(A_k/2)\delta(f-f_k) \qquad (A-6)$$

Ignoring phase angle, the amplitude of the signal should be:

$$P = A_k/2, \quad at \quad f=f_k, \quad k=1,2,3,4$$
$$= 0, \quad elsewhere \qquad (A-7)$$

The sound pressure level (SPL) is defined as:

$$SPL=10 Log_{10}(P/P_{ref})^2 \qquad (A-8)$$

where SPL is in dB and P_{ref} = 20x10^{-6} Pa. Considering this reference pressure, equation (A-8) becomes:

$$SPL = 100 + 20 Log_{10}(P/2) \qquad (A-9)$$

For the signal given in equation (A-1), the SPL can be calculated by:

$$SPL = 100 + 20 Log_{10}(A_k/4) \qquad (A-10)$$

at $f=f_k$, k=1,2,3,4. This is the analytic solution of SPL after Fourier transform for the signal (A-1) at the 4 peaks. The frequency, amplitude and SPL at the 4 peaks are listed in Table A1 according to the calculation by (A-7) and (A-10).

Table A1. Frequency, Amplitude and SPL at Peaks

Peak #	1	2	3	4
f (Hz)	10	20	40	80
P (Pa)	20	10	5	2.5
SPL (dB)	120	114	108	102

Making Discrete Fourier Transform (DFT) for the signal in equation (A-1) using XMGR with a Hanning window, we get its frequency spectrum, as shown in Figure A2.

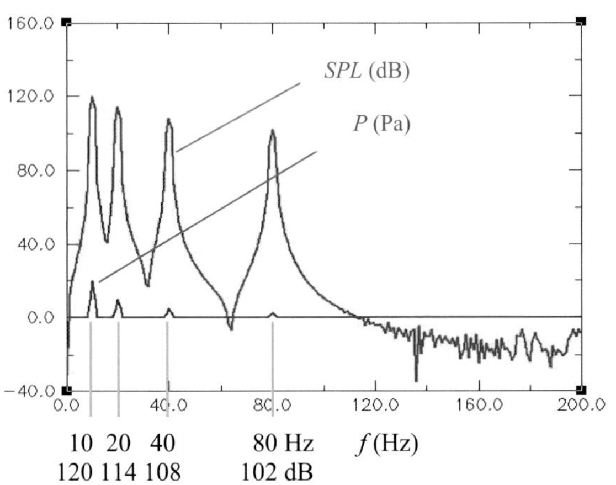

Figure A2. Spectrum of Simple Noise Signal

The values of P and SPL in the locations other than the 4 peaks are of less concern for buffeting. The values at the 4 peaks are exactly the same as the analytic solution (shown on the last two rows of Table A1). In practice, an accurate value of SPL, as well as frequency f, at the first peak in the spectrum curve is good enough for buffeting prediction. Therefore, this simple example gives confidence about the accuracy of XMGR to conduct Fourier transform for buffeting analysis.

ACKNOWLEDGMENTS

The authors are thankful to DaimlerChrysler Corporation for the financial support. They would also like to thank Ms. Dorothy Tekelly at DaimlerChrysler Information Center for her excellent assistance during the course of this study.

REFERENCES

[1] Bodger, W.K., and Jones, C.M., 1965, "Aerodynamic wind throb in passenger cars," SAE Transactions, **73**, pp. 195-206.

[2] Aspinall, D.T., 1966, "An empirical investigation of low frequency wind noise in motor cars," Rept., Motor Industry Research Association (MIRA) #1966/2, Warwickshire, UK.

[3] Hucho, W.-H., 1987, *Aerodynamics of road vehicles*, pp. 250-256, Butterworth, London, UK.

[4] Ota, D.K., Chakravarthy, S.R., Becker, T., and Sturzenegger, T., 1994, "Computational study of resonance suppression of open sunroofs," J. Fluids Eng., **116**, pp. 877-882.

[5] Ukita, T., China, H., and Kanie, K., 1997, "Analysis of vehicle wind throb using CFD and flow visualization," SAE paper 970407.

[6] Karbon, K. J., and Kumarasamy, S., 2001, "Computational aeroacoustics applications in automotive design," First MIT Conference on Computational Fluid and Solid Mechanics, MIT, MA, USA.

[7] Karbon, K. J., and Singh, R., 2002, "Simulation and design of automobile sunroof buffeting noise control", AIAA paper 2002-2550.

[8] Hendriana, D., Sovani, S.D. and Schiemann, M.K., 2003, "On Simulating Passenger Car Side Window Buffeting," SAE Paper 2003-01-1316.

[9] Noelting, S., 2003, "Audi A6 sunroof buffeting study," EXA Aeroacoustics Consortium Conference, Stuttgart, Germany.

[10] Tenstam, A., 2003, "Eliminating unpleasant sunroof noise," SAAB Automobile AB, Sweden, J. STAR-CD Dynamics, issue 20, p. 16.

[11] Muller, J., and Seydell, B., 2003, "Numerical investigation of sunroof buffeting," Rept. Volkswagen AG, Wolfsburg, Germany.

[12] An, C.-F., Alaie, S. M., Sovani, S.D., Scislowicz, M. S., and Singh, K., 2004, "Side window buffeting characteristics of an SUV," SAE Paper 2004-01-0230.

[13] Rayleigh, B., 1945, *The Theory of Sound*, Vol. II, pp. 170-172, Dover Publications, New York, USA.

[14] Pierce, A.D., 1981, *Acoustics*, pp. 330-336, McGraw-Hill Book Company, New York, USA.

[15] Kinsler, L.E., Frey, A.R., Coppens, A.B., and Sanders, J.V., 1982, *Fundamentals of Acoustics*, pp. 225-228, John Wiley & Sons, New York, USA

[16] Blevins, R.D., 1979, *Formulas for Natural Frequency and More Shape*, pp. 353-358, Van Nostrand Reinhold Company, New York, USA.

[17] Alster, M., 1972, "Improved calculation of resonant frequencies of Helmholtz resonators," J. Sound & Vibration, **24**(1), pp. 63-85.

[18] Roshko, A., 1955, "Some measurements of flow in a rectangular cutout," NACA TN 3488.

[19] Block, P.J.W., 1976, "Noise response of cavities of varying dimensions at subsonic speeds," NASA TN D-8351.

[20] Tam, C.K.W., 1976, "The acoustic modes of a two-dimensional rectangular cavity," J. Sound & Vibration, **49**(3), pp. 353-364.
[21] Rockwell, D. and Naudascher, E., 1978, "Review - self-sustaining oscillations of flow past cavities," J. Fluids Eng., **100**, pp. 152-165.
[22] Rockwell, D., and Knisely, C., 1979, "The organized nature of flow impingement upon a corner," J. Fluid Mech., **93**, Part 3, pp. 413-432.
[23] Blake, W.K., 1986, *Mechanics of Flow-Induced Sound and Vibration*, Vol. I, pp. 130-149, Academic Press, Inc., New York, USA.
[24] Blevins, R.D., 1990, *Flow-Induced Vibration*, 2nd ed., pp. 375-383, Van Nostrand Reinhold, New York, USA.
[25] Mongeau, L., Bezemek, J., and Danforth, R., 1997, "Pressure fluctuations in a flow-excited door gap cavity model," SAE Paper 971923.
[26] Henderson, B., 2000, "Automotive noise involving feedback – sound generation by low speed cavity flows," NASA/CP 2000-209790, pp. 95-100.
[27] Ashcroft, G.B., Takeda, K., and Zhang, X., 2000, "Computations of self-induced oscillatory flow in an automobile door cavity," NASA/CP 2000-209790, pp. 355-361.
[28] Shieh, C.M., and Morris, P.J., 2000, "A parallel numerical simulation of automobile noise involving feedback," NASA/CP 2000-209790, pp. 363-370.
[29] Henderson, B., 2003, "Sound generation by flow over a cavity," 4th Computational Aeroacoustics Workshop on Benchmark Problems, p. 26, Cleveland, OH, USA.
[30] Ashcroft, G.B., Takeda, K., and Zhang, X., 2003, "A numerical investigation of the noise radiated by a turbulent flow over a cavity," J. Sound & Vibration, **265**, pp. 43-60.
[31] Lamb, H., 1932, *Hydrodynamics*, 6th ed., pp. 373-375, Cambridge Univ. Press, London, UK.
[32] An, C.-F., Goodman, R.H., Brown, H.M., Clavelle, E.J., and Barron, R.M., 1997, "Oil-water interfacial phenomena behind a boom on flowing water," Proceedings, 10th Int'l Symp. on Transport Phenomena, K. Suzuki, ed., Kyoto, Japan, Vol. 1, pp. 13-18.
[33] Esch, R.E., 1957, "The instability of a shear layer between two parallel streams," J. Fluid Mech., **3**, pp. 289-303.
[34] Michalke, A., 1964, "On the inviscid instability of the hyperbolic-tangent velocity profile," J. Fluid Mech., **19**, pp. 543-556.
[35] Michalke, A., 1965, "Vortex formation in a free boundary layer according to stability theory," J. Fluid Mech., **22**, part 2, pp. 371-383.
[36] Michalke, A., 1965, "On spatially growing disturbances in an inviscid shear layer," J. Fluid Mech., **23**, part 3, pp. 521-544.
[37] Browand, F.K., 1966, "An experimental investigation of the instability of an incompressible, separated shear layer," J. Fluid Mech., **26**, part 2, pp. 281-307.
[38] Lin, C.C., 1955, *The Theory of Hydrodynamics Stability*, Cambridge Univ. Press, London, UK.
[39] Betchov, R., and Criminale, W.O. Jr., 1967, *Stability of Parallel flows*, Academic Press, New York, USA.
[40] Abernathy, F.H., and Kronauer, R.E., 1962, "The formation of vortex sheets," J. Fluid Mech., **13**, pp. 1-20.
[41] Boldman, D.R., Brinich, P.F., and Goldstein, M.E., 1976, "Vortex shedding from a blunt trailing edge with equal and unequal external mean velocities," J. Fluid Mech., **75**, part 4, pp. 721-735.
[42] Nakamura, Y., and Nakashima. M., 1986, "Vortex excitation of prisms with elongated rectangular, H and T cross-sections," J. Fluid Mech., **163**, pp. 149-169.
[43] Fluent 6.0 User's Guide, 2001, Fluent Inc., Lebanon, NH, USA.
[44] Papoulis, A., 1962, *The Fourier Integral and Its Applications*, pp. 269-282, McGraw-Hill Book Company, Inc., New York, USA.
[45] Brigham, E.O., *The Fast Fourier Transform*, pp. 19-21, Prentice-Hall, Inc., Englewood Cliff, NJ, USA.

NASA/CP—2004-212954

Fourth Computational Aeroacoustics (CAA) Workshop on Benchmark Problems

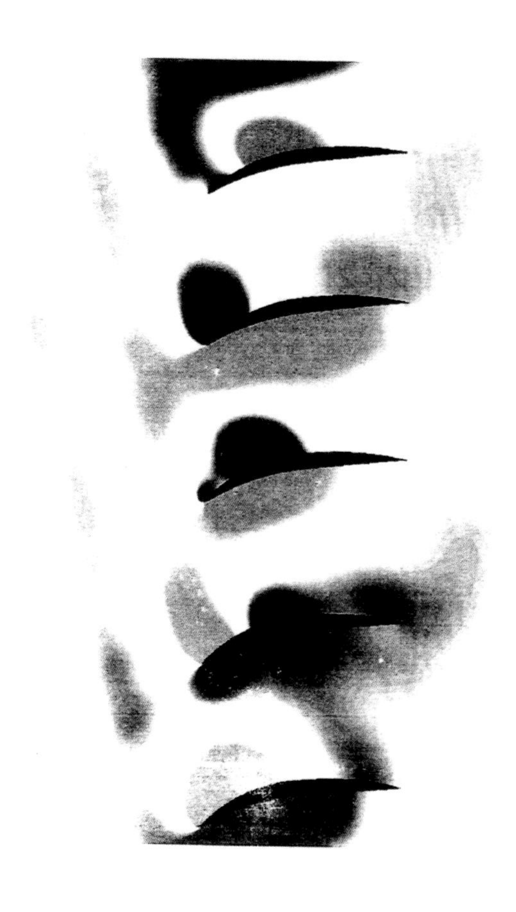

September 2004

SPECTRAL ANALYSIS FOR AIR FLOW OVER A CAVITY

Z. Zhang, R. Barron* and C.-F. An
Fluid Dynamics Research Institute
University of Windsor
Windsor, ON, Canada N9B 3P4

ABSTRACT

A procedure has been developed to perform spectral analysis for air flow over a cavity. The air flow is simulated numerically using the commercial CFD software Fluent and the pressure fluctuations are acoustically post-processing using FFT in the mathematical tool Matlab. A parametric study is undertaken that illustrates the importance of achieving a stable and accurate CFD solution in order to obtain reliable noise predictions. Simulation results are compared to experimental data presented at the Computational Aeroacoustics Workshop.

INTRODUCTION

Air flowing over a cavity may cause pressure fluctuation and noise radiation, which is of great concern in industries such as aeronautical and automotive (refs. 1 to 3). In fact, one source of aerodynamic noise of a road vehicle comes from the opening of a window or sunroof, which can be considered as flow over a cavity. The self-sustaining oscillations of flow over a cavity can be categorized into three groups (ref. 4): fluid-dynamic, fluid-resonant and fluid-elastic. Fluid-dynamic oscillations are excited by the amplification of an unstable cavity shear layer, and enhanced by a feedback mechanism. Fluid-resonant oscillations are strongly coupled with the resonant wave at the cavity. Fluid-elastic oscillations are coupled with the motion or vibration of a solid boundary.

Due to the technical importance, flow over a deep cavity with a lip at the cavity entrance was selected as one of the benchmark problems in the Third Computational Aeroacoustics (CAA) Workshop in 1999. In that workshop, Henderson (ref. 5) provided experimental data for a particular cavity. Some participants (refs. 6 to 9) presented their computational results and compared with the data. A similar problem was selected as a new benchmark in the Fourth Computational Aeroacoustics (CAA) Workshop in 2003. The present paper includes results that the authors presented at this workshop and additional results based on more recent simulations.

In this paper, we present the results of our computation for the prediction of peak frequencies and sound pressure levels in a cavity, and discuss the effects of parameters used in the process of spectral analysis.

PROBLEM DESCRIPTION AND SOLUTION METHODOLOGY

Problem description

The geometry of the cavity for the benchmark problem is shown in Fig.1. The turbulent flow over the cavity has a mean velocity of 50 m/s. In the experimental setup, the thickness of the boundary layer developed over the flat plate is 0.14m at the entrance of the cavity. The objective is to calculate the power spectra at the center of each cavity wall and the center of the cavity floor and to compare the computational results with the experimental data.

Solution methodology

The computational domain includes the cavity and a rectangle above it, as shown in Fig. 1. The rectangle is 0.3m high and 0.3m downstream from the trailing edge of the cavity. For the purpose of investigating the effect of boundary layer thickness, upstream lengths of 0.3m, 0.5m and 0.682m are considered. The lower edge of the rectangle coincides with the upper edge of the cavity. Since the geometry of the computational domain is rectangular and all boundaries are parallel to the coordinate axes, a non-uniform Cartesian grid system is utilized. Generally, a uniform mesh is used in the cavity neck area where the noise source is located. Then, the grid is smoothly enlarged from the cavity neck outwards towards all far boundaries of the rectangle and clustered near the walls within the cavity. In order to investigate the effect of mesh size, several grid number distributions are also chosen, as listed in Table 1. In this table, each column depicts a case of mesh size arrangement and the capital letters in the first column represent the segments of the boundaries in Fig. 1.

---

* Corresponding author: address as above, email: az3@uwindsor.ca; fax: (519) 971-3667; phone: (519) 253-3000,x2110

The flow velocity on the left side of the rectangle (inlet boundary) is uniform and equal to 50 m/s. The right side of the computational domain is set to be a pressure-outlet boundary (p=0). The top boundary of the rectangle is set to be either a frictionless wall (symmetry boundary) or a pressure-outlet boundary. The rest of the boundaries are no-slip solid walls. The rectangle and the cavity are connected through the cavity neck.

In the simulation of the cavity flow, unsteady compressible Reynolds Averaged Navier-Stokes equations are solved using the commercial CFD software, Fluent (ref. 10). To capture the high frequency fluctuating features of the cavity flow, a second order implicit scheme is used for time marching and a second order upwind scheme is used for the convection terms in the momentum, energy and turbulence equations. Ideal gas is selected as the air model to consider density change along with pressure fluctuations. Both RNG k-ε and one-equation turbulence models are considered in this study.

All computations are started from a uniform initial flow field of 50m/s. Relaxation factors for various parameters are adjusted as required to get convergence. The residual of the pressure equation is found to be the most critical residual for convergence. In our simulations the residual of the pressure equation is set at 10^{-5} to 10^{-4} depending on the particular case.

In most cases, the time step is set to be 1×10^{-5}sec. This time step provides sufficient accuracy and stability to capture high frequency signals (over 1 kHz). To study the effect of time step, other time steps from 1×10^{-6} to 1×10^{-3} sec were also employed in the computation. To reach the prescribed residual of the pressure equation, about 20 - 30 sub-iterations are carried out in each time step. This is especially important during the first thousand time steps.

The computation is performed step-by-step until a stably fluctuating pressure signal at a given location is obtained. During the computation, the pressure signals at three points in the cavity, the centers of the two sidewalls and the center of the cavity floor, are monitored.

The recorded signals are then acoustically post-processed using the mathematical tool Matlab that has an option of Fast Fourier Transform (FFT). Using Matlab, a special procedure has been designed and used for acoustic post-processing, as shown in the plots in Fig. 2. Figure 2(a) shows a typical noise signal, i.e. a time history of pressure fluctuation. After application of the FFT, the signal is converted from the format of time history to the format of power density expressed in terms of frequency, as shown in Fig. 2(b). Then, the amplitude of the signal can be calculated and expressed as a function of frequency, as shown in Fig. 2(c). Finally, the sound pressure level (SPL) is calculated from $SPL = 10Log_{10}(P/P_{ref})^2$, where P is the amplitude of the signal and $P_{ref} = 2x10^{-5}$ Pa, as shown in Fig. 2(d).

RESULTS AND DISCUSSION

Comparison with experimental data and other simulations

The measurements for the benchmark problem provided by Henderson in the 3rd CAA workshop are chosen for comparison, since the geometry and operating condition are quite close to the current study.

Figure 3 shows the comparison of SPL-frequency spectra between the computational results and the experimental data for both thin and thick boundary layers considered by Henderson (ref. 5). Here, three cases from our computations are included. This figure demonstrates that: (1) the spectral patterns of our results are quite similar to the experimental data. The frequencies of the first peaks for all cases are around 1.9kHz with the deviation of about 50Hz; (2) the SPL of the first peak is also very close to one another with the deviation of about 5-10dB; (3) the spectra obtained from our computations show a harmonic pattern. The frequency of the second peak is almost twice the frequency of the first peak. In turn, the frequency of the third peak is about three times the frequency of the first peak, etc. This demonstrates the resonant characteristics of the flow over a cavity. However, the experimental results seem to have only three harmonics, including two smeared high frequency peaks; (4) all peaks in the spectral curves from our computations are very sharp. This means that the computational results have low background noise, whereas the experimental spectrum appears to contain more background noise.

Figure 4 shows the peak frequency and SPL from our computational results for Grid 3, i.e., 30×30 cells in the neck area, using time steps 5×10^{-6} and 1×10^{-5} sec, two turbulence models and different boundary layer thicknesses at the entrance of the cavity. In addition, some computed peak frequency and SPL from other researchers (refs. 6 to 9) are included to compare with the experimental data. This figure indicates that our models can predict at least the first two peak frequencies and SPL quite well. The detailed differences between our simulations are discussed below.

Parameter study for flow over a cavity

Certain parameters have significant influence on the spectral characteristics of the flow over a cavity, especially time step and mesh size. Therefore, special attention has been paid to both time step and mesh size in this study.

Effect of time step

In general, a large time step ($\Delta t \geq 5\times10^{-5}$ sec) tends to activate and enhance the low frequency components and reduce or smear the high frequency components of the fluctuation. As a matter of fact, in the early stage of our study, where $\Delta t \geq 5\times10^{-5}$ sec was used, only the low frequency peak was captured, as shown in Table 2. Moreover, an increase in time step may cause further decrease in peak frequency. On the other hand, although a decrease in time step may give a better signal to involve high frequency components, very small time steps may cause difficulty with convergence and high computational cost. From our experience, a time step in the range $\Delta t = 5\times10^{-6}$ to 2×10^{-5}sec is an appropriate choice for this problem. Therefore, most of the computations in this study were performed at a time step $\Delta t = 1\times10^{-5}$sec.

Effect of mesh size

Using an upstream length $l_f = 300$mm and $\Delta t = 1\times10^{-5}$ sec, the computed values of peak frequency and SPL for various mesh size, represented by the grid number in the cavity neck, are listed in Table 3. Computations were also carried out for a fine mesh of 100×100 grids in the cavity neck area, but convergent results could not be achieved. It can be seen from this table that the high frequency first peak can be captured even for the coarsest mesh, 15×15, with a small time step $\Delta t = 1\times10^{-5}$sec. However, the predicted frequency of the first peak on this coarse mesh may not be accurate enough. Discarding the results of the 15×15 mesh, we can see from the table that refining mesh size leads to a lower first peak frequency, but seems to have very little effect on the SPL.

Effect of boundary layer thickness

In the computations, the upstream length l_f, from inlet to the leading edge of the cavity neck, is adjusted to establish boundary layer thickness δ_f at the leading edge of the cavity neck. This thickness is measured in the plot of velocity magnitude contour. If the number of the contour curves from the bottom to the top of the rectangle is set to be 100, the location of the 99th contour curve at the leading edge of the cavity gives boundary layer thickness. Therefore, δ_f can be measured from this plot. In Table 4, three upstream lengths $l_f = 300$, 500 and 682mm are used, and two conditions at the top boundary, symmetry (frictionless wall) and pressure-outlet, are applied to each of the three upstream lengths to give six different boundary layer thickness, as listed in the table. These thicknesses represent their actual values in the flow. However, the boundary layer thickness for a flow over a flat plate should be thicker than that for the flow over a cavity. For example, the boundary layer thickness for a plate at 682mm from the leading edge is about 14mm. This value is higher than the 12.5mm indicated on the bottom line of Table 4.

In order to reduce computational time for this study, and based on the results presented in Table 3, a coarse mesh of 20×20 in the cavity neck is used, the time step is $\Delta t = 1\times10^{-5}$sec and the RNG k-ε turbulence model is invoked. From Table 4 we can see that as the boundary layer thickness at the entrance of the cavity neck increases, the frequency at the first peak usually decreases, but the SPL of the first peak remains almost constant. It appears that both the symmetry and pressure-outlet boundary conditions are adequate when the boundary layer is thin. However, as the boundary layer thickens, the symmetry condition becomes less reliable. Convergence could not be obtained for the 12mm thick boundary layer with the symmetry condition on the top boundary.

Comparison of spectra at different locations in the cavity

As shown in Fig. 5, the spectra at different locations (center of the floor, center of the front wall and center of the rear wall) are quite close to one another. The peak frequencies are the same for all the locations. However, the SPL at the cavity floor is higher than that on the two sides of the cavity. This indicates that the air near the cavity floor has the strongest pressure fluctuations.

CONCLUSION

In this paper, a procedure has been established and implemented to perform spectral analysis for air flow over a cavity. The air flow is simulated numerically by solving transient flow using the commercial CFD software Fluent and the pressure fluctuations are acoustically post-processing using FFT in the mathematical tool Matlab.

The computational results for the spectra of pressure fluctuations are obtained at three locations in the cavity, centers of the cavity walls and center of the cavity floor. The numerical predictions compare favorably with the corresponding experimental data provided by Henderson at the 3rd CAA Workshop.

The computed frequency at the first peaks in the spectral curves, for various combinations of computational parameters, are close to one another and to the experimental data within the range of about 50Hz. The computed SPL of the first peaks are also close to the experimental data with a deviation of 5-10dB.

A parametric study demonstrates that time step is the most crucial factor that has a strong influence on the computational convergence and solution accuracy, especially on the capture of the high frequency components of pressure fluctuations. It also shows that mesh size is another important parameter in the spectral analysis. Boundary layer thickness has some effect on the frequency at the first peak but not much effect on the SPL in the spectrum of pressure fluctuations. Two tested turbulence models show little difference in this study.

It is readily apparent that the reliability of the noise predictions is heavily dependent on the accuracy of the underlying CFD simulations. The procedure still needs to be fine-tuned in order to improve its convergence behavior, solution accuracy and computational efficiency.

Acknowledgement: The authors would like to thank Prof. C. Tam for his invitation to participate in the workshop and his encouragement to submit this paper. Also, they are grateful to Dr. J. X. Lou for his assistance in the implementation of Matlab.

REFERENCES

1. Block, P.J.W.: Noise response of cavities of varying dimensions at subsonic speed, NASA TN D-8351, 1976.
2. George, A.G.: Automobile aeroacoustics, AIAA-89-1067, 1989.
3. Mongean, L.; Bezenek, J.; and Danforth, R.: Pressure fluctuations in a flow-excited door gap cavity model, SAE Paper 971923, 1997.
4. Rockwell, D.; and Naudascher, E.: Review - Self-sustaining oscillations of flow past cavities, J. Fluids Eng., Vol. 100, 1978, 152-165.
5. Henderson, B.: Automobile noise involving feedback-sound generation by low speed cavity flows, NASA/CP-2000-209790, 2000, 95-100.
6. Moon, Y.J., et al.: Aeroacoustic computations of the unsteady flows over a rectangular cavity with a lip, NASA/CP-2000-209790, 2000, 347-353.
7. Aschcroft, G.B.; Takeda, K.; and Zhang, X.: Computations of self-induced oscillatory flow in an automobile door cavity, NASA/CP-2000-209790, 2000, 355-361.
8. Shieh, C.M.; and Morris, P.J.: A parallel numerical simulation of automobile noise involving feedback, NASA/CP-2000-209790, 2000, 363-370.
9. Kurbatskii, K.K.; and Tam, C.K.W.: Direct numerical simulation of automobile cavity tones, NASA/CP-2000-209790, 2000, 371-383.
10. Fluent User's Manual, version 6.0, Fluent Inc., Lebanon, NH, USA, 2001.

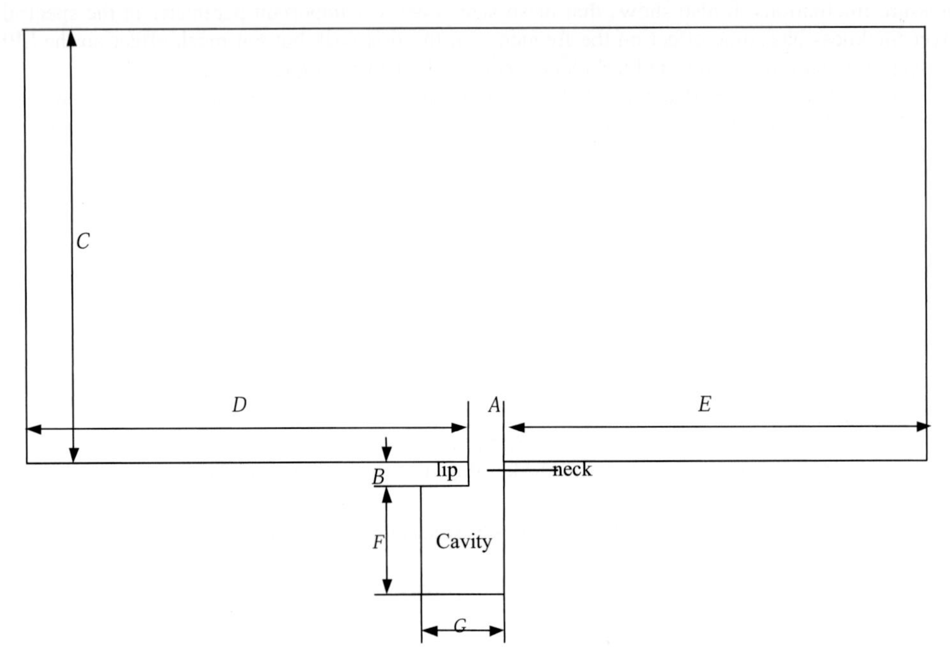

Figure 1. Cavity geometry and computational domain

Table 1. Grid number in various cases

	Grid 1	Grid 2	Grid 3	Grid 4
A	15	20	30	60
B	15	20	30	60
C	40	60	90	150
D	30	40	60	120
E	30	40	60	120
F	30	40	60	120
G	30	40	60	120

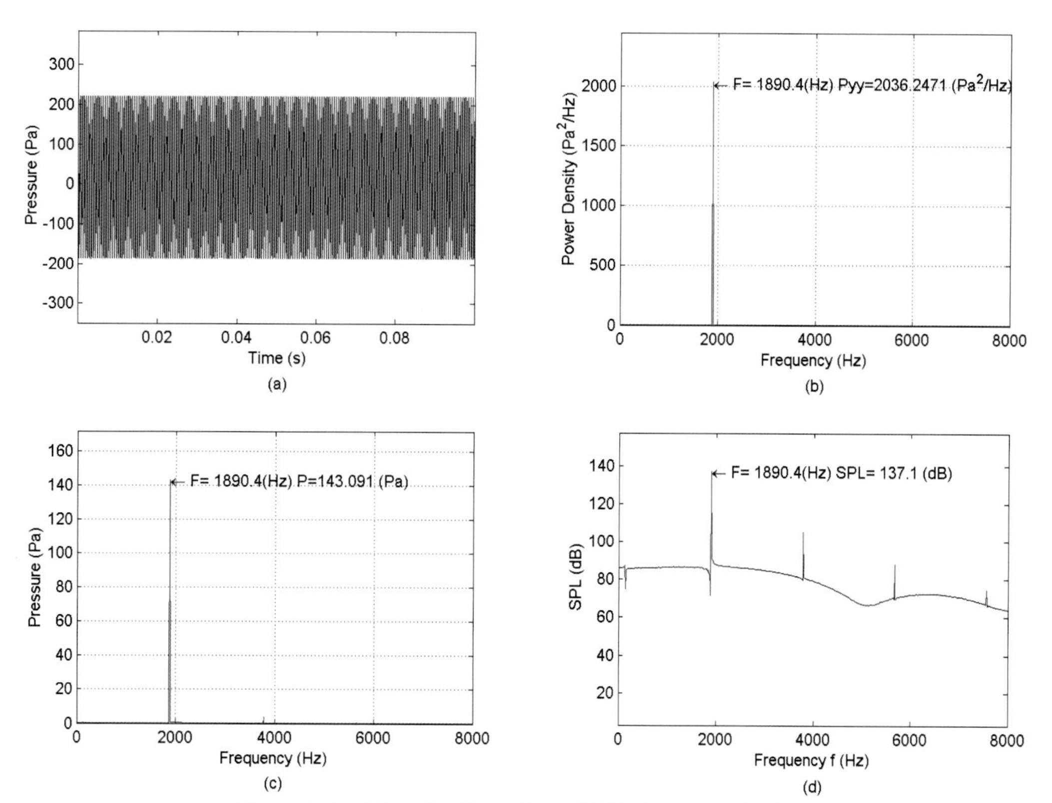

Figure 2. Fast Fourier Transform (FFT) for a sample signal

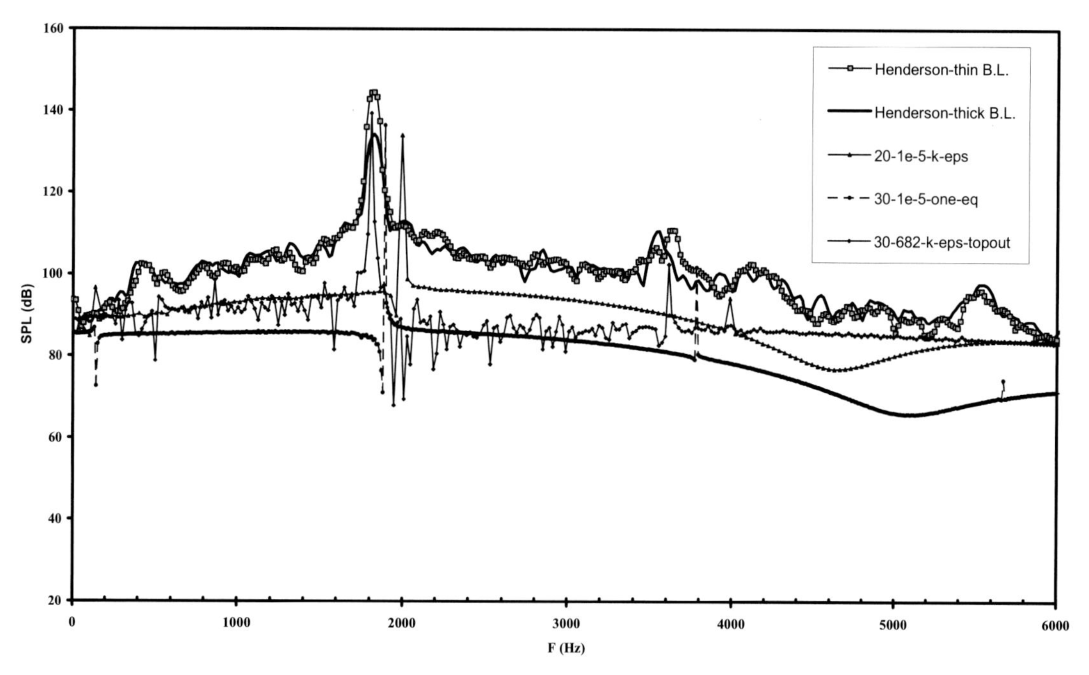

Figure 3. Comparison of spectra between computational and experimental data provided by Henderson in 3rd CAA Workshop (ref. 5)

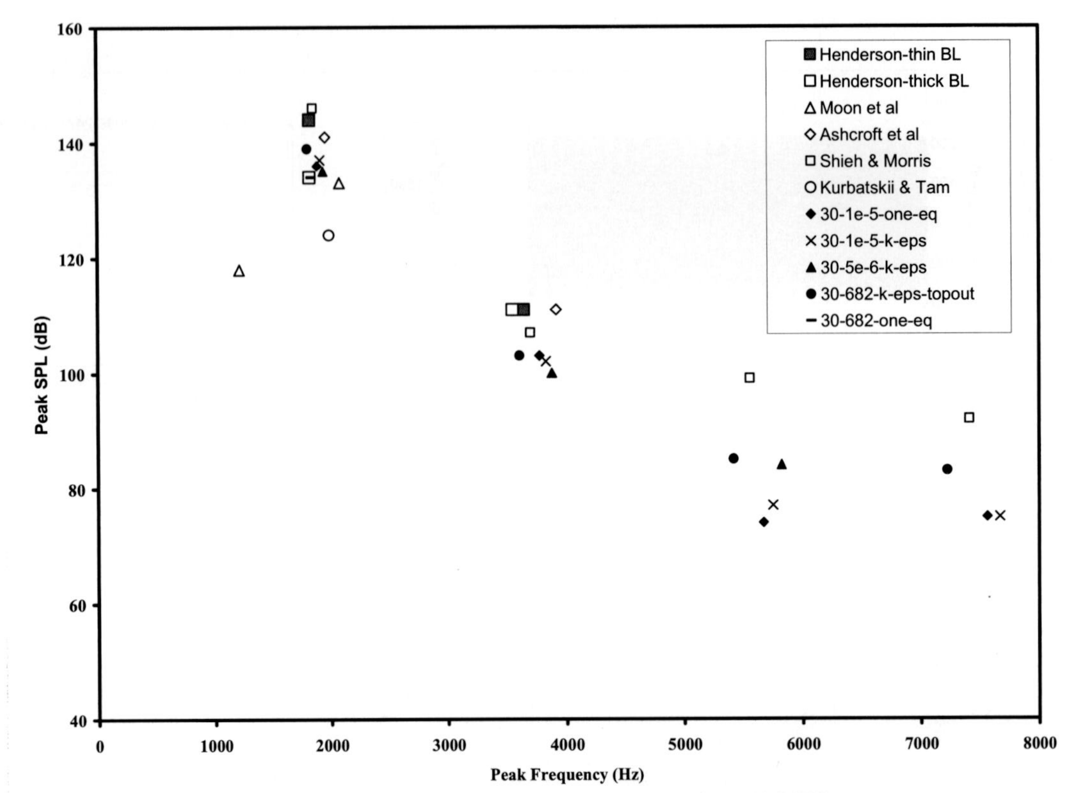

Figure 4. Comparison of peak frequency and SPL

Table 2. Peak frequency and SPL for various mesh size and time step

Mesh	Δt (sec)	F (Hz)	SPL (dB)
10×10	5×10^{-5}	125.2	110.1
	1×10^{-4}	104.7	105.5
	2×10^{-4}	83.6	79.3
15×15	2×10^{-5}	129.1	109.2
	5×10^{-5}	120.0	100.7
	2×10^{-4}	88.9	93.4
	5×10^{-4}	62.6	94.1
	1×10^{-3}	45.3	90.3
20×20	2×10^{-5}	125.0	110.0
	5×10^{-5}	112.4	102.1
	1×10^{-4}	98.5	95.4
30×30	5×10^{-5}	84.7	119.6
	1×10^{-4}	67.7	121.3

Table 3. Effect of mesh size on first peak frequency and SPL (l_f= 300mm, Δt=1×10^{-5} sec)

	F (k-ε)	F (one-eq)	SP L (k-ε)	SPL (one-eq)
15×15	1963.5	/	127.9	/
20×20	1993.0	1974.8	135.0	137.0
30×30	1917.7	1890.4	138.2	137.1
60×60	1897.4	1915.3	131.6	136.0

Table 4. Effect of boundary layer thickness (20×20 mesh, Δt=1×10^{-5} sec)

l_f (mm)	Top B.C.	δ_f *(mm)*	F_{peak} (Hz)	SPL_{peak} (dB)
300	Symmetry	7.0	1974.8	137.0
300	Pressure-outl	7.5	2004.0	126.0
500	Symmetry	10.0	1938.8	124.7
500	Pressure-outl	10.5	1935.7	126.1
682	Symmetry	12.0	Not available	
682	Pressure-outl	12.5	1882.6	126.3

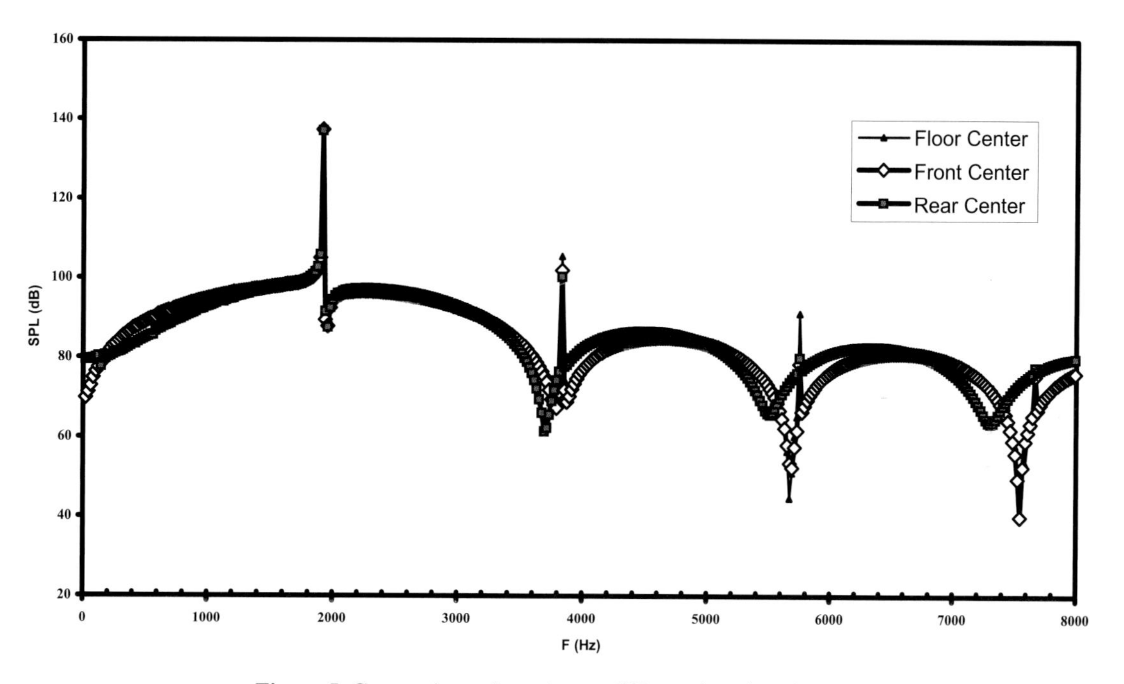

Figure 5. Comparison of spectra at different locations in the cavity

第IV部分　任职于美国克莱斯勒汽车公司时期

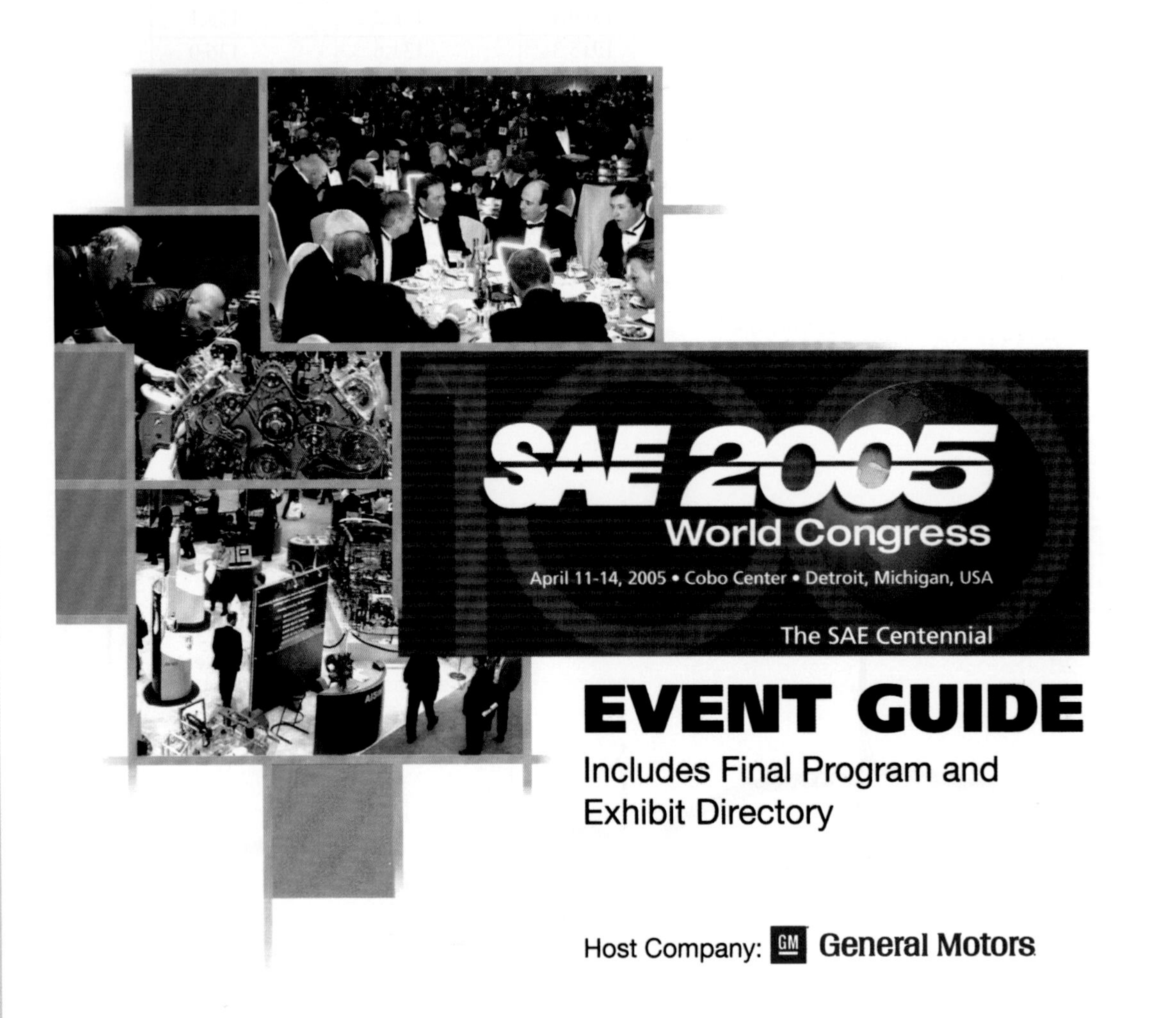

EVENT GUIDE

Includes Final Program and Exhibit Directory

Host Company: GM General Motors

Published by:
automotive engineering international
The Official Publication of SAE 2005 World Congress

Detroit, MI, USA
April 11, 2005

050298

Vehicle Aerodynamics 2005

SP-1931

Changfa An
Detroit, MI, USA
April 11, 2005

第IV部分　任职于美国克莱斯勒汽车公司时期

SAE International™

Published by:
Society of Automotive Engineers, Inc.
400 Commonwealth Drive
Warrendale, PA 15096-0001
USA
Phone: (724) 776-4841
Fax: (724) 776-5760
April 2005

2005-01-0603

Attempts for Reduction of Rear Window Buffeting Using CFD

Chang-Fa An
Exa Corporation, Auburn Hills, MI
Mitchell Puskarz
DaimlerChrysler Corporation, Auburn Hills, MI
Kanwerdip Singh
DaimlerChrysler Corporation, Auburn Hills, MI
Mark E. Gleason
DaimlerChrysler Corporation, Auburn Hills, MI

ABSTRACT

This paper summarizes the major activities of CFD study on rear window buffeting of production vehicles during the past two years at DaimlerChrysler. The focus of the paper is the attempt to find suitable solutions for buffeting suppression using a developed procedure of CFD simulation with commercial software plus FFT acoustic post-processing. The analysis procedure has been validated using three representative production vehicles and good correlation with wind tunnel tests has been attained which has gained the confidence in solving the buffeting problem. Several attempts have been proposed and tried to find solution for buffeting reduction. Some of them are promising, but feasibility and manufacturability still need discussion. In order to find suitable solution for buffeting reduction, more basic research is necessary, more ideas should be collected, and more joint efforts of CFD and testing are imperative.

INTRODUCTION

In the current automobile market buffeting noise is one of the customer complaints on luxury cars and SUVs. Buffeting is a low frequency, high amplitude wind noise, which may occur on a moving vehicle when a side window or sunroof is open. This noise makes people inside the passenger compartment uncomfortable if it lasts for a long period of time. The physics of buffeting is a complex aeroacoustic phenomenon. At the upstream portion of the opening, there exists a shear layer between the fast moving air outside the opening and the relatively stationary air inside the passenger compartment. The shear layer is unstable in nature which causes vortex shedding periodically. If the frequency of vortex shedding happens to coincide with the natural frequency of the passenger compartment, the so-called Helmholtz resonance will be excited and buffeting will take place. The frequency and the sound pressure level (SPL) of the buffeting depend on a number of parameters such as vehicle speed, yaw angle, shape and size of the opening, volume of passenger compartment and its aeroacoustic characteristics.

CFD simulations of automotive sunroof buffeting were started in the last century using simplified 2D or 3D car models [1-2]. Since the new millennium, more realistic vehicle models have been used for CFD simulations of sunroof and side window buffeting in car companies [3-6]. Also, CFD packages have been used in benchmark studies in collaboration with the automotive industry, such as PowerFLOW by Exa [7-8], STAR-CD by Adapco [9-10] and PAM-FLOW by ESI [11].

From a literature review, the methods of buffeting control are usually categorized into two types: passive and active. The passive methods are more popular at the present time. A deflector installed in the upstream of the sunroof opening can "deflect" the vortex away from the opening and reduce buffeting. The method of "glass comfort position" is based on the buffeting mechanism that the natural frequency of passenger compartment can be changed by adjusting the size of the opening so that the excitation of resonance can be avoided. The method of venting is an approach based on the observation that buffeting can be "vented" by a second opening. The active control methods include mass injection at the upstream of the opening, an oscillating device with opposite phase of the buffeting noise, a loud speaker with an electronic control system and so on [12]. In practice, the buffeting control methods can work only under certain circumstances and many of them are still not well understood and not effectively utilized.

The present paper is the continuation of the previous work for side window buffeting using CFD technology [5]. Firstly in this paper, the analysis procedure is briefly outlined and updated. Secondly, three validation cases are reported on rear window buffeting for SUV1, SUV2 and a minivan. The emphasis is to try to find solutions for buffeting reduction. Several attempts have been made, but only those that are more or less successful are included in the paper. The feasibility and manufacturability are also briefly mentioned whenever they are needed.

ANALYSIS PROCEDURE

The procedure of buffeting analysis is described in detail in the previous paper [5] and briefly outlined here for convenience. The flow chart is shown in Figure 1. The main steps are mesh generation, problem set-up, solving and acoustic post-processing.

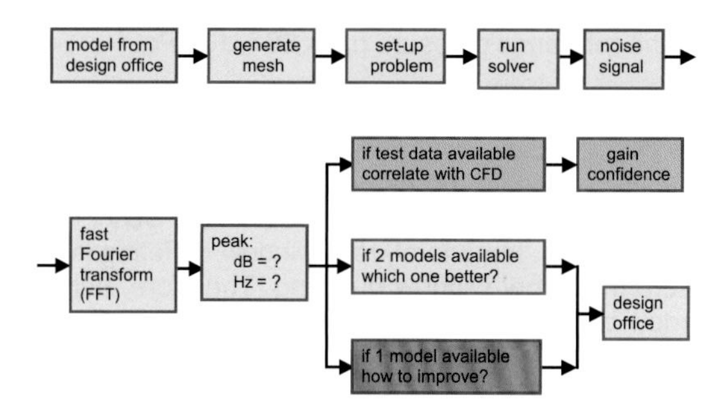

Figure 1. Flow chart of CFD analysis for buffeting

MESH GENERATION

The vehicle model is provided from the design office in CATIA format. Then, ANSA is used to clean-up geometry model and generate surface mesh. The mesh in flow sensitive areas, such as window opening, side view mirror and A-pillar, is finer (5-8mm) while it is coarser on the rest of the vehicle surface (20-30mm). Typically the total number of triangles on the vehicle surface (outside and inside) is about 300,000 - 400,000.

The meshed vehicle model is placed into a virtual wind tunnel of 50x20x10m^3. The surface mesh of the virtual wind tunnel is coarser far from the vehicle and finer near the vehicle. The entire flow domain contains the external region outside the vehicle and the passenger compartment that is connected to the external region through the window. Within the passenger compartment, ceiling, carpet, seats, dummies, instrument panel are included. All climate control duct openings are closed and the passenger compartment is assumed to be perfectly sealed. In addition, all underhood and underbody components are omitted because their influence on buffeting is insignificant.

The surface mesh of the vehicle and virtual wind tunnel is transferred to TGRID for volume mesh generation and refinement. Four levels of volume mesh refinement are utilized to ensure that the mesh is sufficiently fine in the regions of predominant flow unsteadiness. The final volume mesh typically contains 3.2 - 3.5 million tetrahedral cells.

It should be mentioned at this point that due to the limitation of computer resources and the tight timeline for most projects, the meshing strategy is to optimize the final mesh in order to achieve an acceptable level of accuracy. From the previous mesh study [5] and practice on a couple of vehicles, the above mesh size is a good compromise and is used for each new project.

PROBLEM SET-UP

The commercial CFD software Fluent 6.1 [13] is used to simulate unsteady flow around the vehicle and the problem set-up is listed in Table 1.

Table 1. Problem Set-Up

Function	Setting
Solver	Segregated Unsteady
Time marching	2nd Order Implicit
Pressure discretization	2nd Order
Momentum discretization	2nd Order Upwind
Pressure-velocity coupling	SIMPLE
Energy discretization	2nd Order Upwind
Turbulence modeling	LES
Fluid model	Air (Ideal Gas)

It should be pointed out that the compressibility of the air is important in buffeting study. Theoretically, a sound wave is the longitudinal propagation of a small disturbance in the compressible medium (air). If the medium were incompressible, the speed of sound would be infinite. That is, whenever a disturbance starts at a source point, it would be heard immediately at infinity. This is not physically true. Only if the compressibility of the medium is considered, the speed of sound stays finite. In this case, the sound can be heard at distance shortly after the disturbance starts. As shown in [5], the incompressible model leads to a non-acceptable result. The peak SPL is over-predicted by about 10dB and peak frequency is over-predicted by a factor of 2, compared to the wind tunnel measurements. Therefore, the ideal gas, a simple compressible model, is used in all cases.

Boundary conditions for the case of zero yaw angles are described as below. The inlet of the virtual wind tunnel has constant velocity whose magnitude is equal to the vehicle speed. The outlet of virtual wind tunnel is set at a constant (atmospheric) pressure. The top, two sides and tunnel ground are frictionless walls. The vehicle surface is frictional wall. In contrast to conventional external flow simulation, the effect of ground friction is small on buffeting characteristics and, therefore, is neglected.

For the case of non-zero yaw angle, boundary conditions at left and right sides of the tunnel are different. If a left window is open, for a positive yaw angle the left side of the tunnel is specified as a velocity inlet, just like the tunnel inlet. The right side of the tunnel is specified as a pressure outlet just like the tunnel outlet. Likewise, for a negative yaw angle the right side of the tunnel is specified as a velocity inlet and the left side of the tunnel is specified as a pressure outlet. If a right window is open, the settings are opposite to the case of left window open.

SOLUTION

At the beginning of each case, a steady state mode is run for about 100-200 iterations using the standard k-ε turbulence model. The steady state flow is used as an initial condition to start unsteady simulation and the turbulence model is switched to LES. A time step of 0.002 seconds is chosen for unsteady simulation. For buffeting noise, the representative frequency is about 20Hz and time period is about 0.05 seconds. Hence, there are about 25 sampling points within a time period. This is enough to capture the feature of the buffeting wave form. Higher frequency components may not be caught by this setting, but it is alright as they are not the objective in this study. Within each time step, 15 sub-iterations are set and this is enough for most cases. The residual convergence criteria are: continuity 1e-4, momentum and turbulence 1e-3 and energy 1e-7. Relaxation parameters are set based on experience. Typical running time is 3-4 days with 16 processors to reach 1200 time steps. This is based on an SGI Origin 2000 machine with 64 300-MHz CPU and 64 GB memory using the IRIX 6.5 operating system.

ACOUSTIC POST-PROCESSING

The procedure of acoustic post-processing (fast Fourier transform - FFT) is shown in Figure 2.

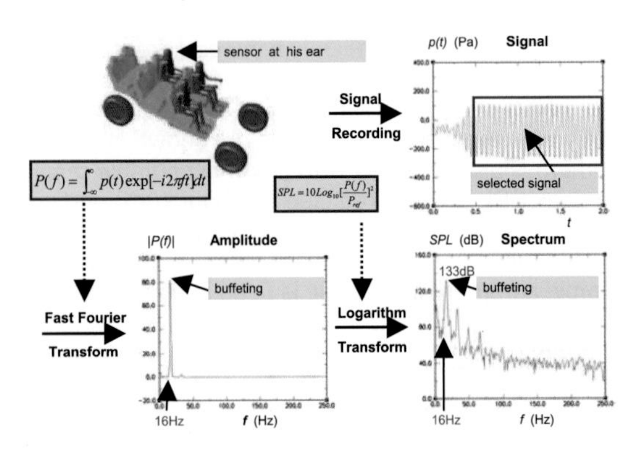

Figure 2. Procedure of acoustic post-processing

In the upper left part of the figure, sensors (microphones) are installed at dummy ears. Static pressure will be recorded at the sensors in each time step. After an irregular initial progress in the first 200-300 time steps, the pressure signal reaches a status of dynamically stable periodic fluctuation. The pressure fluctuation at each sensor is recorded for acoustic post-processing. The recorded signal is treated by cutting off the irregular part of initial process (see upper right part) and the selected signal is used for Fourier transform. The Fourier transform with a Hanning window is available in XMGR and it can transform the selected signal from the format of time history to the format of spectrum. After Fourier transform, the amplitude of pressure fluctuation is expressed as a function of frequency (see lower left part). Finally, the amplitude of pressure fluctuation is converted to the format of sound pressure level (SPL) in dB (see lower right part) using the conventional formula:

$$SPL(dB) = 10\log_{10}\left(\frac{p}{p_{ref}}\right)^2$$

where p is the amplitude of pressure fluctuation in Pa and the reference pressure $p_{ref} = 20 \times 10^{-6}$ Pa.

VALIDATION CASES

In order to validate the procedure of CFD simulation for buffeting, three vehicle models are chosen: SUV1, SUV2 and a minivan. In general, rear window buffeting is more severe than front window buffeting. Therefore, only rear window buffeting at zero yaw angle is considered in this paper.

MODEL 1 – SUV1

The first validation case is the right rear window buffeting of SUV1 at 60mph whose result was published in [5]. For the purpose of procedure validation it is cited here.

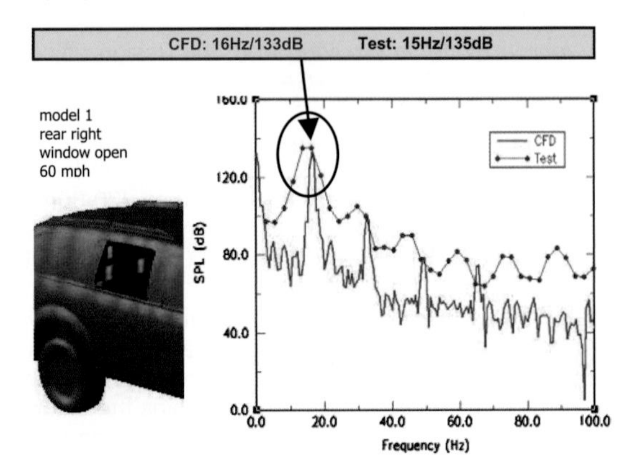

Figure 3. Validation case for model 1

It can be seen from Figure 3 that CFD simulation predicts the first peak SPL of 133dB at 16Hz. The wind tunnel test for the same vehicle measures the first peak SPL of 135dB at 15 Hz. The agreement between CFD and wind tunnel test is pretty good. Moreover, higher harmonic peaks in buffeting spectral curve from CFD simulation are clearly visible from the figure.

MODEL 2 – SUV2

The second validation case is the left rear window buffeting of SUV2 at speeds 60 - 90mph with increments 10mph. As shown in Figure 4, the correlation between CFD and wind tunnel test is very good. The difference in peak buffeting SPL is about 2dB and the difference in peak buffeting frequency is about 2Hz. Figures 5 and 6 show the buffeting spectra at 60-90mph from wind tunnel measurements and CFD simulation, respectively. The trend of changes in first peak SPL, as well as frequency, is the same. The trend of the second harmonic peaks is also the same.

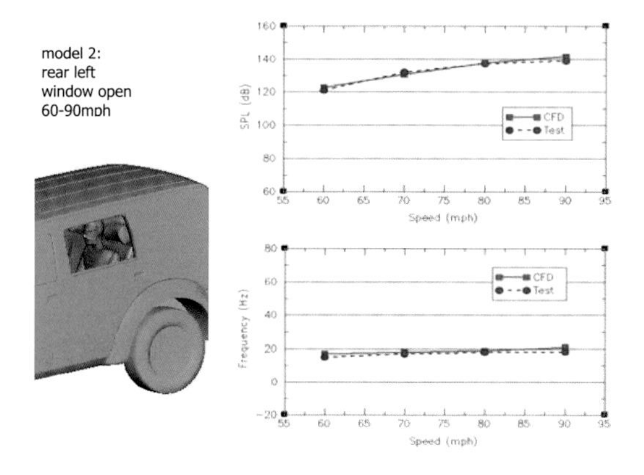

Figure 4. Validation case for model 2

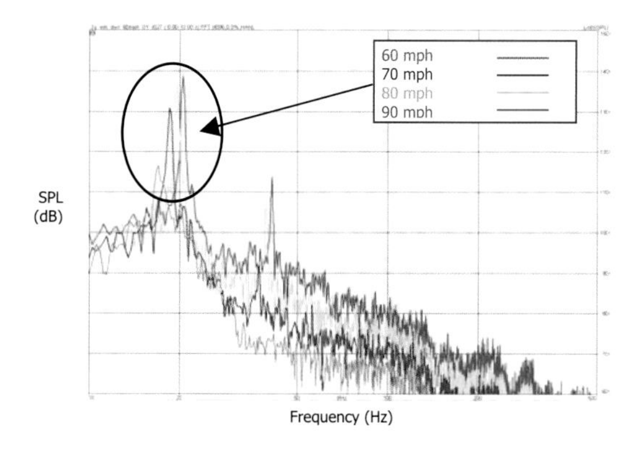

Figure 5. Buffeting spectrum for model 2 - TEST

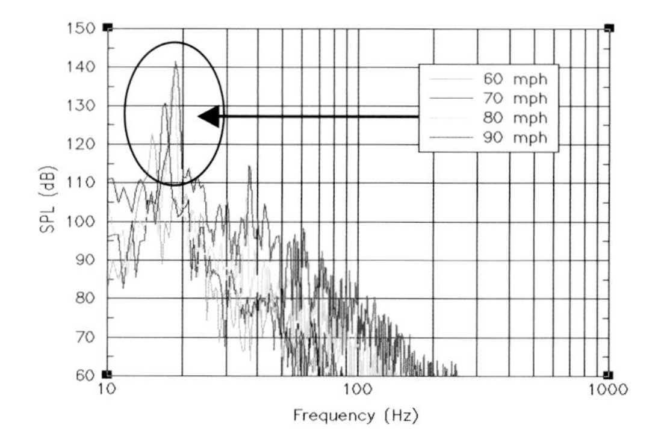

Figure 6. Buffeting spectrum for model 2 - CFD

MODEL 3 - MINIVAN

The third validation case is the left rear window buffeting of a minivan at 60mph. Figure 7 shows the spectra of buffeting signals at 4 locations within passenger compartment: left and right ears of driver and passenger at the seat of second row. Although there exist differences in noise SPL at higher frequencies, the first peaks of SPL curves are quite close, i.e. 125-126dB at 15-16Hz. This phenomenon was reported in [4] and [5] without reasoning details. Here, a brief explanation is given. The dominant frequency of buffeting is roughly 20Hz and the time period of buffeting is about 0.05sec. Considering the sound speed of about 340m/s, the wave length of buffeting should be 340m/s times 0.05sec = 17m. This is about four times longer than the dimension of a typical vehicle, 4m. So, the difference in buffeting characteristics within passenger compartment is negligibly small. Therefore, it is not necessary to distinguish sensor location within passenger compartment. To be consistent, only the signal at the left ear of the second row passenger is taken. In the wind tunnel test of this model, the corresponding measurement of buffeting spectrum gives the dominant peak SPL of 127dB at 13Hz. The correlation is also good.

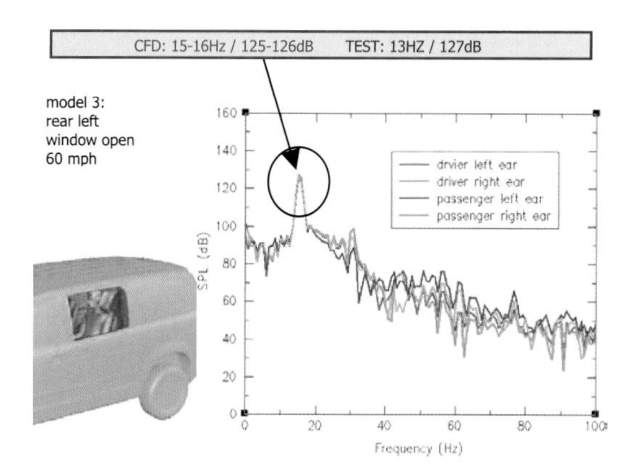

Figure 7. Validation case for model 3

ATTEMPTS FOR BUFFETING REDUCTION

With good correlation of these cases, more confidence was gained and more attention was paid to "solutions" of buffeting reduction. At the beginning of 2004, Gleason [14] initiated a discussion on passenger compartment buffeting and distributed a categorized list of possible solutions. The information sources for the list are from theory of unsteady flow, physical mechanisms of buffeting, principle of energy conservation and energy loss in fluid flow. The information sources are also from wind tunnel experiments, from on-road experience and from a literature survey. It is impossible to try all the possible solutions. Instead, only those that are most likely to give a favorable trend are selected for examination using the vehicle models which have been validated. Only those trials that are more or less successful are included in this paper.

As stated before, the buffeting mechanism is the excitation of passenger compartment, as a Helmholtz resonator, by the fast sweeping flow outside the open window in a certain resonance range of vehicle speed. If, in some way, the resonance condition is destroyed or, in other words, the natural frequency of passenger compartment is shifted to be outside of the resonance range of vehicle speed, the excitation may be avoided. As a result, buffeting may be reduced or even completely eliminated. Passenger compartment venting is one of the approaches to change the resonance condition. The following three attempts are based on this justification.

ATTEMPT 1: QUARTER GLASS VENTING

As shown in the left part of Figure 8, while rear window is open, the quarter glass is swung outward for some angle (5°). In this circumstance, the natural frequency of passenger compartment will be shifted and the venting effect will take place.

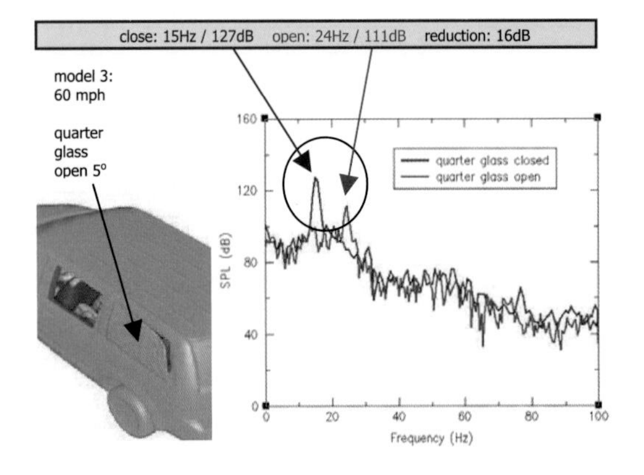

Figure 8. Attempt 1: Quarter glass venting

Figure 8 (right part) shows the buffeting spectral curves for the above-validated model 3 with quarter glass closed (blue) and open (red). The peak frequency is shifted from 15Hz, if quarter glass is closed, to 24Hz, if quarter glass is open. The first peak SPL is reduced from 127dB, if quarter glass is closed, to 111dB, if quarter glass is open. The venting effect of open quarter glass leads to a buffeting reduction by about 16dB.

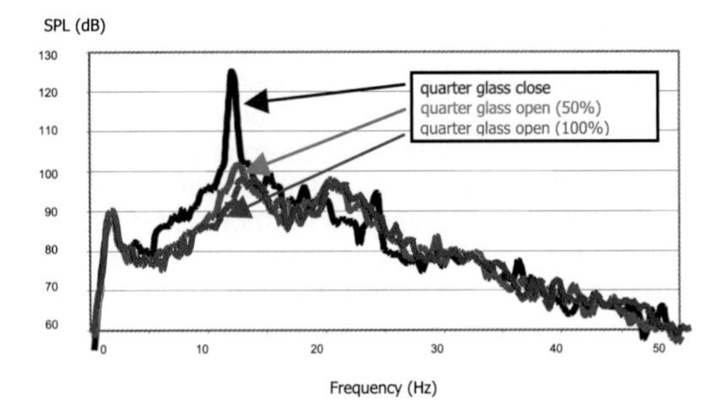

Figure 8a. Experiment for quarter glass venting

In the same time frame, a wind tunnel test confirmed this prediction and claimed a buffeting reduction by more than 20dB for the same vehicle model (see Figure 8a). As additional evidence of venting effect, many subjective observations have been made by people who are driving on road and swinging the quarter glass outward.

In order to apply this idea to the real design, a motorized mechanism may be manufactured and installed at C-pillar to adjust the angle and gap of the quarter glass with a limited cost addition. If automatic deployment is desired additional complexity is incurred.

ATTEMPT 2: DIVIDING POST AT OPEN WINDOW

During the discussion of buffeting solution list, the idea of dividing post was proposed. A CFD simulation was done for models 3. A dividing post is added in the rear left window at 65% length from B-pillar and it is of 75mm wide (see left part of Figure 9).

The mechanism of buffeting reduction in this case is also "venting effect", similar to attempt 1. The front part of the window serves as the main opening in buffeting phenomenon, while the rear part of the window plays a venting role, just like the quarter glass gap in attempt 1. The mechanism of buffeting reduction in this case may be explained using the principle of energy dissipation. This needs some more work in the future.

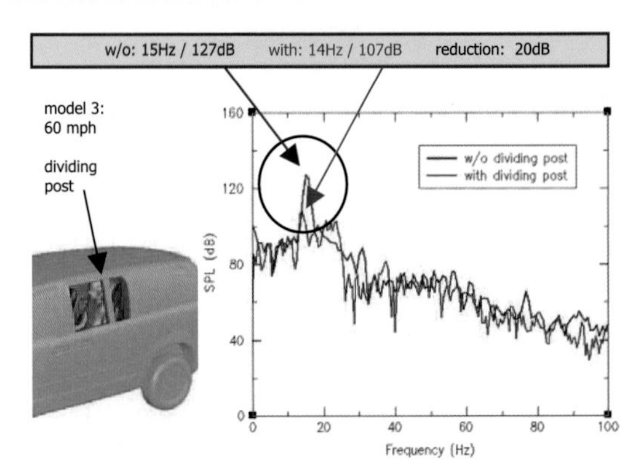

Figure 9. Attempt 2: dividing post at window

As shown in the right part of Figure 9, the peak buffeting SPL is reduced from 127dB to 107dB by adding a dividing post, i.e. 20dB reduction. Further optimizing the structure may lead to more buffeting reduction.

As a matter of fact, carefully checking the spectral curve (red) of the dividing post case, one may find out that there is no obvious peak at all! Probably one can conclude that the dividing post structure could eliminate buffeting completely!

Almost at the same time, a wind tunnel test confirmed the prediction and claimed a buffeting reduction by more than 20dB. If the two pieces of glasses can be moved up and down with synchronization and customers do not mind a post at the open window, it could be another ideal solution.

ATTEMPT 3: C-PILLAR VENTING HOLE

The favorable results of attempts 1 and 2 stimulated a new idea in this direction. A narrow vertical hole at the C-pillar can be considered as an alternative venting element (see left part of Figure 10).

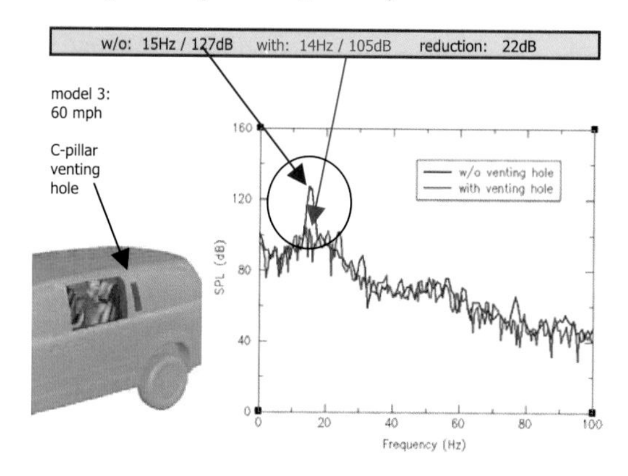

Figure 10. Attempt 3: C-pillar hole venting

The spectral curves show that the peak buffeting SPL can be reduced by as much as 22dB due to the addition of a venting hole at C-pillar. Similar to attempt 2, the spectral curve (red) for the case with a venting hole does not have an obvious peak at all! So, one can conclude that the C-pillar venting hole structure may also eliminate buffeting completely!

Practically, if a movable lip can be tactfully designed and installed at the C-pillar such that it can move up and down with window glass with synchronization and customers do not mind the styling; it could be another good choice for buffeting suppression.

ATTEMPT 4: B-PILLAR MASS INJECTION

As mentioned before, buffeting control methods can be classified into two categories: passive and active. All the above three attempts belong to passive type. Hardin [12] summarized active control methods in computational aeroacoustics. The next attempt is one of the active control methods, mass injection at B-pillar. In this attempt, a very narrow vertical gap, parallel to the leading edge of the window, is constructed and connected to a pipe within B-pillar (see close-up of Figure 11). A certain amount of air flux is supplied by a blower through the pipe and, then, injected into the main air flow outside the vehicle.

The mechanism of mass injection to reduce buffeting is that the mass flow can thicken the boundary layer at the leading edge of the opening and dissipate the energy in the boundary layer and, eventually, reduce buffeting.

The preliminary CFD simulation gives a promising result as shown in the right part of Figure 11. The peak buffeting SPL is reduced by about 6dB if mass flux is 320cfm. By optimizing flow rate of mass injection, geometry and dimension of the gap, buffeting level may be further reduced.

The disadvantage of this attempt is obvious. Adding the duct and blower will increase the cost and tooling the gap may have negative influence on styling. However, keen demands of customers for more quiet luxury vehicles could make it a prospective candidate of active means of buffeting reduction.

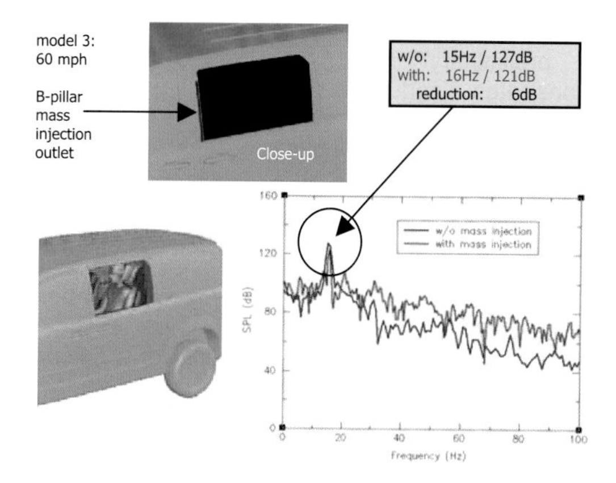

Figure 11. Attempt 4: B-pillar mass injection

ATTEMPT 5: B-PILLAR CAVITY

Encouraged by the 2D study of the impact of a cavity on sunroof buffeting [15], a trial of a B-pillar cavity is made for vehicle model 1 at 50mph. As shown in the close-up of Figure 12, a groove (structural term), or a cavity (aerodynamic term), is constructed at B-pillar, just upstream of the open window.

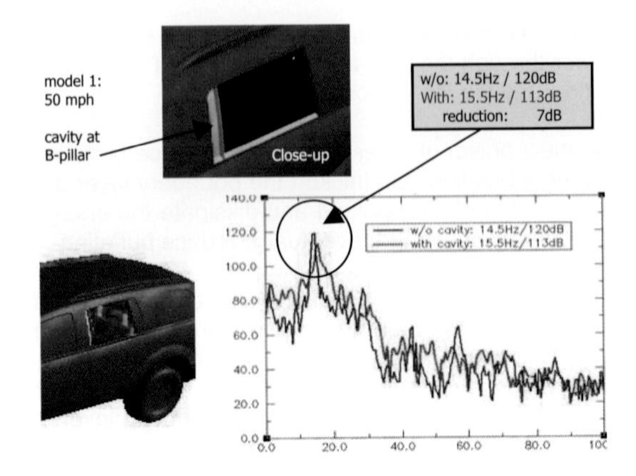

Figure 12. Attempt 5. B-pillar cavity

The functionality of the cavity is to hold the vortex, shed from the shear layer at B-pillar, to retain the vortex there, to keep the vortex from escaping and to avoid the vortex intrusion into the passenger compartment. In this way, it may have a favorable impact on buffeting control.

The preliminary result of the CFD simulation demonstrates a reduction of the buffeting peak SPL by about 6dB, as shown in the spectral curve in the right part of Figure 12. By optimizing the geometry and dimension of the cavity, it is possible to get more buffeting reduction.

Additional proposals were tried with less success, so they are omitted in this paper. Some of the attempts were not tested due to difficulties in tooling and manufacturing, or due to tight schedule of wind runnel. These attempts may be examined in the future if more interests come from vehicle development or design studio.

CONCLUSIONS

Rear window buffeting characteristics of production vehicles can be analyzed using an analysis procedure of CFD simulation plus FFT acoustic post-processing.

The analysis procedure has been validated with three representative types of production vehicles and more confidence has been gained through good correlation with wind tunnel experiments.

Deeper understanding of the physical mechanism of buffeting helps answer questions: why and when buffeting occurs, how severe the noise is and how to reduce it.

Several attempts have been tried to find possible solutions for buffeting reduction and some of them are promising.

For those attempts that have favorable trend of buffeting control, feasibility and manufacturability should be investigated by design and styling engineers.

Future work includes:

1) more basic research is needed to gain insight into buffeting phenomenon and to accumulate more knowledge of buffeting control;

2) more ideas should be collected and open discussions should be made for buffeting reduction;

3) more joint work of CFD simulation and wind tunnel test is necessary to explore possible solutions.

ACKNOWLEDGEMENTS

The authors would like to thank Xijia Zhu and Deep Chakraborty of DaimlerChrysler Corporation for providing wind tunnel experimental data of model 2 and model 3, respectively.

REFERENCES

[1] Ota, D.K., Chakravarthy, S.R., Becker, T. and Sturzenegger, T., Computational study of resonance suppression of open sunroofs, J. Fluids Eng., Vol. 116, pp. 877-882 (1994)
[2] Ukita, T., China, H. and Kanie, K., Analysis of vehicle wind throb using CFD and flow visualization, SAE paper 970407 (1997)
[3] Karbon, K., Singh, R., Simulation and design of automobile sunroof buffeting noise control, 8th AIAA/CEAS Aeroacoustics Conference & Exhibit, June 17-19 (2002)
[4] Hendriana, D., Sovani, S.D. and Schiemann, M.K., On simulating passenger car side window buffeting, SAE Paper 2003-01-1316 (2003)
[5] An, C.-F., Alaie, S.M., Sovani, S.D., Scislowicz, M.S. and Singh, K., Side window buffeting characteristics of an SUV, SAE Paper 2004-01-0230 (2004)
[6] Inagaki, M., Murata, O., Horinouchi, N., Takeda, I. and Kakamu, T., Numerical prediction of wind throb noise in vehicle cabin, Proc. of 6th World Congress on Computational Mechanics, Abstract Vol. I, p. 105, Beijing, China, Sept. 5-10 (2004)
[7] Noelting, S., Audi-A6 sunroof buffeting study, Exa Aeroacoustics Consortium Conference, Stuttgart, Germany, June 3, 2003
[8] Santhooran, S., Elasis LanciaY sunroof buffeting, Exa Aeroacoustics Consortium Conference, Detroit, MI, USA, May 25-26, 2004
[9] Tenstam, A., Eliminating unpleasant sunroof noise, SAAB Automobile AB, Sweden, J. STAR-CD Dynamics, Issue 20, p. 16 (2003)
[10] Muller, J. and Seydell, B., Numerical investigation of sunroof buffeting, Report of Volkswagen AG, Wolfsburg, Germany (2003)
[11] Gardner, B., Zhu, M., Shorter, P. and Tabbal, A., Aero-acoustic applications in the automotive CFD

analysis, ASME 5th International Symposium on Computational Technologies with Industrial Applications, San Diego, CA, USA, July 28 (2004)
[12] Hardin, J.C., Computational aeroacoustics applied to active noise control, SAE paper 951645 (1995)
[13] Fluent 6.1 User's Guide, Fluent Inc., Lebanon, NH, USA (2003)
[14] Gleason M.E., Passenger compartment buffeting solutions, Private communication, March 24 (2004)
[15] An, C.-F., Alaie, S.M. and Scislowicz, M.S., Impact of cavity on sunroof buffeting – a two dimensional CFD study, Proc of ASME 5th International Symposium on Computational Technologies with Industrial Applications, Vol. 1, pp 133-144, (2004)

CONTACT

Chang-Fa An, ca56@daimlerchrysler.com

13th Annual Conference of the
Computational Fluid Dynamics
Society of Canada

CFD 2005

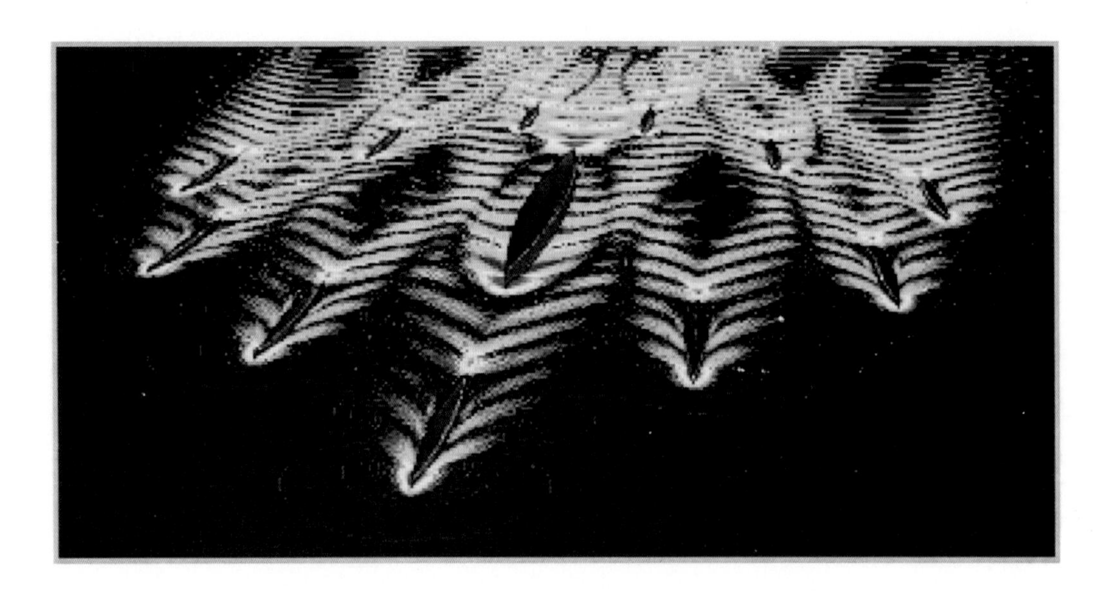

(Image courtesy of Dr. Yang Chi, George Mason University, U.S.A.)

ABSTRACTS and PROGRAMME

St. John's, NL, Canada

July 31 – August 3, 2005

Chaofa An
St. John's, NL, Canada
Aug. 1, 2005

Hosted By:

NRC · CNRC

Memorial
University of Newfoundland

安长发博士科研论文集

Click on CFDSC Logo to Start

CFD 2005

13th Annual Conference
of the
Computational Fluid Dynamics Society of Canada

PROCEEDINGS

Survey of CFD Studies on Automotive Buffeting

Chang-Fa An

DaimlerChrysler Technology Center, 481-33-01
Auburn Hills, MI, USA 48326-2757

Email: *ca56@daimlerchrysler.com*

安长发博士科研论文集

ABSTRACT

In the current automobile market buffeting is one of the customer frequent complaints on luxury cars and SUVs. Buffeting is a low frequency but high level wind noise and makes people inside the vehicle uncomfortable if it lasts for a long period of time. The physical mechanism of buffeting is a complicated phenomenon of aeroacoustic resonance. The aeroacoustic characteristics of buffeting depend on vehicle features and operating conditions. In this paper, a survey of CFD studies on the automotive buffeting is presented. Firstly, several buffeting-related concepts, such as Helmholtz resonator, flow over a cavity, shear layer instability and vortex shedding, are reviewed and relevant references are listed. Then, a historic survey of the buffeting investigation is made with emphasis on computational studies. As an example, the buffeting studies at DaimlerChrysler are selected to demonstrate the procedure of CFD simulation for automotive buffeting. The procedure is then validated by the correlation with wind tunnel testing. After that the validated procedure is applied to find solutions for buffeting reduction. Finally, some comments on buffeting studies are addressed.

1. INTRODUCTION

The terminology of buffeting was originated from aeronautical and architectural industries. In aeronautical industry buffeting, or flutter, is referred to as the elastic response of an airplane component, such as a wing, a tail or a fin, to the temporal load of the air flow [1]. In architectural industry, buffeting is referred to as the vibrating response of a construction structure, such as a long bridge [2] or a high building [3], to the wind gust. In automotive industry, however, buffeting is referred to as the aeroacoustic response of the air bulk inside the passenger compartment to the transient flow outside the vehicle. More precisely, automotive buffeting is a low frequency (<20Hz) but high level (>100dB) wind noise perceived within the passenger compartment of a road vehicle which is moving at a certain speed range (30-80mph) and a side window or sunroof is open. In this paper, only automotive buffeting is considered. Figure 1 shows the cases of (a) sunroof (b) front window and (c) rear window buffeting.

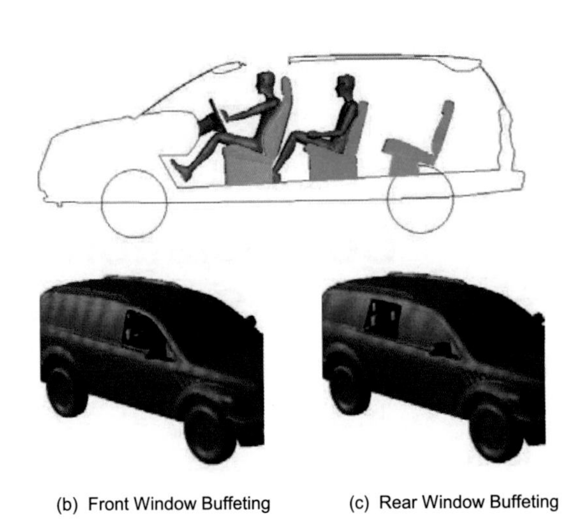

Figure 1. Automotive Buffeting

The physical mechanism of buffeting is a complicated phenomenon of aeroacoustic resonance. At the upstream of the sunroof or side window opening, there exists a shear layer between the moving air flow (outside) and the relatively stationary air bulk (inside). The shear layer becomes unstable when the relative velocity between the outside and inside air exceeds a critical value. As a result, vortices will be formed and shed periodically. If the frequency of the vortex shedding happens to coincide with the natural frequency of the interior air bulk, the so-called Helmholtz resonance will occur and buffeting will take place. The aeroacoustic characteristics of buffeting depend on vehicle features, e.g. windshield, A- and B-pillars, side view mirrors, volume of the interior air bulk, shape and size of the opening, and the driving conditions, such as vehicle speed, wind direction and strength, etc.

In this paper, a survey of CFD studies on the automotive buffeting is presented. To begin with, several buffeting-related concepts are discussed to get some ideas of the phenomenon. Secondly, a historic review of buffeting investigation is made with emphasis on computational work. As a sample, the CFD studies on buffeting at DaimlerChrysler are selected to demonstrate the procedure, validation and applications. In the context of the paper, relevant references are cited as wide as possible from the author's knowledge to provide a literature clue to the readers who are interested in the subject. Finally, concluding remarks on buffeting studies will be given and the future prospect will be overlooked.

2. BUFFETING-RELATED CONCEPTS

In order to get better understandings of buffeting mechanism several buffeting-related concepts are discussed in this section, including (a) Helmholtz resonator (b) flow over a cavity (c) shear layer instability and (d) vortex shedding, as shown in Figure 2.

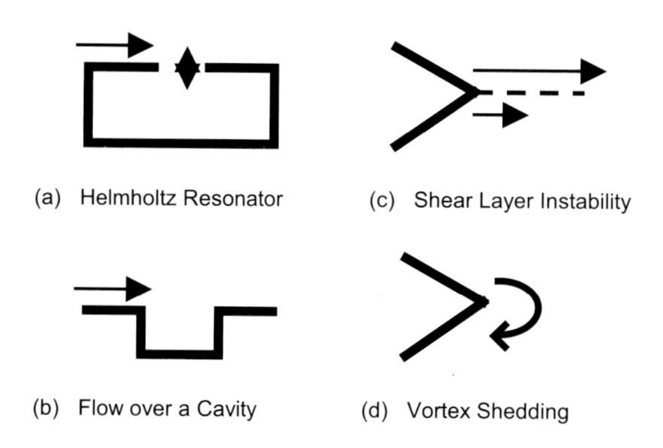

Figure 2. Buffeting-Related Concepts

2.1 Helmholtz Resonator

A Helmholtz resonator is an air bulk chamber that is connected to the air flow outside the chamber through a narrow neck (Figure 2a). Communicating with the outside flow, the air inside the chamber is forced to move-in and -out through the neck. In this way aeroacoustic noise is made with a certain natural frequency. Rayleigh [4], Pierce [5] and Kinsler et al [6] derived a formula for natural frequency of a Helmholtz resonator as $f_n = c/(2\pi)\sqrt{A/(VL)}$ where f_n is natural frequency, c is speed of sound, V is volume of the Helmholtz resonator, L and A are length and cross-area of the neck, respectively. Blevins [7] collected natural frequency formulas for many structural and fluid systems, including the above formula for a Helmholtz resonator. However, this formula usually over-predicts the natural frequency if the length of the neck is short compared to the scale of the cross-area of the neck. Alster [8] derived a modified formula for the natural frequency of a Helmholtz resonator which is a lengthy expression depending on specific geometry of the chamber. In the automotive buffeting study, the passenger compartment can be considered as a Helmholtz resonator, but the formulas are too simple to calculate the natural frequency due to the geometrical complexity of the realistic vehicle interior. Nevertheless, these formulas still can be used to estimate the trend of natural frequency.

2.2 Flow over a Cavity

Another concept that is related to the buffeting phenomenon is the flow over a cavity (Figure 2b). As a matter of fact, the vehicle interior in buffeting study can be considered as a cavity with a big volume and a small opening (see Figure 1). This is because the air flow passing the open sunroof or an open side window has a similar feature as the flow over a cavity. Most early work on flow over a cavity was done in aerospace industry. Roshko [9] measured fluctuating pressure and velocity on the walls of a cavity with various ratios of length to depth. Block [10] investigated noise response of cavities to the flow with subsonic speeds. Tam [11] conducted a theoretical study of flow over a rectangular cavity with the interest in aeroacoustic noise. In 1978, Rockwell and Naudascher [12] published a review paper of self-sustaining oscillations of flow past cavities. They grouped unstable flow over a cavity into three categories: fluid-dynamic, fluid-resonant and fluid-elastic. Each category has its specific feature of pressure fluctuation and noise generation. Rockwell and Knisely [13] observed the vortex impingement upon the trailing edge of a cavity using laser anemometry and hydrogen bubble technique in a water tunnel. Their visualization of vortex movement demonstrated the characteristics of the cavity flow. In mid-1980s, Blake [14] reviewed the flow over cavities with some complex details of the shear layer impingement and feedback. Some years later, Blevins [15] summarized the state-of-the-art of the studies on the sound excited due to flow over cavities. Since 1990s, the study of noise generation from flow over cavities has been applied to automotive industry. Mongeau et al [16] investigated flow-induced pressure fluctuations in a door gap cavity in a quiet wind tunnel. The automotive application of cavity flow gained so much interest that the door gap cavity flow was adopted as one of the benchmark problems in the 3th computational aeroacoustics (CAA) workshop in 1999 [17-19] and the 4th CAA workshop in 2003 [20-23] in NASA.

2.3 Shear Layer Instability

Shear layer instability (Figure 2c) is a classical subject in fluid mechanics and it is believed to be the source of pressure fluctuations in the flow over a cavity. Early work on shear layer instability for a step velocity profile, described in Lamb's book *Hydrodynamics* [24], was named Kelvin-Helmholtz instability. Applying this theory, An et al [25] conducted the analysis of Kelvin-Helmholtz instability with the application to oil-water interfacial phenomena. Esch [26] examined the shear layer instability for a linear velocity profile between upper and lower uniform streams using an eigenvalue technique and determined the region of instability. In mid-1960's, Michalke adopted the linear stability theory for the case of temporally growing disturbances to a hyperbolic-tangent velocity profile [27], analyzed and explained the vortex formation, development and rolling up [28], and finally extended the analysis to the case of spatially growing disturbances [29]. Experimentally, Browand [30] undertook an investigation of shear layer instability in a small wind tunnel. The shear layer instability for a boundary layer velocity profile was extensively investigated and summarized by Lin [31] and Bechov and Criminale [32] in their monographs.

2.4 Vortex Shedding

Vortex shedding (Figure 2d) is an immediate outcome of shear layer instability. Abernathy and Kronauer [33] proposed a vortex formation theory using the concept of point vortices and vortex sheets. The formation of discrete vortices in the wake of a blunt body can be interpreted as non-linear interaction of two vortex sheets with opposite directions. Boldman et al. [34] made a flow visualization of vortex shedding from a trailing edge of a plate in a wind tunnel. They found an interesting fact that strong von Karman vortices develop behind the trailing edge when the velocities on upper and lower surfaces are equal while the vortices tend to disappear when the velocities are unequal. Their observation was in good agreement with the theory of vortex formation in [33]. Nakamura and Nakashima [35] conducted an experimental investigation of vortex excitation where two shear layers were separated from a bluff body in a water tank for visualization purpose. They concluded that the impinging-shear-layer is responsible for vortex excitation.

There are a lot of references in the literature, but it is hard to cite more due to the length limitation of the paper. Instead, the interested readers can use this list as a clue to build-up their own library of the buffeting-related concepts.

3. SURVEY OF BUFFETING STUDIES

The phenomenon of automotive buffeting was first observed and investigated in mid-1960's. Bodger and Jones [36] reported the results of their study on side window "wind throb", the early terminology for buffeting. Aspinall [37] investigated rear window wind throb and proposed some means to reduce the noise. Hucho [38] utilized other terminology "booming" to describe this type of wind noise in his book *Aerodynamics of Road Vehicles*. The automotive buffeting can be divided into two types, sunroof buffeting and side window buffeting. For realistic vehicles, sunroof buffeting may appear at a range of lower cruise speed (30-50mph) while side window buffeting may appear at a range of higher cruise speed (40-80mph). The CFD simulation of sunroof buffeting was done earlier with 2D model because the flow is approximately symmetric and the central section is representative. In contrast, the CFD simulation of side window buffeting was rather late. Unless the two corresponding windows on the opposite sides are open synchronically, which is not a typical scenario, the flow is generally non-symmetric and 3D simulation must be invoked. In the following subsections, representative CFD studies on automotive buffeting will be cited and comments on these studies will be given.

3.1 2D Sunroof Buffeting

With fast development of computer technology in 1990s, CFD simulation for automotive buffeting became possible. Ota et al [39] reported an early simulation for a 2D simplified car model to investigate the sunroof "boom" phenomenon using the CFD code Golde. An et al [40] conducted a 2D study of the impact of cavity on sunroof buffeting. Barron et al [41] and Zhang et al [42] analyzed a 2D cavity flow with applications to sunroof buffeting. The details of these two papers will be presented at the same conference. In general, it is not surprised that the accuracy of buffeting analysis in 2D study cannot be very high because of the difference between 2D model and the realistic vehicle. To obtain more accurate results of buffeting prediction, 3D simulation must be developed.

3.2 3D Sunroof Buffeting

In 1997, Mitsubishi Motors [43] conducted a CFD study of sunroof "wind throb" for a simplified 3D car model using a finite difference code NAGARE.

The calculation was well compared with the corresponding water tunnel testing. Since the new millennium, more CFD studies have been undertaken and more realistic road vehicles have been taken under consideration. In the mean time, the phenomenon has been called "buffeting". In 2002, General Motors [44] reported a study of sunroof buffeting for a 3D realistic car model using a finite element code PAM-FLOW. The simulated results were in good agreement with wind tunnel experiments. They also verified that wind deflector and comfort position work well to reduce buffeting.

3.3 3D Side Window Buffeting

For more complicated side window buffeting, DaimlerChrysler [45] conducted CFD simulations for a 3D passenger car model using a finite volume code Fluent. The results compared well with wind tunnel testing. During the same period of time, several other car companies also took actions towards the sunroof buffeting studies under the collaboration with CFD software vendors. These include Audi's work with PowerFLOW [46-47], SAAB's work with STAR-CD [48], Volkswagen's work also with PowerFLOW [49] and Toyota's work with their own code [50].

4. BUFFETING AT DAIMLERCHRYSLER

Since the year 2002, intensive efforts on buffeting studies at DaimlerChrysler have been made. Following an early work of side window buffeting [45], a series of buffeting studies [51-53] have been done. This includes procedure establishment for buffeting analysis, validation of the procedure by correlation with wind tunnel testing, attempts to find potential solutions for buffeting reduction and parametric optimization for buffeting reducers. These efforts will be outlined in the next subsections.

4.1 Methodology

The procedure of buffeting analysis was described in [51] and [52]. Figure 3 shows the flow chart of buffeting analysis. The regular CFD steps include mesh generation, problem set-up, solver running and post-processing. A new step in buffeting analysis is acoustic post-processing through Fourier transform.

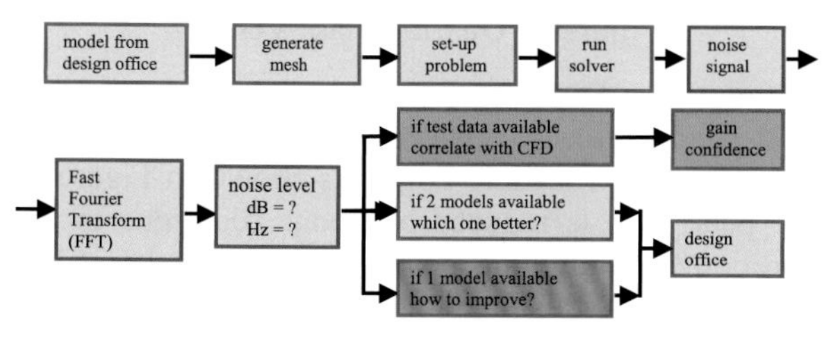

Figure 3. Flow Chart of Buffeting Analysis

The vehicle model is provided from design office in CATIA format. ANSA is used to clean-up geometry and to generate surface mesh. The vehicle surface mesh is placed into a virtual wind tunnel of $50x20x10m^3$. The surface mesh of vehicle and wind tunnel is imported into TGRID for volume mesh generation and local mesh refinement. The final volume mesh contains about 3.5 million tetrahedral cells with 4 levels of local mesh refinement. The commercial CFD code Fluent 6.1 [54] is used to simulate unsteady flow around the vehicle. The air is compressible and assumed to be ideal gas. The solver is segregated transient. The time marching is 2nd order implicit. The space discretization is 2nd order upwind and the P-V coupling is SIMPLE. The inlet boundary condition is a constant velocity of the vehicle speed. The outlet boundary condition is a constant pressure of atmosphere. The top, sides and ground are frictionless walls and the vehicle surface is frictional wall. The effect of ground friction on buffeting is assumed to be small and hence neglected. At the beginning of each simulation, a steady state mode is run for several hundred iterations using the standard k-ε turbulence model. The steady state flow is then used as an initial condition to start unsteady simulation and the turbulence model is switched to LES. A time step of 0.002 seconds is chosen for unsteady simulation in most cases. Within each time step, 15-20 sub-iterations are set and this is enough for most cases. Typical running time is 3-4 days to reach 1200 time steps on an SGI Origin 2000 machine with 64 300-MHz CPU and 64 GB memory using the IRIX 6.5 operating system with 16 processors.

During the unsteady simulation, static pressure is recorded at a sensor in the passenger compartment for acoustic post-processing (Figure 4).

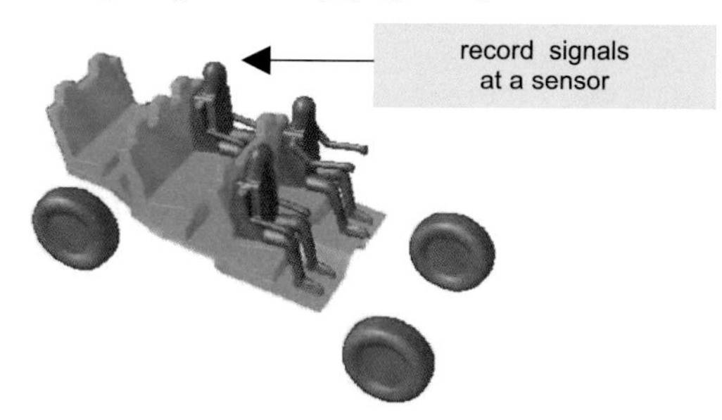

Figure 4. Record Signals During Simulation

As shown in Figure 5 on the next page, after an irregular initial process in the unsteady simulation the pressure signal reaches a status of dynamically stable periodic fluctuation and the selected pressure fluctuation signal (within the red block) can be picked-up for acoustic post-processing, i.e. Fourier transform.

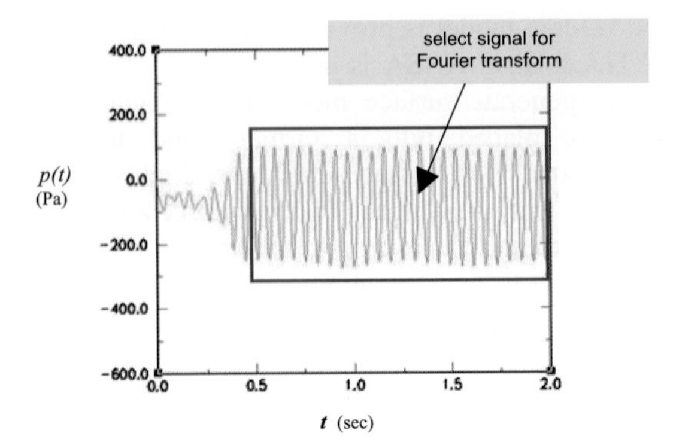

Figure 5. Select Signals for Fourier Transform

Fourier transform with a Hanning window is available in the software XMGR. After Fourier transform, the amplitude of pressure fluctuation is expressed as a function of frequency. Finally, the amplitude of pressure fluctuation is converted to the format of frequency spectrum of buffeting (Figure 6). That is, the sound pressure level (SPL) in dB is expressed as a function of frequency using the conventional formula $SPL=10\log_{10}(P/P_{ref})^2$ where P is the amplitude of pressure fluctuation in Pa and the reference pressure is $P_{ref}=20x10^{-6}$Pa.

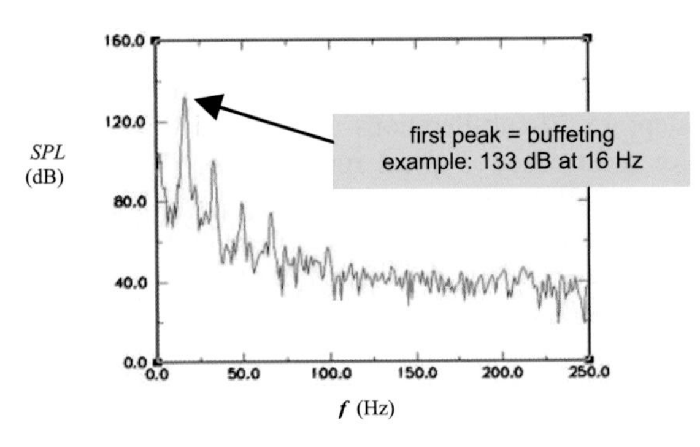

Figure 6. Frequency Spectrum of Buffeting

4.2 Validation Cases

In order to validate the procedure of CFD analysis for buffeting, several realistic vehicles are chosen to correlate the simulated results with wind tunnel testing. They are listed in Table 1.

Table 1. Buffeting Validation Cases

Vehicle (60mph)	Window Open	CFD (dB/Hz)	Test (dB/Hz)
Passenger Car	Front	115/18	116/17
Passenger Car	Rear	129/17	130/17
SUV1	Front	112/18	113/14
SUV1	Rear	133/16	135/15
SUV2	Rear	121/17	122/15
Minivan	Rear	125/15	127/13

From this validation list we can see that for most cases the accuracy of the CFD simulation is that the resonant frequency is within 2 Hz and the peak SPL is within 2dB.

With good correlation of these validation cases, more confidence was gained and then more attention was paid to the possible solutions for buffeting reduction. Among the attempts which have ever been tried to find solutions, only those that seem to be more or less successful are included in this paper.

4.3 Attempts to Reduce Buffeting

As stated before, the buffeting mechanism is the excitation of passenger compartment, as a Helmholtz resonator, by the fast sweeping flow outside the opening in a certain resonant range of the vehicle speed. If the resonance condition is changed in some way or, in other words, if the natural frequency of the Helmholtz resonator is shifted to be out of the resonant range of the vehicle speed, the resonance excitation may be avoided. As a result, buffeting can be reduced or even completely eliminated. Passenger compartment venting is one of the approaches to change resonance condition of the passenger compartment and the first 3 attempts below for buffeting reduction fall into this category.

Attempt 1 - Quarter glass venting: As shown in Figure 7, while the rear window is open, the quarter glass in a minivan is swung out to 5°. Compared to the case of quarter glass closed, the resonant frequency changes and the peak SPL reduces by about 16dB due to quarter glass venting. A wind tunnel test has confirmed the result of CFD prediction.

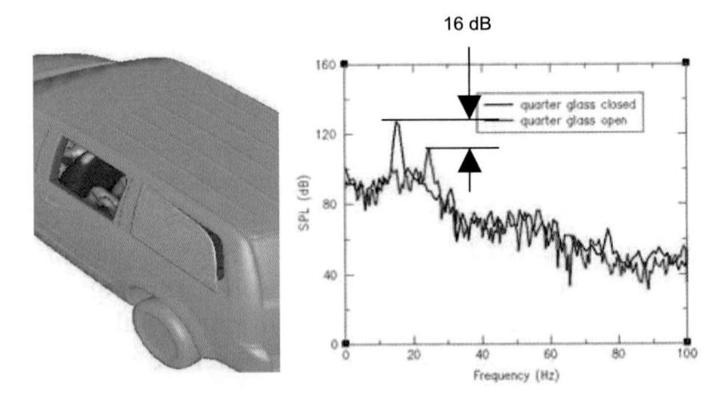

Figure 7. Quarter Glass Venting

Attempt 2 - Dividing post at window: A dividing post is added to the open rear window at about 75% length from its leading edge. As shown in Figure 8, the peak SPL is reduced by about 20dB due to the addition of a dividing post. A wind tunnel test has also confirmed this prediction.

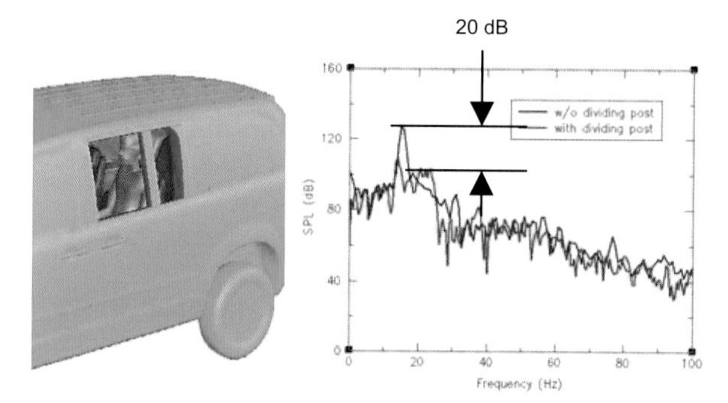

Figure 8. Dividing Post at Window

Attempt 3 – Venting hole at C-pillar: The favorable results of attempts 1 and 2 stimulate a new idea that a narrow vertical hole at C-pillar may play the same role as an alternative venting element. As shown in Figure 9, the peak SPL can be reduced by about 22 dB due to the addition of a venting hole at C-pillar.

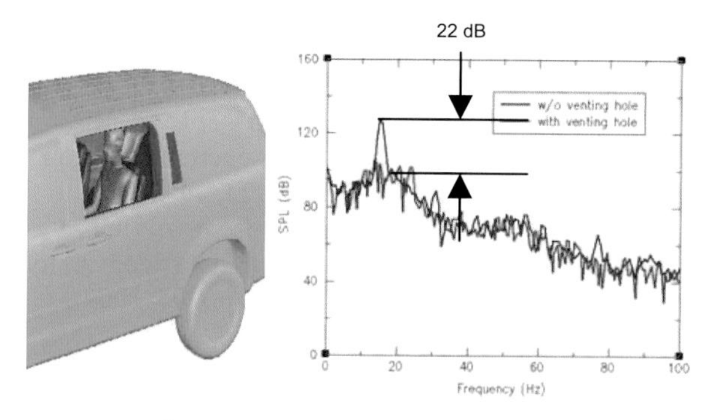

Figure 9. Venting Hole at C-Pillar

Attempt 4 – Mass injection at B-pillar: The buffeting control methods can be divided into two categories: passive and active. All the above attempts belong to passive control. Hardin [55] summarized the active control methods in aeroacoustics. One of the active control methods is mass injection. It is shown in Figure 10(a). The peak SPL can be reduced by about 6dB if mass injection is applied at B-pillar. Of course, the active control is more complex and more costly compared to the passive control.

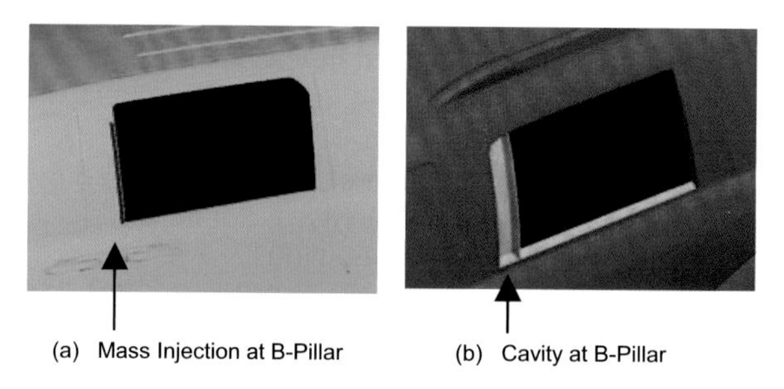

(a) Mass Injection at B-Pillar (b) Cavity at B-Pillar

Figure 10. Mass Injection & Cavity at B-Pillar

Attempt 5 – Cavity at B-pillar: Encouraged by the 2D study for impact of a cavity on sunroof buffeting [40], an attempt of a B-pillar cavity was tried for the rear window buffeting case, as shown in Figure 10(b). The peak SPL can be reduced by about 7dB due to the addition of a cavity at B-pillar.

4.4 Optimization of Buffeting Reducers

With the above achievements as a basis, a systematic investigation with CFD technology is proposed as a long-term plan at DaimlerChrysler. The emphasis will be focused on sunroof buffeting. A matrix of sensitivity study for parametric optimization is made and will be completed within a year. Some of the items in the matrix are listed in Figure 11.

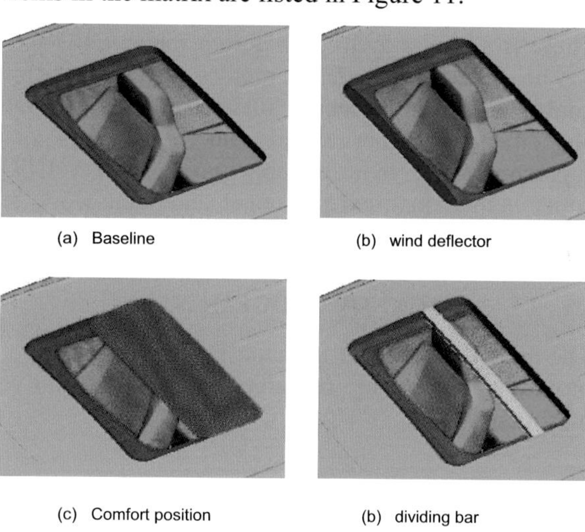

(a) Baseline (b) wind deflector

(c) Comfort position (b) dividing bar

Figure 11. Sunroof Buffeting Reducers

Starting from a baseline design (Figure 11a), a wind deflector is added and its angle, length, width and shape will be changed to find the optimal design (Figure 11b). The sunroof glass position will be slipped back and forth to change the area of the opening for the comfort position (Figure 11c). Reminded by the dividing post at an open window as a buffeting reducer, a dividing bar is put on the sunroof structure (Figure 11d). By changing its location, width and thickness, the optimal combination of these parameters will be found to minimize buffeting. In addition, combinations of these buffeting reducers for sunroof case and side window case will be also investigated.

5. CONCLUDING REMARKS

Buffeting noise is one of the customer frequent complaints in the current automobile market. There is a need to solve such a problem before launching high quality quiet luxury vehicles.

Abundant knowledge of buffeting-related concepts can help understand buffeting mechanism, that is, why it occurs and how to reduce it.

During the past decade, remarkable advancement for buffeting study has been achieved with fast development of computer technology and improved CFD algorithm.

By joint efforts of car companies and CFD software vendors, the methodology of buffeting analysis has been validated with wind tunnel testing for several realistic vehicles.

Some buffeting reducers have been recognized and parametric optimization study has been proposed and will be conducted in the near future.

There is a long way to go before a robust CFD tool can be utilized for buffeting prediction. To reach this goal we need deeper understanding of the phenomenon, higher speed and bigger memory supercomputers, more accurate and efficient CFD code, highly qualified personnel and joint work of computational and experimental studies.

ACKNOWLEDGEMENTS

The author would like to thank DaimlerChrysler for providing computing resources and administrative management.

REFERENCES

[1] Lee, B. H. K., Vertical tail buffeting of fighter aircraft, Progress in Aerospace Sciences, **36**, pp. 193-279, 2000.

[2] Minh, N. N., Miyata, T., Yamada, H., and Sanada, Y., Numerical simulation of wind turbulence and buffeting analysis of long-span bridges, **83**, pp. 301-315, 1999.

[3] Zhou, Y., Kareem, A., and Gu, M., Equivalent static buffeting loads on structures, J. Structural Eng. August 2000, pp. 989-992, 2000.

[4] Rayleigh, B., *The Theory of Sound*, Vol. II, pp. 170-172, Dover Publications, New York, USA, 1945.

[5] Pierce, A.D., *Acoustics*, pp. 330-336, McGraw-Hill Book Company, New York, USA, 1981.

[6] Kinsler, L.E., Frey, A.R., Coppens, A.B., and Sanders, J.V., *Fundamentals of Acoustics*, pp. 225-228, John Wiley & Sons, New York, USA, 1982.

[7] Blevins, R.D., *Formulas for Natural Frequency and Mode Shape*, pp. 353-358, Van Nostrand Reinhold Company, New York, USA, 1979.

[8] Alster, M., Improved calculation of resonant frequencies of Helmholtz resonators, J. Sound & Vibration, **24**(1), pp. 63-85, 1972.

[9] Roshko, A., Some measurements of flow in a rectangular cutout, NACA TN 3488, 1955.

[10] Block, P.J.W., Noise response of cavities of varying dimensions at subsonic speeds, NASA TN D-8351, 1976.

[11] Tam, C.K.W., The acoustic modes of a two-dimensional rectangular cavity, J. Sound & Vibration, **49**(3), pp. 353-364, 1976.

[12] Rockwell, D. and Naudascher, E., Review - self-sustaining oscillations of flow past cavities, J. Fluids Eng., **100**, pp. 152-165, 1978.

[13] Rockwell, D., and Knisely, C., The organized nature of flow impingement upon a corner, J. Fluid Mech., **93**, Part 3, pp. 413-432, 1979.

[14] Blake, W.K., *Mechanics of Flow-Induced Sound and Vibration*, Vol. I, pp. 130-149, Academic Press, Inc., New York, USA, 1986.

[15] Blevins, R.D., *Flow-Induced Vibration*, 2nd ed., pp. 375-383, Van Nostrand Reinhold, New York, USA, 1990.

[16] Mongeau, L., Bezemek, J., and Danforth, R., Pressure fluctuations in a flow-excited door gap cavity model, SAE Paper 971923, 1997.

[17] Henderson, B., Automotive noise involving feedback – sound generation by low speed cavity flows, NASA/CP 2000-209790, pp. 95-100, 2000.

[18] Ashcroft, G.B., Takeda, K., and Zhang, X., Computations of self-induced oscillatory flow in an automobile door cavity, NASA/CP 2000-209790, pp. 355-361, 2000.

[19] Shieh, C.M., and Morris, P.J., A parallel numerical simulation of automobile noise involving feedback, NASA/CP 2000-209790, pp. 363-370, 2000.

[20] Henderson, B., Sound generation by flow over a cavity, NASA/CP 2004-212954, pp. 71-77, 2004.

[21] Zhang, Z., Barron, R.M., and An, C.-F., Spectral analysis for air flow over a cavity, NASA/CP 2004-212954, pp. 197-204, 2004.

[22] Loh, C.Y., and Jorgenson, P.C.E., Computation of tone noises generated in viscous flows, NASA/CP 2004-212954, pp. 213-228, 2004.

[23] Ashcroft, G.B., Takeda, K., and Zhang, X., A numerical investigation of the noise radiated by a turbulent flow over a cavity, J. Sound & Vibration, **265**, pp. 43-60, 2003.

[24] Lamb, H., *Hydrodynamics*, 6th ed., pp. 373-375, Cambridge Univ. Press, London, UK, 1932.

[25] An, C.-F., Goodman, R.H., Brown, H.M, Clavelle, E.J., and Barron, R.M., Oil-water

interfacial phenomena behind a boom on flowing water, Proc. 10th Int'l Symp. on Transport Phenomena, Vol. 1, pp. 13-18, Kyoto, Japan, Nov. 30-Dec 4, 1997.
[26] Esch, R.E., The instability of a shear layer between two parallel streams, J. Fluid Mech., **3**, pp. 289-303, 1957.
[27] Michalke, A., On the inviscid instability of the hyperbolic-tangent velocity profile, J. Fluid Mech., **19**, pp. 543-556, 1964.
[28] Michalke, A., Vortex formation in a free boundary layer according to stability theory, J. Fluid Mech., **22**, part 2, pp. 371-383, 1965.
[29] Michalke, A., On spatially growing disturbances in an inviscid shear layer, J. Fluid Mech., **23**, part 3, pp. 521-544, 1965.
[30] Browand, F.K., An experimental investigation of the instability of an incompressible, separated shear layer, J. Fluid Mech., **26**, part 2, pp. 281-307, 1966.
[31] Lin, C.C., *The Theory of Hydrodynamics Stability*, Cambridge Univ. Press, London, UK, 1955.
[32] Betchov, R., and Criminale, W.O. Jr., *Stability of Parallel flows*, Academic Press, New York, USA, 1967.
[33] Abernathy, F.H., and Kronauer, R.E., The formation of vortex sheets, J. Fluid Mech., **13**, pp. 1-20, 1962.
[34] Boldman, D.R., Brinich, P.F., and Goldstein, M.E., Vortex shedding from a blunt trailing edge with equal and unequal external mean velocities, J. Fluid Mech., **75**, part 4, pp. 721-735, 1976.
[35] Nakamura, Y., and Nakashima. M., Vortex excitation of prisms with elongated rectangular, H and T cross-sections, J. Fluid Mech., **163**, pp. 149-169, 1986.
[36] Bodger, W.K., and Jones, C.M., Aerodynamic wind throb in passenger cars, SAE Transactions, **73**, pp. 195-206, 1965.
[37] Aspinall, D.T., An empirical investigation of low frequency wind noise in motor cars, Rept., Motor Industry Research Association (MIRA) #1966/2, Warwickshire, UK, 1966.
[38] Hucho, W.-H., *Aerodynamics of road vehicles*, pp. 250-256, Butterworth, London, UK, 1987.
[39] Ota, D.K., Chakravarthy, S.R., Becker, T., and Sturzenegger, T., Computational study of resonance suppression of open sunroofs, J. Fluids Eng., **116**, pp. 877-882, 1994.
[40] An, C.-F., Alaie, S.M., and Scislowicz, M.S., Impact of cavity on sunroof buffeting – a two dimensional study, Proc. 5th Int'l ASME/JSME Symp. on Comp. Tech., Vol. 1, pp. 133-144, San Diego, CA, USA, July 25-29, 2004.
[41] Barron, R.M., Zhang, Z., and An, C.-F., Computational aero-acoustic study of airflow over a cavity, 13th Annual Conference of CFD Society of Canada, St. John's, NF, Canada, July 31-Aug. 3, 2005.
[42] Zhang, Z., Barron, R.M., and An, C.-F., Post-processing tools in computational aero-acoustics, 13th Annual Conference of CFD Society of Canada, St. John's, NF, Canada, July 31-Aug. 3, 2005.
[43] Ukita, T., China, H., and Kanie, K., Analysis of vehicle wind throb using CFD and flow visualization, SAE paper 970407, 1997.
[44] Karbon, K. J., and Singh, R., 2002, "Simulation and design of automobile sunroof buffeting noise control", AIAA paper 2002-2550.
[45] Hendriana, D., Sovani, S.D. and Schiemann, M.K., On simulating passenger car side window buffeting, SAE Paper 2003-01-1316, 2003.
[46] Noelting, S., Audi A6 sunroof buffeting study, Exa Aeroacoustics Consortium Conference, Stuttgart, Germany, 2003.
[47] Crouse, B., Senthooran, S., Balasubramanian, G., Freed, D., Nolting, S., Mongeau, L., and Hong, J.-S., Sunroof buffeting of a simplified car model: simulations of the acoustic and flow-induced responses, SAE paper 2005-01-2498, 2005.
[48] Tenstam, A., Eliminating unpleasant sunroof noise, SAAB Automobile AB, Sweden, J. STAR-CD Dynamics, issue 20, p. 16, 2003.
[49] Muller, J., and Seydell, B., Numerical investigation of sunroof buffeting, Rept. Volkswagen AG, Wolfsburg, Germany, 2003.
[50] Inagaki, M., Murata, O., Horinouchi, N., Takeda, I, and Kakamu, T., Numerical prediction of wind-throb noise in vehicle cabin, Proc. 6th World Congress on Comp. Mech., p. 105, Beijing, China, Sept. 5-10, 2004.
[51] An, C.-F., Alaie, S. M., Sovani, S.D., Scislowicz, M. S., and Singh, K., Side window buffeting characteristics of an SUV, SAE Paper 2004-01-0230, 2004.
[52] An, C.-F., Puskarz, M., Singh, K., and Gleason, M. E., Attempts for reduction of rear window buffeting using CFD, SAE Paper 2005-01-0603, 2005.
[53] An, C.-F., and Singh, K., Optimization study for buffeting reduction with CFD technology, will appear at SAE World Congress, 2006.
[54] Fluent 6.1 User's Guide, Fluent Inc., Lebanon, NH, USA, 2002.
[55] Hardin, J. C., Computational aeroacoustics applied to active noise control, SAE paper 951645, 1995.

Post-processing in Computational Aeroacoustic Studies

Z. Zhang[1,2], R. Barron[2] and C.-F. An[3]

[1]College of Chemical Engineering, Beijing University of Chemical Technology, Beijing, China 100029
[2]Fluid Dynamics Research Institute, University of Windsor, Windsor, ON, Canada, N9B 3P4
[3] DaimlerChrysler Technology Center, Auburn Hills, MI, USA 48326-2757

Email: *zzhang@uwindsor.ca*

ABSTRACT

This paper discusses the results of cases studied in the previous paper [1], showing the detailed computed 2D-fields of pressure contours, velocity vectors and vorticity contours, obtained from the airflow over a cavity, at some specific selected moments in a flow fluctuation period through snapshots and animations. These figures help us to achieve a better understanding of aeroacoustic phenomena in these types of problems. They are particularly useful to understand the buffeting mechanism of airflows over an open window or sunroof of a moving road vehicle, and are helpful for suggesting new designs for the local structures. In addition, digital soundtracks are made to simulate the actual feeling of the buffeting noise.

1. INTRODUCTION AND METHODOLOGY

Introduction: Several tools that can be used to interpret and more fully understand the complex physical processes involved in aeroacoustics are discussed in this paper. For an aeroacoustic problem, the flow field is often periodically changing and it is essential to accurately capture the time history of the flow fields, including pressure, velocity and vorticity. To evaluate the acoustic characteristics of the flow induced noise, such as noise frequency and Sound Pressure Level (*SPL*), from the steadily changing air flow, the calculated fluctuating pressure histories at certain monitoring points are spectra-analyzed using a Fast Fourier Transform, which has been presented in our previous paper [1]. In this paper we discuss in detail some results of the cases studied in [1] by showing the pressure contours, velocity vectors and vorticity contours at selected moments in a fluctuation period, using some particular snapshots to represent the full animation. In addition, to hear the noise created from the flow, the calculated pressure histories are transformed to a sound signal and played back by audio equipment.

Methodology: Before discussing the specific cases, a few words should be said about the methodology of creating the flow animation and the digital sound track of a fluctuating pressure history.

Animated flow visualization is a powerful tool in the analysis of unsteady flow fields and can be used to study the cause of flow induced noise. In order to make an animation from a series of time-sequential flow fields, the following steps are required: (i) the series of time-sequential 2D-flow fields, including a selected view of pressure contours, flow velocity vectors and vorticity contours, which can be computed by a CFD simulation package such as FLUENT [2], should to be displayed and hard-copied in a relatively simple graphic form, such as .jpg or .tiff, at specific moments of the fluctuating flow fields; (ii) the .jpg or .tiff figures are further modified and enhanced, such as selecting a partial view of pictures to be shown, adding some text, sharpening the images etc., to obtain a series of improved flow field figures for animation. In this work, the graphic software produced by ACDSee, FotoCanvas [3] has been used; (iii) MATLAB [4] can be adopted to create an .avi movie file, which can be played back by a standard video device, such as Windows Media Player. However, we have chosen to use ACDSee to play the animation show of the time-sequential 2D-fields. A convenient feature of ACDSee is that the movie can be stopped at any frame and the operator can go forward or backward step-by-step to examine particular interests at any time.

Finally, MATLAB is used to transform the fluctuating pressure history into an actual sound-signal file, which can be played back by Windows Media Player. This allows the researcher to hear the noise while observing the animated flow field features.

2. ANIMATION

The flow animation shows a sequential series of frames of flow diagrams, such as pressure and vorticity contours, velocity vectors, etc. Obviously, only selected still frames can be shown in the paper. The full animation of the whole series of frames may be seen as a movie in the presentation at the conference. Since the flow pattern for a particular aeroacoustic problem changes periodically with time, it is necessary to choose some typical frames in a period to show some extreme situations at special moments. Here, five snapshots are chosen for each particular flow diagram to represent the evolution of the flow.

In general, four kinds of flow diagram are presented in this paper. They are (i) pressure contours in the whole calculation domain (see Fig.1 in [1]), (ii) pressure contours around the throat area; (iii) velocity vectors around the throat area; and (iv) vorticity contours around the throat area. In order to show the values of parameters in the figures, a color map is used. Figure 1 gives the color scales of the three flow parameters for all subsequent figures.

2.1 Basic Case: Case A, 22m/s, Cavity x2y2 and 2mm Grid

Figure 2 shows the snapshots of the pressure contours in the whole domain for the basic case (Case A, 22m/s, cavity x2y2 and 2mm grid). The first frame, Fig 2(a), shows the pressure contours at a moment when the surrounding pressure only changes slightly, in other words, the inlet pressure is quite close to the outlet pressure. This moment may be set as a starting point, 0ms in a period. After 9.5ms, the pressure of the outflow surrounding the cavity experiences a large drop, as seen in the second snapshot Fig. 2(b). At 18.5ms, the pressure drop at the outflow returns to a lower value, the third snapshot is taken and shown in Fig. 2(c). Figure 2(d) shows the surrounding outflow at 27.5ms, when the largest negative pressure difference appears. Finally, the pressure contours at 37ms are quite close to those in Fig. 2(a), as seen in Fig. 2(e). This means that the flow pattern at this moment has returned to the pattern at the starting moment, 0ms. Therefore, a full period has elapsed, and 37ms is the period of this flow, which gives the fluctuating frequency as 1/0.037 = 27Hz. This coincides with the value 26.9 obtained from the spectrum analysis in [1].

The same moments as in Fig. 2 are used for the snapshots of the other parameters, to illustrate and analyze the changes of the other parameters during a period. Figures 3(a-e) give the close-up snapshots of these same pressure contours around the throat area at all selected moments in a period. At 0ms, the pressure around the front end of the throat is larger than around the rear end of the throat, see Fig. 3(a). At 9.5ms, the pressure across the throat does not change very much. However we can observe some lower pressure points that appear near the front end of the throat, shown in Fig. 3(b). At the third moment, 18.5ms, pressure near the lower part of the front end of the throat becomes very low (less than -1000 Pa), see Fig. 3(c). This occurs near monitor #2. At 27.5ms, the low pressure region has moved down into the cavity, travels towards the rear of the throat and gradually becomes smaller, as seen in Fig. 3(d). Finally, at 37ms, Fig. 3(e) shows that the contours return to the similar pattern as in Fig. 3(a). These figures tell us that the amplitude of the pressure fluctuation in this case is quite high around the throat region. In particular, the pressure may fluctuate by more than 1000Pa near monitor #2, giving a very high sound pressure level, i.e., buffeting, as predicted in [1].

In order to show the flow pattern around the throat area, Figs. 4(a-e) give the chosen snapshots of the velocity vectors in this region. At 0ms, air escapes from the cavity with higher velocity (>20m/s) around the rear end of the throat as well as in the middle region, as seen in Fig. 4(a). At 9.5ms, a large volume of air enters the cavity with high velocity (>20 m/s) through the frontal region of the throat, see Fig. 4(b). At 18.5ms, the incoming flow moves from the front end towards the middle region, and a strong vortex grows near the front end, as illustrated in Fig. 4(c). At 27.5ms, the flow vortex grows larger and moves to the middle region, and penetrates deeply into the cavity, see Fig. 4(d). Finally, at 37ms, the velocity vectors become similar to those in Fig. 4(a), as shown in Fig. 4(e). At this moment another positive vortex appears near the rear end, which enlarges and carries the flow out of the cavity. These snapshots show that the air at a very large flowrate with high velocity periodically comes in and goes out of the cavity, and penetrates deeply into the cavity. Mechanically, this large flowrate air entering and leaving the cavity is the source of the flow pressure fluctuations and the aeroacoustic noise. The higher fluctuating flowrate will produce the higher sound level or noise.

The vorticity contours Figs. 5(a-e) give further evidence of the flow pattern periodicity, and demonstrate the oscillatory nature of the flow that enters and exits the main cavity.

2.2 Other Cases

In order to further discuss the flow mechanisms that lead to buffeting for the various cases studied in [1], animations of pressure and vorticity contours and velocity vectors for other cases have been made, and will be presented in the conference. For each case to be discussed in this paper, the related snapshots are selected in a similar way as they were for the basic case.

Since all of the other cases studied are based on and compared to the basic case, each particular case is identified by only one distinct feature. For example, the case with a mainstream flow speed of 10m/s represents the case in which the speed is the only feature different from the basic case. Each case is discussed separately in the following sub-sections.

2.2.1 10m/s flow speed

The pressure contours of this case, examined at five specific moments, 0, 23, 46, 68.5 and 91.5ms cover a single period of the flow. Therefore, 91.5ms is the period of the pressure fluctuations, which gives a frequency of 1/91.5 = 10.93Hz, coinciding with the value predicted in [1]. The pressure fluctuations in this case are quite low, with some small changes seen around the throat area. It can be observed that the main flow passes smoothly over the throat and there is very little flow entering or leaving the cavity. Similarly, the shear flow coming from the front end of the throat passes over the throat and mostly runs off the rear wall. There is only some low vorticity flow near the rear end of the throat that moves into the cavity. The sound level generated by this flow is very low compared to the basic case.

2.2.2 40m/s flow speed (time step 0.1ms)

It has been shown in [1] that a time step of 0.5ms is too large and may cause serious computational errors at this flow speed. It is recommended that a smaller time step of 0.1ms should be used. Figures 6(a-e) show the pressure contours around the throat in this case. The period is found to be 7ms, which gives the frequency of the pressure fluctuations 1/7ms = 143Hz, which is quite close the frequency of 141.5Hz predicted in [1]. These figures show that the pressure fluctuation level is quite low compared to the basic case with speed 22m/s shown in Figs. 3(a-e). Figures 7(a-e) illustrate that the high speed outer flow simply passes over the throat. The strong flow fluctuation is tempered and a relatively large but weak vortex sits in the middle of the throat and blocks the exchange of air between inside and outside the cavity. Therefore, the high intensity and low frequency buffeting is suppressed in this case. Figures 8(a-e) show the vorticity contours of this case. These figures also demonstrate that the oncoming shear flow from the front end generally passes over the throat, being prevented from entering the cavity by the large vortex, and only a small amount of air moves down into the cavity near the end of the throat.

2.2.3 5mm grid size

The effect of grid size has been discussed in [1]. Figures 9(a-e), 10(a-e) and 11(a-e) show the pressure contours, velocity vectors and vorticity contours at the same selected moments as for the basic case discussed in section 2.1, but with the calculation performed on a coarser 5mm grid. It can be clearly seen that the flow fluctuation is significantly smeared out by using a coarse grid. As discussed in [1], the 5mm grid may give a poor prediction of the *SPL*, and therefore, it is necessary to carefully choose a correct grid size for accurate computational aeroacoustic analysis.

2.2.4 Cavity x1y1

Figures 12(a-e), 13(a-e) and 14(a-e) give the pressure contours, velocity vectors and vorticity contours at five selected moments for the case with a main cavity which is quarter the size (half length in each direction) of the basic cavity, referred to as case x1y1. It can be seen from the pressure contours in Figs. 12(a-e) that under these conditions the pressure fluctuation level is much lower than in the basic case. The period for this case is 29ms, corresponding to a frequency of 1/29 = 34.5Hz, which coincides with the predicted value in [1]. Figures 13(a-e) show that the flow pattern in the throat area is similar to the basic case, see Figs. 4(a-e), but the vortices are weaker and the outside flow cannot penetrate into the cavity as deeply as in the basic case. The vorticity contours shown in Figs. 14(a-e) also help to explain the flow pattern and are consistent with the lower *SPL* prediction.

The following figures are used to illustrate the throat structure effect in this study.

2.2.5 Case B

Case B is a wider throat case, in which the throat width is taken as in Case A with the sub-cavity cut off. Figures 15(a-e) show the pressure contours for this case at five selected moments, where a low pressure centre is created and grows near the front end of the throat, then moves towards the rear end of the throat. As this low pressure centre moves, a high pressure centre is formed at the front end and moves forward. A new low pressure centre is again formed near the front end, and this pattern repeats itself periodically. The period for this case is 36.5ms, which gives a frequency of 27.4Hz, close to the value given by Case A. However, the pressure fluctuation level, and hence the *SPL*, is somewhat lower than Case A, as predicted in [1]. This is supported by comparing Figs. 4 and 5 with Figs. 16 and 17, which show that the outside air does not penetrate as deeply into the cavity as in Case A.

2.2.6 Case C

Case C has a narrow throat, in which the sub-cavity in Case A is fully filled with solid, see Fig. 4(c) in [1]. Figures 18, 19 and 20 give the pressure contours, velocity vectors and vorticity contours around the throat area. These figures show very similar pressure distributions, flow pattern and vortex formation in this case as their counterparts for Case A, except for some slight difference at moment (a), 0ms. These demonstrate that Case C has almost the same fluctuation frequency and pressure level as for Case A. This indicates that a sub-cavity (with high rear wall) before the front end of the throat does not have any significant affect on the flow pattern and aeroacoustic results.

2.2.7 Case D

Case D is very similar to Case A, but with a full rear wall of the sub-cavity as shown in Fig. 4(d) of [1]. From a close examination of the pressure contours, velocity vectors and vorticity contours for this case, it can be seen that all of the figures obtained for case D are almost identical to their counterparts in Figs. 18, 19 and 20 for Case C, except for some small circulation in the sub-cavity of case D which ultimately does not affect the flow pattern around the throat area. This illustrates that Cases D and C should have the same aeroacoustic characteristics, as predicted in [1].

2.2.8 Case E

Figures 21, 22 and 23 give corresponding contours and vectors for Case E, a case with a short rear wall of the sub-cavity, see Fig. 4(e) of [1]. It can be seen that the flow pattern for this case is somewhat between Cases A and B. Therefore, as predicted, the aeroacoustic characteristics for Case E are between these two cases.

3. SOUNDTRACK EFFECTIVENESS

The monitored pressure fluctuation histories at some particular monitoring locations have been transformed to digital soundtracks using MATLAB [4]. These will be played back using a sound device during the presentation.

4. CONCLUSIONS

In computational aeroacoustic studies, the post-processing of the time-accurate CFD simulations is very important to an enhanced understanding of the flow mechanisms that create noise. In this paper, and in [1], we have demonstrated that the post-processing should include: (i) a Fast Fourier Transform (FFT), which is essential for processing the pressure fluctuation histories to transform the information to a sound spectrum which identifies the fluctuation frequencies and sound pressure levels (*SPL*), as discussed in [1]; (ii) the animation and/or the series of snapshots of the pressure contours, velocity vectors and vorticity contours of the whole domain or in a specific region of interest to clearly illustrate the flow pattern and how it changes with time. Comparing these animations/snapshots for different cases provides additional useful information for understanding the aeroacoustic phenomena; (iii) the digital soundtracks of the pressure fluctuation histories give the audience some realistic feeling about the noise generated by the air flow.

REFERENCES

[1] Barron, R., Zhang, Z. and An, C.-F., Computational Aeroacoustic Study of Airflow over a Cavity, Proceedings of the 13th Annual Conference of the CFD Society of Canada, CFD 2005, St. John's, NL, July 31-Aug. 2, 2005.

[2] FLUENT 6.1 User's Guide.

[3] FotoCanvas 1.1 Help File.

[4] MATLAB Manuals.

Pressure (Pa)	Velocity (m/s)	Vorticity (s^{-1})
1000	20	2000
920	19	1920
840	18	1840
760	18	1760
680	17	1680
600	16	1600
520	15	1520
440	14	1440
360	14	1360
280	13	1280
200	12	1200
120	11	1120
40	10	1040
-40	10	960
-120	9	880
-200	8	800
-280	7	720
-360	6	640
-440	6	560
-520	5	480
-600	4	400
-680	3	320
-760	2	240
-840	2	160
-920	1	80
-1000	0	0

Fig. 1 Color map scales

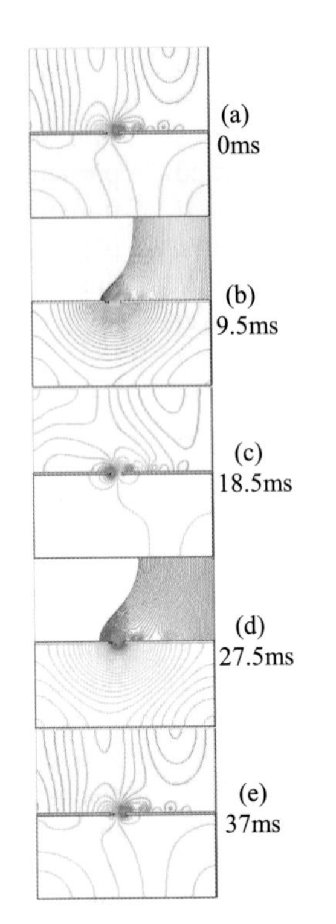

Fig. 2 Pressure contours in whole domain (Basic case)

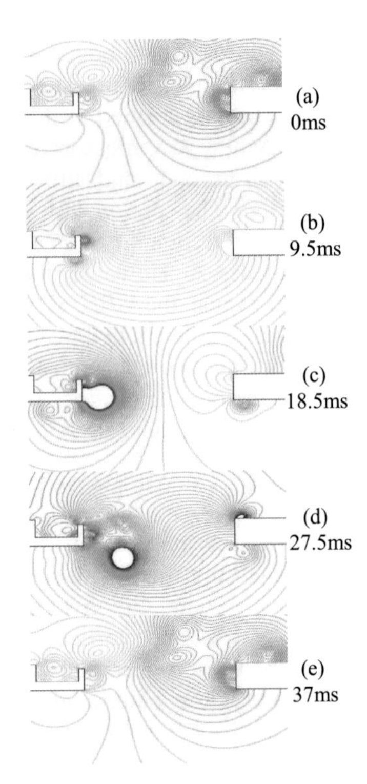

Fig. 3 Pressure contours around the throat (Basic case)

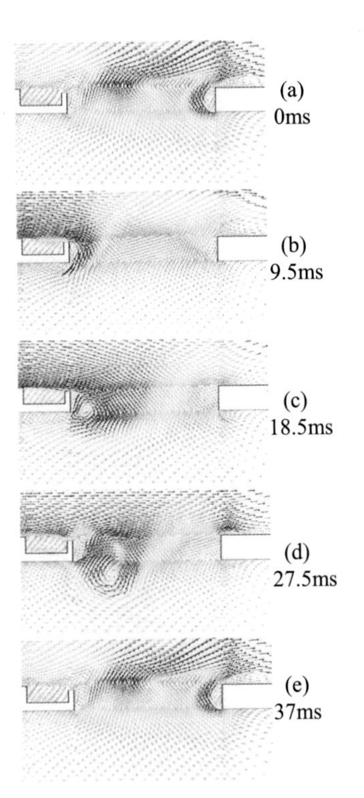

Fig. 4 Velocity vectors around the throat (Basic case)

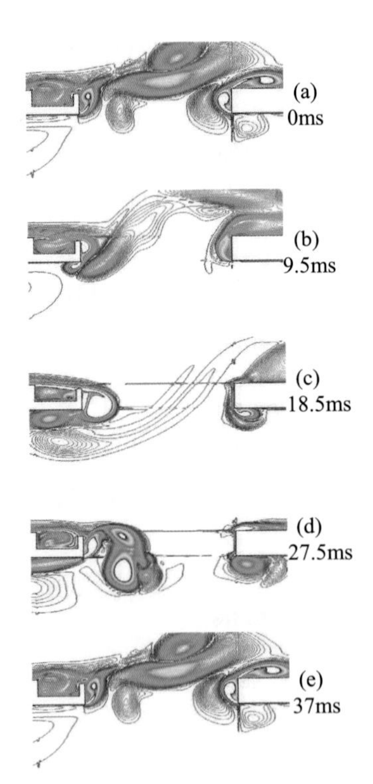

Fig. 5 Vorticity contours around the throat (Basic case)

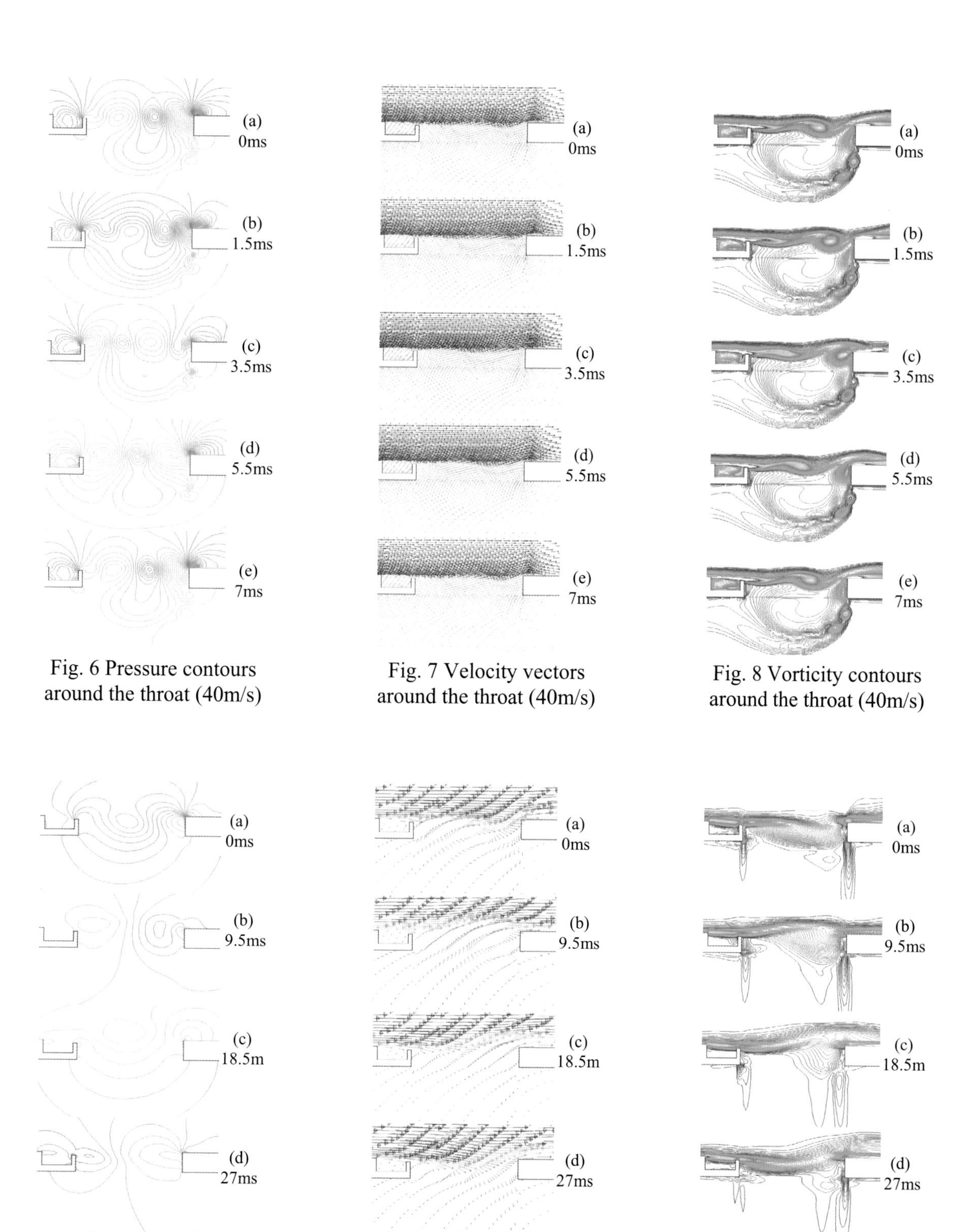

Fig. 6 Pressure contours around the throat (40m/s)

Fig. 7 Velocity vectors around the throat (40m/s)

Fig. 8 Vorticity contours around the throat (40m/s)

Fig. 9 Pressure contours around throat (5mm grid)

Fig. 10 Velocity vectors around throat (5mm grid)

Fig. 11 Vorticity contours around throat (5mm grid)

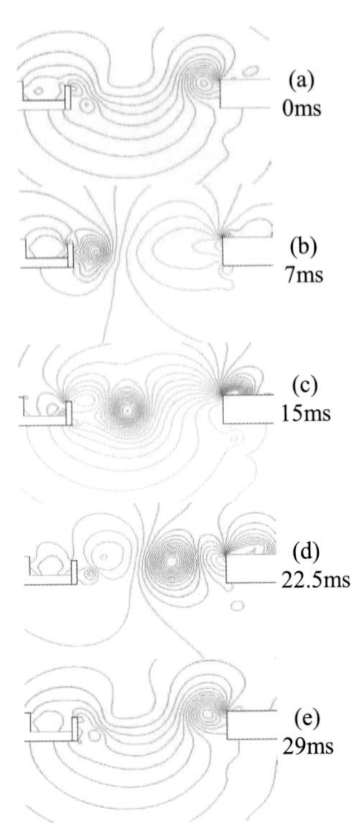

Fig. 12 Pressure contours around throat (cavity x1y1)

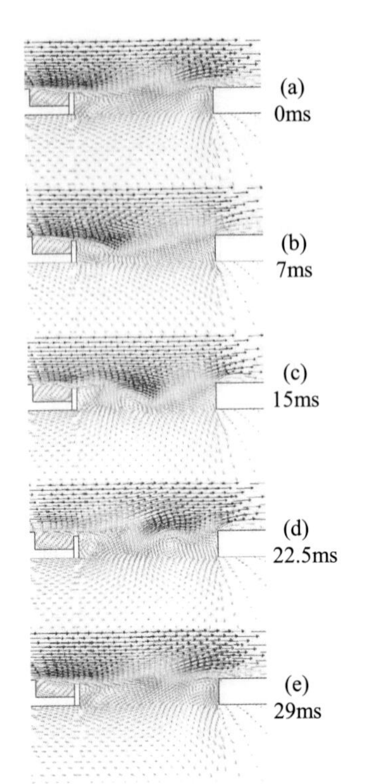

Fig. 13 Velocity vectors around throat (cavity x1y1)

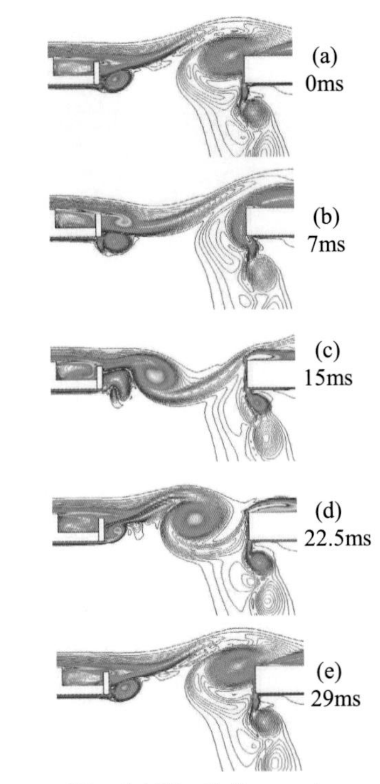

Fig. 14 Vorticity contours around throat (cavity x1y1)

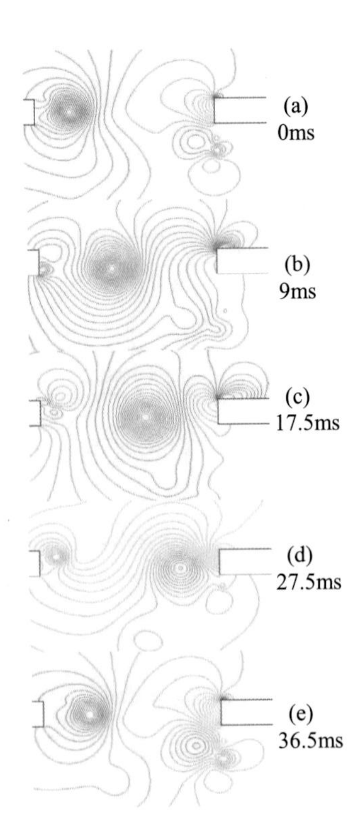

Fig. 15 Pressure contours around throat (Case B)

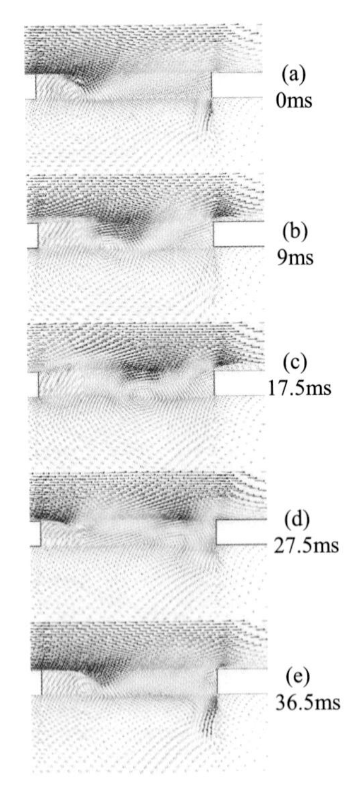

Fig. 16 Velocity vectors around throat (Case B)

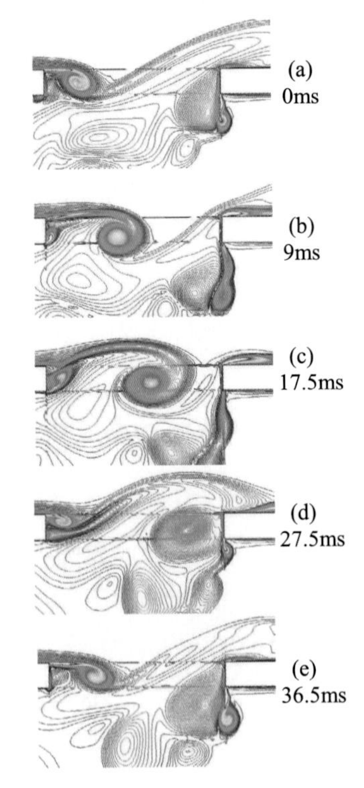

Fig. 17 Vorticity contours around throat (Case B)

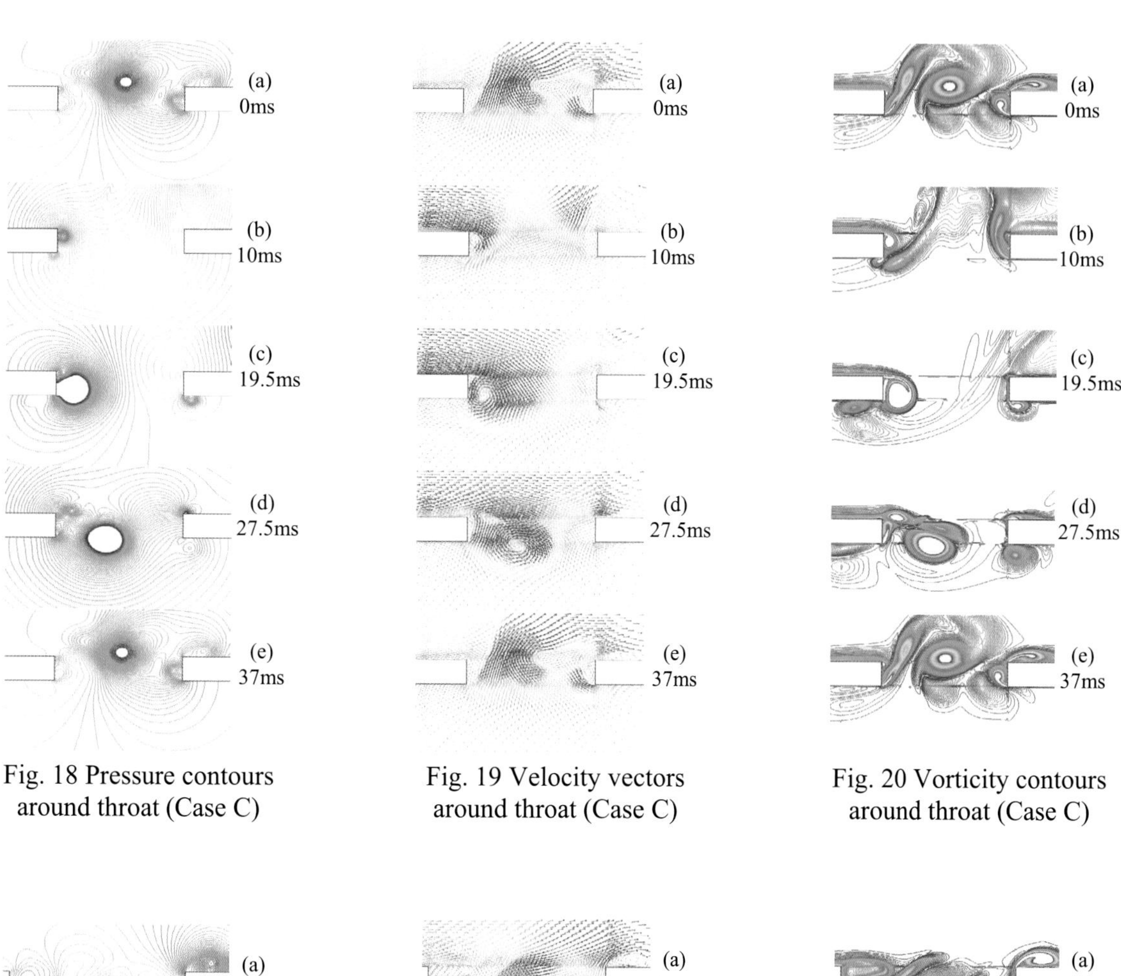

Fig. 18 Pressure contours around throat (Case C)

Fig. 19 Velocity vectors around throat (Case C)

Fig. 20 Vorticity contours around throat (Case C)

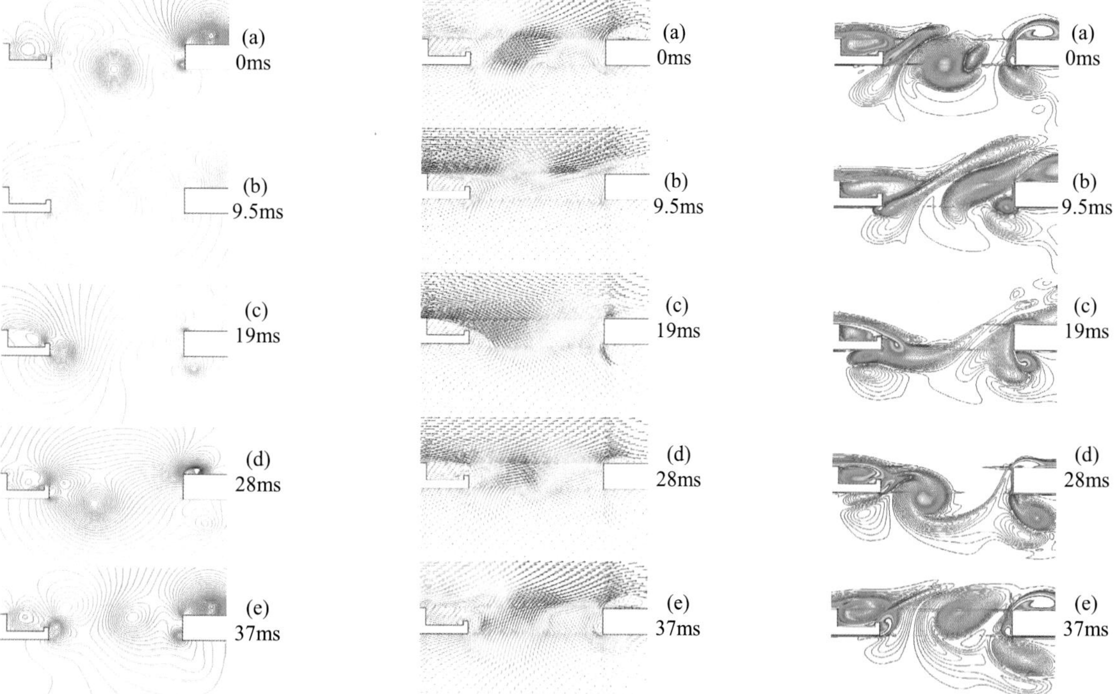

Fig. 21 Pressure contours around throat (Case E)

Fig. 22 Velocity vectors around throat (Case E)

Fig. 23 Vorticity contours around throat (Case E)

Computational Aeroacoustic Study of Airflow over a Cavity

R. Barron[1], Z. Zhang[1,2] and C.-F. An[3]

[1]*Fluid Dynamics Research Institute, University of Windsor, Windsor, ON, Canada, N9B 3P4*
[2]*College of Chemical Engineering, Beijing University of Chemical Technology, Beijing, China 100029*
[3] *DaimlerChrysler Technology Center, Auburn Hills, MI, USA 48326-2757*

Email: *az3@uwindsor.ca*

ABSTRACT

This paper presents a detailed investigation of the airflow over a cavity and the flow induced noise resulting from the pressure fluctuations inside the cavity. This simplified flow model is representative of the air flowing over an open window or sunroof of a road vehicle. The unsteady air flow is simulated using Fluent and the noise characteristics, i.e., main resonance frequency and sound pressure level (*SPL*), are processed using a Fast Fourier Transform technique. The effect of geometry modification, monitoring location, cavity dimension and mainstream flow speed are analyzed. The influence of the numerical accuracy of the CFD simulation on the frequency and *SPL* is also discussed.

安长发博士科研论文集

1. INTRODUCTION

Flow induced noise has been of significant concern to both the aerospace and ground vehicle industries [1-3]. Noise and vibration may be induced by pressure fluctuations when air flows over an obstacle. A major source of aerodynamic noise experienced by automobile passengers is due to the air flowing over an open window or sunroof. The automotive industry refers to the low frequency high amplitude noise as buffeting. Field tests clearly demonstrate the occurrence of buffeting in a certain range of vehicle speeds, and buffeting disappears for most vehicles traveling at very high speeds. Automotive manufacturers are interested in developing strategies to reduce noise levels to enhance passenger comfort and, in particular, to eliminate buffeting since it can cause passengers to experience a feeling of illness. Recent studies have demonstrated the utility of CFD in noise prediction, and in the reduction of noise levels by guiding redesign of component geometry [4-7].

The main features of the airflow over an open window or sunroof can be simulated by modeling the airflow over a cavity. Recognizing the importance of this problem, the organizers of the Third Computational Aeroacoustics (CAA) Workshop selected flow over a deep cavity with a lip as one of its benchmark problems in 1999, and again for the Fourth CAA Workshop in 2003 [8]. CFD can serve as an effective tool to provide a time-accurate solution for the pressure fluctuations which are the source of the aerodynamic noise. The induced noise frequency and sound pressure level are strongly dependent on the parameters affecting the flow characteristics, such as flow speed, cavity structure and dimensions, and accuracy of the CFD simulations is critical to achieving accurate noise predictions [9]. The effects of these parameters are investigated in this research.

2. PROBLEM DESCRIPTION AND SOLUTION METHODOLOGY

The general geometry of the 2-dimensional flow domain of interest in this study is illustrated in Fig. 1. The domain is comprised of two sections, a lower cavity and a rectangular region above it representing the acoustic cavity of a moving vehicle and its surrounding environment, respectively. There is a throat between the two sections representing an open window or sunroof of the vehicle, with a special local structure in front of the throat.

As the airflow passes over the cavity, depending on the throat width and inlet flow speed, some air may move down through the throat into the cavity and then escape back out through the throat. This "up and down" movement causes pressure fluctuations inside the cavity. This motion creates wind noise which, in some situations, may cause discomfort to the passengers inside the vehicle. In the automotive industry, this type of wind noise, if it reaches a certain frequency and sound pressure level, is usually referred to as buffeting.

These pressure fluctuations can be simulated using computational fluid dynamics (CFD) by solving unsteady compressible RANS equations. Since the pressure fluctuations are of small amplitude, the ideal gas model can be used to treat the density change along with the pressure fluctuations. In order to solve this problem, the commercial CFD software, Fluent, is adopted. For the computational domain, a non-uniform Cartesian grid system is created using Gambit. Generally, a finer uniform grid is used in the throat area. The grid in this region ultimately determines how large the mesh of the whole domain will be. The grid size is then gradually expanded from the throat area towards the outer boundaries of the rectangle and the cavity. In order to capture the high frequency fluctuations, a second order implicit scheme is used for time marching and a second order upwind scheme is used for space terms in the momentum, energy and turbulence equations. The RNG k-ε model is adopted to capture the turbulence.

For convenience in modeling the flow, the simulated vehicle (the cavity) is assumed to be stationary and the surrounding air passes over it. In this case, the uniform airflow enters the flow (computational) domain from the left side of the rectangle, as a velocity inlet boundary, see Fig. 1. The upper boundary of the rectangle (far field) is taken as a frictionless wall and the right side of the rectangle is set as an outlet boundary with constant pressure (p = 0). The lower boundary of the rectangle, except the throat, and all sides of the cavity are no-slip walls (u,v = 0). In this study, the upper rectangular region is taken to be 2m high, 2m in length downstream and 2m long upstream from the rear edge of the throat. In the basic case, the cavity has the same dimension as the rectangle above it. The effect of the cavity volume is investigated later by reducing the dimensions of the cavity in both x and y directions.

Tests were performed to verify that the initial field for all variables does not affect the final results; hence the computation may be started from either a uniform flow field or from any simulated flow field under a similar condition. Optimal relaxation factors for each of the variables vary significantly in different cases, and are strongly problem dependent. Normally, we have used relaxation factors $\lambda_{u,v}$ = 0.6-0.85, λ_p = 0.4-0.65, λ_ρ = 0.4-1.0, λ_k = 0.9-1.0, λ_ε = 0.8-1.0, λ_ω = 0.4-1.0. Here λ is relaxation factor, u and v are velocity components in x and y directions, respectively, p is pressure, ρ is density, k is turbulence kinetic energy, ε is turbulence dissipation rate and ω is vorticity. Generally, it is impossible to find a universal set of optimal relaxation factors for all flow conditions, and some trial and error is needed for each particular problem. The number of sub-iterations within a time step is taken to be 15-30 for most simulations performed in this study.

The unsteady flow simulation is performed at a prescribed time step until pressure fluctuations at all pre-set monitor locations (see Fig. 1) are stable, as illustrated in Fig. 2. The recorded pressure fluctuations, or noise signals, are then acoustically post-processed by the Fast Fourier Transform (FFT), using the mathematical tool Matlab. A specially designed procedure for spectral analysis is shown in Fig. 3. Figure 3(a) shows a typical noise signal, i.e. a time history of the pressure fluctuations. A specific time period of the signal is selected for spectral analysis as shown in Fig. 3(b). Then, the power spectrum is obtained after FFT as shown in Fig. 3(c). Finally, the sound pressure level (*SPL*), shown in Fig. 3(d), is calculated:

$$SPL = 10\log_{10}\left(\frac{p}{p_{ref}}\right)^2$$

where p is the amplitude of the pressure fluctuations and the reference pressure is taken to be $p_{ref} = 2\times10^{-5}$ Pa.

Six variations of the local structure around the throat area are shown in Fig. 4. Case A, which represents a sunroof opening, contains a sub-cavity at the upstream side of the throat. Case B has a wider throat and no sub-cavity. Case C is the case with a narrower throat. Cases D-F are variants of case A with different heights of the sub-cavity rear wall. In case D, the rear wall has the same height as the throat. In case E, the rear wall has height equal to one-half that of the throat, and in case F, there is no sub-cavity rear wall.

3. COMPUTATIONAL ISSUES: MESH SIZE AND TIME STEP

3.1. Determination of Mesh Size

A typical mesh is shown in Fig. 5. In the throat area, the grid spacing in both x and y directions are uniform. From this region towards all of the outer boundaries, the grid spacing is gradually expanded with a uniform ratio of the neighboring cells. For most of the results presented here, a grid size of 2mm is used in the throat area, and the total number of cells in each direction is 200. In this case, 2mm is used to identify the grid fineness. To compare the effect of the mesh size, and ensure the solutions are grid independent, simulations were also carried out with a grid fineness of 1mm and 5mm.

Table 1 documents the effect of the grid fineness in the throat region for cases A-E, at an inlet velocity of 22m/s, for 5mm and 2mm grids. The main frequency and *SPL* at four monitoring locations are recorded. In most cases the main frequency, but not the *SPL*, is accurately predicted on both the 5mm and 2mm grids. This is critical information regarding accuracy, since the flow field solutions on the 5mm and 2mm grids may appear to be quite similar.

For case A, at a flow speed of 22m/s, the simulation on a 1mm grid predicts a main frequency of 27.1Hz at monitor #2, and corresponding *SPL* of 142.4dB, only 4dB lower than the 2mm grid result. At 40m/s, the *SPL* on the 1mm and 2mm grids is almost identical, at about 129dB. Based on comparison of the 1mm, 2mm and 5mm grid results, and considering the significant increase in computational time required to achieve the 1mm grid results, all subsequent simulations were carried out on a 2mm grid.

3.2. Determination of Time Step

Since the flow variables, such as pressure, velocity and vorticity, may undergo very rapid changes with time in an aeroacoustic problem, the main resonant frequency is strongly dependent on the characteristics of the flow. A large time step used in the computation may smooth out the rapidly changing flow characteristics. Therefore, for a problem with higher resonant frequency, a smaller time step may be necessary in the simulation. In general, a suitable time step for a particular problem must be carefully justified.

In case A, with a 2mm grid and inlet velocity 30m/s, the main resonant frequency of the noise inside the cavity is about 27Hz and a relatively large time step, 5×10^{-4}sec, appears to be good enough. Figure 6 shows the noise spectra obtained at monitor #2 from simulations using two different time steps, 1×10^{-4}sec and 5×10^{-4}sec. The two spectral curves are quite close and the calculated main resonant frequency (27Hz) and *SPL* (146.8dB) are almost identical. Therefore, one can conclude that $\Delta t = 5\times10^{-4}$sec is good enough and a smaller time step is not necessary. Furthermore, it can be shown that $\Delta t = 5\times10^{-4}$sec is also sufficiently small for other cases with even smaller main resonant frequency, corresponding to upstream velocities less than 30 m/s.

On the other hand, it is found that when the mainstream velocity increases to 35m/s, the time step $\Delta t = 5\times10^{-4}$sec is not small enough. At this speed, the low frequency resonance disappears. Instead, a higher resonant frequency and lower sound pressure level are obtained. In this case, a smaller time step is needed to capture the higher frequency resonance. Figure 7 shows the comparison of the results at 40m/s for several time steps: $\Delta t = 5\times10^{-4}$sec, $\Delta t = 2\times10^{-4}$sec, $\Delta t = 1\times10^{-4}$sec and $\Delta t = 5\times10^{-5}$sec. From the spectral analysis, the main resonant frequencies and *SPL* at 40m/s for the above time steps are calculated as: f_{main} = 105.2Hz, *SPL* = 76.3dB for $\Delta t = 5\times10^{-4}$sec; f_{main} = 125.3Hz, *SPL* = 139.3dB for $\Delta t = 2\times10^{-4}$sec; f_{main} = 141.5Hz, *SPL* = 128.7dB for $\Delta t = 1\times10^{-4}$sec. Since the noise spectrum for $\Delta t = 1\times10^{-4}$sec is different from that for $\Delta t = 2\times10^{-4}$sec, one more run for time step $\Delta t = 5\times10^{-5}$sec is also performed. For $\Delta t = 5\times10^{-5}$ sec, the main resonant frequency is f_{main} = 141Hz and *SPL* = 126dB. This is quite close to the case of $\Delta t = 1\times10^{-4}$sec, and the spectral curves for the two time steps are almost the same. Therefore, the captured values of f_{main} and *SPL* for time step $\Delta t = 1\times10^{-4}$sec are considered acceptable. Clearly, a time step no larger than $\Delta t = 1\times10^{-4}$sec is required.

4. Structure and Flow Parameter Effects

4.1. Effect of the Throat Structure

The calculated main resonant frequency and *SPL* at monitor #2 at speed 22m/s for all six cases are listed in Table 2, and illustrated in Fig. 8.

From Table 2 and Fig. 8 we can observe the trend of the main resonant frequency and *SPL* of the noise inside the cavity. The main resonant frequencies for all cases, but case B, are almost the same (~27Hz). In fact, even for the wider throat case B, the main resonant frequency is 27.3Hz, only 1% higher than the other cases. According to theoretical analysis [10], the main resonant frequency is directly proportional to the square root of the throat width. This means that the noise frequency increases with the increase in throat width. The computational results indicate that the difference in the structure details of the sub-cavity at the front end of the throat may not have as strong an effect on the main resonant frequency as the theoretical analysis suggests. On the other hand, the difference in the structure details may have a noticeable effect on the intensity of the noise, i.e. *SPL*. Among all the six cases, the intensities for cases A, C and D are close to one another and stronger than the other cases. Cases E and F are second and case B is the weakest. One can conclude that the difference in the structure details at the front end of the throat may not have a strong influence on the noise intensity provided the rear wall of the sub-cavity is high enough. The height of this rear wall affects the behaviour of the mainstream flow into and out of the main cavity.

4.2. Effect of Flow Speed

The study of flow speed effect is based on case A. The simulated results are plotted in Fig. 9 to show the effect of mainstream speed. For a low speed of 10m/s, both the main resonant frequency and sound pressure level are quite low, i.e. 118dB at 10.9Hz. There is no buffeting at this speed. When the speed increases to 15m/s and higher, the main resonant frequency increases to about 27Hz and remains at this level until a speed of 30m/s. At the same time, the sound pressure level increases to 126dB for 15m/s and 146.4dB for 22m/s, and then remains the same until a speed of 30m/s. This demonstrates that the obvious buffeting starts around a speed of 15m/s and the buffeting intensity increases continuously until a speed of 22m/s is reached. The buffeting intensity between speeds of 22m/s and 30m/s is quite high, i.e. around 146dB. However, when the speed increases to 35m/s, the low frequency buffeting noise disappears. At a mainstream flow speed of 40m/s, the main resonant frequency is raised to 141.5Hz with the sound pressure level of 128.7dB.

4.3. Effect of Cavity Dimension

All the discussion and analysis thus far relates to the main cavity with dimensions 2m by 2m, referred to as x2y2 (x=2m, y=2m), as shown in Fig. 1. Three additional cases with different dimensions, referred to as x2y1 (x=2m, y=1m), x1y2 (x=1m, y=2m) and x1y1 (x=1m, y=1m), have also been investigated. All the other parameters are the same as those used in the above analysis for case A.

The noise spectra for the four cases with different cavity dimensions are compared in Fig. 10. The sound pressure level for case x2y2 is the highest among the four cases. The *SPL* for case x1y1 is the lowest. The sound pressure levels for cases x2y1 and x1y2 are close to each other and are located between cases x1y1 and x2y2. On the contrary, the main resonant frequency for case x2y2 is the lowest, and it is the highest for case x1y1. For cases x2y1 and x1y2, the main resonant frequencies are almost the same, lying between cases x2y2 and x1y1. These results clearly demonstrate that under our particular parameters, the volume of the cavity, rather than the actual cavity dimensions, seems to have stronger influence on the noise spectrum. The larger cavity volume causes more severe buffeting. The main resonant frequencies and *SPL* for the four cases are listed in Table 3. It is worthwhile to note that the trend of the change in main resonant frequency with the volume of the cavity is consistent with the theoretical formula in which the main resonant frequency is inversely proportional to the square root of the volume of the cavity [10].

4.4. Effect of Monitor Location in Cavity

Passengers sitting at different locations inside a vehicle may perceive the noise differently. Therefore, the effect of the monitor location has also been considered in this work.

Figure 11 shows the noise spectral curves at six different monitor locations inside the cavity for case A. The dimension of the cavity is x2y2. The grid at the throat is 2mm and the speed is 22m/s.

As shown in Fig. 11, the frequencies of the main resonance, as well as higher harmonics, for different monitor locations match very well, but the sound pressure levels (*SPL*) are different. The *SPL* in the throat area (monitors #1 and #2) and at the bottom of the cavity (monitor #4) is quite high (>140dB), while the *SPL* at locations below the throat area (monitors #5, #6 and #7) is considerably lower (<125dB).

5. CONCLUSIONS

CFD analysis has been performed to investigate the buffeting noise inside a cavity induced by air flowing over the cavity. Examination of the effects of grid size and time step on the predicted acoustic properties, i.e., main frequency and sound pressure level, illustrates the importance of achieving an accurate numerical simulation of the flow field.

Several cavity openings have been studied to determine the effect of the local physical structure on buffeting. It appears that the *SPL* does not change if the rear wall of the sub-cavity at the front of the throat is high enough to deflect the mainstream shearing flow and cause it to impact on the rear end of the throat opening. Consistent with theoretical analysis, a wider throat results in a higher main resonance frequency. In this case, the noise intensity inside the cavity is reduced. Is has also been shown that the larger main cavity (i.e., larger volume) has higher *SPL* and lower frequency.

It is shown that buffeting begins to occur at a mainstream speed of about 15m/s, and intensifies up to about 30m/s. beyond 35m/s, the main frequency rises sharply and the buffeting noise disappears. This behavior is consistent with field observations and wind tunnel tests.

Finally, the frequency of the pressure oscillations is independent of the location at which the pressure is being monitored. However, the *SPL* does depend on the monitoring position.

REFERENCES

[1] Block, P.J.W., Noise Response of Cavities of Varying Dimensions at Subsonic Speeds, NASA TN D-8351, 1976.

[2] George, A.G., Automobile Aeroacoustics, AIAA-89-1067, 1989.

[3] Mongean, L., Bezenek, J. and Danforth, R., Pressure Fluctuations in a Flow-excited Door Gap Cavity Model, SAE Paper 971923, 1997.

[4] Karbon, K. and Singh, R., Simulation and Design of Automobile Sunroof Buffeting Noise Control, AIAA Paper 2002-2550, 8th AIAA/CAES Aeroacoustics Conference and Exhibit, June 2002.

[5] Hendriana, D., Sovani, S.D. and Schiemann, M.K., On Simulating Passenger Car Side Window Buffeting, SAE Paper 2003-01-1316, 2003.

[6] An, C.-F., Alaie, S.M., Sovani, S.D., Scislowicz, M.S. and Singh, K., Side Window Buffeting Characteristics of an SUV, SAE Paper 2004-01-0230, 2004.

[7] Karbon, K., Computational Analysis and Design to Minimize Vehicle Roof Rack Noise, SAE Paper 2005-05B-98, 2005.

[8] Proceedings of the Fourth Computational Aeroacoustics (CAA) Workshop on Benchmark Problems, NASA/CP-2004-212954, September 2004.

[9] Zhang, Z., Barron, R. and An, C.-F., Spectral Analysis for Airflow over a Cavity, Proceedings of the Fourth Computational Aeroacoustics (CAA) Workshop on Benchmark Problems, NASA/CP-2004-212954, September 2004, p. 197-204.

[10] Rayleigh, J.W.S., “The Theory of Sound”, Vol. II, Chapter XVI. Theory of Resonators, First American Edition, 1945, Dover Publications, Inc. New York, NY, USA.

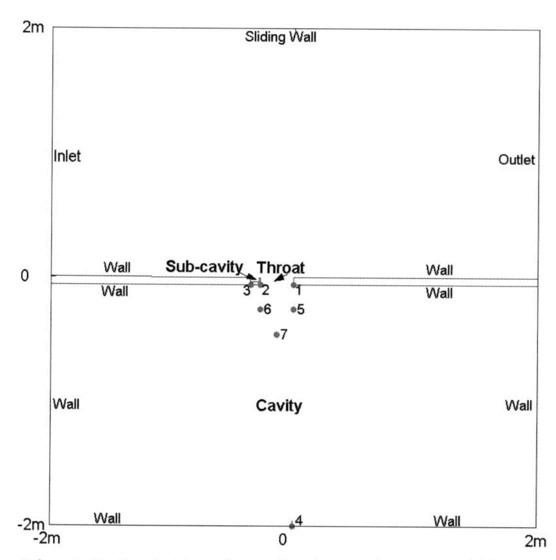

Fig. 1 Calculation domain, boundary conditions and monitor locations

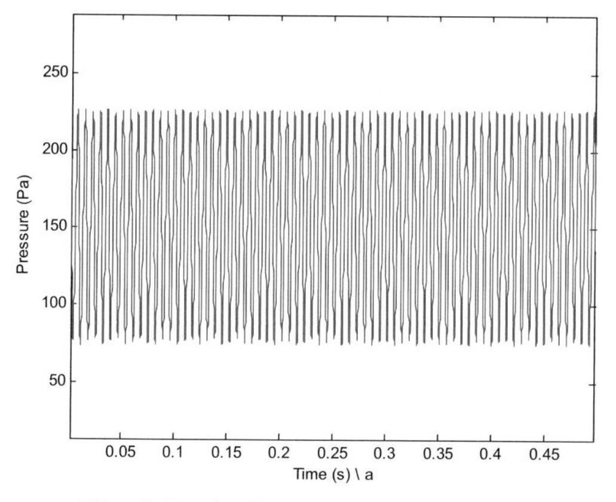

Fig. 2 Typical pressure fluctuations

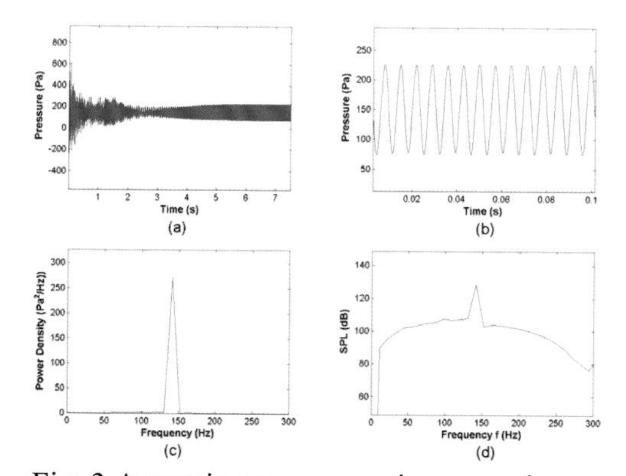

Fig. 3 Acoustic post-processing procedure

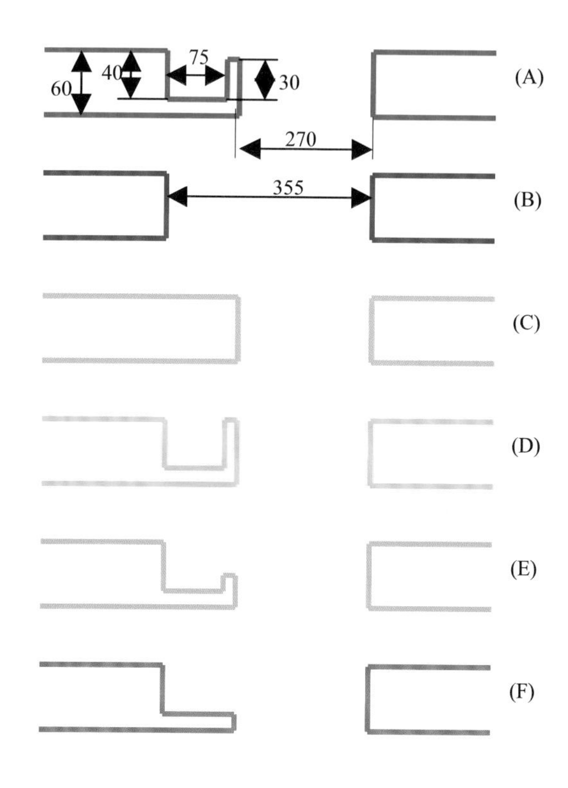

Fig. 4 Different throat structures

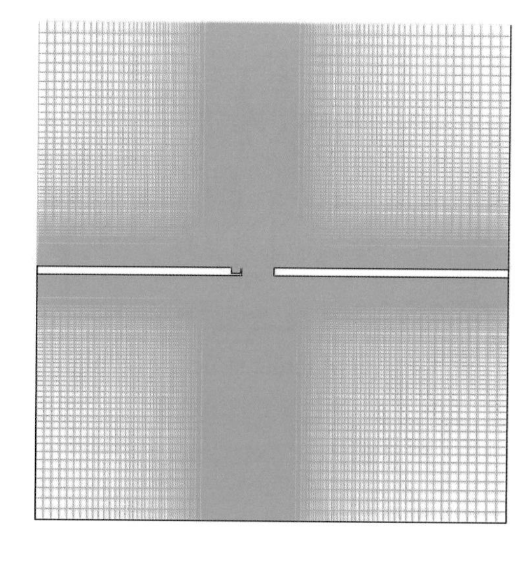

Fig. 5 Mesh for geometry in case A (2mm grid)

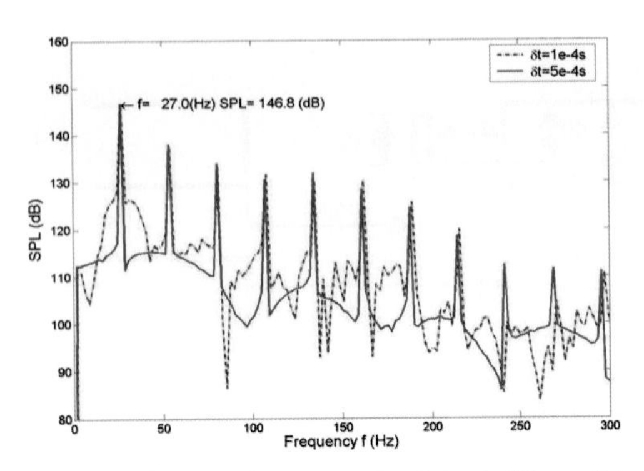

Fig. 6 Effect of time step for 30m/s (Case A)

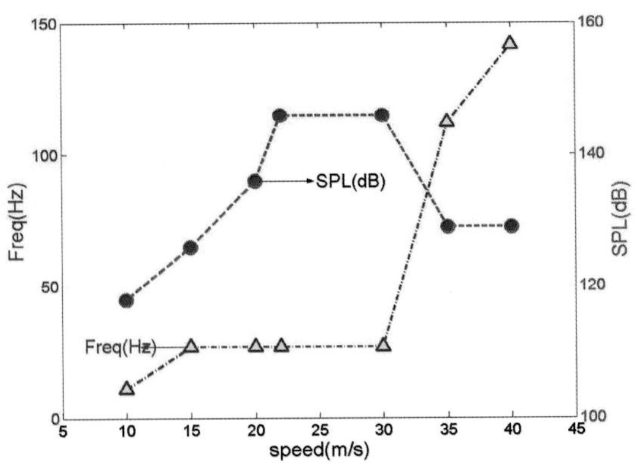

Fig. 9 Effect of flow speed

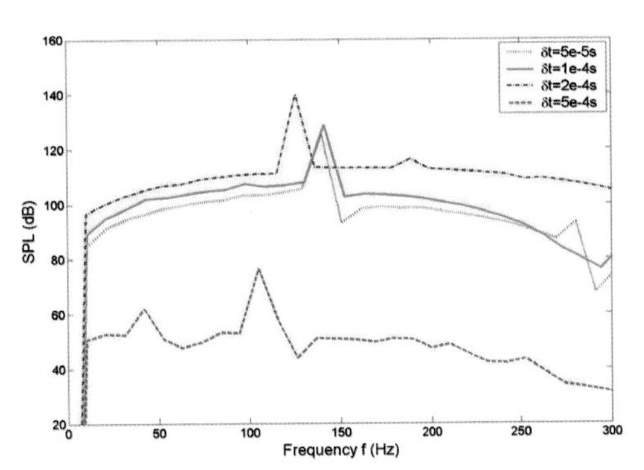

Fig. 7 Effect of time step for 40m/s (Case A)

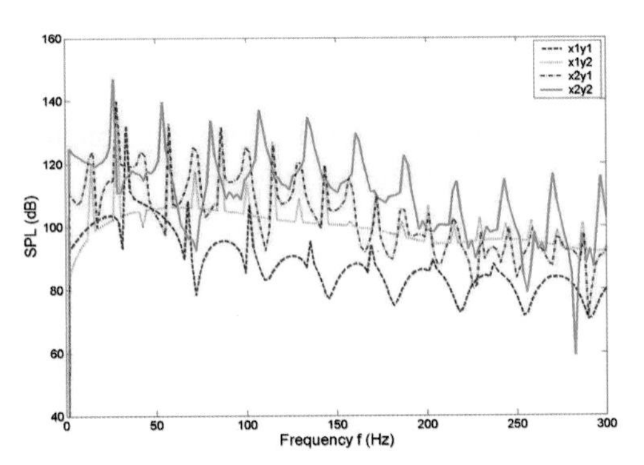

Fig. 10 Effect of cavity dimension (Case A, 22m/s)

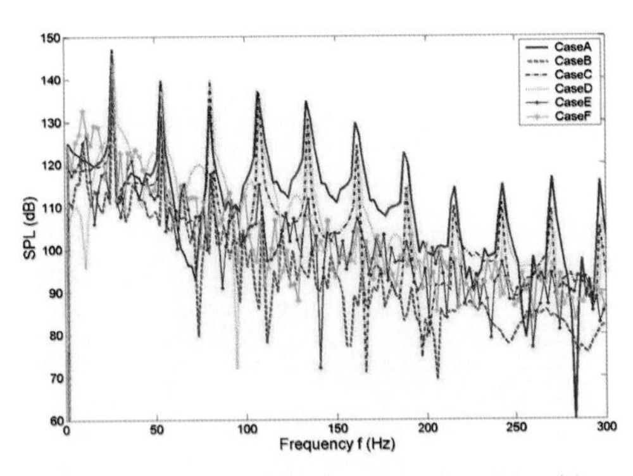

Fig. 8 Effect of throat structure (22m/s)

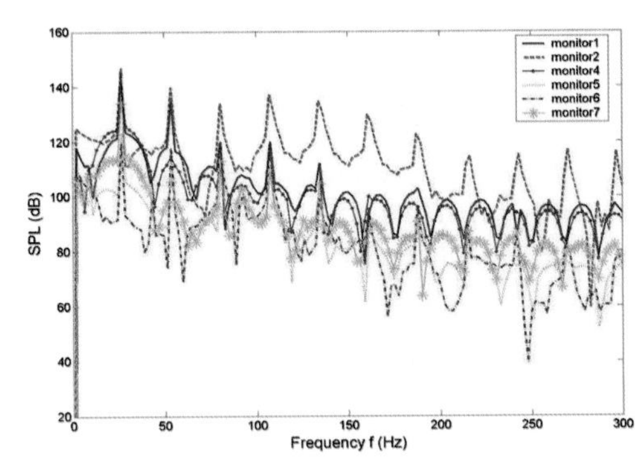

Fig. 11 Effect of monitor location in cavity (Case A, 22m/s)

Monitor	Case		A	B	C	D	E
1	F_{main}(Hz)	2mm	26.9	11.2	27.1	27.0	27.1
		5mm	27.3	27.2	27.9	27.2	26.9
	SPL(dB)	2mm	142.6	125.7	140.6	139.5	135.5
		5mm	121.8	142.1	119.6	140.3	133.7
2	F_{main}(Hz)	2mm	26.9	27.3	27.0	27.1	27.0
		5mm	27.2	26.9	27.8	26.9	26.9
	SPL(dB)	2mm	146.4	137.8	146.5	146.3	140.1
		5mm	125.7	146.3	137.8	145.7	139.3
3	F_{main}(Hz)	2mm	26.9	27.3	27.0	27.1	27.0
		5mm	27.3	43.1	27.4	26.9	43.1
	SPL(dB)	2mm	130.3	133.3	129.0	129.9	127.9
		5mm	114.3	128.1	132.3	129.1	127.1
4	F_{main}(Hz)	2mm	26.9	27.3	27.1	27.1	27.0
		5mm	27.2	27.3	27.1	27.4	26.9
	SPL(dB)	2mm	144.6	133.6	143.7	143.4	141.2
		5mm	129.1	145.6	134.1	144.2	140.2

Table 1: Grid Size Effect (Case A, 22m/s)

Case	A	B	C	D	E	F
F_{main} (Hz)	26.9	27.3	27.0	27.1	26.9	26.9
SPL (dB)	146.4	137.8	146.5	146.3	140.9	141.9

Table 2: Effect of Throat Structure (22m/s, 2mm grid)

Cavity Dimension	x1y1	x2y1	x1y2	x2y2
F_{main} (Hz)	34	28.7	28.7	26.9
SPL (dB)	132	139	138.5	146.4

Table 3: Effect of Cavity Dimension (Case A, 22m/s, 2mm grid)

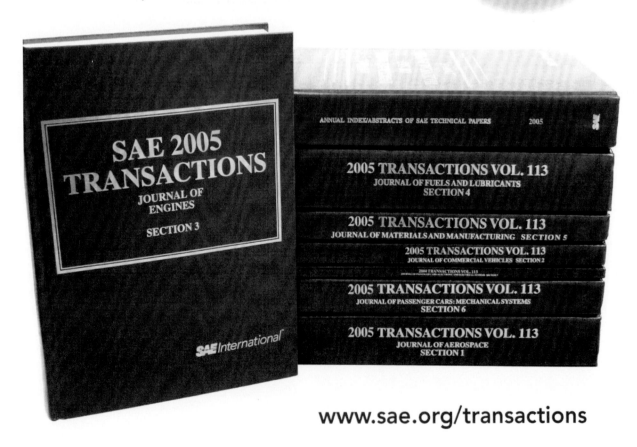

2005-01-0551 **Development of Direct Electromagnetic Clutch System** 625
Rikiya Kunii, Akihiro Iwazaki, Shigenobu Sekiya, Yutaka Kanno, Katsuhiro Kitamura, and Yasuji Shibahata

2005-01-0553 **Torque Vectoring Driveline: SUV-Based Demonstrator and Practical Actuation Technologies** 631
Jonathan C. Wheals

2005-01-0602 **Computational Analysis and Design to Minimize Vehicle Roof Rack Wind Noise** 649
Kenneth J. Karbon and Urs D. Dietschi

2005-01-0603 **Attempts for Reduction of Rear Window Buffeting Using CFD** 657
Chang-Fa An, Mitchell Puskarz, Kanwerdip Singh, and Mark E. Gleason

2005-01-0604 **Experimental Study of Unsteady Wakes Behind an Oscillating Car Model** 665
F. Chometon, A. Strzelecki, V. Ferrand, H. Dechipre, P. C. Dufour, M. Gohlke, and V. Herbert

2005-01-0606 **Experimental Study for Applicable Limit of Acoustic Analogy to Predict Aero-Acoustic Noise of Commercial Vehicles** 674
Yukio Matsushima and Itsuhei Kohri

2005-01-0607 **A Numerical Study of High-Lift Single Element Airfoils With Ground Effect For Racing Cars** 682
Wael A. Mokhtar

2005-01-0611 **Studies of Adaptive Finite Element Method for Component Crashworthiness Analysis** 689
Shen R. Wu

2005-01-0702 **Numerical Investigation of Effects of Frame Trigger Hole Location on Crash Behavior** 697
Wayne Li, Tau Tyan, Guofei Chen, Xiao Ming Chen, and Ming F. Shi

2005-01-0707 **Evaluation of Accident Parameters in a Numerical Fleet for Assessing Compatibility** 707
Gijs Kellendonk, Cor Van Der Zweep, and Herman Mooi

2005-01-0735 **A Subsystem Crash Test Methodology for Retention of Convenience Organizer Equipment System in Rear Impact** 718
Pradip K. Syamal and Patrick L. Brown

2005-01-0737 **Early Detection of Rollovers With Associated Test Development** 724
Jerry Jialiang Le, Robert W. McCoy, and Clifford C. Chou

2005-01-0739 **Spatial Encoding of Structured Light for Ranging With Single Camera** 731
Henry Kong, Qin Sun, and William Bauson

2005-01-0740 **Use of Photogrammetry in Extracting 3D Structural Deformation/Dummy Occupant Movement Time History During Vehicle Crashes** 736
Robert V. McClenathan, Said S. Nakhla, Robert W. McCoy, and Clifford C. Chou

The SAE 2005 Transactions Journal of Passenger Car: Mechanical Systems, Volume 113, Section 6, can be found from the library of University of Michigan, Ann Arbor, MI, USA

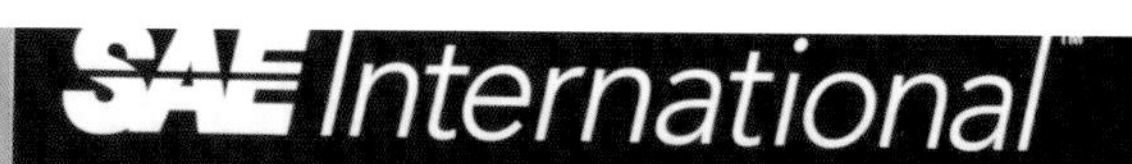

Changfa An
Detroit, MI, USA
April 3, 2006

"The premier society dedicated to advancing mobility engineering worldwide"

SAE 2006 World Congress

April 3-6, 2006 • Cobo Center • Detroit, MI USA
2006 Theme: Driving Innovation through Partnerships.

EVENT GUIDE

Includes Final Program and Exhibit Directory

Host Company:

BMW Group  

Published by:

automotive engineering international

The Official Publication of SAE 2006 World Congress

060209

Vehicle Aerodynamics 2006

SP-1991

Chang-Fa An
Detroit, MI, USA
April 3, 2006

SAE International™

Published by:
Society of Automotive Engineers, Inc.
400 Commonwealth Drive
Warrendale, PA 15096-0001
USA
Phone: (724) 776-4841
Fax: (724) 776-5760
April 2006

2006-01-0138

Optimization Study for Sunroof Buffeting Reduction

Chang-Fa An and Kanwerdip Singh
DaimlerChrysler, Auburn Hills, MI

ABSTRACT

This paper presents the results of optimization study for sunroof buffeting reduction using CFD technology. For an early prototype vehicle as a baseline model in this study a high level of sunroof buffeting 133dB has been found. The CFD simulation shows that the buffeting noise can be reduced by installing a wind deflector at its optimal angle 40 degrees from the upward vertical line. Further optimization study demonstrates that the buffeting peak SPL can be reduced to 97dB if the sunroof glass moves to its optimal position, 50% of the total length of the sunroof from the front edge. For any other vehicles, the optimization procedure is the same to get the optimal parameters. On the other hand, however, this optimization study is only based on fluid dynamics principle without considering manufacturability, styling, cost, etc. Further work is needed to utilize the results in the production design.

INTRODUCTION

The terminology of buffeting has different meaning in different industry. In aeronautical industry buffeting, or flutter, is referred to as the elastic response of an airplane component, such as a wing, a tail or a fin, to a temporal load of the air flow [1]. In architectural industry buffeting is referred to as the vibrating response of a construction structure, such as a long bridge [2] or a high building [3], to the wind gust. In automotive industry, however, buffeting is referred to as the aeroacoustic response of the air bulk inside the vehicle to the outside transient air flow. More precisely, automotive buffeting is a low frequency but high level wind noise perceived in a moving vehicle at a certain speed when a side window or sunroof is open. In this paper only automotive sunroof buffeting is considered.

The physical mechanism of buffeting noise now is well understood. At front edge of the opening, there exists a shear layer between the outside moving air flow and the inside relatively stationary air bulk. The shear layer becomes unstable when the relative velocity between the outside and inside air exceeds a critical value. As a result, vortices will be formed and shed periodically. If the frequency of vortex shedding happens to coincide with the natural frequency of the vehicle interior air bulk, the so-called Helmholtz resonance will occur and buffeting will take place.

Although the buffeting mechanism is well understood, it is not easy to tell how severe the buffeting noise is for a specific vehicle until the physical model of the prototype vehicle is available. This is because the geometry of road vehicles is very complex and no simple formulae exist to calculate buffeting level. Therefore, it is too late when a new vehicle is on the road and the buffeting noise is found to be very severe.

The methods of buffeting control are usually categorized into two types: passive and active. The passive methods are more popular at the present time. One of the passive methods is a deflector, installed at front edge of the opening, which can “deflect” the vortex away from the opening and therefore reduce buffeting. Another passive method is “glass comfort position”. It is based on the buffeting mechanism that the natural frequency of the passenger compartment can be changed by adjusting the size of the opening so that the excitation of resonance can be avoided. The method of “venting” is also a passive method which is realized from the observation that buffeting can be “vented out” through a second opening of the passenger compartment. The active buffeting control methods include mass injection at the upstream of the opening, an oscillating device with opposite phase of the buffeting noise, a loud speaker with electronic control system and so on. The active control methods are more expensive and more difficult to implement and therefore rarely used in the current vehicles. In general, a specific buffeting control method can work only under a certain circumstance. Some of them are still not well understood and not effectively utilized.

With fast development of high speed computer and improvement of CFD technology, buffeting can be predicted whenever the digital model of a new vehicle is available. During the last decade automotive companies, with the help of CFD vendors, conducted CFD simulations of sunroof and side window buffeting [4-7]. Since the year of 2003, intensive efforts of CFD studies on buffeting have been made at DaimlerChrysler [8-11]. These efforts include procedure establishment, correlation with wind tunnel testing, attempts to find solutions for buffeting reduction and parameter optimization to suppress buffeting.

This paper presents the up-to-date results of sunroof buffeting reduction using CFD optimization study. In the next section, several buffeting-related concepts are reviewed to get better understanding of buffeting mechanism. Then the vehicle model, the CFD procedure and its validation are briefly outlined. As the major part of the paper, the parameter optimization study is reported in details in the subsequent section. At the end of the paper concluding remarks are addressed.

BUFFETING-RELATED CONCEPTS

The buffeting-related concepts cited in this section include: (a) Helmholtz resonator, (b) flow over a cavity, (c) shear layer instability and (d) vortex shedding.

HELMHOLTZ RESONATOR

A Helmholtz resonator is an air chamber that is connected to the air outside the chamber through a narrow throat (Figure 1). Communicating with the outside flow, the air inside the chamber is forced to move-in and –out through the throat. In this way aeroacoustic noise is generated with a certain frequency. In early 1940s, Rayleigh [12] derived a formula to calculate natural frequency of a Helmholtz resonator as $f = c/(2\pi)\sqrt{A/(VL)}$ where f is natural frequency, c is speed of sound, V is volume of the Helmholtz resonator, L and A are length and cross-area of the throat, respectively. Blevins [13] collected natural frequency formulas for many structural and fluid systems, including the above formula for a Helmholtz resonator. Alster [14] derived a modified formula for the natural frequency of a Helmholtz resonator which is a lengthy expression. In automotive buffeting study, the passenger compartment can be considered as a Helmholtz resonator, but the formula may be not accurate due to the geometry complexity of the realistic vehicle interior. Nevertheless, these formulas can be used to estimate the trend of natural frequency.

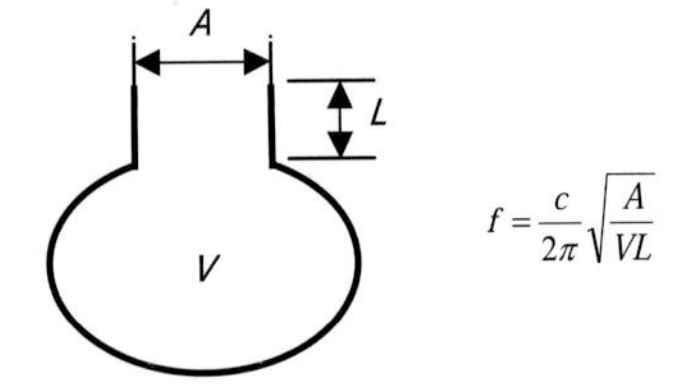

Figure 1. Helmholtz resonator [12]

FLOW OVER A CAVITY

Another buffeting-related concept is the flow over a cavity. As a matter of fact, the vehicle interior in buffeting study can also be considered as a cavity with a big volume and small opening. Most early work on flow over a cavity was done in aeronautical industry. For example, Roshko [15] measured pressure fluctuations in a cavity. Tam [16] conducted a theoretical study of flow over a cavity with the interest in aeroacoustic noise. In 1978, Rockwell and Naudascher [17] published a review paper of self-sustaining oscillations of flow past cavities. In mid-1980s, Blake [18] reviewed the flow over cavities with some complex details of shear layer impingement and feedback. Some years later, Blevins [19] summarized the state-of-the-art studies on the sound excited due to flow over cavities. Since 1990s, the study of noise generation from flow over a cavity has been applied to automotive industry. The cavity flow at the door gap of a road vehicle was adopted as one of the benchmark problems in the 3rd and 4th computational aeroacoustics (CAA) workshop of NASA in 2000 [20] and 2004 [21], respectively. Figure 2 shows an example of the benchmark.

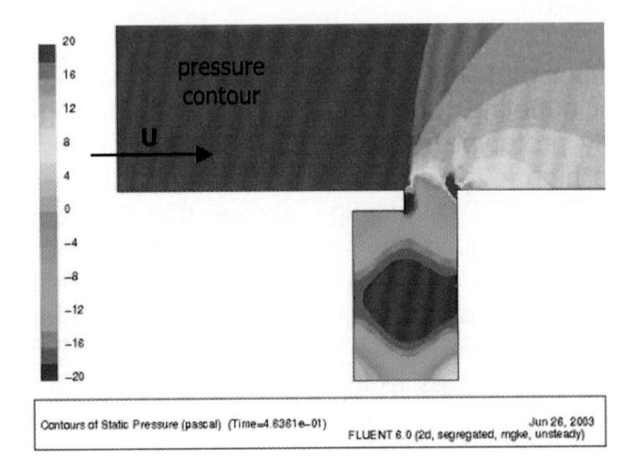

Figure 2. Flow over a cavity [21]

SHEAR LAYER INSTABILITY

The next buffeting-related concept is shear layer instability which is a classical subject in fluid mechanics. Early work on shear layer instability for a step velocity profile was named Kelvin-Helmholtz instability and analyzed by Lamb [22].

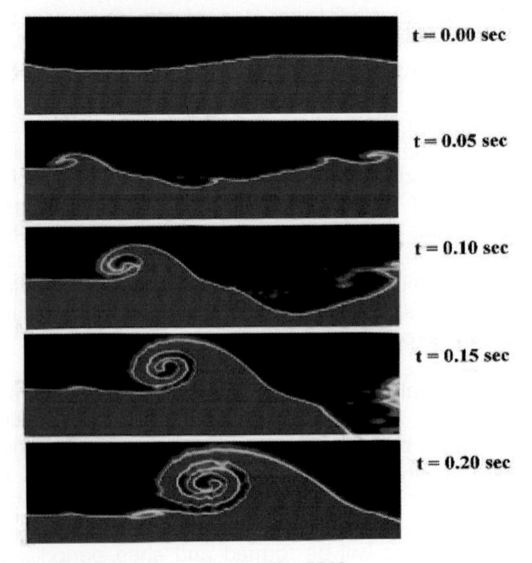

Figure 3. Shear layer instability [23]

An et al [23] conducted a study of Kelvin-Helmholtz instability with application to the oil-water interfacial phenomena. As shown in Figure 3, if relative velocity between oil (blue) and water (red) exceeds a critical value, the shear layer will become unstable. In this circumstance, any small disturbance will be magnified and shear layer will be distorted and vortices will be generated. Esch [24] examined the shear layer instability for a linear velocity profile between upper and lower uniform streams using an eigenvalue technique and determined the region of instability. In mid-1960's, Michalke applied the linear stability theory for the case of temporally growing disturbances to a hyperbolic-tangent velocity profile [25], analyzed and explained the vortex formation, development and rolling up [26], and finally extended the analysis to the case of spatially-growing disturbances [27]. The shear layer instability was widely investigated and summarized in Lin [28] and Bechov and Criminale [29] in their books.

VORTEX SHEDDING

Vortex shedding is the natural outcome of shear layer instability. This subject can be traced back to the famous observation of vortex street by von Karman. Abernathy and Kronauer [30] proposed a theory of vortex formation using the concept of point vortices and vortex sheets. Boldman et al [31] made a flow visualization of vortex shedding from a trailing edge of a plate in a wind tunnel. They found an interesting fact that strong von Karman vortices develop when the velocities are equal while the vortices tend to disappear when the velocities are unequal. Nakamura and Nakashima [32] conducted an experimental investigation of vortex excitation where two layers were separated by a bluff body of various cross sections and concluded that the impinging-shear-layer is responsible for vortex excitation. Figure 4 shows vortex shedding from the sunroof in a 2D study. In the graph, blue color represents vortex in the clockwise direction while red color represents vortex in the counter-clockwise direction. It is clear that the counter-clockwise vortex is dominant.

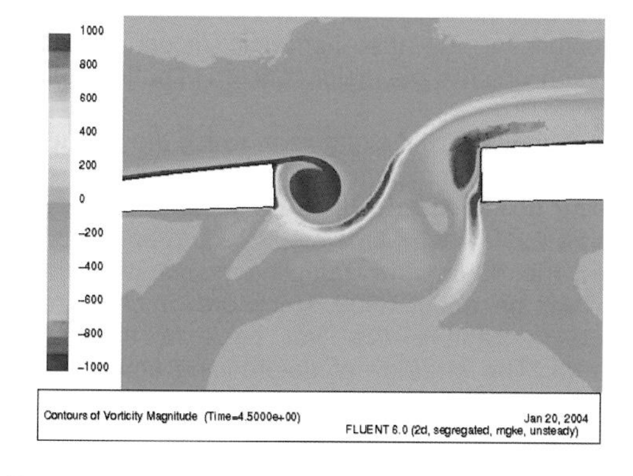

Figure 4. vortex shedding [10]

MODEL, PROCEDURE AND VALIDATION

The procedure of CFD simulation for side window buffeting was described in the previous papers [9, 11]. For sunroof buffeting the procedure is almost the same except a minor difference in local mesh refinement. For side window buffeting the finest locally-refined volume mesh is in the open window region while for sunroof buffeting the finest locally-refined volume mesh is in the open sunroof region. To provide readers an outline a brief description of the procedure is written in this section.

MODEL

The geometry model of an early prototype vehicle which is used for simulation is provided from design office in CATIA format. As shown in Figure 5, the model includes (a) exterior with the open sunroof and (b) interior with dummies, seats, carpet, etc. The software ANSA is used to clean-up geometry and to generate surface mesh. The vehicle model is placed in a virtual wind tunnel of 50m x 20m x 10m, as shown in Figure 6.

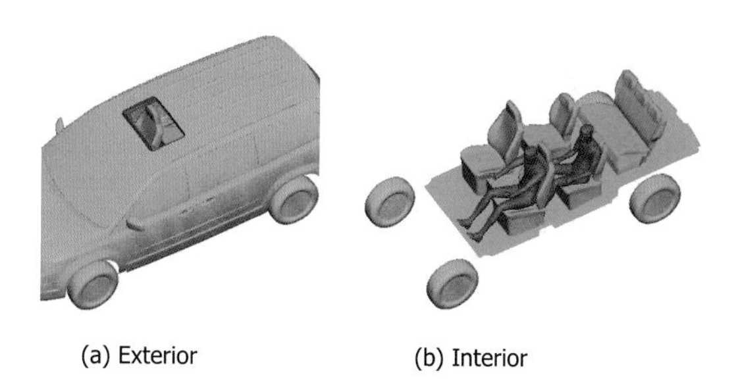

Figure 5. Geometry model of the vehicle

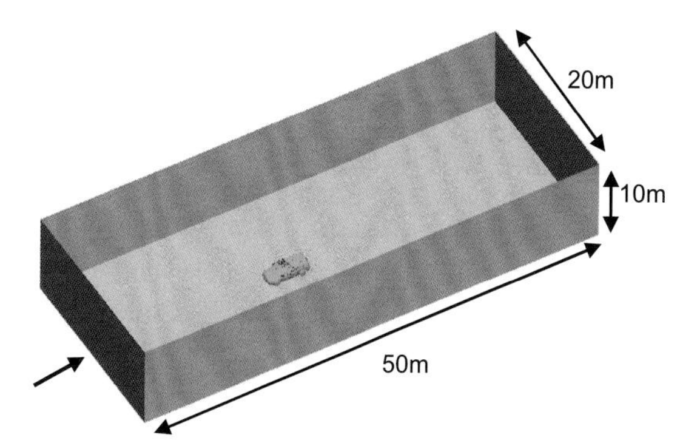

Figure 6. Virtual wind tunnel with the vehicle

The surface mesh of the vehicle and the virtual wind tunnel is then transferred to TGRID for volume mesh generation and local refinement. Four levels of local refinement are utilized to ensure that the mesh is sufficiently fine in the regions of sensitive flow unsteadiness. The final volume mesh typically contains about 3.5 million tetrahedral cells. Figure 7 shows the mesh of the 4 levels of local refinement regions on the central plane. Region 1 and 2 encompass the vehicle exterior, region 3 inscribes the vehicle interior and region 4 covers the sunroof opening.

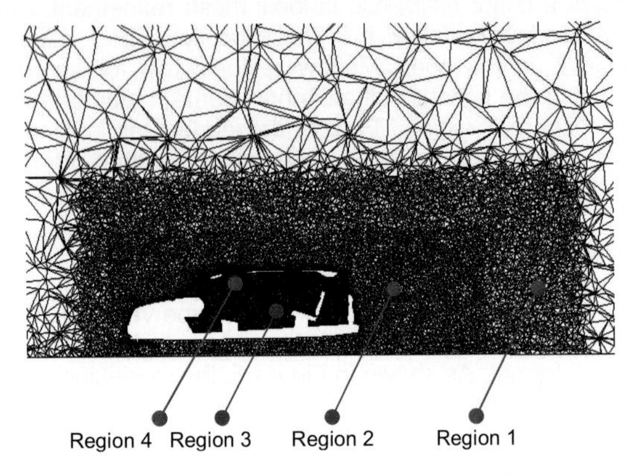

Figure 7. Local refinement of volume mesh

PROCEDURE

The commercial CFD software Fluent 6.2.16 [33] is used to simulate unsteady flow around the vehicle. The air is assumed to be ideal gas – a compressible fluid. The solver is segregated unsteady. Time marching is 2nd order implicit. Space discretization is 2nd order upwind.

The turbulence model is chosen to be large eddy simulation (LES) which is available in Fluent. Turbulent flows are characterized by eddies with a wide range of length scale and time scale. The largest eddies are typically comparable in size to the characteristic length of the mean flow. The smallest scales are responsible for the dissipation of turbulent kinetic energy. It is theoretically possible to directly solve the whole spectrum of turbulent scales using an approach known as direct numerical simulation (DNS). However, DNS is not feasible for practical engineering problems. Theoretical analysis shows that to resolve all turbulent scales, the mesh size in 3D simulation will be proportional to $\mathrm{Re}_t^{9/4}$ where Re_t is the turbulent Reynolds number. This huge mesh size makes it impossible to simulate flows with high Reynolds number using DNS. The conventional flow simulation employs the solution of the Reynolds-Averaged Navier-Stokes (RANS) equations. In RANS, all the turbulent motions are modeled resulting in significant savings in computational effort. Conceptually, the LES model is situated somewhere between DNS and RANS. In LES, large eddies are resolved directly while small eddies are modeled. Typical mesh sizes in LES can be at least one order of magnitude smaller than with DNS. This is affordable for engineering calculations. Furthermore, the time step sizes are proportional to the eddy-turnover time, which is much less restrictive than with DNS. Therefore LES model can be considered as a possibility of engineering calculations and is used in the present buffeting study.

The governing equations, employed for LES, are obtained by filtering the time-dependent Navier-Stokes equations. The filtering process effectively filterers out the eddies whose scales are smaller than the filter width or grid spacing (so-called “subgrid-scale”) used in the computations. In the filtered governing equations, the subgrid-scale stress τ_{ij} requires modeling. The majority of subgrid-scale models are the eddy viscosity model in the following form:

$$\tau_{ij} - \frac{1}{3}\tau_{kk}\delta_{ij} = -2\mu_t \overline{S_{ij}}$$

where μ_t is the subgrid-scale turbulent viscosity and $\overline{S_{ij}}$ is the rate-of-strain tensor for the resolved scale:

$$\overline{S_{ij}} = \frac{1}{2}\left[\frac{\partial \overline{u_i}}{\partial x_j} + \frac{\partial \overline{u_j}}{\partial x_i}\right]$$

Fluent contains two models for μ_t, the Smagorinsky-Lilly model and the RNG-based subgrid-scale model. In the present study, the Smagorinsky-Lilly model is used in which the turbulent eddy viscosity is modeled by

$$\mu_t = \rho L_s^2 |\overline{S}|$$

where L_s is the mixing length for subgrid-scales and

$$|\overline{S}| = \sqrt{2\overline{S_{ij}}\,\overline{S_{ij}}}$$

The time step is chosen to be 2×10^{-3}sec. It should be mentioned at this point that there are some arguments regarding the scale of time step. In some other sunroof buffeting CFD work, smaller time step (~10^{-5}sec) is used. In fact, the frequency of the dominant resonance (the frequency of the first harmonic mode) is about F=20Hz in most buffeting cases. That is, the time period is about T=1/F=5×10^{-2}sec. This implies that if time step is set to be 2×10^{-3}sec there will be about 25 sampling points within a time period. This should be enough to capture the feature of the first harmonic mode of pressure fluctuation which is the major concern in buffeting study.

Figure 8 shows a real example in which the frequency is 17Hz and the time period is 0.059sec. There are 29 sampling points within a time period when time step is set to be 2×10^{-3}sec. If time step is set to be the order of 10^{-5}sec the number of sampling points within a time period will be about 5900 which are much more than enough to capture the first peak in the buffeting spectrum. The strategy of determining time step is to keep it as large as possible for the reason of computational efficiency as long as the solution accuracy is reasonably acceptable. The accuracy of the present set-up is considered to be acceptable based on the validation cases shown in Table 1 in the next sub-section. Making a compromise between solution accuracy and computational efficiency, a time step of $2x10^{-3}$sec is used in buffeting study for all production vehicles.

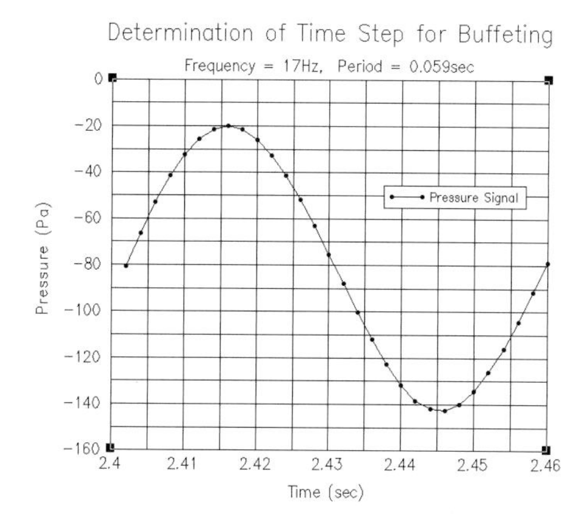

Figure 8. Determination of time step for buffeting

The inlet boundary condition of the virtual wind tunnel is a constant mass flow rate (required by the new version of Fluent 6.2.16) such that the velocity at the inlet is equal to the vehicle speed. The outlet of virtual wind tunnel is atmospheric pressure. The top, two sides and bottom are frictionless walls. The vehicle surface is frictional wall.

At the beginning of each case, a steady state simulation is run for several hundred iterations using the standard k-ε turbulence model. The steady state flow is then used as an initial field to start unsteady simulation and the turbulence model is switched to LES. Typical unsteady run time is 4-5 days with 16 processors to reach 1200 time steps. This is on an SGI Origin 2000 machine with 64 300-MHz CPU and 64GB memory using the IRIX 6.5 operating system.

During the unsteady simulation static pressure is recorded at the sensors which were pre-set at dummy ears. Excluding the irregular initial progress of the first several hundred time steps, the pressure signal reaches a status of dynamically stable periodic fluctuation. Then, the selected signal can be used for Fourier transform using the software XMGR. After Fourier transform, the amplitude of pressure fluctuation is expressed as a function of frequency. At last, the amplitude of pressure fluctuation is converted to the format of sound pressure level (SPL) in dB using the conventional logarithm formula $SPL=10Log_{10}(P/P_{ref})^2$ where P is the amplitude of pressure fluctuation in Pa and the reference pressure $P_{ref} = 20 \times 10^{-6}$ Pa.

VALIDATION

In order to validate the procedure of CFD analysis for buffeting, several realistic vehicles have been used to correlate CFD simulation with wind tunnel testing. Some of them are listed in Table 1. From this table we can see that for most buffeting cases the dominant resonance frequency is within 2Hz and the peak SPL is within 2dB. With reasonable correlation of these buffeting cases, more confidence was gained about the accuracy of the CFD procedure and our attention was turned to solution and optimization. The previous paper [11] was focused on the possible solutions to reduce side window buffeting. The present paper is focused on the optimization study on possible solutions to suppress sunroof buffeting.

Table 1. Validation of CFD procedure for buffeting

Vehicle (60mph)	Open window	CFD (dB@Hz)	Test (dB@Hz)
Passenger Car	Front	115@18	116@17
Passenger Car	Rear	129@17	130@17
SUV1	Front	112@18	113@14
SUV1	Rear	133@16	135@15
SUV2	Rear	121@17	122@15
Minivan	Rear	125@15	127@13

BASELINE CASE

Before the optimization study to minimize buffeting level, a baseline case should be determined as the standard of comparison. Driving experience and wind tunnel test indicate that 40mph is the worst speed at which sunroof buffeting level for the selected vehicle reaches its maximum. Therefore, the model of sunroof glass full open without a deflector at speed of 40mph is chosen as the baseline case for the optimization study. Figure 9 shows the simulated results of buffeting spectrum for the baseline case which tells us severe buffeting occurs at 17Hz with the peak SPL 133dB.

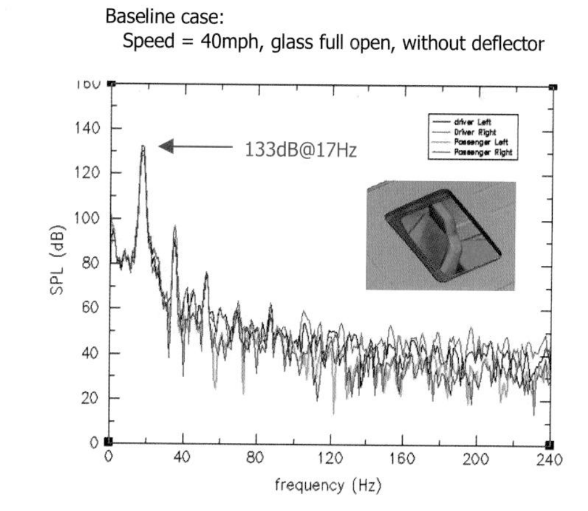

Figure 9. Baseline case for optimization study

Figures 10 shows velocity vector on the central plane at a certain time moment and Figure 11 shows static pressure on the same plane at the same time moment to give a quick glimpse at the flow of the baseline case. From both figures, we can observe a strong vortex, shed

第IV部分 任职于美国克莱斯勒汽车公司时期

from the front edge of the sunroof, which is considered as the source of buffeting noise. For a time-dependent flow, animation is a useful tool to show a sequential series of frames of flow variable, such as static pressure. The flow pattern changes with time can be seen visually and the inherent characteristics of a specific unsteady flow may be discovered. Obviously, only selected still frames can be drawn in the paper. The full animation of the whole series of frames can be watched as a movie only in the presentation at the conference. Since the flow pattern for a particular buffeting case changes periodically with time, it is appropriate to choose typical frames within a time period to show extreme situations at special time moments. In the following, 8 frames are chosen for the baseline case and the subsequent cases.

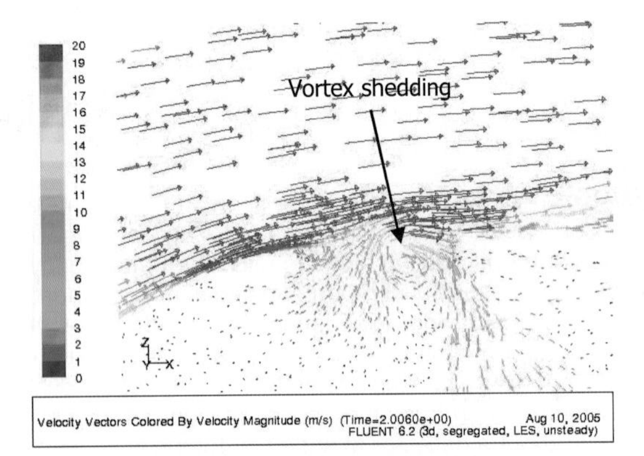

Figure 10. Velocity vector of baseline case

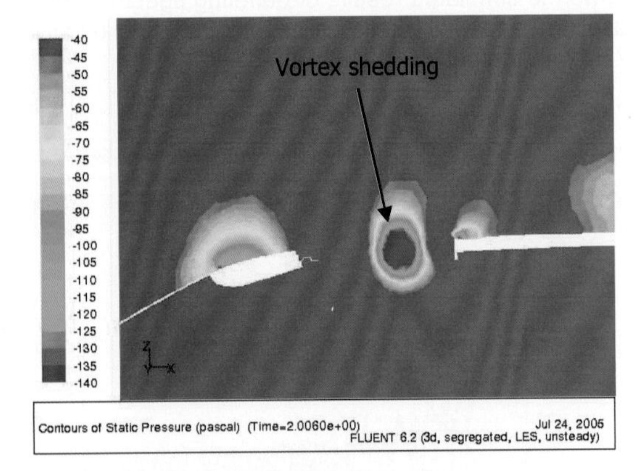

Figure 11. Static pressure of baseline case

Figure 12 shows a sequential series of pressure contour of the baseline case. The color map of pressure scale is the same as Figure 11 in Figure 12 and the following figures 15 and 18. As we have already known that the dominant resonance frequency is F=17Hz and, hence, the time period is T=1/F=0.059sec. Graphs (a), (b), (c), (d), (e), (f), (g) and (h) are pressure contours at time moment t=0, T/8, T/4, 3T/8, T/2, 5T/8, 3T/4 and 7T/8, respectively. At t=T, graph (i) is the same as graph (a) at t=0. From these graphs we can observe pressure fluctuations and vortex movement within a time period so that we can explain why and how buffeting occurs at such a high level of 133dB.

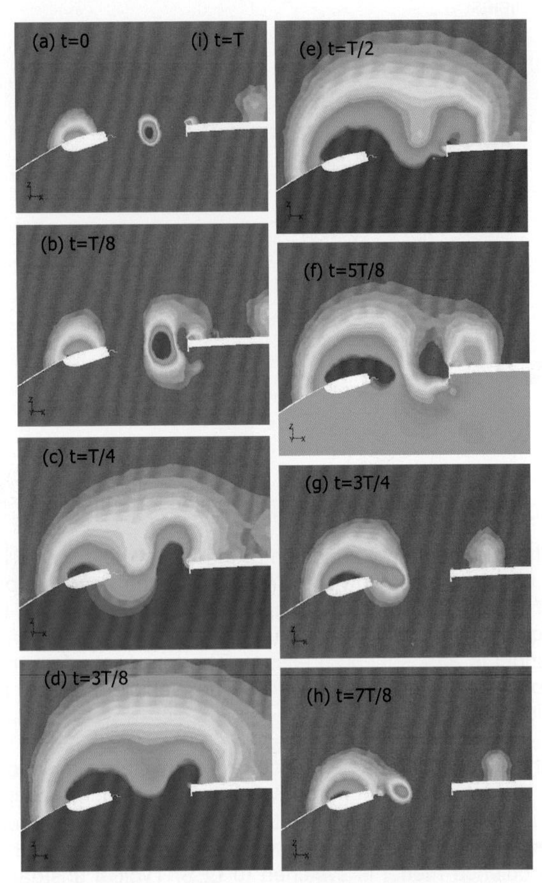

Figure 12. Flow pattern of baseline case
Color map of pressure scale is same as Figure 11

At t=0, pressure in the vehicle interior is very high and a strong vortex with a low pressure core is on the half way of the sunroof opening. At t=T/8, the vortex moves downstream and becomes larger. At t=T/4, the vortex hits the rear edge of the sunroof, breaks-down and intrudes into the interior of the vehicle so that the interior pressure decreases significantly. At t=3T/8, the above process continues. At t=T/2, the interior pressure goes down to its minimum and a secondary vortex is about to appear at the rear edge of the sunroof. At t=5T/8, the interior pressure starts to increase and the secondary vortex leaves the sunroof region. At t=3T/4, the interior pressure continues to increase and a new vortex is about to form at the front edge of the sunroof. At t=7T/8, the interior pressure almost reaches its maximum and the newly-formed vortex is shed from the front edge of the sunroof. At t=T, the period ends and the flow pattern returns to its initial state of t=0.

Next, we will turn our attention to find the means of buffeting reduction. As stated in the introduction, wind deflector and glass comfort position are popular methods for buffeting reduction. Firstly, we will optimize deflector angle and then optimize deflector height while keeping its optimal angle. Lastly, we will try the combination of deflector and glass position. Figure 13 shows the sketch of the wind deflector.

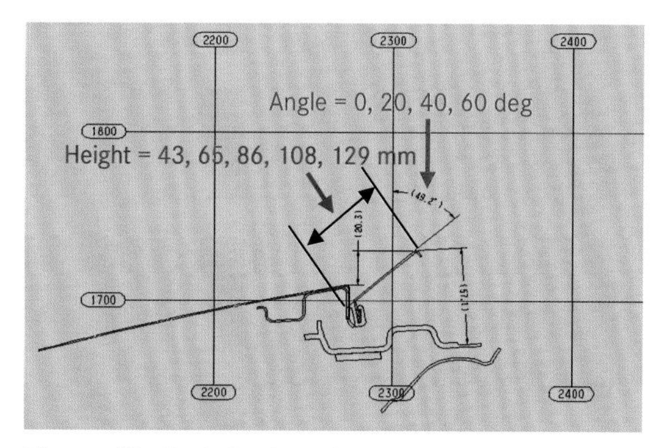

Figure 13. Optimization of wind deflector

OPTIMIZATION OF DEFLECTOR ANGLE

The first step of the optimization study is to optimize the angle of wind deflector. Based on the provided sunroof model with a deflector height of 43mm, various deflector angles, 0, 20, 40, 60 degrees, are simulated to find out what angle is optimal to minimize buffeting. Figure 14 shows the result of this parameter sensitivity study for deflector angle which is measured from the upward vertical line. As shown in the figure, smaller angles do not work well and larger angle is even worse. The optimal angle of deflector is about 40 degrees at which the peak buffeting SPL decreases from 133dB in the baseline case to 124dB.

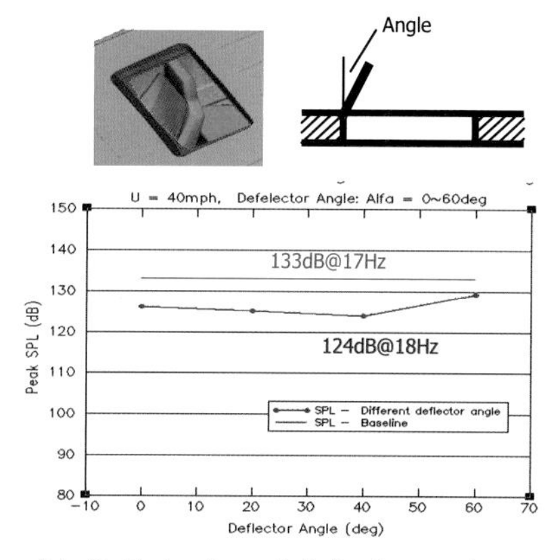

Figure 14. Optimization of deflector angle

Similar to the baseline case, a sequential series of 8 frames of pressure contour within a time period for the case of deflector angle 40 degrees, identified as DA40 case, and are shown in Figure 15. The color map of pressure scale is the same as that in Figure 11.

At t=0, the flow feature is similar to Figure 12(a) but the interior pressure is not as high as that in the baseline case and the vortex is weaker than its counterpart due to the existence of deflector. At t=T/8, the flow feature is also similar to Figure 12(b), but the vortex is weaker and the interior pressure change is milder. At t=T/4, the interior pressure reaches its average and the vortex is approaching to the rear edge of the sunroof. At t=3T/8, the vortex hits the rear edge of the sunroof and breaks-down and the interior pressure continues to decrease. At t=T/2, the interior pressure reaches it's lowest, but still higher than that in the baseline case, c.f. Figure 12(e). At t=5T/8, the interior pressure begins to increase and no secondary vortex to appear, c.f. Figure 12(f). At t=3T/4, the interior pressure reaches its average again and a new vortex is about to form at the front edge of the sunroof. At t=7T/8, the vortex is shed from the front edge of the sunroof and moves downstream and the interior pressure continues to increase. At t=T, the period ends and the flow pattern returns to its initial state t=0. The interior pressure reaches it's highest, but still lower than that in the baseline case. In general, the changes in flow pattern are similar to the baseline case, but slower and milder. These graphs demonstrate the evolution of the unsteady flow and explain why the buffeting noise is weaker in this case, compared to the baseline case.

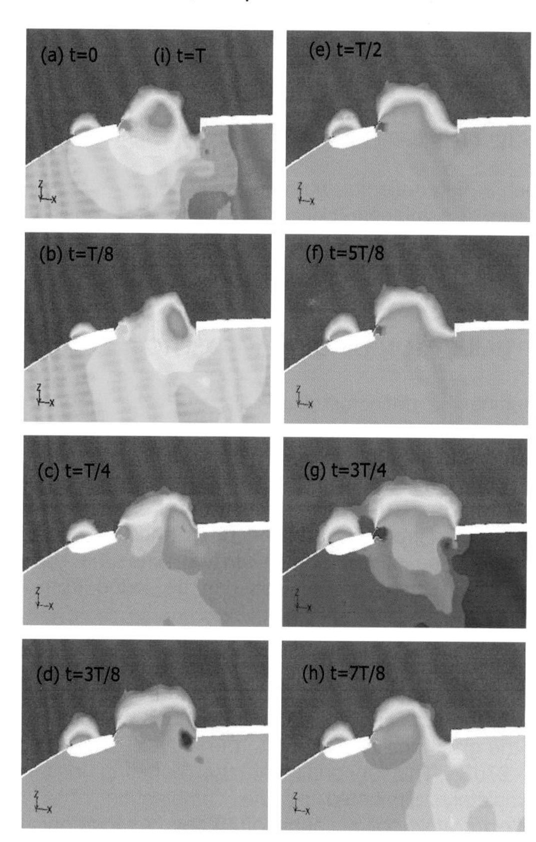

Figure 15. Flow pattern of DA40 case
Color map of pressure scale is same as Figure 11

OPTIMIZATION OF DEFLECTOR HEIGHT

The next step of the study is to optimize the height of the deflector. Based on the deflector model at optimal angle of 40 degrees, various deflector heights, 43, 65, 86, 108, 129mm, are simulated to find out what height is optimal to minimize buffeting. Figure 16 shows the result of this parameter sensitivity study for the deflector height which is measured along the deflector. As shown in the figure, in the range of height 43~129mm, the peak SPL increases slightly and then decreases quickly with increase in the deflector height. The optimal height of deflector in this range is 129mm at which the peak buffeting SPL decreases to 105dB.

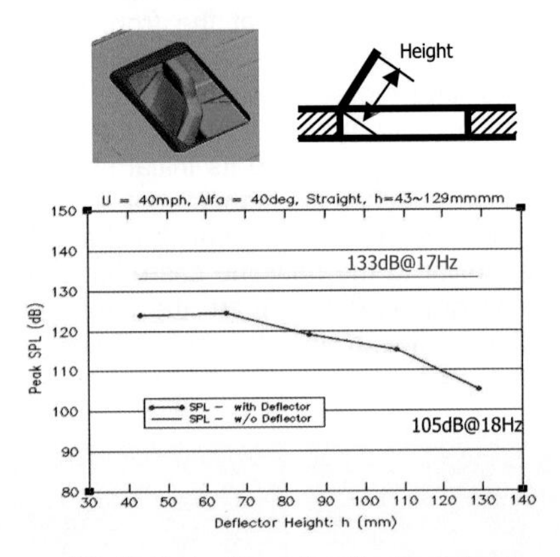

Figure 16. Optimization of deflector height

Such a high deflector, however, may not be accepted in styling. A combination of deflector and glass position is considered to look at more buffeting reduction. The result is presented in the next section.

OPTIMIZATION OF GLASS POSITION

Keeping the deflector at its optimal angle 40 degrees and its low height 43mm, we move the glass to several positions to find out which is optimal position to minimize buffeting. The glass position changes from x/L=50% to 80% where x is the distance between front edge of the sunroof and front edge of the glass and L is the total length of the sunroof. After several runs, the optimal position of the sunroof glass can be found from the simulation.

As shown in Figure 17, with decrease in the distance between front edge of the sunroof and front edge of the glass, the buffeting peak SPL is reduced monotonically. At the glass position x/L=50%, the peak SPL is reduced to 97dB. This means that the buffeting noise is completely eliminated by the combination of a low deflector at its optimal angle of 40 degrees together with the optimal position of the sunroof glass at optimal position x/L=50%. This case is identified as DA40_G050.

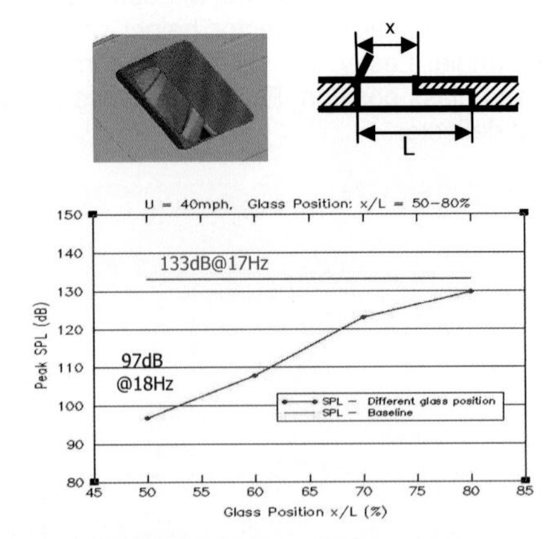

Figure 17. Optimization of glass position

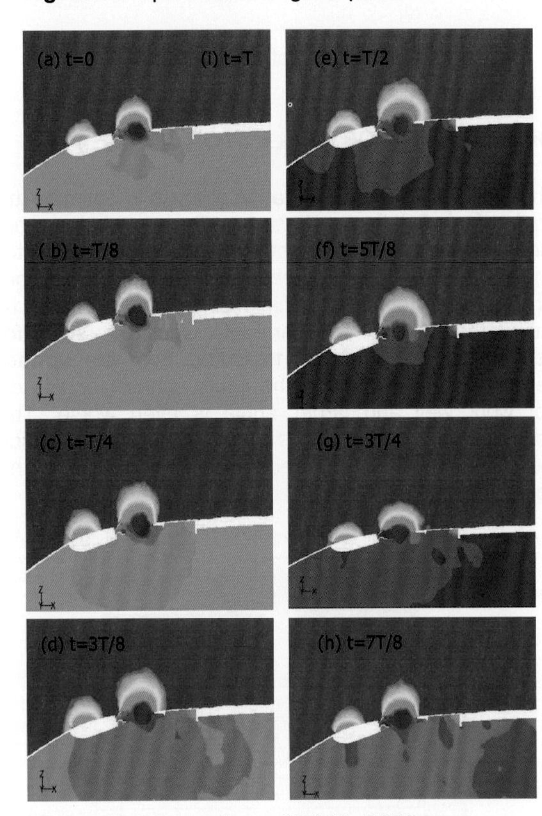

Figure 18. Flow pattern of DA40_G050 case
Color map of pressure scale is same as Figure 11

Figure 18 shows a sequential series of 8 frames of pressure contour within a time period for the case of DA40_G050. The color map of pressure scale is the same as that in Figure 11. Throughout a time period, from (a) t=0 to (h) t=7T/8, the vortex almost stays on the half way of the sunroof opening but moves little. This means that the outside flow is almost steady and the vortex is not shed periodically in this case, unlike in the cases of baseline and DA40. Moreover, the interior pressure changes little within a time period by observation of the interior pressure color from light blue to dark blue. This explains the reason why buffeting is almost eliminated in this case.

CONCLUSIONS

According to the present CFD study, sunroof buffeting can be eliminated using the combination of a deflector at its optimal angle with the partially closed sunroof glass at its optimal position.

For the selected prototype vehicle, the baseline model of sunroof has a high level of buffeting noise with the peak SPL 133dB at 17Hz.

The peak buffeting level can be reduced to 124dB if a deflector is installed at its optimal angle 40 degrees from the upward vertical line.

The peak buffeting level can be further reduced to 105dB if the deflector takes the optimal length of 129mm while keeping its optimal angle of 40 degrees.

The sunroof buffeting can be completely eliminated (the peak buffeting level to 97dB) when a combination of a deflector (at the low height of 43mm and the optimal angle of 40 degrees) with the partially closed sunroof glass (at the optimal position of 50%) is taken. That is, the distance between front edge of the sunroof and front edge of the glass equals to 50% of the total length of the sunroof.

For any other vehicle, the optimization process is the same, but the optimal deflector angle and height and the optimal glass position should be re-calculated accordingly.

The optimization study is only based on fluid dynamics principle without considering manufacturability, styling, cost, etc. Further work is needed to utilize the results in the production design.

ACKNOWLEDGEMENTS

The authors would like to thank Jean Mallebay-Vacqueur, Director of Scientific Labs, Richard Sun, Senior Manager of Aero-Thermal Center of Competency and Core CFD, Mark Gleason, Manager of Aerodynamics and Aeroacoustics Wind Tunnel, Mitchell Puskarz, Senior Aeroacoustics Specialist, for their valuable comments, suggestions and support on this study.

REFERENCES

[1] Lee, B.H.K., Vertical tail buffeting of fighter aircraft, Progress in Aerospace Sciences, Vol. 36, pp.193-279 (2000)

[2] Minh, N.N., Miyata, T., Yamada, H. and Sanada, Y., Numerical simulation of wind turbulence and buffeting analysis of long-span bridge, J. Wind Eng. & Industrial Aerodynamics, Vol. 83, pp. 301-315 (1999)

[3] Zhou, Y., Kareem, A. and Gu, M., Equivalent static buffeting loads on structures, J. Structural Eng., August 2000, pp. 989-992 (2000)

[4] Ota, D.K., Chakravarthy, S.R., Becker, T. and Sturzenegger, T., Computational study of resonance suppression of open sunroofs, J. Fluids Eng., Vol. 116, pp. 877-882 (1994)

[5] Ukita, T., China, H. and Kanie, K., Analysis of vehicle wind throb using CFD and flow visualization, SAE paper 970407 (1997)

[6] Karbon, K.J., Singh, R., Simulation and design of automobile sunroof buffeting noise control, 8th AIAA paper 2002-2550 (2002)

[7] Inagaki, M., Murata, O., Horinouchi, N., Takeda, I. and Kakamu, T., Numerical prediction of wind-throb noise in vehicle cabin, Proc. of 6th World Congress on Computational Mechanics, Abstract Vol. I, p. 105, Beijing, China, Sept. 5-10 (2004)

[8] Hendriana, D., Sovani, S.D. and Schiemann, M.K., On simulating passenger car side window buffeting, SAE Paper 2003-01-1316 (2003)

[9] An, C.-F., Alaie, S.M., Sovani, S.D., Scislowicz, M.S. and Singh, K., Side window buffeting characteristics of an SUV, SAE paper 2004-01-0230 (2004)

[10] An, C.-F., Alaie, S.M. and Scislowicz, M.S., Impact of cavity on sunroof buffeting – a two dimensional CFD study, Proc. ASME 5th Int'l Symp. on Comp. Tech. with Industrial Applications, Vol. 1, pp 133-144, (2004)

[11] An, C.-F., Puskarz, M., Singh, K. and Gleason, M.E., Attempts for reduction of rear window buffeting using CFD, SAE paper 2005-01-0603 (2005)

[12] Rayleigh, B., *The Theory of Sound*, Vol.II, pp.170-172, Dover Publications, New York, USA (1945)

[13] Blevins, R.D., *Formulas for Natural Frequency and Mode Shape*, pp.353-358, Van Nostrand Reinhold Company, New York, USA (1979)

[14] Alster, M., Improved calculation of resonant frequencies of Helmholtz resonator, J. Sound & Vibration, Vol. 24(1), pp. 63-85 (1972)

[15] Roshko, A., Some measurements of flow in a rectangular cutout, NACA TN 3488 (1955)

[16] Tam, C.K.W., The acoustic mode of a two-dimensional rectangular cavity, J. Sound & Vibration, Vol. 49(3), pp. 353-364 (1976)

[17] Rockwell, D. and Naudascher, E., Review – self-sustaining oscillations of flow past cavities, J. Fluids Eng., Vol. 100, pp. 152-165 (1978)
[18] Blake, W.K., *Machanics of Flow-Induced Sound and Vibration*, Vol. I, ppl 130-149, Academic Press, Inc., New York, USA (1986)
[19] Blevins, R.D., *Flow-Induced Vibration*, 2nd Ed., pp. 375-383, Van Nostrand Reinhold, New York, USA (1990)
[20] Henderson, B., Automotive noise involving feedback – sound generation by low speed cavity flows, NASA/CP 2000-209790, pp. 95-100 (2000)
[21] Henderson, B., Sound generation by flow over a cavity, NASA/CP 2004-212954, pp. 71-77 (2004)
[22] Lamb, H., *Hydrodynamics*, 6th ed., pp. 373-375, Cambridge Univ. Press, London, UK (1932)
[23] An, C.-F., Goodman, R.H., Brown, H.M., Clavelle, E.J. and Barron, R.M., Oil-water interfacial phenomena behind a boom on flowing water, Proc. 10th Int'l Symp. on Transport Phenomena, Vol. 1, pp. 13-18, Kyoto, Japan, Nov.30-Dec.4 (1997)
[24] Esch, R.E., The instability of a shear layer between two parallel streams, J. Fluid Mech., Vol. 3, pp. 289-303 (1957)
[25] Michalke, A. On the inviscid instability of the hyperbolic-tangent velocity profile, J. Fluid Mech., Vol. 19, pp. 543-556 (1964)
[26] Michalke, A. Vortex formation in a free boundary layer according to stability theory, J. Fluid Mech., Vol. 22, part 2, pp. 371-383 (1965)
[27] Michalke, A. On spatially growing disturbances in an inviscid shear layer, J. Fluid Mech., Vol. 23, part 3, pp. 521-544 (1965)
[28] Lin, C.C., *The Theory of Hydrodynamics Stability*, Cambridge Univ. Press, London, UK (1955)
[29] Bechov, R., and Criminale, W.O.Jr., *Stability of Parallel Flows*, Academic Press, New York, USA (1967)
[30] Abernathy, F.H., and Kronauer, R.E., The formation of vortex sheets, J. Fluid Mech., Vol. 13, pp. 1-20 (1962)
[31] Boldman, D.R., Brinich, P.F., and Goldstein, M.E., Vortex shedding from a blunt trainling edge with equal and unequal external mean velocities, J. Fluid Mech., Vol. 75, part 4, pp. 721-735 (1976)
[32] Nakamura, Y. and Nakashima, M., Vortex excitation of prisms with elongated rectangular, H and T cross-sections, J. Fluid Mech., Vol. 163, pp. 149-169 (1986)
[33] Fluent 6.2.16 User's Guide, Fluent Inc., Lebanon, NH, USA (2005)

CONTACT

Chang-Fa An, ca56@daimlerchrysler.com

Where the Automotive Industry Comes to Do Business

Event Guide

Includes Final Program and Exhibit Directory

Host Company: **TOYOTA**

Event Theme: Engineering for Global Sustainable Mobility—It's Up to Us

sae.org/congress

Changfa An
Detroit, MI, USA
04/16/2007

See page 32 for details on the SAE 2007 Car Give Away. Automobile funding provided by: DASSAULT SYSTEMES

Published by:

The Official Publication of SAE 2007 World Congress

070235

第IV部分　任职于美国克莱斯勒汽车公司时期

Vehicle Aerodynamics 2007

SP-2066

安长发博士科研论文集

Changfa An
Detroit, MI, USA
April 16, 2007

SAE International™

Published by:
SAE International
400 Commonwealth Drive
Warrendale, PA 15096-0001
USA
Phone: (724) 776-4841
Fax: (724) 776-0790
April 2007

2007-01-1552

Sunroof Buffeting Suppression Using a Dividing Bar

Chang-Fa An and Kanwerdip Singh
DaimlerChrysler, Auburn Hills, MI

ABSTRACT

This paper presents the results of CFD study on sunroof buffeting suppression using a dividing bar. The role of a dividing bar in side window buffeting case was illustrated in a previous study [8]. For the baseline model of the selected vehicle in this study, a very high level of sunroof buffeting, 133dB, has been found. The CFD simulation shows that the buffeting noise can be significantly reduced if a dividing bar is installed at the sunroof. A further optimization study on the dividing bar demonstrates that the peak buffeting level can be reduced to 123dB for the selected vehicle if the dividing bar is installed at its optimal location, 65% of the total length from the front edge of the sunroof. The peak buffeting level can be further reduced to 100dB if the dividing bar takes its optimal width 80mm, 15% of the total length of the sunroof for this vehicle, while staying at its optimal location. The buffeting noise can be completely suppressed and the peak buffeting level can be reduced to 85dB if the dividing bar further takes its optimal thickness 45mm, 8% of the total length of the sunroof for this vehicle, while keeping its optimal width and location. For other vehicles, the optimization procedure would be the same, but the optimal parameters should be re-calculated in general. On the other hand, this study is only based on the fluid dynamic principles without considering manufacturability, styling, cost, etc. Further work is needed to utilize these results in the production environment.

INTRODUCTION

In a recent survey of automotive quality, about 50% of the customer complaints on wind noise relate to buffeting, especially for luxury cars and SUVs (Figure 1).

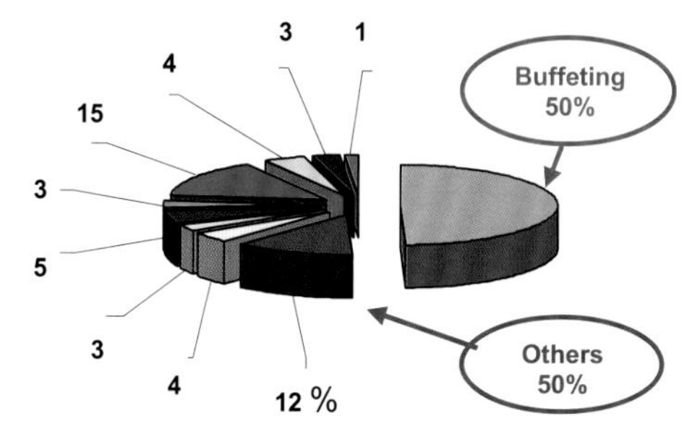

Figure 1. Customer complaints on wind noise

After many years of buffeting investigation, theoretically and experimentally, the mechanism of buffeting is well understood. However, it is not easy to predict how severe the buffeting noise is for a specific vehicle before a physical model of the vehicle is available. This is because of the complexity of the vehicle geometry and the fact that no simple formulas can be used to evaluate buffeting level. Obviously, discovering that a new vehicle has strong buffeting when it is already on the road is too late to make cost-effective changes. With rapid development and improvement of CFD technology, buffeting characteristics can be predicted numerically provided that the digital model of a new vehicle is available. During the last decade, automotive companies, with the help of CFD developers, conducted CFD simulations of sunroof and side window buffeting [1-4]. Since 2002, intensive efforts for buffeting studies with CFD technology have been made at DaimlerChrysler [5-9]. These efforts include procedure establishment, validation with wind tunnel testing, attempts to find practical solutions for buffeting and parameter optimization to minimize buffeting. Among possible solutions for buffeting reduction, a dividing bar has been proven to reduce buffeting for the side window buffeting case [8]. Later, the concept of a dividing bar in the side window case is further extended to the sunroof case. For the sunroof buffeting simulation, the procedure is almost the same except for a minor difference in the local mesh refinement. For the side window case, the finest locally-refined mesh is at the open window area. Likewise, for the sunroof case, the finest locally-refined mesh is at the open sunroof area. Furthermore, in the sunroof buffeting study, optimization studies for dividing bar parameters are also conducted to minimize the buffeting level.

In this paper the most recent results of sunroof buffeting suppression using a dividing bar are reported. The procedure of CFD simulation is described in [6, 8] for side window buffeting and in [9] for sunroof buffeting and will not be repeated here. In the next section, the positive role of a dividing bar in the side window buffeting case is cited and the concept of dividing bar is further extended to the sunroof buffeting case. As the emphasis of the present paper, the details of the parameter optimization for the dividing bar are presented in the subsequent sections. Finally, some concluding remarks are addressed.

POSITIVE ROLE OF DIVIDING BAR

DIVIDING BAR AT SIDE WINDOW

During the progress of the buffeting study several attempts were made to reduce side window buffeting, including an idea of a dividing bar [8]. A CFD simulation of the side window buffeting was done for a selected vehicle model. On the basis of the rear window open case (Figure 2-a), a dividing bar was added in the rear window (Figure 2-b).

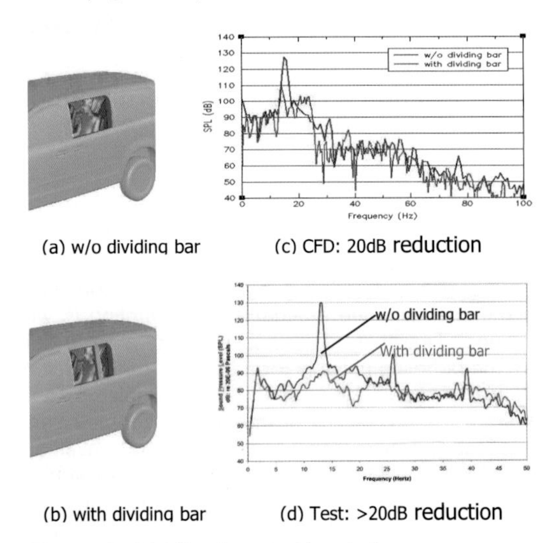

Figure 2. Dividing bar at side window

The mechanism of buffeting reduction in the dividing bar case is the "venting effect". The front portion of the open window serves as the main opening of the resonant chamber in buffeting phenomenon while the rear portion of the open window plays a venting role. The CFD result of buffeting spectrum is shown in Figure 2-c and the peak buffeting level is reduced by about 20dB due to the addition of a dividing bar. A wind tunnel test confirmed the CFD prediction and documented a peak buffeting level reduction by more than 20dB, as shown in Figure 2-d.

DIVIDING BAR AT SUNROOF

The favorable results of buffeting reduction by a dividing bar for the side window buffeting case inspired a further effort in this direction for the sunroof buffeting case. On the baseline model of the sunroof with glass full open (Figure 3-a), a dividing bar is installed at a certain distance from the front edge of the sunroof (Figure 3-b). As we can see in the subsequent sections, the peak buffeting level can be suppressed completely if a dividing bar is installed and properly optimized.

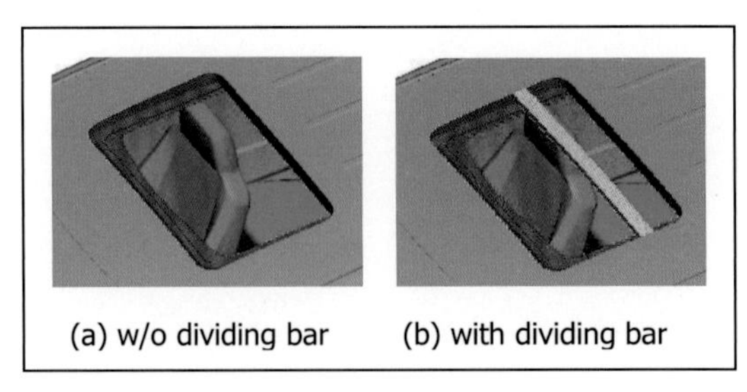

Figure 3. Dividing bar at sunroof

DETERMINATION OF BASELINE CASE

Before the optimization study of a dividing bar to minimize buffeting, a baseline case should be determined as the standard of comparison.

At the beginning of the sunroof buffeting project, two models were presented for simulation. One is a sunroof model with glass full open and a small deflector (blue color), as shown at the upper left corner of Figure 4. Another is a sunroof model of glass half open with the same deflector, as shown at the upper right corner of the same figure. Four vehicle speeds, 20, 30, 40 and 50mph, were required to run in order to determine which speed causes the highest buffeting level.

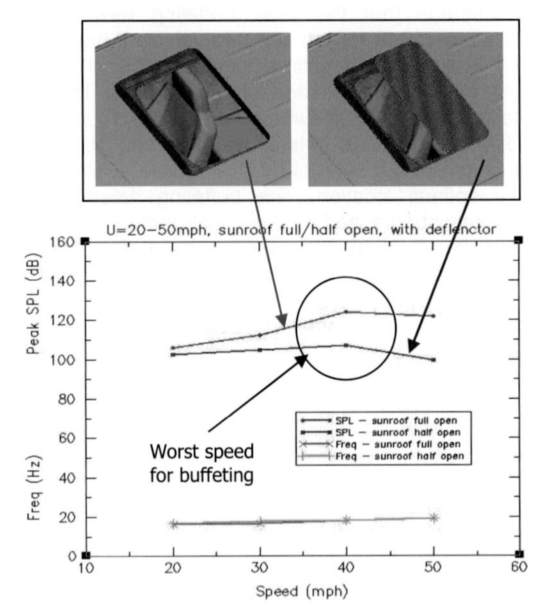

Figure 4. Effect of vehicle speed on buffeting

Figure 4 (lower portion) shows the effect of vehicle speed on the buffeting spectrum, i.e. on the dominant resonance frequency (2 lower curves) and on the peak SPL (2 upper curves). From this figure we can see that with increase in vehicle speed the dominant resonance frequency changes little (<2Hz), while the peak SPL changes considerably (up to 10dB) and reaches its maximum value at about 40mph for both cases. This indicates that the worst case for this vehicle regarding sunroof buffeting is at speed 40mph.

Moreover, in order to investigate the influence of "dividing bar only" on buffeting characteristics, the small deflector should be removed from the model and sunroof glass should be full open. Therefore, the model of glass full open without a deflector should be chosen as the baseline case for the dividing bar optimization study.

Figure 5 shows the buffeting spectrum results for the baseline case which tells us that buffeting occurs at the dominant resonance frequency of 17Hz and the peak buffeting SPL is 133dB.

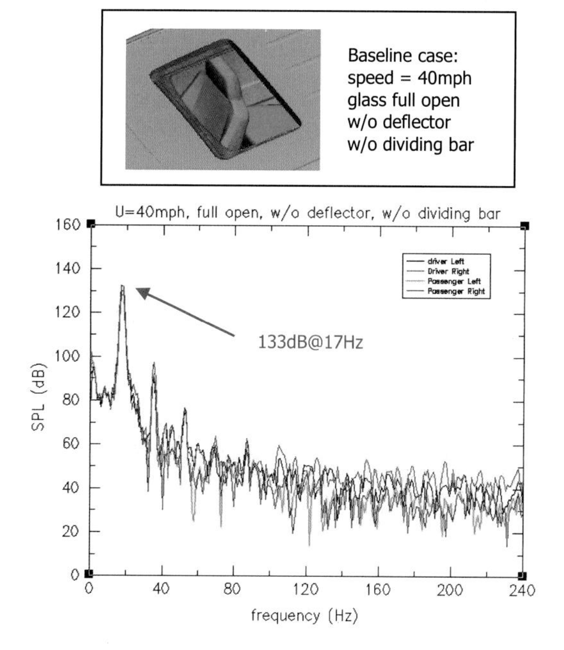

Figure 5. Baseline case for optimization study

Figures 6 and 7 demonstrate the velocity vectors and the static pressure on the central plane at a certain time moment to give a general impression on the flow feature for the baseline case. From both figures we can see that a strong vortex is shed from the front edge of the sunroof and it is intruding into the vehicle interior, which is usually considered as the physical mechanism of buffeting.

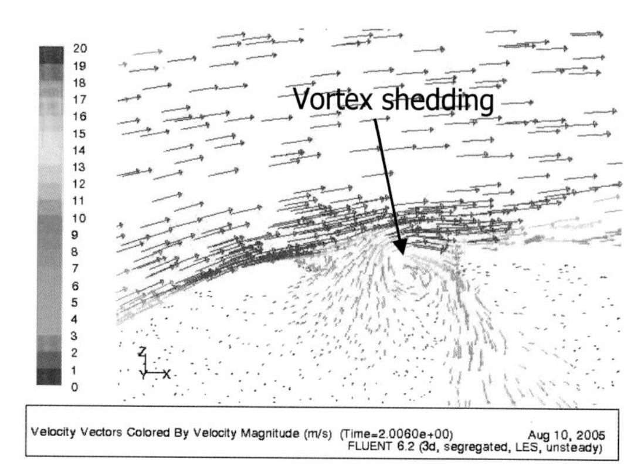

Figure 6. Velocity vector of baseline case

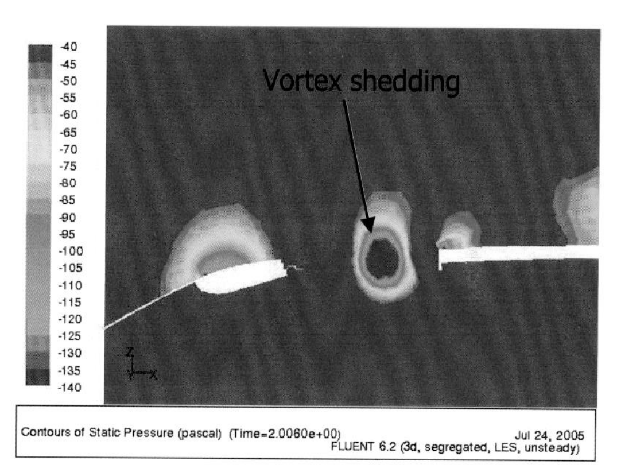

Figure 7. Static pressure of baseline case

For a transient flow, animation is a useful tool to show a sequential series of frames of flow variables, such as velocity, pressure, etc. The flow pattern changes with time can be seen and the inherent feature of a specific flow may be discovered from the animation (of course, only still frames can be shown in the paper). The full animation of the entire series of frames can only be watched as a movie in presentation. Since the flow pattern for a buffeting case changes periodically with time, it is proper to choose typical frames within a time period to show representative situations at special time moments. In the following context, 8 frames are chosen for the baseline case in Figure 8 and for the case of L065 in Figure 11, while 4 frames are chosen for the cases of L065-w80 and L065-w80-t45 in Figure 13. The meaning of the last 3 cases will be identified in the next paragraph.

Figure 8 shows a sequential series of pressure contours for the baseline case. The color map of static pressure scale (-140 ~ -40Pa) is the same as in Figure 7 as well as Figures 11 and 13. As we have already known that the dominant resonance frequency is F=17Hz and, hence, the time period is about T=0.059s. Graphs (a), (b), (c), (d), (e), (f), (g) and (h) are pressure contours at time moment t=0, T/8, T/4, 3T/8, T/2, 5T/8, 3T/4 and

7T/8, respectively. At t=T, the graph (i) is the same as graph (a) at t=0. From these graphs we can see pressure fluctuations and vortex movement. Therefore, we can explain why and how buffeting occurs at such a high level of 133dB.

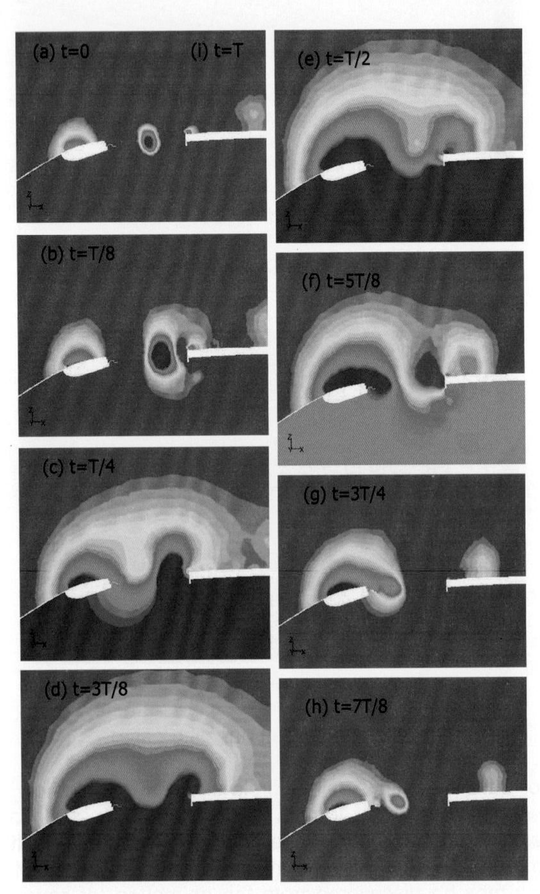

Figure 8. Flow pattern of baseline case

At t=0, pressure in the vehicle interior is very high (pure red) and a low pressure vortex has traversed half the longitudinal distance of the sunroof opening. At t=T/8, the vortex moves downstream and becomes larger. At t=T/4, the vortex hits the rear edge of the sunroof, breaks-down and intrudes into the interior of the vehicle. As a result, the interior pressure decreases significantly (dark blue). At t=3T/8, the above process continues. At t=T/2, the interior pressure goes down to its minimum value and a secondary vortex is about to appear at the rear edge of the sunroof. At t=5T/8, the interior pressure starts to increase and the secondary vortex leaves the sunroof region. At t=3T/4, the interior pressure continues to increase and another vortex is about to form at the front edge of the sunroof. At t=7T/8, the interior pressure almost reaches its maximum value and the vortex is shed from the front edge of the sunroof. At t=T, the period ends and the flow pattern returns to its initial state of t=0.

OPTIMIZATION OF DIVIDING BAR

As mentioned before, installation of a dividing bar at a certain location in the sunroof is expected to reduce buffeting. What is the optimal location and what is the optimal dimension? To answer these questions an optimization study on dividing bar is undertaken for the parameter range shown in Figure 9. The optimization study takes the following three steps.

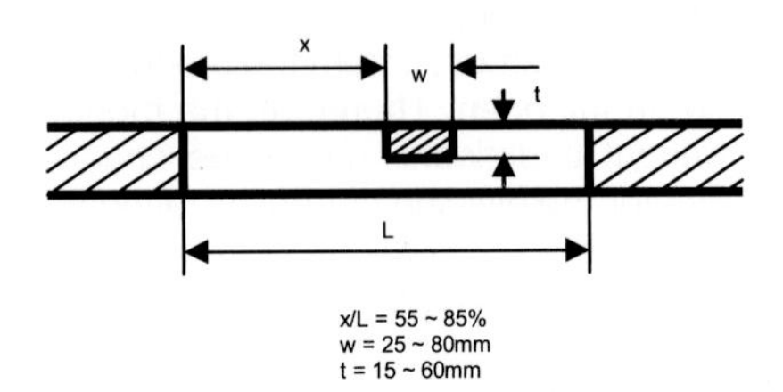

Figure 9. Optimization of sunroof dividing bar

Step 1: The location of a dividing bar of dimension 50x30mm^2 moves back and forth from x/L= 55%~85% to find the optimal location. In Figure 9, w and t are the width and thickness of the dividing bar and x and L are the location of the dividing bar and the total length of the sunroof.

Step 2: The width of the dividing bar varies from w=25~80mm at the optimal location (found at step 1) to find the optimal width of the dividing bar.

Step 3: The thickness of the dividing bar varies from t=15~60mm for the optimal width (found in Step 2) at the optimal location (found in Step 1) to find the optimal thickness of the dividing bar.

OPTIMAL LOCATION

As shown in Figure 10, after 6 runs the optimal location is found to be x/L=65%, identified as the case of L065. At this location the SPL, defined in [8], is reduced to about 123dB, 10dB reduction from the baseline case.

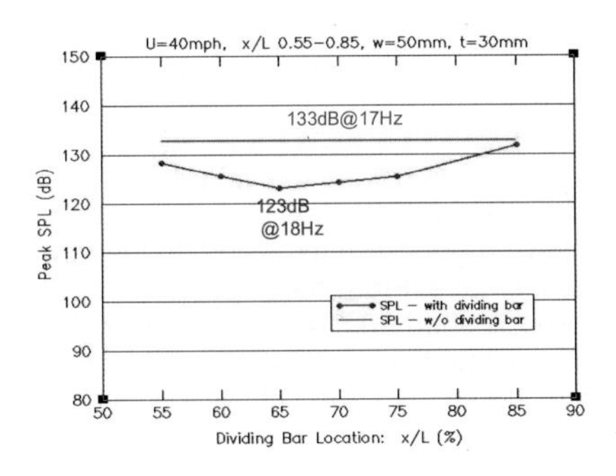

Figure 10. Optimal location: x/L = 65%

In order to understand the reason of buffeting reduction, a similar sequential series of pressure frames is drawn in Figure 11 to discuss the flow pattern change within a time period.

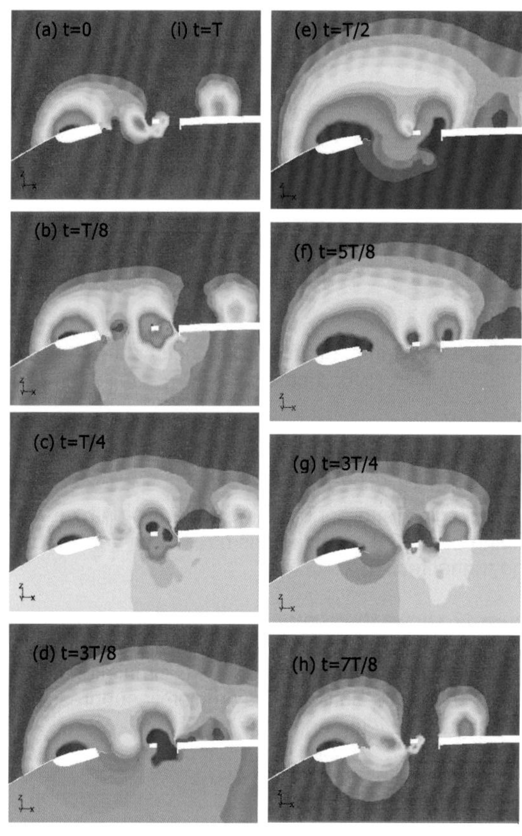

Figure 11. Flow pattern of L065 case

At t=0, the interior pressure is high and a vortex, which is weaker than its counterpart in the baseline case, is just approaching the dividing bar. At t=T/8, the interior pressure decreases. The vortex hits the dividing bar and wraps around it. At t=T/4, the interior pressure reaches its average and the vortex is broken into 2 pieces. At t=3T/8, the interior pressure continues to decrease, but higher than that in the baseline case. At t=T/2, the interior pressure reaches it's lowest value, but still higher than that in the baseline case. At t=5T/8, the interior pressure begins to increase and a secondary vortex is about to appear at the rear edge of the sunroof. At t=3T/4, the interior pressure reaches its average again and the secondary vortex is leaving. At t=7T/8, a new vortex is shed from the front edge of the sunroof and moves towards the dividing bar. The interior pressure continues to increase. At t=T, the period ends. The interior pressure reaches its highest value, but still lower than that in the baseline case. In general, the changes in flow pattern are similar to the baseline case, but slower and milder. The vortex is broken-down by the dividing bar and is considerably weakened. These graphs show the evolution of the unsteady flow and explain why and how the buffeting noise in this case is weaker than that in the baseline case.

OPTIMAL WIDTH

Keeping the dividing bar at its optimal location x/L=65% and varying its width w=25~80mm (see Figure 9), the optimal width is found to be w=80mm (15% of the total sunroof length) after 5 runs and the peak SPL is further reduced to 100dB, as shown in Figure 12.

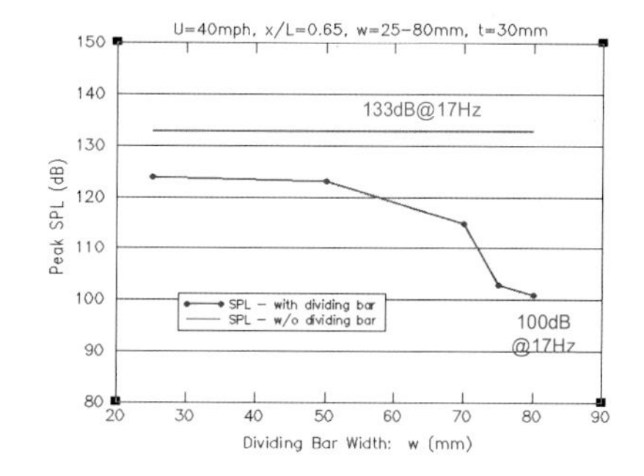

Figure 12. Optimal width: w = 80mm

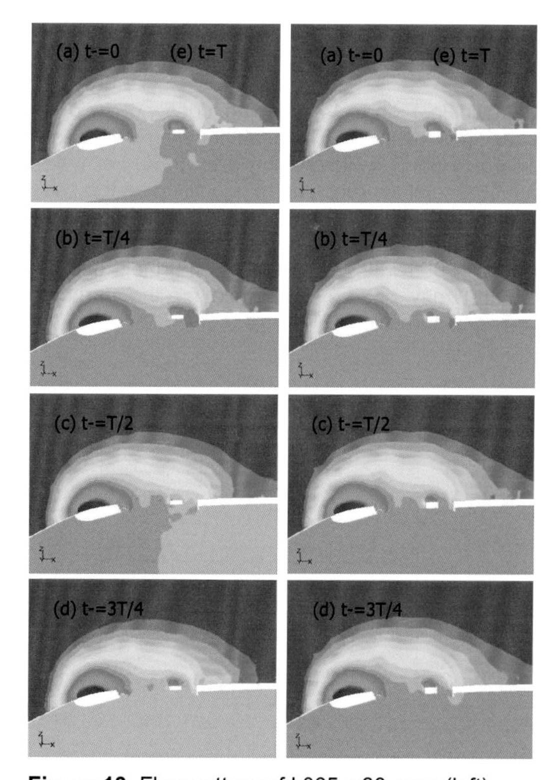

Figure 13. Flow pattern of L065-w80 case (left) and L065-w80-t45 case (right)

Further increasing the width of the dividing bar might decrease the SPL to some extent, but we stopped here due to the consideration of material cost and the resultant reduction in sunroof opening area. For the same purpose, a sequential series of pressure frames within a time period is given in Figure 13, left column, identified as the case of L065-w80. From graphs (a) to (d), the interior pressure changes very little. No obvious vortex is shed from the front edge of the sunroof. This is a convincing interpretation of buffeting suppression.

OPTIMAL THICKNESS

To complete the optimization study, 4 more runs were done by varying the thickness of the dividing bar from 15~60mm while keeping its optimal width w=80mm and its optimal location x/L=65%.

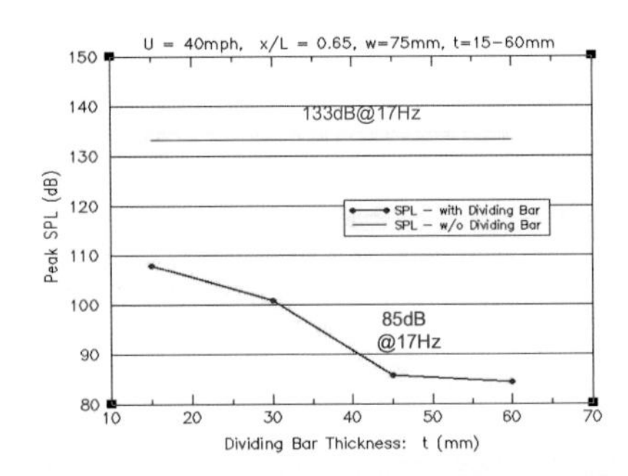

Figure 14. Optimal thickness: t = 45mm

As shown in Figure 14, thinning does not help while thickening is favorable. The peak buffeting SPL is further reduced to 85dB by thickening it by 50%, i.e. t=45mm (8% of the total sunroof length), identified as the case of L065-w80-t45. The flow pattern does not change much, as shown in Figure 13, right column, compared to the case of L065-w80, left column. Further thickening (t=60mm) may not be worthwhile for the purpose of buffeting suppression.

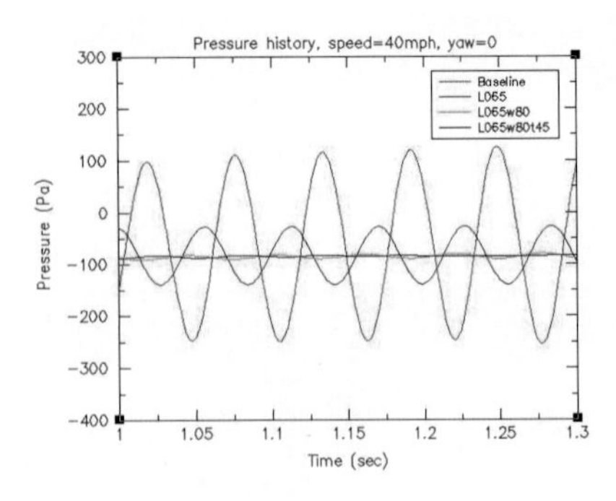

Figure 15. Comparison of pressure history for 4 cases

Figure 15 shows the pressure history and Figure 16 shows the buffeting spectra for the 4 cases. These figures clearly indicate that the optimized dividing bar can suppress buffeting completely.

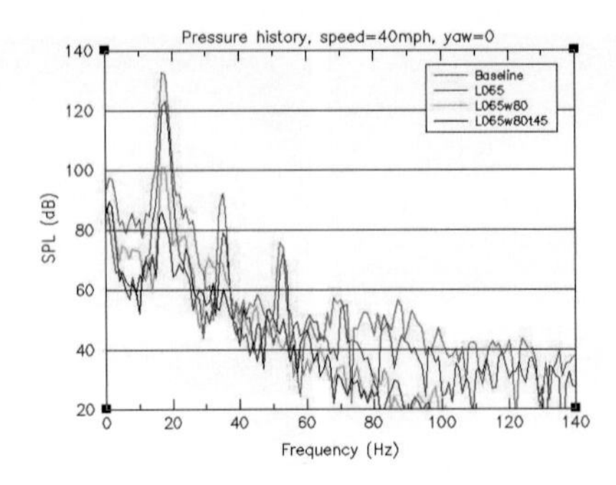

Figure 16. Comparison of buffeting spectra for 4 cases

WIND TUNNEL VERIFICATION

The findings of sunroof buffeting suppression by a dividing bar in CFD simulations have been verified by wind tunnel testing.

Figure 17 shows the comparison of buffeting spectra between the sunroof models with and without a dividing bar for the selected vehicle from CFD simulations. As discussed before, buffeting is completely suppressed by an optimized dividing bar.

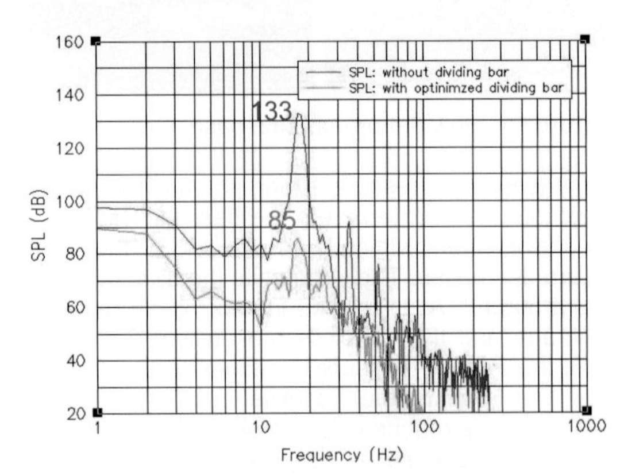

Figure 17. Comparison of sunroof buffeting – CFD for the selected vehicle

In order to verify the idea of buffeting suppression by a dividing bar a wind tunnel test of sunroof buffeting was done for a different vehicle. The reason to choose a different vehicle is that the physical model of the selected vehicle was not available at the time of CFD study. The key point is to qualitatively verify if the idea of buffeting suppression by a dividing bar works.

Figure 18 shows the test result of sunroof buffeting for that vehicle. It is clear that the dividing bar works well to suppress buffeting. Although the baseline model of that vehicle is not too severe (110dB) regarding buffeting, a dividing bar can still reduce buffeting by about 18dB.

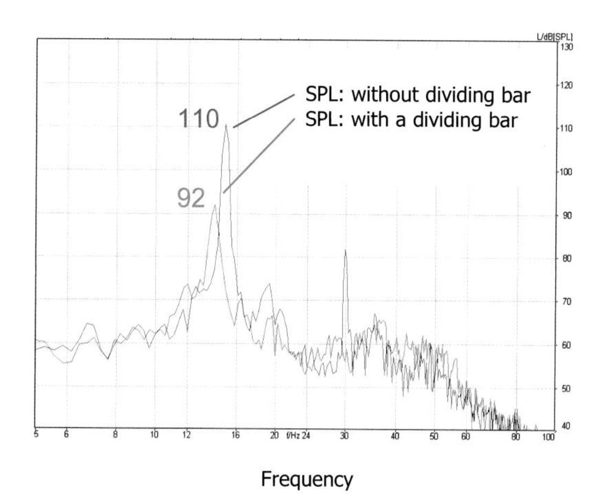

Figure 18. Comparison of sunroof buffeting - Test for a different vehicle

CONCLUSIONS

According to the present CFD study, sunroof buffeting can be suppressed using an optimized dividing bar installed at the sunroof.

For the selected vehicle, the baseline sunroof model has high level of buffeting noise with peak SPL of 133dB.

The peak buffeting level can be reduced to 123dB if a dividing bar is installed at its optimal location, 65% of the total length of the sunroof measured from its front edge.

The peak buffeting level can be further reduced to 100dB if the dividing bar takes its optimal width, 80mm or 15% of the total sunroof length for this vehicle, while staying at its optimal location.

The buffeting noise can be completely suppressed if the dividing bar further takes its optimal thickness, 45mm or 8% of the total sunroof length for this specific vehicle, while keeping its optimal width and its optimal location.

For any other vehicles, the optimization process is the same, but the optimal dimension and optimal location should be re-calculated accordingly.

Other related issues, such as manufacturability, styling, and cost and so on, are still under discussion.

ACKNOWLEDGEMENTS

The authors would like to thank Jean P Mallebay-Vacqueur, Director of Scientific Labs & Proving Grounds, Richard Sun, Senior Manager of Aero-Thermal Center of Competency and Core CFD, Mark Gleason, Leader of Aerodynamics and Aeroacoustics Wind Tunnel, Mitchell Puskarz, Senior Aeroacoustics Specialist, for their valuable discussions, comments, suggestions and support on this project. The authors also appreciate David Tao and Xijia Zhu for the wind tunnel test to verify the findings in CFD simulation. Finally, thanks should be given to Mark Gleason for his careful review and revision of the manuscript.

REFERENCES

[1] Ota, D.K., Chakravarthy, S.R., Becker, T. and Sturzenegger, T., Computational study of resonance suppression of open sunroofs, J. Fluids Eng., Vol. 116, pp. 877-882 (1994)

[2] Ukita, T., China, H. and Kanie, K., Analysis of vehicle wind throb using CFD and flow visualization, SAE paper 970407 (1997)

[3] Karbon, K., Singh, R., Simulation and design of automobile sunroof buffeting noise control, 8th AIAA/CEAS Aeroacoustics Conference & Exhibit, June 17-19 (2002)

[4] Inagaki, M., Murata, O., Horinouchi, N., Takeda, I. and Kakamu, T., Numerical prediction of wind throb noise in vehicle cabin, Proc. of 6th World Congress on Computational Mechanics, Abstract Vol. I, p. 105, Beijing, China, Sept. 5-10 (2004)

[5] Hendriana, D., Sovani, S.D. and Schiemann, M.K., On simulating passenger car side window buffeting, SAE Paper 2003-01-1316 (2003)

[6] An, C.-F., Alaie, S.M., Sovani, S.D., Scislowicz, M.S. and Singh, K., Side window buffeting characteristics of an SUV, SAE Paper 2004-01-0230 (2004)

[7] An, C.-F., Alaie, S.M. and Scislowicz, M.S., Impact of cavity on sunroof buffeting – a two dimensional CFD study, Proc of ASME 5th Int'l Symp on Comp Tech with Industrial Appls, Vol. 1, pp 133-144, (2004)

[8] An, C.-F., Puskarz, M., Singh, K. and Gleason, M.E., Attempts for reduction of rear window buffeting using CFD, SAE Paper 2005-01-0603 (2005)

[9] An, C.-F., Singh, K., Optimization study for sunroof buffeting reduction, SAE Paper 2006-01-0138 (2006)

CONTACT

Chang-Fa An, ca56@daimlerchrysler.com

SAE 2007 TRANSACTIONS

JOURNAL OF PASSENGER CAR: MECHANICAL SYSTEMS

Section 6 - Volume 116

PUBLISHED BY: SAE International
400 Commonwealth Dr., Warrendale, PA 15096-0001
Phone (724)776-4841 Fax (724)776-5760
www.sae.org

All SAE papers, standards, and selected books are abstracted and indexed in the Global Mobility Database.

2007-01-1527 **Experimental and Theoretical Studies of Rubber Properties; The Application to Tire Tread Rubber** 1453
Hiroshi Yokohama, Stephen D. Hall, and Robert B. Randall

2007-01-1548 **On Various Aspects of the Unsteady Aerodynamic Effects on Cars Under Crosswind Conditions** 1463
Jochen Mayer, Michael Schrefl, and Rainer Demuth

2007-01-1549 **Laminar Flow Whistle on a Vehicle Side Mirror** 1475
Todd H. Lounsberry, Mark E. Gleason, and Mitchell M. Puskarz

2007-01-1551 **Aeroacoustic Measurements in Turbulent Flow on the Road and in the Wind Tunnel** 1481
N. Lindener, H. Miehling, A. Cogotti, F. Cogotti, and M. Maffei

2007-01-1552 **Sunroof Buffeting Suppression Using a Dividing Bar** 1501
Chang-Fa An and Kanwerdip Singh

2007-01-1655 **Modelling Computer Experiments with Multiple Responses** 1508
Zhe Zhang, Runze Li, and Agus Sudjianto

2007-01-1656 **Data Fusion and Modelling for Fatigue Crack Growth Prediction** 1515
D. Gary Harlow

2007-01-1660 **One-Factor-at-a-Time Screening Designs for Computer Experiments** 1521
Aijun Zhang

2007-01-1766 **The Virtual Driver: Integrating Task Planning and Cognitive Simulation with Human Movement Models** 1525
Omer Tsimhoni and Matthew P. Reed

2007-01-1780 **Experimental and Numerical Study of Underbody Drive and Soak Thermal Conditions on the Basis of a Heat Shield Test Rig** 1532
Ed Bendell, Luka Gorlato, and Michael Hauenstein

2007-01-2128 **FMVSS 201: Appropriate Injected Rib Design for Trim Panels** 1549
G. Spingler

2007-01-2131 **The Effects of Loading Devices on the Stability of Ratio-Changing Characteristics in a Half-Toroidal CVT after Quick Torque Changes** 1555
Shinji Miyata

2007-01-2135 **Simplified Modeling of Cross Members in Vehicle Design** 1570
Yucheng Liu and Michael L. Day

2007-01-2140 **Mathematical Methodology: Incremental Harmonic Balance Method and Its Application in Automotive Crashes** 1579
Ahmed Elmarakbi

2007-01-2160 **Application of Computational Models for Interior Noise in a Vehicle Development Process** 1597
Hyo-Sig Kim and Seong-Ho Yoon

2007-01-2179 **Errors Associated with Transfer Path Analysis When Rotations are not Measured** 1605
Akira Inoue and Rajendra Singh

The SAE 2007 Transactions Journal of Passenger Car: Mechanical Systems, Volume 116, Section 6, can be found from the library of University of Michigan, Ann Arbor, MI, USA